T0230675

A TEXTBOOK ON

ATM
Telecommunications

Principles and Implementation

A TEXTBOOK ON

ATM
Telecommunications

Principles and Implementation

P. S. NEELAKANTA

CRC Press

Boca Raton   London   New York   Washington, D.C.

Library of Congress Cataloging-in-Publication Data

Neelakanta, Perambur S.
  A textbook on ATM telecommunications : principles and implementation / P.S. Neelakanta.
      p.   cm.
  Includes bibliographical references and index.
  ISBN 0-8493-1805-X
  1. Asynchronous transfer mode. I. Title.
TK5105.35. N44 2000
621.382′16—dc21

00-029431
CIP

Visit the CRC Web site at www.crcpress.com

No claim to original U.S. Government works
International Standard Book Number 0-8493-1805-X
Library of Congress Card Number 00-029431

PREFACE

There has been extraordinary growth in telecommunication engineering in the last few decades. The progress of current years faces a second revolution in telecommunication as the world marches ahead into the new century along the information superhighway. There is, however, a gamut of engineering challenges strewn along this path while facilitating the socioeconomic needs posed in the telecommunication area.

The pace of evolving telecommunications is incomprehensibly fast. Therefore, it has become an impetus at all levels — whether in educating the technicians or creating a new breed of electrical communication engineers — to become proficient and conversant of the past profile of telecommunications as well as to gain a comprehensive knowledge and hands-on perspective about the next generation of telecommunication technology.

In the panoramic realm of telecommunications and in the associated networking of the present and future interests exists a subset called *broadband technology*, which includes multiple applications of electrical communication technology, namely, voice, video, and data in an integrated fashion.

Among the broadband technologies in practice, there is a specific state-of-the-art strategy that prevails known as the *asynchronous transfer mode* (ATM), and the pertinent engineering commitments associated with it involve a plethora of users, service providers, and equipment vendors on a global scale.

Asynchronous transfer mode (ATM) is conceived as a terminus of the evolutions and developments that had taken place in switching and transmission technology in the last two to three decades. It makes the highly sought after *broadband integrated services of digital networks* (B-ISDN), a reality and it makes B-ISDN accessible to subscribers.

The modern applications of telecommunications, which require high-bit rates for the transfer of voice, video and data information, are hopefully turning to the B-ISDN protocols supported by the ATM. Asynchronous transfer mode has a constant cell-size for the packet-structure that is used for the transmission of associated information. Further, it accommodates multiservice telecommunication strategies as needed by current technology. It uses the broadband network schemes namely, the *synchronous optical network* (SONET) and the *synchronous digital hierarchy* (SDH), which have a formattable framing structure consistent with ATM protocols.

In essence, ATM is a high-performance, cell-oriented switching and multiplexing technology that utilizes fixed-length packets ("cells") to carry different types of low-bit to high-bit rate (or, synonymously, low bandwidth to high bandwidth) traffics supporting heterogeneous applications such as data, voice, and video. High *quality of service* (QOS) being the mantra to market telecommunication network services of today, the aforesaid ATM supplemented by SONET or SDH architectures provides an appropriate and flexible infrastructure to the emerging B-ISDN technology. Use of optical fiber in SONET/SDH as the physical layer for transporting the packetized information to its destination has enhanced the realizable broadband attributes to ATM networking and increased the reliability of transmission due to reduced bit errors.

Further, ATM is envisioned towards *local area networking* (LAN), *wide-area networking* (WAN), as well as public networks of national and global communications. Therefore, it is governed by certain protocol issues, unique engineering considerations, and networking principles pertinent to the technology and switching/services constrained by a prescribed set of universal standards.

ATM is a technology that enables service providers to capitalize on a number of revenue opportunities through multiple ATM classes of services, high-speed local area network (LAN) interconnections, voice, video, and data transmission as well as multimedia applications. Relevant ventures in business markets enclave short-term perspectives and stretches into long-term strategies of commercial and residential markets.

In addition to providing such revenue opportunities, ATM also reduces infrastructure costs through efficient bandwidth management, operational simplicity, and by consolidating the overlay networks. From a tariff point of view, carriers may no longer afford to shoulder the financial burden and spend undue time required in deploying an individual network for each version of new service requirement (for example, dedicating a network for a single service such as transparent LAN or frame-relay). However, by pursuing ATM technology, it will allow a stability in the core network and simultaneously permit the service interfaces and other equipment to be developed at a rapid pace. Thus, ATM is a catalyst for robust telecommunications, which is commensurate with the choices imposed by the consumers, capabilities of the vendors, and the potentials of service providers in implementing the services required.

Technically, ATM can be viewed as an evolution of packet-switching. Like packet-switching for data (for example, *X.25, frame-relay, TCP/IP*), ATM also integrates the multiplexing and switching functions. It is well-suited for bursty traffics (in contrast to circuit-switching), and allows communications between devices that operate at different speeds. Unlike the conventional packet-switching, ATM is designed for high-performance multiple-applications networking. In its current emerging phase, ATM technology is implemented in a spectrum of networking devices. It is a capability that can be offered as an end-user service by service-providers (as a specially tariffed service). Also, it may refer to a unique networking infrastructure at the telco level. The basic service building block of ATM networking is the *virtual circuit*, which is an end-to-end logical connection with defined end-points and preset routes but does not have a specific bandwidth dedicated to it. Bandwidth is, rather, allocated on demand by the network when the users define the type of traffic to be transmitted with a prescribed quality of service. Hence, to meet the broad range of application needs, ATM also defines various classes of service involved.

ATM systems are designed in conformance with a set of international interfaces and signaling standards defined by the *International Telecommunications Union Telecommunications* (ITU) *Standards Sector* (formerly the CCITT) as recommended by the *ATM Forum*. The ATM Forum is an international voluntary organization composed of vendors, service providers, research organizations, and users. It has played a pivotal role in the ATM market since its inception in 1991. Its purpose is set to "accelerate the use of ATM products and services through the rapid convergence of interoperability specifications, promotion of industry cooperation and other activities." And, ATM Forum works towards this goal by developing appropriate multi-vendor implementation agreements.

Recently, there is also a considerable interest shown on the topic of *wireless ATM*, a companion system to the emerging wireline B-ISDN/ATM networking. Relevant feasibility and implementation issues and engineering perspectives on wireless ATM have become, therefore, a viable part posing adequate contents for a section in a conceivable book on ATM.

About this book ...

The primary goal of this book is to cast the salient aspects of ATM telecommunications technology in the perspective of the associated engineering. It is written to suit classroom presentations (for adoption as a textbook) as well as a book of guidance for practicing engineers. It outlines the necessary underlying principles and implementation considerations of ATM in a lucid manner for students, network planners, and engineers employed in telecommunications industry. It is designed to educate the students and practicing engineers of the industry on the changing trends in the telecommunication area. The goal of the book, in a nut-shell, is to make its audience appreciate the query "Why ATM at all?"

There are a number of books that have emerged in the recent past, offering excellent descriptions on various aspects of ATM as a broadband switching technology. The associated methods of transmission, protocols involved, and signaling strategies adopted with relevant descriptions on hardware and/or software requirements, etc., have been portrayed comprehensively in those books.

A modern approach to teaching engineering subjects (as encouraged by various accreditation bodies) is to blend design considerations along with the theoretical contents. Such design examples of practical interest and implementation are presented explicitly in this book. Therefore, the pedagogy of this book is conceived to meet the following objectives consistent with the audience profile and its requirements:

- Present the subject-matter on ATM cohesively and logically in a chapterwise format compatible for a semester-based set of lectures
- Each chapter is provided with a set of (solved) example problems, and assignments. Relevant practical design exercises in those chapters are also presented

- Most of all, in each chapter, some topics, which do not directly come under the scope of ATM, but are required as background information, are highlighted within "key concept boxes" along with the contents of the text deliberated in the chapters
- Further, each chapter begins with an overview of its contents and offers, as necessary, the background details. A glossary of terms and networking concepts is also presented at the end of the book
- Adequate examples on issues concerned with the *migration to ATM* and *ATM-enabling* are presented to illustrate the associated principles and techniques. Each example has a balanced emphasis on the concept and calculations involved. The basics on *signaling* considerations are also presented in a systematic manner: First, the signaling in non-ATM situations is discussed and details on ATM signaling are presented along with examples and illustrations
- The book will meet the needs of the engineers of the telecommunications industry. For many of the existing industries, ATM is a new concept. The engineers associated with them should, therefore, learn the basics and prepare themselves for the trends in broadband networking via ATM. This book will be a companion to their conceivable efforts.

The author's experience, as an instructor of this course at the 6000 level, provides insight into the students' needs on this subject. Further, the ample feedback received from students (regular graduate students and/or practicing engineers from various telecommunication industries who participated in this course) has been used judiciously to make the book reader friendly.

This book also covers the basics of electrical communications and networking as the background materials so as to enable the readers to understand and grasp the intricacies of the engineering and science behind the ATM technology.

In short, the book as it is written will aptly match the students at upper undergraduate level, graduate level, as well as ATM professionals belonging to telecommunication service-providers, equipment manufacturers, and research and development sectors.

The book is organized in eight chapters. The contents of each chapter are preceded by a chapter-opener portraying a preview on the real-world aspects and application considerations pertinent to the contextual motive of the chapter. As stated earlier, a comprehensive glossary of terms and their brief definitions is presented at the end of the book. Effort will be made to include any necessary updated information when the book is to be revised in future editions. Also, relevant bibliography is appended to each chapter.

Perambur S. Neelakanta

Boca Raton
2000

ACKNOWLEDGEMENTS

Books are written, bought, read, resold and may be forgotten! But in the heart of the author always linger the hardships strived and the help that was received in making the book.

My foremost thanks are due to all my students who patiently listened to my lectures on the subject-matter of this book and contributed their abundant wisdom, constructive criticisms, and above all, splendid solutions to my homework assignments!

I am also sincerely thankful to those practicing engineers who enrolled in my course and shared their knowledge, which has been profusely inculcated in this book.

My specific thanks are due to my Ph.D student Mr. Wichai Deecharoenkul, M.S.E.E. (a Telecommunication Engineer on deputation from the Communications Authority of Thailand), who helped me immensely in preparing the manuscript of this book. Without his help, a timely release of this work would have been impossible.

I extend my thanks to the publisher and staff of CRC Press for providing me with an opportunity to write and publish this book. Their support and help are greatly appreciated.

I have learned a lot by reading the books written by other authors on this subject. The knowledge I gained thereof helped me to cast this book in the right perspective. I thank all those authors.

My words of gratitude will be incomplete, if I fail to place on record my heartfelt "thanks" to my wife Manju and my children Mahesh and Sabarish for being always beside me in all my endeavors.

Lastly, in all humility I dedicate this book to Thirumurugan of Tiruporur.

ABOUT THE AUTHOR

 **Perambur S. Neelakanta** is a professor of electrical engineering at Florida Atlantic University in Boca Raton, Florida, in the United States. He received his Ph.D. in electrical engineering in 1975 from the Indian Institute of Technology, Madras (India). He was also a research fellow at the Technical University, Aachen, Germany. Career-wise, he served as a member of the faculty at the Indian Institute of Science, Bangalore (India), Indian Institute of Technology, Madras (India), University of Science, Penang (Malaysia), National University of Singapore, and the University of South Alabama, Mobile, Alabama. He was also director of research at RIT Research Corporation, Rochester, New York.

Concomitant to his pedagogical pursuits, he has a wide range of research interests: Electromagnetics, antennas, microwaves and RF engineering, neural networks, radar systems and telecommunications. His current research efforts are directed at ATM telecommunications, mobile communications, ADSL and economics of telecommunications network planning.

Dr. Neelakanta has published extensively, in excess of 130 papers in journals and about 60 in conference proceedings. He has authored a book-chapter and published three books in addition to the present one. He is also working on a new book on wireless communication antennas and currently supervises several Ph.D. dissertations and a couple of M.S.E.E. theses.

He has received university-wide awards, for excellence in undergraduate teaching and for being the distinguished teacher of the college of engineering.

TABLE OF CONTENTS

Chapter 6   ATM Switching and Network Products

Chapter 7   ATM: Operations, Administration, Maintenance, and Provisioning

1

INTRODUCTION

A Perspective of Telecommunications: What is new?

> *"If there is a cliché that hounds our lives unmercifully, it is that we live in the Information Age, a time when ever more information, wanted and unwanted, pours in on us from every side. So relentless and insistent has this flood become that it often seems impossible to escape. Marshall Mcluhan's once fanciful 'global village' has emerged with astonishing swiftness in the form of a planet interconnected by elaborate media networks that transfer data and images almost instantly...."*
>
> Maury Klein
> *Invention & Technology*, Spring, 1993

1.1   The Tale of Telecommunication

One way industrial societies measure progress (and productivity) is the speed with which their technologies compete, cherish, and develop and mutate [1.1]. In this lasting marathon, an impressive record has been set by communication technology ever since its birth and stretching all the way into "the modern maze of telephones, radio, television, computers, fax machines, and satellites". The motivation of deploying telecommunication was spelt even at its inception through telegraphy "as a prophetic foreshadowing of the important part which electrical communications were to play in bringing nations into a better understanding and closer cooperation". An informal telegraphic transmission over a ten-mile wire by Samuel F.B. Morse on January 24, 1838 at the New York University was the imprint and utterance of this prophecy - "Attention, the Universe! By kingdoms, right wheel" [1.2].

Scores of years have passed since communicating a message by electrical signal was conceived; and, as prophesized in 1838 there has been stupendous growth, radical advances and outstanding applications mushrooming in the telecommunication arena shrinking the world within the yard-stick of mass-communication and linking the nations via information superhighways. Specifically, the on-course of current years has faced a more furious pace of changes in the associated technology as the world marches ahead into the twenty-first century on the set tracks of information passage. But, there is a gamut of engineering challenges strewn along this path while facilitating the socioeconomic needs posed in the telecommunication sector which is inevitably entwined with and progressively surfing along the computer environment.

Such challenges are, however, the part of any societal transformations dictated by technological innovations. Even at the birth of telecommunication conceived through telegraphy, the man who ushered that system into society, Samuel Morse, "knew too little to realize the obstacles that lay ahead". Nevertheless, the obsessing quest of Morse, namely, "the instantaneous transmission of intelligence by electricity to any distance" has sprawled across the past and present centuries and spurred technology and engineering not only to overcome the intervening obstacles but also to inspire inventions and installations of newer marvels of telecommunication. It has been an effort comparable to that of Hannibal trying to find his way into Europe across the Alps. When he could not find a route, he rather chose to make one! So has been the aggressiveness that has molded a stage on which the story of telecommunication can be well enacted.

1.1.1 Just what is telecommunication?

It is an art — a technology and a science of *communicating a message at a distance* by electrical means. It follows the basic notion of transporting an *electrical signal* — an entity that contains *information.* The concomitant of a *message* depicts the presence of useful information-content.

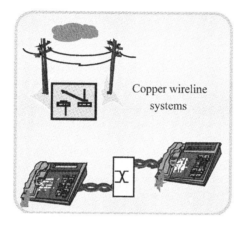

Fig. 1.1: Wireline telecommunications systems

The message is intended for transportation between communicating end-points — namely the point of origin and the point of destination. The link that establishes the connection between these end-points could be *wireline* or *wireless.*

In the modern context of telecommunications, the wireline connection is conceived via electrical conduction through a copper conductor or by optical transmission in a fiber. Classical telegraphic and telephone transmissions were supported by open copper-wire transmission systems. Subsequent versions adopted are largely *twisted pair of copper-lines* and to some extent are the so-called *coaxial lines.*

Optical fiber transmission is a state-of-the-art method used widely in high-speed telecommunication links.

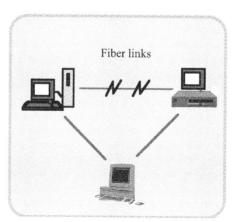

Fig. 1.2: Optical fiber links

Wireless communications refers to a system realized with a propagating electromagnetic wave as the via *media* between the transmitter and the receiver. The information transmitted over telecommunications systems could be simple telegraphic messages, voice signals over telephone, facsimile transmissions, data transfer between machines (computers), an integrated passage of voice, video and data, or a collection of cohesive, multimedia presentations. These information types have dictated the dawn of comprehensive telecommunication engineering and its subsequent growth as a technology. Further, the need for a single framework that can support a large number of diverse services of kaleidoscopic information has paved an alley to explore the varieties in networks and modes of information transmission.

Fig. 1.3: Wireless telecommunication

The following subsections trace the history and conjectural necessities that placed telecommunication in its present status.

1.1.2   Telecommunication -an inception from telegraphy

Morse telegraphy
The genesis of telecommunication can be traced to the invention of wireline telegraphy by Samuel Morse in 1837. The performance exhibition of telegraphy in 1838 was followed by an official operational system envisaged between the railroad depot on Pratt Street in Baltimore and the Supreme Court Chamber in Washington D.C. on Friday May 24th 1844. In the following year, telegraph lines began to be built over other routes. It was a "landmark in human development from which there could be no retreat. For the first time messages could routinely travel great distances faster than man or beast could carry them" [1.1]. Ironically, this technological breakthrough was an epiphany of electromagnetism, of which Samuel Morse knew very little. Yet his intuitive innovation helped him to develop "a system that could reduce the way of writing the letters of the alphabet to a very simple form so that they could be written by an electrically-controlled instrument" [1.3], the necessary electrical signals being produced by switching on and off a d.c. circuit. This "very simple form" of depicting the alphabets corresponds to the on and off states of an electrical switch (known as the *Morse key*).

Two distinguishable durations of keeping the switch closed yield the so-called "dots and dashes", a combination of which is used to represent an alphabet. Shorter and simpler combinations are reserved for most frequently used letters, thus reducing the average length of time required to signal a message. Hence emerged the first variable length code - the *Morse code*.

The first official message telegraphed by Samuel Morse from Baltimore to Washington D.C. was the historic sentence:

What hath God wrought?

"A hundred and fifty years after Morse sent his first message, our information-soaked world still gropes for an answer to the query it contained!" [1.1].

The telegraph, as the first form of modern communication, "burst upon the sensibilities of a people proud of progress but still new to technical leaps of such magnitude". The invention and implementation of telegraphy by Morse in its day "was no less a miracle of its modern offspring. It dazzled and bewildered people the same way computers can today, with feats that seemed magical if not unnatural" [1.1].

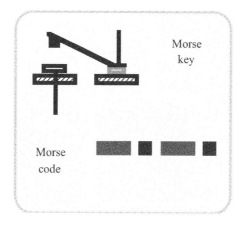

Fig. 1.4: Morse key and Morse code

Morse conceived a few feasible versions of telegraphy and ultimately arrived at a semiautomatic mechanism in which a set of arranged metal contacts (as per a combination of dots and dashes) sent electrical currents, switched on and off in accordance with the encoding of the message envisaged, then the receiver used a pencil to mark the encoded message on paper. Morse also devised repeaters, which facilitated long-distance wireline telegraphic communication. As a result of Morse's invention, the Western Union Telegraph Company was formed in 1845. By the 1860s the telegraphic lines not only spanned the continent, they also became a global telecommunication link with the advent of installing the Trans-Atlantic cable.

Despite the development of automated telegraphic systems, the manual telegraphy was deployed as a means of telecommunication even in the 1960s in many parts of the world. Even today, amateur radio operators (HAMs) enjoy transmitting Morse code on a wireless basis by switching on and off a radio frequency carrier using a Morse key connected to a transmitter. Such a *wireless telegraphy* also played a vital role in the military during the last two world wars.

Apart from the humble, automated on-off telegraphic keying developed by Morse, a more rigorous line of automation of telegraphy refers essentially to a machine operation through the use of what is known as the *Baudot transmitter/receiver* system. Here, unlike the Morse code with varying number of dots and dashes representing each alphanumeric character uniquely, a *fixed length code* (due to Emile Baudot) with distinct five binary (on-off) representations to depict each alphanumeric character was used on a punched tape. The corresponding electrical signals were sensed via contact strips on which the encoded punched tape was enabled to traverse.

Even in the nineteenth century itself, multiplexing several telegraphic messages was also conceived in line with the principle, which in modern terminology refers to the *time-division multiplexing*. The relevant method used a plurality of time division intervals, one for each alphanumeric character.

Many of the telegraphy-based studies were more experimental and practical demonstrations. As Black points out " what seems extraordinary in retrospect is the slow evolution of quantitative expressions concerning the above-described concepts" [1.2]. Nevertheless, telegraphic transmission pointed the direction to derive the so-called Campbell's telegraphic equations that constitute the basic relations of modern transmission-line theory. Further, telegraphy enabled the conceptions behind coding theory, and it threw light on the feasibility of dividing the time-scale for the transfer of a plurality of messages.

The teletypewriter

The extended version of Baudot's arrangement was due to Donald Murray whose patent was adopted by AT&T to develop a teletypewriter with a brand name *Teletype* (TTY). Hence, AT&T's Teletypewriter Exchange (TWX) became operational parallel to a similar service called TELEX offered by Western Union.

Fig. 1.5: Teletypewriter

The TELEX service allowed subscribers to exchange typed messages and it formed the basis for the FAX machines that came into existence in the 1980s. The British Telephone version of teletypewriter is known as the *teleprinter*.

An upgrade of conventional teletypewriting service is known as *teletext*. Such services enable office-to-office document transmission through the use of two interconnecting electronic typewriters. And, in more advanced systems, the electrical typewriters are replaced by computers with wordprocessing capabilities. Teletext also facilitates a background/foreground operation in which transmission/reception of messages proceeds hidden in the background without affecting the machines being deployed for other uses simultaneously. Further, the Teletext interoperates with conventional telex systems compatibly.

Radio telegraphy and radio teletypewriting

The teletypewriting-based wireless transmission was also developed during the 1930s through the 1940s using a short-wave radio frequency band. The concept of *amplitude shift keying* (ASK) with a carrier ON or carrier OFF was replaced by transmitting two distinct frequencies representing the on-off states. This is known as *frequency shift keying* (FSK). Thus the *radio teletype* (RTTY) came into existence and was used for global communication not only by the public and news media (such as UPI and AP) but also by governmental agencies, diplomatic missions, and the military all over the world.

Though modified by the state-of-the-art electronics, RTTY still finds a place in ship-to-shore telex as well as in aeronautical and marine weather information transmissions. It is also popular among HAMs.

During the deployment of telegraphy, emerged a set of transmission procedures or *protocols*, which prescribed distinct identifications (IDs), for the calling and called stations. They facilitated the establishment of a connection between the stations; also, they provided the means to identify the beginning of a message, the end of a message, the acknowledgement (ACK), and the repeat-request as well as the termination of transmission. These protocols were a part of the operational procedures of the telegraphic system and partially were built within the code itself. In radio telegraphy/telephony, these protocols became known as *RT procedures.*

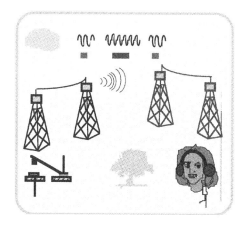

Fig. 1.6: Radio telegraphy

1.1.3 Telecommunication -a lisping voice through analog telephony

Analog telephone

Adjunct to telegraphic communication (which is limited to sending alphanumeric characters on wireline or wireless basis), the transmission of voice signal through a pair of wires became a reality with the invention of the telephone by Graham Bell. Hence sprouted a network of copper conductors interconnecting the subscribers leading to the most familiar telecommunications system, namely, the *public telephone network* [1.4].

Fig. 1.7: Classical analog telephone

Commenced by the Bell Telephone Company in 1877 (later known as AT&T: American Telephone and Telegraph) using the patent rights of Graham Bell, the networking of telephone equipment has proliferated since then across the community and expanded on a world-wide basis constituting a web of global public telecommunication systems.

The most basic utility of telephony since its inception has not changed — to facilitate a conversation between end-users; however, the scope of the analog version of *plain old telephone system* (POTS) has grown grotesquely to accommodate the use of telephony "anywhere and at any time" with the introduction of modern digital telephony.

The general principle of a simple voice telephony circuit is as follows: The telephone instrument converts the changes in the sound pressure due to the talker's voice exerted on its diaphragm into electrical current variations. These variations represent electrical analogs of the time-varying voice (acoustic) signal. The transmitter in the telephone handset thus converts the user's voice into a time-varying current. Most commonly, the transmitter houses tightly packed granules of carbon that are energized by a d.c. voltage. The diaphragm compresses or rarefies this carbon-packing in proportion to the sound pressure due to the impinging voice. Thus, the carbon granule resistance changes and hence the d.c. current through the microphone fluctuates in proportion to the strength of the voice signal. At the receiving end, the fluctuating electric current (that corresponds to the voice waveform) actuates a diaphragm with the help of an electromagnet so that the acoustic (sound) wave arising from the vibrations of the diaphragm maps the voice-induced pressures impressed into the microphone at the sending end.

Some definitions

Analog voice signal

An analog signal is a continuous function of time. It takes values in continuum over time (in contrast with the digital signal, which has discrete values as a function of time).

The voice signal appearing at the output of a microphone, for example, is an electrical analog of the sound/acoustic (pressure) wave upon the diaphragm. This waveform represents a signal voltage that varies continuously with time over the duration of the speech.

The analog signal, in general, can either be a periodic *repetition of the same waveform envelope, or it can be* aperiodic *as in the case of a typical speech signal waveform.*

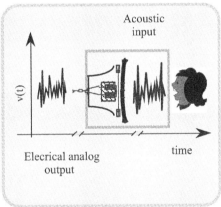

Fig. 1.8: Speech signal

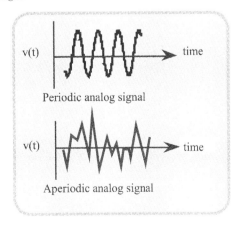

Fig. 1.9: Periodic and aperiodic analog signals

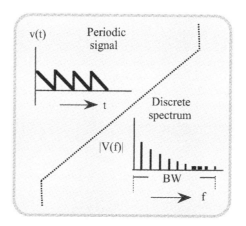

Fig. 1.10: Discrete spectrum of a nonsinusoidal periodic
waveform

The periodic (nonsinusoidal) waveform has a discrete set of amplitude coefficients (known as the Fourier coefficients) of a fundamental frequency component and its harmonics. A plot of these coefficients versus frequency depicts the discrete frequency spectrum of the waveform.

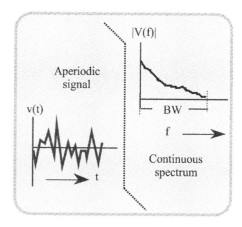

Fig. 1.11: Continuous spectrum of an
aperiodic signal

An aperiodic waveform, on the other hand, has a continuous amplitude spectrum. In both cases however, the coefficients degenerate in their values (as mentioned earlier) and because of this, the frequency spectra of the waveforms can be limited (to an approximate extent) up to a certain maximum value.

Analog baseband signal
The physical, analog waveform (such as speech waveform) converted into electrical format represents an analog baseband signal. This represents the signal of primary interest and bears the useful message (and hence, is also known as a message signal).

Bandwidth concept
Bandwidth (BW), in a broad sense, denotes the extent of significant frequency (f) components present in a signal. In practice, it corresponds to the limited range of the truncated, discrete and/or continuous frequency spectra (Figs 1.10 and 1.11).

In summary, a signal waveform v(t), which is aperiodic or periodic (but non-sinusoidal), can be shown to have a superposition of a large number of (theoretically infinite) sinusoidal (and/or cosinusoidal) components through Fourier analysis. And, the amplitude coefficients of such components degenerate along the frequency scale. As a result, it is possible to truncate the Fourier spectrum V(f) at a frequency beyond which the amplitude coefficients are insignificantly small. The width of the significant frequency spectrum bears the associated information and is termed as the bandwidth. For example, a voice signal has a truncated bandwidth of about 4,000 Hz. This is the baseband *width of the voice signal. Here, the term "baseband" is used to explicitly designate the band of frequencies representing the message signal.*

Analog baseband transmission
This refers to a transmission wherein the band of transmission frequencies supported by the channel matches closely the band of frequencies occupied by the message signal.

Analog telephone switching network
The internetworking between end-users (or, the so-called *subscribers*) of a telephone system was facilitated originally through an operator located at a switchboard. If the telephone switching is confined between a group of users within a business office, switching is done at a switchboard located in the premises of the building and it is known as a *private branch exchange* (PBX). For public connections, the switchboards of telephone companies (*telcos*) are located at a *central office* (CO) also known as the *telephone exchange*s. (More specific definitions pertinent to telephone networks are indicated in later sections.)

The operator-and-switchboard approach soon became totally inadequate, especially in the public telephone systems with growing trends in the number of subscribers. As a result, an automated telephone switching was conceived by Almon Strowger in 1889 in which a step-by-step search of a dialed line was made feasible with the use of a rotary-switch and a set of relays, thus eliminating the need for a human operator. With automated switching, telephony offered a convenient, fast communication means and its evolution was set in pace with the relevant societal inclinations towards aspiring for a variety of telecommunication services.

Apart from civilian interests towards long-distance communication, which promoted the telegraphy and telephony since their inceptions, the two world wars set an unprecedented research and development effort, by all nations concerned, to realize telecommunication transmissions both inland and across the oceans. The wireline techniques implementing advanced teleprinters and automated switches for telephony emerged with a variety of dynamic changes.

Side-by-side, the wireless communication involving radio telegraphy and radio telephony also saw phenomenal growth due to the concurrent developments in vacuum-tube technology, the associated electronics and better understanding of radio wave propagation. In short, telecommunication became the true victor of the wars of this century.

The need for expanded activity in telephony, almost from the beginning of the 1960s, took a different direction that stemmed from the demand placed on conventional analog telephone networks to provide new and different services. New venues of services emerged from the data communication set forth by the dawn of the computer era. Further, the competitive corporation trying to corner the public market together with the intervention of regulatory agencies placing restrictions on monopolistic service-providers accelerated the telecommunication technology globally and specifically in the United States.

The expansion efforts on telephony, which resulted in coping with the societal demand, commercialization strategies, and regulatory prescriptions, can be cast in technological and engineering perspectives:

    ✓   Emergence of and developments in digital telephony
    ✓   Proliferation of telephone networks through hierarchical interconnection of switching offices.

Telephone switching

Regardless of the variety in the service feasibility in telecommunication, the associated telephone switching is concerned with these three major functions:

- *Signaling*: This refers to monitoring the activity of subscribers' lines so as to alert the control elements and enable *a connection setup or release*

- *Control:* This is concerned with functions controlling a connection setup or a release

- *Switching*: This action provides for facilitating the electrical/physical connection between the selected calling and called subscribers.

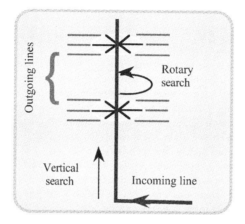

Fig. 1.12: Strowger switching

Classical versions of automated telephone switching are electromechanical types in which the dial-pulses of signaling from the calling subscriber (entering the switch through an incoming line) is used to move a mechanical wiper/contact assembly step-by-step so as to search and connect to the outgoing line (of the called subscriber).

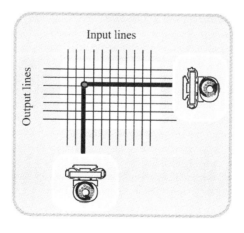

Fig. 1.13: Crossbar switching

The step-by-step movement in response to each dial pulse is made feasible (as a vertical and a rotary motion in succession) by an electromagnetic rotary switch called the *Strowger switch.*

This method was later replaced by *crossbar system* wherein the dial pulses from the calling subscriber are used to search and locate the called subscriber line at the node of a matrix constituted by the vertical set of incoming lines and the horizontal set of outgoing lines.

The interconnection between end-users of telephony, enabled by a set of switching offices in between, constitutes a *telephone network*. The analog telephone network hierarchy, (popularly known as the *Bell system hierarchy*) was conceived to interconnect the subscribers through a hierarchical set of switching offices in a tree-like fashion. Each subscriber line (called a *local loop*) of the local service area was terminated at a local *end-office* (designated as a *central office* or CO). The COs were interconnected via a second level of switching office called a *toll office*. Thus, the tree-like hierarchy of telephone switch offices was continued through three more levels: *Primary centers, sectional centers,* and *regional centers* respectively.

Classically two (open), overhead copper wires were used in the early days of analog telephony. Later, the transmission line used in telephone interconnections consisted of a cable housing over a pair of copper wires. Typically 26-19 AWG paired cables are deployed. Each pair provides for the transmissions both ways at the subscriber loops. However, for long-distance services (over trunk-lines between COs), four wires are used, one pair for each direction. The reason is to enable independent handling of transmission in each direction with respect to facilitating amplification of signals etc. The two-wire pair at the local loop level is transitioned into a four-wire system at the trunk level by means of a balancing, hybrid circuit situated at the end-office.

Multiplexing of analog telephone signals

The term *multiplexing*, in general, refers to realizing a transmission of several signals originating from independent sources on a single channel. In the context of analog telephony, multiplexing implies transmitting a number of voice signals from different subscribers over a single pair of copper wires.

Each pair of copper wire represents a transmission line with inherently associated inductive and capacitive characteristics. As a result, the transmission line acts like a low-pass filter and supports only a low-pass (band-limited) spectrum of electrical energy associated with the analog waveforms of the signals. The bandwidth that a line can support is, however, dependent on its physical length (the longer the line, the more capacitive and inductive effects come into play and limit the cut-off frequency of the low-pass band — thus limiting the bandwidth available for electrical transmissions). Thus, the passband characteristics of a transmission line is decided by the (*bandwidth × distance*) product.

The pair of copper-lines used in analog telephony conforms to frequency spectrum capability in supporting bandwidths much larger than the baseband of a single voice channel (of 4 kHz) over the prescribed transmission distances.

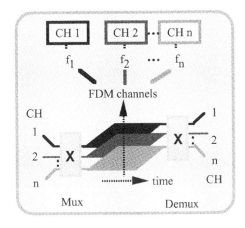

Fig. 1.14: Frequency-division multiplexing: (*See PLATE 1*)

This enabled several voice signal transmissions over a single pair of copper-lines: the *multichannel voice frequency* (MCVF) system. This required dividing the available bandwidth of the cable into a number of narrower bands, each exclusively reserved for an individual voice signal.

Allocation of a number of voice channels along the frequency scale refers to a multiplexing (MUX) strategy called *frequency division multiplexing* (FDM). In this technique, each voice signal is "amplitude modulated" over a distinct *carrier*, so as to form a *sub-band* or *sub-channel* centered around the carrier frequency. Each sub-band is set as 4 kHz of bandwidth corresponding to one side-band of the amplitude modulated spectrum. (More details on FDM are furnished in the next chapter).

Thus, for example, a FDM hierarchy of multiplexed voice telephony transmission was built by forming a group of 12 channels; further multiplexing is done to constitute a supergroup of 5 groups (or 60 channels). The multiplexing could proceed further to realize a jumbogroup of 10,800 channels.

However, in order to accommodate the transmission of a large number of channels constituted by the higher groups, the copper pair cannot meet with the required bandwidth, especially over long-haul transmissions. Therefore, *coaxial lines* for wireline applications and *microwave links* for wireless transmissions were ushered into analog telephony. Thereby, a system of long-haul, wideband multiplexed analog transmissions of voice signals was developed successfully. Intermediate repeaters were included in the transmission path so as to boost up the signal level as necessary. At the receiver end, the multiplexed signals were segregated with appropriate filters and "demodulated" to recover the baseband voice signals. This is called *demultiplexing* (DEMUX).

In the Bell network, coaxial lines (designated as L1-L5 system) supported a FDM traffic of 1,800-108,00 channels. Further microwave links called *analog microwave radios* were used by the Bell system to support wideband FDM transmissions of 600 to 6000 channels in the frequency bands of: 4 GHz (TD system), 6 GHz (TH/TM systems) and 11 GHz (TJ/TL systems).

Both coaxial and microwave systems were used for short- as well as long-haul transmissions with intermediate repeaters as required.

In reference to FDM-based transmission of multiplexed channels, the transmission band of each channel is centered around a frequency much higher than the highest frequency component of the message signal. Each signal transmitted as a part of the multiplexed scheme is referred to as *passband signal*, the generation of which (as indicated above) is accomplished in the transmitter using a modulation process.

Impairments to signal transmissions in the telephony

The scope of voice transmission faces limitations due to inevitable impairments resulting from signal transmission characteristics of the channels. These impairments eventually restrict the quality of the signals received and perceived by the listeners. The causative factors of these impairments stem from cabling and other subsystems used in the transmission technology. Typical telephone signal impairments and their implications are summarized below:

Telephone signal impairments and the contributing factors:

Electrical noise

Electrical noise refers to inevitable and unintentional electrical voltage (or current) fluctuations occurring randomly in the system. This results mostly from inherent thermal agitation of electrical charges in the electrical components and devices used. The electrical noise gets superimposed on the signal under transmission. A parameter called signal-to-noise ratio (SNR) is used to quantify the extent of noise prevailing with the signal.

Attenuation of signal level

　　　　The telephone signal invariably suffers a reduction in its amplitude along the interconnection path between the sending and receiving ends. This is known as channel attenuation. The attenuation in wireline transmissions can be attributed mainly to: (i) The finite, ohmic resistance (lossy) characteristics of the transmission medium and (ii) the imperfections of the hardware/subsystems interposed in the transmission path. For example, in a 24 AWG copper cable, the attenuation per kilometer increases from 0 dB at DC to about 3 dB at 5 kHz. (The unit dB or decibel refers to the ratio of two powers (P_1 and P_2) expressed as $10log(P_1/P_2)$. More details are presented in Chapter 2.

　　　　The design objective of telephone systems is to keep the attenuation impairment within a tolerable level. If necessary, regenerative repeaters are used along the transmission path to increase the signal level so as to counteract the path attenuation. Similar to attenuation of signal in wirelines, the transmission channels of wireless systems also cause degradation of signal intensity.

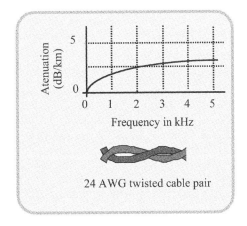

Fig.1.15: Attenuation versus frequency characteristics of a typical twisted pair of copper cables

Radio frequency interference and electromagnetic interference

　　　　The radio frequency interference (RFI) refers to unwanted coupling of radio frequency energy from external sources such as radio transmissions, and the electromagnetic interference (EMI) represents undesirable electromagnetic coupling of electromagnetic emissions from electrical relays, ignition systems, lightning etc.

Crosstalk

　　　　Crosstalk is unintentional coupling from other interconnection lines. A near-end crosstalk (NEXT) implies coupling from a transmitter into a receiver at a common locale. A far-end crosstalk (FEXT) refers to unwanted coupling into a signal from a transmitter at a distant location. Since the extent of unwanted coupling in respect to NEXT is large, the associated crosstalk impairment is prominent in the telephone systems.

Signal distortion

　　　　The waveform of a telephone signal under transmission may get distorted for three reasons: (i) The superimposed noise and interference components would alter the envelope of the signal randomly. Such an envelope or amplitude distortion is nondeterministic due to the associated randomness. Alternatively, the amplitude (or envelope) of the signal may also get distorted whenever the signal is subjected to a nonlinear process (such as modulation). Such distortions are, however, deterministic inasmuch as the nonlinear process involved can be analyzed and the governing process can be predicted. (ii) The second kind of distortion refers to frequency distortion that results from the dependence of signal attenuation on frequency. That is, the different frequency components of the baseband spectrum suffer different extents of attenuation due to the electrical characteristics of the line; and (iii) the third type of distortion is known as the phase distortion. This refers to different frequency components of the baseband spectrum undergoing different extents of phase shift while the signal is under transmission. This is also known as delay distortion and is dictated by the reactive (inductive and capacitive) electric characteristics of the line.

Echo

The transmission lines always encounter electrical mismatches at any transitions introduced. For example, the two-wire to four-wire transition would cause a mismatch. The result is that a transmitted signal would be reflected at a mismatch and may get coupled into the return path; this echo is eventually fedback to its source. If a single reflection occurs, it is called a talker echo; *the second reflection, if it takes place, is known as* listener echo. *Normally, the talker echo is prominant and is a nuisance in telephony. Repeatedly coupling of echo on the forward path would cause a "singing" or an oscillatory condition annoying to both listener and talker.*

More details on the signal impairment aspects of telecommunication are presented in the next chapter (Chapter 2).

1.1.4   Telecommunication - baby talk via digital telephony

The *digital telephony* was conceived as a result of studies on signal-sampling and reconstruction of a signal from the sampled data [1.5-1.7]. Relevant digital circuit implementation became a reality with the advent of phenomenal developments in solid-state technology that incurred since the 1950s. As digital technologies evolved, and as the economies of large-scale integration began to be achieved, an interest was focused on efficient methods for digitally encoding and transmitting speech [1.4]. Hence, "the field of digital speech rapidly (became) an attractive viable technology for communication and man-machine interaction" [1.4].

Fig. 1.16: Digital telephone

What is digital communication?

Digital communication refers to the transmission of a discrete time-signal that can take only a specific set of values over the transmission time. *Signal*, in a general sense, is the bearer of a message being communicated and the message contains *useful information*. A set of signal elements constituted by an ON-OFF pattern of light, for example, can be regarded as a digital message, and when the on-off pattern is decoded, the information pertinent to the message can be extracted.

The on-off pattern is *discrete* in the sense that, at any instant, the "value" of the signal element is either "ON" or "OFF". The signal elements in this case, are restricted to two-values and are designated as *binary elements*.

If numerical attributions are given to the ON and OFF states, they can be denoted by logical 0 and 1 respectively. In electrical format, 0 and 1 may represent the state of a switch that allows the electric current to flow (when ON) in its 0-state and cuts off the current (when OFF) in its 1-state. For example, the telegraphic transmission in essence refers to a digital transmission constituted by dots and dashes generated via on-off Morse keying of a d.c. circuit.

Digital representation is not limited to binary values of signal elements. For example, a message can be constructed in an encoded form by a collection of signal elements formed by switching on/off three lights of different colors in a sequence. Signal elements now are constituted by a permutation of three distinct and discrete states governed by: Any one light on, the other two being off, any two lights on and the other being off, all lights on and all lights off.

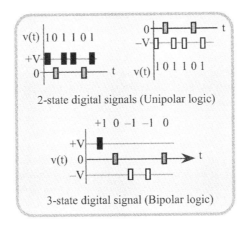

Fig. 1.17: Digital signals

The digital communication in the context of electrical communication (that is, transmitting and receiving signals in the electrical format) signifies sampling an analog electrical waveform (with a continuum of values over a time) and transmitting the discrete sampled-data, typically in a binary-encoded format; and, at the receiving end, the information is retrieved from the received, binary-encoded sampled-data. For example, an analog signal sampled as four distinct levels {1,2,3,4} can be depicted as four binary combinations (1 and 0) as: {00,01,10,11}. Likewise, ternary or more generally, m-ary representations of sampled-data are also feasible.

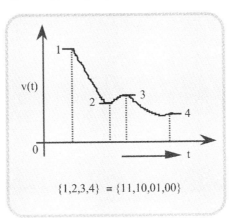

Fig. 1.18: Signal sampling and binary representation of discrete analog values

Digital communication is rendered by transmissions of a series of zeroes and ones. That is, binary format is used in most instances to represent the information flow in digital communications. The signal elements used, namely, "ones" and "zeros" are referred to as *bits* (binary digits). A set of 4 bits is termed a *nimble* and 8 bits constitute a *byte* (or an *octet*).

A collection of bits can be used, for example, to represent a voltage level of a sampled-data (Fig.1.18). Different combinations of zeroes and ones, therefore, allow distinct representations of different such voltage levels. These combinations are called *binary codes*. There are different versions of binary codes, the details of which will be discussed in Chapter 2.

The transmission of encoded zeroes and ones in the electrical format as the states of a switch namely, ON and OFF (alternatively setting the voltage on the line as positive or negative) is known as *digital baseband transmission*.

Voice digitization and its transmission in an encoded-binary format became feasible [1.3,1.4] in the late 1950s with developments in solid-state electronics. This digitization process involves, as the first step, sampling the analog baseband waveform of speech. In a simple form, such sampling is done at uniform intervals. The sampling rate is set at least 8 kHz. This is twice the baseband width of the voice signal. This criterion (known as *Nyquist rate*) is set forth to preserve the intelligibility of the voice in the sampled format, by retaining most of the essential information content of the baseband signal.

As indicated before, the sampling process amounts to quantizing the analog waveform so that each sampled data corresponds to a discrete level. Hence, the sampled-data is a discrete set of values representing the envelope/amplitudes of the (analog) signal. Each sampled-data can be assigned with a distinct set of binary digits. That is, the sampled-data values can be encoded into a set of bits known as *words* and transmission of these words constitutes a digital communication. The analog signal in its encoded digital format constituted by the binary set of words is called a *pulse code modulated* (PCM) signal.

Time-division multiplexing (TDM)

This is a multiplexing strategy, which accommodates several digitized channels on a single channel. For example, TDM can be designed to support digitized, multiple voice signals on a single pair of wires. Suppose each digitized voice channel refers to a PCM channel. The multiplexing of several of these channels refers to accommodating each channel at a designated slot on the time-scale in a synchronous manner.

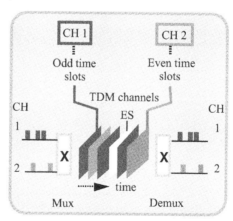

Fig. 1.19: Time-division multiplexing: (*See PLATE 1*)
ES: Empty slot

Two incoming channels (1 and 2) to be multiplexed can be specified as odd and even channels respectively. Upon multiplexing, the incoming bit-streams of a specified length (words) of signal 1 are placed in the odd numbered time-slots; likewise, the bit-streams of a given length (words) of signal 2 are accommodated in the even slots. Thus, the bearer-line or the trunk carries the time-division multiplexed signals 1 and 2 in their bit-encoded forms synchronously. The receiver maintains a time-synchronism with reference to the transmitter, so that the signals 1 and 2 are segregated at the receiver from the odd and even slots synchronously.

T-carrier hierarchy

The TDM-based multiplexing of voice signals is done at many levels, each time increasing the number of channels on the multiplexed line. This is called *T-carrier system* [1.7]. In this system, each sampled level of voice signal is encoded into 8 bits. Thus, each individual TDM channel is assigned 8 bits per time-slot. The lowest multiplexing in the T-hierarchy corresponds to 12 channels. It is called T1 or DS-1 (*digital signal level* 1). Such a multiplexing carried further leads to higher DS levels. For example, at DS-4, there are 4,032 multiplexed channels. The first T-carrier system was designed for exchange trunks (interoffice trunks) over distances in the range of 15 to 80 km.

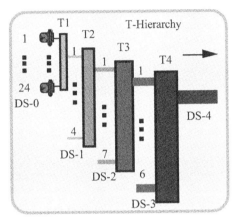

Fig. 1.20: T-Hierarchy

Regenerative repeaters at about 2 km spacing were used at intermediate points to restore digital bit stream to its original form that existed prior to its identity obliterated by transmission impairments. The repeaters included perform:

- Equalization to compensate for intersymbol interference arising from phase and amplitude distortions caused by electrical characteristics of the wire pair used

- Clock recovery to establish a timing-signal to sample the incoming pulses and to transmit outgoing pulses at the same rate as at the input to the line — that is maintaining a synchronism between the transmitting and receiving ends

- Detection of incoming pulses and restore them in good wave shape for onward transmission.

In 1972, the T-2 (or DS-2) system was introduced for telephone transmissions.

The European counterpart of T-hierarchy is called the *CEPT (Conference Europeanne des Postes et Telecommunications)* system. It designates rates commencing at the E1 levels. The E-1 level is derived from PCM coding on 30 channels (instead of the 24 channels of its T-1 counterpart). The E-1 rate is 2.048 Mbps. Table 1.1 lists the T and CEPT digital hierarchies.

The rates of transmission for various signal levels indicated in Table 1.1 will be discussed in Chapter 2.

Table 1.1: T-carrier and CEPT digital hierarchies

T-carrier system (Australia, Canada, Japan and USA)				CEPT-carrier system (Europe and other CCITT countries)			
Signal level designation	Type of carrier system	Transmission rate in Mbps	Number of channels	Signal level designation	Type of carrier system	Transmission rate in Mbps	Number of channels
DS0	-	000.064	1	CEPT0	-	000.064	1
DS1	T-1	001.544	24	CEPT1	E-1	002.048	30
DSC1	T-1C	003.152	48	CEPT2	E-2	008.448	120
DS2	T-2	006.312	96	CEPT3	E-3	034.368	480
DS3	T-3	044.736	672	CEPT4	E-4	139.264	1920
DS4	T-4	274.176	4032	CEPT5	E-5	565.148	7680

Digital telephone network

As early as 1962, in the Chicago area, Bell System implemented a commercial transmission of digitized voice signal facilitated via the time-division multiplexed T-carrier system [1.5].

In the wireless domain, NEC setup a commercial digital microwave radio system in 1968 and in the U.S digital microwave communication started emerging in the 1970s.

The concept of supporting digitized information was extended to fiber optic communication in the 1980s with a short-haul fiber optic route established between Atlanta and Smyrna, Georgia in 1980 [1.6]. Subsequently, long-distance hauling of digital information via fiber optic transmissions came into being on a wider scale.

Digital subscriber loop (DSL)

Digital telephone equipment developed in the beginning (for digital TDM-based local-loop distributions), supported a small group of subscribers and extended them to the nearest CO. This community switch was called a *community dial office* (CDO), which unattended, was controlled remotely from the nearby CO.

Due to the maintenance problems encountered with CDOs, the Bell System introduced the *subscriber loop multiplex* (SLM) and later the *subscriber loop carrier* (SLC-40) system replacing the CDOs.

SLM was designed to serve up to 80 subscribers. It was a *pair-gain* system: It had a *concentrator* which could switch the active users (among the 80 subscribers) into a smaller number (24) of shared output lines, and these 24 lines were multiplexed and transmitted as a T-1 system. At the other end (in the end-office), *deconcentration* (or, *expansion*) was done by switching from the concentrated traffic on shared lines to the original number of stations (namely, 80) by means of a 24 × 80 crossbar matrix switch.

In SLC-40, there was no concentration. It served 40 subscribers and all 40 subscriber lines were multiplexed and transmitted as 40 channels on a T-1 line. Both SLM and SLC-40 used the so-called *delta modulation* for voice digitization *in lieu of* the PCM technique due to the relative simplicity in the associated electronics. However, advances in IC technology by 1979 enabled the realization of the SLC-96 switch designed to cater 96 subscribers. It was equivalent to a (4 × T-1) support and it used PCM instead of delta modulation for voice digitization.

Early versions of simultaneous digital transmission of voice and data on a single medium

Inasmuch as the T-carrier system allows bit-stream transmission, it was also considered a candidate to support the transmission of encoded bits of alphanumeric characters in data communication. For this purpose, AT&T came up with a system called *dataphone digital service* (DDS). However, this system faced a problem; long-haul transmissions with the T-carrier transmission of digitized voice via copper pairs of wires was limited to local exchange areas and/or for short-distance toll-network trunks. Thus, there was no provision to interconnect digital circuits in different exchange areas to accomplish a long-haul digital transmission pertinent to

DDS. Therefore, AT&T developed a technique called *"data under voice"* (DUV) to extend the service of DDS for long-haul applications. This was done by resorting to microwave transmissions to interconnect exchanges separated by long distances. TD and TH radio systems were used and the digital data information from DDS were accommodated below the passband of multiplexed analog voice signals supported conventionally by these radio systems.

Apart from DDS related microwave transmissions of digital information, digital radios operating with the microwave carrier frequencies to transmit and receive higher levels of time-division multiplexed voice traffics were also developed. The frequency band used spanned from 2.1 GHz to 11.7 GHz supporting different extents of DS levels.

The microwave transmission (with repeaters as necessary) adopted for long-haul digital multiplex transmissions of information was adjoined by high-density, long-haul digital transmissions via fiber optics. Thus, in 1986, AT&T introduced transcontinental fiber optic transmission successfully.

It can be gathered from the discussions above, historically, that the transmission medium has been a limiting factor in the viable implementation of long-distance communication. The wireline and wireless bases of telecommunication rely dominantly on the associated media, a brief outline of which is presented below along with some details on the impairments the signal may suffer while being transported via these media. Extended details on transmission media are furnished in Chapter 2.

Transmission media of telephony ... types and characteristics

Wireline media

The nerves of wireline telecommunication to date have been the copper conductors, though they are contemporarily adjoined by optical fibers. Classically, the simplex telegraphy used a single copper wire between the transmitter (Morse key) and the receiver (sounder box) and the electric circuit between them was completed via an earth return.

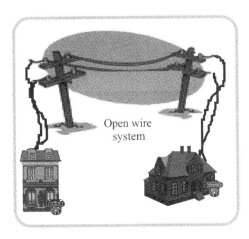

Fig. 1.21: Open-wire transmission medium

Alternatively, two (open), overhead wires were adapted typically in analog telephony, and today, copper wires are still used widely in different categories for telecommunication applications.

Fig. 1.22: Unshielded twisted pair of copper wires

In internetworking that involves computers for data communication purposes, the copper cables are in the form of high quality twisted pairs of copper wires.

Fig. 1.23: Shielded twisted pair of copper wires

The twisted pair can be either shielded and unshielded and is designated as shielded twisted pair (STP) or unshielded shielded pair (UTP) respectively. The twisting and shielding offer the copper lines at least some immunity from external electromagnetic interference (EMI)/ radio frequency interference (RFI) and undesirable crosstalk noise.

Shielding is a brute force technique to combat against EMI/RFI and/or crosstalk. Placing a group of copper wires within a metal braid or foil acts as a barrier and accomplishes the shielding in a simple way. Twisting of copper wires allows mutual cancellation of electromagnetic interfering energy in each wire so that the net effect of interference is minimized.

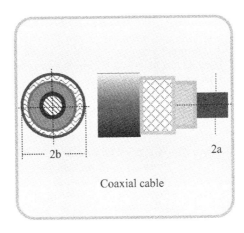

Fig. 1.24: Coaxial line

Another version of electric cable used in telecommunication practices is the coaxial cable. It consists of a central conductor and a concentric conducting tube or a metallic braid or foil. The intervening space between the inner and outer conductor is filled with an insulating (dielectric) material. Both twisted pairs and coaxial cables have protective plastic jackets on the outside.

There are two versions of coaxial cable: Baseband coaxial line and broadband coaxial line. The baseband version can handle up to a bandwidth of 10 MHz whereas the broadband coaxial line can support signals with a bandwidth extending up to 450 MHz.

The inherent electrical characteristics of copper cable in general, dictates the performance in supporting the electrical signal of a specified range of frequency, namely the bandwidth. Twisted copper cables allow transmission of low bandwidth (4 kHz) voice-signal over a long distance. They can also sustain transmissions of higher bandwidths such as 10 MHz or 100 MHz, however, over short distances. The dimensions, geometry of the conductors, and the dielectric properties of the insulation(s) used are the parameters adopted in the design of cables to yield certain bandwidth characteristics.

In general, the bandwidth-distance product limits the desired bandwidth of operation restricted only up to a specified distance of transmission for any transmission media.

The limited bandwidth of operation is also synonymous with the limited speed of electrical pulse transmission (measured in terms of bits per second or bps) used in digital communication. Thus, twisted copper wires do not facilitate a transmission rate beyond 10 Mbps beyond about 180 meters without a repeater. The transmission range gets even limited to meters if the 100 Mbps rate is adopted. Sustaining transmissions beyond such speeds over a few meters becomes impossible with the twisted copper pairs. Therefore, for long-haul transmissions at high bit rates, the use of coaxial lines and/or fibers has become imminent.

The coaxial cables offer larger bandwidths than twisted pairs and they have limited applications in certain ethernet network architectures and TV transmissions. However, the networking industry shuns the use of coaxial cables for local-area networking because for high-speed networks, as well as long-haul communication, use of fiber is more economical.

The electrical cable connections need specific types of connectors for physical lock and snap. The most widely used types of network cable connectors are: BNC connector on coaxial cables, IBM Type 1 connector on shielded twisted pair wiring and an RJ-45 connector on unshielded twisted pair wire. The connectors used have a compatibility not only to provide adequate mechanical connection, but also designed to avoid electrical mismatch between mating units.

Each version of electrical cable has a characteristic or intrinsic impedance dictated by its electrical properties which in turn are dictated by the physical dimensions, geometry, and the associated dielectric properties. The extent of electrical mismatch associated with cabling depends on the ratio of characteristic impedances of the units joined. Such mismatches cause reflection of electrical energy, which would manifest as the echo in the telecommunication link.

A new generation of networking media developed in the last two decades concurrent to the application of copper conductors in networking systems is the optical fiber [1.8, 1.9]. Here, the electrical signal is transducted appropriately into an optical format. The signal, now in the form of light, is guided through the fiber so that the propagating light would not leak out of the fiber.

A variety of fibers capable of supporting few tens of megabits per second to gigabits per second have been developed and deployed. Use of optical fibers has enhanced the scope of conceiving telecommunication transmissions on a broadband scale, that is at high-speed levels. The major advantage of using fiber transmission is its immunity from RFI/EMI. The light wave propagation through the fiber is not affected by any invading electromagnetic wave. The fibers also need specific types of connectors, for example, an ST fiber optic connector.

The extent of signal impairments in reference to a copper wire pair, coaxial cable or an optical fiber may vary significantly and this dictates the appropriate choice of medium to meet the transmission objectives.

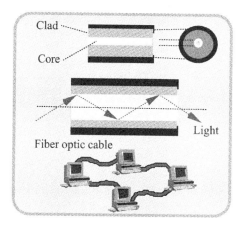

Fig. 1.25: Optical fiber

Wireless media

The wireless telecommunication refers to interconnecting the end-users via radio means [1.10]. Typically the wireless communication involves modulating a high frequency carrier by the baseband signal (or by a group of subcarriers, each modulated by a baseband voice signal) and radiating the modulated carrier as an electromagnetic (EM) wave using a suitable antenna. At the receiving end, an antenna system tuned to the central frequency, such as that of the carrier frequency, receives the EM wave. The baseband voice signal(s) are recovered from the passband of the tuned-in radio frequency spectrum using appropriate demodulation techniques. Further, as needed, regenerative repeaters are interposed between the transmitting and receiving end.

The range of frequencies in the electromagnetic spectrum compatible for radio transmissions with the available technology, in general, spans widely, stretching from almost 100 kHz (termed as long-wave transmissions) up to about 60 GHz (called millimeter waves). This wide range of EM spectrum is divided into specific bands and these bands are designated for specific applications. Exclusively for modern telecommunication purposes, the ultra-high frequency (UHF) and/or microwave bands are used. Classical radio telegraphy and telephony adopted the short-wave band for long-distance communication.

There are unique signal impairment situations associated with wireless telecommunication. The wireless telecommunication is essentially a point-to-point communication, but there could be multiple transmission paths resulting from reflections and scattering of EM waves by physical structures such as buildings etc. or due to refractory effects caused by the atmosphere. The received signal is a vector sum of these multipath-traversed constituents, namely, the primary ray and the delayed secondary rays. The extents of attenuation suffered by these rays would be different and may change. Such changes are significant in mobile and cellular phone applications. The attenuation is controlled by atmospheric conditions as well as by shadowing and other scattering of the EM waves involved. In effect, wireless telecommunication signals face what is known as signal "fading".

To counter the effects of fading, a fade-margin is facilitated. It is done by increasing the transmitted power and/or incorporating frequency- and space-diversity receptions as envisaged in practice. (In frequency diversity systems, the same intelligence is transmitted over more than one carrier. It is expected that, even if one channel fades, the other channels are unlikely to fade. Hence, the information can be extracted from the unfaded channels. The space diversity system uses a single carrier but the reception is done at multiple, spatially dispersed receiving antenna/receiver systems. Again, if a faded reception is perceived at one receiving locale, the other locales may receive unfaded signals. Therefore, the information can be recovered from these unfaded receptions.)

Facilitating reliable communication despite inevitable fading conditions however, poses a challenge in the expansion scenario of modern wireless telecommunication services. Nevertheless, diversity-based system technology, different coding techniques, and spread-spectrum based strategies are adopted to minimize the effects of impairments in such wireless transmissions.

SONET/SDH

In the evolution of fiber-optics based telecommunication, the first generation of networks were proprietary architectures. The equipment, line-codes, multiplexing formats, and operation administration and maintenance (OAM) strategies were uniquely designed and operated by each telco. Only limited commonalties existed between network-to-network.

Soon it became a necessity that a common standard be established so that interoperation with equipment from different vendors could be used. Further, such a common standard became imminent for implementing fiber interfaces between IXCs and LECs. As a result, the Exchange Carriers Standards Association (ECSA) setup a committee to formulate standards towards connecting one fiber system to another at the optical level. In 1985, Bellcore developed the *synchronous optical network* (SONET) which was standardized by the American National Standards Institute (ANSI). SONET defines the rates and formats for the optical transmission of digital information [1.11]. The specifications of SONET indicate a hierarchy of standardized data rates from 51.84 Mbps to 9.953 Gbps. Parallel to SONET, CCITT came up with a single international standard for fiber interconnects between telephone networks of different nations. This is called *synchronous digital hierarchy* (SDH). In SDH, the term "optical" was omitted as it includes both optical and digital microwave interconnection considerations.

The data rates of SONET/SDH are furnished in Table 1.2. The base level of SONET rate is called *synchronous transport signal level - 1* (STS-1) and the higher levels are denoted by STS-N (N = 2, 3, etc.) In terms of optical level specifications, the signal levels of STS-N are designated as *optical carrier level - N* (OC-N).

SONET/SDH was designed to prepare for future sophisticated services that will support the B-ISDN/ATM. These include virtual private networking and time-of-day bandwidth allocation.

Table 1.2: SONET/SDH transmission rates

OC level designation	SONET designation	ITU-T/SDH designation	Data rate (in Mbps)
OC-1	STS-1	-	51.84
OC-3	STS-3	STM-1	155.52
OC-9	STS-9	STM-3	466.56
OC-12	STS-12	STM-4	622.08
OC-18	STS-18	STM-6	933.12
OC-24	STS-24	STM-8	1244.16
OC-36	STS-36	STM-12	1866.24
OC-48	STS-48	STM-16	2488.32
OC-192	STS-192	STM-64	9953.28

The SONET/SDH levels are obtained via systematic multiplexing and framing of basic digital signal levels. SONET frames are formatted to include the information payload plus overheads, which facilitate organized transport of optical signals through a physical connection hierarchy consisting of section-links, line-links and path-links. Each of these links has a dedicated bandwidth for relevant OAM purposes.

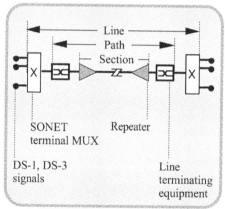

Fig. 1.26: SONET physical hierarchy

Other than the STS rates indicated in Table 1.2 SONET can support lower rate digital signals via virtual tributaries, which can be mapped on STS frames appropriately. Further, SONET/SDH is facilitated with mapping feasibility to support the ATM cells. Details on SONET/SDH are presented in a later chapter (Chapter 3) wherein the distinction between SONET and SDH is also spelt out.

Telephone networks and related organizations

Conceptually, a variety of telephone networks have evolved since the time analog telephone system was patented by Alexander Graham Bell in 1876 and those being deployed today have unique features and are multifunctional. Many of these diverse networks emerged as circuit-switched network configurations with the North American and/or *CCITT* standards supporting voice and the voice band data.

In the North American scenario, the *public switched telephone networks* (PSTNs) have two eras namely, *pre-deregulation/divestiture* and *post-deregulation/divestiture* hierarchies. At the inception of public networks, the aggregate of public-owned companies franchised territorial jurisdictions for the provision of telecommunication services [1.12]. Though the pre-divestiture networks represent the now-defunct configurations, nevertheless, their inception, implementation and deployment have given an insight into modern network planning and the associated cost outlays. These pre-divestiture public-switched networks were based on the following monopolized service strategy: The service was facilitated by a single-provider at a relatively fixed (tariff-invariant) service-cost. This service-cost was based only on the distance between the calling- and called-subscriber. The interconnection technique used the so-called *tandem planning* in which a set of *routing rules* was adopted and these rules, expiration could be programmed at each site on an *ad hoc* basis.

Subsequent to the expiration of AT&T patents in 1983, a number of independent PSTNs mushroomed, especially to cater the telecommunication needs of rural areas, but with extensions to cities as well. This, however, warranted multiple telephones at home so as to access different areas — one telephone for each company serving those areas with no interconnection facilitated by the Bell system.

AT&T started buying up several of these independent companies (the *Independents*) which led the U.S. Congress to enact the Communication Act of 1934 to prevent monopolization of an industry by one company (such as AT &T). This act protects the public from the possibilities of "high-pricing and low-quality service", which may result from monopolization. Further, it renders the FCC (*Federal Communication Commission*) to regulate interstate and international communications rationally.

The major telecommunication equipment designers who supplied the Bell System until 1983 (when the Bell System broke up) were *Bell Telephone Laboratories* and *Western Electric* and local service providers in the rural areas (who were not covered by the Bell system) relied on the *U.S. Independent Telephone Association* (*USITA*, currently known as *USTA -U.S. Telephone Association*) for telephone standards and their implementations.

The divestiture from AT&T of *Bell Operating Companies* (BOCs) led to the formation of the *Exchange Carriers Standards Association* (ECSA) in 1983 to formulate unified standardization common to BOCs and Independents. Hence, the T-1 committee was formed under the sponsorship of ECSA and later was accredited by the American National Standards Institute (ANSI) to formulate a common interconnection set of standards designated as *ANSI T1.nnn-date*. The other North American institutions, which eventually participated in telecommunication-related standards are: Electronic Industries Association (EIA), Institute of Electrical and Electronic Engineers (IEEE) and Bell Communication Research (Bellcore).

At the international level, the telecommunication standards are set forth by the *International Telecommunication Union* (ITU) with the CCITT establishing recommendations on telephony, telegraphy and data communication systems and the *Conference of European Posts and Telecommunications Administration* (CEPT) establishing the CCITT-based standards. The subcommittees of the International Standards Organization (ISO) of the United Nations also work closely with CCITT in formulating telecommunication standards. In regards to the wireless (radio) spectrum-related issues, they are governed by the *International Radio Consultative Committee* (CCIR). In the United States, the FCC manages the radio spectrum.

On the networking side, the introduction of digital telephony and the perpetual growth in local traffic warranted new interoffice transmission fusibilities. The so-called *T-carrier systems* emerged as the cost-effective means to meet the traffic demands. Essentially, a T-carrier system consists of support systems to handle the digitized message. At a lower level, a subsystem either digitizes an analog signal into an encoded binary format (as in digital voice telephony) or generates binary encoded characters as in a computer terminal. The digital bit streams generated at different terminals are time-division multiplexed to form hierarchy of multiplexed digital signal (DS) levels called *T-system*.

The digital form of message transmission also necessitated a compatible *digital switching*. The concurrent developments in solid-state technology since the 1950s enabled the use of semiconductors in digital switching circuits, which are used in realizing the networking of digital transmissions.

1.1.5   *Telecommunication -a cradle of facsimile transmissions*

Facsimile was conceived to replicate a document (with alphanumeric characters and/or pictures, printed matter, drawings, graphics,, hand-written text etc.) at a receiving end of a telecommunication system. Facsimile transmission can be either a document facsimile or a photographic facsimile. The photographic version has provisions to preserve the dark and gray levels of a photo in the transmission process.

Fig. 1.27: Fax machine

The technology of facsimile transmission gained popularity thanks to its business applications especially by the worldwide news services such as Reuters, United Press International, and the Associated Press. The facsimile transmission has grown into the modern *fax* system and has become a viable telecommunication information exchange technique.

Facsimile transmission is facilitated through the electrically sensing (scanning) of the black, gray, and white levels of a document or photograph and sending distinct codes representing these levels encoded in the form of an electrical signal. The codes are similar to the Baudot's fixed length, block code. In modern fax machines, computer block-codes such as ASCII or EBCDIC are used to send the binary (0s and 1s) encoding of the signal-levels obtained from scanning a document or photograph.

Facsimile machines, when introduced in the early 1970s, were grouped into four categories based on their horizontal and vertical resolutions, transmission time per page, and the so-called *compression* scheme adopted. The compression technique refers to "compressing" the scanned data before transmission.

The early versions of facsimile transmissions and the subsequent fax technologies are taking new shape with the advent of integrating computers as a part of the telecommunication systems. A facsimile card attached to a personal computer can now send and receive document/photograph-based messages. Here, the computer prepares the scanned message in the bit format for transmission and reception. Further, through the so-called integrated system of digital networking, high-speed facsimile transmissions have become a reality.

1.1.6   *Telecommunication -a toddler through data communication*

Computer communication

Computers emerged as sophisticated calculators. But, with the increased speed of associated *central processing units* (CPUs) and enormous growth in their memory size, the basic function of a computer namely, "to calculate" took a new turn to include multifunctional applications such as serving as a data-base, word-processing, graphic feasibility and performing multimedia tasks. Further, in today's environment data communication is synonymous with the *distributed networking* of computers. First generation computers were serving the *batch run* users on a centralized basis in which the users brought in their *jobs* to the computer center and the computer after executing the tasks specified in the jobs, returned the results (mostly in the form of hard copy). As the computers grew to accommodate a large number of users and perform more difficult jobs in a shorter time, the strategy was to provide a terminal for each user connected to the computer via private lines. Through time-sharing computers served the users. Essentially, here started the concept of data communication — a feasible strategy to exchange data between a computer and a user workstation [1.13]. Communication was facilitated through a wireline communication and the terminal was used as a teletypewriter.

The genesis of *data communication* is through *computer communication* wherein information is transferred between computers and/or peripherals and computers. Basic computer communication has two forms, namely, *serial* and *parallel*. The so-called ASCII code developed on the basis of Baudot's fixed length code is used to represent each alphanumeric character computer communication in binary formats.

In the serial version of computer communication, data is sent in a serial fashion, one bit after another, on a single pair of wires. The relevant transmission of characters could be either *synchronous* or *asynchronous*. In a synchronous system, both sending and receiving computers are synchronized exactly to the same clock frequency and synchronism is maintained by transmitting a clock pulse from the sender to the receiver. Asynchronous transmission on the other hand, has clocks at either end free-running at approximately the same speed. Each "word" representing a set of serial bits is preceded by a *start bit* and followed by (at least) one *stop bit* to frame a word. The start and stop bits allow the receiver to distinguish each word distinctly in spite of its asynchronous arrival.

Serial data communication follows the so-called *RS-232C Standard Version C* set by *Electronics Industry Association* (EIA). This standard specifies voltage levels, timings and type of connectors for use. Typically, a *DB-25 connector* with two rows of $(12 + 13 = 25)$ pins is used. (For IBM compatible personal computers (PCs), another connector called *DB-9* is more commonly employed.)

RS-232 provides a means to interface a computer with a modem. In modern practice, it is also used to interface a mouse and/or a printer to a PC. Further, the RS-232 standard also specifies the maximum length of cable to be used as 50 feet (so that the associated line capacitance does not exceed 2500 pF).

In certain computer communications, it is also necessary to send each bit of a character (of ASCII) exclusively on one line so that the bits constituting a word are sent via a parallel set of lines. This refers to *parallel computer communication* and facilitates easier data-processing in certain devices like a printer. The serial-to-parallel conversion (or vice versa) can be accomplished by electronic circuits involving gates and shift-registers, etc. The device that performs such a conversion is known as a *universal asynchronous receiver transmitter* (UART). Parallel data transfer has to be done only over a limited short distance, lest the capacitive effects due to parallel wires would impair the fidelity of data transmissions. (In modern context, another parallel interface known as *IEEE-488* is used in computer/instrumentation interface applications).

Data communication involves, in general, the distribution or exchange of data information between machines that process data. In this activity, the bits, which signify the message being transmitted, should have preserved sequences. Data communication is an inevitable infrastructure of computer applications today as it allows computers to communicate with each other. The theoretical foundation that stemmed from the studies on digital communication, the exponential growth in the solid-state technology and the impulsive evolution of computers have decided jointly the diffused permeation of data communication in this information era.

An analog way of digital data transmission: The emergence of modems

The term *modem* is an acronym for modulator-demodulator. As indicated earlier, in telegraphy or in teletypewriting, sending and receiving pulsed direct currents in an encoded fashion is telecommunication. Such communications correspond to low speed transmission and reception of d.c. current changes *vis-à-vis* the encoded message as the generation of encoded characters is dependent on the manual or mechanical operation of key(s). The copper transmission line supported the long-distance transmission of such low speed (that is, low rate of occurrence) pulsed d.c. currents. Transmission of signals as they are (without any frequency translation) is known as *baseband transmission*.

However, with the need that arose to implement transmission of the much higher occurrence rates of binary signals generated electronically (using solid-state devices) in digital communication systems, the copper-lines used for telephone interconnections were incompatible. (The relevant reasons will be detailed in the next chapter.) Therefore it became necessary to "modulate" a voice frequency; that is, convert the d.c. on-off pulses of the encoded binary message into corresponding on-off analog tones (at a voice frequency) so that these modulated voice-frequency pulses can be supported by the voice-grade copper-lines used for telephone applications.

The calling modem, for example, switches between 1,070 Hz to 1,270 Hz (corresponding to logical 1 and 0), and the called modem replies by switching between 2,025 Hz and 2,225 Hz. Thus it became feasible to send and receive over conventional telephone lines, with computer data at speeds much higher than that in telegraphic and/or in teletypewriting transmissions.

The modems (such as the Bell 103) developed in the beginning still operated at an awesomely low speed of 300 bits per second (bps). Increasing the rate of transmission in a modem became a technological pursuit. Thus, a series of modems were developed with speeds of 1,200 bps, 2,400 bps, etc., going up to 56,000 bps with worldwide usage compatibility while assuring ITU-T standards.

Fig. 1.28: Modem

Switching of digital data

Space-division switching involves a spatial layout of incoming lines and a (spatial) set of outgoing lines. The switching simply maps input-line activity to destined output activity on an appropriate outgoing line.

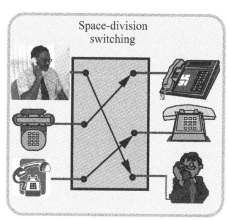

Fig. 1.29: Circuit-switching: Space-division layout: (See PLATE 2)

The time-old manual switching, the automated Strowger, and crossbar switching used in analog telephony were all based on the space-division principle. These classical telephone switching techniques refer to establishing (or releasing) an interconnection between an incoming line (of the calling-subscriber) to an outgoing line (of the called-subscriber) (for example: # 1 ESS of AT&T).

Contrary to space-division switching, the concept of digital switching was developed on the basis of the *time-division principle* circa 1976. AT&T's digital time-division switch (DTDS) version 4 ESS then became operational for toll office applications supporting a line size of 107,000. Subsequently, a number of manufacturers released several types of digital central office switching systems for use at local, toll, and toll/tandem offices. Functionally, a DTDS consists of a matrix which transfers incoming information in a particular time-slot (channel) of a TDM link to a specified time-slot on an outgoing TDM line. This switching process therefore involves both spatial mapping (space-switching) and time translation (time-switching).

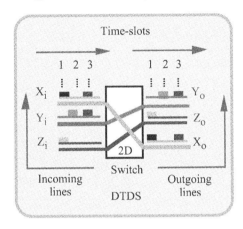

Fig. 1.30: Digital time-division switch: (See PLATE 2)

DTDS is thus a two-dimensional switching. Spatial translation is performed by a digital logic selector circuit and time-switching is carried out by facilitating a temporary storage of incoming messages in a buffer (digital memory/register circuit). The DTDS performs switching on a digital pair-gain system or on a T-carrier interoffice trunk.

Switching of data across a telecommunication network refers to the notion of facilitating connection between one terminal to another terminal in order to perform data communication. Relevant technology has led to realizing a number of network topologies. Such a variety in network topology allows flexible designs that allow sharing of resources and reduce the cost of communication, at the same time offering higher *throughput* of information and decreased delay of services. The way by which the bit-stream representing the data is routed across a network topology can be accomplished via different methods. The following is a summary of relevant strategies and associated definitions:

Routing of binary digits

Circuit-switching

Ideally, switch in a communication system is intended to facilitate transfer of information from one terminal to any other terminal in the network with a prescribed quality of service (QOS).

A classical approach to switching is known as circuit-switching. Here, a pair of end-users have a physically-dedicated connection setup for the length of their communication session with a guaranteed transmission rate. However, this method becomes a colossal undertaking in terms of links required when the number of terminals undertaking becomes large. Therefore, as an alternative to such direct interconnection between terminals, a centralized switching station *is deployed.*

All the terminals are, in turn, connected to this centralized switch, and this switch enables spatial-switching with a physical connection between any two terminals on the network as desired.

The circuit-switching stems from the conventional method of interconnecting telephone subscribers. It was adopted subsequently for end-user interconnections involving modems as well. In classical sense, the interconnection facilitated through circuit-switching in the conventional local-loop arrangement in telephony provides for the establishment and release of three major functions: dialing, ringing and conversation.

Fig. 1.31: Classical circuit- switching of telephones

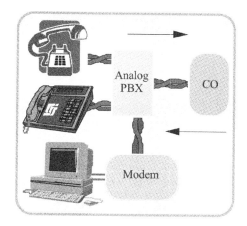

Fig. 1.32: Circuit-switching involving modem

Information flows between the interconnected stations on real-time once the connection is established. Such a flow is transparent to the terminals involved. That is, the terminals are not aware how the data flow is routed. Their main concern is confined to the fact that a connection has been successfully made.

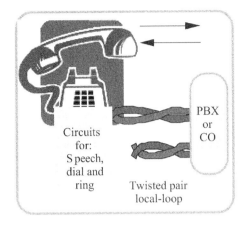

Fig. 1.33: Circuit-switching local- loop level

Circuit-switching in reference to data transmission refers to a connection between two computing systems. Such connections over long distances can be implemented on a wireline or a wireless basis. It can be done via a privately owned transmission service or by leasing data links from a telco service provider (like the telecommunication common carriers in the United States such as AT&T, MCI, Sprint etc.).

In reference to POTS using circuit-switching, the service provider supplies two types of connections:

❑   Leased-line connection — A permanent and dedicated point-to-point circuit. It is always available between two users end-to-end basis

❑   Dial-up connection — A line is accessed by a user to get a connection established to another user by dialing. When the transmission is finished, the call-termination disconnects the users. No dedication of the connection to any specified subscriber prevails.

The physical wiring in circuit-switching is not necessarily dedicated solely to the communicating devices. Node-to-node connections in particular use multiple access techniques. When an established physical channel is in place, other devices may use part of the physical channel, yet reside in a different allocated slot of time or at a different carrier frequency. The path of course, is typically routed through multiple stations within the network, and multiple possible paths should be available for such a network to be reliable.

Circuit-switching involves establishment and disconnection of the circuit. This results in a time-delay prior to the commencement of the actual information transfer session. Another major shortcoming of circuit-switching is that two or more stations may try to establish a connection to a particular (destination) terminal simultaneously. This is known as contention. If contention occurs exactly at the same time, the condition is known as collision. Methods to detect such collisions and providing access to contending stations for a specific terminal in an organized fashion have been developed and will be discussed in detail in the next chapter.

A circuit-switched network sets up a real (as opposed to virtual) connection, with dedicated resources like bandwidth, electrical contacts with a switch and physical wires. Once the connection is set up, it does not matter if the network is used or not; the resources are occupied/consumed in any case. To release a circuit-switched path so that it no longer uses the resources, a specific signaling action is required to tear down the connection.

Store-and-forward switching

     The inefficiency of a dedicated connection becomes burdensome when transfer involves digital data. To avoid this burden, the store-and-forward *technique is adopted. In store-and-forward switching, the nodes store the user information and forwards the same information towards a destination as and when the links become available. This is done by two methods:*

     ❑   *Message-switching*
     ❑   *Packet-switching.*

Message-switching

     Since the direct interconnection is limited by the number of links to be used, and circuit-switching poses a line-contention problem, message-switching was proposed as an alternative strategy. Here, a central switching system accepts the incoming traffic, stores it in a buffer memory and forwards it to appropriate destinations, as lines become free.

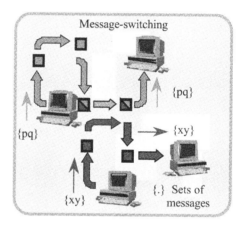

Fig. 1.34: Routing of a set of messages by message-switching: (See PLATE 3)

     The store-and-forward mechanism and message routing are implemented via software. This store-and-forward mechanism does not tie up the resources between the sending and receiving ends on any dedicated criterion. Further, in message switching, once the transmission is initiated, the message is transmitted in its entirety without a break from one node to another. The transmission places statistical attributes on the arrival time at the destination. The relevant attributes are: Arrival rate, variability of arrivals, expected message length, the statistical distribution of the message length and waiting-in-queue delay statistics at each node. In this method, the sending end continuously sends out all the messages without interruption and the messages will reach the destinations unless they experience a delay at the buffers while waiting for a free-line. Thus, the message switching confirms to the store-and-forward technique in which messages in full are sent as unit entities (rather than segmented into packets).

     Messages originating from a source are tagged with an address field and identified by an opening and a closing flag. The message is transmitted across the network to the destinations through several nodes representing a central switching system. At each node, as mentioned before, the message is stored in a buffer and forwarded to an available link.

The node processor in message-switching performs the following functions:

- ❏   *It receives the full user message and stores it in a buffer*
- ❏   *It checks the message for data transmission errors and performs error recovery, if required*
- ❏   *It determines the destination address from the user message. It chooses an appropriate link towards destination based on certain routing criterion.*
- ❏   *It forwards the message to the next node on the chosen link.*

Message-switching has, however, certain drawbacks. For long messages, it becomes important to ensure that there is adequate storage space on the receiving node before the transmission is initiated. Otherwise, the buffer storage may become full, and part of the message may not be stored, thereby requiring retransmission of the entire message. Similarly, if an error occurs during transmission, the entire message may have to be retransmitted. Retransmission of long messages results in large communication overheads in the network. If a high priority short message arrives while a long message is in transmission, it will have to wait until the transmission of the long message ends. That is, when a large message is transmitted through a line, then that specific line or route becomes unavailable over a significant extent of time for those messages waiting in the queue and contending for that line. This undue delay situation can be avoided by chopping the messages into small portions (known as packets*) and transmitting those packets on a store-and-forward basis as described below.*

Packet-switching

*   Paul Baran of Rand Corporation conceptualized the packet switching in 1960. In this switching technique, the messages are split into a number of packets, variable or fixed in size, and the packets are transmitted in a store-and-forward fashion. Messages are split at the source host and reassembled at the destination host. Each packet transmission is independent of the others.*

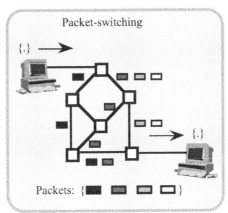

Fig. 1.35: Routing of a packet message via packet-switching: (See PLATE 3)

*   The packets of a single message may travel different routes and arrive at the destination with different delays. This may lead to packets of the same message arriving out of sequence at the destination node. Each packet, however, has to be identified in regards to its origin and destination. Therefore, every packet needs to carry the complete address information, (namely destination identifier, source identifier, message identifier, and packet identifier), in addition to the actual user data.*
In other words, each packet message should carry these sending-end and receiving-end addresses as overheads. Further, inasmuch as the packets of the same message may be stored and forwarded through different routes at the switching nodes, the packets (of the same message) may experience different queueing delay at the switching centers. Hence, the packets of a given message may arrive at the destination at different time instants as a

mixed sequence. Therefore, a reassembling method is required at the receiving end to sequence the packets in the right order before the information is extracted from them.

When packet-switching was introduced, the small, chopped portions of messages became known as datagrams *and the transmission of such datagrams became a unique class of switching system — the packet switched networks. The major advantages of packet switching are the inherent security and privacy of fragmented messages and uninterrupted flow of packets across the network through alternate paths, in the event of a switching node failure. In contrast, such reliability lacked in the hierarchical networks such as AT&T PSTNs, in which the failure of a host amounts to a crash of the entire system.*

In 1967, the Department of Defense (DoD) in the United States started the Advanced Research Projects Agency Network (ARPANET) to interconnect various universities, national laboratories, and governmental establishments across the nation using datagram-based protocol. This constitutes the core and first bed for packet switching. The first two applications of ARPANET were: (i) The TELENET which facilitated a TELENET terminal software supported by a computer so that this terminal could interact with different versions of remote computers, and (ii) the file transport protocol (FTP), which enabled the transfer of files from one computer to another transparently across the network.

Packet transmission has facilitated computer networking specifically to handle data transmissions (rather than voice). For example, the Internet links commercial users, universities, and research organizations throughout the world. Telecommunication service providers also enable access to special computer networks for data transmission. Such networks are known as packet-switched data networks (PSDNs). Examples are SprintNet, Tymnet, and Datapac. The computer network provider facilitates telecommunication circuits or access points (most leased from the conventional telecommunication service providers) to users for necessary interconnections. The devices within the computer network, which enable such interconnections, are called routers.

The packet transmission on wireless media was conceived at the University of Hawaii in the early 1970s leading to the emergence of the so called ALOHA system. This was the precursive effort, which led to the modern versions of cellular digital packet data transmission systems. A variation of the protocol used by ALOHA was adopted by ethernet and developed in the mid 1970s at Xerox PARC.

X.25 protocol

In 1974, CCITT standardized the DTE-to-packet switched-network interface. The associated protocol is known as X.25 [1.14]. While datagrams are constituted out of one message using different routes depending on which route is available, X.25 uses a call-connect *phase to establish a link with the distant DTE. That is, during the call-connect phase on most X.25 implementations, a path is selected through the nodes and links, and all the data during the data transfer phase travels on this given route. This fixed route is called a virtual circuit (VC).*

Thus the routing of packets with X.25 protocol involves a virtual connection wherein a particular path is setup when a session is initiated and maintained during the life of the session. As indicated before, the address in a packet is used to direct the packet to its destination. The path a packet takes refers to a logical connectivity between end-ports. This connectivity is virtual *(not real) in the sense that there is nothing (no specific resources, for example) set aside for the connection on a dedicated basis. The connection is only a permit for packets to pass through the network between the terminal. This permission is "granted" by the routing tables in each node or by an algorithm that lets a switch "decide" if and how arriving packets should be forwarded. If no packets arrive, no switching or transmission resources are used.*

In generic form, a packet is a collection of digital bits that are sent as a unit marked at the beginning and end to a designated receiver.

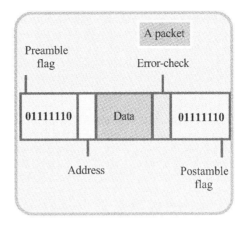

Fig. 1.36: A typical format of a packet

A packet consists of an opening flag, address of the destination (and if required, the sender's), information or payload, a frame check sequence for error detection and a closing flag. The preamble and postamble flags are distinct binary code words. The conventional packets can have variable payload lengths.

Example 1.1
Why is a set of digital bits constituting an ASCII character not a packet?

Solution
ASCII characters consist of a set of digital bits with start-stop bits intended for a conventional asynchronous transmission to mark off the beginning and end of bit groups sent as units.

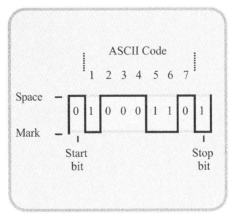

Fig. 1.37: An ASCII encoded character

ASCII characters commonly travel on a cable and only one user can send such characters on each cable; that is, the relevant connection represents one logical channel, depicting a circuit-switched connection. There is no address in a character to control the routing either; therefore, it is not a packet.

Connection-oriented service

This refers to a logical association termed as the "connection" that must be established between the source and destination systems before the data is exchanged between them. Similar to telephone systems, a connection in a computer network depicts either a permanent connection, which is always available (with a leased time), or a dial-up mode connection, which is established as and when data is to be transmitted and, at the end, the line is released.

Connectionless service

Here no connection is established between the source system and the destination system. Each packet transmitted is sent from a source system to a destination system independent of all others without requiring that a connection first be established between them. This is also known as *datagram service*.

Virtual circuits

To support the communication between two end-entities, a VC is established. It appears to the users as if they have an actual point-to-point communication link between them. A connection of hardware and software at end-units enable establishment and release of VCs.

Some defintions ...

DTEs and DCEs

When the computer-to-user distance increased during the expansion era of data communication, the use of private lines had to be substituted by deploying the already existing communication links — the telephone lines. Thus data communication encroached on the telephonic transmission facilities. In other words, the carriers of public telephones became service-providers to data communication users as well. Thus, the data communication system fell within the scope of highly-distributed multilevel switching hierarchy of telephony except that the customer premises equipment (CPE) became a computer (and its peripherals) instead of a telephone.

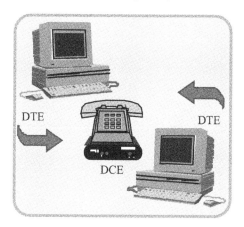

Fig. 1.38: Data terminal equipment and data communication equipment

In the computer-telephony integration, the CPE (in data communication context) became known as data terminal equipment (DTE). Most DTEs are also known simply as terminals. Data information is created, sent, received and interpreted by DTEs. Typically, DTEs include:

□   *Teletypewriters (TTYs)*
□   *Video display units (VDTs), for example, at airline ticketing counters*

❑   *Transaction units such as automatic teller machines, barcode reading cash registers etc.*

❑   *Smart terminals with memory and capability to perform logical operations (Example: Facsimile machines)*

❑   *Intelligent terminals programmed to off load and perform some of the tasks for a host machine (such as screen formatting, data editing etc.)*

❑   *Dumb terminals performing only certain routine functions.*

In a broader sense, DTEs are comprised of: (i) Workstations or personal computers with extensive processing capabilities, and (ii) front-end processors, which perform communication-related functions at a host site. The front-end processor off-loads the communication tasks from the host computer in order to ease out the host computer from laborious communication protocols so that the central processing unit (CPU) of the host system is fully available for computational and other user applications. Facilitating the transmission media to be shared by more than one DTE by means of time-division multiplexers (MUX) further enhances the scope of data communication. Also, there are other components that can be added to enlarge the data communication systems. These include servers that provide support functions such as serving as a database to workstations or servers that facilitate certain intelligent functions that cannot otherwise be performed by a DTE/workstation alone.

The digital (binary) waveform from the DTE modulates an audio frequency tone and the corresponding modulated waveform(s) are transmitted over the voice grade copper-lines just as the voice frequency signals from a telephone pass through. The modulation and (the demodulation) are performed by a modem as mentioned earlier. The modems represent data circuit terminating equipment (or simply, data communication equipment — DCE) which condition the signals received from the DTE for transmission over communication connections, and restore signals received from the network so as to be compatible with the receiving DTEs. In general, a DTE is connected to a DCE by a cable that conforms to a standard. It is further imperative that the DTE-DCE pairs at each end of the communication circuit be synchronized so as to ensure the integrity of the bit stream.

Access networks

The access networks are intended to perform the function of transporting information from core network to the end-user and, in the case of interactive services, also in the reverse direction. It is known popularly as the "last mile" to the end-user.

A variety of access networks used in practice are consistent with much different cabling architectures, which are already installed. The investment these installations represent dictates the necessity to accommodate them in the specifications of any new installation of access network architectures.

Typical access network types are as follows:

✓   Subscriber (local) loop of telephony: End-user telephone access line
✓   Hybrid fiber-coax (HFC)
✓   Asymmetrical digital subscriber line (ADSL)
✓   Very high-speed digital subscriber line (VHSL)
✓   Fiber-to-the-curb (FTTC)
✓   Fiber-to-the-home (FTTH)
✓   Multipoint multichannel distribution system (MMDS)
✓   Local multipoint distribution system (LMDS)
✓   Satellite and terrestrial distribution.

The access network, in general, consists of the local access node connected to the core network via a *local exchange carrier* (LEC) placed at the circumference of the core network, the distribution network, and the network termination (NT) at the end-user.

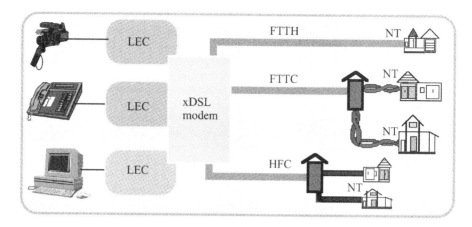

Fig. 1.39: Basic Access network representation

Access node: A connection point to the core-network (such as an ATM network). Conversion of line speed and transmission protocol format (to and from the distribution network) is carried out at the access node.

Distribution network: Transports and distributes signals to and from the end-users. It can be of any physical media - twisted copper-pair, fiber, wireless or any combination of these media.

Network termination (NT): A point of demarcation between the public and private domain. NT can be passive with no built-in functionality and simply a "plug-in the wall" type. It is first the connecting point for the user equipment. Sometimes, the NT may be requested to do a definite functionality (such as signal conversion between media type — example: optical-to-electrical). Then, the NT is called an active-NT.

The access network types and their functions are illustrated in Fig. 1.39

- *Subscriber local loop*: This is end-user access line of telephony. A local telephone network comprised of a central office (CO) supports a number of subscriber access lines, typically of twisted copper pairs. It represents the last-mile line between a subscriber and the exchange (central) office

- *HFC*: Supports wide band transmission such as digital video or interactive video carried via fiber from the CO to Optical Network Units (ONU)/pedestals. At this point, analog signal also is fed in via coaxial lines. HFC structure uses a standard media topology from the ONU to the end-user. The coaxial cable between the ONU and the end-user serves a neighborhood (say, a few hundred domestic end-users)

- *ADSL: Asymmetric digital subscriber line* is based on ANSI standard T1.413. Refers to a modem technology developed by Bellcore in 1989. It converts the existing twisted pair telephone lines into access paths for multimedia and high-speed data communication. ADSL transmits more than 6 Mbps to a subscriber in one direction and as much as 640 kbps in both directions. Such rates expand existing access capacity of telephone lines by a factor of 50 or more without new cabling.

  ADSL is based on advanced digital signal processing and creative algorithms, which squeeze large amounts of information on twisted-pairs of copper cables. As indicated earlier, long twisted-pairs of copper wires, in general, attenuate the signals significantly, almost to an extent of 90 dB around 1 MHz. It means that, in order to implement ADSL, a transmission technology beyond the conventional POTS has to be conceived. A relevant method makes use of the frequency spectrum. It creates three information channels: A high-speed down-stream channel (1.5 to 6.1 Mbps), a medium speed duplex channel (16 to 640 kbps), and a POTS (0 to 4 kHz) channel. The POTS is segregated from the digital section by means of filters.

ADSL will play a crucial role over the next decade as telcos venture into new markets offering services that support information in video and multimedia format. Another technology, known as *very high-speed DSL* (VDSL) is quite similar to ADSL. It can support higher bandwidth from 10 Mbps and possibly up to 50 Mbps, and cover a shorter distance of about 500 ft.

- *FTTC*: This is similar to HFC, but goes closer to the end-user. It uses fibers from the CO connected to a B-ISDN core network to the curbs where the ONUs are located. The type of information supported belongs to digital services such as VoD

- *FTTH*: This is intended to offer highest bandwidth to users. The user has direct access to fiber medium for transfer of information. Intended to support connections at 155 Mbps or 622 Mbps

- *MMDS*: This refers to a short-range technical broadcasting strategy in the down stream direction (frequency 2.5 to 2.7 GHz) carrying 100-200 channels compressed digital video with QAM and other modulation types. Alternatively about 30 analog video are supported.

Transmit antenna covers a small "cell" of end-users within 50 km radius. Line of sight propagation is adopted. POTS/ISDN wire lines are needed for feedback from interactive users.
Local MDS covers a much smaller reach (a few kilometers). Also LMDS supports upstream microwave transmission in lieu of POTS/ISDN wireless usage for interactive feedback. Operating frequency is 10 GHz and the bandwidth (BW) is 1 to 2 MHz.

Satellite and terrestrial distribution systems are used for distribution of analog video satellite and can be used for MPEG compressed signals as well (for example, ETSI ETS-300-421). Interactive application is not yet implemented via satellite.

Terrestrial distribution systems have a lesser geographical coverage than the satellite system (up to 100 km in diameter per transmitting station). The system uses the VHF/UHF (40 - 800 MHz) band. And the channel bandwidth is: 8 MHz (Europe) and 6 MHz (United States).

Local area networks

With the trend towards computer interoperation, transferring of large-sized data over a significant distance became operationally difficult. Hence, the concept of interconnecting a set of DTEs sprouted a *local area network* (LAN) confined within a building or extended to a campus/enterprise geographically limited to a radius of 5-10 km. This interoperation of machines via "in-house" data communication principles can be defined as one of (possibly) many access or switching units (linked by lines) of an existing system which is used to transport information between the attached data terminals (terminals, PCs, hosts).

LANs are characterized by a small spatial coverage (at most a few kilometers: A complex building, a campus, an industrial estate), a relatively simple topology (ring, bus, star, tree), and a private operator (company, organizational unit) [1.7]. Thus came into existence the *ethernet* in the late 1970s (developed by Xerox Corporation). Concurrently, Datapoint Corporation came out with *ARCnet*. Subsequently, IBM released a major networking technology, namely the *token-ring*.

Fig. 1.40: Interconnected set of DTEs in a local area environment

Initial LANs had severe constraints in regards to their protocol descriptions, copper wirelines, associated connectors, and software functions. Eventually, however, these systems became more flexible to embrace various protocols, cables and connectors though "LANs come in different shapes and sizes". Still, there was a need for an open set of protocols, which does not restrict the LAN implementation with products of one vendor.

Open system interconnect (OSI) model

The *open system protocol* standards enable networks to interoperate with no constraints. Mixes of different computers (say, for example, Macintosh and IBM PC) on the same network and their interoperation became a reality. Supporting programs emerged which have made sharing of files and resources (such as modems and printers) easy. In the 1990s workgroup productivity software surfaced and facilitated the search for organized and linked data from documents, spreadsheets, and databases on a shared framework of interoperation.

For interoperational computers, there are a number of complex issues, which have to be resolved so that different computers, which can be totally dissimilar, will be able to communicate with each other. This is accomplished through open-system interface, based on a set of layered protocols. In this approach, each host computer has to follow a chain of hierarchical processes in deliberating the information flow to be received by another host computer. The organization of this protocol structure is as follows: At each computer, the highest layer of protocol is application-specific, designed to handle an application running on a network. The next layer below refers to the physical display or presentation of information associated with the running application. There is a third layer, which enables the establishment/release of a session between the host computers. Beneath this third layer is a transport layer, which takes care of the quality of information exchanged in terms of acceptable errors, error rates, etc. The information is then subjected to a routing assessment by a network layer in order to figure out the best way to exchange information with the remote host. Based on the routing strategy established, a data layer prepares the message for transmission so as not to get lost or misdirected. The message is now ready for transmission over the designated route and is processed in the form of electrical signals or optical signals for transmission in a physical medium.

The family of LAN architectures is a subset of the aforesaid OSI (*open systems interconnection*) reference model which is set forth as a standard by ISO for networking between hosts made by any vendor. The legacy members of LAN architecture(s) which are subjected to the OSI reference model are the following multiaccess networks: *Ethernet, token-ring, and token-bus.*

Ethernet/ IEEE 802.3 standard

In *ethernet,* the stations are connected to one cable called a *bus* [1.15]. Essentially, it is a 10 Mbps baseband LAN. Its specifications were developed jointly by Xerox, Intel, and Digital Equipment. It consists of a collection of end-devices on a single linear bus. These devices (stations)

are required to share the network's transmission capacity. Some means of controlling access to the transmission medium is therefore needed to provide an orderly and efficient use of that capacity. In ethernet this medium access control (MAC) strategy is a low level protocol and is known as *carrier-sense multiple access with collision detection* (CSMA/CD).

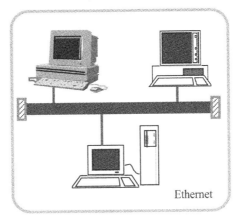

Fig. 1.41: An ethernet

When a station wants to transmit, it listens to the medium (or *carrier senses*) to check whether another transmission is in progress. If the medium is idle, the station may transmit; that is, an ethernet station waits for a clear line and transmits a packet, checking that the other end has received it. However, there could be more than one station trying to transmit at the same time. This condition refers to a *collision*. The MAC protocol not only provides for carrier-sensing, it also detects collision. If there is a collision, the station backs off and is allowed to transmit at a randomly chosen later instant, at which time the medium is likely free. Thus, the MAC protocol specifies what a station should do if the medium is busy and when there is a collision. For lightly used networks, the throughput of an ethernet increases with workstation load, but its throughput performance falls off at larger loads.

The IEEE 802.3 offers several media options compatible for 10 Mbps operation. They are: Unshielded twisted copper wire pair (10 Base-T), coaxial cables (10Base5, 10BASE2, 10BROAD36) and 850-nm optical fiber pair (10BASE-FP).

Token-ring/IEEE 802.5 standard
The IEEE 802.5 *token-ring* standard emerged from IBM's commercial token-ring product. A *token-ring network* (TRN) is a LAN topology and access method is controlled by passing a digital token along a wiring ring formed between connected devices (stations). Its operation is based on the digital token which is a small frame (of a specific bit pattern).

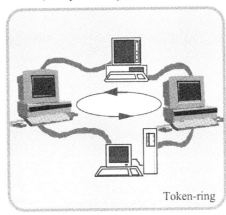

Fig. 1.42: Token-ring

The token circulates when all stations are idle. A station that wishes to transmit waits until a token passes by. It then captures the token and changes one bit in it. Now, the token becomes a start of frame sequence for the data ready for transmission at that station. The station also appends necessary postambles to form a complete frame.

Once a station avails of a token and commences its transmission, no other token is available for the rest of the stations. That is, while a particular station is transmitting, the other stations must wait until the token is re-issued before they can transmit. The transmitted frame makes one round-trip and is absorbed ultimately by the transmitting station. Thus, the nodes (stations) in the ring are offered service once at a time in a round-robin fashion. The IEEE 802.5 specifies the use of shielded twisted copper pairs of wires in the token-ring so as to support 4 and 16 Mbps.

Token-bus/IEEE 802.4 standard

Token-bus refers to a medium access control technique for bus and tree LANs. Stations form a "logical" ring around which a token is passed. A station on the bus receiving the token may transmit data then pass the token onto the next station in the logical ring as an invitation to transmit. This standard is intended largely for factory automation applications.

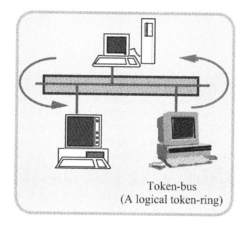

Token-bus
(A logical token-ring)

Fig. 1.43: Token-bus arrangement

ISO-OSI protocols ...

As mentioned earlier, the LANs were built on the ISO-OSI architecture which is layered. Layering is an implicit choice for signal flow in communication architectures. There are seven layers in this protocol as shown in Fig.1.44.

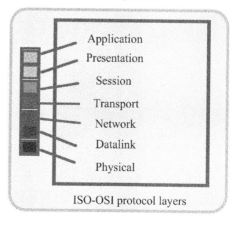

Application
Presentation
Session
Transport
Network
Datalink
Physical

ISO-OSI protocol layers

Fig. 1.44: ISO-OSI protocol architecture

The functional attributes of the ISO-OSI protocol layers are:

❑   *The application layer is, in essence, an application-specific programming interface*

❑   *The presentation layer performs terminal emulation and display format*

❑   *The session layer enables logical connection between applications*

❑   *The transport layer facilitates packet assembly/deassembly and oversees error corrections*

❑   *The network layer does the packet routing and decides a network topology mapping thereof*

❑   *The data-link layer takes care of the transfer of data between adjacent nodes*

❑   *The physical layer places bits on and takes them off the physical medium.*

The characteristic features of a layered protocol of communication is as follows:

❑   *The communication process is set up in a multilayer structure*

❑   *A layer is composed of subsystems or independent smaller autonomous computers and their associated software, peripherals and users wherein information processing and/or transfer may take place*

❑   *Communication from an upper layer [N^{th} layer] to the lower layer [$(N-1)^{th}$ layer] strictly follows a designated protocol*

❑   *Entities in the same layer (peer entities) may share and exchange information called peer protocols*

❑   *Data exchanged between peer entities are known as protocol data units (PDU) and data exchanged between adjacent layers is termed interface data units (IDUs)*

❑   *A layer may obtain services from a lower layer and may provide services to an upper layer; that is, a layer acts as a user as well as a service provider.*

In reference to LANs, the interface between attached terminal devices and the packet switching network, in general, has the task of coordinating the packet generation (and reassembing) at the device side and transporting such packet data across the network. The relevant protocol is imbedded in the universally-used standard namely, X.25.

In packet switching, X.25 identifies the user-network interface (UNI) and it influences the network design itself imposing an excessive overhead in order to separate the device-side and network-side functionality of the UNI. Therefore, a technology known as the frame relay facilitating a streamlined, higher throughput packet transmission was developed.

Frame-relay transmission

    Frame-relay refers to a variable length packet transmission. It is intended for cost effective applications in metropolitan LAN-to-LAN services. It maintains the cost-effective bandwidth and demand schemes of X.25, while providing a tremendous increase in throughput.

    The traditional approach to packet switching makes use of X.25, which not only determines the user-network interface (UNI) but also influences the internal design of the network. This approach calls for considerable overhead. Much of this overhead pertains to error-checking on a link-by-link basis. Such an error-correction is, however, unwarranted due to the fact that the available technology at the physical transport media level enables minimal bit errors.

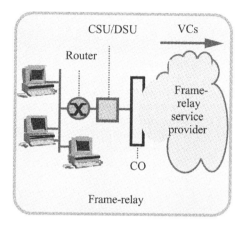

Fig. 1.45: Frame-relay system

The frame-relay is designed to eliminate much of the overhead that X.25 imposes on the end-user systems and on the packet-switching network. It involves a streamlined packet transmission with certain associated functions. These functions refer to eliminating error-checking and flow-control to realize greater throughput. It supports data rates up to 1.544 Mbps (DS-1 rate) which is much higher than the rate available with X.25 packet networks. A frame-relay transmission consists of a customer-premises network routed into a frame-relay service provider cloud through a local access line. Also, a conversion of unipolar bits into bipolar bits by means of a channel service unit/data service unit (CSU/DSU) (so as to be compatible for local access line) is done.

Metropolitan area networks (MANs)

With the flexible data network planning in vogue, the concept of a localized, in-house data communication through interoperation of computers was stretched to go beyond the restricted geographical distance — hence, what are known as *metropolitan area networks* (MANs) and *wide area networks* (WANs) came into operation.

Fig. 1.46: Metropolitan area network

MAN refers to a conglomeration of terminals distributed across a single city (for example teller-machines) whereas WANs serve many locations distributed over a large geographical area possibly extending across several cities. MAN operation is normally confined to an area of ten to a few hundred kilometers radius. It supports 1.5 Mbps to 150 Mbps.

Typically the links required for interconnections in MAN are digital telephone lines leased from the local telephone company. As necessary, microwave radio connections (wireless interLAN) and/or fiber distributed data interface (FDDI) network are also deployed. The IEEE 802.6 MAN is designed as a metropolitan utility serving a large number of organizations across an area of many miles. (The IEEE 802.6 Committee studies MANs and establishes relevant standards.) Its network topology is termed as *distributed queue dual bus* (DQDB) which includes two parallel runs of cable (typically fiber optic) linking each node constituted by a router of a LAN segment.

Wide area networks (WANs)

A WAN may extend through several PSTN clouds over a nationwide geographical span in the order of hundred to a few thousand kilometers. Transmission rates on WAN range from 1.5 Mbps to 2.4 Gbps. The interconnection is facilitated via PSTN leased lines.

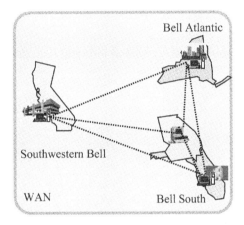

Fig. 1.47: Wide area network

Global area networks (GANs)

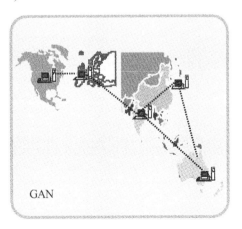

Fig. 1.48: Global area network

Global area networks are designed to operate between countries across the world. The transmission speed in GAN ranges from 1.5 Mbps to 100 Gbps. GANs are designed to serve growing business communication needs. Their interfaces are usually adapted to the regional and/or international standards so as to facilitate interoperation of different versions of equipment.

Transmission control protocol/Internet protocol (TCP/IP)

Transmission control protocol/Internet protocol refers to a set of protocols developed in the late 1970s by DoD/ARPA in order to facilitate interoperability among equipment manufacturers. In contrast to OSI, TCP/IP is not, per se, an international standard, through it is an open standard used worldwide. The Internet Engineering Task Force (IETF), which recommends suggestions through Internet Request for Comments open to the public, informally administers it.

The current Internet is a morphology of original APRANET. It emerged in 1989 and consisted of a collection of independent packet-switched networks which are interconnected as a coordinated system. No single body has overall control over the Internet, through it is used widely by educational institutions, governmental agencies, defense establishments, and commercial enterprises.

Basically, the Internet is used for:

✓ Electronic mail service
✓ File transfer support between hosts
✓ Permission granting facility to log on remote computers
✓ A means to provide user-access to information databases.

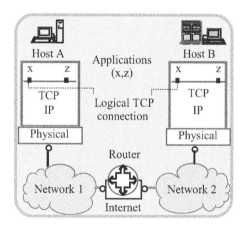

Fig. 1.49: TCP/IP protocol architecture

The layered architecture of TCP/IP provides a combined logic for routing through an Internet with end-to-end control. The associated protocol design, therefore, includes an IP for Internet routing and delivery, and TCP for reliable end-to-end transport. For applications where reliable delivery is not essential, an efficient *user datagram protocol* (UDP) is deployed on the top of IP.

Further, for real-time traffic, a new protocol called *real-time transport protocol* (RTP) is used on top of IP. The application services include *file transfer protocol* (FTP), *simple mail transfer protocol* (SMPT) and the TELENET which allows users to log on a remote computer over the network and operate as if they were directly connected.

An *address resolution protocol* (ARP) and a *reverse address resolution protocol* (RARP) facilitate address resolutions. The TCP/ID architecture further has the data link layer to accomplish the logical link control (LLC) and packet access. The physical layer supports ethernet, token-ring, fiber and the serial link. Ethernet is responsible in large part for the success of the TCP/IP protocol suite.

A number of new protocols have been added to the TCP/IP family (such as IPv6, resource reservation protocol (RSVP), and multicast routing protocols) so as to realize an integrated services architecture (ISA) for the internet to handle both traditional bursty traffic as well as multimedia and real time traffic.

High-speed LANs, MANs and WANs

Initial deployment of LANs had limited options on the speed of data transmission up to 1.54 Mbps. This soon became too small to handle large texts of file transfers not only in MAN and WAN environments but also in LANs. Therefore, high-speed LAN development became imminent and subsequently such high-speed transmissions were adopted in MANs and WANs as well. The high-speed LAN stretched the rate of transmission to beyond 100 Mbps with a trend set towards applications at gigabits per second. Fiber distributed data interface (FDDI) and distributed queue dual bus (DQDB) indicated before are high-speed systems. For similar transmission applications at Gbps rate, *high performance parallel interface* (HIPPI) has been developed.

Fiber-distributed data interface (FDDI)

FDDI is an ANSI prescribed 100 Mbps standard (ANSIX3T95) for fiber-optic networks. Essentially, FDDI is a token-ring composed of two counter rotating rings — a primary ring and a secondary ring that passes tokens and messages in opposite direction to that of the primary ring. FDDI is a high performance, general-purpose multi-station network. It is deployed both in LAN and WAN applications.

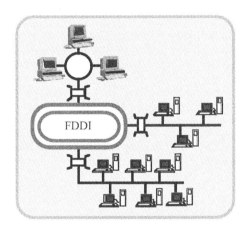

Fig. 1.50: Fiber-distributed data interface

FDDI is an ANSI prescribed 100 Mbps standard (ANSIX3T95) for fiber-optic networks. Essentially, FDDI is a token-ring composed of two counter rotating rings — a primary ring and a secondary ring that passes tokens and messages in opposite direction to that of the primary ring. FDDI is a high performance, general-purpose multi-station network. It is deployed both in LAN and WAN applications.

Distributed queue dual bus (DQQB): IEEE 802.6 protocol

Distributed queue dual bus metropolitan networks contain two parallel, optical transmission paths (Bus A and Bus B) operating in opposite directions. Again, this is a shared medium system governed by a medium access protocol. The two unidirectional DQDB buses to which the access points are connected have a head station (or head of bus station) called *slot/frame generator*. Each node is connected to each bus using a read and a write line. The reading of the bus does not affect the incoming data. That is, the information can pass through the nodes transparently.

The MAC mechanism here is based on distributed queueing. That is, the system (meaning any one of the buses) has logically one queue for data transfer in one direction. However, this queue is distributed physically over the nodes. The service discipline of the queue relies on *first-in first-out* (FIFO). The logical queues manage the traffic, one for each direction/bus. Each bus coordinates the queueing discipline for information transfer over the other bus. This effort involves collecting the requests of the nodes when these nodes have information to transmit.

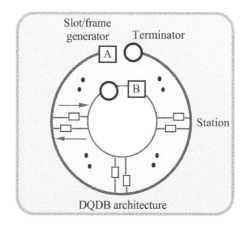

Fig. 1.51: Distributed queue dual bus metropolitan network

The double ring configuration allows reliable operation. Should any one of the buses fail, the nodes at the failure become frame generators. The data does not flow through the nodes, so the failed node can be removed from the bus without disturbing the communication.

Switched multi-megabit data service (SMDS)

Switched multi-megabit data service (SMDS) is a high-speed, connectionless (datagram type) public network deployed by LECs that operates similar to LAN linking applications in the MAN and WAN ambients. It supports 56 kbps to 45 Mbps in 1996 implementations. Bellcore on behalf of Bell operating companies developed it. It was introduced in 1991 and is widely operational in the United States.

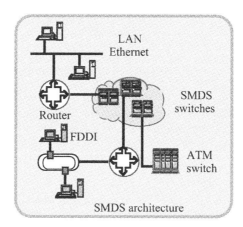

Fig. 1.52: Switched multi-megabit data service system

SMDS defines a service (not a protocol in itself) at the user-network interface (UNI) based on IEEE 802.6 standards. This UNI supports the data and physical layers. It can use any transport mechanism including ATM.

The goal of SMDS is to provide any-to-any connectivity for MANs, FDDI subnetworks, and private LANs. Sharing data is rendered not harder than making a phone call. It is intended to support multiple services (voice, data and video). SMDS is intended to facilitate an easy migration to B-ISDN and SONET. It is considered the "first ATM service", though it can be a stand-alone system sans ATM SONET. It differs from frame relay and ATM in that it supports no permanent virtual circuits. SMDS architecture is a layered one. These layers are SMDS interface protocols (SIP). SMDS is designed to handle bursty traffic. It offers short delays for widely spaced independent data bursts. Further, SMDS accommodates broadcasting. (ATM: *Asynchronous transfer mode* — A transfer mode adopted in the broadband integrated system of digital networks (B-ISDN). Later sections and chapters of this book elaborate on this transfer mode.

Typically SMDS connects together multiple LANs at the branch offices and factories of a single enterprise. It can act as a high-speed LAN backbone, allowing packets from any LAN to flow to any other LAN. The packet format of SMDS has three fields: Destination address, source address and variable user data payload up to 9188 bytes.

High-performance parallel interface system (HIPPI)

The *high-performance parallel interface system* (HIPPI) was developed to interconnect supercomputers and establish ways to interface peripherals with such computers. The objective on the data rate was to achieve 1600 Mbps.

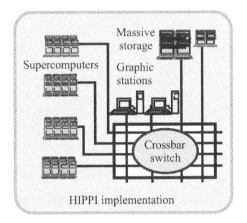

Fig. 1.53: High-performance parallel interface system

HIPPI was conceived as a *data channel*, that is, to operate between point-to-point or from a master computer to slave entities (such as peripherals) on dedicated connectivity with no intervening switching. The standardization committee (ANSI X3T9.3), subsequently offered standards on HIPPI covering the essentials on physical and data link layers. The networks so conceived largely meet the applications of distributed computing.

The supercomputer interconnections shown in Fig. 1.53 are enabled via SONET backbone. A typical HIPPI implementation is on the CASA network testbed established in 1990 to link research centers such as the Jet Propulsion Laboratory, the Los Alamos Laboratory and Caltech.

Fiber/Fibre channels

The *fibre channel* is a successor to HIPPI in high-speed family of LANs. It handles both data as well as network connections. It supports high-speed data channels such as HIPPI, SCSI (small computer system interface; SCSI — pronounced "scuzzy[*]") and multiplexer channels of IBM mainframes. It is also intended to support the network packets including IEEE 802, IP, and ATM. Basically, the fibre channel is a crossbar switch interconnecting the incoming and outgoing lines. Such interconnections can be done on a packet-by-packet basis or over a specified length of time.

---

[*]*"Scuzzy"* or *SCSI* is a high-speed parallel interface defined by the ANSI X3T9.2 Committee. It is deployed to interconnect a computer to peripheral devices using a single port. The devices so connected are said to be "daisy chained". SCSI has been a standard on Mac Plus, IBM RS/6000, IBM PS/2, and some higher computers. Further, it is often used to connect hard-disks, tape-drivers, CD-ROM drivers, other mass storage media as well as scanners and printers.

There are three classes of service designated for fibre channel: (i) Circuit-switched for data traffic with a guaranteed delivery in order, (ii) Packet switching with a guaranteed delivery; and, (iii) Packet-switching with no guarantee on delivery.

The protocol architecture of fibre channel is sectioned into two parts: Data channel section and network service section. Further, it has 5 layers spanning two physical layers at the bottom and three data layers on top. The first physical layer supports data rates of 100, 200, 400 and 800 Mbps and the second physical layer does bit encoding. The lowest data layer performs framing protocol and header layouts. The layer above this makes service primitives available to the interfacing layer above it to support computers and peripherals.

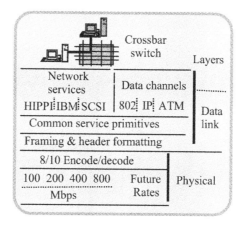

Fig. 1.54: The fibre channel system

Satellite networks

Communication satellite based WANs have multiple channel access feasibility. In general, communication satellites have a footprint of coverage on the earth ranging from 250 km (spot-beam coverage) to 10,000 km (wide-beam coverage). WAN stations within the beam coverage area send frames up-linked to the satellites; the satellites rebroadcast these frames down-linked towards their footprint areas. The uplink and downlink frequencies of transmission are distinct. The satellites are just "bent pipes" in the sense that they are passive relays; that is, they do not perform any on-board processing of the frames.

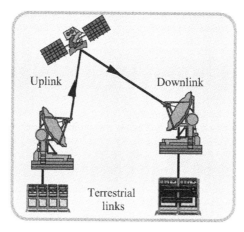

Fig. 1.55: Satellite communication

The antenna beam of the transponder in the satellite illuminates section-by-section of the coverage area and dwells at each of the sections over a brief time so as to facilitate access by the WAN stations in the illuminated section.

The communication-satellite based WAN stations lean towards multiple access; however, they cannot be subjected to CSMA/CD protocols. This is because a station can sense the status of a downlink channel (in order to detect collision within the duration of a few bits and withdraw its transmission thereafter) only after a propagation delay (of 270 ms) involved in the satellite-to-earth communication. Five classes of protocols are adopted in multiple access uplink transmissions. They are: Polling, ALOHA, frequency division multiple access (FDMA), time-division multiple access (TDMA) and code-division multiple-access (CDMA). They correspond to distinct channel allocation methods as described below.

Polling

Allocating a channel to one of the competing ground stations for uplink transmission to the satellite can be done via *polling*. A poll is an invitation from a centralized access point to a secondary (slave) station to transmit. The satellite can poll each ground station in turn subject to the 270-msec delay constraint involved.

An alternate approach is as follows: Suppose all the ground stations are tied together to a low bandwidth packet-switching network and the stations are arranged in a logical ring such that each station knows its neighbors in the ring. Now, a token is circulated among these stations. The satellite itself, however, does not play any role in the circulation of the token. A ground station gets the option to transmit uplink when it accesses the token under circulation. This method is efficient when the number of stations is not excessive and remains constant and the duration of message burst sent uplink is much longer than the circulation time of the token.

Slotted-ALOHA

This is a technique in which every ground station is aware of the available time-slot for uplink transmission. A reference ground-station periodically broadcasts a time-reference to which the time-slots are locked. Therefore, all the ground stations know synchronously when each time-slot begins.

FDMA

In this strategy, the transmissions pertinent to a ground-station are assigned specific uplink and downlink frequency bands within the allotted satellite bandwidth. The allotment frequency bands can either be *preassigned* or *demand-assigned.* The transmission separation in the frequency domain enables collision avoidance. Each width of the divided frequency band is known as subdivision and each subdivision has its own carrier frequency. A control mechanism is used to prevent any two ground stations from using the same subdivision simultaneously for the transmissions. In the demand-assignment technique, the control mechanism also establishes and terminates the links between the source and destination stations. As a result, the subdivisions are utilized only on an ad hoc basis. The transmission powers are carefully managed so that spill over of energy into adjacent channels via sidebands of the transmission spectrum and causing adjacent-channel interference is avoided. A typical FDMA arrangement involving N stations is illustrated in Fig. 1.56. Each FDMA transmission occupies distinct bandwidth in the frequency domain. Hence, the set of all transmissions shares the total system bandwidth as well as the total transponder power.

The FDMA scheme is used for a single voice channel (of 4 kHz bandwidth) per carrier (SCPC) as well as for multiple-channel per carrier (MCPC) supporting a group of voice channels. An example of the FDMA system is the SPADE DAMA (an acronym for *single-channel per carrier PCM multiple access demand assignment equipment*). This satellite system supports 794 simplex PCM voice channels (64 kbps) together with a 128 kbps common signaling channel.

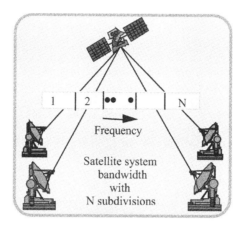

Fig. 1.56: FDMA scheme

Note: The system bandwidth is divided into N distinct satellite channel subdivisions

Comsat developed the first demand-assignment FDMA satellite system for use on the *Intelsat* series *IVA* and *V* satellites.

TDMA

This is the most popular version of multiple access technique. TDMA refers to time-division multiplexing digitally modulated carriers between participating earth stations within a satellite network through a common satellite transponder. Each ground station transmits a burst of information for a short duration in a specific time-slot (epoch) assigned within a TDMA frame as illustrated in Fig. 1.57. The transmitted bursts are synchronized so that the burst corresponding to each ground station arrives at the satellite at a different time. It can be noted that, each TDMA transmission refers to a brief duration in the time-domain, but it is spread entirely over the transponder bandwidth. Correspondingly, the entire power of the transponder is committed to each transmission but over a short duration only.

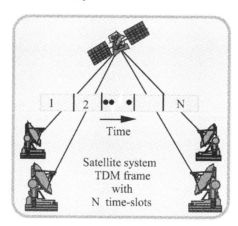

Fig. 1.57: TDMA arrangement

Note: Every TDM frame of the satellite system is divided into N station epochs/time-slots.

CDMA

CDMA is also known as *spread-spectrum multiple access*. It refers to a communication system in which multiplexing of more than one signal within a single transmission medium is done. The choice of the signal is dictated by a distinct code (called a *chip code*), only available at the sender and the receiver stations. That is, the communication process uses specialized codes as the basis of channelization and the transmission of signals through each channel is distinguished by the encoding associated with it. For this purpose, envelope encryption and decryption techniques are adopted in separating the transmissions.

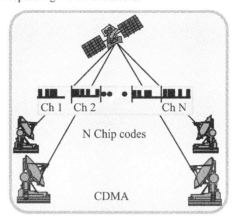

Fig. 1.58: CDMA scheme for N stations

In CDMA scheme, there are no bandwidth or time-slot allocations for the transmissions involved. All stations use the entire satellite transponder bandwidth on a continuous basis. That is, each signal transmission spreads throughout the entire allocated bandwidth.

A note on spread-spectrum ...

Spread-spectrum is a method in which a modulated waveform is modulated (spread) again in such a way that an expanded-bandwidth wideband signal results; however, this wideband signal does not significantly interfere with other signals. The implementation of spread-spectrum (SS) is independent of the type of information involved. Also, SS does not guarantee any immunity against the interference of additive white Gaussian noise on the signal. But, it offers reduced inter-channel interference, allows CDMA implementation, and facilitates a secured communication.

Scope of telecommunication support: Beyond LANs

Computer systems and terminal networks of data communication systems have come together to operate as a part of a total telecommunication system. This fusion has augmented the scope of telecommunication so as to deal with a heterogeneous set of traffic streams constituted by voice, data, and video. As a result, the evolution and growth of LAN technology as an in-house data communication facility, and the associated larger networks namely, MAN, WAN, and GAN intended to support mixed traffics of voice, data, and video, rely upon the flexibility that a network designer enjoys in developing a telecommunication network with multiple degrees of freedom. The design options, which go with relevant telecommunication technology, in this state-of-the-art scenario refer to:

Type of traffic supported:

      ✓   Voice
      ✓   Data
      ✓   Video.

Type of voice signal:

      ✓   Conversational/speech
      ✓   High-fidelity version (music etc.)

Type of video signal:

- ✓ Basic TV quality
- ✓ High definition TV quality
- ✓ MPEG-specified quality.

Type of data — message formats of data transmission characterized as:

- ✓ High density text data
- ✓ Low density text data.

Type of networking associated with:

- ✓ Switching centers (PBX family)
- ✓ Private data switching units
- ✓ Networks of mainframes
- ✓ Local area networks
- ✓ MANs/WANs/GANs.

Application — specifics of networking intended for:

- ✓ Office environment
- ✓ Technological applications
- ✓ Scientific uses.

Traffic characterization — nature of data transmission pertinent to:

- ✓ Continuous and deterministic nature of message under transmission
- ✓ Statistical aspects of the message.

Connection type — type of data transportation specified as:

- ✓ Connection-oriented
- ✓ Connectionless.

Nature of usage — description of the networking as a:

- ✓ Dedicated facility
- ✓ Shared facility with controllable access.

Topology — the way the networking is arranged:

- ✓ Star, ring, bus etc.
- ✓ Hierarchical networks.

Transmission media — the physical media of transmission constituted by:

- ✓ Copper wires, optical fibers, and wireless means.

Transmission attributes — characteristic parameters of the channels specified by:

- ✓ Rate of transmission, bandwidth, and error tolerance.

Standards — standardization in reference to:

- ✓ Architectures (manufacturer-specified)

✓   Protocols (network-specified)
✓   Technology details (organizations such as ISO, IEEE etc. -specified).

The scope of designing and implementing digital telecommunication systems in an expanded profile in supporting voice, video and data culminates in realizing an *integrated approach*, the concept of which is as follows.

1.1.7   Telecommunication- adolescence into integrated services

With the advent of developments in digital communication and experience gained through packet-switching networks across private and public domains, a new protocol system was developed [1.16] to support data, image, video, and voice communications efficiently and in a unified manner. This is known as the *integrated services digital network* (ISDN).

Fig. 1.59: Narrowband integrated services digital network

The concept of integrating the services came from the kaleidoscopic spectrum of information that telecommunication has to cope with. Such a spectrum ranges from voice over the telephone to multimedia transmissions and within this span are multitudes of information pertinent to data/text, graphics, low and high resolution fax, still-pictures, and video presentations.

During the 1980s, data transmission across the world grew at a rate of about 15-25% per year and voice traffic grew 3-5% per year. Total volume of data traffic surpassed that of voice in the 1990s thanks to interoperative computers. Integrating voice and data via ISDN was the original solution to cope up with the competing data and voice. Though ISDN was proposed in the 1970s, its real need was felt only in the 1980s and 1990s.

Each one of these needs a distinct transmission speed and occupies a minimum extent of frequency spectrum specified as the bandwidth. Such speed and/or bandwidth differences are concerns of importance only in the engineering and technological aspects of telecommunication. For the user such technical differences are not of importance. The subscriber is satisfied as long as:

✓   The message passages through the intervening medium to the destination in its natural form
✓   Such a transmission is in real-time; or, at least with only a tolerable delay
✓   Regardless of the type of message (voice, video, text etc.) the communication is facilitated by the same set of procedures.

In other words, despite the vagaries in applications and in the related messages involved, the telecommunication system is required to provide a *transparent service* to the users. The users are not to be involved in the technical aspects in facilitating a *semantic transparency* (that is, the message being retained in its original form in the transmission) and a *temporal transparency* (that is, the transmission approaching the real-time basis). Further, the procedural tasks performed in transporting any type of message have a uniform protocol structure. These can be achieved by the ISDN. The constituent parts and goals of ISDN are as follows.

ISDN ...

*I*ntegration of services ...

 This refers to integrating the different user-functions in a single transport medium — that is, whether be it a telephone dialing by a subscriber (for a conversation with another person), or, a user invoking a multimedia transmission to access a remote destination, the "integration" aspect of ISDN facilitates a single procedure to establish (or release) a call, and finds an appropriate route for the transmission of the message(s) involved. Only a service identification is needed to distinguish the individual services involved. Further, despite the integrated transport of message transfer, the semantic and temporal transparencies are preserved.

*S*ervices integrated ...

 As indicated earlier, the ISDN is intended to support a gamut of services with distinct information attributes. These services when handled in an integrated manner, are to be managed by the associated networking and the terminals involved cooperatively. Further, these coordinated efforts should remain transparent to the user while seeking a connection regardless of the end-application involved (say, voice transmission from a telephone or data transmission from a computer).

*D*igitized transmission ...

 The method of transmitting different types of information/message in an unified manner over a single transporting network calls for translating the different messages into a common format. This is achieved by digitizing the message of any type into a stream of binary digits (bits). The electrical and/or optical versions of these bits from different sources (applications) are time-multiplexed and transported over the medium. Further, the bit stream of each source (or chunks of it) carries an identification so as to be distinguished and segregated as necessary. The segregated bits of a given information-transmission can be converted back to its original format that can be understood by the end-equipment. Modern microelectronics allows this digitization applicable to any form of message involved and has made the ISDN a success.

*N*etworking ...

 ISDN implementation relies on the emergence of networking feasibility which can support different rates of (low-speed or high-speed) information transmission. Also, the developments in prescribing a unified and organized set of protocols which are "open" to include different applications, the associated end-equipment and network services rendered by different operators have made the ISDN a reality.

 The need for ISDN can be perceived by considering the multiple networking status that the telcos faced: POTS and Telex use the classical circuit-switched networks. The data communication such as frame-relay and SMDS warranted its own packet-switching strategies. The so-called DQDB required yet another network and the internal telco call management had to be done by another network called SSN 7. Maintenance of these different networks is a formidable effort and, as a result the single networking concept found its way to support all kinds of information transfer. As stated before, this single networking on an integrated basis allows enormous transmission rates and also has permitted a variety of new services.

 Initially, the ISDN implementation was a narrowband system and hence was called N-ISDN. Its primary goal was to support voice and nonvoice information on an integrated basis. For this purpose, the classical analog telephony was turned into a digital telephony. Next, a worldwide acceptance of an interface standard on ISDN devices/components and equipment came into being, though at a slow pace. The main reason for the slow adoption of ISDN was that the multiple applications to be handled in an integrated fashion, required a much larger transmission rate than that conceived at N-ISDN level. The original ISDN bit-pipe was standardized to carry three standard rates called basic rate, primary rate *and* hybrid rate *interleaved by time-division multiplexing. It was soon realized that these three classes of transmission rates were insufficient to support high-speed transmission of data (such as 10 Mbps and 100 Mbps rates of LANs) as well as even higher rates required in video transmissions. Thus, CCITT came out with a broadband system known as B-ISDN. The broadband service implies transmission channels capable of supporting rates greater than 1.5 Mbps, or a primary rate in ISDN or T1/DS1 in digital technology.*

ISDN uses three types of connections:

- ❑ *Circuit-switched connection*
- ❑ *Packet-switched connection*
- ❑ *Semi-permanent connection equivalent to a leased line. When it is established, it remains available.*

ISDN also defines three categories of end-to-end digital telecommunication services to support voice and nonvoice information transfer:

- ❑ *Bearer service intended for basic transfer of information between users. It offers services in packet-switched mode for data networks and in circuit-mode to substitute the conventional telephone circuits.*
- ❑ *Teleservices depicting higher level functions such as facsimile, videotext, telex, and teletext*
- ❑ *Supplementary services corresponding to additional functions such as call waiting, conference call, caller ID etc. as would be required in both bearer and teleservices.*

1.1.8   Telecommunication - maturity towards broadband integration

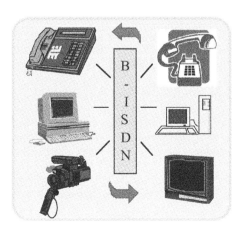

Fig. 1.60: Broadband ISDN

B-ISDN is basically a digital "virtual" circuit that can transport fixed-size chunks or packets of a message (called *cells*) from source to destination at rates of transmission higher than the primary rates so as to support services like video which have invariably high bandwidth requirements. In essence, it was the outcome of a motivation to develop a single type of network capable of supporting narrowband as well as broadband traffics.

B-ISDN refers to a second generation ISDN offering very high data rates to end-users. Building an integrated telecommunication system supporting such high data rates is the answer to the problem. At the same time it unifies those "too many networks" and "too many services" in vogue. B-ISDN was, therefore, conceived to be a *service-independent* system meaning that it is capable of supporting any of the telecommunication services of the past, present, and future. It is further intended to share all the resources that exist between services. It is supposed to be a flexible system that adapts itself to changing trends in communication and circuit technologies. Also, it has to follow a set of well-defined standards and specifications.

The lethargy that was perceived at the inception of ISDN intended to support only those primary rates and narrowband services in the 1970s, and has not affected the proliferation of B-ISDN due to the following reasons:

- ✓ The ubiquitous computer usage and tying several of them together in day-to-day applications have become a part of today's life-style
- ✓ The variety features of telecommunication messages in the form of voice, data, and video have necessitated the need for broader bandwidths
- ✓ The volume and richness of data, not just limited to alphanumeric transfer, but also enclaving the colorful and high-resolution versions of graphics have become a commodity of teleinformation transfers
- ✓ Computer application awareness and multimedia feasibility have turned the business community towards accepting the best that telecommunication can offer
- ✓ The exponential growth in microelectronics and fiber optic technology have encouraged implementation of B-ISDN
- ✓ The milestones reached by the information highway in recent times have become the pinnacles of coordinated developments in communication and computers. There is, indeed, a computer-communication fusion that has established the realization of B-ISDN
- ✓ The crowning aspect of B-ISDN is its inherent digital technology. The art of processing digital information has a wealth of theoretical base, electronic support, and implementable feasibility. Therefore, B-ISDN has seldom faced a bottleneck in blooming as a viable strategy.

In short, service independence, flexibility, efficiency, and standardization have promoted the B-ISDN as a future-safe telecommunication system. But the question is: *What is the target architecture to be pursued?* In 1990, CCITT selected the so-called *asynchronous transfer mode* (ATM) as the designated architecture for B-ISDN [1.17-1.50]. And now, *what is ATM?*

1.1.9   *Telecommunication -a vitalization through ATM*

ATM is a switching technology that uses small, fixed-size *cells* towards a fast, packet-oriented transfer mode for asynchronously multiplexed heterogeneous traffics. In 1988, the CCITT designated ATM as the transport method (or "mode") for the B-ISDN services. ATM "has the potential for replacing the many conflicting technologies that must now be integrated into a cohesive whole" [1.37].

The goal of ATM is to carry all forms of traffic (data, voice, and video) over one switching fabric based on the cells. ATM multiplexes and places the cell-streams over a single, bearer-line or carrier facility. Multiplexing is done with the cell addresses. That is, each user or function is assigned a unique cell address on each link. Further ATM depicts a single communication technology that can operate over short distances using typical LAN technology and over long distances by WAN technology.

How did ATM come into being? As discussed earlier, in the 1960s, a global effort was envisaged to upgrade the PSTN by converting the then all-analog telephones to digital systems, which can support both analog and digital transmissions. As indicated before, the North American effort resulted in the 64 kbps 24 channels multiplexed on a single trunk supporting 1.544 Mbps (T-1 carrier system). Correspondingly, a digital signal hierarchy was devised with 32×64 kbps = 2.048 Mbps in Europe. The availability of low-loss fibers in the 1970s led to fiber-optics based telecommunication facilitating high-speed transmissions. Thus emerged the international standards on optical signal levels in 1984.

An immediate result of the availability of broadband physical transmission media was ITU's series of recommendations for a broadband integrated services digital network. The optical data rates, synchronization and framing formats chosen for the B-ISDN became known as the SDH in Europe and SONET in the United States.

With the availability of broadband physical media plus the associated set of world-wide standards, it became easier to conceive a universal multiplexing and switching mechanism to support integrated transport of multirate traffic. That is, feasible aspects of supporting the diverse requirements of multiple-bit-rate traffic sources and providing services in a cost-effective and efficient manner began to appear. The result was the ATM technology.

ATM is thus not a stand-alone support. It runs on top of highly scalable physical layer protocols namely, fiber channel and wide-area based protocols of SONET and SDH. In the context of present-time telecommunication, ATM is also intended to support the copper-wire (as well as wireless) based information transfers at the existing rates.

Each ATM cell (fixed-length packets) consists of 48 octets (bytes) for information field and 5 bytes for *header*. The header is used to identify cells, which belong to the same source and route them through the same *virtual channel* (VC). Thus, the cell sequence integrity is preserved per virtual channel at all times. ATM relies on frame technology to encapsulate its cells. SONET and SDH are the most popular framing methods adopted.

ATM, in short, is a breakthrough in the wireline network technology that is capable of providing interchangeable support for distinctive services and it is also being inculcated into wireless systems. Further, the success of B-ISDN through ATM has reduced the severity of distinction between isochronous (voice and video) services and asynchronous (data) services in the transport system envisaged.

ATM is being accepted universally as the standardized transport vehicle of B-ISDN; that is, as the logical transfer mode of choice for B-ISDN. As stated before, it can support any kind of information, namely voice, data, image, text and/or video in an integrated manner. It also provides good bandwidth flexibility and can be used efficiently at desktop computers as well as in LAN and/or in wide area network environments.

A factor that led to a rapid acceptance of ATM technology was that the concept behind ATM had already been envisioned in data communications when the cell-based DQDB protocol was standardized for the IEEE 802.6 MAN protocol and its corresponding service SMDS.

ATM is a *connection-oriented packet-switching technique*. No processing related error-control is done on the information field of ATM cells inside the network; rather such efforts are carried out transparently in the network.

Legacy LANs like ethernet, token-ring, and token-bus are limited in speed (10Mbps) and are therefore, limited to specific types of applications, mostly data. For applications such as video transmissions and multimedia applications, the bandwidth requirement is high. Further, the pertinent traffic is a composite of information posed by the heterogeneous presence of voice, video, and data. As a result, it requires a unique transfer mode capable of transporting and switching these different types of information across the network. ATM satisfies this requirement.

Strictly speaking, if all devices communicate using ATM, only one network will be needed for all forms of information. At physical transmission levels, this however, amounts to a global wind up of copper wirelines and an eventual transition to fibers in conjunction with wireless methods, if needed.

In summary, ATM technology facilitates B-ISDN implementation with the following technical considerations. (More details are furnished in Section 1.3, and the rest of the chapters elaborate ATM telecommunication comprehensively.)

✓ It supports all existing services as well as other emerging services by means of an ATM adaptation layer (AAL) that encapsulates various services, thereby offering service-specific utility

    ✓   It makes use of network resources very efficiently. That is, ATM is connection-oriented and the establishment of the connections includes allocation of a *virtual channel identifier* (VCI) and/or a *virtual path identifier* (VPI). An associated task also involves allocating the required resources on the user access and inside the network. These resources, expressed in terms of *throughput* and *quality of service* (QOS), can be negotiated between the user and the network either before the call-set up or during the call in progress

    ✓   It minimizes the switching complexity. The header values, that is, VCI, VPI assignments etc. are done during the connection set-up phase and translated when switched from one section to the other

    ✓   Signaling information is carried on a virtual channel separate from the user information

    ✓   In routing efforts, there are two types of connections involved: Switching on cells is first done on virtual path connection VPC and then on virtual circuit connection VCC

    ✓   ATM minimizes the processing time at the intermediate nodes and supports very high transmission speeds. This is chiefly because ATM employs the fixed-packet size scheme. Therefore, all the overhead that occurs in handling variable-packet sizes is eliminated

    ✓   ATM minimizes the number of buffers required at the intermediate nodes to bound the delay within a negotiated limit and the complexity of buffer management is reduced

    ✓   ATM further guarantees performance requirements of existing and emerging applications.

In short, ATM can be defined as a fast, fixed-sized, packet-oriented transfer mode based on asynchronous switching and multiplexing of composite traffics. It is an emerging broadband technology and is targeted for global implementation including wireless strategies. It has vitalized telecommunication and has forged the information technology to meet the user demands of the forthcoming century.

1.1.10   *Telecommunication comes of age*

Summarizing the past and present tale of telecommunication, it had always been "the best of times" with telecommunications, which has perennially enjoyed the snuggling of technical advances and achievements, and there had never been any "worst of times" either while trading telecommunication as a service commodity. In all, telecommunication technology has been dwelling in the Wonderland of Alice, where the Walrus always mutters " the time has come" for better changes.

The evolution of modern telecommunication to what it is today, is a logical consequence when viewed in terms of relevant information transmissions implemented. These transmissions in their bits and pieces can be associated with various need-based teleservices catered to the customers. Therefore, the growth of telecommunication in the past and the technological pursuits undertaken in the modern information highway can be viewed in three phases of associated service categories:

    ✓   From telegraphy to digital telephony
    ✓   From digital telephony to integrated services digital networks (ISDN)
    ✓   From ISDN to ATM and beyond.

Essentially, concurrent to the telegraphic era and in the subsequent proliferation of telecommunication, the associated technology can be regarded as a step-response to the following services extended chronologically to the public as utilities of mass communication:

Telegraphy to digital telephony ...

The enthusiastic prelude to telecommunication namely, the "magnetic telegraph" was indeed a masterstroke in mass communication. It led people to realize the following: "After all, the man whose horse trots a mile in a minute does not (very likely) carry the most important message" [1.2]. Hence, the telegraphy when introduced in the 1840s soon became superior to mail-carriers who relied on horses and mules. With the concept of electromagnetism, when extended further to telephony, people witnessed a miracle messenger who carried their "voice" on the galloping stallion of electricity. The astonishing speed of such electromagnetic-based information transfer culminated into a viable technology. Thus, the speed of message transfer became the seed of modern telecommunication. A summary of infantile evolution concerning the speedy way of sending a message via electrical means is as follows:

- Transportation of telegraphic messages in the form of Morse code words at a very low speed as decided by the keying ability of an operator

- Transportation of messages in the form of machine-encoded alpha-numeric characters at a low speed (conforming to a humble transmission rate of information of 300 bps or even less!)

- Two-way voice communication of plain old telephone service (POTS) via public switched telephone networks in analog and/or in digital formats.

While the speed of communication through principles of electricity and magnetism brought the world closer together, the concept of computer-based communication enhanced the scope of telecommunication beyond speech-signal transmission. Specifically, the later-day computers that emerged have powerful applications far beyond their calculating abilities. They perform interactive real-time simulations, do graphics, store and process data and build images. Therefore, they posed a need to redefine and enhance the scope of electrical communication — as a means of interaction between machine-to-machine and, between machines-to-persons. Such a communication is imminent to make best use of the application-potentials of computers by the users. Hence emerged a blend of electrical communication methodology that includes conventional voice telephony and computer-based information transfer. It has given the telecommunication system a profile of mixed traffic — voice, data, and video.

As discussed earlier, the status of telecommunication in the 1970s and during the early part of the 1980s, refers to the introduction of digital communication technology and large-scale implementation of T-1/E-1 and X.25 legacies. Of these, T-1/E-1 carrier hierarchy still plays a role in telecommunication and will continue to enjoy the patronage of service providers. Its support capability on transfer rates from 1.544 Mbps to 2.048 Mbps is sufficient for most user applications.

The data traffic of the 1970s and 1980s was managed effectively with the protocol strategies of X.25. With the changing trends in computers to support more powerful applications, however, warranted transmission capacities beyond T-1/E-1 rates and protocol architectures involved more than X.25. More so, mixed traffic handling posed its own technical intricacies. Therefore, in the 1980s and later high-speed telecommunication systems and integrated systems of digital networking came into being.

Digital telephony to ISDN ...

The extension of digital transfer of information to include voice plus data (and video) set the need to expand the telecommunication technology in respect to:

- ✓ Rate of transmission (in terms of bits per second)
- ✓ Networking considerations (switching technique and geographical outlays)
- ✓ Integrated transfer of heterogeneous bit streams belonging to voice, data, and video traffics.

Speed became a vital issue when the advent of high capacity LANs became operational in the 1980s. The conceived strategy included faster bit transfers (such as 45 Mbps and 145 Mbps) facilitated through compatible switching and physical transport media; further, the networks were interconnected to share resources judiciously. Also, the telcos facilitated stretching the high-speed/large bandwidth connectivity across the service cloud. In essence, the relevant developments perceived refer to:

✓ Computer data transmission in private and public domains through circuit-switching and/or by packet-switched data networks
✓ Interconnectivity of LANs for sharing the resources
✓ Implementation of high-speed physical media such as fibers
✓ An integrated approach in rendering networks to support a heterogeneous set of traffic streams constituted by data, voice and video information. (This, as indicated earlier, refers to the ISDN).

These developments in the 1980s, however, led to compatibility hardships with growing user applications, vagaries in vendor equipment, and diverse service providers. Therefore, in the late 1980s, efforts to address these hardships and propose future strategies emerged.

ISDN to ATM and beyond ...

ISDN services through broadband networking was conceived towards:

■ Expanding the scope of N-ISDN by enhancing the rates of transmissions of the constituent service categories; that is, realization of broadband ISDN

■ Conceiving a new mode of information transfer for B-ISDN namely, the asynchronous transfer mode (ATM)

■ Adopting a new generation of wideband personal communication service and realizing a variety of wireless computer networks.

Consistent with the evolutionary aspects of telecommunication as above, the promise of a technologically superior information-highway is synonymous with realizing a high-speed, ubiquitous and seamless integrated services information infrastructure. The associated strategy is to overcome the inadequacy of classical telecommunications and computer network paradigms.

The classical telecommunication architecture is a simple paradigm of POTS. Its access is managed by a dumb telephone apparatus (powered by the network) which controlled the dual function, namely, interchanging of signaling messages between the end-switches and routing the transport of voice information. The service-provider and the subscriber enter into a contract so that the subscriber pays the service-provider for the calls on the basis of call-duration and the geographical separation between the calling and called subscribers.

Improvements and standardization on classical telecommunication architecture aimed at providing for interoperability of end-systems, network servers, and network switching nodes in an open environment. Further, information access and transport technologies (pertinent to digital transmission and switching) have uttered a single *mantra* in unison: *Enhance the bandwidth and facilitate high bit rates.* This slogan can be tied to the quest of integrating services and enabling the users to have a desired bandwidth on demand and bit rate of their choice.

Compared with the twisted pair of copper wirelines, fiber optic channels have enabled enormous bandwidths at reasonably attractive costs. Almost a thousand-fold enhancement in bandwidth realized through fibers has totally changed the associated technology trade-off in telecommunication. But concurrently, such changes have warranted radically different network designs consistent with:

■ *Speed of transmission*: State-of-the-art data networks are based on the promise that *transit time*, namely, the time of transit of a given bit across the physical medium between the end entities, is small in relation to the times expended for processing and queueing. However, data can be placed onto a gigabit network so fast that transit time may become comparable and even larger than the processing and/or queueing times. For example, a megabyte-sized file may be queued in a gigabit network to an extent of ten milliseconds, and if this file has to be transferred coast-to-coast, the transit delay would be about twice as long

- *"On-the-fly" paradigm*: Conventional networks operate slowly on the incoming bit-streams while accommodating them in buffer storage (though temporarily) to examine "on-the-fly". Relevant examinations are pertinent to underlying features concerning dynamic route computation. That is, the precise path that a given message follows in a network is decided at intermediate nodes. But in gigabit transmission endeavors, the volume of data is so large that the time available to perform the paradigm of "on-the-fly" calculations (through store-and-forward operation and dynamic routing) is insufficiently small. Therefore, alternative design options on message routing has become inevitable

- *Economics of telecommunication networking*: Realizing a high-speed transmission with optical fibers burden the gigabit networks economically in regard to implementing the channel sharing among many users (that is, multiplexing). While such multiplexing efforts can be handled comfortably via electrical, microelectronics-based switching hardware, the technology for switching light pulses is in an immature stage. Hence, the associated technology in the present time, poses an economically challenging research agenda.

Yet, realization of gigabit networking with a background support of optical fiber transmissions has paved a path that refers to the emergence of ATM. Further, burgeoning societal demands and market prospects have pushed corporations and carriers to deploy broadband networks profusely. In addition to the traditional data needs, enterprise networks are surged with bandwidth consuming multimedia applications (such as imaging, video, digital voice collaborative computing, and Internet access). Reducing, if not eliminating the congestion in the networks plus improving the reliability and self-healing have become design constraints in facilitating transmission demands of digital voice, image, and data. High-speed networks offering a desired quality of service and bandwidth on demand have become the target architectures in telecommunication.

1.1.11   Telecommunication - the technology of wireless services untied by wires

Wireless telecommunication is a top-notch technology of modern times. *Cellular telephones, paging, mobile radios*, and *personal communication systems* (PCS) are constituents of the state-of-the-art wireless system which are growing into "easy-to-use" communication networks and have been proliferating profusely across the public. Wireless telecommunication in a free sense refers to a global ubiquitous wireless network which can permit its users to communicate with anyone, anywhere and at any time. Wireless access points can connect wandering users to wireline networks as well as to other wireless users. The access to wired infrastructure, in general, is provided for wireless/mobile network users via centralized access points.

The evolution of wireless telecommunication can be traced on the basis of its stratified generations. The classical era (termed as the pioneer phase) from 1921 to 1927 set the gears in motion to facilitate land mobile communication. The first experimental study refers to using mobile radios in police cars in Detroit (in the 1920s). That system used the 2 MHz RF band. In 1934 several municipal police radios were placed in use serving more than 5,000 police cars. The FCC assigned 29 channels in the EM spectrum exclusively for police mobile radios. Until the early 1930s these mobile radios operated on the amplitude modulation (AM) principle. Later, frequency modulation (FM) mobile radios were found to be more resistant to electromagnetic (EM) propagation problems, and by the 1940s all police mobile radio systems in the United States became FM-based.

The scope of mobile radio applications was dramatically enhanced thanks to implementation of such systems on a large-scale basis across the world during World War II for military purposes. Strides in performance, achieving reliability, and realizing cost-effectiveness that were attempted in those war-time developments led ultimately to a very successful mobile communication system market in the post-war period.

Subsequent to World War II, the technological pursuits of mobile communication (during 1946 through 1968) refer to the first commercial phase. During this period, the demand for a mobile wireless system grew beyond the previous, constricted use of such systems in the police and in the military. Many civilian applications came into existence; as a result, there was inevitable congestion posed in utilizing the available EM spectrum. Therefore, efforts were concentrated in multiplexing the channels and adopting a network-based centralized routing of messages, akin to the wireline systems, which were then in vogue. By 1949, mobile radio became a new class of telecommunication service, and in the two decades that followed, the mobile telecommunication user population in the United States alone exploded at least by an order of magnitude.

As a result, the mobile telephone service (which became a part of PSTN even in the 1940s) took the identity as a commercial enterprise with AT&T as the service-provider in specified locales in the United States under a license from the FCC. These services were operated in the VHF band (at 150 MHz). However, these were eventually operating in the UHF band (at 890 MHz) using the FM technology in the middle of the 1950s accommodating a FM bandwidth of 30 kHz for each voice transmission. Again, in order to serve multiple users, multiplexing (trunking) of a group of radio systems was the strategy that was used.

The original systems of wireless telecommunication were based on the radio broadcasting model (with a high-power transmitter placed at an elevated location so as to serve the mobile units over a large area. The concept of having several stations (each of smaller RF power transmission capability and designated to serve only a small area called a *cell*) became popular eventually. In this "*cellular*" system, the same frequencies used for a set of channels are reused in other cells as well. This *frequency reuse* strategy is done with minimal channel interference across the cells.

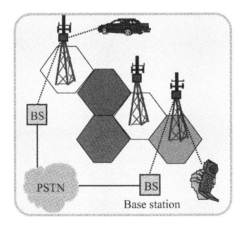

Fig. 1.61: Cellular telephone system

Further, the cells can be split judiciously into smaller cells in the event of increases in the user population per cell. When a mobile user goes from one cell region into another, the service responsibility is shifted from the first cell to the next one by means of a central *base station* control. This is called *hand-off*.

Thus emerged an organized cellular (analog as well as digital) telephone technology. In the early 1980s, Bell started operating a high-capacity mobile telephone system (HCMTS) in the FCC allotted band of 40 MHz in the 850 MHz spectral region. This became the forerunner to the so-called advanced mobile phone service (AMPS) of the 1980s through the 1990s constituting the first generation of commercial analog cellular telephone systems. Parallel developments in Japan and Europe emerged in the span of 1978 through 1986. The UHF bands adopted are 870-960 MHz and 453-468 MHz and the number of channels serviced range from a few hundreds to a couple of thousands.

The second generation of mobile telephone system corresponds to digital cellular systems that came into being to meet the popularity of cellular telephones in the 1990s. Going in for digital cellular systems has definite advantages. For example:

- State-of-the-art advances in digital modulation techniques facilitate high-performance (in terms of spectrum utilization) of cellular telephones

- More voice channels on a single carrier can be accommodated thanks to developments in lower bit-rate digital voice encoders

- The digital technique allows reduction in overheads required for signaling (call set-up etc.)

- To meet the challenges of harsh, EM wave propagation environments faced by mobile systems, robust schemes have been developed for digital source and channel encoding strategies

- Digital techniques have also been developed to reduce the co-channel and adjacent channel interference encountered in cellular telephone systems

- Digital schemes can be devised to accommodate flexible bandwidths

- Access and hand-off techniques are also handled efficiently through digital methods.

The first generation AMPS was confined to a narrowband standard. The second generation of digital cellular phones conforms to three major standards, namely:

- ✓ *Group Special Mobile* (GSM): (also known as *global mobile system*). This is European and international standard and the mobile unit to the base station link operates at 890-915 MHz band
- ✓ *IS-54*: North American Digital Cellulars (NADC) standard operating at: 824-849 MHz (mobile-to-base) and 869-894 MHz (base-to-mobile)
- ✓ *Japanese Digital Cellular* (JDC) standard operating at: 810-915 MHz (mobile-to-base) and 940-960 MHz (base-to-mobile).

The third generation of cellular telephone system refers to a contemplated service that stretches into the century in implementations and operation. It was conceived to include the state-of-the-art advances in FDMA, TDMA, CDMA, and collision sense multiple access (CSMA). Further, *spread-spectrum* considerations will play a significant role in the technology envisaged.

Wireless telecommunication systems other than cellular telephones ...

These refer to: (i) paging; (ii) private mobile radio (PMR); (iii) satellite mobile systems and (iv) personal communication systems.

Paging

A simple pager is a wireless system that notifies a called-party via an alarm (beep) or a defined voice plus an alphanumeric display that someone (the calling-party) is interested in talking. It is essentially a one-way traffic. It could be a private (that is, local premises-oriented, as in hospital paging systems), or it could be a public (wide area) paging system. The first generation of paging system was introduced in the late 1950s and early 1960s. The windows of frequency band adopted for paging lies in the range 80-1000 MHz. Paging service is offered to several million users.

A modern version of paging refers to *smart pagers* (for example Motorola Pagewriter[TM] 2000). This allows two-way communication to send and receive word messages. It is claimed to be smart enough to let one communicate with most anyone else having a pager, Internet, e-mail address, or fax. It is also claimed that it is easy enough to create messages on the go with a full keyboard. Another claim is that, it is direct enough to respond from almost anywhere making it the most personal form of two-way communication.

Private mobile radios

These refer to a fixed base-station serving a number of mobile units for private user applications (for example, to direct taxies to a customer). These operate in the VHF/UHF bands. The transmitter power could be in the range of 5 to 25 watts. Both FM and AM strategies are used.

Satellite mobile systems

These are useful means of communication for long-distance travelers. Also, they are adopted in aircraft and ship navigational purposes. National Aeronautics and Space Administration (NASA) through ATS-6 satellite undertook the premier efforts on mobile satellite communication. In 1979, the INMARSAT (International Maritime Satellite) organization was setup to establish worldwide aeronautical satellite communication standards governing the telephone and telex services (Standard A), ISDN (Standard B), and low data rate services (Standard C). Since 1983, the International Civil Aviation Organization (ICAO) is also involved in the related activities.

The other satellite-based mobile systems include:

    ✓   *Radio determination satellite system* (RDSS) of Geostar: This is a radio navigation/radio location system managed via a single satellite for U.S.-based domestic applications

    ✓   *OmniTracs* system of Qualcom Inc.: This is a two way mobile satellite communication system intended for vehicle positioning and is adopted in the United States and Europe

    ✓   *MobileSat*: This is an Australian system which supports services such as circuit-switched voice, and data/packet-switched data transfers for land, aeronautical, and maritime users

    ✓   Telesat Mobile Inc. (TMI)/Americal Mobile Satellite Corporation (AMSC) systems: These systems provide mobile satellite services in the United States and in Canada.

Personal communication systems (PCS)/Personal communication networks (PCN) ...

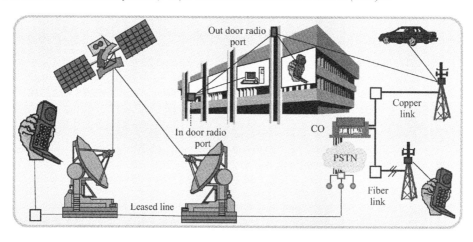

Fig. 1.62: Generic PCS framework and its global connectivity aspects

These are *location-independent* communication systems as illustrated in Fig. 1.61. PSN/PCN systems allow freedom of communication for any type of information between any two points. The locales of end-entities can be indoor/outdoor, in a mobile unit, rural areas with sparse population, crowded metropolitan ambient in an airplane, or at sea. The end entities can be standstill or be moving at jet-speed. The separation between the end-entities could be arbitrary. PCS/PCN is an emerging global system with international connectivity.

A standing committee on ITU, namely the World Administrative Radio Conference (WARC) coordinates the future public land mobile telecommunication (FPLMTS) systems so as to develop a global system of PCS/PCN for aggressive deployment in the upcoming century. Some of the challenging aspects of the associated efforts are:

         ✓   Judicious use of available EM spectrum
         ✓   Choosing of appropriate technology taking into account of spectrum scarcity
         ✓   Poising the usage demands and services to be offered.

TDMA techniques together with SS methods and CDMA strategies are targeted for use in PCS implementations. For example, in order to enhance the traffic capacity of PCS, digital multiple access methods such as time-division duplexed TDMA, frequency-division duplexed TDMA, slow frequency hopped TDMA/CDMA etc. are being considered.

The general framework criterion towards PCS implementation is to render the system interconnected and internetworked so that the users are free of wirleline tether (tetherless) and/or cord (cordless). PCS should enable voice as well as fax and computer data transfers on a personal basis across an exhaustive internetwork of terrestrial (wireline and/or wireless) and satellite links with global coverage. Further, multiuser out-door radio-ports would permit the users in a building to get access to remote end-entities. Likewise, an indoor radio-port would interconnect the users within a building.

Thus, regardless of the locale of a person and the type of end-entity, communication is to be facilitated in the envisioned global PCS. Eventually, adjunct to telephone service, PCS will include end-entities comprised of mobile personal computers — the laptop/palmtop version, handhelds, subnotebooks, and personal assistants (personal organizer plus pager and cellular telephone).

Impairments to wireless communication ...

In the wireless system, the physical medium corresponds to free-space in which the electromagnetic (EM) wave propagates. As in any electrical/electronic system, the wireless signals are subjected to corruption by the inevitable presence of noise at the transmission and reception ends as well as along the transmission medium. Apart from the device/system based electronic noise (such as thermal noise) corrupting the signal, certain characteristics of EM propagation in the mobile environment would also impair the signals under transmission. Such impairments arise from signal fading due to scattering, reflection, and refraction effects that the EM wave may face during propagation and attenuation of EM energy as a result of absorption by rain, snow etc. With the result, for robust implementation of wireless communication, the receiver technology is being trimmed continuously through available techniques and devices.

In short, the "genesis" of noisy wireless communication scriptures is as follows: "In the beginning of wireless, there was just 'noise', and life was simple. Engineers soon called this noise additive white Gaussian noise (AWGN), but even this title was not enough to capture all the idiosyncrasies of a real world RF channel. Then, engineers, scientists, and mathematicians got involved, and they soon gave names to more subtle effects, such as Rayleigh fading, Ricean fading, impulse noise, cyclostationary noise, and inter-symbol interference, to name just a few of the newer noise-family members. All these noises were in addition to basic channel difficulties of signal attenuation and loss".

1.2   A Narration on Networking

1.2.1.   Hierarchical telegraphic and telephone networks

Networking is a scheme that facilitates interconnection between end-users of a telecommunication system [1.51-1.79]. It was a necessary part of telecommunication extended as a utility service. Its main objective is to link one end-user to another reliably and economically with the lowest possible delay. Connecting each end-user to every other end-user by a dedicated set of lines seems to be an ideal (theoretical) plausibility of networking. But, it becomes a formidable web of interconnections when the number of end-users increases. Therefore, the classical notion that emerged in telephony to interconnect any two subscribers refers to the so-called *hierarchical networks* in which a group of users is connected to a *central office* (CO) and a group of central offices in turn is connected to a *zonal office* and so on. Thus, a hierarchy-based tree-like combining of offices at different levels offers a practical scheme of networking across which any two end-users can be interconnected.

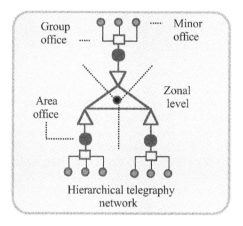

Fig. 1.63: Classical network of telegraphy

The hierarchical interconnection was conceived even in the telegraphic era of telecommunication. *Minor offices* (such as post offices and railroad stations) served as the telegram originating locales where customers (public) delivered their messages to be sent. A few telegraphic offices were grouped to form a *group center* and each group center was tied to a parent *zone center*. The zone centers are again interconnected facilitating a telegram originating at a minor office to reach a destined minor office.

Thus a point-to-point telegraphic interconnection was established essentially through a hierarchical system of interconnections [1.3]. The telephony system borrowed liberally this tree-structure of interconnections in its networking outlays.

The classical telephone terminal equipment (due to Bell) supplemented by the Strowger system of step-by-step switching plus an organized transmission over copper-lines facilitated remarkably myriad of analog voice telephone connections reliably through the public switched networks for decades in all parts of the world. The success of analog voice telephony stems from the standardized interfaces and well-defined functional hierarchy.

Traditionally, in reference to a basic telephone connection, the telco provides a two-wire (*tip* and *ring*) service to each subscriber. This connection is done via a twisted pair of copper wires. It is called a *subscriber loop* or a *local loop* (or as an *end-user access line*) and runs up to a few miles in the case of a CO or within a building complex in the case of a PBX. The transmission refers to a voice grade quality spanning a frequency range up to 4000Hz.

In basic telephony, a number of subscribers (customer telephone sets) are connected to a CO. In certain cases, the voice signals from a number of subscribers may be multiplexed on a single pair of lines and it is known as the *subscriber carrier system*. The multiplexed number denotes a *pair gain*. The CO is a telephone (local) switch. (As mentioned earlier, it is also known by other names such as: *Local office, end-office or Class 5 office.*) CO completes essentially a *local call* by connecting the calling and called subscriber-loops together, and calls destined to distant telephones are switched out of the CO through *interoffice trunks*.

Though the original telephone networks were conceived for voice transmission (constituted by an analog waveform), the emergence of computer interoperation warranted a digital data communication adjunct to the analog transmission of voice. Therefore, it became necessary that the voice grade telephone local-loop be adopted to carry digital waveforms by using the modems.

The transmission between end entities (like a telephone or a computer) in a network, in general, could be *simplex*, *half duplex* or *full duplex*. The simplex transmission is confined to one direction and supports a unidirectional communication. The half duplex permits two-way communication but only one direction at a time. Supporting simultaneous two-way communication is accomplished by a full duplex arrangement.

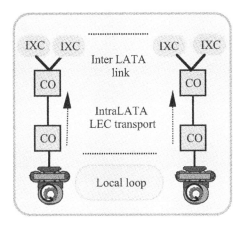

Fig. 1.64: Basic POTS hierarchy

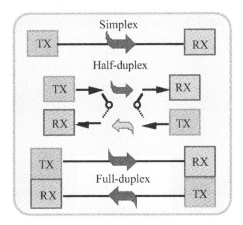

Fig. 1.65: One-way and two-way communications

Some definitions

PSTN --Public switched telephone network: This is the formal name for a public network, which is not restricted to private use. It constitutes the direct distant dialing (DDD) cloud. When the PSTN is stretched to global service, it becomes an international DDD (IDDD) network.

Private network — A telephone network within a building managed privately.

PBX — Private branch exchange: A switching system owned or leased by a business or organization to provide both internal switching functions and access to the PSTN. It can be an operator-assisted manual exchange; if automated, it is known as a private automatic branch exchange (PABX).

Central office (CO): An end-office that provides service to subscribers in a local area whose lines are directly connected to that office. Such lines are referred to as subscriber lines or loop lines. The maximum distance between COs is 80 km and a repeater is required at about every 5 km. When a central office is connected only to similar offices of the same class, then the arrangement is known as local tandem.

Toll center: A toll-exchange containing only trunk circuits and facilitate long-distance calls. It is also responsible for toll-call billing. Distance between toll offices could be in excess of 80 km.

Interoffice calls: Calls between subscribers connected to different end-offices.

Intraoffice calls: Calls between subscribers connected to the same end office.

A tree-like system illustrated in Fig.1.63 was developed in telephony to accommodate the growing subscriber population that desired to have long-distance connections. The subscribers are connected to a central office (CO), (also known as *end-office*), which forms the lowest hierarchical level 5. The COs are in turn connected to toll offices (level 4) and the Bell System contained three more levels of switching in the hierarchical networking as illustrated in Fig. 1.66.

The hierarchical system as described above was conceived to save a number of trunks in comparison to a *fully-connected mesh* arrangement (Fig. 1.67) where all the switching nodes are interconnected at the first-level switches. In such a fully-connected mesh network, the total number of trunks (M) needed to interconnect all the N switching nodes is given by: $M = N(N - 1)/2$. Though a mesh arrangement can provide alternate routes of interconnection (which may be needed in the event of traffic congestion or equipment failure), several trunks may remain idle if the traffic is not intense and shared significantly by each link (trunk).

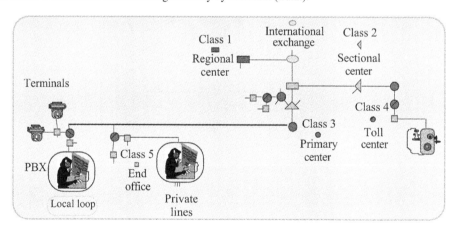

Fig. 1.66: U.S. national telephone network hierarchy in the pre-divestiture era

Further, the trunks needed for interconnections in a fully-interconnected system, grows exponentially with the number of nodes to be interconnected. On the other hand, the hierarchical setup requires fewer trunks but has the demerit of not enabling a connection in the event of an equipment failure or traffic overload at any level of interconnection.

Instead of having a single, fully-connected mesh topology, or a basic backbone hierarchical network, a mixed arrangement that interconnects those COs within an exchange area (having sufficient interoffice traffic volumes to justify direct trunks) was developed. The traffic was routed via the lowest available level of network.

The backbone hierarchical network was also augmented with additional switching facilities called *tandem switches* further to deploying direct interoffice trunks depicting high-usage links. Hence a *mixed tandem switching plus hierarchical connections* came into being. This refers to a hybrid of vertical hierarchy and partial interconnection of nodes (Figs.1.68 and 1.69).

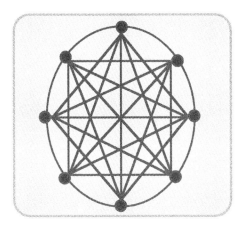

Fig. 1.67: A fully-connected network

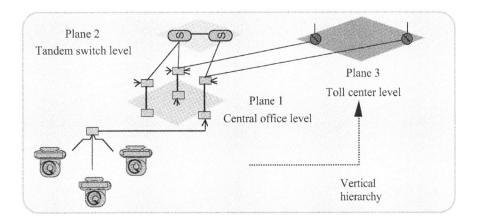

Fig. 1.68: Mixed tandem and hierarchical connections

In the arrangement shown in Fig.1.69, line-to-line, line-to-trunk, and trunk-to-line connections could be established via alternative paths. The *tandem switching* refers to an intermediate switching within the *exchange area*. (In general, an exchange area means a local area within which the calls handled are local and toll-free.)

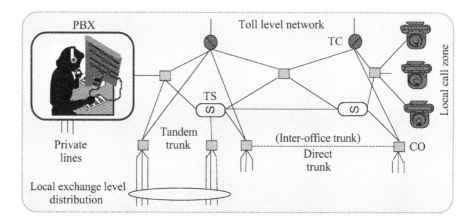

Fig. 1.69: Hierarchical telephony with the inclusion of tandem offices

Post-divestiture era

The telephone networks in the United States took a new structural layout on January 1, 1984 as a result of the break up of the AT&T monopoly leading to the divestiture of Bell Operating Companies (BOCs) from AT&T. The partitioned topology is illustrated in Fig. 1.70.

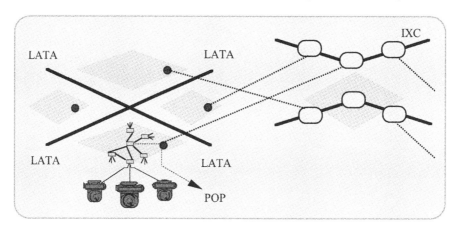

Fig. 1.70: The post-divestiture U.S. telephone network scenario

The technological mutations that adjoined the divestiture include:

- ✓ Massive use of digital switching machines
- ✓ Deploying a single computer to control multiple switching functions; that is, integrating the switching efforts at the end-office, tandem office and/or at the toll office
- ✓ Using fiber-optic and/or wireless transmissions extensively on the cross-sections of the traffic.

The result of these implementations shrunk the number of switching offices in the post-divestiture era but widened the network area considerably. The concurrent change in the administrative profile of telecommunication management was partitioning the U.S. telephone network as illustrated in Fig. 1.70.

1.2.2   Switching and call-routing: An overview

In reference to generic switching and call-routing, a *line* refers to a connection between an end-entity (such as a telephone) and a network. A *switch* is a system of electronic devices (sometimes supplemented by appropriate software) that mediates a connection between two lines or trunks. (A *trunk* is a connection between any two switches.)

A call in telephone systems is routed appropriately between the calling-subscriber and the called-subscriber with intervention of switching offices and service-providers. Some related definitions and considerations are as follows.

Example 1.2

There are two towns A and B far from each other and have two different area codes: 6xx and 4yy. Both towns are served by the same telco. Does a call between A and B require the services of an IXC? Why?

Solution

(i)   No. It is an intra-LATA call.

There are two other neighboring cities C and D each having the same area code 5zz. Two different LECs render the local telephone services for C and D. Does a call between C and D require the services of an IXC? Why?

(ii) Yes. It is an inter-LATA call.

Some definitions

Interexchange carriers (IXC)...

> *These are several competing long-distance interexchange carriers (such as AT&T, MCI, and US Sprint of the post-divestiture U.S. telephone network).*

Local access transport areas (LATAs)

> *These are exclusive service domains of local exchange carriers (LECs) — also known as telcos (telephone companies). IXCs are forbidden to carry intra-LATA traffic. LECs are not allowed to carry traffic between two LATAs even if these two are service domains under a single BOC. Only IXCs can carry inter-LATA traffic. Network design for intra-LATA traffic is left to the discretion of each LEC.*

Point of presence (POP)

> *This is used to interface IXC networks with those of LEC. Existence of POP in every LATA ensures nationwide IXC coverage and handling of inter-LATA calls.*

Evolution of digital telephone network

The conversion of analog telephone networks into digital networks commenced in the 1960s in the United States. It involved installation of T-1 systems on short-haul inter-office trunks within the exchange area.

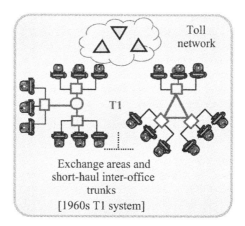

Fig. 1.71: The T-1 system of the 1960s in the United States

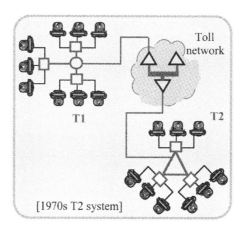

Fig. 1.72: Introduction of T-2 systems in the 1970s

In the early 1970s, the digital network was introduced via T-2 systems between toll networks for short route traffics. The late 1970s saw T carrier strategy being employed extensively through *digital loop carriers* (DLCs), and digital switching became available for use at the CO levels. Hence came into being *digital PBXs* (DPBXs), *digital end-offices* (DEOs), *digital tandem offices* (DTmOs), and *digital toll offices* (DToOs). Further *microwave digital radios* (MDRs) became useful links in both exchange as well as short toll network areas enabling easier and economical digital interfacing. A comprehensively interconnected digital network for integrated transmission of voice and data became feasible in the 1980s with the introduction of optical fiber transmission media. Subsequently, the customer premises equipment (CPE) also became digital thanks to the developments in microelectronics. This made the entire end-to-end connectivity a digital T-hierarchy.

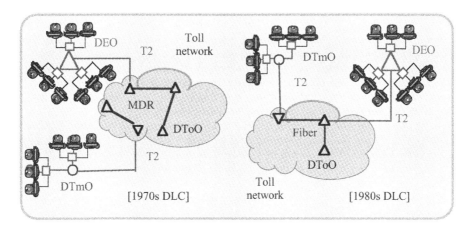

Fig. 1.73: Digital loop carriers of the 1970s and 1980s

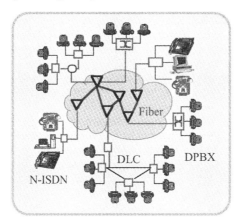

Fig. 1.74: DLC supporting N-ISDN (1980s)

Access networks

In summary, the telecommunication network evolved essentially on the basis of various service categories and networking strategies which have unique features of their own playing distinct roles in the proliferation of telecommunication-based information transfer. The growth of telecommunication networks, in short, rested upon the underpinning efforts associated with POTS and rose to the modern web of information super-highway via the ladder of the following constituent systems:

✓ Private PBX and public switched telephone systems
✓ Transportation of private domain computer data via LAN (local area networks such as ethernet, token bus, and token ring categorized as IEEE 802 Series)
✓ Transportation of computer data via MANs and WANs through public and/or private packet-switched networks
✓ An expanded version of computer data transmission at high-speeds via LANs and/or WANs (For example, FDDI, SMDS and HIPPI)
✓ Wireless networks for voice and personal communications and for LAN/MAN/WAN applications
✓ Satellite-based telecommunication networks
✓ Fiber-optics based interconnections including SONET and SDH
✓ Integrated efforts of transmitting heterogeneous traffic via ISDN and emergence of broadband capabilities through B-ISDN/ATM.

1.3   ATM - A March towards High Performance Telecommunication

Asynchronous Mode Transfer (ATM) is the culmination of the evolution and developments in switching and transmission aspects of telecommunication engineering in the last two to three decades. It has made the *broadband integrated service of digital network* a reality. It is a technology that allows the total flexibility and efficiency required for high-speed multi-service/multi-media networks. It promises high-speed integrated services and a choice of connectivity through universal switching and multiplexing. It is conceived to enable high-bandwidth time-critical applications to reach the tabletop computers.

Now, what exactly is asynchronous transfer mode vis-a-vis telecommunication? How does it differ from the existing network technologies? In what way will ATM support the application programs? What are feasible ventures of new applications facilitated by ATM?

ATM is originally described as a computer-networking paradigm that will bring high-speed communication to the desktop. But, in today's perspective it is more than that. It is envisioned to support a heterogeneous traffic of data from computers, voice from the telephone and video from television/image-processing related equipment. The eventual aspect of ATM is to bring networked multimedia capabilities to desktop level (if possible to the palm-top level as well). ATM offers the possibility of unifying communication paradigms — the unification of services, operating environments, and administrative domain interconnection.

ATM can provide both circuit and packet-switching services in an integrated fashion with the same protocol. As a result, it offers scalability, statistical multiplexing, traffic integration, and network simplicity. However, ATM in the course of enabling these benefits makes certain compromises, which will be discussed later.

Many telecommunication applications requiring high bit rates for the transfer of voice, video, and data information are turning to the B-ISDN protocols supported by ATM in which constant cell-size is used for the packet-structure. Further, in order to accommodate multiservice (voice, video, and data) telecommunication strategies, broadband schemes (SONET and SDH) are adopted. These schemes have a formattable framing structure compatible for ATM.

In essence, ATM is a high-performance, cell-oriented switching and asynchronous multiplexing/switching technology that utilizes fixed-length packets ("cells") to carry different types of traffics — of low bit to high bit rate (or, synonymously, low bandwidth to high bandwidth) *vis-à-vis* applications such as data, voice, and video. High *quality of service* (QOS) is the modern strategy to market telecommunication network services, and the ATM supplemented by SONET or SDH provides an appropriate infrastructure to the emerging B-ISDN technology with the flexibility to offer a QOS on demand. Use of optical fiber in SONET/SDH as the physical layer for transporting the packetized information to its destination has facilitated further realizable broadband attributes to ATM networking.

What have been the objectives of ATM technology? ATM is conceived to be cost-effective and scalable. It should support applications with diverse traffic characteristics (such as multimedia presentations) and it must support multiplexed bit-streams (from different applications, namely, data, voice, and video flowing at different rates) with acceptable (guaranteed) delay-bounds. ATM is also expected to be an efficient multicasting system inasmuch as many applications would often need this requirement. Above all, ATM should have the interoperability capabilities and is envisioned for applications in existing LANs, MANs, and WANs, as well. ATM with interoperable support structures of public networks should facilitate regional and global communication. And, it should adopt the existing standards and protocols as much as possible (especially, in the interoperation situations).

In view of the above objectives concerning its implementation and the scope of its universal deployment, ATM can be illustrated by relevant considerations pertinent to technology, services, and standards. Fig. 1.74 shows the concept that ATM can support a variety of services through switched virtual circuit connectivity.

ATM service - The wide range of service supported by a single transfer mechanism

> ✓   Operating environment: This refers to the wide range of BW and locales (desk top to wide area), wireless and wireline in public and private networking

✓   Administrative domain interconnection: Other networking functions (such as routing) segregated from the basic information transfer etc.

In facilitating multiple classes of services, ATM faces a different physical media of transport. For example, typical T-1 carrier information that flows through a twisted pair of copper wires may converge on to ATM traffic. Fiber-based high-speed data is another class of service that the ATM may support. There are other transmission-media such as coaxial lines and wireless links, which can become a part of an ATM system (Figs. 1.75 and 1.76).

In view of these different end-side transmission media, appropriate user-network interfaces (UNIs) are warranted in ATM implementations. Further, the pipeline through which the logical virtual paths (VPs) and virtual circuits (VCs) are realized through network clouds (mostly fiber-optics based) require appropriate network-network interfaces (NNIs). The ATM connectivity between different end-entities and multiplexing of the heterogeneous traffic are done with ATM switches and multiplexers respectively. ATM switches (to which the end applications are connected) offer a convergence to the incoming bit-streams and adapt them (regardless of the type of application sending in the message) into a common ATM cell structure. Thus the ATM adaptation renders all the bit-streams (belonging to the heterogeneous applications) encapsulated as uniform ATM cell format, ready for transmission via selected VPs and VCs.

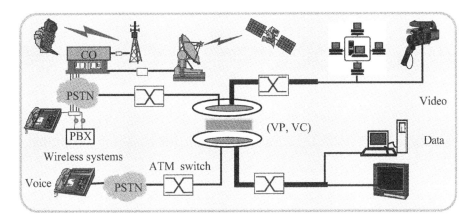

Fig. 1.75: ATM — A switched VC system supporting multiple classes of services

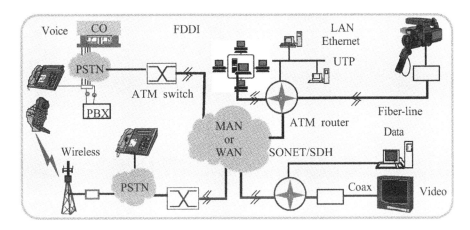

Fig. 1.76: ATM — A system that supports networks using different physical media

ATM is expected to reduce the infrastructure costs through efficient bandwidth management, operational simplicity, and consolidation of overlay networks. The reason is that the carriers can no longer afford to go through the financial burden and time required to deploy a separate network for each new service requirement (for example, dedication of a network for a single service such as transparent LAN or frame-relay). Implementing ATM technology, however, allows stability in the core network, at the same time, permitting service interfaces and other equipment to be evolved rapidly commensurate with customer needs, vendor potentials, and capabilities of the service providers.

ATM standardization is therefore, designed to include specifically the interface considerations, signaling requirements under heterogeneous traffic-handling and management of such multiple service traffics.

ATM is a technology which can enable carriers to capitalize on a number of revenue opportunities through a portfolio of multiple ATM classes of services, high-speed feasibility in LAN interconnections, integrated transportation of voice, video, data, and facilitation of multimedia applications. Such revenue opportunities enclave short-term perspectives and may span to extended strategies in both business as well as community or residential markets. Learning about ATM telecommunication, in short, refers to comprehending the subject-matter pertinent to:

- Concepts of heterogeneous signal-transfer and the associated hardware means of implementation

- Performance goals and practically attainable service-parameters *vis-a-vis* the multiple service signal transmission(s) envisaged

- Management of desirable traffic characteristics of ATM transmissions.

In the following paragraphs further considerations on ATM principle and related issues are outlined briefly as a prologue to this text-book:

1.3.1 *ATM - a formal definition*

ATM is defined as: "A transfer mode in which information is organized into cells; it is asynchronous in the sense that the recurrence of cells containing information from an individual user is not necessarily periodic".

In CCITT Draft Recommendation I. 113 section 2.2, Geneva, May 1990, ATM has been adopted as the universal transfer mode by CCITT for B-ISDN and it is characterized as follows:

- ATM transmission refers to the transfer (asynchronously) over the communication channel of a stream of packetized cells (of constant size), derived by grouping digitized pulses (bits) corresponding to voice, video, and data signals

- It is an accepted means of unifying the communication paradigm for applications in the public wide area networks in the form of B-ISDN as well as a feasible local area networking (LAN) technology

- It not only accommodates the service for multiple types of information transfer across LANs and WANs, but it also permits a straightforward internetworking for such transfers.

There are three major technological advancements which have led to the feasibility of realizing the integrated packet switched network to carry all three fundamental communication traffics namely, data, voice, and video. They are:

- Decades of achievements in very large-scale integrated (VLSI) circuits — Growths in semiconductor technology

- High speed switching: 100 Mbps to a few Gbps (per port) — User demand on a variety of telecommunication applications

- Fiber optics communication — Feasibility of information transmission with light.

Targeted deployments of ATM technology are (and not restricted to):

- ✓ Wide area public networks of telephone companies
- ✓ Campus and local networking (ATM LAN)
- ✓ Video distribution
- ✓ Computer imaging
- ✓ Distributed scientific computation and visualization
- ✓ Distributed file and procedure access
- ✓ Multimedia conferencing
- ✓ Wireless ATM.

ATM concept has emerged from two well-known electrical communication methodologies, namely,

- Time-division multiplexing

- Packet-switching.

Technically, ATM can be viewed as an evolution of *packet-switching*. Like packet-switching for data (for example, X. 25, frame relay, and TCP/IP), ATM integrates the multiplexing and switching functions, is well-suited for bursty traffics (in contrast to circuit-switching), and allows communication between devices which operate at different speeds. However, unlike packet-switching, ATM can be designed for high-performance multiservice networking.

ATM technology has been implemented in a plethora of networking devices such as:

- ✓ PC, workstation, and server network interface cards
- ✓ Switched-ethernet and token-ring workgroup hubs
- ✓ Workgroup and campus ATM switches
- ✓ ATM enterprise network switches
- ✓ ATM multiplexers
- ✓ ATM-edge switches
- ✓ ATM-backbone switches.

ATM is also a capability extended to an end-user by service-providers constituting a basis for tariffed services or for network infrastructuring for the tariffed as well as other services. The simplest service building block is the ATM *virtual circuit,* which is an end-to-end connection that has a set of defined end-points and routes but does not have bandwidth dedicated to it. Bandwidth is rather allocated by the network on demand basis *vis-à-vis* meeting the users' request (or traffic contract) pertinent to the type of traffic being transmitted; the type of traffic for which the bandwidth is negotiated, falls within various classes of service prescribed for a broad range of applications.

ATM has a set of international interface and signaling standards defined by the *International Telecommunications Union Telecommunications* (ITU) *Standards Sector* (formerly the CCITT). Further, *ATM Forum,* an international voluntary organization composed of vendors, service providers, research organizations, and users plays a pivotal role in the ATM market since its formation in 1991. Its objective is to "accelerate the use of ATM products and services through the rapid convergence of interoperability specifications, promotion of industry cooperation, and other activities", and, it implements its objective by developing necessary multi-vendor agreements on an *ad hoc* basis.

1.3.2 Merits of ATM implementation

The benefits that can be achieved by deploying ATM technology can be summarized as follows:

    ✓   Hardware switching (with terabit switches on the horizon) enables high performance

    ✓   Realization of dynamic bandwidth for bursty traffic meeting application needs and delivering high utilization of networking resources

    ✓   Support for multimedia traffic is based on class of service, which allows applications with varying throughput and latency requirements to be met on a single network

    ✓   Scalability in speed and network size at the current link speeds of T-I/E-1 to OC-12 (622 Mbps) extending into the multi-Gbps range

    ✓   LAN/WAN architecture to allow ATM to be used consistently from one desktop to another

    ✓   Simplification of network operation via switched VC architecture

    ✓   International standards compliance in central-office and customer-premise environments permitting for multivendor operation.

1.3.3 ATM technology: A perspective
ATM cell

As indicated earlier, in ATM networks, all information is formatted into a fixed-length cell consisting of 48 bytes (8 bits per byte) of payload and 5 bytes of cell header (Fig 1.77). The fixed cell size ensures that time-critical information such as voice or video is not adversely affected by long data frames or packets. The header is organized for efficient switching in high-speed hardware implementations and carries payload-type information, virtual-circuit identifiers, and header error check.

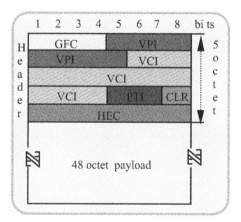

Fig. 1.77: ATM cell

ATM is connection-oriented. By organizing different streams of traffic in separate cells, the user can specify the resources required. This organization allows the network to allocate resources based on these needs. Multiplexing multiple streams of traffic on each physical facility (between the end-user and the network or between network switches) — combined with the ability to send the streams to many different destinations — enables cost-savings through a reduction in the number of interfaces and facilities required for constructing a network.

Standards on ATM define two types of ATM connections: *Virtual path connections* (VPC) which contain *virtual channel connections* (VCC), as shown in Fig. 1.78.

A *virtual channel connection* (or virtual circuit) is the basic unit, which carries a single stream of cells, in a sequential order, from user-to-user. A collection of virtual circuits can be bundled together into a *virtual path connection*. A virtual path connection can be created from end-to-end across an ATM network. In this case, the ATM network does not route cells belonging to a particular virtual circuit. Since all cells belonging to a particular virtual path are routed the same way through the ATM network, it results in faster recovery in case of major failures.

An ATM network also deploys virtual paths internally for purposes of bundling virtual circuits together between switches. A pair of ATM switches can have many different virtual channel connections between them, belonging to different users. These can be bundled by the two ATM switches into a virtual path connection, which can serve the purpose of a virtual trunk between the two switches. This virtual trunk can then be handled as a single entity by, perhaps, multiple intermediate virtual path cross connects between the two virtual circuit switches.

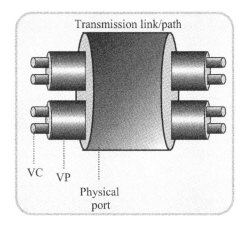

Fig. 1.78: ATM connections

Virtual circuits can be statically configured as permanent virtual circuits (PVCS) or dynamically controlled via signaling as switched virtual circuits (SVCs). They can also be point-to-point or point-to-multipoint, thus offering a rich set of service capabilities. SVCs are the preferred mode of operation because they can be dynamically established, thereby minimizing the reconfiguration complexity.

Statistical multiplexing

In fast packet-switching, statistically multiplexing several connections on the same link based on their traffic characteristics is used as an attempt to solve the so-called *unused bucket problem* of synchronous transfer mode (STM). In other words, if a large number of connections are very bursty (i.e., their peak/average ratio is 10:1 or higher), then all of them may be assigned to the same link assuming that they will not all burst at the same time. And, if some of them do burst simultaneously, then there should be sufficient flexibility so that the burst can be buffered up and placed in subsequently available free buckets. This is called statistical multiplexing, and it allows the sum of the peak bandwidth requirement of all connections on a link to even exceed the aggregate available bandwidth of the link under certain special conditions. This feature is rather impossible in an STM network, and it is the main distinction of an ATM network from other transfer modes.

Essentially ATM can be regarded as an offshoot of time-division multiplexing (TDM). That is, ATM can be regarded as a shared, time-division switch depicted as in Fig. 1.79. As the samples are received in the TDM switch, they are entered into the buffer. The output from the buffer is selected in the same order that the switching is sequenced to receive the inputs.

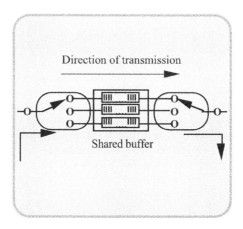

Fig. 1.79: TDM: Concept of a shared buffer

Time-division switching fabrics are efficient for multicasting (that is, for transmissions to all destinations). This is because the transmitted bits go past all the outputs. Instead of only one port being signaled to copy the bits, multiple ports can be signaled to copy them. Further, with time-division switching, instead of all ports operating at the same speed, bandwidth-on-demand can be accessed dynamically by the ports permitting variable input rates.

Fig. 1.80 shows how the TDM preassigns every input channel its slice of available time. This is also called *synchronous transmission mode* (STM). A receiving terminal at the demultiplexer knows the bit of data that belongs to each designated time-slot. The three different traffics (voice, data, and video) shown in Fig. 1.80 are accommodated in sequenced time-slots (1, 2, 3 respectively) on the multiplexed line. If a particular traffic at its designated time-slot on the input side is absent, the corresponding time-slot on the multiplexed line will go empty as shown. Since the receiver works in synchronization with the transmitting end and is aware of the designated time-slot sequence, it can segregate the traffics as they arrive in their respective assigned time-slots.

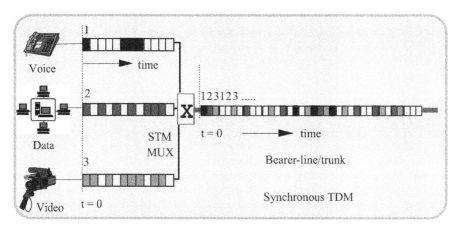

Fig. 1.80: STDM (Each channel is allotted a designated portion of the available time): (*See PLATE 4*)

Contrary to STM, the incoming traffic can also be multiplexed on first-in/first-out basis. This refers to statistical or asynchronous multiplexing as illustrated in Fig.1.81. Since there are no designated time-slot reservations on the trunk line the empty slots are absent. This strategy enables a better utilization of trunk line capacity.

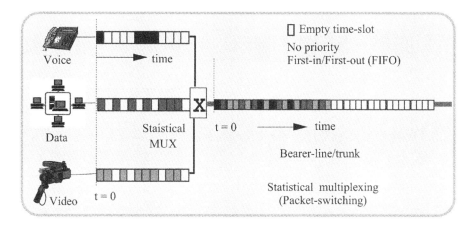

Fig. 1.81: Statistical (asynchronous) multiplexing based on first-in/first-out principle: (*See PLATE 4*)

However, in order to demultiplex and segregate the traffic at the receiver, each of the traffics should bear an ID as its overhead. Based on this information, a traffic will be separated out at the demultiplex (DEMUX) from the stream of incoming traffic (at the DEMUX).

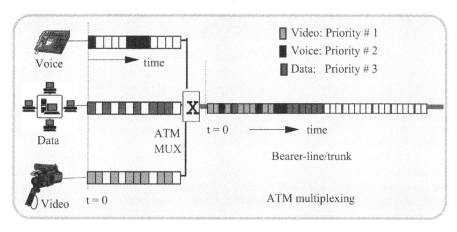

Fig. 1.82: ATM multiplexing based on statistical multiplexing with priority schedule: (*See PLATE 4*)

ATM multiplexed follows the strategy of statistical multiplexing but with a difference. In order to allow the isochronous traffics to flow without being delayed at the multiplexer, the traffics are prioritized. For example, in Fig. 1.82, the video traffic has been given a priority #1, voice the second priority, and data the third. These three traffics are multiplexed on a FIFO basis as long as there is no contention. In the event of contention, the higher priority traffics are serviced first (in their order of priority) and the least priority traffic will be delayed at the buffer and sent later on the trunk.

As distinct from STM, part of ATM's design is to accommodate *isochronous* traffic. Isochronous traffic means traffic that is sensitive to any latency or delay. When a response from a host to a terminal is slow in being received, it can be tolerated if it corresponds to a data set (which is not isochronous). However, when a voice conversation is being chopped into digital packets, they must arrive at the listener's end within a specified amount of time or else the conversation would become unnatural. Voice and video signals are typical examples of isochronous traffic.

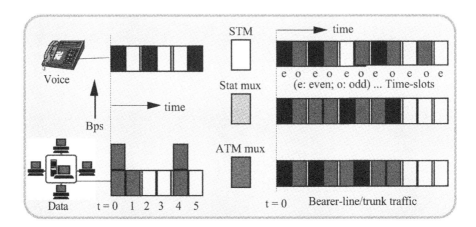

Fig. 1.83: ATM multiplexing: (*See PLATE 5*)

Suppose there are two sources, data and voice, to be carried on an ATM link. Referring to Fig. 1.83, a bursty data traffic is present at t = 0 for transmission. The two overlaid units (2 slabs clustered as shown) denote the strength of the burst. That is, at any time-slot, the number of vertically placed slabs represent bits per second and the horizontal spacing represents time units. Three voice signals are located at t = 0, 2 and 5 while the data is presumed to be transmitted at t = 0, 1 and 4. Further, data at t = 0 and t = 4 are assumed to be twice of what is sent by voice at t = 0, 2, or 5.

If synchronous TDM technology is used (as done in T-1 multiplexers), each source would have to occupy its assigned time-slot in the frames that are transmitted. Suppose voice channel is assigned the even time-slots and the data channel is assigned the odd time-slots. Accordingly, the receiver is synchronized to interpret all even time-slots as voice and odd time-slots as data. This synchronization requirement results in the waste of two slots as indicated; therefore, this is not efficient towards bandwidth utilization and the delays (avoidance of which are rather crucial for isochronous signals) in TDM seem to be rather inevitable.

For the packet-switching, instead of altering between two sources, each transmission is sent as one unit rather than breaking them up into time frames. Voice (of slot 1) is sent first, followed by data (slots 1 through 3) as one unit. Because of this packetized data, the second voice frame does not go through until t = 4. Similarly, the third voice frame is delayed by two time-slots.

The packet switching such as X.25 and frame relay/(FR), in general, uses the bandwidth quite effectively. In the example shown, the entire data is transmitted three slots earlier than it was with TDM. However, the voice is delayed by only one time-slot. If the data at t = 0 had been a lengthy file-transfer, the corresponding delay on the voice at the second slot would have, however, been considerable. In other words, packet switching is favorable for data transmission, whereas TDM is more appropriate for isochronous traffic.

With ATM, isochronous traffic (such as voice) is accommodated with zero delay and the contending data is inserted in sequence as illustrated in Fig. 1.83. There are no wasted time-slots, and the bandwidth is also used effectively. While in TDM, the even/odd slots identify the locations of voice and data signals; in ATM, a header is provided towards identifying the cell's destination. Hence, ATM is a type of *label of multiplexing*.

The *asynchronous* aspect of ATM is that the rate at which information is fed is not equal to the outgoing capacity of the network (though, the integrity of the sequence of information is maintained).

Layered architecture

The open system interconnection (OSI) models constitutes the functional aspects and layering architecture of ATM as illustrated in Figs.1.84 (as specified by International Organization for Standardization, ISO).

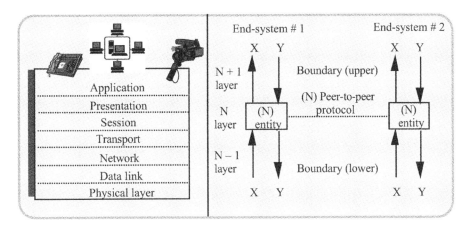

Fig. 1.84: Layered architecture of the ISO-OSI reference model and the associated primitives
Note: X: services; Y: Interface control information

The relevant layering concepts are as follows: A logical grouping of functions is identified as (N)-layer with an upper boundary (N+I) and a lower boundary (N-1).

(N)-service: A capability of the (N)-layer and the layers beneath which is provided to (N + 1)-entities at the boundary between the (N)-layer and the (N + 1)-layer.

(N)-entity is an active element, which provides (N)-functions.

(N)-entities communicate using (N)-protocol and entities in the same layer are peer entities, (N)-protocol information and (N)-user data is transferred between (N)-entities using (N – 1)-connection.

Interface control information: Information transferred between two adjacent layer entities to coordinate their joint operation.

ATM, in essence, is connection-oriented and allows the user to specify dynamically the resources required on a per-connection basis (per SVC). There are five classes of service defined for ATM (as per ATM Forum UNI 4.0 specification). The QOS parameters for these service classes are summarized in Table 1.3.

Table 1.3: <u>ATM service classes</u>

Service classes	Description
Constant bit rate (CBR)	This class is used for emulating circuit switching. The cell rate is constant with time. CBR applications are quite sensitive to cell-delay variation. Examples of applications that can use CBR are telephone traffic (namely, $64 \times n$ kbps), videoconferencing, and television.
Variable bit rate-nonreal time (VBR-nrt)	This class allows users to send traffic at a rate that varies with time depending on the availability of user information. Statistical multiplexing is provided to make optimum use of network resources. Multimedia e-mail is an example of VBR-nrt.
Variable bit rate-real time (VBR-rt)	This class is similar to VBR-nrt but is designed for applications while being sensitive to cell-delay variations. Examples for real-time VBR are voice with *speech activity detection* (SAD) and interactive compressed video.

Available bit rate (ABR)	This class of ATM services provides a rate-based flow control and is aimed at data traffic such as file-transfer and e-mail. Although the standard does not require the cell transfer delay and cell-loss ratio to be guaranteed or minimized, it is desirable for switches to minimize the delay and the loss as much as possible. Depending upon the state of congestion in the network, the source is required to control its rate. The users are allowed to declare a minimum cell rate, which is guaranteed to the connection by the network.
Unspecified bit rate (UBR)	This class is the catch-all "other" class, and is widely used today for TCP/IP.

The ATM Forum has identified the following technical parameters to be associated with a connection. These terms are outlined in Table 1.4.

Table 1.4: ATM technical parameters

Technical parameter	Definition
Cell-loss ratio (CLR)	Cell loss ratio is the percentage of cells not delivered at their destination because they were lost in the network due to congestion and buffer overflow.
Cell transfer delay (CTD)	The delay experienced by a cell between network entry and exit points is called the cell transfer delay. It includes propagation delays, queueing delays at various intermediate switches, and service times at queueing points.
Cell delay variation (CDV)	A measure of the variance of the cell transfer delay. High variation implies larger buffering for delay sensitive traffic such as voice and video.
Peak cell rate (PCR)	The maximum cell rate at which the user will transmit. PCR is the inverse of the minimum cell inter-arrival time.
Sustained cell rate (SCR)	This is the average rate, as measured over a long interval, in the order of the connection lifetime.
Burst tolerance (BT)	This parameter determines the maximum burst that can be sent at the peak rate. This is the bucket-size parameter for the enforcement algorithm that is used to control the traffic entering the network.

In reference to the ATM technical parameters, their association with relevant service classes is depicted in Fig. 1. 85.

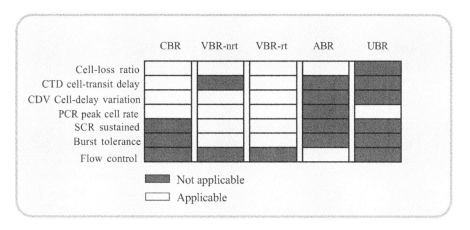

Fig. 1.85: ATM classes of services: Applicability of traffic parameters

Its extensive class of service capabilities as indicated alone makes ATM the technology of choice for multimedia communications. The services supported in a fully evolved multiple service network such as B-ISDN can be expected to produce a wide range of traffic flow characteristics, as well as to have a wide string of performance requirements. Fig. 1.86 illustrates the approximate ranges of the maximum bit-rate and the utilization of a channel at this rate pertinent to common, multiple service categories. It should be noted that a network carries not only a heterogeneous mix of traffic (which could range from narrowband to broadband and from continuous to bursty), but with the advent of multimedia terminals and services (for example, combined voice and image applications), individual traffic sources may themselves be heterogeneous as well.

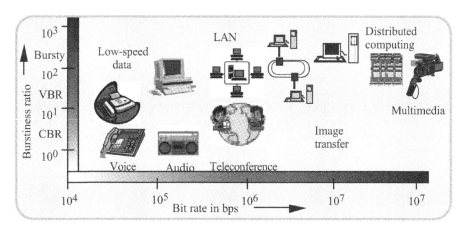

Fig. 1.86: Multiple service traffic characteristics

In an integrated service environment, the underlying (integrated) transport is required to provide a simple information transfer function between user terminals or end-systems for all the supported services. Higher layer service-related functions such as end-to-end flow control and error recovery would be functions of the user terminals or end systems.

Such a transport should:

  ✓ Support a range of traffic flow types
  ✓ Sustain a guaranteed QOS on information transfer performance of all the supported services.

As indicated earlier, ATM is compatible as a B-ISDN connection-oriented telecommunication link. It does synchronous cell operation with minimized per-node processing with no link-by-link error recovery. However, considering the end-to-end performance, a major traffic quality parameter refers to two major considerations, namely:

✓   Maximum cell-delay variation (CDV)

✓   Cell loss characteristics.

1.4   Telecommunications: Quo vadis?

Telecommunication has made an unprecedented progress decade after decade since the 1960s. What was realized in each decade, was only a figment of conception in the previous decade. With the result, the humble beginning of telecommunication through telegraphy of the last century became versatile to support voice, data, and video in unlimited ways and with a variety of QOS to choose. Advances in electronics and optical technologies added extreme flexibility, reduction in physical size and a total restructuring of the telecommunication system. As a result, provisioning any type of service at any speed with any extent of bandwidth at any place at any time and destined to any locale has become the slogan of modern telecommunication. *Has there ever been a customer demand or societal interest to have such an unbounded service type and access?*

As late as 1991, Noll [1.4] has expressed rather a skeptical response: "The utopian view of communication technology and all it could do is cloaked in a sense of euphoria that makes it difficult to separate hyperbole from reality". His apprehension has stemmed from the fact that certain services, though made available, may not fly due to users' limited interest *vis-a-vis* commercial prospects involved.

But, as the new century has emerged, telecommunication is becoming a vital gateway for an information highway of different perspective and prospects. The skepticism on users' preferences is dwindling, the push for new technologies and methods has become a way of life. As a result, developments have been directed at realizing a "high-speed, ubiquitous, and seamless integrated services information infrastructure". This is ambitious, yet conceivable. This is an emancipated effort from the legacy technology and classical paradigms of telecommunication networking and service capabilities.

In the past, the tale of telecommunication was a simple theme built on the strata of telegraphy, analog telephony, and digital telephony. When high-speed computation and interactive machines knocked at the doors of telecommunication technology, a new vista leading to a variety of ventures left the door ajar. This set the climax to the story of electrical modes of message transfer. Telecommunications had since then been cornered to meet user demands for more functions in the combined computer-communication operations and in the related applications excogitating thereof producing utilities and powerful information transfer systems.

What is the base structure of existing and the contemplated efforts in telecommunication strategies? The answer is simple and assertive: *Digital technology*. And, the associated functional aspects of telecommunication are: *Information transport, switching functions, and network management and control.* The superstructure to be built towards any envisioned strategies in telecommunication is to expand the scope of existing services in respect to the aforesaid information transfer and other functional considerations. As rightly stated by Black [1.2], as far as telecommunication technology is concerned, "the future has become the present", it is a beginning of an end and its epilogue of the past has become the prologue of the present. *Based on the review of the growth of telecommunication and its present status as indicated in the previous sections, what are possible avenues of expansion in service control, management architectures, and in related technologies?*

The imminent scope to be emphasized is to make the B-ISDN a full-fledged technology through ATM support. ATM, as envisioned, offers a technology for a connection-oriented, high-speed telecommunication in the regimes of LAN, MAN, WAN, and possibly, in GAN. Conceptually, ATM distinguishes the network controlling (signaling) functions from the information transport. It allows a variety of traffic flows heterogeneously in a single network with different QOS demands, and a gamut of distinct end-entities pertinent to different application profiles can be interfaced to this single network. Yet, the system operates cost-effectively.

What is the future of ATM? The answer is to come out with solutions to harness and exploit the ATM's flexibility, capability, and efficiency by focusing relevant developments in regard to:

      ✓   Applying complex resource allocation schemes
      ✓   Employing advanced congestion and traffic smoothing techniques
      ✓   Enhancing traffic-management capabilities.

In addition, ATM implementation should be considered aggressively in certain domains. Specifically, the following could be the relevant focus areas:

Wireless ATM

      ✓   Interface structures at wireless ATM and fiber-optic based terrestrial transport
      ✓   *Ad hoc* ATM networks.

IP plus ATM

      ✓   IP over ATM
      ✓   Integrating IP and ATM
      ✓   Hybridization of IP and ATM switching.

ATM-based home-access networks

      ✓   ADSL.

1.5   Concluding Remarks

Today's knowledge of telecommunication and associated networking is an ensemble of details gathered through years of experience — from simple POTS through megabit networks via copper-wireline and/or fiber-based transmissions. The question is how to comprehend the technological aspects of modern gigabit environments from the first principles and evolutionary engineering experienced and digested so far. The possible answer is to reengineer the following considerations of telecommunication technology vis-à-vis gigabit networks and their extensions:

    ■   *Ultra large-scale network connections*: The growth of telecommunication customers has emphasized the need for a strategy to engineer an ultra large-scale, global networking which can offer a reliable and consistent operation at a reasonable cost. In the United States alone, the magnitude of networks (such as Internets) sized by 5 to 6 orders to an order of 8 is inevitably imminent. Such mammoth growths in telecommunication should see a parallel resizing and reduction of cost-outlays faced by the customers

    ■   *Criticality of high-speed network performance*: The high-speed traffic through a network representing interconnected units can face a stability problem. That is, a high-speed cell or a packet experiencing a traumatic traffic condition at any unspecified node may trigger technical implications at any other node or at the originating end itself. Such (delayed) feedback issues are still open questions even in low-speed networks. Therefore, it can be expected that the gigabit networks need comprehensive research inputs to resolve the network stability and allied implications

- *Distributed network response*: Fast transmission of data permits dispersed computing tasks. That is, with the advent of interconnection feasibility, the computer technology is becoming increasingly decentralized. Data-flow is passed down a hierarchical set of machines, and the hierarchy itself is flattening. To facilitate the associated connectivity among the distributed parts, computer networks have proliferated as a complex web within the tiers of the telecommunication practice. Therefore, designing a high-speed network topology and implementing it as a telecommunication facility has become more comprehensive and to a certain extent, flexible. The overall network response under such an ambient refers to a performance-study issue of gigabit networks

- *Overhead information and management considerations*: Realization of gigabit networks imposes extraordinary overhead information requirements to manage the colossal extent of messages processed and routed across the network. Therefore, fast network protocols and management algorithms have to be revisited in order to shape them as efficient paradigms.

Bibliography

[1.1]   M. Klein: What hath God Wrought? *Invention and Technology*, Spring 1993, 34-42.

[1.2]   H.S. Black: *Modulation Theory* (D. Van Nostrand Co., Inc., Princeton, NJ: 1966)

[1.3]   E.H. Jolley: *Introduction to Telephony and Telegraphy* (Hart Publishing, Inc., New York, NY: 1970)

[1.4]   A.M. Noll: *Introduction to Telephones and Telephone Systems* (Artech House, Inc., Norwood, MA: 1991)

[1.5]   J. Bellamy: *Digital Telephony* (John Wiley & Sons, Inc., New York, NY: 1991)

[1.6]   R. E. Crochierer and J.L. Flanagan: Current perspectives in digital speech. *IEEE Communications Magazine,* Vol. 21(1), January 1983, 32-40

[1.7]   K.E. Fultz and D.B. Perick: The T1 carrier system. *Bell System Technical Journal*, Vol. XLIV (7), September 1965, 1405-1451

[1.8]   J.C. Palais: Fiber Optic Communications (Prentice-Hall, Inc., Upper Saddle River, NJ: 1998)

[1.9]   G. Keiser: *Optical Fiber Communications* (McGraw-Hill, Inc., New York, NY: 1991)

[1.10]   T.S. Rappaport: *Wireless Communications Principle and Practices* (Prentice Hall PTR, Upper Saddle River, NJ: 1996)

[1.11]   U. Black and S. Waters: *SONET & T-1: Architectures for Digital Transport Networks* (Prentice Hall PTR, Upper Saddle River, NJ: 1997)

[1.12]   T. Ramteke: *Network* (Prentice Hall Education, Career and Technology, Englewood Cliffs, NJ: 1994)

[1.13]   E. Ramos and A. Schröder: *Contemporary Data Communications: A Practical Approach* (Macmillan Publishing Co., New York, N.Y.: 1994)

[1.14]   P.J. Fortier (Editor): *Handbook of LAN Technology* (McGraw-Hill Publishing Co., New York, NY: 1992)

[1.15]   H.G. Hegering and A. Läpple: *Ethernet* (Addison-Wesley Publishing. Co., Workingham, England: 1995)

[1.16]   W. Stallings: *ISDN and Broadband ISDN with Frame Relay and ATM* (Prentice-Hall, Inc., Englewood Cliffs, NJ: 1995)

[1.17]   U.D. Black: *ATM Foundation for Broadband Networks* (Prentice Hall PTR, Englewood Cliffs, NJ: 1995)

[1.18]   L.G. Cuthbert and J.C. Sapanel: *ATM – The Broadband Telecommunications Solutions* (The Institution of Electrical Engineers, London, England: 1994)

[1.19]   M. de Prycker: *Asynchronous Transfer Mode – Solution for Broadband ISDN* (Prentice-Hall, London, England: 1995)

[1.20]   R.P. Davidson: *Broadband Networking ABCs for Managers*: ATM, BISDN, Cell/Frame Relay to SONET (John Wiley & Sons, Inc., New York, NY: 1994)

[1.21]   R.P. Davidson: *Broadband Networks Manager's Guide* (John Wiley & Sons, Inc., New York, NY: 1996)

[1.22]   I.J. "Duffy" Hines: *ATM: The Key to High-Speed Broadband Networking* (M & T Books, New York, NY: 1996)

[1.23]   H.J.R. Dutton: *High-Speed Networking Technology: An Introductory Survey* (Prentice Hall PTR, Upper Saddle River, NJ: 1995)

[1.24]   H.J.R. Dutton and P. Lenhard: *Asynchronous Transfer Mode (ATM) – Technical Overview* (Prentice Hall PTR, Upper Saddle River, NJ: 1995)

[1.25]   A. Edmonds Jr.: *ATM Planning and Implementation* (International Thomson Computer Press, Boston, MA: 1997)

[1.26]   W.A. Flanagan: *ATM – User's Guide* (Flatiron Publishing Co., Chelsea, MI: 1994)

[1.27]   C. Gadecki and C. Heckart: *ATM for Dummies* (IDG Books Worldwide Inc., Foster City, CA: 1997)

[1.28]   L. Gasman: *Broadband Networking.* (Van Nostrand Reinhold, New York, NY: 1994)

[1.29]   W.J. Goralski: *Introduction to ATM Networking* (McGraw-Hill, Inc., New York, NY: 1996)

[1.30]   D. Ginsberg: *ATM – Solution for Enterprise Internetworking* (McGraw-Hill, Inc., New York, NY: 1999)

[1.31]   R. Händel, M.N. Huber and S. Schröder: *ATM Networks Concepts, Protocols, Applications* (Addison-Wesley, Workingham, England: 1994)

[1.32]   O.C. Ibe: *Essentials of ATM Networks and Services* (Addison-Wesley Longman Inc., Reading, MA: 1997)

[1.33]   B. Kercheval: *TCP/IP over ATM: A No-Nonsense Internetworking Guide* (Prentice Hall PTR, Upper Saddle River, NJ: 1998)

[1.34]   S. Keshav: *An Engineering Approach to Computer Networking – ATM Network, the Internet and the Telephone Network* (Addison-Wesley, Reading, MA: 1997)

[1.35]   B. Kumar: *Broadband Communications A Professional Guide to ATM, Frame Relay, SMDS, SONET and B-ISDN* (McGraw-Hill, Inc., NY: 1994)

[1.36]   B.G. Lee, M. Kang and J. Lee: *Broadband Telecommunications Technology* (Artech House, Inc., Boston, MA: 1996)

[1.37]   J. Martin, K.K Chapman and J. Leben: *Asynchronous Transfer Mode: ATM Architecture and Implementation* (Prentice Hall PTR, Upper Saddle River, NJ: 1997)

[1.38]   D.L. McDysan and D.L. Spohn: *ATM Theory and Applications* (McGraw-Hill, Inc., New York, NY: 1998)

[1.39]   D.E. McDysan and D.L. Spohn: *Hands-on ATM* (McGraw-Hill Co., New York, NY: 1998)

[1.40]   D. Minoli, T. Golway and N. Smith: *Planning and Managing ATM Networks* (Maning Publications Co., Greenwich, CT: 1997)

[1.41]   R.O. Onvural: *Asynchronous Transfer Mode Network: Performance Issues* (Artech House, Inc., Boston, MA: 1995)

[1.42]   R.O. Onvural and R. Cherukuri: *Signaling in ATM Networks* (Artech House, Inc., Boston, MA: 1997)

[1.43]   M. Orzessek and P. Sommer: *ATM & MPEG-2: Integrating Digital Video into Broadband Networks* (Prentice Hall PTR, Upper Saddle River, NJ: 1998)

[1.44]   A.S. Pandya and E. Sen: *ATM Technology for Broadband Telecommunications Network* (CRC Press, Boca Raton, FL: 1998)

[1.45]   S. Schalt: *Understanding ATM* (McGraw-Hill, Inc., New York, NY: 1996)

[1.46]   O. Spaniol, A. Danthine and Effelsberg (Editors): *Architecture and Protocols for High-Speed Networks* (Kluwer Academic Press, Boston, MA: 1994)

[1.47]   W. Stallings: *High-Speed Networking – TCP/IP and ATM Design Principle* (Prentice-Hall, Inc., Upper Saddle River, NJ: 1998)

[1.48]   W. Stallings: *Advances in ISDN and Broadband ISDN* (IEEE Computer Society Press, Los Alamitos, CA: 1992)

[1.49]   M. Toy (Editor): *ATM Development and Applications – Selected Readings* (IEEE Inc., Piscataway, NJ: 1996)

[1.50]   T. Wu and N. Yoshikai: *ATM Transport and Network Integrating* (Academic Press, San Diego, CA: 1997)

[1.51]   D. Bertsekas and R. Gallager: *Data Networks* (Prentice-Hall, Inc., Englewood Cliffs, NJ: 1992)

[1.52]   U.D. Black: *Data Networks Concepts – Theory and Practice* (Prentice-Hall, Inc., Englewood Cliffs, NJ: 1989)

[1.53]   U.D. Black: *Engineering Communications Technologies* (Prentice Hall PTR, Upper Saddle River, NJ: 1997)

[1.54]   U.D. Black: *Data Communications and Distributed Networks* (Prentice-Hall, Inc., Englewood Cliffs, NJ: 1993)

[1.55]   U.D. Black: *Data Networks Theory and Practice* (Prentice-Hall, Inc., Englewood Cliffs, NJ: 1989)

[1.56]   R.L. Brewster: *Telecommunications Technology* (Ellis Horwood Ltd., Chichester, England: 1996)

[1.57]   B. Carne: *Telecommunications Primer* (Prentice Hall PTR, Upper Saddle River, NJ: 1995)

[1.58]   M. Cole: *Telecommunications* (Prentice Hall PTR, Upper Saddle River, NJ: 1999)

[1.59]   D.E. Comer: *Computer Networks and Internets* (Prentice-Hall Inc., Upper Saddle River, NJ: 1999)

[1.60]   J.H. Green: *The Irwin Handbook of Telecommunications* (Irwin Professional Publishing Chicago, IL: 1997)

[1.61]   J.K. Hardy: *Inside Networks* (Prentice-Hall Inc., Upper Saddle River, NJ: 1995)

[1.62]   G. Held: *Local Area Network Performance – Issues and Answers* (John Wiley & Sons, Chichester, England: 1994)

[1.63]   W. Hioki: *Telecommunications* (Prentice-Hall, Inc., Upper Saddle River, NJ: 1998)

[1.64]   H. Kern, R. Johnson, M. Hawkins, H. Lyke, W. Kennedy and M. Cappel: *Networking – the New Enterprise* (Sun Microsystems Press, Mountain View, CA: 1997)

[1.65]   M. Khader and W.E. Barnes: *Telecommunications Systems and Technology* (Prentice-Hall Inc., Upper Saddle River, NJ: 2000)

[1.66]   S.H. Rowe II: *Telecommunications for Manager* (Prentice-Hall, Inc., Upper Saddle River, NJ: 1999)

[1.67]   M. Schwartz: *Telecommunication Networks – Protocols Modeling and Analysis* (Addison-Wesley Publishing Co., Reading, MA: 1988)

[1.68]   W.A. Shay: *Understanding Data Communications and Networks* (PWS Publishing Co., Boston, MA: 1995)

[1.69]   J.D. Spragins, J.L. Hammond and K. Pawlikowski: *Telecommunications Protocols and Design* (Addison-Wesley Publishing Co., Reading, MA: 1991)

[1.70]   W. Stallings: *Local and Metropolitan Area Networks* (Prentice-Hall, Inc., Upper Saddle River, NJ: 1997)

[1.71]   A.S. Tannenbaum: *Computer Networks* (Prentice Hall PTR, Upper Saddle River, NJ: 1996)

[1.72]   D. Teare: *Designing Cisco Networks* (Cisco Press, Indianapolis, IN: 1999)

[1.73]   J. Van Duuren, P. Kastelein and F.C. Schoute: *Telecommunications Networks and Services* (Addison-Wesley Publishing Co., Workingham, England: 1992)

[1.74]   J. Walrand: *Communications Networks: A First Course* (WCB/McGraw-Hill Inc., Boston, MA: 1998)

[1.75]   J. Walrand and P. Varaiya: *High Performance Communication Networks* (Morgan Kaufmann Publishers Inc., San Francisco, CA: 1996)

[1.76]   R.G. Winch: *Telecommunication Transmission Systems* (McGraw-Hill Inc., New York, NY: 1993)

[1.79]   K. Ziegler Jr.: *Integrating NetWare with the Enterprise Network* (John Wiley & Sons, Inc., New York, NY: 1994)

2

BASICS OF ELECTRICAL COMMUNICATION SYSTEMS

What is communication?

> *"Boy winks at girl. Girl smiles. Here is an example of communication wherein messages are sent and received ..."*
>
> H.S. Black: *Modulation Theory*

2.1   Communication Systems

2.1.1   *Communication — A mode of transferring information*

Communication[*] is a mode of information transmission from one point to another [2.1]. The sending-end is termed as the *source of information* or a *transmitter* and the receiving end is called the *sink* or the *receiver*. The intervening medium over which the transmission takes place is termed as the *channel*.

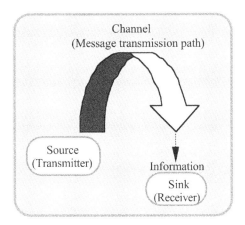

Fig. 2.1: A basic communication system

The transferred entity between the source and the sink that bears useful information is known as the *message*. A message is composed of signals constituted by a set of *signal elements*. Some specific examples of signal elements are presented in Fig. 2.2.

When does a set of signal-elements constitute a message?

A message is constructed when the signal elements are arranged in a predetermined manner so that the source and the sink have a prior knowledge of the type of arrangement. Also the arrangement of the signal elements should confirm to a meaningful message. In other words, the

---

[*] The term "communication" is derived from the Latin word *communicare* meaning "to share".

permutation of signal elements should attribute a semantism to the stretch of such elements transmitted. For example,

- When the dashes and dots of the telegraphic code are sent in a manner known a priori (as decided by the Morse code combinations), the sequence of such dashes and dots will be an "information-bearing message". The encoding and decoding of the message is feasible only if knowledge on the format of the code exists a priori between the sending and receiving ends. The decoding then renders the extraction of useful information from the signal elements transmitted

- Likewise, when the alphabets of a language are arranged in a preset (known) sequence as words and sentences, it will correspond to a message (with a meaningful information content in it), and presumably the language is known a priori to the users.

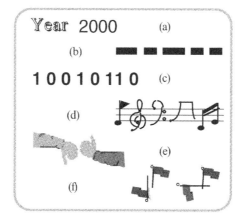

Fig. 2.2: Examples of signal elements
(a) Alphanumeric characters; (b) Dots and dashes of Morse code; (c) Ones and zeroes of a binary code;
(d) Musical notes; (e) Sign language; and (f) Semaphore (flag signs used by the navy)

Communication, in living systems, is achieved or realized through perceptions such as audio, visual, and tactile sensations. Such perceptive interactions between the species via hearing, seeing or touching, lead to a communication process. Each of these modes of communication uses a set of distinct signal-elements. The verbal communication through speech and hearing among human beings is rendered feasible by means of a language constituted by a set of words uttered. Similar communication among animals could be, for example, by means of growls and in birds, through birdcalls. Other typical audio perceptive communications include the public address like soapbox oration, a tom-tom message, etc.

A simple visual communication could be (in the perspectives of Black [2.1]) "the boy winking at a girl". Reading of a newspaper and browsing through a Web site are also day-to-day examples of visual communication. Corresponding sets of gestures such as "winking" or actions such as viewing the gamut of news prints or surfing along the scores of Web pages portray the set of visual signal-elements. Likewise, the on-off switching of a lamp in an encoded fashion or the semaphore signaling with a flag are other examples of a visual communication. Similarly, the sign language provides the signal elements for the disabled persons to exchange information visually. Further, the smoke signal of American Indians is again a known example of visual mode of communication.

The tactile basis of message transfer among living systems is well known. The withering response of a *mimosa* (touch-me-not) plant to a touch is a typical communication that the plant receives so as to protect itself against any possible predatorial encroachment. Reading the Braille characters, for example, constitutes a touch-based communication.

Thus, communication in the most general format refers to how message from a source is delivered to a receptor. In the viewpoint of mass media it describes what is heard, read, seen, or felt. With the advent of the technological breakthroughs that the world has witnessed since the nineteenth century, a new class of mass media has emerged in which the communication is enabled via electrical methods. As indicated in the previous chapter, this unique mass media of communication is termed as the *electrical communication* [2.2 –2.10].

2.1.2 *Electrical communication systems*

Electrical communication refers to the mode of transmission of messages by electrical means, with the ultimate destination being human perception or a machine (such as computer or any other data terminal equipment). The following can be considered the engineering infrastructure that governs a practical electrical communication system:

✓ Types of signal sources
✓ Characteristics of electrical signals
✓ Information-bearing ability of signals
✓ Useful and redundant information content in the electrical signal
✓ Source and mode of interference of signal-obscuring impairments such as noise
✓ Processing of signals for optimal transmission through a medium
✓ Characteristics of transmission media
✓ Detection and extraction of information from the received signal.

Electrical communication as a technology uses a plethora of hardware commencing from a humble piece of copper wire. It is based on all the intriguing concepts behind the equations of Maxwell. In the modern context, electrical communication may refer to a simple telegraphic communication between two remote stations. Or it can be a wireless toy-control from the hands of a child, an endless telephonic conversation between a boy and a girl (who preferred to skip Black's strategy!), thousands of videos cabled through, and zillions of bits crunched across computers. The panoply of modern electrical communication in an enigmatic scale, could even be the much anticipated radio astronomical replies from "something out there" in space. In short, electrical communication embraces any electrical process that represents, transforms, interprets, or processes information among persons or machines situated at different locales.

To understand the underlying concepts, discover the associated engineering and explore the methods of electrical communication deployed in the electrical communication endeavors, it becomes crucial to study the fundamentals of electrical engineering principles as applied to transference of information. Hence, the present chapter highlights the basics of electrical communication and the associated concepts of signals and systems.

The scope of electrical communication covers in a global sense the totality of techniques and electrical/electronic means necessary for optimal transfer of information via a given transmission medium in the presence of possible impairments such as electrical noise. Though the generic art of telecommunication implies electrically communicating "at a distance", in a restricted sense, it refers to the technology and procedures adopted to facilitate optimum information transfer by means of public network utilities (such as PSTN) and/or private facilities such as LAN.

To meet the scope of this textbook poised towards this restricted meaning of telecommunication, relevant basics on telecommunication networks beyond the detailed discussions on electrical communication are also presented in this chapter.

2.1.3 *Characteristics of electrical signals*

Consistent with the evolution of electrical communication systems discussed in Chapter 1, indicated below are brief descriptions of the classical electrical means of communication developed, which eventually paved the way to the modern information highway. Since electrical communication involves transference of the message-bearing signals across an electrical system, it is necessary to understand the interactive aspects of electrical signal and the communication system that handles it. The interaction of electrical signal (elements) and the system (of electrical network) refer to the associated communication process. It can be ascertained by analyzing the response of the system to the signal input. In Fig. 2.3, the response of a linear system is denoted by g(.) in time domain.

For the purpose of analysis, a *signal* can be characterized as follows [2.3, 2.4]:

- ✓ In general, a signal is complex-valued; that is, $f(t) \equiv |f(t)| e^{j\theta(t)}$
- ✓ A signal f(t) is a *single-valued function* of time
- ✓ The energy associated with a signal is finite-valued. That is,

$$\int_{-\infty}^{+\infty} |f(t)|^2 \, dt \leq \infty \qquad (2.1)$$

- ✓ A signal is *periodic* if it repeats itself exactly after a fixed length of time. That is, f(t +T) = f(t) for all T which denotes the *signal period*
- ✓ A signal f(t) for which there is no specific value of T that defines the periodicity, is termed as *aperiodic* or *nonperiodic*
- ✓ A signal is considered as random when there is some degree of uncertainty before it actually occurs. A signal is nonrandom or deterministic when there is no uncertainty in its value at any instant of time.

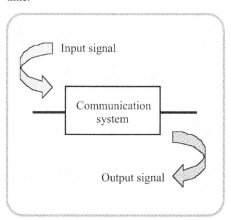

Fig. 2.3: A linear system representation of a communication link

The system depicted in Fig. 2.3 responding to a signal can be characterized as follows:

- ✓ *Linear or nonlinear*: If a system is linear, then the superposition principle applies. That is, if $f(t) = [a_1 f_1(t) + a_2 f_2(t) + \dots]$, then $g[a_1 f_1(t) + a_2 f_2(t) + \dots] = a_1 g[f_1(t)] + a_2 g(f_2(t) + \dots]$ where a_1, a_2, ... are constants
- ✓ *Time-invariant or time-varying*: A system is time-invariant, if a time-shift in the input results in a corresponding time-shift in the output. That is, if $f_{in}(t) = f(t - \tau)$, then $f_{out}(t) = g[f(t - \tau)]$ where τ is a constant time-shift
- ✓ *Realizable or nonrealizable*: A physically realizable system cannot have an output response before an arbitrary input function is applied. That is, the output of a physical system at $t = t_o$, namely, $g[f(t_o)]$ must depend only on values of the input f(t) for $t \leq t_o$. A system having this property is known as a *realizable* or a *causal* system.

2.1.4. Types of electrical signals

The electrical signals can be broadly classified as *digital* and *analog signals*. They are defined as follows:

Digital signals

When the transmission of information is done by a sequence of discrete electrical voltage levels, these discrete electric levels constitute a set of digital signals. For example, computers transmit data using digital signals [2.7]. Considering a repetitive sequence of digital signals, it can be of two types as illustrated in Fig. 2.4.

    ✓   Unipolar pulses
    ✓   Bipolar pulses.

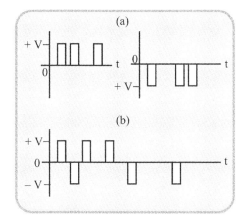

Fig. 2.4: Digital signals: (a) Unipolar signals and (b) bipolar signal

Analog signals

In contrast to the digital format of signals, the analog signals are constituted by a continuous variation of voltage or current level.

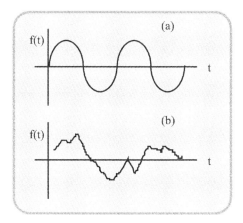

Fig. 2.5: Analog signals

Continuous signal can be of two categories as shown in Fig. 2.5.

    ✓   Periodic signal with a period, T (second) and corresponding frequency, $f = 1/T$ Hz
    ✓   Aperiodic signal that has no periodicity.

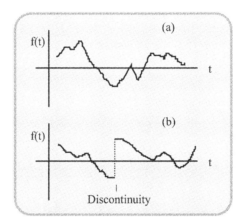

Fig. 2.6: (a) Continuous and (b) piecewise-continuous signals

The analog signal can also be *piecewise* or *continuous*. That is, when a discontinuity of the function with respect to time is observed, the signal can be treated as continuous only on a piecewise basis as depicted in Fig. 2.6.

2.1.5   Spectral characteristics of signals
The spectrum of a signal refers to the description of a (signal) waveform in the frequency domain, which has a correspondence to the description of the signal in the time-domain. The spectral characteristics of a signal can be ascertained by the following mathematical analyses:

Spectral analysis of a periodic function
In reference to a periodic signal f(t), let T_0 be the fundamental period. Then, f(t) can be expanded as an infinite sum of sinusoidal and/or cosinusoidal functions provided that :

    ✓   f(t) is periodic
    ✓   f(t) is single-valued
    ✓   f(t) is continuous or piecewise-continuous in the period T_0
    ✓   Energy associated with the signal is finite-valued. That is,

$$\int_{0}^{T_0} |f(t)|^2 \, dt \leq \infty \tag{2.2}$$

    ✓   f(t) has only a finite number of maxima and minima over, T_0
    ✓   f(t) is integrable. That is,

$$\int_{0}^{T_0} |f(t)| \, dt < \infty . \tag{2.3}$$

The above conditions are known *as Dirichlet's conditions*. The expansion of a real valued f(t) is given by a *trigonometric Fourier expansion* as follows:

$$f(t) = A_0 + \sum_{n=1}^{\infty} A_n \cos\left(\frac{2\pi nt}{T_0}\right) + \sum_{n=1}^{\infty} B_n \sin\left(\frac{2\pi nt}{T_0}\right) \tag{2.4a}$$

where A_0 denotes the average value of f(t), and A_0, A_n, and B_n are known as *Fourier coefficients*. They are explicitly given by:

$$A_0 = \frac{1}{T_0} \int\limits_{-T_0/2}^{+T_0/2} f(t)dt$$

$$A_n = \frac{2}{T_0} \int\limits_{-T_0/2}^{+T_0/2} f(t)\cos(n\omega_0 t)dt$$

$$B_n = \frac{2}{T_0} \int\limits_{-T_0/2}^{+T_0/2} f(t)\sin(n\omega_0 t)dt \qquad (2.4b)$$

where $\omega_0 = 2\pi f_0$, $f_o = 1/T_o$ and $n = 1, 2, \ldots \infty$. The entity f_0 is known as the *fundamental frequency component* and nf_0 is the nth *harmonic frequency* of the signal.

___

Example 2.1

A given rectangular function is defined over a time interval $(0, 2)$. That is,

$$f(t) = \begin{cases} +1 & 0 < t < 1 \\ -1 & 1 < t < 2 \end{cases}$$

as illustrated in Fig. 2.7.
Find the Fourier series of the function.

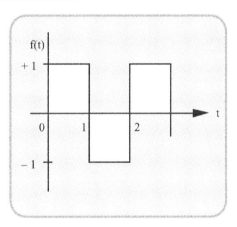

Fig. 2.7: A rectangular periodic function of time

Solution

The given function satisfies all the *Dirichlet's conditions*. Using Eqn.(2.4),

$A_0 = 0$

$A_n = \dfrac{4}{m\pi}\sin(m\pi t)$, where $m = 1, 3, 5, \ldots \infty$

$B_n = 0$

Hence the Fourier expansion of f(t) is given by

$$f(t) = \frac{4}{\pi}[\sin(\pi t) + \frac{1}{3}\sin(3\pi t) + \frac{1}{5}\sin(5\pi t) + \ldots \infty]$$

___

From the example shown above, the function illustrated in Fig. 2.7 can be obtained by the superposition of a weighted sum of sine functions as illustrated in Fig. 2.8.

An alternative form of the trigonometric Fourier expansion of a periodic function is given by:

$$f(t) = C_0 + \sum_{n=1}^{\infty} C_n \cos[2\pi nt/T_0 - f_n] \tag{2.5}$$

where $C_0 = A_0$, $C_n = [A_n^2 + B_n^2]^{1/2}$ and $\phi_n = \arctan[B_n/A_n]$.

Yet another Fourier expansion of a periodic function can be specified, in general, in exponential form as:

$$f(t) = \sum_{-\infty}^{\infty} D_n \exp[j2\pi nt/T_0] \tag{2.6a}$$

where

$$D_n = \frac{1}{T_0} \int_{-T_0/2}^{T_0/2} f(t) \exp[-j2\pi nt/T_0] dt$$

$$\equiv \frac{C_n}{2} \exp(-j\phi_n) \tag{2.6b}$$

and $D_0 \equiv C_0$.

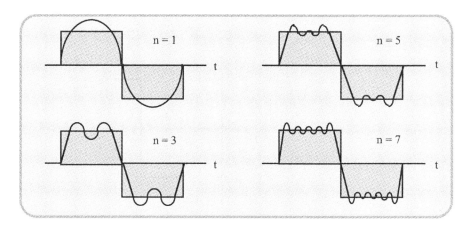

Fig. 2.8: Approximation of a rectangular waveform by superposition of its Fourier sine components

---

Example 2.2
Determine the Fourier series of an impulse train shown in Fig. 2.9.

Solution
The function shown in Fig. 2.9 can be explicitly written in terms of delta-Dirac function $\delta(t)$ as follows:

$$f(t) = \sum_{-\infty}^{\infty} \delta(t - nT_0)$$

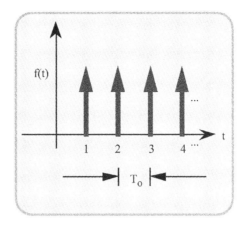

Fig. 2.9: Impulse train function

In terms of the exponential form of Fourier series,

$$f(t) = \sum_{-\infty}^{\infty} D_n \exp[j2\pi nt / T_0]$$

where

$$D_n = \frac{1}{T_0} \int_{-T_0/2}^{T_0/2} f(t) \exp[-j2\pi nt / T_0] dt = \frac{1}{T_0} \int_{-T_0/2}^{T_0/2} \delta(t) \exp[-j2\pi nt / T_0] dt = \frac{1}{T_0}.$$

Therefore,

$$f(t) = \sum_{n=-\infty}^{\infty} \delta(t - nT_0) \equiv \frac{1}{T_0} \sum_{n=-\infty}^{\infty} \exp(j2\pi nt / T_0)$$

Odd and even symmetry signals

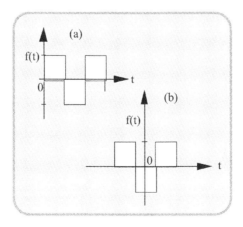

Fig. 2.10: Periodic functions
(a) Odd and (b) even

103

The real-valued signal f(t) expanded via trigonometric Fourier series may correspond to either *even* or *odd* type. The even function is symmetric about the ordinate and the odd function is symmetric about abscissa as illustrated (Fig. 2.10).

For even functions, the Fourier coefficients $B_n \equiv 0$, and for odd function, $A_n \equiv 0$. Further, for even harmonic signals, $f(t+T_0/2) \equiv f(t)$ and for odd harmonic functions, $f(t+T_0/2) \equiv -f(t)$.

Problem 2.1

Identify whether the function illustrated in Fig. 2.11 is odd or even. Hence determine its Fourier series expansion.

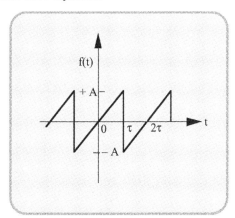

Fig. 2.11: The saw-tooth function

Spectral analysis of an aperiodic signal

Suppose the signal f(t) is nonperiodic as illustrated in Fig. 2.12. Its spectral characteristics can be specified via Fourier transform, $\mathcal{F}(f)$ as follows:

$$f(t) = \frac{1}{T_0} \int_{-\infty}^{\infty} F(f) \exp(j2\pi ft) df \qquad (2.7)$$

where $\mathcal{F}(f) = \int_{-\infty}^{\infty} f(t) \exp(-j2\pi ft) dt$.

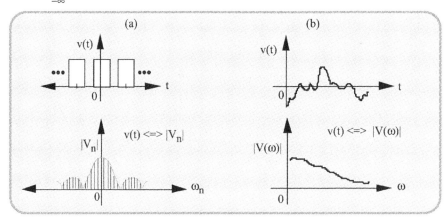

Fig. 2.12: Time and frequency domain representation of: (a) A periodic and (b) an aperiodic signal

The corresponding Dirichlet's conditions are as follows:

- f(t) is integrable. That is, $\int_{-\infty}^{\infty} |f(t)|dt < \infty$

- f(t) in any time interval can have only finite number of discontinuity

- f(t) can have only finite number of maxima and minima in any time interval.

A Fourier transform relation of a pair $v(t) \Leftrightarrow V(f)$ denoted by, $V(f) = \mathcal{F}[v(t)]$ or $v(t) = \mathcal{F}^{-1}[V(f)]$ where $\mathcal{F}$ and $\mathcal{F}^{-1}$ denote the Fourier transform and inverse Fourier transform operators respectively. In general, V(f) can be complex. Its magnitude depicts the continuous spectrum of a typical aperiodic signal (in contrast to the Fourier series-based discrete spectrum of a periodic signal).

Example 2.3
Determine the Fourier spectrum of the delta function.

Solution
The δ-function is as illustrated in Fig. 2.13:
The Fourier transform of $v(t) = \delta(t)$ is given by:

$$V(f) = \mathcal{F}[\delta(t)] = \int_{-\infty}^{\infty} \delta(t)\exp(-j2\pi ft)dt \equiv 1 \text{ [shifting property of } \delta(t)] \text{ (See Appendix 2.1)}$$

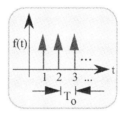

Fig. 2.13: The delta-function

Hence, the spectrum of δ(t) is flat and continuous (Fig. 2.14).

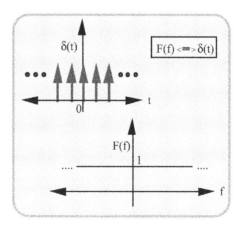

Fig. 2.14: The continuous flat spectrum of δ(t)

2.1.6   Concept of signal bandwidth

Consider the rectangular pulse train shown in Fig. 2.15 and its discrete spectrum obtained via the exponential form of Fourier series expansion.

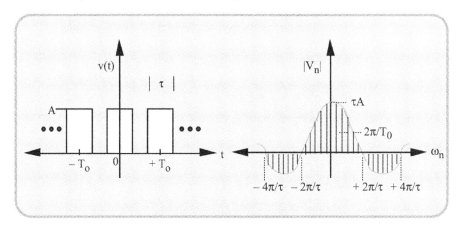

Fig. 2.15: Rectangular pulse train and its discrete spectrum

Using the exponential form of Fourier expansion, the spectral components of the pulse train waveform, v(t) illustrated is given by

$$V_n = \int_{-\tau/2}^{+\tau/2} Ae^{-2\pi f_n t} dt$$

$$= A\left[e^{j2\pi f_n \tau/2} - e^{-j2\pi f_n \tau/2} \right]/(j2\pi f_n)$$

$$= \tau A \sin(w_n \tau/2)/(w_n \tau/2) \qquad (2.8)$$

where $\omega_n = 2\pi f_n = 2\pi n f_0$ and $f_0 = 1/T_0$.

For each characteristic (distinct) value of n, the amplitudes of the spectrum, namely, $|V_n|$s are discretely specified. That is, the value of $|V_n|$s are eigen (discrete)-valued as illustrated in Fig. 2.15. In reference to the discrete spectrum presented in Fig. 2.15, the following inferences can be made:

    ✓   As T_0 decreases (meaning more pulses per second), the spacing between the discrete spectral components increases

    ✓   As T_0 increases (meaning less pulses per second), the spacing of spectral components decreases. That is, a slow variation in time domain, refers to crowding of spectral components at the low frequency region. Conversely, more rapid variations in the time-domain would indicate more high-frequency components in the frequency-domain

    ✓   As the pulse-width (τ) decreases, the frequency content of the signal extends out over a larger frequency range. The first zero-crossing is at $\omega = 2\pi/\tau$. That is, the frequency spread is inversely proportional to the pulse-width

    ✓   Bandwidth of periodic pulses: For periodic impulses of period T_0, the spectral components are discrete and spaced by $2\pi/T_0$. As the pulse width $\tau \to 0$, the bandwidth (namely, the effective extent of significant frequency components of the Fourier spectrum) $\to \infty$.

Problem 2.2

Shown in Fig. 2.16 are two distinct single pulses with different amplitudes and pulse widths.

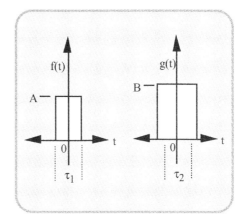

Fig. 2.16: Rectangular pulses

Suppose $h_2 > h_1$ and $\tau_2 < \tau_1$.

- Determine and sketch the continuous spectra of the waveforms
- Justify the following statements:

"The shorter the pulse, the broader its spectrum"
"Spiky pulse shape has a spectrum rich in high frequencies"

Assume that the pulses depicted in Fig. 2.16 are periodically repetitive. Assume the period as T_1 for A and as T_2 for B. Discuss the implications on the bandwidth of the waveform in each case separately, when the pulse repetition rate (in each case) becomes very small or very large.

2.1.7   Bandwidth and data rate

Bandwidth is (an effective) range representing the constituent frequency components (of a signal), which contain most of the energy in the signal. In practice, any transmission medium would permit only a limited spectral band of the signal. In order for signal transmission to take place effectively over that medium, most of the energy associated with the signal should be transported across the medium. For this to occur, the following condition should be met: The transmission band of the medium should be greater than or equal to the effective bandwidth of the signal.

For different telecommunication services this requirement is fulfilled as described below:

Analog voice signal considerations

Consider a continuous (analog) voice signal, $v(t)$. Its spectrum is specified by the Fourier transform, $V(f) = \mathcal{F}[v(t)]$. This time-domain/frequency-domain Fourier pair is denoted by $v(t) \Leftrightarrow V(f)$. Fig. 2.17 illustrates that the spectrum $V(f)$ can be truncated at $f \approx 3.5$ to 4 kHz beyond which the spectral amplitude is negligible. Hence, the effective bandwidth (BW) of the voice signal can be 4 kHz at most.

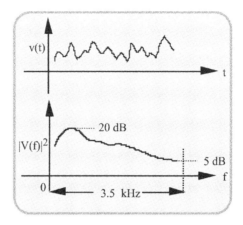

Fig. 2.17: Time and frequency domain representation of an analog voice signal

Suppose a copper cable is used to transport this voice signal (Fig. 2.17) as illustrated in 2.18. It represents an iterated network of distributed L, C, R, G components. A signal applied on this line will be subjected to lowpass filtering with its high frequency components curtailed. The extent of such filtering is determined by the total reactive (capacitive and inductive) influence that is exerted by the transmission line on the signal. This influence will depend on the length of the line. That is, the total distributed L, C, R, G values increase with the increase in the length of the line. Hence the line will pose a length-dependent bandwidth to the signal applied to it. This bandwidth refers to the passband of the transmission channel. Thus, the integrity of signal transmission along the line is decided by the (bandwidth × length) product.

Again considering voice transmission, the cable as a transmission medium should have passband characteristics equal to at least 3.4 to 4 kHz width in order for the received voice signal to almost retain its waveform integrity. (*Note*: Speech spectrum normally has a spectral range of about 20 Hz to 20 kHz. However, most of the energy of the speech signal is at the lower end of the spectrum up to about 4 kHz. Hence, even if the baseband of the speech signal is restricted to 4 kHz, as a result of the passband characteristics of the medium, the received signal will still be "intelligible". That is, it will maintain its *articulation efficiency* or the *understandability* characteristics.)

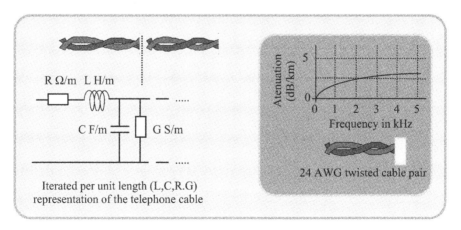

Fig. 2.18: Copper-cable as the transmission medium for a voice signal

Digital signal considerations

Consider a bipolar binary signal v(t) illustrated in Fig. 2.19.

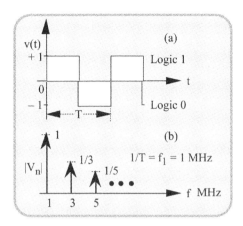

Fig. 2.19: A bipolar signal: (a) Waveform and (b) its discrete spectrum

Suppose a binary data is sent by means of a sequence of bipolar pulses illustrated in Fig. 2.19. Then the data rate is equal to $2f_1$ bps where f_1 is the pulse repetition frequency and $f_1 = 1/T$. Since v(t) is an odd function, its Fourier series is given by (see Example 2.1):

$$v(t) = \sum_{k=1,3,5,...}^{\infty} (1/k)\sin(2\pi f_1 t) \tag{2.9a}$$

For example, when $f_1 = 1$ MHz,

$$v(t) = \sin(2\pi \times 10^6 t) + \frac{\sin(2\pi \times 3 \times 10^6 t)}{3} + \frac{\sin(2\pi \times 5 \times 10^6 t)}{5} + ... \infty \tag{2.9b}$$

The discrete spectrum of v(t) is also sketched in Fig. 2.19.

Suppose a binary data is sent by means of a sequence of pulses. If the spectrum of this pulse stream is truncated, the approximated waveforms are illustrated in Fig. 2.20 depending on the extent of truncation of the harmonics carried out.

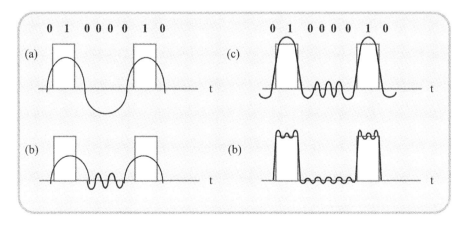

Fig. 2.20: Time-domain approximation of a pulse train supporting a data rate of 2000 bps with different extents of spectral truncation at: (a) 500 Hz, (b) 1300 Hz, (c) 1700 Hz, and (d) 2500 Hz

From Fig. 2.20, the following can be inferred:

- The waveform with its frequency band truncated even at 1700 Hz is still a fair representation of the bits transmitted

- That is, despite curtailing or limiting the bandwidth of a digital signal (corresponding to a data stream at a rate of 2000 bps) to 1700 to 2500 Hz, the time-domain integrity of the bits remains good. Hence, it can be concluded that: Under noise-free conditions, if the data rate of a digital signal is W bps, the signal can be represented in time-domain fairly in good shape, if the effective bandwidth, B is taken such that $W < B < 2W$

- The example illustrated in Fig. 2.8 indicates similar effects on the waveform of a bipolar signal when the spectrum of the signal is truncated to different extents.

In practice, pulses are not transmitted as they are. Several modulation techniques are used for pulse/data transmission. In such modulated systems, the bandwidth requirement will be less than 2W as will be discussed later.

Video signal considerations

As a third case consider a video signal. A TV picture denotes an ensemble of pixels on the screen obtained by scanning an image. The scanning process yields a set of scan and trace lines. As illustrated in Fig. 2.21, there are scan lines and horizontal retraces and a vertical retrace line across a screen.

In order to realize good resolution, the U.S. standard specifies 483 horizontal scan lines to be scanned at a rate of 30 screen scans per second. Normally, 525 lines are considered so that 43 are blanked out during the vertical trace intervals. Hence, the scanning frequency corresponds to [525/(1/30)] scans per second. The time involved thereof is, 63.5 μs per line, which is split as 52.5 μs per video line, and 11 μs for the horizontal retrace.

Considering (4 × 3) as the width-to-height ratio of a TV screen, the horizontal resolution is (4/3) × 483 lines. Of this, only about 70 percent is considered in order to account for the subjectiveness of resolution perceived. That is, the horizontal resolution is: (4/3 × 338) = 450 lines.

Considering the alternate black and white aspects of the pixels, 450/2 = 225 cycles occur in 52.5 μs. Or, 1 cycle refers to a period of 52.5/ 225 μs. Hence, the corresponding frequency f = 1/T = 4.3 MHz is the occurrence rate of the signal elements (pixels) of a TV picture.

Thus, a TV signal requires at least a bandwidth of 4.3 MHz.

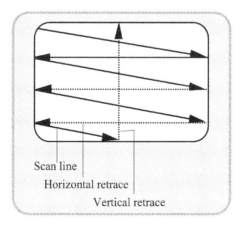

Fig. 2.21: A TV screen composition of scan and trace lines

Problem 2.3

In manual telegraphy, the signal is made of dots and dashes following the Morse code. The length of a dot in Morse code is taken as one base time unit. The duration of a dash is three base time units. The space between dots and dashes in one letter is one base unit, the time between letters is three base units and the spacing between words is set as seven base units. Determine the approximate bandwidth required to transmit a manual telegraphic message in terms of the base time units.

2.2 Impairments to Electrical Signal Transmission

2.2.1 Signal distortions

The electrical signal traversing a communication link normally undergoes impairments. In other words, the integrity of the signal may suffer and the received waveform will be a "distorted" version of its original counterpart. There are three possible distortions that a signal may face. They are as follows:

✓ Amplitude distortion
✓ Frequency distortion
✓ Phase distortion.

Amplitude distortion

The *amplitude distortion* refers to a variable attenuation with the envelope of the signal waveform. It happens when the transmission system poses a nonlinear transfer function to the incoming signal as illustrated in Fig. 2.22.

The nonlinear transformation (namely, $v_{out} = g(v_{in})$) across the transmission path renders the output waveform envelope visibly changed as illustrated. (Thus, the amplitude distortion and *envelope distortion* are synonymous.) A Fourier analysis would indicate the presence of harmonic frequency components in this distorted output signal, which are not present in the input signal. Thus, the amplitude distortion can be identified by the presence of harmonic components not originally present in the signal. As a result, the amplitude distortion is also known as *nonlinear* or *harmonic distortion*. The causative mechanism of amplitude distortion is the presence of nonlinear device or components in the transmission system across which the signal may pass through.

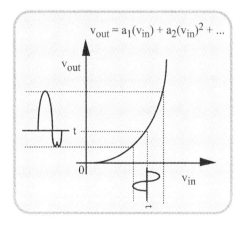

Fig. 2.22: Amplitude distortion due to nonlinearity in the transmission path of the signal

Frequency distortion

The *frequency distortion* occurs when the transmission system attenuates or amplifies different frequency components of a signal to different extents. That is, the transfer function is frequency-dependent. For example, if the system transfer function is given by $g(t) \Leftrightarrow G(f) = k/(1 + jf/f_c)$ where k is a constant and f_c is the cut-off frequency, the signal will be subjected to a low-pass filtering with f_c at the 3 dB cut-off frequency as illustrated in Fig. 2.23.

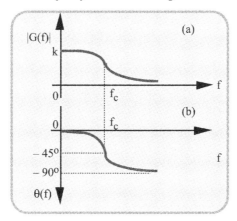

Fig. 2.23: Transfer function of a transmission system: (a) Frequency response and (b) phase response

The frequency-dependent of G arises as a result of reactive components being present in the transmission system. For example, the distributed inductance and capacitance of a transmission line will make the transfer function frequency dependent. In general G(f) is complex. That is, $G(f) = |G(f)|\angle\theta(f)$.

Phase distortion

Inasmuch as the phase angle $\theta(f)$ associated with the transfer function of the transmission system is also frequency-dependent, the phase angles of each frequency component of the signal will undergo different extents of phase change across the system with a transfer function, G(f).

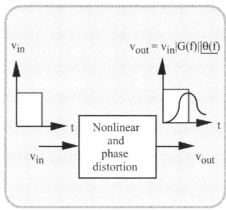

Fig. 2.24: A rectangular waveform subjected to amplitude and phase distortions

Corresponding to the low-pass filter function $G(f) = k/(1 + jf/f_c)$ considered above, the phase response θ versus f can be plotted as shown in Fig. 2.23. The implication of the frequency-dependent phase shift $\theta(f)$ on a signal can be understood by studying the phase-distorted output signal function in time domain. A phase-shift $\theta(f)$ in a sinusoid corresponds to a value $2\pi f\tau = \omega\tau$

where τ is the delay imposed by the phase-lag suffered by the sinusoid across the transmission system. A linear relation between θ and f leads to a constant delay on the sinusoids of all frequencies. However, a nonlinear phase-response will introduce varying extents of delays on different frequency components of the signal.

In practice, the signal waveform may undergo both amplitude distortion (due to the presence of nonlinear active elements in the circuit) as well as frequency/phase distortions as a result of passive elements and distributed reactive components that prevail in the transmission system. Typically, a rectangular waveform subjected to both amplitude and phase distortions will appear as depicted in Fig. 2.24.

2.2.2 *Signal attenuation*

Attenuation represents a fall in the signal strength as the signal is transported along a medium (channel) over a distance. It is a characteristic of the type of medium. The typical channel (wireline or wireless) which attenuates the signal by absorbing a part of energy in it, is known as a lossy channel or a lossy transmission system [2.3].

Consider the signal-strength $v_1(t)$ at a point A where the signal enters the medium. At the point B where the signal emerges, $v_2(t) < v_1(t)$ due to the attenuation suffered by the signal along the length ℓ of the transmission medium. This attenuation occurs due to finite losses the signal may face in the intervening medium. (For example, the finite conductivity of the copper wire will attenuate the signal along its path.) Likewise, rain can attenuate wireless signal in certain frequency ranges.

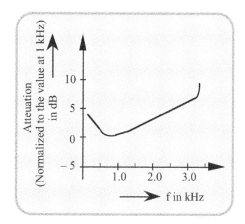

Fig. 2.25: Attenuation of a signal along the length of a lossy medium

Suppose corresponding to the voltages v_1 and v_2, the signal power levels are P_1 and P_2 respectively at A and B separated by a distance ℓ meters. Then the *insertion loss* (IL) due to the transmission line between A and B is given by:

$$IL = \text{Insertion loss} = \log_e(v_1/v_2) \qquad \text{nepers} \qquad (2.10)$$

Defining α as the attenuation coefficient of the transmission line under consideration, the power loss between A and B in *decibel* (dB) units is given by

$$\alpha\ell = 10 \log_{10}(P_2/P_1)$$
$$= 10 \log_{10}(|v_2/v_1|^2)$$
$$= 20 \log_{10}(|v_2/v_1|). \qquad (2.11)$$

In terms of IL, the attenuation coefficient α is equal to IL/ℓ nepers/meter or kft. (The unit 'neper' is named after the mathematician John Napier and "decibel" is used to honor Alexander Graham Bell; 1 neper = 8.868 dB.)

In general, α is a function of frequency. As such, different frequency components of a signal suffer different extents of attenuation. For a typical voice signal, α versus f characteristics is shown in Fig. 2.25. The attenuation factor α(f) represents the transfer function of the transmission line.

Repeaters/amplifiers are commonly required along the transmission path of a signal to compensate for the signal attenuation suffered.

2.2.3   Delay distortion

In addition to a signal experiencing an attenuation of its envelope (amplitude) level across a medium (channel), the phase angles associated with the signal components also would change as a function of frequency.

The signal may face phase distortion as indicated earlier. That is, different frequency components of a signal may undergo different extents of delay while being transmitted across a medium inasmuch as α is a function of frequency. Typical delay versus frequency characteristics of a voice signal is as shown in Fig. 2.26.

Historically, equalization of delay distortion was done on telegraph and telephone lines using inclusion of inductive (loading) coils at different points along the line. This compensates for the capacitive effects of the line and equalizes the delay. In modern telephone transmission systems delay equalization is done electronically, as will be indicated in Chapter 3.

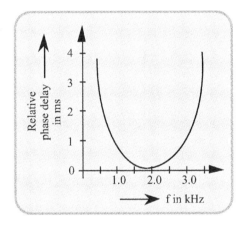

Fig. 2.26: Envelope delay distortion

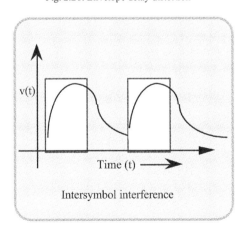

Fig. 2.27: Pulse (symbol) to pulse (symbol) interference due to phase distortion

Without delay equalization of analog voice signals, the received signal would be unnatural due to the presence of delayed echoes. In digital (pulse) transmissions delay distortion would cause inter-symbol interference. That is, in time-domain, excessive delay distortion corresponds to a transient overlap of a pulse onto a succeeding pulse causing pulse-to-pulse interference as illustrated in Fig. 2.27. More details on attenuation and delay distortion will be indicated in Chapter 3.

2.2.4   Noise considerations

Noise is an undesired electrical entity interfering and corrupting a signal. This interfering entity corresponds to random voltage or current variations in a circuit. In an electromagnetic field point of view, noise refers to random electric and magnetic field intensities pervading the signal of interest. Noise is a major limiting factor in a communication systems performance. The classes of noise, which are of concern in telecommunication systems, are as follows:

      ✓   Component/device noise
      ✓   Intermodulation noise
      ✓   Crosstalk
      ✓   Impulse noise.

These undesirable entities in the communication system can be broadly categorized into:

      ▪   *Noise*: System generated or system-inherent undesirable voltage/or currents caused by passive components and/or active elements

      ▪   *Interference*: Undesirable voltage/current induced into the system by external causes.

Types of component/device noise

Thermal noise or Johnson noise: This is due to random agitation of charges in a conducting medium as a result of acquired thermodynamic energy. Hence, the noise voltage v(t) that appears across a resistance R is given by its mean squared value as follows:

$$E[v^2(t)] = \frac{2(\pi k_B T)^2}{3h} R \quad (volt^2)$$ (2.12)

where,

k_B: Boltzmann's constant $= 1.37 \times 10^{-23}$ J/K
h: Planck's costant $= 6.62 \times 10^{-34}$ J s
T: Temperature in °K.

     The amplitude distribution of this random thermal noise is Gaussian with,

      ✓   Mean value (μ) = 0
      ✓   Variance (σ^2) $\equiv$ E $[v^2(t)]$.

Further, its power spectral density, G(f) is equal to: $\dfrac{2Rh|f|}{\exp(h|f|/k_B T) - 1}$ volt²/Hz. This corresponds to what is known as *white noise* spectral characteristics with flat (power spectral) values versus frequency as illustrated in Fig. 2.28. However, if this white noise is passed through a system with specific filter characteristics, its bandwidth gets restricted. This band-limited noise is called *colored noise* or *pink noise*. In summary, the characteristics of thermal noise are as follows:

      ✓   Its spectrum is flat. That is, its frequency components are of the same extent along the frequency spectrum. As a result, the thermal noise is regarded as a "white noise" due to its flat spectral characteristics
      ✓   Thermal noise power density: $N_o = k_B T$ watts/Hz

      ✓   Total noise power $= \int_0^\infty N_0 df = N$.

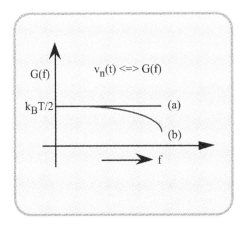

Fig. 2.28: Spectral characteristic of: (a) White and (b) colored noise

In a system with a finite passband (namely, the bandwidth, BW), the total noise power in watts is given by:

$$N = \int_0^{BW} N_0 df = (k_B TBW) \qquad W \qquad (2.13a)$$

$$= [-228.6 + 10 \log_{10}(T) + 10 \log_{10} (BW)] \qquad dBW \qquad (2.13b)$$

where dBW is the decibel representation of N (in watts) in reference to 1 W. That is, $N_{dBW} = 10 \log$ [(N in watts/1 watt)]

Equivalent thermal noise voltage
The thermal noise power, N associated with a resistor R can be regarded as if it is caused by an root-mean-squared (rms) voltage source of V_n. In terms of R and N, V_n is given by:

$$V_n = (4NR)^{1/2} = (4k_B TBWR)^{1/2} \qquad V \qquad (2.14)$$

Example 2.4
Express the noise power generated by a 10 KΩ resistor at 27 °C in dBm over a passband of 10 kHz. (b) Determine the noise voltage.

Solution
(a) Noise Power $= k_B TBW = 1.38 \times 10^{-3} J/K \times (27+273) K \times 10^4 Hz$
$$= 4.14 \times 10^{-17} \qquad W$$
Noise power (in dBm) $= 10 \log (4.14 \times 10^{-17}/1 \times 10^{-3})$
$$= -133.83 \qquad dBm$$

(b) Noise voltage $= [4(k_B TBW)R]^{1/2} \qquad V$
$$= [4 \times 4.14 \times 10^{-17} \times 10^3]^{1/2} \qquad V$$
$$= 0.407 \, \mu V$$

Other types of device-based noises

- *Shot Noise (Bursty noise)*: The bursts of electron emissions from a cathode surface are invariably random and each burst has unique stochastical characteristics. Also, such bursts occur at random intervals. These random bursts represent shot noise in electronic devices

- *Transistor noise (1/f noise)*: The electron/hole emission and its transport through a transistor/diode change randomly. Rigorous analysis and measurements indicate that the spectrum of this noise has decreasing amplitude of spectral components at the high frequency region. (Hence, it is named as *1/f noise.*)

Intermodulation noise

When signals at different frequencies (say f_1 and f_2) share a common medium, (in the transmission system) intermodulation may occur due to nonlinearity. This causes mixing of two signals. With the result, new components of frequencies $2f_1$, $2f_2$, $(f_1 \pm f_2)$ and $(2f_2 \pm f_1)$ etc., (which were not present in the original signal) would appear in the output.

Of these frequency components, $2f_1$, $2f_2$ and $(f_1 \pm f_2)$ are second order intermodulation products that may lie outside the passband of the system and therefore, may not interfere (as unwanted noise) with the desired signal under transmission. However, the third order products, namely, $(f_1 \pm 2f_2)$ and $(2f_2 \pm f_1)$ represent components that appear in the system passband. Therefore, they are regarded as significant intermodulation noise (IM) components. Third order IM products are of prime concern in the RF stages of wireless communication receivers.

2.2.5   Electromagnetic interference (EMI)

The EMI refers to undesirable entry of electromagnetic energy into the signal-bearing media. There are two versions of EMI: The first type is known as the *radiated mode EMI* and the second is called *conduction mode EMI.*

The *radiated mode EMI*, as the name implies, is the direct pick up of undesirable electromagnetic energy radiated by external sources. Typically, the radio frequency (RF) sources (such as AM/FM transmitters) can invade the telecommunication equipment or cable in the vicinity causing RF interference (RFI).

The conduction mode represents the coupling of unwanted signal into the equipment through via ducts, back-plane, etc.

- *Cross-talk*: This refers to the unwanted coupling of electrical energy associated with a signal propagating in a transmission line (say, telephone wires) to a nearby line or a set of lines. It means that a signal propagating in a particular line may undesirably interfere with the signals in the nearby lines. The type of lines, proximity of lines, terminal impedance, and the strength of signals involved decide the extent of crosstalk.

- *Impulse noise*: These are spiky, irregular, or sporadic fluctuations of electrical voltages occurring for a short duration; their peak values are of considerable magnitude. In effect, they cause unpleasant clicks and crackles in analog transmissions such as in voice telephony. They may also constitute the primary source of error in digital communication systems.

    For example, a sharp spike over 0.01s duration may cause about 50 bits of data erased in a 4800 bps transmission.

There are several mechanisms responsible for impulse noise: Lighting, power supply transients, and interference due to EMP (electromagnetic pulses) etc. would induce unpredictable spiky noise in telecommunication lines.

Transmission impairment in telephony

In general, the transmission objectives in a telephone system are challenged by impairments on:

✓ Semantic transparency
✓ Temporal transparency

with the information flow across end-to-end connectivity in telecommunication systems.

Maintenance of semantic transparency refers to realizing a transmission objective despite prevailing impairments masking the information carried over the transmission system. The contributing facts and phenomenology of corruption introduced on the signal as indicated earlier, are signal attenuation, interference, and noise.

The temporal transparency is affected when the signal suffers excessive delays and delay variations while on transit across the network.

Measurement of noise

Normally, any disturbing entity on the signal such as noise or interference can be measured in root-mean-square (rms) scale. Such a measurement signifies the power level of the noise under measurement. Specific to voice channels, there are certain frequencies in the passband at which the invading noise could be subjectively more annoying than at other frequencies. These subjective effects are considered in the telephone circuit noise measurements via (i) *C-message weighting curve* or (ii) *psophometric weighting curve*. These curves essentially represent a filter response that weights the frequency spectrum of noise in relation to the annoying effects of noise perceived by the listeners [2.11].

- C-message weighting: This applies to a 500 type telephone set. It is used in North America

- Psophometric weighting: This is a European (CCITT) standard.

Studies conducted by Western Electric Company indicate that the voice transmission when corrupted by noise is most annoying when the noise invades the spectrum in the range of 600 to 3000 Hz. Considering this annoyance, the so-called *C-message filters* have been developed. These filters increase attenuation of the noise power at frequencies below 600 Hz and above 3000 Hz. These filters when used in noise measuring sets replicate closely the human-hearing response to noise levels as per the C-message weighting curve (shown in Fig. 2.29). In essence, C-message weighting is a selective attenuation of voice band noise in accordance with subjective effects perceived as a function of frequency.

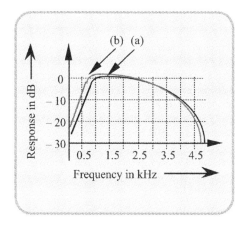

Fig. 2.29: C-message weighting curve

C-message noise measurement determines the continuous root-mean square (rms) noise power. This background noise is characterized by additive white Gaussian noise (AWGN) statistics. Being an additive noise, C-message noise is line-length dependent.

C-message noise measurements could be affected in the presence of frequency-dependent gain characteristics of equipment like compander and other automatic gain control devices. As such, a different technique known as *C-notched noise measurement* is used in practice. Here, a holding tone (1004 Hz or 2804 Hz) is applied at the transmitter. It emulates a "loaded" voice or data transmission in the sense that it indicates the presence of a signal power level comparable to that in actual message transmission. This holding tone is notched out prior to C-message filtering in the measurement set up. The noise power is then measured in the usable passband on C-message basis. The noise reading reflects the loaded circuit characteristics of a normal voice band.

Noise levels

A standard noise reference used in telephony is 1 pW. In decibel units, it refers to:

$$1 \text{ pW} \quad = \quad 10^{-12} \text{W} \quad \Rightarrow \quad 10 \log_{10}(\frac{10^{-12}}{10^{-3}}) \quad = -90 \text{ dBm} \qquad (2.15)$$

where dBm is the power-level in decibels relative to a mW. Any noise measured relative to the above reference of 1 pW is expressed in decibels as dBrn.

Example 2.5

Determine the absolute power level of a noise specified as 30 dBrn.

Solution

Given that, $\quad (P_n)_{dBm} = 30$ dBrn

$$\equiv 10 \log(\frac{P_n}{P_{rn}})$$

where P_m is the reference power level of 10^{-12} w. Therefore, $P_n = 10^{-9}$ w.

It may also be noted that 30 dBm $\equiv (-90 \text{ dBm}) + (30 \text{ dBm}) = -60$ dBm $(= 10^{-9} \text{ W})$.

When power measurement is done on a C-message weighted scale, the power level is expressed as dBrnC. In psophometric weighting, the weighted picowatts of noise is expressed as pWp. The various conversions can be summarized as in Table 2.1.

Table 2.1: Noise levels in decibel notations

Weighting Type	Noise level		Noise level relative to 1 pW
-	N dBm	⇔	(N + 90) dBm
C-message	N dBm 3 kHz flat	⇔	(N + 90) – 2 dBrn
Psophometric	N dBp 3 kHz flat	⇔	(N – 90) + 2.5 dBrnC
C-message	N dBrn 3 kHz flat	⇔	(N – 2) dBrnC

Other useful relations are:

✓ Reference noise power = 1 pW
✓ N dBC ⇔ (N – 0.5) dBp
✓ Pn pW (3KHz flat) ⇔ (0.562 × Pn) pWp.

Example 2.6

Determine the absolute power levels (in pW, mW and W) of a measured C-message weighted noise of 33 dBrnc

Solution

$$[(Pn)_{dBrnC} = 33] \equiv 35 \text{ dBrn (see Table 2.1)}$$

$$35 \text{ dBrn} \equiv (35 - 90) = 55 \text{ dBm} = 10\log[\frac{Pn(mW)}{1mW}]$$

$$\therefore (Pn)_{mw} \Rightarrow 2\times 10^{-6} \text{ mWC} = 2000 \text{ pWpWC}$$

Problem 2.4

Determine the dBrnC and dBp values of a 30 pW noise power with flat spectrum in the range of 0 to 3 kHz

Problem 2.5

Expressing the reference noise power as P_r watts, the C-message weighted power level as P_{nC} and the psophonetrically weighted power level as P_{np}, establish relevant conversion algorithms using the data in Table 2.1.

Transmission level point

A transmission level point (TLP) is a reference point on a transmission circuit. In reference to testing a circuit, the tone level at a TLP is set as 0 dBm. This is known as 0-TLP. It is a hypothetical point not necessarily accessible. Given a 0-TLP, the power level at any other point in the circuit is directly specified relative to 0 dBm. For example, an absolute noise power of 20 dBrn at a − 6 dB TLP corresponds to 26 dBrn0. The 0 is affixed to explicitly denote the 0-TLP condition. Under C-message weighting 0-referenced power level is expressed as dBrnC0, and for psophometric weighting the corresponding unit is dBm0p or pWp0.

Example 2.7

A transmission system requires a certain parameter measurement at − 15 dBm0. The input point is specified as − 15 TLP.

What is the absolute power level required to be applied at the input for this measurement?

Solution

Input reference level: − 15 dB TLP. Therefore, to realize the −15 dBm0 level, the required power level (p) at the input is: (− 15 − 15) = − 30 dBm

$$- 30 = 10\log[\frac{P(mW)}{1mW}] \Rightarrow P = 10^{-6} \text{ W}.$$

Problem 2.6

What is the absolute noise power level indicated as 27 dBrnC0 at a − 5 dB-TLP?

Problem 2.7

The following are the noise measurement data on a transmission line. Determine the absolute noise added at the input in performing the measurements:

Absolute noise power at the input test point:

15 dBm at – 10 dB TLP

Absolute noise power at the output test point:

30 dBmm at – 2 dB TLP

Electromagnetic interference on access lines

Accessing digital broadband services at home or in an office is an objective of modern telecommunications. Such services are rendered by high-speed digital systems and a promising mode for delivering such broadband services at home or office environment is, still, largely copper-based. Relevant wiring, therefore, encounters challenges posed by EMI and warrants trade-off between data rates to be supported and compliance considerations being imposed. An overview on the deployment of appropriate EM considerations in "future-proofing" of residential and/or commercial buildings is as follows [2.12].

Home-wiring refers to the task of network cabling to facilitate telecommunication access into a building. It consists of a pair of copper-wires running from the curb (where the subscriber-loop carrier terminal is located) to a pedestal terminal. From the pedestal it runs into the house or office building where a standard jack-and-plug arrangement (such as RJ-11 or RJ-45) is used for connecting a telephone or a data-terminal to the incoming copper wiring.

When voice telephony was the major telecommunication service, the story of network cabling within a building was simple and square. Through a variety of premises the wiring system has existed as dictated by different interfaces needed by the host of service providers. Using a mix of varying quality and wire sizes (such as 2 pair, 3 pair, 4 pair, 6 pair, 12 pair, and 25 pair) was also not uncommon. Further, in such voice-alone types of communication service facilitated, any impairments on the quality of transmission arising from the vagaries and variety of the wiring adopted was tolerantly accepted. However, with the advent of data communication, the quality of relevant transmission over copper-wires requires a new outlook on the standards for a structured wiring system. These standards are aimed at careful choice of materials, proper manufacturing, and high-quality installations of the cables so that high data rates can be supported within the conceivable distances encountered in premise wiring. For example, Table 2.2 lists various EIA/TIA Standards on telecommunication wiring pertinent to building environment.

These standards were developed to address the structured wiring that can support both voice and data. Further, the associated wiring basically refers to using shielded and unshielded twisted pair of copper-wire (known as UTP and STP respectively), but can include other media such as optical fibers and/or coaxial lines, besides the twisted-pair.

Table 2.2: EIA/TIA wiring standards

Standard	Refers to:	Premise-type
1. EIA/TIA-568	Cabling standard	Commercial building
2. EIA/TIA-569	Pathways and spaces	Commercial building
3. EIA/TIA-606	Administrative infrastructure	Commercial building
4. EIA/TIA-607	Grounding and bounding	Commercial building
5. EIA/TIA-570	Wiring standard	Residential building

While the original standards as indicated above on building-related telecommunication wiring were conceived and conditioned for the bandwidth of 4 kHz (for telephone and/or voice-band modem communications), a new application of such wiring has, however, been indicated in the recent times. A question which is relevant is as follows: whether such residential cabling (essentially based on UTP) can serve as access networks within the building to offer a variety of new communication services to subscribers at home and other consumers at business enterprises.

Such services include high-speed data for computers, interactive television, video-on-demand, video telephony, and other services focused on TV and video capabilities.

In essence, all these services are based on accessing high-speed digital information from the curb into the building; as such, the existing UTP is required to cope with the associated broad-bandwidths of these new services in contrast to the classical voice grade, 4 kHz bandwidth attributions.

Further in addition to the aforesaid services, access to the Internet and on-line services, both national and local, is becoming a highly promising application for residential telecommunication supported by socioeconomic demands.

A viable and pragmatic technology to realize and support a high-speed digital information access into UTP-based premise wiring is, for example, asymmetric digital subscriber lines (ADSL). This system has the capability to increase significantly the (asymmetric) transport capacity of the embedded twisted-pair cable infrastructure. It is asymmetric in the sense that it supports a high bit rate in the down-stream (from the curb to the building) and lower bit rate in the up-stream transmission. These up and down-stream transmissions with asymmetrically poised data rates of digital information are done without any impairment to the conventional voice transmission (POTS) supported by the twisted-pair cable.

The classical concept of high-speed data transmission via voice-band modem technology has reached a limit with modems such as the V.90. However, the demand on still greater communication capabilities from the end-users is imminent and is increasing. Under such prevailing demands, the strategy to make best use of the existing UTP cabling in the building premises also coexists. As a result, many operators and subscribers are pinning their hopes on technology such as ADSL, which analysts predict will be installed by the millions in coming years.

In addition to high-speed digital subscriber lines (DSLs), the cohabited use of voice and high-speed/broadband transmissions may also be conceived with installations that support the B-ISDN protocol, namely, the asynchronous transfer mode (ATM).

In the context of home-based high-speed broadband access plus the deployment of prevailing UTP cabling, the following queries can be posed:

- To what extent would the existing structured cabling support high-speed/broadband information access from the PSTN into the building premises?

- Is rewiring of existing networking needed with the replacement of prevailing cables by enhanced categories and related connectivity hardware?

- What is the EMI consideration vis-à-vis high-speed/broadband support via UTP wiring?
  (It is believed that allowing competitors to use ADSL technology in the different part of the network may create a significant interference problem on an existing service if one copper pair in the same binder group, or wire bundle, is designated to deliver the ADSL service.)

- What are the strategies towards mitigating such EMI environments so as to meet the compliance requirements?

- If rewiring is done towards "future-proofing" of the structured wiring, how will EM material technology considerations be helpful?

- In implementing the "future-proofing" type of shielding required, will the prevailing art of shielding be sufficient?

- What are the cabling products to which EMI-proof material technology can be applied?

■ How will the compliance requirements on thermal, mechanical, and electrical characteristics of the compatible materials of cabling products be traded off against telecommunication performance or quality of expected service under voice plus high bit rate environments?

In essence, to study the mitigating considerations of UTP-based premise wiring under high speed DSL implementation as well as to decide on "future-proofing" strategies for the structured wiring intended to support broadband/high-speed services, it is necessary first to identify the EMI problems. Then, the compliance aspects can be studied and the role of novel EM-shielding schemes to achieve the required telecommunication performance can be considered under the constraints posed by thermal, mechanical, and electrical insulation performance requirements.

Typically, a building-wiring environment consists of:

✓ Power cables supporting 60/50 Hz domestic/office a.c. power distribution
✓ UTP analog voice-grade telephone wiring
✓ Cable-TV wiring and/or wiring to a satellite-dish
✓ Home-office LAN wiring requiring high-speed access to Internet
✓ Wiring for central alarm and security system including surveillance camera plus a dial-up service to a monitoring station
✓ Wiring for home automation.

Apart from the power cabling, the rest as enumerated above constitutes telecommunication wiring. There are two grades of information for residential applications specified in the TIA/EIA 570A: Residential Cabling Standard, (which awaits final approval). Grade I is for basic telephone and video service. One 4-pair Category 3 (or better such as Category 5) UTP and one RG-6 coaxial cable to each information outlet are suggested in the above standard. And, Grade 2 is specified to provide enhanced voice, video, and data service to the residence. Two 4-wire Category 5 UTP cables and two RG- coaxial cables are recommended for Grade 2 services. One of the Category 5 cable is intended for voice and other is for data; one of the RG-6 cables is for satellite and the other is for local programming via a roof-top antenna or cable-TV connection.

Notwithstanding the new set of standards as above which are intended to offer a future-proof wiring, the concept of high bit rate traffic (such as ADSL technology) is to make use of the UTP for broadband/high-speed access, despite inevitable impairments to copper-based transmissions which may be expected.

In copper-based DSLs, the sources of signal impairments are:

✓ Cross-talk interference
✓ Impulse noise
✓ RF interference.

Cross-talk interference on DSLs

There are two versions of crosstalk of concern in digital subscriber lines (DSLs). The so-called *near-end cross-talk* (NEXT) refers to the crosstalk effect between transmit and receive pairs at the same end of a cable section. The other version of crosstalk is the *far-end cross-talk* (FEXT) defined as the effect of crosstalk due to adjacent transmitters perceived by a line at the remote-end from the input section of the line where the signal is injected. The *crosstalk loss parameter* is defined as the attenuation for a disturbing signal to pass through the coupling mechanism arriving at a disturbed receiver.

Impulse noise interference on DSLs

This is generated from a number of diverse sources such as lightning, switching equipment at the CO, power-line transients, and motors and other devices in the building. Impulse noise is a major concern inasmuch as the received signal on the high-speed data line may be weak due to heavy subscriber loop losses. Typically, impulse noise may occur with the following characteristics:

- ✓ Occurrence rate of 1 to 5 per minute
- ✓ Peak amplitudes in the range, 5-20 mV
- ✓ Spectral energy is concentrated within 40 kHz
- ✓ Duration of the impulse falls in the range, 30-150 μs.

Unlike the crosstalk interference, the origin of impulse noise is more difficult to locate. It is often characterized by a random pulse waveform whose amplitude is much larger than the system noise. Further impulse noise would cause spiky, irregular or sporadic change in electrical voltage on the transmission lines. Relative to signal level on the line, impulse noise will be significantly higher in magnitude.

RF noise interference (RFI) on DSLs
    This refers to the direct pick-up of RF energy by the copper lines. The source of RF energy could be the radio transmission operating in the vicinity of the lines. Typically, high-power AM stations may appear as noise interference at the receiver front-end of DSL receivers.

Thermal and other system noises effects on DSLs
    These are the electronic noises associated with the system, especially at the front-end of the receiver. It is essentially a thermal noise approximately – 140 dBm/Hz. It is also known as *background noise.*

2.2.6 Electromagnetic compatibility(EMC) issues
    When the telephone lines leave the telcos, they are packaged together in 50 wire pair bundles or binder groups; therefore, each line may encounter cross-talk from the other 49 lines. At high frequencies, more cross-talk is injected into the adjacent lines. Thus, NEXT and/or FEXT could seriously impair the high-speed information access via UTP wiring.
    With the inevitable EMI interference (such as cross-talk) manifesting as the noise and appearing at the receiver front-end along with the background noise, it becomes essential to "guard" the telecommunication wiring and cable housing units within the building premises. The performance of the services (such as ADSL) supported by the copper-lines in the domestic and/or commercial buildings, largely depends on realizing a fair signal-to-noise ratio at the receiver front-end. To decrease the effect of EMI-based noise, a variety of cabling products have emerged to place the wiring in the building in such a way that minimal EM interference is observed.

Cabling products vis-a-vis EMI
    In reference to the essential cabling products and shielding accessories used in the indoor copper-based telecommunication wiring, the choice of relevant materials are based on:

- ✓ Electrical conductivity
- ✓ Electrical insulation properties
- ✓ Complex permittivity versus frequency range of operation
- ✓ Thermal durability (plenum rating)
- ✓ Mechanical properties (tensile and compression strengths)
- ✓ EMI shielding, as required in the shielding parts.

In general, the following gamut of materials has been deployed in realizing viable, cabling products:

- ▪ Electrical materials
  Conducting material: Copper (annealed), Be-Cu
  Insulating materials: PVC, PE, PEP, Teflon, polyolefin, fluropolymer
  Shielding materials: Copper and aluminum

- Mechanical/structural materials:
  Steel
  Aluminum
  Thermoplastics.

The host of cable products that house the cable in the building environment and concurrently provide EMI immunity, are largely based on metals and insulators modified to offer desired electrical conductivity so that they pose lossy dielectric relaxation to the EMI/RFI energy. While the metallic materials (such as Cu, Al, Ag, and Ni) are chosen on the basis of their electrical conductivity and cost-effectiveness, the insulating materials, which are rendered lossy, are largely proprietary. A number of polymeric compounds are available which can be either surface-treated or volume-loaded to realize lossy dielectric characteristics.

Synthesizing appropriate lossy material also refers to designing a composite EM material as outlined by the author in [2.13]. In reference to the discussion and information furnished in [2.13], one can see that making of composite EM materials could be a practical solution to realize shielding materials required for EMI suppression over a select frequency band. The scope synthesizing for composite EM materials is wide. For example, one can think of "smart" EM composites, as narrated by the author in [2.14]. Especially, with the incoming demand posed by high-speed/broadband transmissions across the building telecommunication wiring [2.15], the "future-proofing" of such cabling heavily depends on new EM material technology, specifically the composite versions.

Quantitative representation of noise

- Signal-to-noise ratio (SNR or S/N) parameter: This represents a relative measure of desired signal power to (undesired) noise power at a specific measurement point (such as at the input or the output) of a communication system. Expressing the signal power as S and noise power as N, the SNR in decibels is given by:

$$SNR|_{dB} = 10 \log10 \ (S/N). \tag{2.16}$$

- Noise figure (NF): This is a figure of merit depicting how noisy a device or a system is. It is specified as the ratio of SNR at the input to the SNR at the output of the device or a system. In decibel measure, it is specified as:

$$NF|_{dB} = 10 \log10[(SNR)input/(SNR)output] \tag{2.17}$$

2.2.7   *Semantic transparency and statistics of bit errors*

In digital communication, the binary signal waveforms may become corrupted as a result of the noise introduced. In essence, the noise may cause the following types of errors in the transmission of bits:

- *Single-bit error*: This refers to a logic 1 being changed to logic 0 or vice versa causing a corresponding loss of information

- *Multiple-bit error*: A bursty noise can wipe out a stretch of bits in the bit-stream causing a significant extent of semantic loss in the message transmitted

- *Logic upsets*: These refer to single and/or multiple bit errors which impair the operation of a control circuits which critically perform the timing and synchronization of pulse transmission across the network.

The statistics of occurrence of each of the above category of bit errors are normally quite different. Further, the occurrence probabilities of 1s and 0s can also be distinct. If they are same, then the channel is called the *binary symmetric channel*.

Bit error rate (BER)

The corrupted transmission of bits under noisy conditions leads to erroneous reception and semantic misinterpretation of the message transferred. Occurrence of bit errors are random in nature and, as such, a probability of error P(e) can be specified. It refers to the expectation of the *bit error rate* (BER) for a given system. The BER itself is a measurable entity and it depicts the observed bit error performance of the system. Suppose the BER is indicated as 10^{-6}. It means that in the test system, one bit error per every 10^6 bits transmitted has been observed. The corresponding mathematical specification of this system performance corresponds to $P(e) = 10^{-6}$.

The probability of error is a function of the average signal energy per bit to noise power density responsible for the bit error in the system. Suppose the signal energy per bit is E_b and the noise power density (noise power per bandwidth) is N_0. Then,

$$E_b = (\text{Signal power}) \times (\text{Time of transmission of 1 bit}) \qquad (2.18a)$$
$$= S\, \tau_b \text{ Watt-s or J}$$

where $R_o = 1/\tau_b$, refers to the rate of transmission of bits, and τ_b is the duration one bit. Therefore,

$$\frac{E_b}{N_0} = \frac{S\tau_b}{k_B T} = \frac{S}{k_B TR} \qquad (2.18b)$$

Expressing in decibels,

$$(E_b/N_o)_{dB} = (S)_{dB} - 10 \log_{10}(R_o) + 228.6 \text{ dBW} - 10 \log_{10}(T). \qquad (2.18c)$$

Natural bit rate

This represents the natural information rate. In order to characterize a service in a generic fashion, the concept of *generated bit rate* (pertinent to a service under consideration) should be replaced by a *natural information rate*: This is the rate at which the source is generating information if no limitations on the functionality and/or cost of the telecommunication network are imposed. The natural information rate of each source (such as voice and video) is very much dependent on the coding and compression techniques used. As a result, natural information rate relies on state-of-the-art signal-processing and technology along with the related economical consideration.

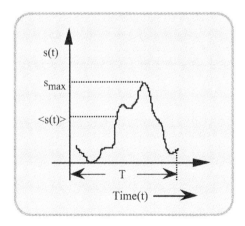

Fig. 2.30: The stochastical process of bit rate fluctuation
(The peak natural bit rate s_{max} and the average natural bit rate $<s(t)>$ over the duration T.)

This natural information rate can be represented by a stochastic process s(t) (Fig. 2.30). Suppose this stochastic process is observed over a duration T of the information transfer; for example, this can be the duration T of a telephone conversation, the duration of a computer session for computer-to-computer data communication, or the duration of a video conference session, etc. Two important values can be obtained from the stochastic process representing the natural bit rate fluctuation with respect to time.

The duration over which the peak and average is calculated is an important parameter to characterize a service. In Fig. 2.30, $s_{max} = \max s(t)$ and $<[s(t)> = \dfrac{1}{T} \int s(t)dt$.

The ratio between the maximum and the average natural information rate is a measure of burstiness B.

$$B = \frac{s_{max}}{<s(t)>} \qquad\qquad (2.19)$$

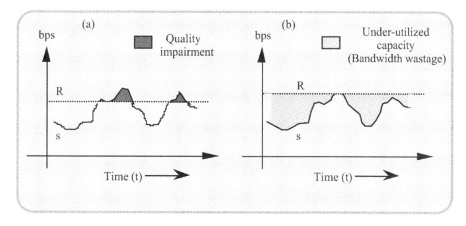

Fig. 2.31: Fluctuation of natural bit rate with respect to time: (a) Quality and (b) bandwidth versus peak cell rate considerations

It can be observed that for every session of a service, the stochastic process s(t) will have a distinct behavior, but the average and peak values will be typical for a service. In Table 2.3, some typical values are presented for a number of services. The burstiness for the voice basically comes from the talkspurts and silence periods, which each typically last about 50 percent of the duration of a consideration; for bulk data transfer, the very high burstiness is mainly due to the fact that some contiguous sectors of a disk are read before the magnetic head of the disk has to be moved and then a next set of contiguous sectors is read; for video one can imagine that a coding technique only generates bits if there is non-redundant information available (for example, some motions in an image).

Table 2.3: Typical telecommunication services and their bit flow rate characteristics

Service	$<s(t)>$	B
Voice communication	32 kbps	2
Data communication		
Interactive data	1-100 kbps	10
Bulk data	1-10 Mbps	1-10
Video communication		
Standard quality video	1.5-15 Mbps	2-3
High definition TV (HDTV)	15-150 Mbps	1-2
High quality video telephony	0.2-2 Mbps	5

Major inferences that can be gathered from Table 2.3 as follows:

- A "typical" service cannot be uniquely specified. All services have different characteristics both for their average bit rate and burstiness factor

- None of the services has a burstiness equal to one. By source coding, one can always transform the natural bit rate to one fixed value, but that is either at the expense of lower quality (if the peak bit rate is reduced, Fig. 2.31a) or at the expense of a lower efficiency (Fig. 2.31b) because idle information is transported, wasting resources in the network.

In Fig. 2.31a the transfer rate (R) is smaller than the peak of the natural bit rate. This results in a quality reduction, since during the periods that the natural rate is larger than the transfer rate, some bits will have to be discarded to limit the natural rate to the acceptable transfer rate. In Fig. 2.31b on the contrary, the transfer rate is always larger than or equal to the natural bit rate. Only redundant information can be used to fill up the difference between natural bit rate and transfer rate. Therefore, resources will be wasted in the network.

Ideally, a transfer mode intended for services (such as those described in Table 2.3), must be quite flexible in the sense that it can transport a wide range of natural bit rates. Further, it should be able to cope with services, which have a fluctuating natural bit rate in time. Hence, an optimal transfer mode should support communication bearing a variety of uses of information via an integrated access, especially one which places few or no constraints on the way in which the customer wants to use such access. Ideally, the transfer mode must provide a capability to transport information, regardless of type of information through the network.

Example 2.8

Given that $(E_b/N_o)_{dB} = 8.4$ dB at a probability of error $= 10^{-4}$ in transmitting bits over a certain communication channel at a rate, $R_o = 2400$ bps.

Assuming the operating temperature as $290°K$, determine the signal power in watts.

Solution

$(E_b/N_0)_{dB} = 8.4 = (S)_{dBW} - 10\log_{10}(2400) + (228.6)_{dBW}$
$- 10\log_{10}(290)$

$\therefore S|_{dBW} = 161.8$ dBW $\equiv 10\log_{10}[(S) \text{ watts}/1 \text{ watt}]$
$\Rightarrow s = 1.514$ W

(1 Watt = Reference level of 0 dBW)

Suppose M-ary encoding conditions are used and the average carrier power used (M-ary refers to the number of encoded bits used to modulate the carrier frequency). Then, the carrier-to-noise power ratio can be specified as C/N where C is the carrier power in watts and $N = (N_o \times BW)$ watts. Further, $E_b = (C \times \tau_b)$ watt-second/bit (or J/bit). In terms of the bit rate $R_o = 1/\tau_b$, E_b can be specified as C/R_o J/bit. In essence, $E_b/N_o = (C/N) \times (BW/R_o)$.

The entity E_b/N_0 is commonly used to compare the various digital modulation systems or an M-ary aspect of encoding schemes. Since E_b/N_0 ratio specifies the energy per bit referenced with respect to the noise power present per 1 Hz of bandwidth, it offers a normalized common base to compare the modulation schemes. Such a comparison refers to the relative error performance of the modulation schemes subjected to common noise bandwidth conditions.

Example 2.9

Suppose a particular digital modulation scheme refers to the following parameters: $C = 1 \text{ pW}$; $N = 0.012 \text{ pW}$; $R_o = 60$ Kbps and $BW = 120$ KHz.
Determine (a) N_o in dB, (b) E_b in dBJ, (c) C/N in dB and (d) E_b/N_o in dB.

Solution

(a) $\qquad N_o = \dfrac{N}{BW} \text{ W/Hz} = \left[\dfrac{0.012 \times 10^{-12}}{120 \times 10^3} \right] \qquad \text{W/Hz}$

$\qquad\qquad \Rightarrow N_{dBm} - 10\log_{10}(BW) = -160 \text{ dBm}$

(b) $\qquad E_b = C/R_o \text{ J/bit} = \left[\dfrac{1 \times 10^{-12}}{120 \times 10^3} \right] \qquad \text{J/bit}$

$\qquad\qquad \Rightarrow 10\log_{10}(C) - 10\log 10(R_o) = -167.8 \text{ dBJ}$

(c) $\qquad C/N = \left[\dfrac{1 \times 10^{-12}}{0.012 \times 10^{-12}} \right] \quad \Rightarrow 10\log_{10}(1/0.012) \text{ dB} = 19.2 \text{ dB}$

(d) $\qquad E_b/N_o = (C/N) \times (B/R_o) \quad \Rightarrow 10\log_{10}(C/N) + 10\log_{10}(B/R_o) \text{ dB} = 22.2 \text{ dB}$

Table 2.4: Error performance of different modulation schemes

Modulation type	Parameters	Probability of error P(e)	Remarks
1. FSK			
a) Noncoherent or asynchronous	(E_b/N_o)	$(1/2)\exp(-E_b/2N_o)$	In the noncoherent modulation, the transmitter and the receiver are neither frequency nor phase synchronized.
b) Coherent or synchronous	(E_b/N_o)	$\text{erfc}[(E_b/N_o)^{1/2}]$	In the coherent system, the local receiver reference signals are frequency and phase locked with the transmitted signal.
2. PSK	M-phase system (E_b/N_o)	$\dfrac{1}{\log_2(M)}\text{erf}(Z)$ $Z = \sin(\pi/M)$ $\times [\log_2(M)]^{1/2}$ $\times [E_b/N_o]^{1/2}$	$P(e)|_{QPSK} = P(e)|_{BPSK}$
3. QAM	L: Number of levels on each axis of constellation (E_b/N_o)	$\dfrac{1}{\log_2(L)}\left(\dfrac{L-1}{L}\right)\text{erf}(Z)$ $Z = \dfrac{[\log_2(L)]^{1/2}}{(L-1)} \times$ $\left(\dfrac{E_b}{N_o}\right)^{1/2}$	

Error performance of modulation schemes

The following is a summary of error performance of the FSK, PSK, and QAM schemes often used in digital communication.

Problem 2.8

Construct a table indicating the C/N (dB) and E_b/N_o (dB) requirements under a common BER performance of 10^{-6} for the following modulation schemes: BPSK, QPSK, 4-QAM, 8-QAM, 8-PSK, 16-PSK, 16-QAM, 32-QAM, and 64-QAM.

Problem 2.9

Write a computer program to calculate the minimum bandwidth required to achieve a specified P(e) for different modulation schemes operating at a given R_o and C/N ratio.

Problem 2.10

Compare the following systems operation under identical bandwidth conditions in respect of their BER performance:

System A: QPSK, C/N = 20 dB, R_o = 50 Mbps
System B: 8 PSK, C/N = 10 dB, R_o = 60 Mbps

Error control

BER is the most significant quality parameter in digital telecommunication systems. The structure of the noise involved in digital communication decides the extent of BER it causes. The two factors which can be changed to control the bit errors are: (i) S/N ratio at the receiver and (ii) rate of transmission of bits. The strategies adopted in practice towards bit error control are:

✓   Error control coding (Channel coding/forward error correction (FEC))
✓   Error control incorporated modulation process (Trellis coded modulation, TCM)
✓   Automatic repeat for request (ARQ).

Error control coding

This involves encoding the bit stream prior to modulation by a process of adding extra bits to the stream as per certain specified rules. This process deliberately introduces redundancy. That is, an improvement in BER performance is achieved at the expense of transmission rate or *throughput*. It is a strategy of forward error correction, which does not need a return path. The details on specific encoding methods are presented in a later section.

Trellis coded modulation(TCM)

The FEC involves improving the BER performance by increasing the bit rate or utilizing more bandwidth. TCM on the other hand reduces bit errors without resorting to increasing the bit rate. The preamble of Trellis code implementation on different modulation schemes will be presented in a later section.

Automatic request for repeat (ARQ)

This is based on positive and negative acknowledgments (for ARQ). The positive acknowledgment version refers to confirming the receipt of error-free data. Negative acknowledgment specifies that an error has occurred and hence a retransmission is solicited. The negative acknowledgment scheme is less redundant and follows a "Go back N" procedure. That is, when an error is observed, a request is made to repeat the last N blocks of the data (with N = 4 or 6). Alternatively, a request can be made for retransmission of a selective block of data as needed.

2.2.8   Temporal transparency

As indicated earlier, transmission of a signal may suffer delay distortion due to the presence of distributed L, C, R, and G elements in the transmission line. This delay is termed as *the propagation delay*. In addition to this delay, the signal may suffer delay due to various processing and transfer considerations across the hardware (such as multiplexers and switches) as listed below.

In certain services (such as voice and video transmission) excessive delay is not tolerated, because the quality of signal would suffer. An unduly delayed voice may sound unnatural to a listener and a delayed video signal would cause a blurred image on the screen. These delay sensitive signals are called *isochronous signals*. Their transmissions warrant maintenance of a guaranteed temporal transparency.

Transfer-delay

Transfer delay is defined as the elapsed time between a signal (represented in terms of a bit, a packet or an instantaneous part of an analog waveform) exit event at a measurement point-1 (for example, at the source user-network interface, UNI), and the corresponding entry event at measurement point-2 (for example, the destination UNI) for a particular connection. Thus, the transfer-delay depicts the sum of the total inter-node transmission delay and the total node processing delay(s) between the two measurement points.

There are several contributing factors which decide ultimately the extent of total transfer delay. Briefly they can be summarized as follows:

- *Coding delay*: It refers to the time required to convert a nondigital signal to digital bit patterns. It may also include the data compression delay, if any, in addition to the delays arising from analog-to-digital conversion

- *Packetization delay*: This delay occurs while accumulating the required number of bits to constitute a packet

- *Propagation delay*: This is due to the finite speed of signal transmission between the source and destination. It depends on the electrical properties of the line

- *Transmission delay*: This is dependent on the speed of the link and becomes negligible as the transmission speed increases

- *Switching delay*: It refers to the total delay incurred by a packet to transverse a switch. It depends on the interval switch speed and the amount of overhead added to the packet for routing within the switch

- *Queueing delay*: Switches may have buffers at the input ports, at the output ports, or a combination of input, internal, and output buffers depending on the switch fabric. The resulting traffic-handling at the buffers would cause congestion and resulting queueing delays

- *Reassembly delay*: Depending on application (service), several packets of a frame are collected at the receiver before they are passed to the application. For example, to provide a continuous 64 kbps constant rate service, a number of voice packets are collected before the frames are to be played out.

The delays listed above are encountered as a part of random traffics involved and are largely stochastical in nature. And none would follow any repetitive pattern. Hence it is rather justified to assume the time interval between the reception and the transmission of data across the UNI can be regarded as a stochastical variable.

Delay variation: Another traffic parameter used in defining the quality of service (QOS) associated with virtual channel connections in an ATM network. It refers to the upper bound of variability in temporal pattern of an ATM packet (cell) arrival observed at a single measurement point with reference to the peak cell rate of an ATM connection.

The end-to-end delay of the ith cell can be specified as $(D + W_i)$ where D is a constant and W_i is a stochastical variable representing the *cell-delay jitter*. That is, W_i represents the random delay arising out of buffer overflows within the network caused by asynchronous multiplexing as well as due to cell-losses arising from uncorrectable bit-errors. Writing in terms of interarrival times of the cells,

$$(D + W_{i+1}) - (D + W_i) = \delta_i. \tag{2.20}$$

This delay δ_i will be zero when $W_i = W_{i+1}$, which corresponds to the interarrival times being equal to the interexit times of the cells.

However, inasmuch as randomness always persists due to multiplexing and bit-error based cell-losses, it is rather inevitable that $W_i = W_{i+1}$. Therefore, δ_i has a non-zero value and represents a random variable as decided by the stochasticity of the *cell delay variation* (CDV) or jitter. This CDV is perceived in the cell delay whenever a difference between the values of the transit delay of the cells of a connection exceeds a limiting value, that is, in terms of probabilistic attributes.

$$\text{Probability } (W_{i+1} - W_i) > w_i \tag{2.21a}$$

where w_i is a designated upper limit and $(W_{i+1} - W_i)$ is regarded as a measure of CDV. Alternatively, the instant delay variation from the mean can also be regarded as a measure of CDV with

$$\text{Probability } \{W_{i+1} - E[W_i]\} > w_2 \tag{2.21b}$$

where w_2 is a specified upper limit and $E[.]$ represents the expected value of the variables considered. Likewise, it is also possible to specify the variations of transmission delay in a connection in terms of the variance of the transmission.

Upper bound on transfer delay
In reference to the transfer delay and associated delay variations, there are two end-to-end delay parameter objectives, which can be specified. They are:

    ✓   Maximum transfer delay
    ✓   Peak-to-Peak transfer delay.

More discussions on the above considerations are presented in Chapter 4 in reference to ATM transmission.

Table 2.5: <u>Examples of permissible limit on transfer delay and delay variations in typical applications supported by ATM transmissions.</u>

Applications	Rate	Delay (ms)	Jitter (ms)
Video conference	64 kbps	300.00	130.00
MPEG NTSC video	5 Mbps	5.0	6.5
HDTV video	20 Mbps	0.8	1.0
Compressed voice	16 kbps	30.00	130.0
MPEG voice	256 kbps	7.0	9.1

Delay performance and delay variations in typical telecommunication services supported by a transfer mode such as ATM are listed in Table 2.5. Normally, the state-of-the-art telecommunication systems have to cope with the aforesaid delay and/or delay variation considerations as well as the variety in the transmission rates. Telecommunication networks are required to facilitate the transmissions of information pertinent to a variety of services which can range from a very low-speed data to a very high-speed information transfer. Examples are as follows:

- ✓ *Low speed services*: Telemetry, telecontrol, telealarm, voice, telefax, and low speed data
- ✓ *Medium speed services*: Hi-fi sound, video telephony, high speed data
- ✓ *(Very) high speed services*: High quality video distribution, video library, video education etc.

The rate of transmission of bits in the previous services would range from few bits to gigabits per second in practice. Consistent with the side range of transmission rates, the information transfer mode of the telecommunication network should be such that it must transport a wide range of natural bit rates, and it should be able to cope with services which have a fluctuating character in time (in respect of natural bit rate) as discussed earlier.

2.3   Modulation and Demodulation

Modulation, in general, is a process of changing a signal so that the signal is conducive for transmission over a medium, and *demodulation* is an inverse process of recovering the signal (information) at the destination from the modulated waveform received [2.2].

2.3.1   Analog signal modulation

A typical modulation, for example, is changing the pulsed data from a personal computer, (which is in the form of digital pulse) into "voice-like" analog form and transmitted over a pair of telephone wires. This is done because telephone wires are not an efficient medium to transport digital pulses. That is, the digital pulses correspond to a large baseband Fourier spectrum extending to a few MHz. The copper-wire cannot support this large bandwidth over a long distance as indicated earlier. In essence, the copper-wire transmission line is a low-pass filter and hence, the high frequency spectral components of digital pulses will be filtered out when the pulses are transmitted across these wires.

Therefore, the modulation of digital waveforms representing the bits of information from a personal computer is done by a modem and at the receiving end another modem converts the analog format of the signal into digital pulses (which is then sent to the receiving host computer via the modem port). Modem is an acronym for *modulator-demodulator*. In Bell system (AT&T) terminology, a modem is known as a *data set*.

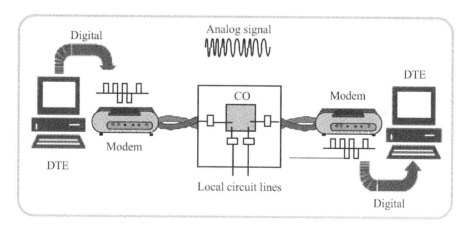

Fig. 2.32: Deploying a modem-based data communication across a PSTN

The conversion of digital waveform into an analog form in a modem enables the converted signal to be treated as a voice-like entity, so that it can pass through the telephone line like a speech signal. The modem-based voice band data transmission is described below:

2.3.2   Modems

The *modem* in essence, is a *data communication equipment* (DCE) which is used to interface *data terminal equipment*, DTE, (such as a personal computer) to an analog local-loop and line circuit on the PSTN, by converting the digital signals into modulations of an analog carrier. Each modem is attached to a computer or terminal via an RS-232 cable.

Originally, as is known, the local telephone networks were designed to handle the telephone conversation and hence were intended to support baseband of voice signals extending at most to 4 kHz. When data pulses originating from a DTE were to be supported by the local loops, the data pulses had to be converted into analog waveforms in the 300 to 3000 Hz band. The physical channels (namely, the copper-lines) can then accommodate them within their pass bandwidth over the distance involved between the DTEs.

The modems are used in pairs with one located at the sending end and the other at the receiving end. The modem at the receiving end converts the received analog signal back into appropriate digital form compatible for the computer on the receiving end. In short, modems allow the connection of a data circuit to the PSTN as illustrated in Fig. 2.32.

Indirect modem

Historically, the version of modem developed (in 1967) was an *indirect type*. It was exactly based on the same principle as how the telephones at the ends of two subscribers are interconnected. Hence, it was then known as an *acoustically coupled modem* or simply, an *acoustic coupler*. Its operation is as follows: The digital pulses from a DTE modulate voice frequency carriers. The resulting voice frequency tones make audible sounds at the mouth-piece of a conventional telephone via a speaker. Hence, the telephone mouth-piece (transmitter) sends the corresponding tones electrically across the network (as in conventional telephone conversation mode). At the receiving end, the received data is acoustically coupled from the telephone's (handset's) speaker on to a microphone. Hence, the modulated tones are available at the receiving end in the electrical format. Demodulation of these tones leads to the digital pulses at the receiver end DTE. The above principle is illustrated in Fig. 2.33.

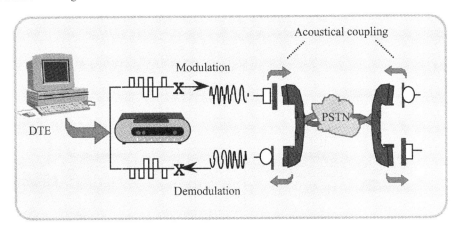

Fig. 2.33: Acoustically coupled duplex-modem

Since the arrangement as illustrated in Fig. 2.33 involves the use of conventional telephone network, call setup is done via dialing in the usual manner. Once ringing is heard at the receiver, the handset is taken off the hook and placed in a position so that the modems at the ends can "talk" to each other. That is, the data flow between the DTE's is facilitated. The acoustical coupler can be operated for speeds up to 1200 bps. It is, however, prone to ambient (acoustical) noise. A pair of rubber cups that fit over the mouthpiece and ear-piece of the handset are used in the acoustical coupler so as to reduce the external sound noise picked up.

Direct connected modems

Direct connected modems refer to the DTEs communicating across the PSTN directly as illustrated in Fig. 2.33. The DTE is connected to the modem via an RS-232 cable. The digital pulses from a computer modulated on a voice frequency carrier are allowed into the PSTN circuit via a standard telephone wall jack (such as RJ 11 C) with a limited amplitude up – 9 dBm. Direct-connect modems are placed on the telephone network subject to FCC Part 68 Registration and FCC Part 15 Approval.

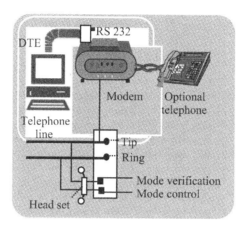

Fig. 2.34: Direct-connect modem on the PSTN

Modes of transmission in modems

Like telephone connections, modems can operate in simplex, half-duplex, and full-duplex modes. Simplex operation allows data to be sent or received in one direction only. In the half-duplex mode, the channel is shared so that the modems at the two ends send and receive the data alternately. The time interval over which a modem is on the sending (or receiving) mode, and returns to the sending (or receiving) mode is known as turn around time.

In the full-duplex mode, data transmissions between DTEs can take place simultaneously. On common two-wire telephone circuits, two distinct carriers are used for the two directions. These are known as *low-band* and *high-band frequencies*, both of which fall in the pass-band of the telephone lines. Alternately, full-duplex operations of modems can be done in private or leased lines where four-wire circuits are commonly used without splitting the available frequency band.

Echo canceling method

This allows the modems to transmit simultaneously on the same frequency. In such single channel full-duplex operations, the two end-stations decide, a priori, the channel to be used to transmit and receive the data. In this operation bidirectional communication can take place only if echo canceling is used.

In the full-duplex communication a modem is configured for "*originate mode*", if the station originates the transmission and this transmission occurs on the low band of frequencies. The modem that answers the call operates in the " *answer mode*". In this mode the transmission occurs on the high band.

Modem protocols

The DCEs (such as modems) at either end of the communication link should share and operate on the same protocol. In reference to modems, a protocol defines the following:

✓   The maximum bit rate at which the modem can exchange the data. This bit rate is decided by the type of transmission line (for example, the category of copper pairs of wires)

✓   The maximum bit rate versus the electrical characteristics of the line which decides the method of encoding the data bits on the channel. The amplitude, frequency and/or the phase of the carrier is normally changed (modulated) towards the encoding. These modulation methods (known as amplitude shift-keying, frequency shifting-keying, phase shift-keying etc.) will be detailed later in this chapter

✓   Establishment of common voltage reference level between the modems via equalization. The process of establishing a commonness of the voltage between the modems refers to equalization. It compensates for any differences in signal amplitude versus frequency. The equalization is essential for high bit rate transmissions

✓   Type of transmission, namely, synchronous or asynchronous. The synchronous and asynchronous transmissions refer to the following: In synchronous transmission, the modems on either side of transmission, would first establish a strict interval time reference between them and then bursts of bits of a certain feed length are sent when the connection is established. The transmission and reception are constrained within the synchronized timetable originally established. Staying in synchronism is done by placing one or zero every so often

✓   In asynchronous mode, the data bits are not sent on any strict synchronized timetable. The start of a character is preceded by a start prompting bit and the end of the character is followed by a stop bit. Therefore, at any time along the time-scale, the receiving modem will recognize the arrival and end of a character. Normally, the modems remain synchronized during the length of the character interposed between the start-stop bits. (Even if the clocks of the modems slightly go off synchronism, data transfer would still be successful)

✓   Type of mode of transmission: Simplex, half-duplex, or full-duplex.

The mode of transmission is facilitated by the type of hard-wired transmission connectivity that prevails between the modems through the central office (CO). The subscribers are normally connected to the CO by subscriber loop-lines made of twisted copper-pairs known as *switched* or *dial-up lines*.

A company other than a communication carrier can also own a telecommunication line. Or, a line can be leased by an organization from a common carrier. This class of dedicated lines can be either two-wired or four-wired.

A half-duplex transmission involving one-way at a time communication takes place on the two-pairs of wires, one-pair per each direction. The full-duplex allows two way communication simultaneously, again one on each pair of wires.

Modem family

Historically, the evolution of modem technology can be summarized by considering the family tree of modem, which emerged and proliferated the data communication system:

■   *Indirect modems*

    ✓   103 compatible device
    ✓   202 compatible device
    ✓   VA 3400 compatible coupler (1200 bps full-duplex).

■   *Direct modems*

The Bell Systems historically commenced the robust operation of modems and set their operational specifications. Eventually those specifications became the de facto standards in the modem technology as specified by the ITU-TS. A summary of the various modems with the specifications at different stages of their development is indicated in Table 2.6.

Table 2.6a: The modem family: Bell103/113 modem

Modem type	Specification				
	Data type	Transfer rate	Modulation	Transmit level	Receive level
Bell 103/113 Modem	Serial Binary Asynchronous Full-duplex	0-300	FSK-FM Origination Answering end end Transmit Space 1070 Hz 2025 Hz Mark 1270 Hz 2225 Hz Receive Space 2025 Hz 1070 Hz Mark 2225 Hz 1270 Hz BW: 3000-34000 Hz	0 to – 12 dBm	0 to – 50 dBm

Remarks:
Receive level is specified assuming the adjacent channel transmitter can be at 0 dBm

Table 2.6b: The modem family: Bell 202 modem

Modem type	Specification				
	Data type	Transfer rate	Modulation	Transmit level	Receive level
Bell 202 Modem	Serial Binary Asynchronous Half-duplex on two wire lines	0-1200 on switched network 0-1800 on leased lines with C2 conditioning	FSK-FM Space: 1200 Hz Mark: 2200 Hz BW: 300-3400 Hz	0 to – 12 dBm	0 to – 50 dBm on switched network 0 to – 40 dBm on leased network

Remarks:
1. Pseudo full-duplex; A 387 carrier is used with ASK modulation at 5 bps on reverse channel. The low bps prevents spectral spill over on the FDM main channel.
2. In the full-duplex operation the echo suppressors must be disabled.

Table 2.6c: The modem family: Bell 212A modem

Modem type	Specification			
	Data type	Transfer rate	Modulation	Remark
Bell 212A Modem	Serial Binary Full-duplex Asynchronous or synchronous (switched lines)	Low speed 300 bps High speed 1200 bps	−FSK-FM/BELL 103 Specification (Asynchronous) −(Synchronous or asynchronous) Four-phase DPSK (Quadrature PSK) Origination phase: 1200 Hz Answer phase: 2400 Hz	▪ Enclosed bits: Gray codes ▪ Consecutive 2 bits of the serial binary data sent to 212A modem are encoded into a single phase change of the carrier. The encoded two bits are called: *dibits*

Table 2.6d: The modem family: Bell 201 B/C modem

Modem type	Specification			
	Data type	Transfer rate	Modulation	Remark
Bell 201 B/C Modem	▪ Half-duplex on switched line ▪ Full-duplex on four wire private line	Fixed 2400 bps on two or four wire lines	201A: Obsolete 201B: For private/leased line 201C: Switched/leased line Four phase DPSK using dibits (11, 01, 00, 10)	- -

Table 2.6e: The modem family: Bell 208 A/B modem

Modem type	Specification			
	Data type	Transfer rate	Modulation	Remark
Bell 208 A/B Modem	Synchronous	4800 bps on four wire private/ leased switched lines	8-phase DPSK on 1800 Hz carrier	Consecutive 3 bits of serial binary data (called *tribits*) are encoded into a single phase change on the carrier

Table 2.6f: The modem family: Bell 209A modem

Modem type	Specification			
	Data type	Transfer rate	Modulation	Remark
Bell 209A Modem	Full-duplex Synchronous	9600 bps	Quadrature amplitude modulation (QAM)	Uses the quad bits M-ary: $M = 4$

Table 2.6g: The modem family: ISDN modem

Modem type	Specification			
	Data type	Transfer rate	Modulation	Remark
ISDN Modem (Digital modem/ Terminal Adapter)	Asynchronous	300 to 230,400 bps	Digital lines: ▪ Basic rate interface (BRI) ▪ Primary rate interface (PRI) • BRI delivers ISDN services to subscribers via two B (bearer) channels for voice/data/video and one D (delta) channel for controlling and packet switching • PRI delivers 23 B channels and 1 D channel for ISDN services at PBXs, host computers, and LA systems	▪ Use: Data, voice, video transmissions on a single digital connection between the CO and the users

Table 2.6h: The modem family: Cable modem

Modem type	Specification			
	Data type	Transfer rate	Modulation	Remark
Cable Modem	Asynchronous	Down stream: 10-30 Mbps Up stream: 19.2 Kbps – 3.0 Mbps 9600 bps	Down-stream data channel: 250-850 MHz Up-stream data channel: 5 MHz – 40 MHz Modulations: QPSK, QAM, VSB BW: 250 kHz to 6 MHz *Note*: Down stream modem is a CATV head-end modem Up stream modem is home-based modem	Uses high-speed Internet access and video services via cable TV network

ITU-TS modem family – V series

These modems are in conformity with data transmission standards outside the United States. The V-series recommendations are set by the ITU-TS (formerly known as CCITT). Listed in Table 2.7 are V-series modems and their operation features.

Table 2.7: V-series modems

Series type	Description			Remarks
	Line speed (bps)	Modulation rate/type bps	Carrier frequency(Hz)	
V.21	300	300/FS	1080 & 1750	Switched line Similar to Bell 103 FDX
V.22	1200	600/PS	1200 & 2400	FDX Switched/Leased lines
V.22 bis* (*Second revision in French)	2400/1200	600/QAM	1200 & 2400	FDX Switched/Leased networks
V.23	600/1200	600/FM	1300 & 1700	Similar to Bell 202 HDX Switched network
V.26	2400	1200/PS	1800	4 wire FDX Similar to Bell 201B
V.26 bis	2400/1200	1200/PS	1800	Similar to Bell 201C HDX Switched lines
V.26 ter* (* Third revision in French)	2400/1200	1200/PS	1800	Switched line using echo canceller
V.27	4800	1600/PS	1800	Similar to Bell 208A Leased circuit with manual equalizer
V.27 bis	4800/2400	1600/PS	1800	Leased lines with auto-equalizers
V.27 ter	4800/2400	1600/1200/PS	1800	HDX Switched lines
V.29	9600/7200 /4800	2400/QAM/Ps	1700	HDX & FDX Similar to Bell 209
V.32	9600/4800	2400/QAM/TCM	1800	FDX
V.32 bis	14,400			FDX
V.35	48,000	AM/FM	100,000	FDX

Note: FDX: Full duplex   HDX: Half duplex   Modulation rate: Baud rate
               TCM: Trellis coded modulation   Switched lines: Dial-up lines

Hayes modem family

These are low and medium speed modems (developed by Hayes Microcomputers Products Inc.) and have become subsequently de facto standards. They are available for FDX and HDX operations with a number of other options. Tables 2.8 summarizes typical versions.

Table 2.8: The Hayes modems

Type	Line speed (bps)	Modulation	Remarks Comparable to:
Smart Modem ™ 300	300	FSK/PSK 1070-1270 Hz 2025-2225 Hz	Bell 103
Smart Modem ™ 1200	1. 300 2. 1200	PSK 1200 and 2400 Hz	Bell 212A Bell 103
Smart Modem ™ 2400	3. 300 4. 1200 5. 2400		Bell 103 Bell 212/V.22 V.22 bis
Smart Modem ™ 9600	6. 300 7. 1200 8. 2400 9. 4800		Bell 103 Bell 212A/V.22 V.22 bis V.32 FDX

ISDN modems

These are also known as *terminal adapters* (TA) and refer to *digital modems*. Unlike the analog modems, TAs allow voice, video, and data communication over a single digital connection between CO and the users.

Typically, the associated data rate is 300 to 230,400 bps. Asynchronous multichannel (8) protocols can be supported with provisions for point-to-point and point-to-multipoint connection feasibility. The ISDN modems conform to ISDN specifications in regard to ISDN rates (namely, the basic rate and primary rate). (An example of an ISDN modem is: BitSURFR ProEZ™ of Motorola Inc.)

Cable modem

This has been developed to facilitate high-speed Internet access and video services. Also included is the picture-phone service. This *cable modem* is available to those who subscribe for cable TV. The cable modem allows asymmetric transmission in the following manner:

- *Upstream operation*
  Subscriber to head-end of cable TV

  ✓ Transmit frequency: 5 to 40 MHz
  ✓ Receive frequency: 250-850 MHz
- *Down stream operation*
  Head-end of cable TV to subscriber

  ✓ Transmit frequency: 250-850 MHz
  ✓ Receive frequency: 5 to 40 MHz.

Types of modulations adopted in cable modems are: QPSK, QAM, and vestigial sideband (VSB).

File transfers via modems

Data transmission via modems across the PSTN is prone to bit errors. However, the integrity of transmission is expected when text transfers are made on an interactive basis between PCs, mainframes, and bulletin boards. Hence, a number of protocols have been developed to enable efficient file transfers and are commercially available. Examples are: X modem™, Y modem™, Z modem™, Kermit™ etc.

Baud rate concept

Modems use complex modulation methods to enhance their data transfer capabilities. The way the modulations (or waveform changes) support the bits constituting the data leads to the so-called *baud-rate transmission concept*.

In reference to a DTE, the baud-rate refers to the rate at which the characters are transmitted per second. As indicated earlier, it is the number of cycles or reversals of the digital waveform, in a channel serving a DTE accommodate. That is, baud rate refers to the rate at which signal elements are generated so that the channel will be capable of accommodating the reversals (or changes) of signal elements, as they come by.

Normally, the characters generated by a DTE consists of a string of n bits. A typical character generated by a terminal is depicted in Fig. 2.35.

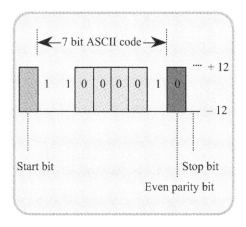

Fig. 2.35: An example of character generated at a DTE (ASCII character 'a')

The character shown in Fig. 2.35 includes a start-bit, a seven-bit data word, a parity bit, and a stop-bit. The bit-time (that is, the duration of each bit) is obtained by taking the reciprocal of the baud rate. Suppose the baud rate is 1200; bit-time is then equal to 833 μs (1/12000). Standardized baud rates at DTEs are as follows: 75, 110, 150, 300, 600, 1200, 2400, 4800 etc.

Bit rate and baud rate

Bit rate refers to the actual number of binary bits transmitted per second (bps) by the communication channel. It is the *transmission rate* or *data rate*.

The *baud rate* described before needs not be equal to the bit rate because, the baud rate, of the signal as mentioned earlier, is the number of times the signal changes per unit time. In essence, it is a *signaling rate*. It depends on the modulation adopted and how the modulation alters the waveform corresponding to a bit. For modems, it is the actual modulation rate of the carrier frequency as it is encoded in terms of the bits of information it carries.

Baud rate is measured as the inverse of the duration of the shortest signal-element transmitted.

The change in signal waveform in reference to specifying the baud rate corresponds to a change in the amplitude, frequency, or phase of the signal. Baud is the number of changes occurring in a signal.

Examples 2.10

Suppose a transmission involves sending signals at 1800 Hz or 1900 Hz and the change in the frequency occurs 2400 times per second.
What is the baud rate? If each change in frequency represents 1 bit, what is the bit rate?

Solution

By definition, the change in the waveform (via change in frequency) occurs at a rate of 2400 per second. Hence, the baud rate is 2400 baud.
Since each change corresponds to 1 bit, the bit rate is 2400 bps.

Example 2.11

The Bell 212A modem operates in two modes. In its low speed mode, it uses frequency-shift keying (FSK) modulation in which the carrier frequency is shifted at the same rate as the binary serial stream at 300 bps. In its high speed mode, at a bit rate of 1200 bps, two bits of the serial binary data are encoded into a single phase change in the carrier. (The encoded two bits are called dibits.) The modulation is known as four-phase differential phase-shift keying (DPSK).

What is the baud rate in the low speed mode?

What is the baud rate in the high speed mode?

Solution

(a) In the low speed mode, the low bit rate of transmission (300 bps) is same as the rate at which FSK takes place. Therefore, the baud rate is 300 baud.

(b) In the high-speed mode, the high bit rate 1200 bps is halved so that a phase-shift occurs at every two bits (dibits). Therefore, the baud rate is 1200/2 = 600 baud.

Example 2.12

Suppose a DTE generates a character made up of: 1 start bit + 5 data bits + 1 stop bit. Each data is of duration 18 ms and the start/stop bits are of 35 ms duration each.

What is the rate of transmission in bps?

What is the signaling rate?

Solution

(a) Total duration of transmission of the character

$$= (5 \times 18) + 2 \times 35 \text{ ms} = 160 \text{ ms}$$

Since the total number of bits transmitted per character is 7, the transmission rate is equal to:

$$\frac{7 \times 10^{-3}}{160} = 43.75 \text{ bps.}$$

(b) Signaling speed corresponds to the shortest duration, namely 18 ms. Therefore, the signal rate is equal to $1/(18 \times 10^{-3})$ baud = 55.56 baud.

Example 2.13

Suppose a modem is devised to transmit 16-level signal elements. Each signal element requires 0.8333 ms for transmission.

Determine the baud rate, and the transmission rate.

Solution

(a) The baud rate, by definition corresponds to the duration shortest signal element. That is,

$$\text{Baud rate} = (1/0.8333 \times 10^{-3})$$
$$= 1200 \text{ baud.}$$

(b) The possible (distinguishable) levels or signal elements for each word is 16. In binary format, this refers to 4 binary strings per word, ($2^4 = 16$). It means 4 bits are transmitted per every 0.8333 ms. Hence,

$$\text{Bit rate} \equiv \text{Number of bits/time}$$
$$= 4/0.8333 = 4.8 \text{ bps.}$$

Note: In general, corresponding to L signal elements that can be generated with n number of bits, $L = 2^n$. Hence, it follows that:

$$n = \log_2(L); \text{ Data or bit rate} = R \text{ bits/sec}$$
$$\therefore D = R/n = R/\log_2(L) \text{ baud.}$$

Baud rate

$R = D \log_2(L)$ bps [that is, bit (data) rate = baud rate $\times$ n]

For example, using only 2 frequencies f_o and f_1 means that the signal change sends 1 bit of data.

That is, $R = D \log_2(2)$, $D = R$.

Baud rate $\equiv$ bit rate $= D \times 1$data

However, it is also possible to use more than 2 frequencies. For example, each frequency change may represent 2 bits of data. In this case, baud rate $= R/2$ [$R = D \log_2(2^2) = 2 D \log_2(2) = 2D$].

In general, n bits can yield 2^n signal elements each one of which can be assigned a frequency (in FSK scheme). Therefore, $R = D \log_2 2^n = (n \times D)$ bps.

Analog waveforms of modem-based transmissions

Digital-to-analog conversion: As indicated earlier, a modem converts a bit stream into a voice-band analog form. This requires a digital-to-analog (D/A) conversion. This conversion or modulation technique is done essentially by three methods as follows:

    ✓   FSK   ⇒   Frequency shift keying
    ✓   ASK   ⇒   Amplitude shift keying
    ✓   PSK   ⇒   Phase shift keying.

FSK

This is essentially a *frequency modulation* technique. Relevant to two digital levels, namely, 0 and 1, two analog carriers with distinct frequencies are allotted. The analog frequency carriers chosen are such that they pass through a voice-grad transmission medium, and when received, they are distinguished in terms of their frequency distinction. Consider for example, the waveforms illustrated in Fig. 2.36.

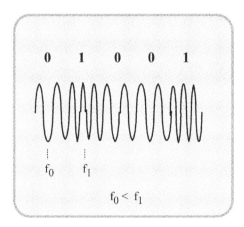

Fig. 2.36: Frequency shift keying assignment to represent two distinct analog levels

    In Fig. 2.36, the digital "0" is designated by analog signal of frequency f_o, and digital "1" corresponds to an analog signal of frequency f_1 and $f_1 \neq f_o$ (in the above case $f_1 < f_o$). The illustration in Fig. 2.36 is a FSK format of a bit string 01001.

Note: Along the time scale, since $f_1 \neq f_o$, the corresponding period occupied by a "0" is not equal to that of "1".

FSK is used in the modems for low speed asynchronous transmissions at 300-1800 bps. Among the two discrete frequencies chosen to represent the logic 1 and 0, the lower frequency is used to depict the logic 0 (called *space*), and the higher frequency is used for denoting logic 1 (called *mark*). The mark and space frequencies lie within 300 to 3400 Hz passband of the PSTN. FSK is used in the Bell System 100 Series (such as 103/113 type) modems. The relevant frequency shifting corresponds to:

- In one direction, 1 and 0 are represented by 1170 Hz with a shift of 100 Hz on each side

- In the other direction, 1 and 0 are represented by 2125 Hz with a shift of 100 Hz on each side.

The resulting spectral assignment is shown in Fig. 2.37. In reference to Fig. 2.37, it can be observed that:

✓ There is a mirror overlap and a possible two way interference
✓ The FSK modulation used was compatible for voice-grade pass band (300-3400 Hz) transmission via twisted copper pair of wires.

Asynchronous modems offering a rate of transmission up to 1800 bps were used in the 1970s with FSK modulation.

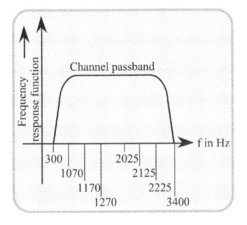

Fig. 2.37: Spectral constituents of the FSK signals in Bell system 100 series

ASK

This is a form of *amplitude modulation* with a carrier waveform taking an on-off pattern. That is, amplitude shift keying is essentially an on-off keying (OOK). In modem applications ASK is restricted to speeds limited to 5 bps due to bandwidth constraints and noise problems. For example, Bell 202 types of modem use the ASK for error control.

In the amplitude-modulating scheme of shift keying, the digital "0" and "1" are allotted two different voltage levels of an analog signal. For example, suppose the signal elements are 00, 01, 10 and 01. To each one of them, an amplitude level, namely, A_1, A_2, A_3, and A_4 are assigned. That is, $n = 2 r = D \log_2(2)^2 = 2D$. That is, the signal corresponds to 2 bits/baud.

PSK

This refers to the *phase shift keying* modulation technique of an analog carrier with its frequency kept constant. Any two digital signal sets are distinguished by their phase shifts (in reference to a fixed phase reference). Typically, a phase of a signal is measured relative to the previous signal, in which case, the technique is called differential PSK (DPSK), where n bits are assigned to a signal element, the signal can assume any on of 2^n possible phase - shift values ($R = D \times \log_2 2^n = nD$).

The DPSK is employed in modems supporting data rates in excess of 1200 bps. Normally, multiphases are used to encode groups of bits. For example, a two-bit set as mentioned before is called a *dibits*; a three-bit set is a *tribits;* a four-bit set is known as a *quadbits* and so on.

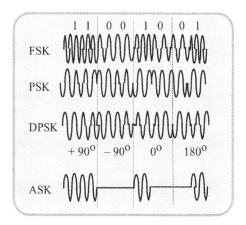

Fig. 2.38: FSK, PSK, and DPSK modulation schemes

The merit of DPSK over the PSK is that an absolute phase reference is not necessary for demodulation of the data. DPSK allows efficient utilization of the bandwidth — several bits can be encoded into a single phase change. Typical modems that use DPSK are:

✓   Bell 212A System: 4-phase DPSK - 1200 bps
✓   Bell 208 System: 8-phase PSK - 4800 bps.

Generally, DPSK is more expensive to implement than PSK systems. In Fig. 2.0, ASK, FSK, PSK, and DPSK schemes are illustrated for comparison.

M-ary concept

M-ary refers to the number of encoded bits used to modulate the carrier frequency. Suppose n is the number of encoded bits used to represent a carrier and M is the number of state changes that the carrier may undergo in representing the n bits. Then, the M-ary relation is given by:
$$n = \log_2(M)$$

Example 2.14
The number of bits in an encoded set in a modem is 3. The modulation technique used is DPSK. Determine the number of phases associated with the DPSK.

Solution
Given that M-ary = 3 bits = n
If M is the number of (phase) changes in the DPSK, then
$$n = \log_2 (M) = 3.$$
Therefore, M = 8.
Thus, the modulation is an eight-phase DPSK. This is typically used in the Bell 208 modem.

Example 2.15

Establish the M-ary relations in respect of Bell 201B/C and Bell 208A/B modems described in Table 2.6. Draw phasor diagrams in each case.

Solution

Modem type	n	Type of modulation and phasor diagram	$n = \log_2(M)$ bits
Bell 201B/C	2	Four-phase DPSK	$2 = \log_2(4)$
Bell 208A/B	3	Eight-phase DPSK	$3 = \log_2(8)$

Bell 201B/C phasor diagram:

```
      01          11
     135°       45°
        ↖  ↗
        ×
        ↙  ↘
     225°       315°
      00          10
```

Bell 208A/B phasor diagram:

```
              010
      010    90°    011
     135°         45°
        ↖   ↗
110 180° ×——— 0° 000
        ↙   ↘
     225°   0|0  315°
     111   270°  100
              101
```

It is technologically more viable to use a combined phase, amplitude, or differential PSK schemes. Examples are: 2000 and 2400 bps modems (DPSK versions).

The first example refers to a 2400 bps Bell 201B/C modem, which uses a signal rate of 1200 bauds/s; in each baud time the phase of the carrier signal is shifted by either 45, 135, 225, or 315° from its previous level. With four possible phase shifts all possible combinations of two-bit pairs (dibit pairs) can be represented (00, 01, 10, 11). The other example refers to Bell 208A/B with a possibility to use 0, 90, 180, and 270° phase changes for the same purpose. The former arrangement has the detection advantage that some measurable phase change will occur in every baud. Demodulation involves the receiving modem comparing the phases of two consecutive signal samples and inferring which pair of bits was sent over the line by the observed magnitude of this phase difference.

Quadrature amplitude modulation

QAM is a technique in which a set of bits is assigned a distinct phase and amplitude designating it as a signal element identifiable by means of the assigned phase and amplitude values.

Fig. 2.39 shown below illustrates changing signal elements pertinent to a string of bits: 001, 010, 100, 011, 101, 000, 111, and 110.

Each signal is characterized by is distinct value of amplitude element A and phase value ϕ shift as shown in Fig. 2.39.

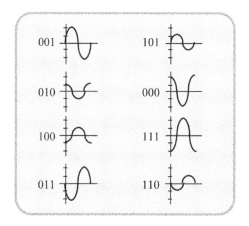

Fig. 2.39: QAM amplitude elements

Applications of QAM

At speeds of 4800 bps and faster, commonly a combination of phase and amplitude modulation systems are employed. These may be illustrated by the mean of a vector diagram. The carrier signal may be regarded as a vector rotating about the origin whose length at any instant corresponds to the maximum amplitude of the encoded signal and whose phase is the angular displacement from the x-axis. For example, Fig. 2.39 denotes a four-phase, two-amplitude modem, giving rise to eight discrete signal elements (points) in phase-amplitude space. A modem adopting this type of encoding could, therefore use these eight different signal elements (points) to send all possible combinations of three-bit groups {000, 001, 010, 011, 100, 101, 110, 111}. For example, 4800 bps operation could be achieved using a baud (line signaling) rate of 1600 signal elements/sec with three bits in each baud. In Fig. 2.39 a signaling rate of 1600 baud would imply that the receiving modem samples the incoming carrier signal amplitude and phase 1600 times per second. Each sample produces one of the eight discernible points of amplitude and phase change from the prior sample, enabling the set of data bits to be reconstructed.

Example 2.16

In reference to the Bell 209A modem described in Table 2.6,

(i)   Establish the M-ary relation
(ii) Determine the baud rate
(iii) Construct the phase diagram

Solution

(i)   The Bell 209A modem supports data in quadbits. Therefore, $n = 4$.
      The modulation used in QAM allows each state represented by one of 16 possible quadbits

(ii) The M-ary relation is: $n = 4 \equiv \log_2(16)$
      The bit rate of transmission in Bell 209A is 9600 bps. The transmission is done on quadbits basis. Therefore, baud rate = 9600/4 = 2400 baud.

(iii)

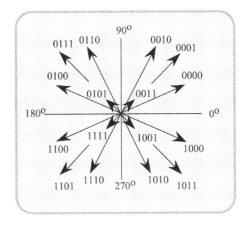

Fig. 2.40: Phasor diagram of the Bell 209B

Nonvoice transmissions on voice band-limited lines

When voice-grade copper-lines are adopted for nonvoice signal (such as data) transmissions, there are two specific considerations. They are as follows:

✓ The d.c. and low frequency components could be blocked by some parts of the telecommunication equipment (such as two-wire to four-wire hybrid transformers, FDM separation filters etc.) used in the voice telephony circuits. Therefore, if supporting nonvoice applications on voice telephone band is considered, these (nonvoice) signals should be modulated appropriately for voice band transmission. That is, the modulated passband should include all the frequency contents of the baseband down to d.c. For example, the data pertinent to nonvoice signals (such as facsimile) normally have low (almost d.c.) components in them

✓ The second consideration refers to supporting modulated carriers of high speed data. The phase response of the voice band channel as indicated earlier would introduce envelope delay and in high-speed transmission such delays may cause intersymbol interference. Therefore, the voice grade lines need " conditioning" so as to impose the quality of transmitted data.

Standard PSTN lines are classified as 3002 unconditioned voice grade category. The bandwidth of such line is limited appropriately from 300 Hz to 3400 Hz. The transfer function, $\alpha(f)$ of such lines is governed by the distributed (L, C, R, G) parameters of the line. The size and geometrical factors of the copper-wire and the dielectric properties of the insulation determine the values of these parameters. The magnitude of $\alpha(f)$ as a function of frequency (f) was indicated in Fig. 2.37 and the envelope delay (due to phase distortion) versus frequency was presented in Fig. 2.26. Shown in Fig. 2.41 is an approximation of envelope delay characteristics of a voice signal across a telephone line as a function of frequency.

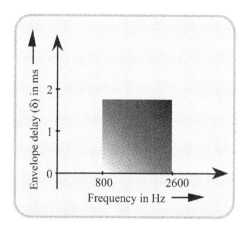

Fig. 2.41: Envelope delay δ in ms versus frequency in Hz

The public networks are normally unconditioned and are guaranteed for data transmission from 300 bps to 9600 bps. However, in permanent connection of the private lines, the telcos can *condition the lines* so as to compensate, to some extent, for the distortion indicated above. For this purpose, *phase* and *amplitude equalizers* are incorporated on leased lines. These circuits counteract the lossy and reactive effects of the (L, C, R, G) elements of the line and enable flat attenuation and linear phase delay characteristics over the voice band spectrum.

The types of conditioning are known as *C-type* and *D-type*. These are offered by telcos at an extra monthly tariff. These two conditionings can be summarized as follows:

- *C-type conditioning:* This conditioning refers to limiting the maximum attenuation distortion on the line. Measured at a reference frequency of 1004 Hz, FCC Tariff No. 60 stipulates five versions of C-type conditioning as indicated in Table 2. The types of modulations adopted under C-conditioning are FSK and PSK

- *D-type conditioning:* The *D-type conditioning* (introduced by AT&T) for 9600 bps transmissions sets the limit on the extent of signal-to-C notched noise ratio and harmonic distortion perceived on the line.

C-notched noise refers to the standard of measured channel background noise. The D-type conditioning specifications are:

- Signal-to-C-notched noise ratio = 28 dB

- Signal-to-harmonic ratios are as follows:

  ✓ For second harmonic: 35 dB
  ✓ For third harmonic: 40 dB.

Unlike C-type conditioning which is facilitated by compensations on the line (such as equalization), D-type conditioning usually does not involve special treatment of any particular line. Instead, the telco facilitates transmission along those lines on which the measurements indicate compatible quality. Specifically, paths devoid of impulse noise are designated for data pulse transmissions (which are otherwise prone to such impulse noise effects).

Sometimes echo suppressors on long distance lines are disabled by means of in-channel control signals sent from the DTEs.

Table 2.9: A summary of C1-C5 line conditioning for private leased lines

Conditional type	Frequency Hz	Attenuation distortion limit (dB)	Frequency Hz	Envelop delay limit (µs)
C_1	300-2700	− 2 to + 6	1000-2400	1000
	1000-2400	− 1 to + 3		
C_2	300-3000	− 2 to + 6	500-2800	3000
	500-2800	− 1 to + 3	600-2600	1500
			1000-2600	500
C_3		− 0.8 to +3(A), +2(B)	500-2800	650(A), 500(B)
A: Access lines	300-3000	− 0.5 to +1.5(A), +1(B)	600-2600	300(A), 260(B)
B: Trunk lines	500-2800		1000-2600	100(A), 260(B)
C_4	300-3200	− 2 to + 6	500-3000	300
	500-3000	− 2 to + 3	600-3000	1500
			800-2800	500
C_5	300-3000	− 1 to + 3	500-2800	600
	500-2800	− 0.5 to + 1.5	600-2600	300
			1000-2600	100

Equalization techniques

As mentioned earlier, equalization refers to compensating for the amplitude and phase distortions. This is accomplished within the modems by equalizer circuits. In essence, it "sharpens" the received waveform so that waveform becomes much akin to the original waveform.

The so-called *compromise equalizers* are contained in the transmit section of the modem. It prequalizes (or predistorts) the shape of the transmitted signal by changing the phase and amplitude characteristics of the signal so that the impairments due to the line are cancelled out. The compromise or predistortion is set to optimize the bit error rate (BER). The compromise equalization can further be of four types, namely: *Amplitude only type compromise, delay only type compromise, amplitude and delay type compromise* and *neither delay nor amplitude type compromise*. And, the corresponding settings can be applied to low, high, or medium ranges of voice band either symmetrically or asymmetrically. The settings can be changed manually. The choice of setting versus BER specifications eventually depends on the line length of the circuit.

Another version of equalizer is known as the *adaptive equalizer* located at the receiver section of the modem. It does post equalization to the received analog waveform. It involves automatic adjustment of the gain and phase characteristics of the equalizer circuit to offset the impairments suffered by the signal on the line. The adaptive aspect of the circuit is based on the information derived from the received signal within the equalizer circuit itself. Or such information can be supplied to the equalizer by the support circuit such as *descramblers* and/or *demodulators*. The adaptive equalizer continuously varies its settings to offer flat amplitude and phase transfer characteristics over the voice band.

A note on envelope delay and phase delay

Ideally a linear phase versus frequency relation is required for error-free data transmission. Since relevant measurement needs a phase reference (which is difficult to specify), normally the delay suffered by a signal as it propagates on the line from the source to the destination is ascertained in lieu of the phase-shift. The envelope delay implicitly indicates that the transmitted phase versus frequency characteristics $\phi_T(f)$ will not correspond to the received characteristics, namely $\phi_R(f)$. The absolute phase delay is the actual time of propagation relevant to a specified frequency. It means different frequency components would experience different extents of delays. Normally, the phase delay versus frequency is a nonlinear curve.

Envelope delay is closely estimated by the change rate of phase with frequency. The envelope delay distortion (EDD) is measured in terms of the phase difference at the different carrier frequencies. For the voice band, typically the reference carrier is taken as 1800 Hz and the EDD is specified as the envelope delay at different carrier frequencies relevant to 1800 Hz. EDD is only a close approximation of the actual phase response of the line. The EDD limit of a basic 3002 line is 1750 µs in the 800 Hz to 2600 Hz band. It means the envelope delay between any two carriers in this band cannot exceed the 1750 µs limit. Suppose the absolute delay at 1800 Hz is 400 µs. It means the absolute delay experiences by any carrier over 800 Hz to 2600 Hz cannot exceed (1750 + 400) = 2150 µs since 1800 Hz is the reference datum.

Example 2.17
The propagation delays measured on a line at different frequencies are as follows:

Carrier (Hz)	Delay (ms)
400	4
1000	3
1800	2.5

What are the relative phase delays of these signals?

Solution
The reference carrier is 1800 Hz. Hence, the relative delays are:

Carrier (Hz)	Delay (ms)
400	1.5
1000	0.5

2.3.3   Digital carrier technology

The classical transmissions of voice (POTS) over a twisted pair of copper-wires refers to an analog carrier system in which the spectrum of the voice waveform generated at the telephone is accommodated by the bandwidth of the lines. This refers to the *baseband analog transmission*.

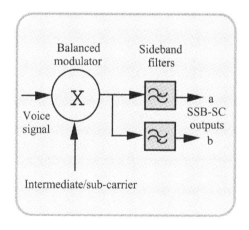

Fig. 2.42: Single sideband suppressed carrier modulation

Multiplexing of several voice transmissions in the analog transmission era (1940-1960) was done via frequency division multiplexing (FDM). That is, each voice waveform is frequency modulated over a subcarrier. Hence the baseband spectrum of the signal gets shifted around this subcarrier frequency as illustrated in Fig. 2.42. Each voice signal resides as a single-sideband at the subcarrier frequency. Several of such sidebands at distinct subcarrier frequencies are generated and group modulated on a single carrier for transmission over a coaxial line. This multi-channel voice frequency (MCVF) system based on single-sideband suppressed carrier (SSBSC) modulation leads to an L-multiplex FDM as illustrated in Fig. 2.43.

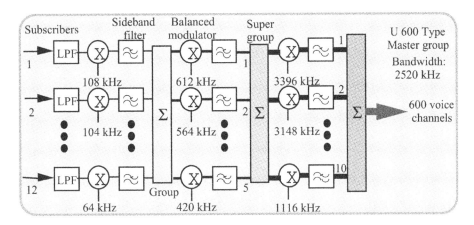

Fig. 2.43: U 600-multiplex MCVF transmission

The analog transmission system was also logically adapted for the transmission of nonvoice data using appropriate modem technology as discussed earlier.

With the developments that took place in semiconductor technology, digital circuit technology sprouted and paved the way for the digital telephony and subsequent digital carrier technology. Modern communication technology, hence enabled the development of better transmission line systems which can carry digital signals with prescribed integrity. An example of such a transmission system is the fiber optic lines. In using such transmission systems, it is required to modulate the analog signal (such as voice) into a predetermined (or encoded) set of pulse trains. Upon reception, such pulse trains can be demodulated, that is, decoded back as the analog waveform (constituted by the original voice signal).

Here, the task of modulation-demodulation involved refers to an encoding-decoding process. "*Codec*" is an acronym for coding-decoding and refers to a device which translates the analog signal into a digital equivalent or vice versa.

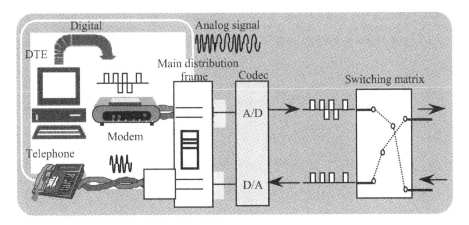

Fig. 2.44: Codec-based transmission of voice

The coder portion of a codec changes analog voice signal into a digital signal (known as DS-0 — digital signal level 0). In essence, coder is an analog-to-digital (A/D) converter. Its counter part, namely the decoder, is a digital-to-analog converter (D/A). Illustrated in Fig. 2.44 is the method of implementing a codec-based transmission of analog voice on the PSTN.

The method of converting analog waveform (data) into digital data is based on two methods known as:

- Pulse code modulation (PCM)

- Delta modulation.

2.3.4   Pulse code modulation (PCM) and its variations

Sampling theorem...

"*If a signal f(t) is sampled at regular intervals of time and at a rate greater than twice the highest significant signal frequency, then the sampled data contains all the information of the original signal*".

Consider a voice signal f(t). The Fourier transform of this aperiodic waveform is F(f). That is, f(t) ⟺ F(f). Suppose f_{max} is the highest significant frequency component of f(t). Then, if sampling is done at a rate equal or greater than 2 f_{max} the original waveform can be reconstructed from the sample data. This is known as the Nyquist-sampling criterion.

Nyquist rate

If f_c is the maximum frequency a physical medium can transmit, the receiver can completely reconstruct the signal, when the signal is sent in samples taken at a rate $2 \times f_c$ per second. (This is true for a noiseless channel.) That is, the receiver can reconstruct a signal sampled at the intervals of $1/2f_c$ second or twice each period, T_c, (since $f_c = 1/T_c$). This sampling rate of 2f is known as the *Nyquist rate*. For example, suppose a voice signal is considered. The truncated bandwidth (BW) of the voice signal is approximately 4000 Hz. That is, the highest frequency of importance in its spectrum corresponds to 4000 Hz. Therefore, the Nyquist sampling rate = 2 × 4000 = 8000 samples/second. In other words, if the voice signal is sampled at every $(1/8000)^{th}$ of a second, it can be recovered from the sampled data. The sampling interval is (1/800)s.

For voice, f_m = 4000 Hz. Therefore, the samples/second should be at least 8000/second. Suppose the sampled data corresponds to eight quantized levels as shown in Fig. 2.45.

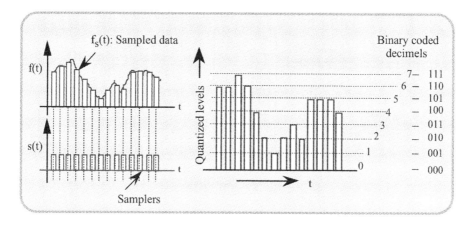

Fig. 2.45: Voice sampling and bit code word assignments to 8 levels of the sampled voice data

Each of the 8 levels in Fig. 2.45, can be represented by a 3 bit code word as illustrated. In the quantized form, the levels of the sampled data denote only an approximate signal envelope. The difference between actual analog signal and its quantized value is known as *quantization noise* (N_q).

The signal-to-quantization noise ratio (S/N_q) is given by (6n-a) dB where 2^n corresponds to the number of the quantized levels (L) and $a \approx 0$ to 1.

Example 2.18
Suppose 256 levels are used for the voice signal.
Determine the rate of transmission of the bits generated.

Solution
$\qquad$ Number of levels, $L = 256 \equiv 2^8$
$\qquad$ That is, each level is represented by an 8 bit code word.
Therefore, bit rate = (8000 samples/second) $\times$ 8 bits/sample.
$\qquad\qquad$ = 64 kbps.
Note: This rate is called the standard digital signal level 0 or DS-0

Example 2.19
Consider color TV bandwidth: $(f_{max}) = 4.6$ MHz.
Determine the rate of transmission of the bits, if the sampled data is encoded in 10 bit code words.

Solution
Number of samples/second = 2×4.6
$\qquad\qquad\qquad$ = 9.2
Since 10 bit code is used, bit rate = (9.2 samples/second $\times$ 10 bits/sample
$\qquad\qquad\qquad$ = 92 Mbps

Companding: Compression - expanding technique

$\qquad$ This refers to a nonlinear encoding technique. That is, in constructing a sampled data of an analog waveform, the quantization levels are not equally spaced. This is done on the following considerations: A larger set of quantizing steps is used at low signal amplitudes and a smaller number of quantizing steps are adopted at high signal amplitudes. By this scheme the quantization process does less distortion to the signal envelope.

$\qquad$ Thus, in the companding technique the encoded signal is compressed to divide low-level signals into more steps, and the high-level signal divided into fewer steps leads to an expanded encoding of the signal. At the receiver, a decoding process does reversals on compression and expansion adopted at the encoding stage.

$\qquad$ The United States follows a companding law known as *μ-law coding*. In Europe, a slightly different nonlinearity is adopted (*A-law*). The μ-law and A-law are known as *companding functions* of nonlinearity.

μ-law companding

$\qquad$ This is defined by the following relation:

$$|y| = \log[1+\mu|x|]/\log[1+\mu(x)] \qquad\qquad (2.22)$$

where x is the input to the compander and y is the output. The μ law is approximately linear at low input levels corresponding to $\mu|x| \le 1$ and approximately logarithmic at high input levels corresponding to $\mu|x| \ge 1$.

A-law companding

This compression algorithm can be specified as follows:

$$y = \begin{cases} \dfrac{A|x|}{1+\log(A)} & 0 \le |x| \le \dfrac{1}{A} \\ \dfrac{1+\log(A|x|)}{1+\log(A)} & \dfrac{1}{A} \le |x| \le 1 \end{cases} \tag{2.23}$$

when $A \to 1$, y is linearly related to x. When $\mu \to 0$ (or $A \to 0$), it corresponds to the uniform quantizing technique. Normally, μ or A is in the vicinity of 100. In the practical companding circuitry used in actual PCM, the compressions law refers to a piecewise linear approximation of the μ- or A-law curve.

The companding on voice signals facilitate a relatively constant S/N_q ratio performance over a wide dynamic range. The parameter μ defines the amount of compression to achieve a given performance. For the same intended dynamic range, A-law companding has a slightly flatter S/N_q ratio than the μ-law.

Voice transmission requires a minimum dynamic range of about 40 dB and a 7-bit PCM code. Hence, the Bell System digital transmissions prescribed a 7-bit PCM code with $\mu \approx 100$. In the recent digital systems, 8-bit PCM codes and $\mu = 255$ are adopted.

The A-law transmissions suffer from *idle-channel noise* at small signal levels in comparison to μ-law companded voice transmissions.

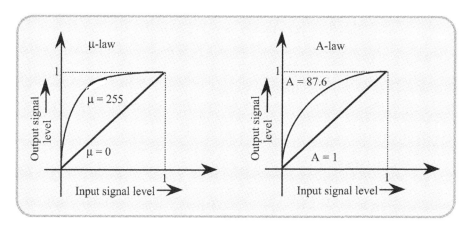

Fig. 2.46: Compander characteristics

The PCM encoding of analog signals via μ- or A-law is known as *log-PCM encoding*. Typical 13 segment compander characteristics (as per CCITT Recommendation G.711) are illustrated in Fig. 2.46. They corresponds to $A = 87.6$ and offers about 26 dB improvement in S/N_q ratio for 8 bit coding.

Example 2.20

Relevant to the A-law characteristics depicted in Fig. 2.46, determine the following:

The gain of the compander for the input values of:
(a) V_{max}; (b) 0.75 V_{max}; (c) 0.5 V_{max}; and (d) 0.01 V_{max}.

Solution

For A-law companding,

$$V_{out} = \frac{V_{max} \times A|V_{in}/V_{max}|}{1+\log(A)}, \quad 0 \le \left|\frac{V_{in}}{V_{max}}\right| \le \frac{1}{A}$$

$$= V_{max}\left\{\frac{1+\log(A|V_{in}/V_{max}|)}{1+\log(A)}\right\}, \quad \frac{1}{A} \le \left|\frac{V_{in}}{V_{max}}\right| \le 1$$

Given that = 87.6.   ∴ 1/A = 0.0114155

(a) $\dfrac{V_{in}}{V_{max}} = 1 > 1/A$

∴ $V_{out} = V_{max}\left\{\dfrac{1+\log(A)}{1+\log(A)}\right\} = V_{max}$

∴ Gain = 1

(b) $\dfrac{V_{in}}{V_{max}} = 0.75 > 1/A$

∴ $V_{out} = V_{max}\left\{\dfrac{1+\log(0.75A)}{1+\log(A)}\right\}$

$= V_{max} \times \left\{\dfrac{5.185}{5.472}\right\}$

∴ Gain = 1.0553

(c) $\dfrac{V_{in}}{V_{max}} = 0.50 > 1/A$

∴ $V_{out} = V_{max}\left\{\dfrac{1+\log(6.50A)}{1+\log(A)}\right\}$

$= V_{max} \times \left\{\dfrac{4.7796}{5.472}\right\}$

∴ Gain = 1.145

(d) $\dfrac{V_{in}}{V_{max}} = 0.01 > 1/A$

∴ $V_{out} = V_{max}\left\{\dfrac{A(V_{in}/V_{max})}{1+\log(A)}\right\}$

$= V_{max} \times [0.160]$

∴ Gain = 6.25

RMS value of quantization error

For 8 bit coding, the A-law used gives 26 dB improvement on S/N_q performance. Consider uniform quantization: The step approximation due to quantization is illustrated in Fig. 2.46.

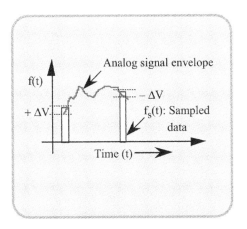

Fig. 2.47: Quantization error

The maximum quantization error voltage per quantization step is equal to: $\pm$ $\Delta V/2$. Assuming all error values up to this maximum can occur randomly and equally likely (that is with a uniform distribution), the RMS value of the quantization error (ΔV_{rms}) can be written as:

$$\Delta V_{rms} = \Delta V / 2\sqrt{3} \qquad (2.24)$$

Further, peak-to-peak amplitude of the quantizied signal = $(\Delta V) \times 2^n$. Hence, the RMS signal amplitude $= (2^n \times \Delta V) \times (1/2) / \sqrt{2}$

$$\therefore \qquad \left(\frac{S}{N_q}\right)_{voltage} = \left(\frac{2^n \times \Delta V}{2\sqrt{2}}\right) \times \frac{1}{(\Delta V / 2\sqrt{3})}$$

$$\therefore \left(\frac{S}{N_q}\right)_{dB} = 20\log[(S/N_Q)_{voltage}]$$

$$= [20n \times \log_{10}(2)] + 10\log_{10}(3) - 10\log_{10}(2)$$
$$= (6n + 1.76). \qquad (2.25)$$

Suppose it is given that:

$$(S/N)_A = \left.\left(\frac{S}{N_q}\right)_{dB}\right|_{\substack{\text{with A-law} \\ \text{companding}}} = 75.6\text{dB}$$

$$(S/N)_U = \left.\left(\frac{S}{N_q}\right)_{dB}\right|_{\substack{\text{without} \\ \text{companding}}} = (6n + 1.76) = 49.6\text{dB}.$$

Assuming signal power is unchanged:

$$(S/N)_A \text{ ratio} = 36.3 \times 10^6$$
$$(S/N)_U \text{ ratio} = 9.1 \times 10^4$$

$$\therefore \left(\frac{S}{N_1}\right)_A \times \left(\frac{N_2}{S}\right)_U = \frac{36.3 \times 10^6}{9.1 \times 10^4} = 3.98 \times 10^2$$

$$\therefore \left(\frac{N_2}{N_1}\right)_{voltage} \approx 20 \text{ or } (N_1)_{voltage} = (1/20)(N_2)_{voltage}$$

$\therefore$ RMS error voltage with A-law companding will be $(1/20)^{th}$ of the corresponding error with uniform quantization scheme.

Problem 2.11

Develop an algorithm that enables a compression ratio of 63.5:1 in a compander having a dynamic range of voltage equal to 60 dB.

Merits and demerits of PCM

With the advent of codec-technique, PCM has sprouted in telecommunication applications. Its major advantages are as follows:

- ✓ Rugged performance under noisy and interference conditions
- ✓ Permits efficient regeneration of the coded signal at the repeaters
- ✓ Leads to an encoded format of information (in bits) regardless of the type of the source
- ✓ The resulting encoded bits allow (time division) multiplexing of several messages
- ✓ Security can be implemented in the encoding adopted
- ✓ Offers a significant S/N ratio.

The limitations of PCM are:

- ✓ System complexity with corresponding enhanced cost
- ✓ Increased bandwidth requirements.

Developments in VLSI technology, availability of wideband communication channels (via fiber) and techniques like data compression, however, pose reduced concerns on the limitations indicated above.

Digital companding

It is a technique in which compression is done at the transmitting end after the input sample is converted to a linear PCM code, and expanding is done at the receiving end prior to PCM decoding. Typically, 12-bit linear code and 8-bit compressed code have been tried. This digitally companded PCM performance is close to that of μ-law analog system with μ ≈ 255.

Vocoders

These are special encoders/decoders specifically designed for speech signals. In order that a speech signal be communicated with an acceptable performance, the following can be take into account:

- ✓ Preservation of the short-term power spectrum of the speech signal
- ✓ Phase distortion is not of major concern.

The above considerations are compatible with human audio perception. Consistent with the above allowances, vocoders are designed with fewer bits by the techniques as described below:

Channel vocoders

Digital vocoders split the voice bandwidth into sub-bands. Each sub-band is digitally encoded/decoded with limited bits. About 200-2400 bps rates are used.

Format vocders

These are based on extracting the short-term, salient power spectra at format frequencies of the voice signal. Encoding/decoding is centered around these short-term components. The system thus operates at lower speed/lower bandwidth.

Line codes

As indicated in the above section PCM, in general, allows information transmission in the form of bits. The bit sequences are comprised of logical 1s and 0s.

In reference to the physical layer of transmission (that is, across a copper wireline or fiber optic medium), these 1s and 0s should be formatted in electrical and/or optical signals compatible for transmission across the physical channel in question. Representation of PCM encoded logical 1s and 0s in pulse forms (electrical and/or optical) at the physical layer level is known as *line coding*.

Construction of a line code in pulse form depends on the following considerations relevant to digital pulses:

    ✓   Unipolar or bipolar digital pulses
    ✓   Non-return-zero (NRZ) and return-zero (RZ) pulses.

The bipolar and unipolar pulses are illustrated in Fig. 2.48. A unipolar pulse has a value 0 to +V or 0 to −V and a bipolar pulse has values +V to −V as indicated

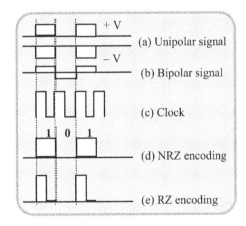

Fig. 2.48: Unipolar and bipolar a digital signals and non-return to zero a and return to zero encoding

The NRZ pulse refers to the state that the pulse amplitude return to 0 only at the end of one full clock period; whereas, the RZ signal has the pulse value returning to 0 level even during the clock period as illustrated in Fig. 2.48.

The following are considerations relevant to the design of line coding schemes:

▪   PCM equipment needs a code converter to realize a specific line code format

▪   The line code should not average out to give a large d.c. level. If such a d.c. level is present, it may get blocked at the capacitive and transformer sections of the communication equipment

▪   The energy at low frequency part of the spectrum of the code should be small. Large low frequency content requires low-pass action and hence requires the use of large-sized components such as capacitors

- High d.c. levels may also lead to d.c. drift. Such drifts will cause performance impairments via bias variations at the receiving end. For example, in fiber optic communication, the d.c. drift will affect the APD (avalanche photo diode) performance significantly due to bias-drift induced thermal instability.

More details on line coding schemes are elaborated in a later section of this chapter.

Delta modulation

This is an alternative to PCM technique. Here, an analog waveform is approximated by a staircase function that moves up or down by one quantization level (δ) at each sampling interval (T_s) as illustrated in Fig. 2.49.

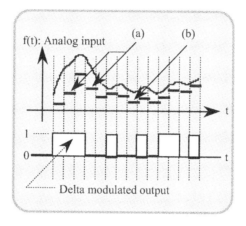

Fig. 2.49: Delta modulation

In Fig. 2.49, the staircase function is overlapped on the original signal waveform. It can be noted that the staircase function has a binary behavior. Since the function moves up or down by a constant amount δ at each sampling instant, the output of the delta modulation process can be represented as a single binary digit for each sample. In other words, approximating the derivative of the analog signal (that is, the slope of the analog signal), rather than its amplitude, produces the bit stream.

When the staircase function goes upward (during the next interval), then a logic 1 is set; a logic 0 is set otherwise. The up/down transition the occurs at each sampling interval is so chosen that the staircase function tracks the original analog waveform as closely as possible. A low-pass integrating circuit can smooth out the staircase function decoded at the receiver, and the resulting waveform would closely approximate the original signal waveform. The performance parameters of delta modulation are:

✓ δ: Size of the step assigned to each binary digit
✓ T_s: Sampling rate.

In delta modulation technique there are two types of noise encountered. They are:

✓   *Quantization noise* (*granular noise*) when the analog waveform changes slowly. This increases as δ is increased
✓   *Slope overload noise*: When analog waveform changes rapidly enough, the staircase would not follow the slope changes closely. This noise increases as δ is decreased.

Delta modulation is a variation of the so-called differential PCM wherein the analog input signal is converted to a continuous serial data stream of 1s and 0s at a rate determined by a sampling clock.

Delta-sigma modulation

The performance of delta modulation largely depends on the derivative of the incoming signal. This causes an accumulative error in the demodulated signal. Integrating the signal prior to delta modulation can compensate this drawback. The beneficial aspects of such integration are as follows:

- ✓ Preemphasis of the low-frequency components in the signal
- ✓ The adjacent samples of the delta modulator input get closely correlated and reduce thereby the variance of the error signal at the quantizer input
- ✓ Simpler architectures are feasible.

Adaptive delta modulation (ADM)

In this technique, the step-size of D/A converter is varied automatically as per the amplitude changes of an analog input signal. The algorithm that accomplishes ADM operation is as follows: Whenever three consecutive 1s (or 0s) occur, the step-size of the D/A converter is increased (or decreased) by a factor 3/2. When alternate 1s and 0s are present, it corresponds to a large granular noise. Hence, the D/A converter adaptively resorts to minimum step-size.

Differential PCM

In practical analog waveforms such a speech signals, the successive samples are quite often identical. Therefore, transmission of such identical successive samples in PCM encoded format is rather information-redundant. Hence, in the differential PCM, the difference in amplitudes of successive samples is encoded and transmitted. This technique requires fewer bits than the conventional PCM.

On the receiver side, subsequent to D/A conversion, the difference amplitude obtained is stored and added to the next sample received. This operation is cumulative and leads to a close approximation of the analog envelope transmitted as the output.

2.4 Concepts of Information Theory

2.4.1 *Entropy and information*

Entropy is a measure of uncertainty or ignorance. It refers to a negative measure of information content [2.1, 2.3, and 2.16]. That is, the uncertainty of the signal, which is measurable in terms of the "unpredictable signal variation changing continuously in time" is an implicit measure of information borne by the signal.

"A continuous whistling of one note conveys no information". However, if the note is varied, that is, changed continuously in a manner that one can interpret, the note begins to carry meaningful music. Or it bears useful information. It is then said to have a *semantic* attribute.

2.4.2 *Hartley-Shannon law*

Consider a time-varying signal envelope shown in Fig. 2.50: Let the amplitude of the signal $f(t)$ be divided into discrete levels. On the time scale, suppose the signal duration, T is divided into m subdivisions each of width τ. That is, $m = T/\tau$. Further, the total amplitude A of the signal is divided into n levels as shown in Fig. 2.50.

Over time T, the second, the unpredictably varying signal, can represent n^m possible distinct values. Therefore, the corresponding information content, I should be functionally related to n^m. That is,

$$I = \text{Function of } (n^m)$$
$$= \text{Function of } [n^{t/\tau}] \tag{2.26a}$$

Intuitively, I should be directly proportional to the total time T, or as T increases, I should monotonically increase.

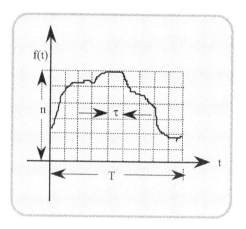

Fig. 2.50: A time-varying signal

Hence, in the above relation, the function must corresponds to the following valid identity: $I \equiv [k_1(T/\tau)\log n]$ where k_1 is a constant of proportionality. Eqn. (2.26a) implies the following:

✓   I increases with time of the signal, T

✓   I increases with $1/\tau$. Therefore, by decreasing τ, I would increase. But the minimum value of τ is rather limited by the rise-time characteristics of the system

✓   I increases with n. The relevant implications are:

⇒   If the number of subdivisions (n) are increased, I would increase. But the minimum value of the subdivision $\Delta v = A/n$ is limited by noise level. That is, the smallest extent of signal level can be limited up to the noise level.

⇒   n can be increased by increasing the signal level, A. However, power limitations will set the upper limit on A. In other words, the maximum value of A is set by the power (proportional to A^2) the system can handle.

In the relation, $I = k_1 \, t/\tau \, \log(n)$, k_1 can be so chosen, by appropriate choice of the base of the logarithm as indicated below:

$$I = (T/\tau)\log_e(n) \text{ nats (natural units)}$$
$$= (T/\tau)\log_2(n) \text{ bits (binary digits)} \qquad (2.26b)$$

Therefore, it follows that, the maximum channel capacity C_{max} is given by:

$$C_{max} = (I/T)_{max} = 1/\tau \log_2(n)|_{(\tau \to \text{ minimum value})} \qquad \text{bits/second (bps).} \qquad (2.27a)$$

That is, the minimum value of τ leads to a maximum rate of transmission. In a physical system, the minimum τ corresponds to the finite rise-time, t_r. That is $t_r = (1/\tau)_{min}$ and $(1/t_r)$ approximately corresponds to the bandwidth (BW) of the system. Hence,

$$C_{max} = BW \times \log_2(n) \quad \text{bps.} \qquad (2.27b)$$

The maximum number of n (number of quantized levels) of the signal is decided by the amplitude level A and by the minimum (quantized) division Δv. Since A is limited by the power-handling constraints of the system, n can be maximized by choosing a minimum value for Δv. However, as stated earlier, the minimum Δv is limited to the existing noise level. Therefore,

$$n = (S/N) + (\text{zero level}) = (S/N + 1) \qquad (2.28)$$

where S/N is the signal-to-noise power ratio. Therefore,

$$C_{max} = BW \times \log_2 (S/N + 1) \text{ bps.} \qquad (2.29)$$

Eqn. (2.30) is known as *Hartley-Shannon's law*

2.4.3   Measure of information

Consider a binary digit (0, 1) transmission of the quantized levels of an analog voltage discussed above. There are $m = T/\tau$ possibilities of 0 or 1 occurring in the transmission. On an average, 0 appears mp times and 1 appears mq times where p and q are a priori probabilities of 0s and 1s. Corresponding to n levels (1/n) refers to the probability of occurrence of each level. Therefore,

$$\begin{aligned} I &= m \log(n) \\ &= - m \log (\text{probability of occurrence of each level}) \end{aligned} \qquad (2.30)$$

By defining, H_{ave} as the *average information* over the interval τ, it follows that,

$$H_{ave} = I_{ave}/m \text{ bits/interval} \qquad (2.31)$$

and in reference to the binary states 0 and 1,

$$I_{ave} = - mp \log_2(p) - mq \log_2(q) \text{ bits} \qquad (2.32a)$$

or

$$I_{ave}/m = [- p \log_2 (p) - q \log_2(q)] \equiv H_{ave}. \qquad (2.32b)$$

The corresponding, average channel capacity is given by

$$C_{ave} = H_{ave}/\tau = - B\Sigma p_i \log p_i \text{ bits/sec.} \qquad (2.33)$$

The average information per interval with two possible signals is as indicated in Fig. 2.51.

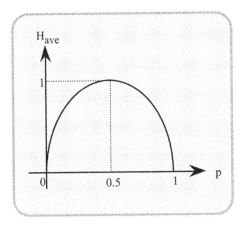

Fig. 2.51: H_{ave} versus occurrence probability of two possible signals

Example 2.21

Consider a standard 4 kHz telephone. Determine its channel capacity, if S/N ratio is 32 dB.

Solution

Given that

$$SNR = 32 \text{ dB}$$
$$32 \text{ dB} \Rightarrow \text{Ratio } 1585$$
$$\therefore C = B \log_2 (1 + S/N)$$
$$= 400 \times 10.63 = 42520 \text{ bps}$$

$$32 \text{ dB} = 10 \log_{10} [(S/N)_{Ratio}]$$
$$\therefore (S/N)_{Ratio} = 1585.$$

Example 2.22

A system has a bandwidth of 4 kHz and a SNR of 28 dB at the input to a receiver.
(i) Calculate its information-carrying capacity.
(ii) Calculate the capacity of the channel, if its bandwidth is doubled, while the transmitted signal power remains constant.

Solution

(i) $(S/N)_{Ratio}$ = antilog $(28/10) = 631$
$$\therefore C_1 = 4000 \log_2 (1 + S/N) = 4000 \log_2 (1 + 631)$$
$$= 37,216 \text{ bps}$$

(ii) Noise power, $N = (k_B T) \text{ BW}$

When BW is doubled, the noise power is doubled. Given that the original $(S/N) = 631$, after BW is doubled, $(S/N) = 631/2$ since N is doubled.

$$\therefore C_2 = (2 \times 4000) \log_2 (1 + 631/2) = 66,448 \text{ bps}$$

Trade-off between SNR and bandwidth

Harltey-Shannon's law, namely, $C = BW \times \log_2(1 + S/N)$ bits/sec, specifies the maximum rate of transmission in a channel. Suppose the bandwidth is increased to infinity. Will the channel capacity increase to infinity? The answer is "no". The reason is as follows:

It is known that noise is directly proportional to BW. That is, $N = \eta \text{ BW}$, where $\eta = k_B T$

$$\therefore C = BW \times \log_2 [1 + S/(\eta \text{ BW})] \text{ bps}$$

$$= (S/\eta) (\eta BW/S) \times \log_2 [1 + S/(\eta BW)] \text{ bps}$$

$$= (S/\eta) (1/x) \times \log_2 (1 + x) = (S/\eta) \times \log_2 (1 + x)^{1/x} \qquad (2.34a)$$

where $1/x = \eta BW/S$.
Under the limit BW $\to \infty$ or x $\to 0$, $(1 + x)^{1/x} \to e$.

$$\therefore \lim_{B \to \infty} C = (S/\eta) \times \log_2 (e) = 1.44 (S/\eta) \text{ bps.} \qquad (2.35)$$

Therefore, in order to achieve a given C, bandwidth and SNR have to be traded off.

Problem 2.12

Suppose a discrete information source emits a binary digit, either 0 or 1, with equal probability every T s. The rate of binary traffic from the source is $R = (1/T)$ bps. The information content of each output form the source is:

$$I(x_i) = -\log_2[P(x_i)] \qquad x_i = 0,1$$
$$= -\log_2(1/2) = 1 \text{ bit}$$

Now suppose successive outputs from the source are statistically independent. Consider a block of k binary digits from the source which occurs in a time interval nT. There are $M = 2^n$ possible n-bit blocks, each of which equally probable with probability $1/M = (1/2^n)$. The self-information of a k-bit block is

$$I(x_i) = -\log_2(1/2^n) = n \text{ bits}$$

emitted in a time interval nT. Thus the logarithmic measure for information content possesses the desired additivity property when a number of source outputs is considered as a block.

Note: A useful conversion: $\log_e(a) = [\log_2(a)] [\log_e(2)] = 0.69315 \times \log_2(a)$

Example 2.23

Find the entropy of a source that emits each one of third symbols A, B, and C in a statistically independent sequence with probabilities 1/2, 1/4, and 1/4 respectively.

Solution:
$H \cong (1/2)\log_2(2) + (1/4)\log_2(4) + (1/4)\log_2(4)$ bit/symbol
$= (1/2) + 2/4 = 1.5$ bits per symbol.

Example 2.24

In Example 2.23, if the symbols are emitted each in every millisecond, find the information rate.

Solution
$H = 1.5$ bits per symbol

$\therefore$ Information rate $= 1.5$ bits/symbol $\times 1$ symbol/1ms
$$= 1.5 \times 1000 \text{ bps}$$
$$= 1500 \text{ bps}$$

Example 2.25

A binary symmetric channel is shown in Fig. 2.52. Find the rate of information transmission over this channel when $p = 0.9$, 0.8, and 0.6. Assume that the bit rate is 1000 per second.

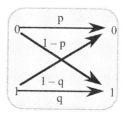

Fig. 2.52: Binary symmetric channel

Generated (source) entropy $\cong$ H(x)

$$H(x) = - p(0) \log_2 [p(0)] - p(1) \log_2 [p(1)]$$

In a BSC p(0) = p(1)

$$\therefore \; H(x) = (1/2)\log_2(2 + (1/2) \log_2 2 = 1 \text{ bit/symbol}.$$

Example 2.26

A computer monitor displays alphanumeric data. It is connected to the computer through a voice grade telephone line having a usable bandwidth of 3000 Hz and an output S/N of 10dB. Assure the terminal has 128 characters and that the data sent from the terminal consist of independent equiprobable characters.

(i) Find the capacity of the channel.
(ii) Find the maximum (theoretical) rate at which data can be sent from the terminal to the computer without errors.

Solution

$$C = B \log_2 (1 + S/N)$$
$$= 3000 \log_2 (11) = 10{,}378 \text{ bits/sec}.$$

Average information per character = H = $\log_2$ (128)
$$= 7 \text{ bits/character}$$

For errorless transmission, rate of transmission should be R such that:

R (characters/sec) $\times$ H (bits/character) $\leq$ channel capacity.
That is, R $\times$ 7 $\leq$ 10,378.

Example 2.27

A typical ASCII code is shown in Fig 2.53. Its baud rate is 110. Determine:

(i) Width of each bit interval
(ii) The maximum number of characters that can be sent per second
(iii) The maximum information rate at maximum signaling speed.

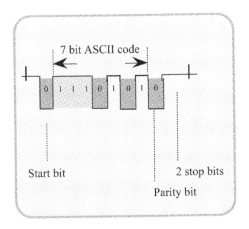

Fig. 2.53: ASCII code

Solution

(i) Baud rate $\Rightarrow 1/\tau$, τ = Width of the shortest element

Given that: $1/\tau = 110$

$$\therefore \tau = 1/110 = 9.01 \text{ ms}$$

(ii) The given word has 7 data bits

+ 1	start bit
+ 1	parity bit
+ 2	stop bit

Total: 11 bits

Hence, the total length $= 11 \times 9.09 = 100$ ms

$\therefore$ Number of characters that can be transmitted per second

$$= 1/100 \text{ m s} = 10 \text{ characters/s.}$$

(iii) At the maximum signaling speed the difference between the baud rate and the information rate is the stop interval equal to 2 stop bits (in the present case) which is considered equal to 1 information bit. Therefore, the effective number of bits per character = 10 since there are 10 characters sent per second, the information rate is 100 bps.

Example 2.28

Determine the average information content of a hypothetical language with 25 alphabets, assuming each alphabet is equally likely to occur.

Solution

Since all the alphabets are equally likely to occur, the corresponding probability of occurrence of each alphabet is 1/25. Hence the average information content is:

$$H_{ave} = -\sum_{i=1}^{25} p_i \log_2(p_i) \qquad \text{bits}$$

$$= -\sum_{i=1}^{25} \left(\frac{1}{25}\right) \log_2\left(\frac{1}{25}\right) \qquad \text{bits}$$

$$= \log_2(25) = 4.644 \qquad \text{bits/letter}$$

Problem 2.13

Considering 26 English alphabets, $\log_2(26)$ corresponds to 4.7 bits/letter. However, the actual information content of English is only about 4.15 bits per letter. Justify.

Problem 2.14

A standard voice-band communications channel with a S/N ratio of 1000 and a bandwidth 2.7 kHz indicates the Hartlay-Shannon information capacity as $2700\log_2(2700) = 26.9$ kbps. Does it mean that 26.9 Kbps can be transmitted through the 2.7 kHz channel using a binary system? Justify your answer.

2.5   Encoding and Decoding

2.5.1   Principle of coding

A salient aspect of telecommunications is representing the data generated by a discrete source efficiently so as to be most compatible for transmission. The process by which this representation is enabled is called *source encoding*. Relevant to digital communication, source coding refers to bit representation in a specific format of the information transmitted. The unit that performs the required representation is called a *source encoder*. For the source encoder to be efficient, a statistic knowledge of the source is required. For example, if some source symbols are known to be more probable than the others, then one may exploit this feature in the generation of a source code by assigning short code words to frequent source symbols, and large code words to rare source symbols. Such a source code is known as a *variable-length code*. As indicated earlier, the Morse code is an example of a variable-length code. In the Morse code, the letters of the alphabet and numerals are encoded into streams of marks and spaces, denoted as dots ("·") and dashes ("-"), respectively. Since in English, the letter E occurs more frequently than the letter Q for example, the Morse code encodes E into a single dot ".", the shortest code word in the code, and it encodes Q into "---", a longer code word.

Suppose the number of messages under transmission is M. Let each message be coded into N bits. Therefore, $M = 2^N$. If each message is equally likely to occur, then $p_i = 1/M$ and average information/message $H = -\log_2(1/M) = \log_2(M)$. Further, $M = 2^N$. Therefore, the average information carried by individual bit $= H/N = 1$ bit.

This average information content can be improved by encoding the message. That is, the number of bits assigned for each message can be variable instead of being the same for each message. This is possible due to the fact that each message has its own probability of occurrence. The message most likely to occur is given the fewer number of bits, and likewise, the least-likely to occur message is given the largest number of bits. This is the basis of Morse code as indicated above. A complete list of *International Morse Code* is listed in Fig. 2.54.

International Morse Code

A	0 —	1	0 — — — —	. Period	0 — 0 — 0 — 0
B	— 000	2	00 — — —	; Semicolon	— 0— 0— 0 —
C	— 0 — 0	3	000 — —	, Comma	— — 00 — —
D	— 00	4	0000 —	: Colon	— — — — 0 0 0
E	0	5	00000	? Query	00 — — — 00
F	00 — 0	6	— 0000	'Apostrophe	0 — — — — — 0
G	— — 0	7	— — 000	- Dash (Hyphen)	— 0000 —
H	0 0 0 0	8	— — — — 00	/ Slash	— 00 — 0
I	00	9	— — — — — 0	" Quotes	0 — 00 — 0
J	0 — — — —	0	— — — — —	_ Underscore	00 — — — 0 —
K	— 0 —			= BT	— 000 —
L	0 — 00	Å	0 — 0 —	SOS/Distress	000 — — — — 000
M	— —			Attention	— 0 — 0 —
N	— 0	A'		CQ	— 0 — 0 — — 0 —
O	— — —	or Á	0 — — 0 —	DE	— 0 0 0
P	0 — — 0	Ė	00 – 00	Go ahead	— 0 —
Q	— — 0 —			Wait AS	0 — 000
R	0 — 0	CH	— — — —	Break BK	— 000 — 0 —
S	000			Understand	000 — 0
T	—	Ñ	— — 0 — —	Error	00000000
U	00 —			Okay	0 — 0
V	000 —	Ö	— — — — 0	End of message AR	0 — 0 — 0
W	0 — —			End of work SK	000 — 0 —
X	— 00 —	Ü	00 — —	$	0 — — — 0 — 0 —
Y	— 0 — —			More to follow	— 000
Z	— — 00	(or)	— 0 — — 0 —	ASAR	0 — 0000 — 0 — 0

Fig. 2.54: The International Morse Code

Example 2.29

Consider a signal transmission of 4 symbols α, β, γ, and μ.

Determine the average information per symbol if a coding is done on the basis that each symbol is equally likely to occur.

Solution

All the symbols are equally likely to occur. Therefore, $p_\alpha = p_\beta = p_\gamma = p_\mu = 1/4$. Hence, the average information in bits/message is given by

$$H_a = [-\sum_{i=1}^{4} p_i \log_2(p_i)] \times 4 = [-\sum_{i=1}^{4} (4)\log_2(4)] \times 4 = 2 \text{ bits.}$$

Example 2.30

Suppose the probabilities of occurrence of a set of symbols are as follows: $p_\alpha = 0.5$; $p_\beta = 0.25$; $p_\gamma = 0.125$ and $p_\mu = 0.125$. Devise an unequal length code for the transmission of α, β, γ, and μ. Determine the average information in bits/message.

Solution

Message (Signal element)	Probability	Code scheme			Number of bits per message
α	1/2	0			1
ß	1/4	1	0		2
γ	1/8	1	1	0	3
μ	1/8	1	1	1	3

Average number of bits/message (H)

$$H = -\sum p_i \log_2 p_i = (1/2)\log_2(2) + (1/4)\log_2(2^2) +$$
$$(1/8)\log_2(2^3) + (1/8)\log_2(2^3)$$
$$= (1/2) + (1/2) + (3/8) + (3/8) = 1.75 \text{ bits/message}$$

Using the code devised as above:

$$H = 1(1/2) + 2(1/4) + 3(1/8) + 3(1/8)$$
$$= 1.75 \text{ bits / message}$$

That is, source coding does not change the source entropy. Suppose each symbol was assigned the same length of bits, and the average bits/message = 2 as indicated in the previous example. This is 12.5% larger than 1.75. As a result, this code will have the following characteristics:

❑ It will consume more time per bit (meaning reduced channel capacity)
❑ If same transmission time is maintained, it will require more bandwidth as per Hartley-Shannon's law
❑ The increased bandwidth would reduce the noise immunity.

Problem 2.15

Indicated in below is the relative occurrence frequency of roman characters adopted in Bahasa Malaysia (the *lingua franca* of Malaysia) [2.17].

Letter	Relative frequency	Letter	Relative frequency
A	1.00000	P	0.15731
N	0.58625	H	0.12686
I	0.42889	B	0.10655
E, $\overline{E}$	0.38828	Y	0.07355
T	0.31218	O	0.06595
U	0.30203	J	0.04820
K	0.26902	W	0.02535
S	0.24362	V	0.02280
G	0.23347	C	0.01264
D	0.22587	Z	0.01264
L	0.18526	F	0.00760
M	0.18526	X	0.00000
R	0.18526	Q	0.00000

Devise a compatible variable bit length code and compare it with the International Morse Code (prescribed to roman characters on the basis of English language). (Hint: See [2.17])

As indicated above a variable bit length code refers to allocating the less probable messages, fewer bits (as they are likely to be transmitted most often). This makes the coding scheme "*efficient*". The "*coding efficiency*" is defined as follows: Considering a source emitting statistically independent (m) symbols, the associated entropy Shannon information content is given by,

$$H = \sum_{m=1}^{M} p(m) \log_2[1/p(m)] \quad \text{bit/symbol}. \qquad (2.36a)$$

The maximum possible entropy, however, is realized, if all these (m) symbols were equiprobable. That is, when $p(m) = 1/m$. Therefore, it follows that

$$H_{max} = \log_2(m) \quad \text{bit/symbol}. \qquad (2.36b)$$

Hence , the coding efficiency (η_{code}) can be defined as:

$$\eta_{code} = \left(\frac{H}{H_{max}} \right) \times 100\%. \qquad (2.36c)$$

Problem 2.16
Determine the coding efficiency of Captain Kidd's message concerning the whereabouts of a hidden treasure. The message was written with typographical signs with a relative frequency of occurrence indicated below [2.18].

Symbol/Character	8	;	4	±	(	*	5	6	+	1	0	g	2	I	3	?	Π	–	•
Relative frequency of occurrence	33	26	19	16	16	13	12	11	8	8	6	5	5	4	4	3	2	1	1

2.5.2 Huffman coding

This is an improved version of the simple variable bit length coding discussed above. It also relies on the concept of entropy. Huffman code offers an organized technique to find the best possible variable length code for a given set of messages. The procedure to construct a Huffman code involves two steps, namely *reduction and splitting*.

The reduction procedure refers to the following: The symbols are first listed in their descending order of probability. The two least probable symbols are then reduced to one with probability equal to their combined probability. As a subsequent step, reordering is done in descending order of probability. This procedure is continued until only two symbols are left out.

The second part is called splitting and it involves assigning 0 and 1 to the two final symbols and working backwards. To cope with each successive split, the code is lengthened at each stage; at the same time, two split symbols are distinguished by adding another 0 and 1 respectively to the code word. It has been proved that the Huffman code will always require less than one binary digit per symbol more than entropy.

In reference to the reduction and splitting procedure indicated above, the underlying concept can be better understood via the following example: With the Huffman code, the most commonly used characters are assigned fewest bits and the less commonly used characters are given the most bits. For example, suppose the characters α, β, γ, and ε constitute a specified character set. Suppose, on an average, α account for 52% of the total transmitted traffic, β accounts for 25%, γ accounts for 14% and ε accounts for 10%. This skewed character distribution is illustrated in Fig. 2.55. The rules in developing the tree are:

✓ The character set is arranged in a column with the most frequently occurring character first in the list on down to the least frequently occurring character

✓ Lines are drawn out horizontally from the bottom of the column and merge the two lowest frequencies to obtain a new summation of the two relative frequencies of occurrence. A line is drawn from this line with the new summed frequency entered

✓ This process is continued until all lines are merged

✓ 0 and 1 are placed at the end of each nodal line

✓ The tree is traced back to the origin to obtain the code for the character.

In reference to a set of four transmitted messages, {a, b, g, e} with the occurrence probabilities and Huffman encoding as indicated in Fig. 2.55, the decoding can be done via a decision chart illustrated in Fig. 2.56. In the relevant procedure, the data are examined and decoded by reading the bit stream left to right and following the decision chart down the proper logic path. Thus, the Huffman code can be decoded as the individual bits arrive at the receiver, without waiting for the entire user data stream to arrive. The average bits per symbol for this example is given by: $H = (1 \times 0.52 + 2 \times 0.23 + 3 \times 0.14 + 3 \times 0.11) = 1.73$ bit/symbol.

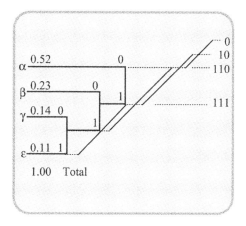

Fig. 2.55: Construction of Huffman code tree

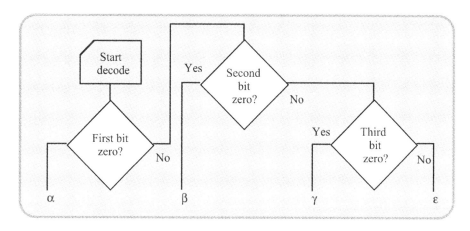

Fig. 2.56: Huffman decision chart to construct the code tree

Note: Follow the tree from right to left in order to build the code.

The number of bits (β) needed to encode a character using the Huffman scheme, is given by $B = INT(- \log_2 P)$ where INT is the function rounded up to the nearest integer, and P is the probability of the occurrence of the letter in the character set.

Problem 2.17

The letter "a" has a probability of occurrence equal to 0.15731 in the romanized text of Bahasa Malaysia (see Problem 2.15).

Determine the bits required to represent the letter "a".

Huffman codes have a characteristic feature called *a prefix property*. This means no short code set will be interpreted as the beginning of a large set. For example, if the bit stream 111 is a code for a symbol, then 11101 cannot be a code for another symbol since a left to right scan would detect a 111 symbol, followed by 01.

The prefix property also implies that it is not necessary to know where a message commences so as to actually decode it. The only requirement is that the decoding should be started by searching for the shortest sequence of bits that represents a known symbol.

Example 2.30

Consider the set of four characters (α, β, γ, ε) depicted in Fig. 2.55 and their corresponding Huffman codes constructed. Suppose the following bit stream refers to the binary format (written in terms of the Huffman code developed): 10 0 111 110 0 111. Decode the message.

Solution

Now, following the tree shown in Fig. 2.55 from right to left, the binary stream can be decoded to $\beta\alpha\varepsilon\gamma\alpha\varepsilon$.

Problem 2.18

Indicated below is a set of words {A, B, C, D, E, F, G, H} and their probabilities of occurrence in a text. Suppose the following message is sent.

"A B H A E C A B A G D C A B"

Word	Relative frequency of occurrence
A	0.50
C	0.15
D	0.12
B	0.10
E	0.04
G	0.04
F	0.03
H	0.02

(i)   Construct the Huffman encoding tree.

(ii) Show how decoding can be done ensuring the prefix property.

In terms of code words (number of binary digits) assigned to m = 1, 2, ..., M symbols, the coding efficiency defined earlier is determined as follows:

$$\eta_{code} = \frac{H}{H_{max}} \times 100\% \qquad (2.37)$$

where $H = - \sum_{m=1}^{M} p(m) \log_2 [p(m)]$ bit/symbol with p(m) being the probability of occurrence of mth symbol, and H_{max} denotes the maximum information that can be conveyed per B digit binary code word. This average code word byte is given by

$$H_{max} = \sum_{m=1}^{M} p(m) b_m \qquad \text{bit/code word} \qquad (2.38)$$

where b_m is the binary digit of the code word representing the mth symbol.

For the coding Huffman coding scheme indicated in Fig. 2.55, the coding efficiency is 86.5 %

Problem 2.19

Construct a Huffman code for the Captain Kidd message source specified in Problem 2.16. Hence, determine the corresponding coding efficiency. Compare this value to that obtained in Problem 2.16. Justify the results.

The so-called Shannon-Fano code is similar to Huffman code. The major difference is that in the Shannon-Fano code, the operations are performed in a forward, rather than backward, direction. Therefore, the storage requirements are simplified and the code is easier to implement. The average lengths are the same in both Shannon-Fano and Huffman coding. The characteristic and performance of these two codes are comparable.

Example 2.31

Construct a Shannon-Fano code for a set of words {A, B, C, D, E,} with the set of probabilities {$P_A = 0.06$, $P_B = 0.12$, $P_C = 0.28$, $P_D = 0.04$, $P_E = 0.50$}

Solution

Step 1: Arrange the words in their decreasing order of probabilities

Word	Probabilities
E	0.50
C	0.28
B	0.12
A	0.06
D	0.04

Step 2: Partition the words into the most probable subsets. Assign 1 or 0 as indicated

Subsets	Word	Probabilities	0/1 Assignment		
I	E	0.50	0	or	1
II	C	0.28	1		0
	B	0.12	1		0
	A	0.06	1		0
	D	0.04	1		0

Step 3: Repeat step 2 on subset II

Subset	Word	Probabilities	0/1 Assignment	
III	C	0.28	10 or	01
	B	0.12	11	00
IV	A	0.06	11	00
	D	0.04	11	01

Step 4: Repeat step 2 on subset IV

Subset	Word	Probabilities	0/1 Assignment	
V	B	0.12	110	001
VI	A	0.06	111	000
	D	0.04	111	000

Step 5: Repeat step 2 on subset VI

Subset	Word	Probabilities	0/1 Assignment	
VII	A	0.06	1110	0001
VIII	D	0.04	1111	0000

Hence, the resulting codes are: (A: 1110 or 0001), (B: 110 or 011), (C: 10 or 01), (D: 1111 or 0000), and (E: 0 or 1).

Problem 2.20

Construct a Huffman code for Example 2.31 and show that the resulting code(s) will be the same as in the Shannon-Fano procedure.

2.5.3 Practical source codes

Source coding implies digital representation of reformation in bit forms. The relevant coding schemes indicated above aim at reducing the average number of symbols required to transmit a given message. The source coding does not alter the source entropy, namely, the average number of information bits per source symbol. But it may alter the entropy of the source coded symbols.

Practical examples of source coding schemes are as follows:

✓ Morse code
✓ Huffman code used in Group 3 fax transmission (ITU-T standard)
✓ Source coding adopted in speech signal transmissions.
✓ Source coding used in image compression in MPEG video transmissions
✓ Creation of zip files in computers.

The Morse code was discussed earlier in detail and is not, therefore repeated here.

The Group 3 fax transmission is based on the statistics of the relative presence of black letters (or lines) and white areas in fax transmissions. Appropriate (relative) extents of black and white regions occurring along the run length of typical fax messages are indicated in Table 2.10. Hence, the Huffman coding for Group 3 fax is done by allocating the shortest codes to the most common run lengths. Commonly in a fax, the scanning of the A4 sheet is done at a rate of 3.85 scan lines per mm vertically with 1728 pixels across each scan line. Each pixel is then binary quantized into black or white to produce 2 Mbps of data per A4 page. Transmission is done over a telephone modem at 4.8 kbps. At this rate, it requires about 7 minutes per page. (Faster modems such as the V.17 data modem can transmit fax signals at 14.4 kbps.)

Table 2.10: <u>Relative occurrence extents of black and white runs in an average fax (A4) transmission</u>

Run lengths	Runs black	Runs white
2.5	1	0.9
5.0	1	0.9
7.5	0.9	0.9
10.0	0.1	0.9
50.0	0.1	0.5
100.0	0.1	0.3
500.0	0.1	0.2
1000.0	0.1	0.1
2000.0	0.1	0.3

Source coding removes redundancy and enables the efficient information transmission. Source codes are *uniquely decodable*, meaning that each code word has a designated single meaning. The code is also *instantaneously decodable* in the sense that each code word will not form a prefix of any other code words. Source code is *lossless* inasmuch as it is precisely reversible in the all sense of symbol errors. Group 3 fax code (Huffman code) is an example of lossless, unique, and instantaneous code.

The speech coding and source coding adopted in image compression (video) techniques are generally lossy. That is, the information is first mapped into a domain wherein only those parts which essentially contain information are presented and the rest are rejected. For example, in speech vocoders, the parts of speech waveform, which contain the major part of energy, are conserved, and, among these conserved parts the bit assignment is done in Huffman coding fashion, in the sense that the most frequent parts are given a minimum number of bits.

The zip file compression in computers is based on the Lempel-Ziv-Welsh (LZW) algorithm. It is an improved version of source coding. It gives a better compression ratio. But it is more difficult to implement than Huffman coding. The procedure of LZW coding involves constructing a table of frequently occurring strings and represent new strings by joining their prefixes in the table.

Normally, the use of source coding techniques like Huffman coding reduces the line of data to 30-50% of its original size.

2.5.4 Error control coding

While source coding is non-redundant, deliberate introduction of redundancy in the coding scheme can lead to error detection and sometimes enable correction of errors in data that may occur during transmission or storage. A basic format of such codes refers to *linear block encoding*. This is an error control coding scheme which provides an option on the signal structure such that errors can be recognized at the receiver.

Some examples of linear block codes are as follows:

✓ *Baudot code*: This is a 5-bit code developed to represent the alphanumerics for the teletype machines
✓ *ASCII code*: This code was developed by ANSI to represent the alphanumeric symbols each by a 7-bit code. *ASCII-77* is the corresponding international standard (ASCII: American Standard Code for Information Interchange)
✓ *EBCDIC code*: This is an 8-bit alphanumeric code (EBCDIC: Extended Binary-Coded Decimal Interchange Code)
✓ *Gray code*: This is a numeric code that represents the decimal values from 0 to 9. It is used in slow data transmission systems (such as telemetry).

The linear block codes are redundant codes due to the following reason. Suppose a block code is designed to represent L information-bearing messages. Let each message be represented by a code word of k bits. Then, it follows that $L = 2^k$. In the block-coding scheme, however, redundant bits are added to k, so that $(r + k) = n$. Hence, with n bits, $2^n (> 2^k)$ possible code words can be constructed. However, of these 2^n code words, only $2^k (= L)$ code words are needed. The balance, namely, $(2^n - 2^k)$, therefore, refers to redundant combinations. This scheme is known as *systematic block code* or *(n,k) code*.

The principle of choosing 2^k out of 2^n code words to represent L symbols is as follows: The 2^k code words chosen should be distinctly dissimilar to each other. In the event of a specific code word getting corrupted, the corrupted code word will not resemble any of the other code words used. Hence, the corrupted code word can be isolated (that is, detected).

Now, the norm of logic in choosing 2^k out of 2^n combinations, as appropriate, leads to the following error control coding schemes.

Parity check code

As discussed above, the (n, k) coding refers to (n – k) redundancy in the constructed code words. The redundant bits are termed *parity bits*. The use of parity bits in error detection can be illustrated by considering a *single-bit error detection* method.

Suppose k bit code words are assigned to represent $L = 2^k$ symbols and an extra bit is added to the code word. The addition of the extra bit (parity bit) can conform to two rules:

- The added (parity) bit to the code words can render the sum of all 1s in each code word an even number. This is known as *even parity*

- If, however, the sum refers to an odd number, then it is known as *odd parity*.

With the inclusion of a parity bit in a code word, should an error (single bit error) occur on the code word, it will lead to a count on 1s as an odd number (if even parity had been adopted), at the receiver. Similarly, the count on 1s will be an even number (if odd parity had been adopted). A violated count on 1s when observed at the receiver indicates that an error has occurred. That is, a single bit error detection is accomplished.

Error detection, is thus a mechanism of maintaining data integrity in bit transmissions. Essentially, it is concerned with sensing the inadvertent change of logic 1 to 0 or vice versa occurring in digital transmissions.

In practice, the error in the transmission of bit streams can be either a *single-bit error* or a *multiple-bit error*. The latter refers to contiguous bits having many errors resulting from a bursty noise.

> *"A message with content and clarity*
> *Has gotten to be quite a rarity*
> *To combat the terror*
> *Of serious error*
> *Use bits of appropriate parity"*

> — Solomon Golomb

Error control

The error control options adopted in practice are as follows:

✓   Automatic repeat request (ARQ)
✓   Error detection (ED)
✓   Forward error control (FEC).

ARQ

This procedure involves the receiving end making a request to the transmitter for a retransmission whenever a corrupted message in a block of data is received and identified.

ED

Error detection techniques play a key role in telecommunication assuring a reliable information transmission. The commonly used ED techniques are as follows:

✓   Parity checking
✓   Cyclic redundancy coding (CRC).

Parity checking

As indicated earlier, parity checking is used for detecting single-bit errors.

Example 2.32

For the data set indicated, include parity bits for even parity checking.
Demonstrate the error detection, assuming a single-bit error occurs in a parity-added data.

	Data set
A	010011
B	101110
C	010101
D	010001

Solution

		Parity	(Even) number of 1s
A	010011	1	$3 + 1 = 4$
B	101110	0	$4 + 0 = 4$
C	010101	1	$3 + 1 = 4$
D	010001	0	$2 + 0 = 2$

Suppose a single-bit error occurs in A and C. It leads to odd parity as indicated below:

(i) In A: 0100011 ⇒ odd parity
(ii) In C: 0001011 ⇒ odd parity

The corruption (shown underlined) in each case leads to odd parity. Hence error detection is affirmed.

Multiple bit errors

Multiple bit errors are more common than single bit errors in wireline transmissions. For example, a bursty noise interference on a cable may cause a set of bits to be corrupted. Attempting to locate the occurrences of (or detecting) bursty errors in arbitrary positions is rather difficult. Therefore, systematic error-detection methods for bursty error have been developed.

Normally, long messages are divided into smaller and manageable "frames". Instead of sending the entire bits collectively, they are sent separately, frame-by-frame. This involves disassembling the bits frame-by-frame and reassembling them at the receiver.

The construction of message frames and transmitting the bits in columns involves the following steps in reference to the illustration presented in Fig. 2.57.

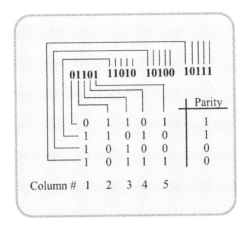

Fig. 2.57: Parity-check procedure for multiple bit errors

Step 1: Creation of a two dimensional bit array.

Step 2: In the array, the first column contains the first bits from each frame; the second column contains the second bits from each frame and so on.

Step 3: For each frame, a parity bit is added and an extra column is therefore created.

Step 4: Now, instead of sending each frame with its parity bit, each column is sent separately.

Step 5: Suppose a burst error occurs and destroys some (or all) of the bits in, say, column 4. (For example, all the bits in column 4 may become 0.)

Step 6: Then, the received pattern of columns will be as shown below (with asterisks depicting errored rows):

$$
\begin{array}{ll}
01101 & 1 \\
11000 & 1^* \\
10100 & 0 \\
10101 & 0^*
\end{array}
$$

Step 7: The receiver performing the parity check will find those frames indicated by asterisks have failed.

Step 8: However, the receiver would not known which column has failed. Therefore, it will request the retransmission of the entire column.

The drawback of the above scheme is that assembling and reassembling of the frames would require extra overhead.

Error correction

When errors are detected, there are two choices, namely,

✓   Retransmit the original frame (ARQ)

✓   Fix the damaged frame. This refers to pinpointing and correcting the errors at the receiving node without requesting additional information from the transmitter (that is, without any ARQ). This process is called *forward error correction* (FEC) because there is no need for any transmission in the reverse direction such as a request for retransmission.

Simple parity checking cannot accomplish error correction. Hence, a method due to Hamming involves creating special code words from the data sent which enable error correction. Hamming codes are constructed by inserting multiple parity bits to each bit string. Error detection and correction are done by computing the Hamming distance between the code words received erroneously and the transmitted code word patterns.

Hamming distance concept

Consider two encoded words :

C_i 110001

C_j 011000

C_j differs from C_i at : 1st, 3rd, and 6th positions from the left. Calling the number of differing positions as d_{ij} ($= 3$, presently), represents the extent of dissimilarity between C_i and C_j. It is known as *Hamming distance*. The lesser the value of d_{min}, the larger will be the confusion of distinguishing between C_i and C_j. Therefore, d_{min} is a measure of the distinguishability of coded messages. Hamming distance can be ascertained by modulo-2 addition between the code words. Normally, Hamming code is designed such that

$d_{min} = e + 1$ (for error detection)

$= 2e + 1$ (for error correction)

where e is the errors per word to be detected, or errors per word to be corrected.

Example 2.33

A given message consists of the following code words: 0000000, 0011110, 01101101, 0111000, 1001100, 1011001, 1101010, 1110100. If 1011011 is received, what is the transmitted code word (based on Hamming distance criterion) which has been erroneously received?

Solution

First, the receiver identifies that the received code word does not belong to the set of code words assigned a priori for transmission. Hence, it performs the modulo-2 (XOR) operation between the erroneous code word with all the code words designated for transmission as shown in following table. The transmitted code, which yields minimum Hamming distance, is declared as the correct code word sent.

Received code A	Transmitted code B	$A \oplus B$	d
1011011	0000000	1011011	5
1011011	0011110	1000101	3
⋮	⋮	⋮	⋮
1011011	**1011001**	0000010	**1**
			$\Rightarrow$
			Minimum value

In the example, the correct transmitted code word is: 1011001. This code word, upon modulo-2 operation with the received code word yields the minimum Hamming distance as indicated in the table.

Algebraic codes

The one-error code (single-parity-bit-check code) is a simple form of an *algebraic code* discussed below. The parity-check based detection of an error and rejecting the corresponding code word is known as *erasure*. When erasure occurs, the retransmission request of the code word is pursued. Thus, a single-parity check allows detection and not correction as mentioned earlier.

Suppose a two-bit error occurs. Then the parity check would lead to the deception of incorrectly decoding the word.

Example 2.34

A set of 3-bit code words are encoded with a fourth parity-check bit. Suppose a BER equal to 5×10^{-3} is observed in the transmission of these encoded words. Calculate the probability of an undetected error at the receiver.

Solution

If a single bit or 3-bit errors occur, the parity-check will enable the receiver to detect the error. However, if two-bit or 4-bit error occurs, parity-check will not show the error and the receiver will incorrectly decode the word.

Assuming statistical independence in and between two-bit and four-bit errors, the following results can be obtained. The possible two error bit combinations are as follows: {1^{st} and 2nd bit, 1st and 3rd bit, 1st and 4th bit}, {2^{nd} and 3rd bit, 2nd and 4th bit and 3rd and 4th bit}. Corresponding probability of two bit errors is,

$$P_2 = 6 \times [(5 \times 10^{-3})(5 \times 10^{-3})] \times [(1 - 5 \times 10^{-3})(1 - 5 \times 10^{-3})] = 2.475 \times 10^{-5}.$$

There are four possible ways of 4 bit errors. Hence, the corresponding probability of four bit errors is:

$$P_4 = [(5 \times 10^{-3})(5 \times 10^{-3})(5 \times 10^{-3})(5 \times 10^{-3})] = 6.25 \times 10^{-10}$$

Therefore, the probability of either two or four bit errors is equal to $P_2 \times P_4 \equiv 1.5 \times 10^{-14}$.

Problem 2.21

In Example 2.34, determine the probability of a detected error at the receiver. Hint: The detected error corresponds to one or three bit errors.

In discussing information bits (U_l) supplemented with redundant bits (C_l), a code word was depicted as $C_1 C_2 \ldots C_r \; U_1 U_2 \ldots U_K$ where k refers to a number of message (or information) bits and r denotes the number of check (redundant) bits.

In the so-called *algebraic codes*, a code word constructed is related to the information word via a matrix equation of the form $v = u\,(G)$, where U is a $[1 \times k]$ vector denoting the information, and v is a $[1 \times n]$ vector depicting the code word; [G] is a $[k \times n]$ generating matrix and n is the length of the code. This format of coding refers to a (n, k) linear code with k information bearing bits originally present in the encoded message. For example, Hamming codes indicated earlier are linear block codes.

An (n, k) linear code can be identified as *systematic* if a part of the code bits will exactly match the entire block of the information word. This is achieved via a portion of the generating matrix taken as an *identity matrix*. The following example will illustrate the underlying concepts.

Example 2.35

Consider a set of (1×3) vector representing eight messages (information). It is required to encode these messages into 4-bit code words. Hence, the generating matrix G is a (3×4) matrix. Construct a (4×3) linear code, given that $U_l = (000, 001, 010, 011, 100, 101, 110, 111)$ assuming that

$$[G] = \begin{bmatrix} 1 & 1 & 0 & 0 \\ 1 & 0 & 1 & 0 \\ 1 & 0 & 0 & 1 \end{bmatrix}$$

Solution:

Each of the eight possible 3-bit information words is multiplied by the generating matrix to get the linear code. Thus,

Information words	Linear code word
0 0 0	0 0 0 0
0 0 1	1 0 0 1
0 1 0	1 0 1 0
0 1 1	0 0 1 1
1 0 0	1 1 0 0
1 0 1	0 1 0 1
1 1 0	0 1 1 0
1 1 1	1 1 1 1

Note:

❑   The last three linear code bits exactly make the information word. (Systematic property)

❑   The right-hand portion of G is a 3-D identity matrix

❑   The added bits in the linear code represent parity check bits. (In the present case, it is an even parity.)

Parity check matrix

Given a $[G] = [k \times n]$ generator matrix, an associated $[H] = [(n-k) \times n]$ matrix exists. This is known as the *parity check matrix*. This matrix is used in error correction purposes, using the *syndrome* concept.

The result of multiplying the received vector by the transpose of $[H]$, namely $[H]^T$ yields what is known as *syndrome*.

Consider a transmitted code vector v, which is corrupted. It is equivalent to adding an error vector e to v to represent the received vector, R. Now, R is multiplied by $[H]^T$ to obtain the syndrome. That is,

$$[R][G]^T = [v + e][H]^T = v[H]^T + e[H]^T. \tag{2.39}$$

By the appropriate choice of generating matrix, the corresponding $[H]^T$ is such that $v[H]^T \equiv 0$. Therefore, the syndrome $\equiv e[H^T]$. In a single-bit error situation $e = 1$, therefore $e[H]^T$ would match the row of $[H]^T$ corresponding to the error position. To realize this syndrome-based error detection and correction, in the event of a single error, for example, $[H]$ should meet the following conditions:

    ✓   The columns of $[H]$ must not be repetitive. They must be distinct

    ✓   No column of $[H]$ can consist of all zeros.

Syndrome for the code can be generated using shift registers.

2.5.5   Cyclic redundant code (CRC)

Practical implementation of algebraic codes warrants circuits performing matrix multiplication and comparing the results with binary numbers. An alternative strategy is to use a look-up table. Nevertheless, the circuit complexity persists. As a result *cyclic codes* were developed which are easier to be implemented. A cyclic code has the property that any cyclic shift is a shift (to the right or left) by one position. The end bit cycles around to the starting side. The cyclic codes can be done via using a *generating polynomial*, $g(X)$ and a polynomial representation of a number. That is, a polynomial is used to represent a binary number by setting the coefficients equal to the bits of the number. For example, a binary number 1011101 corresponds to a polynomial $u(x) = 1 + X^2 + X^3 + X^4 + X^6$. Suppose $g(X) = 1 + X + X^3 \ (\Rightarrow 1101)$; a code word can be generated in terms of a polynomial $v(X) = u(x)g(x)$. This cyclic code generation concept is used in constructing a popular error-control code. It follows the *cyclic redundancy check* method, which can be described as follows: Given a k-bit block representing a message, the transmitter generates an n-bit sequence, known as the *frame check sequence (FCS)*, so that the resulting frame, consisting of (k + n) bits, is exactly divisible by some predetermined number. The receiver then divides the incoming frame by that number and, if there is no remainder, it confirms that there was no error.

Modulo-2 arithmetic

Modulo-2 arithmetic uses binary addition with no carries. It is just an exclusive-OR operation:

Exclusively OR:

A	$\oplus$	B	$\Rightarrow$	d
A		B		d
0		0		0
0		1		1
1		0		1
1		1		0

For example:

Addition:	*1111*		*Multiplication*	*11001*
	+1010			*×11*
	0101			*11001*
				11001
				101011

Cyclic redundancy checking (CRC) is a very reliable scheme for error detection. With CRC, almost 99.95% of all transmission errors can be detected. CRC is generally used with 8-bit codes such as EBCDIC or 7-bit codes when parity is not used.

In North American context, the most common version of CRC code used is known as CRC-16, which is identical to the international standard, CCITT V.41. With CRC-16, 16 bits are used for the *block check sequence* (BCS) which is the other name for frame check sequence (FCS). Essentially, as mentioned before, the CRC character is the remainder of a division process. A data message polynomial U(x) is divided by a generator polynomial function G(x); the quotient is discarded, and the remainder is truncated to 16 bits and added to the message as the BCS. With CRC generation, the division is not accomplished with a standard arithmetic division process. Instead of using straight subtraction, the remainder is derived from an XOR operation. At the receiver, the data stream and the BCS are divided by the same generating function P(x). If no transmission errors have occurred, the remainder will be zero.

For example, suppose the generating polynomial for CRC-16 is

$$G(x) = x^{16} + x^{12} + x^5 + 1. \tag{2.40}$$

The number of bits in the CRC code is equal to the highest exponent of the generating polynomial. The exponents identify the bit positions that contain a 1. Therefore, b_{16}, b_{12}, b_5, and b_0 are 1s and all of the other bit positions are 0s.

The cyclic code is classified as systematic or nonsystematic on the basis of the existence or nonexistence of an identity matrix as a part of the generating matrix respectively.

Example 2.36

Suppose the data- and CRC-generating polynomials are specified as:

$$\text{Data } u(x) = x^7 + x^5 + x^4 + x^2 + x^1 + x^0 \quad \text{or } 10110111$$
$$\text{CRC } G(x) = x^5 + x^4 + x^1 + x^0 \qquad\qquad \text{or } 110011$$

Determine the BCS.

Solution

First u(x) is multiplied by the number of bits in the CRC code, namely, 5.

$$x^5 (x^7 + x^5 + x^4 + x^2 + x^1 + x^0) = x^{12} + x^{10} + x^9 + x^7 + x^6 + x^5$$
$$= 1011011100000$$

Then, the result is divided by G(x).

The CRC is appended to the data to give the following transmitted data stream:

u(x)	CRC
10110111	01001

At the receiver, the transmitted data are again divided by P(x).

Problem 2.22

Suppose the data- and CRC-generating polynomials are as follows. Determine the BCS.

$$u(x) = x^7 + x^4 + x^2 + x^0 = 10010101$$
$$G(x) = x^5 + x^4 + x^1 + x^0 = 110011$$

2.5.6   Line coding

The transmission of digital data in binary format (denoted by +5 V – high, 0.0 V – low) is usually done in a serial fashion for short distances such as a computer to a printer interface. For long distance transmissions of digital data, the binary set (0 , 1) should be encoded so as to be compatible for easy identification of the associated high-low logic states without ambiguity when received. Further, inasmuch as the transmission could be synchronous or asynchronous, the relevant timing (sync) information should also be recovered at the receiver without "slipping" out of the synchronism or timing information vis-à-vis the sending end and the receiving end. Commonly, the logical states are encoded in a line code format compatible for transmission across wireline. A *code convert* accomplishes the mapping of PCM codes into line codes [2.11, 2.19, and 2.20].

The common line encoded waveforms are classified as follows:

✓   NRZ (Nonreturn-to-zero)
✓   RZ (Return-to-zero)
✓   Phase-encoded and delay modulation
✓   Multilevel binary (M-ary) code.

These encoding formats are known as the *baseband* type. (A *baseband signal* is one that is not modulated. These waveforms are still in a binary or pseudo-binary format, and as such they are classified as *baseband* versions).

The salient considerations in constructing line codes can be enumerated as follows:

✓ The waveform should not have a significant direct current (d.c.) component. Otherwise, the resulting d.c. current will not pass the transformer or capacitive coupling used in most wire-line transmission systems (to eliminate ground loops)

✓ The energy associated with these baseband waveforms at low frequencies must be small; otherwise large-sized components will be required for the equalization circuitry

✓ A significant number of zero-crossings should be facilitated for timing recovery at the receiving end without ambiguity

✓ The code should not be prone to high error multiplication

✓ A good coding efficiency should be achieved in order to minimize bandwidth

✓ The code should be conducive for error-detection or correction to ensure high-quality performance.

The NRZ family of line codes

The NRZ set of line-codes is a popular method used for encoding binary waveforms. They are also one of the simplest to implement. NRZ codes derive their name inasmuch as the data signal does not return to zero during an interval or clocked time-slot. In other words, NRZ codes remain constant during an interval. Because of non-return zero property the code has an average (d.c.) component in the waveform. For example, a data stream of 1s or 0s will show an average (d.c.) component at the receive side.

Another important aspect of the NRZ codes is that it does not contain any self-synchronizing capability. That is, NRZ codes will require the use of start bits or some kind of synchronizing data pattern to keep the transmitted binary data synchronized. There are three coding schemes in the NRZ family: NRZ-L (level), NRZ-M (mark), and NRZ-S (space) as described in Table 2.11.

Table 2.11: <u>NRZ family of line codes</u>

	Type of code	Logic levels	Line-code waveform
1	NRZ-L (Nonreturn-to-zero – level)	High 1	High level
		Low 0	Low level
2	NRZ-M (Nonreturn-to-zero – mark)	High 1	Transition at the beginning or the interval (clocked time-slot)
		Low 0	No transition
3	NRZ-S (Nonreturn-to-zero – space)	High 1	No transition
		Low 0	Transition at the beginning or the interval (clocked time-slot)

Spectral characteristics of NRZ Codes

NRZ code is cent percent unipolar code. Its circuit implementation is simple. Therefore NRZ code is the most common form of digital signal since all logic circuits operate on the on-off principle. As a result, the NRZ code is used inside the equipment (e.g., multiplexers, optical fiber line terminals, etc.). In reference to the signal depicted in Fig. 2.58, it can be seen that all 1 bits have a positive polarity, and so its spectrum has a d.c. component whose average value depends on the 1/0 ratio of the signal stream. If the signal consists of a 10101010 sequence, the d.c. component will be V/2. Depending on the signal, the d.c. component can assume a value from 0 (all 0) to V volts (all 1). The spectrum of the NRZ code is as indicated in Fig. 2.58. Its fundamental frequency component is at half the clock frequency f. Further only the odd harmonics are present. There is no signal amplitude at the clock frequency; as such it is impossible to extract the clock frequency at the receiving end. Also, during transmission via wire-cable, if the noise peaks are summed up so that a 0 is simulated as 1, it is rather impossible to detect the error. These disadvantages make the use of the NRZ unsuitable for transmission via copper cables.

NRZ is widely used in data communications systems. The asynchronous process uses the NRZ coding scheme. For example, the universal asynchronous receiver transmitter chip (UART) uses NRZ code. The UART has now become an informal industry standard. Though NRZ makes efficient use of bandwidth (a bit is represented for every signal change, or baud), it does not lack self-clocking capabilities. Normally self-clocking is achieved when each successive bit cell undergoes a signal-level transition. A continuous stream of 1s or 0s would not facilitate a level transition until the bit stream changed to the opposite binary number. As a result, the receiver is at a loss to know where to begin the bit sampling or what group of bits constitutes a character (byte). Hence, the NRZ scheme requires an independent clocking mechanism, often a separate transmission. This approach solves the non-self-clocking deficiency of NRZ but incurs additional synchronization problems if the data signal and clocking signal "drift" from each other as they traverse down the communications path. The clocking and synchronization problems with NRZ codes can be diminished through the randomized NRZ code. The NRZ signal is passed through a component that randomizes the bit stream to increase the number of signal-level transitions.

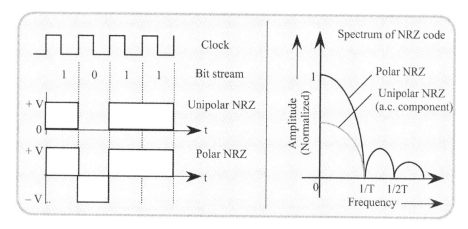

Fig. 2.58: NRZ Code and its spectral characteristics

The RZ family of line codes

The RZ-*unipolar* code is illustrated in Fig. 2.59. It has the same limitations as the NRZ group. It has a d.c. level on the data stream for a series of 1s or 0s. Synchronizing capabilities are also limited. These deficiencies are, however, overcome by modifications in the coding scheme, by including bi-polar signals and alternating pulses. The RZ-*bipolar* code enables a transition at each clock cycle. Further, the bipolar pulse technique minimizes the d.c. component.

Another RZ code is known as *RZ-AMI*. This alternate-mark-inversion (AMI) code provides alternating pulses for the 1s. This technique virtually removes the d.c. component from the data stream. Since a data value of 0 is 0 V, the system poses poor synchronizing capabilities if a continuous series of 0s is transmitted. This deficiency can also be overcome by transmission of the appropriate start, synchronizing, and stop bits. Table 2.12 describes the RZ codes.

Table 2.12: RZ family of line codes

	Type of code	Logic levels	Line-code waveform
1	RZ (unipolar) (Return-to-zero)	High 1	Transition at the beginning or during the clock interval
		Low 0	No transition
2	RZ (bipolar) (Return-to-zero)	High 1	Positive transition in the first half of the clock interval
		Low 0	Negative transition in the first half of the clock interval
3	RZ-AMI (Return-to-zero – alternate-mark inversion)	High 1	Transition within the clock interval alternating in direction
		Low 0	No transition

As illustrated in Fig 2.59, RZ codes provide a transition in every bit time-slot. Therefore, they have good synchronization characteristics. RZ is a self-clocking code because of its bit cell transition properties. However, since RZ experiences two signal transitions within each cell, its information-carrying capacity is not as good as the NRZ code. That is, a trade-off is done to achieve synchronization capabilities. As a result RZ requires a baud that is twice the bit-per-second rate.

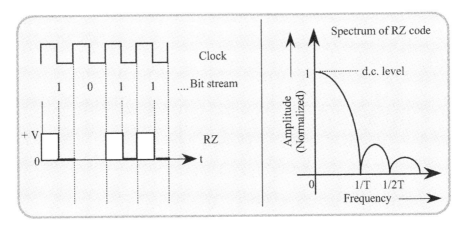

Fig. 2.59: The RZ code and its spectral characteristics

In essence, the RZ code is 50% unipolar. That is, RZ code is similar to the NRZ signal except that the pulse duration is reduced to one-half. RZ code is also convenient for equipment circuitry because all the logic circuits operate on the on-off principle. Thus, the RZ is another code used inside the apparatus. From Fig. 2.59 it can be noted that RZ code also produces a d.c. component in the spectrum. However, the fundamental frequency is now at the frequency of the clock signal with only the odd harmonics existing as in Fig. 2.59. This makes it possible to extract the clock frequency at the receiving end provided no long sequences of 0s are present. However, detection of errors, as explained earlier, is not possible. A unipolar version of RZ code is, therefore rarely used at any stage of the system. But, a bipolar version of the RZ code known as RZ-AMI mentioned earlier and indicated in Table 2.12 is widely used for the reasons indicated below.

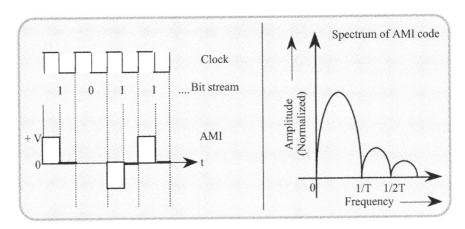

Fig. 2.60: AMI code and its spectral characteristics

Fig. 2.60 illustrates an alternate mark inversion (AMI) code (bipolar scheme). In reference to Fig. 2.60, it can be seen that the 1s are alternately positive and negative. Hence, there is no d.c. component in the spectrum. The apparent absence of the clock frequency in the spectrum can be overcome by simply rectifying the received signal to invert the negative 1s and making the resulting signal similar to the RZ signal. When this is done, since the received AMI signal has already passed through the line, the consequent appearance of d.c. component is of no interest, and the clock frequency can be extracted from the new spectrum of the signal. Another merit of the AMI signal is that it can correct errors.

Suppose during line transmission, noise peaks are summed up to simulate a 1 instead of a 0. There would be a violation of the code which dictates that the 1s are alternately positive and negative. The recovery of the clock frequency is not, however, easy with the AMI coding whenever a long sequence of 0s is present. The AMI code is recommended by the CCITT (G.703) for the 1.544-Mb/s interface. Its modified version, known as alternate digit invention (ADI) code which has 100% unipolar duty cycle and used in 2.048 Mbps transmission is described below.

Alternate digit inversion (ADI) code

In ADI code, every second, or alternate, digit or bit is inverted as illustrated in Fig. 2.61. The 8-bit PCM words are coded in ADI code in Fig. 2.61. For example, the speech code from the A/D converter contains a relatively high number of 0s for speech levels close to zero. On the other hand the occurrence of 1s increased as the level is raised. The probability of the speech signal being near zero is significant. First, when a given channel is not seized at all, the level is zero. Second, when a channel is seized, only one subscriber at a time speaks, which means that the other subscriber has zero level in direction of transmission. Therefore, in general, the probability of a level near zero is significantly large. ADI is very useful because even with large strings of 0s or 1s, the receiver would extract the clock signal when the transmit signal is ADI encoded.

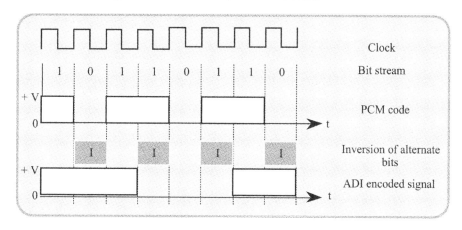

Fig. 2.61: Construction of ADI signal

Bi-phase and Miller family of line codes

The phase-encoded and delay-modulated codes refer to bi-phase and Miller codes. Bi-phase codes are very useful in optical systems, satellite telemetry links, and magnetic recording systems. Bi-phase M is used for encoding SMPTE (*Society of Motion Picture and Television Engineers*) time-code data for recording on videotapes. The bi-phase code is an excellent choice for this type of media because the code does not have a d.c. component to it. Another important benefit of the code is *self-synchronizing*, or *self-clocking*. This feature allows the data stream speed to vary (tape shuttle in fast and slow search modes) while still providing the receiver with clocking information.

The *bi-phase* L code is known as *Manchester coding*. This code is used on the *ethernet* standard IEEE 802.3 for local area networks (LANs). Table 2.13 summarizes the bi-phase/Manchester line code characteristics:

Table 2.13: Phase-Encoded and Delay-Modulation (Miller) line codes

Type of code	Logic levels	Line-code waveform
Bi-phase M (Bi-phase mark)	High 1	Transition in the middle of the clock interval
	Low 0	No transition in the middle of the clock interval
		Note: There is always a transition at the beginning of the clock interval.
Bi-phase L (Bi-phase-level/Manchester)	High 1	Transition from high-to-low in the middle of the clock interval
	Low 0	Transition from low-to-high in the middle of the clock interval
Bi-phase S (Bi-phase space)	High 1	No transition in the middle of the clock interval
	Low 0	Transition in the middle of the clock interval
		Note: There is always a transition at the beginning of the clock interval.
Differential Manchester	High 1	Transition in the middle of the clock interval
	Low 0	Transition at the beginning of the clock interval
Miller/delay modulation	High 1	Transition in the middle of the clock interval
	Low 0	No transition at the end of the clock interval unless followed by a zero

As indicated above, biphase codes have many variations and names; such variations correspond to phase encoding, frequency encoding, and frequency shift encoding. All biphase codes have at least one level transition per bit cell, which is similar to RZ codes. However, most of the biphase codes have the 1s or 0s defined by the direction of the signal-level transition.

Biphase codes are found extensively in magnetic recording and in data communications systems utilizing optical fiber links. The codes are used in applications requiring a high degree of accuracy; the code is self-clocking.

As specified above, one of the variations of biphase is the Manchester code. It is used in several data communications systems, especially the local area network ethernet. In the biphase code known as *differential Manchester coding*, the polarities of the signal are dependent on the last half of the previously transmitted bit cell. For a binary 0, the preceding signal element is the opposite polarity of the first half of the 0 bit cell. The situation is reversed for a binary 1; the polarity of the previous signal element is the same as the first half of the 1 bit cell.

A comparative study of NRZ, RZ, biphase-Manchester, differential Manchester and Bell system PCM code as illustrated in Table 2 can provide a relative performance feature of these codes.

Coded mark inversion (CMI) encoding

The CMI is a line code in which the 1 bits are represented alternately by a positive and a negative state, while the 0 bits are represented by a negative state in the first half of the bit interval and a positive state in the second half of the bit interval. That is, CMI is a two-level NTZ code in which binary zero is encoded so that both amplitudes A1 and A2 are attained one after another, each for unit time slot (T/2). Binary 1 is encoded so that either of the amplitude levels A1 and A2 are attained for one full time-slot (T). The CMI code specifications are given in CCITT Recommendation G.703, as the interface code for fourth-level multiplex signals transmitted at 139.264 Mb/s. An example of CMI encoding is shown in Fig. 2.62.

CMI has evolved similar to digital biphase code. In CMI, there is no d.c. energy in the signal and plentify of transitions exist, similar to diphase. Further, there is no ambiguity between 1s and 0s. Therefore, it suffers from large error performance (3dB worse than diphase when bit-by-bit detection is used). This inefficiency results because for one-half of time-slot, a 1 resembles a 0. However, CMI has inherent redundancy on which the inefficiency can be constructed via the maximum likelihood detection (such as Viterbi) schemes.

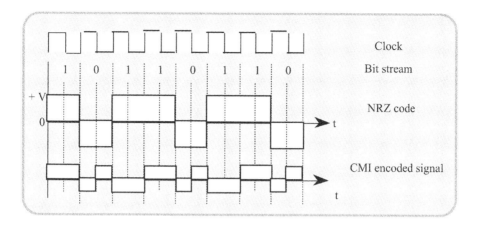

Fig. 2.62: An example of the CMI code

Block codes

The line codes discussed at so far are bit-oriented codes in which the coding principle applies to individual bits. In *block codes* the coding format refers to a block of bits. Block codes require more complex encoding and decoding circuitry. However, because of their high efficiency, block codes allow longer cable lengths between repeaters. Hence, block codes have applications in long-distance telecommunication. In such applications, any extra complexity in the encoding and decoding circuits is justified by the savings in the number of repeaters. Like other line code, block codes are also designed to prevent *baseline wander* (drift in d.c. level) and to have good timing content.

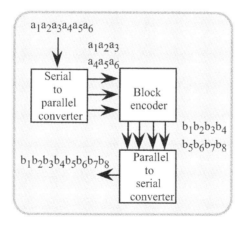

Fig. 2.63: An encoder to construct a block code

The principle of designing block codes is as follows: The blocks of input bits are encoded to give blocks of code symbols. This is different to all the codes looked at so far which involve coding a single bit to give one or more code symbols.

The encoder for a block code segmentizes the incoming data stream into appropriately sized blocks (call words). It then encodes each block of input bits into a block of code symbols (code words). Finally, it transmits the code words serially. Fig. 2.63 is a typical block schematic of constructing block codes.

A binary block code can be specified as an *nBmB code*, where n binary bits are encoded into m binary bits. Hence, in the case of a 3B4B code the serial input data is first of all broken down into blocks 3 wide; the 3-bit blocks are then encoded into 4-bit blocks and the 4-bit blocks are then transmitted as serial data. It is also possible to have a ternary block code which can be specified as an nBmT code, where binary bits are encoded into m ternary symbols.

Many times the incoming data may not have inherent block structure; the location of the initial block boundary chosen by the encoder therefore, is rather arbitrary. However, once the initial block boundary is fixed, the encoder can then take successive sequences of n adjacent symbols.

A major advantage of block codes is that they can be designed to have an efficiency greater than either HDB3 or AMI. It can be noticed that the nature of a block code is such that a block of m binary output bits from an encoder is carrying the information from n binary input bits. The information per symbol used is n/m bits and the information per symbol available for a binary symbol is 1 bit. As a result, the efficiency of a general nBmB code is: $\eta = (n/m) \times 100\%$. A code with $n > m$ means an efficiency greater than 100%. This is not possible because no coded blocks can carry information more than the theoretical limit. Codes with $m = (n + 1)$ are particularly common since they have higher efficiency than any with $m > (n + 1)$.

The efficiency of a block code is determined by the choice of n and m. In general, as $2^m > 2^n$ for a binary code or $3^m > 2^n$ for a ternary code, not all the possible code words are required by the code. This redundancy allows some flexibility in the choice of which code words are used. It is this flexibility that allows block codes to be designed with such good timing content. It also means that block codes can be designed to suit a range of applications. Some typical nBmB codes are as follows:

3B4B code

This is one of the basic types of block code. It retains all of the characteristics of a block code. Block codes are specified in a *coding table*, which shows how each input word is coded. Table 2.16 gives the encoding rules for the 3B4B code. In the table, the code is implemented by positive and negative pulses where (+) indicates a positive pulse and (–) indicates a negative pulse.

Table 2 is constructed so that the encoded words have the following properties:

    ✓   Null mean voltage offset (to prevent baseline wander)
    ✓   Frequent transitions to preserve timing.

The left-hand column lists all the possible input 3-bit binary words. The next three columns contain all the possible output words according to the coding rules. The fifth column contains the code word disparity. The difference between the number of positive and negative pulses is known as disparities or digital sum. It is a normalized measure of the mean voltage of the word. For example, a word with equal numbers of positive and negative pulses has zero disparity and a zero mean voltage and it is called a *balanced word*. Unbalanced words can have positive or negative disparity. For example the code word ++-+ has a disparity of + 2. To ensure that there is no mean voltage offset, balanced code words are used in Table 2.14 for the output wherever possible, that is, for input words 001 to 110. The remaining two input words, however, are to be encoded as unbalanced words. Still a running digital sum can be maintained. That is, in order to maintain zero mean offset despite the presence of unbalanced words, each of the unbalanced words can be alternately set to positive and negative disparity. During encoding, the encoder keeps a record of the running digital sum of each transmitted word.

Table 2.14: Encoding table for 3B4B block code

Input Binary words	Output code words			
	Negative	0	Positive	Disparity = Number of (+) pulses – Number of (–) pulses
001		– – + +		0
010		– + – +		0
100		+ – – +		0
011		– + + –		0
101		+ – + –		0
110		+ + – –		0
000	– – + –		+ + – +	± 2
111	– + – –		+ – + +	± 2

Binary N zero substitution (BNZS) codes

In bipolar coding schemes such as (AMI), the limitation is that a minimum density of 1s in the source code is required so as to maintain the timing at the regenerative repeaters. Otherwise timing jitter will result.

Therefore, BNZS codes have been developed. They augment a basic bipolar code by replacing a string of N zeros with an N length code, which contains pulses which deliberately introduce *bipolar violations*.

Bipolar violation

Bipolar encoding, in general, helps to eliminate the mean (d.c.) level of a code. The presence of such d.c. levels would cause a baseline d.c. drift (known as d.c. wander). Inherently, with bipolar coding, the pulses on the line alternate in polarity. Should two successive pulses have the same polarity, it implies an error. This error condition is called bipolar violation. No single error can occur without introducing bipolar violation. That is, bipolar codes offer a kind of line code parity.

In the BNZS coding scheme, with the introduction of deliberately introduced bipolar violations, the original data bits are recognized upon reception and are replaced by N strings 0s. The BNZS line code specified by CCITT Recommendation G.703 for the 44.736-Mb/s (DS-3) interface is B3ZS code. Another code used in North America is the binary six zero substitution (B6ZS) used by the Bell system at the 6.312-Mb/s rate for *one symmetric pair*. The B8ZS line code is specified by CCITT Recommendation G.703 for *one coaxial pair* at 6.312 Mb/s.

The rules for substitutions can be summarized as follows:

Notations

00V: 2 bit intervals with no pulse (00) followed by a pulse representing a bipolar violation

B0V: A single pulse with bipolar alternation (B) followed by no pulse (0) and ending with a pulse with bipolar violation (V)

B3ZS code: Substitution rules

  ✓ A decision to substitute is made (with 00V or B0V) so that the number of B pulses between V pulses is odd: If an odd number of ones has been transmitted since the last substitution, 00V is chosen to replace three zeros. If an even number of ones has been transmitted since the last substitution, B0V is chosen to replace three zeros

  ✓ Substitution algorithm: A summary.

Table 2.15: B3ZS code: Substitution algorithm

Pulse polarity of the preceding data	Substitution sequence representing number of bipolar pulses (1s) since last substitution	
	Odd	Even
(−)	0 0 −	+ 0 +
(+)	0 0 +	− 0 −

Remarks:

  ✓ Bipolar violations alternate in polarities leading to minimal d.c. wander

  ✓ When channel errors occur, an even number of bipolar pulses between violation would occur

  ✓ Every deliberately introduced violation is immediately preceded by a 0

✓ Continuously strong timing component is offered via increased density of pulses through deliberately introduced violations.

B6ZS code

Here, bipolar violations are introduced in the second and fifth bit positions of a six zero substitution. That is, each block of six successive zeros is replaced by 0VB0VB. The substitution algorithm is as follows:

Table 2.16: B6ZS: substitution algorithm

Pulse polarity of the preceding data	Sequence used for substitution
(–)	$0 - + 0 + -$
(+)	$0 + - 0 - +$

B8ZS code

In this code, each block of eight successive 0s is replaced by 000VB0VB.

Example 2.37

A data sequence is specified as: 001000101000110000001. Construct a B3ZS code.

Solution

Assuming odd number of pulses were transmitted following the previous violation: (see Table 2.15)

Data:	001	000	101	000	11	000	000	1
		Substitution		Substitution		Substitution	Substitution	
Code:	00+	00–	+0–	+0+	–+	–0–	+0+	–

Assuming even number of pulses were transmitted following previous violation: (see Table 2.15)

Data:	001	000	101	000	11	000	000	1
		Substitution		Substitution		Substitution	Substitution	
Code:	00+	00+	–0+	–0–	+–	+0+	–0–	+

Example 2.38

Given that: Data 1000000101100000000000000001. Construct a B6ZS code.

Solution

Suppose polarity of pulse immediately preceding six 0s to be substituted is (–). See Table 2.16

Data:	1	000000	1011	000000	0000000	0001
Code:	+	(0+–0–+)	–0+–	(0–+0+–)	(0–+0+–)	000+
		(Substitutions)				

Suppose polarity of pulse immediately preceding six 0s to be substituted is (+). See Table 2.16

Data:	1	000000	1011	000000	0000000	0001
Code:	–	(0–+0+–)	+0–+	(0+–0–+)	(0+–0–+)	000–
		(Substitutions)				

Problem 2.23

Using the following rules construct a B8ZS substitution code for the data:
10100000000010100000000001

Rules for substitution in B8ZS code

Polarity of preceding pulse	Substitutions
(–)	000–+0+1
(+)	000+–0–+

Like BZNS substitution codes, there is a pair-selected ternary (PST) code, which also helps to improve the timing content of a binary signal. The procedure to construct a PST code is as follows:

✓ The input binary data is paired to form a sequence of 2 bit code words
✓ These 2 bit code words are then translated into two ternary digits (+, – or 0) for transmission across the line
✓ The 2 bit binary set corresponds to {b} = {00, 01, 10, 11} and there are 9 elements in the two digit ternary combinations set, namely, {Ter} = {+–, –+, +0, –0, ++, ––, 00, 0–, 0+}
✓ Therefore, a flexible translation from {b) to {Ter} is feasible so as to ensure a strong timing content. The translation can also be taken into account to ensure no d.c. wander. The binary input to two-digit ternary translation is done by the following encoding scheme:

Positive mode	Negative mode
$00 \Rightarrow$ –+	$00 \Rightarrow$ –+
$01 \Rightarrow$ 0+	$01 \Rightarrow$ 0–
$10 \Rightarrow$ +0	$10 \Rightarrow$ –0
$11 \Rightarrow$ +1	$11 \Rightarrow$ +–

✓ The encoder sends either a positive-mode or a negative-mode ternary digits. When the binary digit transmitted contains a single 1, the next binary digit transmission will be done with the mode reversed. This process is illustrated in the following example:

Example 2.39

Construct a PST code for the following sequence of binary data stream: 10110001010011

Solution

There are two possible solutions. Suppose the encoder is in (+ mode) initially.

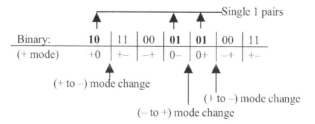

After every single 1 binary pair encountered the mode change takes place as illustrated. Hence the PST code is constructed using the procedure indicated above.

Problem 2.24

In Example 2.39, assuming the encoder is in (– mode) to start with, construct the PST code for the binary data stream indicated.

In PST encoding the pair-wise boundary is maintained in tact. Like B6ZS code, the energy content of PST is also high due to increased pulse density. Hence, relative to bipolar encoding, PST (as well as B6ZS) may cause increased crosstalk. However, the timing performance realized in the coding scheme can offset this demerit.

High density bipolar (HDB) coding

This is another format of BNZS coding recommended by CCITT. It is implemented in CEPT primary digital signal. It replaces strings of four 0s with sequences containing a bipolar violation in the last bit position. Inasmuch as this coding structure precludes strings of zeros greater than three, it is called a HDB3 coding. That is, the purpose of the HDB3 code is to limit the number of zeros in a long sequence of zeros to three. This assures clock extraction in the regenerator of the receiver. This code is recommended by the CCITT (G.703) for the 2-, 8-, and 34-Mb/s systems. An example is shown in Fig. 2.64.

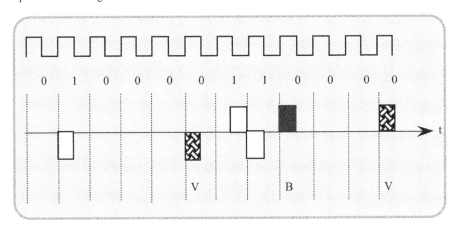

Fig. 2.64: HDB3 code

Longer sequences of more than three zeros are avoided by the replacement of one or two zeros by pulses according to specified rules. The relevant substitution rules (presented in Table 2) ensure that the receiver recognizes that these (substituted) pulses are replacements for zeros and does not confuse them with code pulses. This distinction is achieved by selecting the polarity of the pulse so as to violate the alternate mark inversion polarity of the AMI code. Also, the replacement pulses themselves should not introduce an appreciable d.c. component. The HDB3 coding principle can be summarized as follows:

    ✓   Every second 1 is inverted as long as a maximum of three consecutive zeros appear

    ✓   If the number of consecutive zeros exceeds three, the violation pulse is set in the fourth position. The violation pulse would deliberately violate the AMI rule

    ✓   Alternate violation pulse is set to change the polarity. If this rule cannot be applied, 1 is inserted (as per AMI rule) in the position of the first zero in the sequence.

The algorithmic steps for HDB3 code is as follows:

- Apply the three rules as above, step by step

- 000V and B00V generation:

    ✓   000V is substituted if there is an odd number of 1s since the last violation pulse

    ✓   B00V is substituted if there is an even number of 1s since the last violation pulse. The B-pulse follows the AMI rule for its polarity

- By observing the polarities of the preceding data pulse and the violation pulse, apply the substitution sequence as above.

For the HDB3 code indicated in Fig. 2.64, the spectrum is as follows (Fig. 2.65).

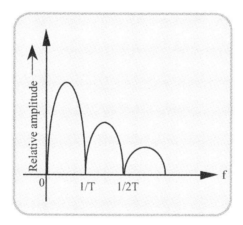

Fig. 2.65: Spectrum of HDB3 code

    In HDB3 code, some violation pulse substitutions may not be identifiable from genuine marks. To avoid this confusion, any violation pulse which is not of opposite polarity to the preceding mark pulse is forced into violation by the insertion of a "parity" pulse substituted for the first zero immediately following the genuine mark. The next mark following a parity pulse is then made of opposite polarity to the parity pulse, irrespective of the polarity of the previous genuine mark. Hence a unique decoding is possible.

Problem 2.25

For the data sequence 01 00000 101 0000 101 1001 10000, construct AMI and HDB3 codes following the appropriate rules.

Multilevel binary family of line code

    Codes with more than two levels representing the data are known as *multilevel binary* codes. Mostly these codes have three levels: Two of these codes belong to the RZ group. They are: RZ (bipolar) and RZ-AMI. Others included in this group are *Dicode NRZ* and *Dicode RZ*.

Table 2.17: <u>Multilevel binary line codes</u>

	Logic levels	Line-code waveform
Dicode NRZ	Multilevel	One-to-zero and zero-to-one data transitions change the signal polarity. If the data remain constant, then a zero-level is output.
Dicode RZ	Multilevel	One-to-zero and zero-to-one data transitions change the signal polarity in half-step voltage increments. If the data do not change, then a zero-voltage level is output.

Whenever bandwidth is limited, but higher data rates are desired, the number of levels can be increased while maintaining the signaling rate unaltered. The data rate that can be achieved by a multilevel system is given by:

$$R = \left(\frac{1}{T}\right)\log_2(L) \tag{2.41}$$

where L is the number of levels chosen during each interval and T is the signaling interval. (*Note*: signaling rate (1/T) refers to the symbol rate in bauds.)

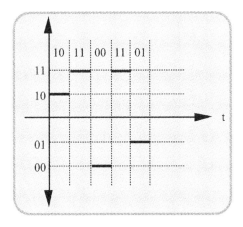

Fig. 2.66: A four-level transmission with 2 bits per signal time-slot

The limitation of multilevel binary code is that it requires greater S/N ratio for a given error rate. But its lower baud rate (for a given data rate) offers crosstalk reductions in wireline transmissions. A typical example of multilevel transmission is the ISDN basic rate on a digital subscriber line (DSL) where a four level transmission at a signaling rate of 80 baud is used to receive a data rate of 160 kbps. Its main advantages are as follows:

✓   Near-end crosstalk elimination
✓   Reduction of intersymbol interference caused by bridged tap reflections.

An example of multilevel transmission with 2 bits per symbol accommodated per signal interval (slot-time) is illustrated in Fig. 2.66.

Example 2.40
Consider the bit pattern of 101101110.

Construct the following line codes: Bipolar NRZ, RZ, biphase Manchester, differential Manchester, and Bell System PCM code

Solution:

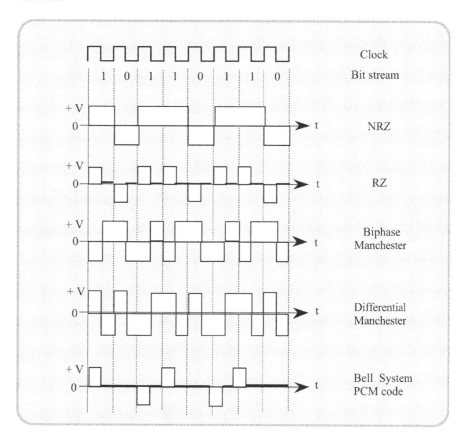

Fig. 2.67: Construction of various line codes

Problem 2.26

For the data stream 1101100000000111, construct the following types of line-codes: Unipolar NRZ, bipolar NRZ, NRZ1, Manchester, differential Manchester, unipolar RZ, bipolar RZ, Miller code, AMI, and CMI.

Problem 2.27

In reference to a bit pattern 1101001100111111 construct the waveforms of the following line encoding schemes: Unipolar NRZ, bipolar NRZ, NRZ1, Manchester, differential Manchester, unipolar RZ, bipolar RZ, Miller code, AMI, and CMI.

A quick reference note on line codes

NRZ family:

> *Unipolar NRZ code: Signal level transition stays (+) and does not return to 0 during binary 1 slot.*

> *Bipolar NRZ code: (+) and (–) levels and does not return to 0 within the slot.*

> *NRZ1 code: No change in the signal level at the beginning of bit ⇒ 1 and vice versa.*

RZ family:

> *Unipolar RZ code: Signal level goes to (+) and returns to 0 from the 1 state during the slot duration.*

> *Bipolar RZ code: Comprised of two non-zero voltage levels: 1 ⇒ (–) and 0 ⇒ (–).*

Manchester code family:

> *Manchester code: Signal level transition occurs at the center of each slot. 1 ⇒ (–) transition and 0 ⇒ (–) transition.*

> *Differential Manchester code: Signal level transition occurs at center of each slot. 1 ⇒ No transition at the beginning of the slot, and 0 ⇒ transition at the beginning of the cell.*

> *Miller code: Signal level transition occurs at the center of the slot for 1s only. No transition for a 0 unless it is followed by another 0, in which case the signal level transition is accommodated at the end of the slot for the first 0.*

> *Alternate mark inversion (AMI) code: This corresponds to a bipolar RZ signal. It carries a pulse density of at least 12.5%. 1s are alternately position and negative. For logic 1, a transition occurs within the clock interval. For logic 0, no transition takes place.*

> *Code mark inversion (CMI) code: 1s are represented alternately by a positive and a negative state. 0s are denoted by negative state in the first half of the time-slot of a bit and positive in the second half.*

Spectral characteristics of line codes

A major required attribute of a line code is that the spectra of the transmitted data should match the line characteristics. In comparing line codes by evaluating the spectral characteristics one may notice that a particular pulse shape invariably dominates the spectral diagram of each code. Fig. 2.68 shows the spectral characteristics of typical line code. The graphs are obtained by using impulses as digits. The spectrum of randomly encoded data using such impulses is constant (flat). Hence for any such data encoded by the normal coding rules the deviation from a flat spectrum is due entirely to the code.

For the straightforward binary code the spectrum is flat and includes d.c. and low-frequency components thereby giving rise to baseline wander. AMI code has some very attractive features. Its spectral diagram shows no d.c. component and very little low-frequency content thereby reducing considerably the effects of baseline wander. The bandwidth required is equal to the transmission rate. It is a very simple code to implement but does have the big disadvantage of poor timing content associated with long runs of binary zeros.

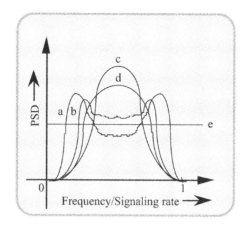

Fig. 2.68: Spectral characteristics of typical line codes

The line codes used in practice as specified by different organizations are summarized in Table 2.18.

Table 2.18: <u>Line codes specified by different organizations for digital transmissions</u>

Carrier	Organizations	Line code	Bit rate × line length Mbps × ft	Physical media
T-1		AMI/B8ZS	1.544 × 6000	TP copper
T-1C		Bipolar	3.152 × 6000	TP copper
T-1D		(+D) Duobinary*	3.152 × 6000	TP copper
T-1G	AT&T	4 level	6.443 × 6000	TP copper
T-2		B6ZS	6.312 × 14,8000	TP copper (Low capacitance)
T-4M		Polar (Binary/NRZ)	274.176 × 5700	Coax
CEPT1	CCITT	HDB3/B4ZS	2.048 × 2000	TP copper

* Note: The following note on band-limiting filters used in digital transmission explains the duobinary concept.

Band-limiting filters for digital transmission systems...

The band-limiting filters used in digital transmission systems constrain the signal bandwidth so that the adjacent symbol interference is avoided. A class of technique in which the system response is trimmed so that the receiver detects the symbols correctly in spite of inevitable spectral spill over causing intersymbol interference is termed as: Duobinary, correlative level encoding, or partial response signaling.

The partial response signaling (PRS) implies the following: The intersymbol interference would cause overlapping of adjacent bipolar waveform of an encoded signal. The resulting output (due to a two level input) would exhibit multiple levels. As such, it becomes difficult to detect such multilevel waveforms correctly.

Now the PRS waveform corresponds equivalently to a superposition of the input with a delayed version of the same input. Denoting D as the delay, the partial response is designated as (1 + D) PRS. To limit the overlap effect arising from (1 + D) PRS, the encoder can perform a correlative level encoding, known as the (1 − D) system. It uses a single cycle of a square wave across two signal intervals to encode each bit. The resulting spectrum "shapes" the bandwidth of the encoded waveform so that with the (1

+ D) PRS occurring along the transmission line, the spectral changes caused thereof, will be counteracted by the (1 – D) encoding done at the input (prior to its passage along the line). The T1D/T-carrier system of AT&T uses precoded (1 + D) level encoding (called duobinary).

Convolution codes

This code is specified by three parameters namely (n, k, m) where n is the length of the output code block, k is the input block, and m describes the memory of the code. For example a (3, 1, 4) convolution code depicts that the encoder yields a 3 bit block code for every single bit input, and the output depends on the "memory" specifying the four previous input blocks.

There are some obvious difference between the block coding and convolution coding. The following table (Table 2.19) indicates salient differences and similarities between those codes.

Table 2.19: <u>Block coding versus convolution coding</u>

Block coding	Convolution coding
The data stream is divided into blocks of bits and each block is transformed into a sequence of bits called the code word.	The data stream is divided into blocks of bits and each block is transformed into a sequence of bits called code words.
The code word is usually longer than the data word.	The code word is smaller than the data word.
The system has no memory between one block and the next. That is, the encoding of any specific block of data bits does not depend on what happened before and after that specific block.	The system bears memory. That is, encoding of a specific block of data bits depends on the prior input blocks.

In the (n, k, m) code, n > k. That is, redundancy is provided in the output. The code rate R is defined as k/n.

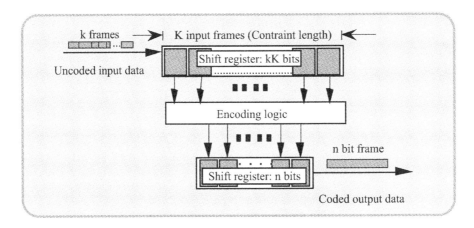

Fig. 2.69: Block encoding implementation

Referring to Fig. 2.69, one input frame (of k bits) is shifted in each time slot; concurrently, one output frame (of k bits) is shifted out. Hence, every k-bit input frame produces an n-bit output frame. There is a *constraint length* K which refers to the number A input frames that are held in the k K-bit shift register. For a specific convolution code to be generated, data from the k K stages of the shift register are called modulo-2 and are adopted to place the bits in the n-stage output register.

Decoding of convolutionally encoded signal

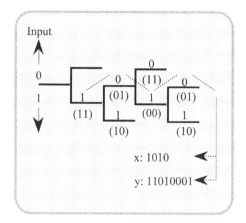

Fig. 2.70: A partial code-tree

The encoded signal when received is matched against the corresponding bit pattern in a code-tree. In the event of a received bit frame being corrupted, the matching and the associated Hamming distance evaluation would show the correct format of the corrupted frame.

The structural properties of a convolution code are normally represented via: (i) Code tree, (ii) trellis, and (iii) state diagram.

A pair of code trees is illustrated in Fig. 2.70. In Fig. 2.70 each branch of the tree denotes an input symbol and the corresponding pair of output binary symbols is shown on the branch. Following the dotted path in Fig. 2.70, the y = 11010001 depicts a received signal for which the closest match of decoded data is x = 1010.

Trellis

The code-tree grows exponentially vis-à-vis number of states. This is avoided in the so-called trellis diagram in which identical states are identified and over laid.

State transition diagram: Here, the input to the encoder is shown on the appropriate branch and the corresponding outputs are shown in brackets beside the input.

Viterbi decoding

This is an optimum decoding algorithm for convolutionally encoded data. It uses a path tracing along a code-tree. It examines the possible paths and selects the best ones based on some conditional probabilities.

Trellis code modulation (TCM)

In terms of Eb/No equal to the energy per bit/noise density ratio, a *coding gain* can be defined as the reduction in Eb/No (in dB) that is realized when appropriate coding is used (as compared to an uncoded case at some specific level of BER).

TCM refers to a multilevel modulation plus convolution encoding so as to achieve coding gain without bandwidth expansion. It is used in the CCITT V.32 modem at 9600 bps rate.

2.6   Concluding Remarks

The basis of modern telecommunication technology is in essence the science of electrical communication in an applied engineering framework. Ever since the birth of telephony, the intricacies of electrical communication were studied and applied to the gamut of communication technology over more than a century. Any evolving telecommunication considerations have, therefore, a basis of solid reinforcements specified in different perspectives: Technology, and engineering trends mark a successful broadening of the relevant scope of telecommunications especially for ATM technology.

Bibliography

[2.1]   M. Schwartz: *Information Transmission Modulation and Noise* (McGraw-Hill, Inc., New York, NY: 1980)

[2.2]   H.S. Black: *Modulation Theory* (D. Van Nostrand Co., Inc., Princeton, NJ: 1966)

[2.3]   W.L. Everitt and G.E. Anner: *Communication Engineering* (McGraw-Hill, Inc., New York, NY: 1956)

[2.4]   L.W. Couch II: *Modern Communication Systems – Principles and Applications* (Prentice-Hall, Inc., Englewood Cliffs, 1994)

[2.5]   G.M. Miller: *Modern Electronic Communication* (Prentice-Hall, Inc., Upper Saddle River, NJ: 1999)

[2.6]   W. Tomasi: *Advanced Electronic Communication Systems* (Prentice-Hall, Inc., Upper Saddle River, NJ: 1998)

[2.7]   M.A. Miller: *Introduction to Digital and Data Communications* (West Publishing Co., St. Paul, MN: 1992)

[2.8]   S. Haykins: *Communication Systems* (John Wiley & Sons, Inc., New York, NY: 1994)

[2.9]   W. Tomasi: *Electronic Communications Systems* (Prentice-Hall, Inc., Upper Saddle River, NJ: 1998)

[2.10] U.D Black: *Emerging Communications Technologies* (Prentice Hall PTR, Upper Saddle River, Inc., NJ: 1997)

[2.11] J. Bellamy: *Digital Telephony* (John Wiley & Sons, Inc., New York, NY: 1991)

[2.12] P.S. Neelakanta and A. Preechayasombon: Mitigating EMI in high-speed digital transmission networks, Part I, *Compliance Engineering*, Vol. XVI, 1999, 36 – 48

[2.13] P.S. Neelakanta and K. Subramaniam: Controlling the properties of electromagnetic composites, *Advanced Materials and Processes*, Vol. 14, 1992, 20 – 25

[2.14] P.S. Neelakanta: "Smart Materials" in the *Electrical Engineering Handbook* (Editor: R.C. Dorf.) (CRC Press, Boca Raton, FL: 1997), 1277 –1307

[2.15] P.J. Kyees, R.C. McConnell and K. Sistanizadeh: ADSL: A new twisted pair access to the information highway, *IEEE Communications Magazine*, 1995, 52 – 59

[2.16] T.M. Cover and J.A. Thomas: *Elements of Information Theory* (John Wiley and Sons, Inc., New York, NY: 1991)

[2.17] S. Gnanaprakasam and P.S. Neelakantaswamy: Romanised non-european languages as information sources for the communication links. *Zeitschrift für elektrische Information und Energietechnik*, Vol. 9, 1979, 456 – 459

[2.18] G. Gamow: *One Two Three ... Infinity* (Dover Publications, Inc., New York, NY: 1974)

[2.19] R.G. Winch: *Telecommunication Transmission Systems* (McGraw-Hill, Inc., New York, NY: 1993)

[2.20] W. Hioki: *Telecommunications* (Prentice-Hall, Inc., Upper Saddle River, NJ: 1998)

Appendix 2.1

Fourier Transform Theorems and Pairs

Table A2.1: Fourier transform theorems

Operation	Function	Fourier Transform
Linearity	$aF_1(t) + a2F2(t)$	$a_1F_1(f) + a_2F_2(f)$
Delay in time	$F(t-\tau)$	$F(f) \, e^{-j2\pi f\tau}$
Scale change	$F(at)$	$\dfrac{1}{|a|}F(f/a)$
Conjugation	$F^*(t)$	$F^*(-f)$
Duality	$F(t)$	$F(-f)$
Differentiation	$\dfrac{d^nF(t)}{dt^n}$	$(j2\pi f)^nF(f)$
Integration	$\displaystyle\int_{-\infty}^{t} F(\lambda)d\lambda$	$(j2\pi f)^{-1}F(f) + (1/2)F(0)\delta(f)$
Convolution	$F_1(t) * F_2(t)$	$F_1(f)F_2(f)$
Multiplication	$F_1(t)F_2(t)$	$F_1(f)*F_2(f)$

Table A2.2: Fourier pairs

Function	Time-domain representation F(t)		Spectrum of the function F(f)
Unit step	$u(t) = \begin{cases} +1 & t>0 \\ 0 & t<0 \end{cases}$	$\Leftrightarrow$	$\dfrac{1}{2}\delta(f) + \dfrac{1}{j2\pi f}$
Signum	$\mathrm{sgn}(t) = \begin{cases} +1 & t>0 \\ 0 & t<0 \end{cases}$	$\Leftrightarrow$	$\dfrac{1}{j2\pi f}$
Constant	k	$\Leftrightarrow$	$k\delta(f)$
Impulse at $t = t_0$	$\delta(t - t_0)$	$\Leftrightarrow$	$e^{-j2\pi ft_0}$
Phasor	$e^{-j(\varpi t+\Phi)}$	$\Leftrightarrow$	$e^{j\Phi}\delta(f - f_0)$
Sinusoid	$\mathrm{Cos}(\omega_0 t + \Phi)$	$\Leftrightarrow$	$\dfrac{1}{2}e^{j\Phi}\delta(f - f_0) + \dfrac{1}{2}e^{j\Phi}\delta(f + f_0)$
Impulse train	$\displaystyle\sum_{k=-\infty}^{+\infty}\delta(t - kt)$	$\Leftrightarrow$	$\dfrac{1}{T}\displaystyle\sum_{\ell=-\infty}^{+\infty}\delta(f - \ell/T)$

Appendix 2.2

Linear System Theory: Concepts

2-A2.1 Signal Response of a Linear System

The signal response of a linear system is specified by:

✓ Transfer function, H(f)
✓ Impulse response function, h(t).

The *Transfer function* H(f) of a system is a complex-valued function which can be determined at a frequency, f by applying a signal to the input of the system at that frequency and noting the output (Fig. 2A2.1).

The *Impulse response function h(t)*, is obtained by recording the output when the input f(t) = δ(t), and the delta-Dirac function at t = t_0.

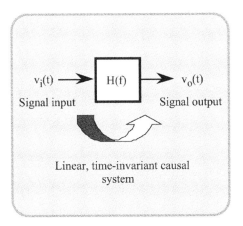

Fig. 2A2.1: Linear system

2-A2.2 Time Domain ⇔ Frequency Domain

In reference to Fig. 2.2A.1, the following time domain /frequency domain relations can be specified: For any time-domain function x(t), its frequency domain counterpart is given by X(f). Explicitly,

$$X(f) = \mathcal{F}[x(t)] = \int_{-\infty}^{+\infty} x(t)e^{-j2\pi ft}dt \Leftrightarrow x(t) = \mathcal{F}^{-1}[X(f)] = \int_{-\infty}^{+\infty} X(f)e^{j2\pi ft}df . \tag{2A2.1}$$

Hence, the input/output relation in Fig. 2.2A.1 can be written in terms of $v_i(t) \Leftrightarrow V_i(f)$ as follows:

$$v_o(t) = \left[\int_{-\infty}^{+\infty} V_i(f)e^{j2\pi ft}df\right]H(f) = \int_{-\infty}^{+\infty} [V_i(f)H(f)]e^{j2\pi ft}df \tag{2A2.2a}$$

$$\therefore V_o(f) \Leftrightarrow V_i(f)H(f) = \mathcal{F}^{-1}[v_0(t)]. \tag{2A2.2b}$$

Suppose $v_i(t) = \delta(t)$. Then, from Eqn. (2A2.2a), it can be ascertained that $v_o(t) = h(t)$. That is, $\mathcal{F}^{-1}[H(f)] = h(t)$.

2-A2.3   Power Spectral Density

Associated with a signal $v(t) \Leftrightarrow V(f)$, a spectral power, $S(f) = V(f)V^*(f)$, where $V^*(f)$ is the complex conjugate of $V(f)$. Then, the *power spectral density* (PSD) is defined as follows:

$$PSD = \frac{dS(f)}{df}.$$

(2A2.3)

Note:

Spectral symmetry: If $v(t)$ is real, that is, when real-valued signals are considered, $V(-f) = V^*(f)$.

Parseval's theorem

This theorem specifies that the energy contained in a signal evaluated in the time-domain is equal to that determined in the frequency domain. (This is in conformity with the principle of conservation of energy.)

$$\int_{-\infty}^{+\infty} v_1(t)v_2{}^*(t)dt \equiv \int_{-\infty}^{+\infty} V_1(f)V_2(f)df$$

(2A2.4a)

If $v_1(t) = v_2(t) = v(t)$ then,

$$\int_{-\infty}^{+\infty} \left| v(t) \right|^2 dt \equiv \int_{-\infty}^{+\infty} \left| V(f) \right|^2 df$$

(2A2.4b)

where $V(f) \Leftrightarrow v(t)$.

PSD and transfer function

When a signal is limited over the time $+ T/2$ to $- T/2$, then

$$PSD = G(f) = \underset{T \to \infty}{\text{Limit}} \left| V(f) \right|^2 / T.$$

(2A2.5)

In reference to Fig. 2.2A.1,

$$\text{Input PSD} = G_i(f) = \left| V_i(f) \right|^2 / T.$$

(2A2.6)

$$\text{Output PSD} = G_o(f) = \left| V_o(f) \right|^2 / T$$

(2A2.7)

$$= \left| V_i(f)H(f) \right|^2 / T.$$

Therefore, it follows that $G_o(f) = G_i(f)H(f)$.

2-A2.4   Convolution

This is a mathematical operation performed in respect to two real-valued functions of the same variable. That is,

$$p(t)*q(t) = \int_{-\infty}^{+\infty} p(\lambda)q(t-\lambda)d\lambda$$

$$= \int_{-\infty}^{+\infty} q(\lambda)p(t-\lambda)d\lambda.$$

(2A2.8)

In reference to the signals $v_1(t)$ and $v_2(t)$, *the convolution integrals* are as follows:

$$v_1(t)*v_2(t) = \int_{-\infty}^{+\infty} v_1(\lambda)v_2(t-\lambda)d\lambda$$

$$= \int_{-\infty}^{+\infty} v_2(\lambda) v_1(t-\lambda) d\lambda. \tag{2A2.9}$$

By taking appropriate Fourier transforms, the following relations can be proved:

$$v_1(t) * v_2(t) \Leftrightarrow V_1(f) V_2(f) \tag{2A2.10a}$$

$$v_1(t) \, v_2(t) \Leftrightarrow V_1(f) * V_2(f) \tag{2A2.10b}$$

2-A2.5 Frequency Translation Property
Consider a product of two signals of the form,

$$y(t) = [x(t) \, [\cos(2\pi f_0)]. \tag{2A2.11a}$$

Here, y(t) represents a *modulated signal*. The Fourier transform of y(t) can be written as follows:

$$Y(f) = \mathcal{F}[y(t)] = \frac{1}{2}[X(f - f_0) + X(f + f_0)]. \tag{2A2.11b}$$

The above relation indicates the frequency translation of baseband signal x(t) when modulated by a *carrier*, $\cos(2\pi f_0 t)$. That is, the *baseband spectrum* X(f) gets shifted as a *passband spectrum* located at f_0 as illustrated in Fig. 2A2.2

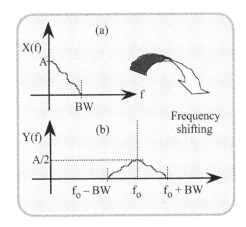

Fig. 2A2.2: Frequency shifting principle

2-A2.6 Sampling Property
Mathematical representation of sampling: A signal v(t) can be represented by a set of sampled information $v_s(t)$:

$$v_s(t) = v(t) \sum_{n=-\infty}^{+\infty} \delta(t - nT_s) v(t). \tag{2A2.12}$$

The above relation (Eqn.2A.12) indicates that the *sampled signal*, $v_s(t)$ is made of a discrete sequence of impulses separated in time by the sampling internal, T_s. The sampling rate or frequency is given by $f_s = 1/T_s$. Taking the Fourier transform of $v_s(t)$,

$$v_s(t) \Leftrightarrow \mathcal{F}[v_s(t)] = V_s(f) = V(f) * [f_s \sum_{n=-\infty}^{+\infty} \delta(f - nf_s)]$$

$$= f_s \sum_{n=-\infty}^{+\infty} V(f - nf_s) \tag{2A2.13}$$

where $v(t) \Leftrightarrow V(f)$. The above considerations can be illustrated as shown in Fig. 2A.2.3.

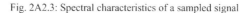

Fig. 2A2.3: Spectral characteristics of a sampled signal

2-A2.7 Correlation

Correlation depicts the similarity (or dissimilarity) between a set of two given signal waveforms. Suppose $v_1(t)$ and $v_2(t)$ are two given waveforms. The average *cross-correlation* between $v_1(t)$ and $v_2(t)$ is defined as follows:

$$R_{12}(\lambda) = \lim_{T \to \infty} \frac{1}{T} \int_{-T/2}^{+T/2} v_1(t) v_2(t + \lambda) dt \qquad (2A2.14a)$$

assuming that the signals are aperiodic. In the event of the signals being periodic with a period T_0, then,

$$R_{12}(\lambda) = \frac{1}{T_0} \int_{-T_0/2}^{+T_0/2} v_1(t) v_2(t + \lambda) dt. \qquad (2A2.14b)$$

In the above relations, τ refers to a searching on a scanning parameter, which can be chosen as a time-shift between the waveforms (in coupling R_{12}) so that the extent of relation between the waveforms will be revealed to a maximum possible extent.

When $R_{12}(\tau) \to 0$, v_1 and v_2 are *totally uncorrelated*. When $R_{12}(\tau) \to 1$, v_1 and v_2 are *totally correlated*.

Autocorrelation

When a signal waveform performs correlation on itself, the resulting function is *autocorrelation*. That is, when $v_1(t) = v_2(t) = v(t)$, then $R_{12}(\tau) = R(\tau)$, which is known as the *autocorrelation coefficient*. It is given by:

$$R(\lambda) = \lim_{T \to \infty} \frac{1}{T} \int_{-T/2}^{+T/2} v(t) v(t + \lambda) dt. \qquad (2A2.15)$$

Correlation function and power relations

Suppose, $v_1(t)$ and $v_2(t)$ are two signals whose power spectra are given by:

$$S_1 = \frac{1}{T} \int_{-T/2}^{+T/2} v_1(t)^2 dt \qquad (2A2.16a)$$

$$S_2 = \frac{1}{T} \int_{-T/2}^{+T/2} v_2(t)^2 dt. \tag{2A2.16b}$$

Let

$$S_{12} \cong \frac{1}{T} \int_{-T/2}^{+T/2} [v_1(t) + v_2(t+\tau)]^2 dt$$

$$= S_1 + S_2 + 2 \int_{-T/2}^{+T/2} [v_1(t) + v_2(t+\tau)] dt$$

$$= S_1 + S_2 + 2R_{12}(\tau). \tag{2A2.16c}$$

Note: When the waveforms are totally uncorrelated then the total power due to v_1 and v_2 is a constant equal $S_1 + S_2$ since $R_{12}(\tau) \equiv 0$.

Wiener-Khintchine theorem

This theorem states that the power spectral density of a signal is the Fourier transform of the autocorrelation function. That is,

$$G(f) = \mathcal{F}[R(\tau)]. \tag{2A2.17}$$

Example 2A2.1

Determine the PSD of an aperiodic signal, whose autocorrelation coefficient is given by

$$R(\lambda) = \exp(-\tau^2 / 2\sigma^2).$$

Solution

By the Wiener-Khintchine theorem,

$$P.S.D = G(f) = \int_{-\infty}^{\infty} \exp(-\tau^2 / 2\sigma^2) \exp(-j2\pi f\tau) d\tau$$

$$= \sqrt{2\pi\sigma^2} \exp[(-2\pi f\sigma)^2 / 2].$$

Problem 2A.1

Determine the normalized average power content (S) of the signal described in Example 2A2.1. *Hint:* $S = R(\tau)|_{\tau \to 0}$

2-A2.8 Nyquist Rate: Sampling Theorem

Nyquist rate is the minimum rate required when the sampling is performed so that the original waveform can be reconstructed from the sampled values of the waveform. Suppose a signal $v(t)$ has a bandwidth BW corresponding to its highest significant frequency component, then the Nyquist rate should be greater than or equal to $2 \times$ BW. When sampling is done as per the Nyquist sampling theorem, the spectrum of the sampled signal, namely $V_s(f) \Leftrightarrow v_s(t)$ contains all the information as in the unsampled signal spectrum $V(f) = v(t)$. In short, any $2 \times$ BW independent samples per second will completely characterize a band-limited signal. Alternatively, any $2 \times$ BW $\times$ T unique (independent) pieces of information are needed to completely specify a signal over an interval T seconds long.

Aliasing

Nyquist sampling indicates that the spectrum of a sampled waveform consists of replicating the spectrum of the unsampled waveform about the harmonics of the sampling frequency, f_s. If $f_s < 2BW$ (where BW is the highest significant frequency component) of the signal, then *aliasing errors* will occur in the sampled waveform. Using a higher sampling frequency, or a presampling low-pass filter, can decrease this aliasing error.

Scaling uncertainty

For a signal v(at), $\mathcal{F}[v(at)] = (1/a)V(f/a)$ with a > 0. This relation implies that a short-term duration signal yields a wider spectrum or vice versa. That is, both time and bandwidth cannot be arbitrarily (and independently) made small without influencing the other. This is known as the *time-bandwidth uncertainty principle*. (This is akin to *Eisenberg's uncertainty principle* in quantum mechanics on momentum and position of a particle.)

Problem 2A2.2

Determine the Nyquist rate required in sampling (i) voice signal and (ii) TV signal waveforms.

3

NETWORKING: CONCEPTS AND TECHNOLOGY

What is a good network?

"A good network should be a network that doesn't look like a network to the end user..."

T. Ramteke
Networks, 1996

3.1   Introduction

The modern telecommunication network is a complex interconnection of a variety of heterogeneous switches. As indicated in earlier chapters, it emerged originally as a means of transporting simple telephone conversations (voice information). Later it evolved as a system to perform various teleservices such as transporting computer data, simple video, facsimile, high-definition video, video-conferencing, higher-speed data transfer, video-telephony, video-library, home-education, video-on-demand, multimedia information, and so on.

The ongoing expansions on new telecommunication services have come into being as a result of the following major considerations:

- Customer needs and the dawn of the information age

- Revolutionary developments in :

   ✓ Semiconductor technology, specifically, the high-speed integrated circuits (ICs)
   ✓ Software vis-á-vis hardware
   ✓ Optical media of transmission involving fiber optics, optical switching etc.
   ✓ System concepts enlarging the superfluous transport functions to the edge of network
   ✓ Commercial trends to meet residential and business subscriber expectations.

This chapter is devoted to describing various phases of such evolution in telecommunication networking and the associated transfer modes of information. The relevant details form the logical foundation to study the asynchronous transfer mode (ATM) in depth.

3.2   Transfer Modes of Information in Telecommunication Networks

This topic can be discussed in two perspectives namely,

- Evolution of various transfer mode technologies in the history of telecommunication and the gist of factors, which necessitated changes in the modes of information transfer

- State-of-the-art aspects of information transfer modes in the modern telecommunications.

3.2.1   *The making of a telecommunication network*

Consistent with the evolution of electrical communication systems discussed in Chapter 1, indicated below are brief descriptions on the networking considerations pertinent to the classical electrical means of communication developed, which eventually paved the way to the modern information highway via a maze of telecommunication networking.

Networking in telegraphy

A network of telegraphic apparatus constitutes telegraphy, which is basically an electromagnet connected to a battery through a switch (called the *Morse key*). It is illustrated in Fig. 3.1. The Morse key is at the sending end and the electromagnet arrangement is located at the receiving end as shown.

How is a telegraphic transmission done across a network?

The answer to this question can be stated in terms of network functions associated with the telegraphy. Referring to Fig. 3.1, when the Morse key is pressed, a closed-loop electric circuit is formed via the ground-return path. Hence, the solenoid is energized. The soft-iron core of the solenoid thereby is magnetized attracting an armature. A bob attached to the armature moves and hits the bell-gong making a sound. If the key is pressed over a long duration, the sound made will be long. A shorter period of keeping the key closed makes a shorter sound. The longer sound corresponds to a "dash" and a shorter sound refers to a "dot". Thus, Morse devised a *variable-length binary code* in which the letters of the English alphabet are represented by a sequence of dots and dashes (code words). When a message is sent by these code words composed of dots and dashes, the telegraphist at the receiving end, by listening to the corresponding sounds at the *sounder box,* deciphers the encoded message [3.1].

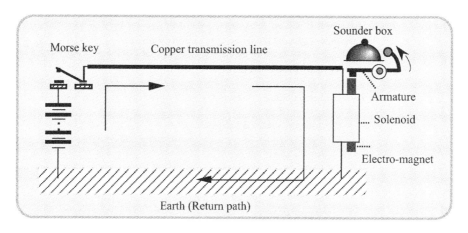

Fig. 3.1: The Morse telegraphic point-to-point networking

Radio/wireless telegraphy

An extension of terrestrial telegraphy is known as *radio or wireless telegraphy* facilitated via a wireless transmission network. Here, the on-off keying (corresponding to dots and dashes) allows an electromagnetic wave at a radio frequency transmitted or cut-off. The on-off state of the carrier is detected and decoded at the receiving end. As necessary, the radio telegraphic message can be relayed from one station to another station via a wireless network.

Automated telegraphy

The networking of manual telegraphy was eventually automated. As a first step towards this endeavor, the original telegraph of Morse was partially automated to send messages and a print mechanism was used to print the dots and dashes received on a long strip of paper. The mechanized system consisted of a notched stick called a *port rule* on which a set of metal pieces were arranged as per the encoding of a message. When the port rule is cranked, the encoded arrangement of the metal pieces facilitates a "make and break" set of electrical contacts (that could be otherwise established via manual operation of a Morse key). At the receiving end, the sounder box is modified with a pencil recorder to make dashes and dots of the transmitted message on a paper tape. This message is read subsequently by a telegraphist and deciphered.

Another version of printing telegraphy was conceived, by Thomas Alva Edison. Subsequently, Emile Baudot developed an automated telegraphic transmission with a *fixed length (block) code* constituted by five on-off (binary) states. When a key is depressed, the Baudot machine creates automatically an appropriate code for the desired alphanumeric character on a perforated tape (Fig. 3.2).

A hole (perforation) on the tape represents a "mark" and the absence of a hole denotes a "space". The mark and space are the two distinguishable states (corresponding to the dot and the dash in the Morse telegraphy respectively). A set of five conducting fingers (brushes) traversing on the perforated tape senses these states at the perforations and transmits the corresponding electrical signals. The network of Baudot machines eventually led to the so-called teleprinter networking in which a typewriter-like keyboard replaces the Morse key.

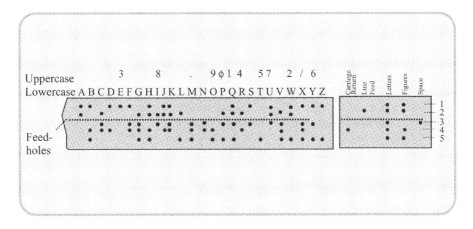

Fig. 3.2: A CCITT-2 code based on Baudot's 5-unit code

Example 3.1

Determine the average transmission time of each word keyed at a rate of 60 words per minute with a 5-unit code Baudot telegraphy. Each coded letter is preceded by a 22 ms space and followed by a 33 ms mark. (In a typical English text, there are on an average six letters per word.)

Solution

Average time of transmission of each word:
$$= 6 \times [22 \text{ ms} + (5 \times 22 \text{ ms}) + 33 \text{ ms}]$$
$$\approx 1 \text{ s}$$

Problem 3.1

Determine the minimum width of the passband required in a telegraphic circuit, which supports machine telegraphic message transmissions. Assume 1 intervals are allowed for each letter group. Normally, 60 words per minute are sent. Each word on an average has 5 letters.

The extended version of Baudot's telegraphy with a typewriter keyboard arrangement was due to Donald Murray whose patent was adopted by AT&T to develop a teletypewriter with a brand name *Teletype* (TTY). Hence, AT&T's Teletypewriter Exchange (TWX) network became operational parallel to a similar service called Telex offered by Western Union. The Telex network service allowed subscribers to exchange typed messages and it formed the basis for the *fax* (facsimile) machines, which came into existence in the 1980s.

Perhaps the network of automated telegraphy can be regarded as the seed which set the pace for the modern network of high-speed information transmission.

The networking system of telephony

The networking system with telephone apparatus and equipment commenced almost with the invention of the telephone itself. The patent rights of Graham Bell on the telephone led to the formation of a monopolized telephone company, the Bell Company (later AT&T), as the service provider and the network operator of telephony in the United States. Hence, came into being the telephone communication networking marking the dawn of analog voice communication facilitated through electrical conduction via a pair of copper wires [3.2 – 3.6]. Essentially, the telephone networking consists of a transmitter (the mouthpiece) called a *microphone* and a receiver the *earphone*. In principle, the mouth-piece has a vibrating plate (diaphragm) and when the sound energy (voice) impinges upon this diaphragm, the mechanical vibration is converted (transducted) into corresponding electrical voltage (or current) variations which are transmitted along the copper wires with their other end connected to an earpiece. The electrical variations received are then used to vibrate a diaphragm in the earphone causing a corresponding sound. That is, the electrical variations are transducted back into audible sound waves at the receiving end.

Graham Bell's telephone transmitter had a membrane backed by a mild acid in a cup. The vibrations due to impinging sound waves on the membrane proportionately alter the electrical resistance of the acid, and hence modulate the d.c. current flowing in the loop proportional to the sound wave as illustrated in Fig. 3.3.

At the receiver, the changing electric current causes a corresponding magnetic induction to vary in an electromagnet, and thereby, a soft-iron diaphragm placed in the magnetic induction field moves in accordance with the changes in the electrical current through the solenoid of the electromagnet. The moving diaphragm at the earpiece thereof, replicates the analog sound (voice) at the transmitting end.

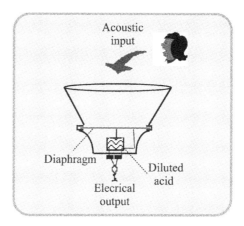

Fig. 3.3: Bell's liquid acid microphone

Analog voice signal

The sound wave generated by voice causes acoustical pressure variations on the diaphragm. The corresponding electrical voltage, v(t) or current i(t) variations with respect to time, t at the output of a microphone depict an analog voice signal, the wave shape of which follows that of the input sound wave.

Ideally, the transformation from sound to electrical energy should be on a one-to-one basis. This would correspond to a linear transduction. However, in practice, some nonlinearity in the mapping is rather inevitable.

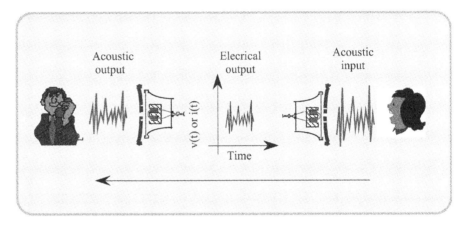

Fig. 3.4: A simple analog voice telephony

Bell's telephone, especially the transmitter part, was later substituted by a patent by Thomas Alva Edison. This substituted unit had carbon granules replacing the unpleasant liquid acid of Bell's invention.

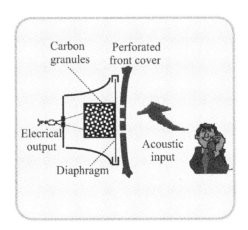

Fig. 3.5: Edison's carbon microphone

The hand set of the modern telephone houses the mouthpiece and the earphone in various aesthetic forms and shapes, which basically work on the same principle as the inventions of Bell and Edison with some technological changes, improvements, and colorful appearance.

Voltage source representation of a microphone...

    The liquid-acid version, the carbon-granule type, or an electromagnetic induction-based microphone can be equivalently represented by a Thevenin's voltage source $v_S(t)$ in series with a low impedance, R_S.

    The voltage source denotes the analog electrical voltage generated (in proportion to the sound input) and the source impedance depicts the nominal internal resistance offered by the liquid acid, the carbon granules, or the induction coil.

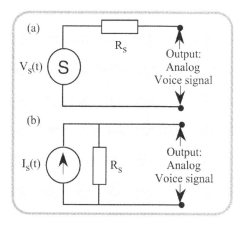

Fig. 3.6: (a) Thevenin's voltage-source equivalent of a low impedance microphone and (b) Norton's current-source equivalent of a high impedance microphone

In EM induction-based microphones, the time-variations of the diaphragm placed in the magnetic field cause analogous electric voltage variations in the solenoid housed in the microphone, yielding a corresponding (analog) voice signal.

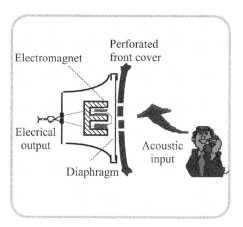

Fig. 3.7: Electromagnetic induction-based microphone

Electromagnetic induction principle ...

The phenomenon that leads to the generation of voice signal voltage at the terminals of an electromagnetic (EM) induction type microphone refers to Faraday's law of EM induction which can be stated as follows:

Whenever a time-varying interaction exists between an electric conductor and a magnetic field, a voltage will be induced in the conductor proportion to the time-varying interaction involved.

High impedance microphones...

The *electret* type microphone is a modern version, which uses an electret material for its diaphragm. The electret material works on the *principle of electrostriction.* The vibration (or the corresponding mechanical stress) caused by the sound wave at the electret diaphragm would alter the extent of electric charge developed on the electret (Fig. 3.8). Hence, associated changes in the electric field across the capacitance between the back plate and the electret diaphragm cause an electric signal variation at the output terminal of the microphone. The electret material is typically a fluorine-based organic compound.

The microphones based on induced electric-charge variation (such a electret microphones) have a high internal impedance. Therefore, they are represented equivalently by a constant current source $i_S(t)$ with a high shunt impedance, $R_S(t)$. This is known as *Norton's equivalent circuit* as depicted in Fig. 3.6.

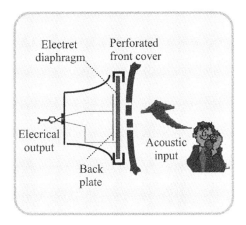

Fig. 3.8: Electret microphone

Analog telephony

The network of analog telephony commenced with the classical version of telephone equipment placed into an analog network, which interconnected the subscribers via a *switching system.* The switching system is a common locale to all the subscribing telephone users. It refers to the *exchange,* or the *switching office,* or a *central office* (CO). Essentially, a switching system enables the following:

- It facilitates the establishment and release of connections between subscribers via signaling (or controlling) functions

- It maintains a reliable connectivity between the end-users during the session of a conversation.

A basic connection between two subscribers is the *simplex telephone circuit* as shown in Fig. 3.9. The simplex circuit allows one-way communication between a speaker and a listener. It refers to a "receive only" or "transmit only" system. An improvement on the circuit of Fig. 3.9 is the *half-duplex communication* illustrated in Fig. 3.10 where two-way communication is made feasible, but not simultaneous, between two subscribers. A push-button allows talking and when talking is over, release of the push-button is done after the statement "over" so that the person on the other end will commence talking. Either of the users can terminate the session with a statement "out". In short, the user at either end will be, at any time, on the "talk mode" or "listen mode".

A modification of the half-duplex circuit refers to the *full-duplex transmission* wherein two subscribers can converse simultaneously across the link. That is, a full-duplex is a two-way simultaneous mode of transmission. An extended version of full-duplex is the *full/full-duplex* in which a user can transmit to another user and also simultaneously receive from a third user.

Fig. 3.9: Basic simplex telephone connection

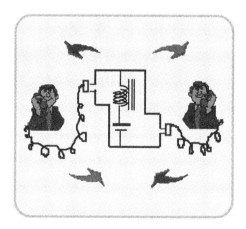

Fig. 3.10: Half-duplex telephone system

An interlink between subscribers is facilitated by a set of switching systems constituting a hierarchical network operated by telcos. The link enabled between any two subscribers can be based on either using *two-wire circuits* or *four-wire circuits*. The two-wire circuit is half-duplex. More practically, in order to allow full-duplex operation as well as to enable amplification of voice signals on long-haul links, a four-wire system is common in the practice of telephony. It is illustrated in Fig. 3.11.

The full-duplex implies a segregated passage of transmitted and received voice signals across the link. This segregation can be accomplished via two distinct pair of physical wires in each direction, or, it can be a single pair of wires, and the bidirectional signals are separated by being carried by two distinct bands of frequencies.

The use of four physical wires is common in interoffice trunks, and the two-wire system enabling full-duplex via a carrier frequency band separation is called a *derived four-wire circuit.*

Hardware of telephone circuits and networking
In reference to the basic duplexed transmissions of voice signals, there are four important hardware components used. They are:

      ✓   Hybrids
      ✓   Echo suppressors and echo cancellers
      ✓   Analog companders.

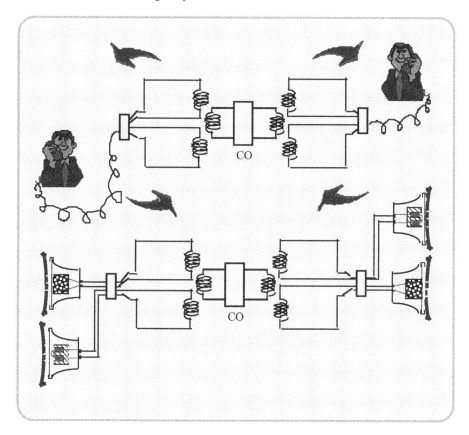

Fig. 3.11: A full-duplex system: The four-wire configuration

Hybrids
      These are used to convert a two-wire subscriber loop (Fig. 3.12) into a four-wire trunk circuit at the CO. This interfacing circuit is known as the *hybrid circuit.* A hybrid is based on the *cross-connected transformer* principle.

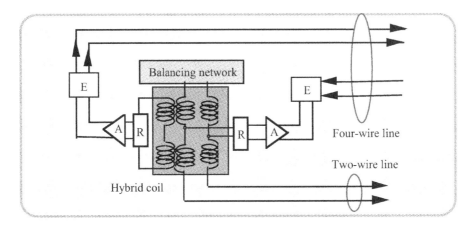

Fig. 3.12: A hybrid circuit used in telephony (A: Amplifier; R: Attenuator pad; E: Equalizer)

A hybrid separates the two-way signals as illustrated in Fig. 3.12. Further, a balancing circuit is included so that the subscriber loop impedance is matched to that of the hybrid. This matching is required for maximum transfer of power between the connected ends. The advantages of four-wire circuits over the two-wire circuits are as follows:

✓   Four-wire circuits provide better isolation between the bidirectional transmissions
✓   A good balancing leads to minimal echo
✓   Four-wire system offers independent bidirectional amplifier gain settings
✓   Facilitated matching allows maximum power transfer across the network.

However, a four-wire system is more expensive than the two-wire counterpart due to the replicated set of hardware, one in each direction used.

A note on maximum power transfer…

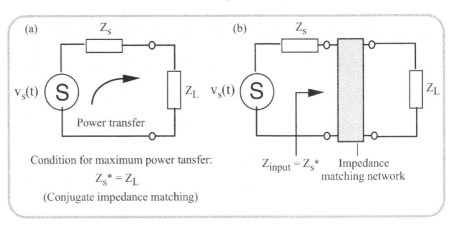

Fig. 3.13: Power transfer from source to load: (a) Condition for maximum power transfer and (b) impedance matching technique

In a given circuit, as illustrated in Fig. 3.13a, the signal source v_S has impedance Z_S associated with it. The source delivers power to a load, Z_L. In general Z_S and Z_L could be complex impedances.

In order for the maximum power transfer from the source to the load to take place, Z_L must be a complex conjugate of Z_S. That is, $Z_L = Z_S^*$. When this matching is perfect, all the energy supplied to the load will be absorbed and none will be reflected.

If the above condition is not satisfied, a matching circuit is interposed as shown in Fig. 3.13b. Conjugate matching at the input and output terminals (of the matching circuit) are achieved via appropriate designs of matching circuit elements.

A transformer can be adopted typically as a matching device. At high frequencies, techniques like stub matching and quarter-wave line matching are used.

Echo suppressors and cancellers

The balancing network used in conjunction with the hybrid circuit may not allow a perfect matching between the coupled impedances. Therefore, some reflected energy due to mismatch would result and this "echo" signal will then return to the sending-end. And, in short-haul trunks, the echo will be received by the sending-end, almost on a real-time basis. This echo would then reinforce the caller's own voice and may serve as the "side-tone"[*] effect deliberately introduced in the telephone system.

An echo arising from mismatches at the hybrid may, however, arrive delayed at the speaker end if the trunk length is excessive. If this time delay (corresponding to the round trip on the line) exceeds 45 ms, the echo received will be unpleasant and annoying to the speaker. Therefore, *echo suppressors* are introduced in long-distance communication links.

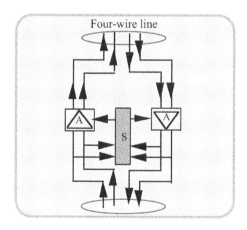

Fig. 3.14: An echo suppression arrangement (A: Amplifier; S: Audio tone sensing device)

An echo suppressor (Fig. 3.14) is a logic circuit which lets the speaker's voice signal be amplified in its forward direction and, at the same time, it cuts off the amplification for any echo that comes into the reverse direction towards the speaker-end. Echo suppression to an extent of 60

---

[*] *Side-tone:* The voice signal heard at the ear-piece of the generating-end is called the *side-tone*. It is deliberately facilitated in telephone equipment for physio-psychological purposes. Since the side-tone offers a feeble feedback of what is spoken, the speaker may automatically adjust the volume of conversation. If the side-tone is absent, the speaker may perceive a feeling of equipment being "dead". However, too high a side-tone should be avoided or the speaker may tend to speak in too low a voice level. Further, the side-tone should be in real-time in concurrence with what is spoken. Otherwise, a delayed echo will constitute an annoying effect to the speaker.

dB is feasible. The speech sensing circuit is a detector, which senses the presence of the signal and enables the amplifier in the appropriate direction and disables the amplifier in the opposite direction as necessary. The logic extended by the echo suppressors between two speakers let the system to operate essentially half-duplex.

If half-duplex mode is not desired, then *echo cancellers* are used in lieu of echo suppressors. The echo canceller does not disable amplification in the reverse direction. It rather subtracts the echo from the original signal electrically. Since the logical cutting off of the reverse direction amplification is absent, the two speakers can engage in full-duplex conversation.

Analog companding

Voice signal, in general, poses a large dynamic range (60 dB or more) in its amplitude variations. In long distance telephone transmissions, where several repeaters are used, this large dynamic variation would cause, en route, the amplifiers to saturate at large signal levels and the low signal level to be concealed by the amplified noise.

The above problem is overcome via a *compander circuit*. It compresses the high amplitude ranges of the voice signal and expands the low amplitude ranges of the signal. This can be accomplished via *logarithmic amplifiers*.

Amplifier saturation – overloading of an amplifier

Typical signal input (v_{in}) versus output (v_{out}) characteristics of an amplifier is illustrated in Fig. 3.15.

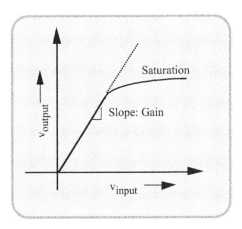

Fig. 3.15: Input versus output characteristics of a signal amplifier

The slope of the curve in Fig. 3.15 depicts the amplifier gain. It can be seen that for large input levels, the signal output is not proportionally amplified. The amplifier becomes saturated. This nonlinear behavior is due to the inherent nonlinear characteristics of the active device (such as junction or field-effect transistors) used in the amplifier circuits. This nonlinear behavior of an amplifier will limit the dynamic range of signal variations in the system.

The telephone set

The telephone set at a subscriber end in conjunction with its connectivity with another subscriber through the CO and the network is designed to perform certain basic functions. They are:

    ✓   Alerting the called-subscriber of an incoming call via ringing
    ✓   Facilitation of voice transmission to and from the subscriber ends

✓ Regulating the constancy of voice signal level regardless of the length of the distance between the subscribers

✓ Enabling a calling-subscriber to contact a called-subscriber via dialing

✓ Gaining the attention of CO by lifting the handset off the hook

✓ Releasing the lines when a subscriber places the handset back on the hook

✓ Energizing the circuit during a conversation session with the d.c. supply from the CO and opening the circuit when the set is not in use

✓ Allowing normal side-tone levels.

The above functions are accomplished in a conventional conversation telephone set by the associated networking that includes telephone lines and switching equipment of the telcos. Indicated below is an outline on the working principle of telephony and the associated networking.

Telephony at its inception stage was a manual system. Its network was built using a pair of copper (or sometimes iron!) wires suspended on open poles connecting the telephone apparatus and the exchange office.

Along with the transmitting and receiving part of the telephone set was a *magneto generator* at each subscriber end. The alternating current generated by the magneto generator (at a frequency of 17-20 Hz) was used to signal the exchange of a subscriber's intention to have a telephone interconnection made to another subscriber. Also, the subscriber equipment was energized locally by a 3V battery. Hence, the system was known as a *local battery system* (LBS).

Subsequent developments in telephony refer to what is known as the *central battery system* (CBS) in which energization of the subscriber's set is facilitated by a battery situated at the exchange.

A manual telephone network using CBS is illustrated in Fig. 3.17. The constituent hardware is as follows:

At the subscriber's end:

✓ Telephone transmitter
✓ Telephone receiver
✓ Telephone switch hook (normally open until, the hand set is lifted)
✓ A balanced hybrid which segregates transmitted and received voice signals
✓ A ringer.

At the manual exchange:

✓ A central battery (common to all the subscribers connected to the exchange)
✓ A lamp (per subscriber) with an on-off relay switch
✓ Ringing circuit generator
✓ Operator's headset
✓ Set of jacks (each connected to a subscriber-line) on a patch-board
✓ Patch cords to interconnect the subscribers.

The operation of a manual exchange-based telephony is as follows (Fig. 3.16): Suppose subscriber A wants to be connected to subscriber B. Lifting the handset off the switch hook at A will let a d.c. circuit close with the energization from the central battery (CB) located at the exchange. Hence, a relay on this circuit at the exchange would close, signaling the operator of an incoming call establishment request from A. The operator would then use the headset connected to the loop and acquire the called party's (B's) information from A. If the called party (B) was free (as could be ascertained from the associated lamp being off on the board), the operator would ring the called party (B) using a ringing circuit generator. This would signal the called party of an incoming call. Once the called party would (B) lift the hand set off the switch hook, the associated circuit closes and enables d.c. energization of the path established. This would close a relay switch at the exchange and a light would be on to inform the operator of the availability of B and hence the connection would be established. The operator would then interconnect the subscribers using the patch-cord jacks at the panel.

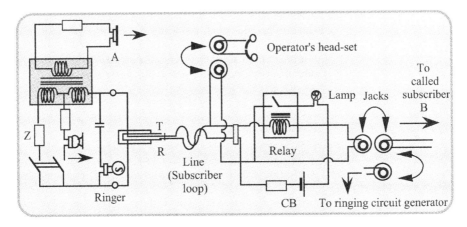

Fig. 3.16: A manual exchange telephone network

A modern automated telephone circuit is illustrated in Fig. 3.17. In this arrangement lifting a handset off the switch hook closes and completes a d.c. path for the central battery (CB) located at the CO through the transformer TR (at the CO). The two loop conductors at the subscriber end are called the *tip* (T) and *ring* (R). The ring takes the negative polarity when the circuit is closed. When the handset is lifted, the corresponding off-hook condition is called the *busy state*. Otherwise, the on-hook condition refers to the *idle state* of the circuit.

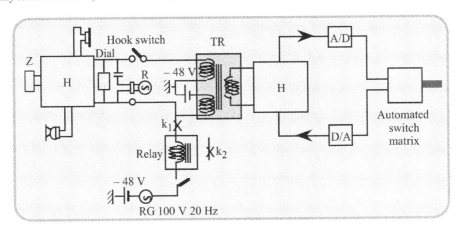

Fig. 3.17: Automated telephone circuit

At the next phase, the calling subscriber (A) dials the number of the called subscriber (B). There are two possible dial circuits, namely, *rotary dial* and *tone dial* used in practice.

The rotary dial generates pulses on the telephone line by opening and closing an electrical circuit when the dial is operated. That is, the rotary dial sends 1 to 10 dial pulses (corresponding to 1 to 10 digits). The rotary dial contacts are placed in series with the loop so as to send pulses via interruption of the circuit current path. Each pulse is 1/20 s long with 1/20 s pause between pulses. The rotary dial process is called *out-pulsing*.

Tone dialing is done via push buttons. It is known as *dual tone multifrequency* (DTMF) dialing. It uses a 4×4 matrix of tones as shown in Fig. 3.18. With each push-button pressed, a unique pair of frequencies per digit dialed is sent and the CO receives these tones. A common control at the CO rings B by closing the relay k1 in the line circuit B. Also, it sends an audible ringing tone to line A. Ringing generator (RG) refers to 100 V a.c. at 20 Hz super-imposed on -48 V.

The circuit of a ring trip relay k2 operates on d.c. When the called party (B) answers (by taking the handset off the switch hook), the d.c. powered by the loop makes k2 operate. This, in turn alerts the common control to disengage ringing and establish a connection between A and B.

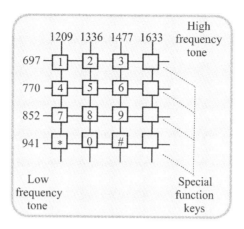

Fig. 3.18: DTMF matrix

3.3   Basics of Telecommunication Networking

As discussed in the above sections, telegraphy was the trend setter of telecommunication networking, but the real story of telecommunication commenced with the invention of the telephone by Alexander Graham Bell. This art of telephony facilitated point-to-point voice communication links, which later extended and proliferated into the telecommunication networking with several options as discussed in earlier chapters.

In such options, suppose *universal connectivity* is aimed at realizing any entity of the network being able to communicate with any other entity of the system as illustrated below (Fig. 3.19a).

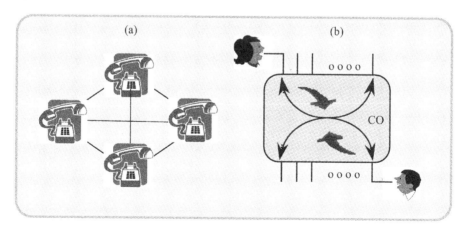

Fig. 3.19: Examples of end-to-end connectivities: (a) Universal connectivity and (b) CO-based connectivity

In reference to Fig. 3.19a,

- The network with point-to-point links indicates a "universal connectivity" among all the subscribers belonging to the network system

- Further, the circles represent the end-entities and the arrows indicate the flow of messages between them.

Networks with point-to-point links among all the entities as illustrated in Fig. 3.19a are known as *fully-connected networks*. To realize a fully-connected networking, $n(n-1)/2$ links are required with n entities or *nodes*. Hence, as n increases, the corresponding number of links that are required for fully interconnecting all the entities grows enormously. This limits the economical feasibility of realizing such networks in practical systems.

An alternative strategy to fully-connected networks is to have a switching office (or an exchange) wherein the entities are not connected to one another directly; instead, they are connected to a switching system, which establishes a connection between two entities (namely, the calling and called subscribers) only when needed. For example, two end-users are linked at the switching system or the exchange in Fig 3.19b.

As indicated before, in implementing the central office (CO) or the exchange-based switching system illustrated in Fig. 3.19b, it is necessary to have a *signaling function* in order to draw the attention of the switching system to establish or release a connection. This function of the switching system facilitating an establishment or releasing a connection is known as *control function*. A historical evolution of switching systems in telecommunication networks can be summarized as follows:

- *Manual switching*: Operator-assisted switching

- *Automated switching*: Classically, this refers to two classes of electromechanical switching (Table 3.1).

Table 3.1: Electromechanical switching

Step-by-step or rotary switch (Strowger system)	Cross-bar switch
Circuits associated with a set of rotary switching elements perform relevant control functions.	Control functions are done by a set of hard-wired control subsystems in the form of a matrix which uses relays and/or latches.

Subsequently, automatic switching was rendered electronic as indicated below (Table 3.2).

Table 3.2: Electronic switching: Stored program control

Space-division switching	Time-division switching
A dedicated path is established between the calling and called subscribers during the entire duration of the call.	Sampled values of speech signal are transferred at fixed time intervals during which a connection exists between the calling and called subscriber.

3.3.1 Circuit-switching concepts

Modern telecommunication networks illustrated in Fig. 3.20 consist of an interconnection of a number of nodes made up of intelligent processors (for example table-top computers, main frames, telephones, video terminals etc.). The primary function of these nodes is to route information through the network. Each node may have one or more stations attached to it; the stations indicated refer to devices wishing to communicate. The network is designed to serve as a shared resource to move the messages exchanged between stations in a robust manner. They also provide a framework to support new applications and services. The telephone network is an example of a telecommunication network in which *circuit-switching* is used to provide a dedicated communication path or circuit between two sections. The circuit consists of a cascade of links from source to destination. The links may consist of time-slots in a time-division multiplexed

(TDM) system or frequency-slots in a frequency-division multiplexed (FDM) system. The circuit, once established, remains uninterrupted for the entire duration of transmission (*session*).

Circuit-switching is usually controlled by a centralized hierarchical control mechanism operating within the purview of a network organization. To establish a circuit-switched connection, an available path through the network is identified and then dedicated to the exclusive use of the two stations wishing to communicate.

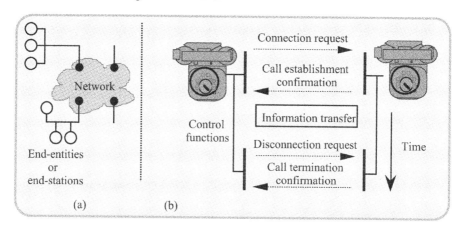

Fig. 3.20: An overview sketch of a modem telecommunication network

In particular, a call-request signaling must be sent all the way to the destination, and be acknowledged, before transmission begins. Thereafter, the network established remains effectively transparent to the end-users. That is, during the connection time, the resources allocated to the circuit are essentially possessed by the two end-entities until the circuit is disconnected. The circuit thus depicts an efficient use of resources only to the extent that the allocated resources are properly utilized.

Whether it is a manual or an automated (mechanical or electronic) type, switching is intended to interconnect two end-users by a sequence of control functions as illustrated in Fig. 3.20 prior to and after the transfer of information.

The story of telecommunication networking is in fact a narration of the details on the signaling (control) and information transfer functions developed over the decades with appropriate cohabitation with the developments in state-of-the-art electronics.

The expansions facilitated in the transfer function of the information are essentially concerned with handling different types of information between the end-entities placed at various geographical dispositions. The growth in signaling-function engineered refers to achieving a reliable and high-speed switching. Jointly, these functions were developed to realize a fast, almost error-free information transfer mode across a complex set of networks between a variety of end-entities in a robust manner.

Although in the current practice a telephone network may be used to transmit computer data, the voice still constitutes the bulk of the traffic involved. The circuit-switching is well suited to the transmission of such voice signals, since voice conversation tends to be of long duration (about 2 to 3 minutes on the average) compared to the time (about 0.1-0.5 seconds) required for the call setup procedure. Moreover, in most voice conversations, the flow of information lasts for a relatively large percentage of connection time, which makes circuit-switching all the more compatible for voice conversation.

In a broad sense, a network offers two services, namely, the information transportation and signaling. These services enclave the following functional considerations as a part of information transfer facilitated by a telecommunication network:

     ✓   Transmission
     ✓   Switching
     ✓   Multiplexing.

Transmission

This refers to the information transportation between end-entities across a network. The transmission of information requires a *medium* (*channel*) between end-stations. As indicated in earlier chapters, this transmission medium can be wire-line or wireless. It represents, in a classical sense, a *physical medium*. In telegraphy, for example, it is a single, overhead copper-wire between the transmitter and the receiver. A closed return path via the conducting earth constitutes the electrical circuit. The electrical transmission along this closed-loop is a baseband transport of current, on or off as per the Morse code used.

Switching

It is a functionality of a telecommunication device that interconnects its incoming line to an appropriate outgoing line.

Multiplexing

This refers to a process of placing several information-bearing transmissions on a single medium. Corresponding demultiplexing processes at the receiving end would segregate these transmissions.

The extension of the single-wire telegraphic line into the picture of telephony refers to the use of a pair of uninsulated copper-wires supported on poles with cross-arms and ceramic/glass insulators. This open-wire system was a familiar scene (along the railway tracks) in the early era of telephony.

Subsequently, multipair buried copper cables and coaxial lines replaced the open-wire system, and the present trend includes concurrent use of optical fibers for telecommunication transmissions.

Though radio means was sparingly used in the telecommunication arena, it has become today an extensively used parallel transmission medium for telecommunication purposes with the advent of *wireless personal communication services*. Described below is an account of the details concerning the types and characteristics of the state-of-the-art wireline transmission media.

3.3.2   Wireline transmission media

Telephone cabling

The proliferation of telephony warranted a criss-cross deployment of wireline on an extensive geographical basis. Home-based wiring, the subscriber local-loop and the trunks therefore, emerged with buried cables [3.3, 3.4]. In modern applications, a cable may consists of 6 to 2700 copper wire pairs, and this cable distribution could be buried cable or underground conduit (cable) based. Typically, the AWG standards of copper wires used are 26, 24, 22, and 19. The corresponding diameter and d.c. resistance per unit length are presented in Table 3.3.

Table 3.3: Types of copper-wire used in paired cables of telephone networking

AWG	Diameter in mils	DC resistance in ohm/kft
26	16	40.810
24	20	25.670
22	25	16.140
19	36	8.051

The resistance, and hence, the attenuation (α) offered by the copper-wire line pairs are functions of frequency in the spectral range of voice signal (up to 4 kHz) supported by the cable as illustrated in Fig. 3.21.

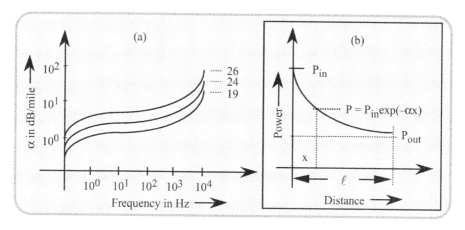

Fig. 3.21: Attenuation versus frequency of copper-cables

The attenuation indicated in Fig. 3.21(b) refers to the loss of power per unit length. Expressing the power at the input of the line of length ℓ meter as P_{in} and the attenuated power at the end of the line as P_{out}, the relation between P_{in} and P_{out} can be depicted as in Fig. 3.21(b).

In Fig. 3.21(b), the attenuation constant α is indicated in the units of nepers per meter (if ℓ is in meter). Since $P_{out}/P_{in} = \exp(-\alpha\ell)$, α is given by:

$$\alpha = (1/\ell)\log_e(P_{in}/P_{out}) \quad \text{neper/meter} \tag{3.1}$$

Alternatively, α can be specified in decibels (dB) units via the following definition:

$$\text{Power-loss (or gain)} = [10\log_{10}(P_{out}/P_{in})]/\ell \quad \text{dB/meter} \tag{3.2}$$

A negative decibel value indicates attenuation and a positive value refers to gain. The corresponding decibel loss (or gain) in terms of voltage (V) or current (I) entities can be expressed as:

$$\text{Power-loss (or gain)} \equiv 10\log_{10}(V_{out}^2/V_{in}^2) = 20\log_{10}(V_{out}/V_{in})$$

$$\equiv 10\log_{10}(I_{out}^2/I_{in}^2) \quad = 20\log_{10}(I_{out}/I_{in}). \tag{3.3}$$

Example 3.2
Suppose a signal of 1 W at the input of a line of length 10 meters is a attenuated to 0.5 W at the end of the line. Determine the power loss per unit length (i) in neper and (ii) in dB.

Solution
Power loss ratio = P_{out}/P_{in} = 0.5 W/ 1.0 W = 1/2
In nepers, $\log_e(1/2) = -0.693$
In dB, $10\log_{10}(1/2) = -3$
Per unit length, the attenuation is:
In neper/m   = -0.0693
In dB/m   = -0.3

Example 3.3
A signal voltage of 1 V impressed at a transmission line suffers a total attenuation of 3 dB along the length of the line. Determine the signal level at the end of the line.

Solution

Signal voltage attenuation $= -3$ dB.

That is, -3 dB $\equiv 20\log_{10}(V_{out} / V_{in})$

or, V_{out} / V_{in} $= 0.707$

hence, V_{out} $= (0.707 \times 1)$ V.

---

dB notation…

While dB is a scale of relative measure of power, voltage, or current, the absolute level of power (or voltage or current) can be expressed in reference to a standard value of power (voltage or current). For example, a power level of P watt can be expressed in decibels with reference to a milliwatt of power (dBm). That is,

$$\text{Absolute P watt of power} \Rightarrow 10\log_{10}\left[\frac{P \times 10^3\, mW}{1mW}\right] \quad dBm$$

$$= [30 + 10\log_{10}(P)] \qquad dBm \qquad (3.4)$$

which implies that P watt of power is $[30 + 10\log_{10}(P)]$ decibels larger than 1 mW.

In telecommunication practice, there are several contexts in which the relative entities are specified in decibels. The following is a summary of decibel reference scales:

❑ *dB (decibel): A relative logarithmic measure of power, voltage, or current ratios*

❑ *dBm: A decibel measure of an absolute power in mW relative to 1 mW*

❑ *dBm(R): A dBm measure of a power in mW across R ohm (R = 50, 75, 600 ohm)*

❑ *dBW: A decibel measure of an absolute power in watts relative to 1 watt reference*

❑ *dBμV: A decibel measure of the voltage in μV relative to 1 μV reference. (Normally wireless receiver RF levels are specified in this unit)*

❑ *dBV: A decibel measure of the voltage in volts relative 1 V reference*

❑ *dB/bit: This refers to the dynamic range of resolution for a PCM. For example, $20\log_{10}(2)$ per bit = 6.02 dB/bit*

❑ *dBc: This is the decibel measure of a signal level with respect to the RF carrier level*

❑ *dB/Hz: The unit to specify relative noise power in 1 Hz bandwidth*

❑ *dBW/K-Hz: A quantity that specifies carrier to thermal noise ratio in decibel scale. For example, the Boltzmann constant $k_B = 1.38 \times 10^{-23}$ W/°K-Hz can be specified as $10\log 10\ (1.38 \times 10^{-23}) = -228$ dBW/K-Hz.*

Telcos closely control the net attenuation (or amplification) of each transmission system in order to maintain the rigid end-to-end power levels of a circuit constituted in practice, by a variety of transmission system components.

For the purpose of administering the net loss of transmission links, the transmission levels at various points in a transmission system are specified in terms of a bench-mark level known as *the reference point*. As per CCITT recommendations, this level is known as *zero-relative level point* (0-RLP), and in the North American standard, it is known as *zero-transmission level point* (0-TLP) as mentioned in Chapter 2. For example, the sending end terminal of a 4-wire switch is specified to be at − 2 dB TLP (in the North American context). It means that this point is at a power level of 2 dB down the hypothetical reference point in the 4-wire circuit. That is, a 0 dBm (or 1 mW) test tone applied at this reference point will indicate a − 2 dBm level at the sending end terminal of the 4-wire switch.

Considering the range of talk levels over a telephone, the output dynamic has a typical high-end of about 0 dBm. The maximum power level can be observed at:

✓   The subscriber end of a local loop
✓   The CO end of a local loop
✓   The CO end of a tandem connecting trunk.

These points, in practice, are recommended to be specified as 0-TLPs. The typical assigned TLPs in analog and digital tandem offices are illustrated in Fig. 3.22.

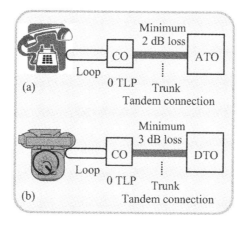

Fig. 3.22: TLPs at the telco offices: (a) Analog system and (b) digital system

Via net loss (VNL) and fixed loss

A power loss plan is deliberately introduced in the telephone transmission system to control the echo [3.3]. That is, in order to combat the annoyance resulting from an echo in the telephone circuits, the toll circuits (Fig. 3.23) are designed with a fixed loss of about 5 dB plus an additional loss proportional to distance. This additional loss is known as via *net loss* (VNL*)*. The 5 dB is split into (2.5 + 2.5) dB on each tandem connecting trunk as illustrated (Fig. 3.23).

The via net loss plan is adopted in the analog network, and the fixed loss plan is specified for digital networks. For desirable performance in digital networks, VNL is avoided and a fixed loss is applied at each tandem connecting trunk as shown.

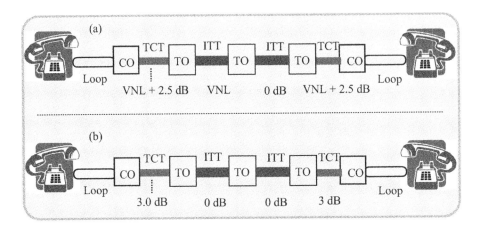

Fig. 3.23: VNL and FL plans: (a) Analog network and (b) digital network

Example 3.4
An analog tandem connecting trunk between two switches implemented on a cable pair is shown in Fig. 3.24 Assume a cable loss as 6 dB. Determine the expected measured loss (EML) due to the insertion.

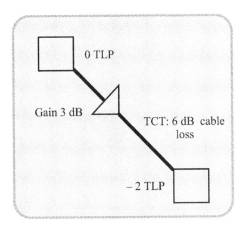

Fig. 3.24: Tandem connected trunk insertion on a cable

Solution

EML = 0 dB (TLP) + 6 dB cable-loss − 3 dB (gain) + 2 dB (TLP)
    = 5 dB

Problem 3.1.

Determine the EML in Fig. 3.25.

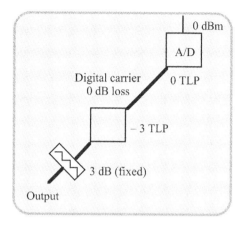

Fig. 3.25: Digital tandem connected trunk

Local loop telephone line loss…

Considering the local loop of a telephone, the associated line loss includes the impedance of in-house wiring plus the impedance of the line from the pedestal (outside the house) to the nearest CO.

Fixed loss loop (FLL) is a factor preset by the local carrier by modifying the existing line loss and adding or subtracting impedance. This is to ensure that the transmitted carrier of the modem reaches the CO at an acceptable level. Normally, the FLL is set at – 9 dB. The transmitter-to-receiver loss is maintained at 16 dB so that for a FLL of – 9 dB, the level at the CO is – 25 dB.

A programmable jack is an alternative to FLL. Here, the modem adjusts its transmit level whenever the phone line impedance changes ensuring a steady level of transmitted voltage level.

Noise power levels in telecommunications

As indicated in Chapter 2, in telephony, 1 pW or – 90 dBm is taken as the standard noise reference. Any noise power level specified against this reference has the unit dBrn. For example, 30 dBrn refers to – 60 dBm or 1 nW of noise power [3.4, 3.7].

The unwanted power (such as noise or interference) in the passband of telephone voice frequency could be subjectively more annoying at certain frequencies. Hence, commonly, the noise is weighted across the speech signal spectral range by a filter to mitigate the annoyance effect. As mentioned in Chapter 2, the North American weighting is known as *C-message weighting* and the corresponding European (CCITT) standard is called *psophometric weighting*.

The dBrn under the C-message weighting is denoted as dBrnC. Similarly, the pW when psophometrically weighted is expressed as pWp. The conversion between the noise level measures described in Chapter 2 is reproduced in Table 3.4 for convenience and quick reference. Also more illustrative examples relevant to the present discussions are presented below.

Table 3.4: <u>Relative noise power units and conversions</u>

Noise level	Spectrum	Conversion	Noise level	Spectrum
x dBm	3 kHz flat	⇔	(x+90) dBrn	3 kHz flat
x dBm	3 kHz flat	⇔	(x+90 – 2) dBrnC	C-message weighted
x dBm	3 kHz flat	⇔	(x+90 – 2.5) dBrnC	Psophometric-weighted
x dBrn	3 kHz flat	⇔	(x – 2) dBrnC	C-message weighted
x dBC	C-message weighted	⇔	(x – 0.5) dBp	Psophometric-weighted
x pW	3 kHz flat	⇔	p Wp	Psophometric-weighted

Problem 3.2

(a) A survey of voice signals in the Bell System indicates the average signal level is –16 dBm0 where "0" indicates that the specification is relative to the 0-TLP. Express this signal as an absolute measurement value at a – 14 dB TLP.

(b) The noise power objective for a 1000-mile analog line is 34 dBrnC0. Express this in dRmC0.

(c) What is the value in dBrn0 of an absolute noise power of 100 pW measured at a – 6 dB TLP?

Example 3.5

A set of TLP values along a cable is indicated in the Fig. 3.26. (i) Determine the signal level to be applied at X that will correspond to TLP levels specified at Y and Z. (ii) What is the loss or gain that a signal will experience from X to Z? (iii) If the absolute noise measured is α dBrnC and the line between Y and Z adds additional noise of N dB, determine the noise level at Z.

Solution

(i) Since point X is at – a dB TLP, a signal level of – a dBm will maintain the TLP values indicated at Y and Z.

(ii) The net loss (or gain) between X and Z is:

$$- a - (- c) = (- a + c) \qquad \text{dB}$$

If c > a, there is a net loss
If c < a, there is a net gain.

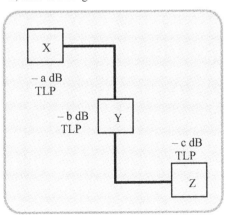

Fig. 3.26: TLP specifications

(iii) Absolute noise measurement at $Y = (\alpha + b)$ dBrnC0

At Z, this value will be: $[(\alpha + b) + N]$ dBrnC0

Absolute noise power measured at Z is:

$$[(\alpha + b) + N - c]$$ dBrnC

Characteristics of telephone cables

The twisted-pair cable used in telephony is classified in terms of the wire gauge used, the sheath material used, the type of protective outer jacketing, and the number of pairs contained within the sheath. Sizes available range from one-or two-pair drop wire to 3,600-pair cable used for central office building entrance. The upper limit of cable size, which depends on wire gauge and the number of pairs, is dictated by the outside diameter of the sheath. Sheath diameter, in turn, is limited by the size that can be pulled through 4-inch conduit. Cables of larger sizes, such as 2,400 and 3,600 pairs, are used primarily for entrance into telephone central offices, which are fed by conduit in urban locations. Normally, wire gauges of 26, 24, 22, and 19 AWG are used in loop plants. Economy considerations dictate the use of the smallest wire gauge possible, consistent with technical requirements. Therefore, the finer gauges are used close to the central office to feed the largest concentrations of users. Coarser gauges are adopted at greater distances from the central office as needed to reduce loop resistance.

Cable sheath materials are predominantly high-durability plastics such as polyethylene and polyvinyl chloride. Cable sheaths guard against damage from lightning, moisture, induction, corrosion, rocks, and rodents. Besides the outer sheath, a layer of metallic tape, which is grounded on each end, shields the cables from induced voltages.

Loop resistance consideration

The loop resistance is essentially decided by the type of cable used. Constructional details of a typical copper cable are shown in Fig. 3.27. Sometimes the entire cable assembly under the outer cable jacket is 100% flooded with a petrolatum-polyethylene gel filling.

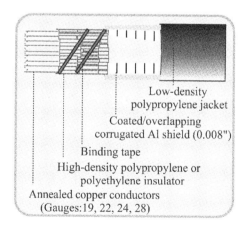

Low-density
polypropylene jacket

Coated/overlapping
corrugated Al shield (0.008")

Binding tape

High-density polypropylene or
polyethylene insulator

Annealed copper conductors
(Gauges:19, 22, 24, 28)

Fig. 3.27: Construction of a typical telephone cable

1. Conductors:	Solid annealed copper, available conductor sizes are 19, 22, 24, and 26 gauge. (*Note*: The larger the gauge size, the smaller the diameter of the conductor)
2. Insulation:	High-density polyethylene or polypropylene
3. Binding tape:	Overlapping high dielectric tape
4. Shield:	0.008" corrugated aluminum tape (coated and overlapping)
5.Outer jacket:	Low density, high molecular weight polyethylene, with sequentially numbered length markers along the outside

The total resistance at the subscriber loop level is controlled by:

- ✓   Battery feed resistance of the switching system (usually 400 ohms)
- ✓   Central office wiring (nominally 10 ohms)
- ✓   Cable pair resistance (variable to achieve the design objective of the central office or PBX)
- ✓   Drop wire resistance (nominally 25 ohms)
- ✓   Station set resistance (nominally 400 ohms).

A further consideration in selecting telephone cable is the capacitance of the pair, expressed in microfarads (μF) per mile. Ordinary subscriber loop cable has a high capacitance of about 0.083 μF per mile. Low-capacitance cable is used for trunks because of its improved frequency response. It has a capacitance of 0.062 μF per mile. Special 25-gauge cable used for T-carrier has a capacitance of 0.039 μF per mile.

In practice, special types of cable are used for cable television, closed-circuit video, local area networks, and other applications. Some types of cable are constructed with internal screens to isolate the transmitting and receiving pairs of a T-carrier system.

Distributed transmission line parameters

A transmission line represents essentially an electrical wireline circuit facilitating flow of electromagnetic energy from the signal source to a load. It is made of two parts. The source-to-load section, which is at a higher potential and the load-to-source current-return path, which is at a lower potential, (normally at a ground potential). The incremental length Δx in Fig. 3.28 (specified as $\Delta x/\lambda$ in terms of the wavelength (λ) of the signal energy transmitted) is comprised of the following: (i) The finite resistance of the wireline $\Delta x \times R$, where R is the resistance per unit length of the line; (ii) The finite inductance of the wireline $\Delta x \times L$, where L is the inductance per unit length of the line; (iii) The finite capacitance of the wireline $\Delta x \times C$, where C is the capacitance per unit length of the line. (This capacitance exists between the lines set at two potential levels, by virtue of the dielectric (insulation) prevailing between the lines); and, (iv) the finite conductance of the wireline $\Delta x \times G$, where G is the conductance per unit length representing the leakage across the insulation due to the lossy nature of the dielectric.

The distributed parameters indicated above are illustrated in Fig. 3.28. These parameters are decided by the geometry and material characteristics of the wire and insulation used. Table 3.5 indicates relevant characteristics of typical wirelines.

A uniform electric transmission-line can be regarded as infinite symmetrical sections of impedance, each of infinitesimal size cascaded in series as an iterated network shown in Fig. 3.28. Each section includes its proportionate share of the distributed inductance, capacitance, resistance, and *leakance* (conductance) per unit or line length.

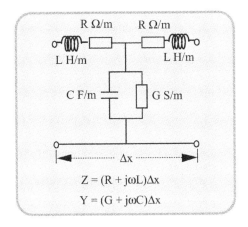

Fig. 3.28: An incremental length of transmission line

Table 3.5: Transmission line parameters

Line	R (ohm/m)	L (H/m)	G (S/m)	C (F/m)	Remarks
Two wire line	$\dfrac{R_S}{\pi a}$ $R_S = \sqrt{\dfrac{\pi f \mu_c}{\sigma_c}}$	$\dfrac{\mu}{\pi} \cosh^{-1}\left(\dfrac{D}{2a}\right)$	$\dfrac{\pi\sigma}{\cosh^{-1}\left(\dfrac{D}{2a}\right)}$	$\dfrac{\pi\varepsilon}{\cosh^{-1}\left(\dfrac{D}{2a}\right)}$	[μ_c: permeability and σ_c: conductivity] of the conductors used in the wirelines
Coaxial line	$\dfrac{R_S}{2\pi}\left(\dfrac{1}{a}+\dfrac{1}{b}\right)$	$\dfrac{\mu}{2\pi} \ln\left(\dfrac{D}{2a}\right)$	$\dfrac{2\pi\sigma}{\ln\left(\dfrac{b}{a}\right)}$	$\dfrac{2\pi\varepsilon}{\ln\left(\dfrac{b}{a}\right)}$	[μ: permeability, σ conductivity and ε permittivity] of the insulation used in the wirelines

$$\text{Characteristic impedance, } Z_o = \sqrt{\frac{R + j\omega L}{G + j\omega C}} \cong \sqrt{L/C} \text{ ohm} \qquad (3.5)$$

(In the above expression R and G are neglected if the line is assumed to be lossless.)

$$\text{Velocity of propagation along the line} = 1/\sqrt{LC} \text{ m/s.} \qquad (3.6)$$

Telephone cables

A summary of the geometrical and constructional features of voice-grade telephone cables, namely, unshielded twisted pair (UTP) and shielded twisted pair (STP) are as follows [3.8, 3.9]:

❑ *Quad: Four non-twisted parallel wires in one cable*
❑ *Silver-satin cable: A flat cable with silver-colored vinyl jacket*
❑ *Key-system cable: Multiconductor telephone cable (not used currently)*
❑ *UTP: Unshielded four pairs of copper wire placed in the same jacket. Each pair has a different degree of twisting to reduce coupling and crosstalk between them*
❑ *STP: Two twisted-pairs of copper wire, in which is each pair shielded individually. It has the advantage of coupling/crosstalk cancellation due to the twist and it is also less prone to EMI as a result of shielding provided.*

Use of copper-pairs for LAN cabling

With the advent of computer communication in the telecommunication arena, the cabling requirements and standards had to be overviewed in order to support the bit rates of data transmission and bandwidth needed thereof. In short, it became evident that "telephone wire ≠ LAN cable".

Specifically, pertinent to local area networking, the cabling standards emerged in terms of two major considerations, namely,

✓ Run length of the cable (in meters or feet)
✓ Data rate to be supported (in bps).

The aforesaid considerations stem from the fact that the performance of an electrical transmission line in supporting a given bit rate relies on the transmission length involved. That is, as indicated in Chapter 2, the cable performance is dictated by (bandwidth × distance) product.

Hence, for various data networking applications, the cabling standards were indicated consistent with the following:

- ✓ Network topology (that is, the interconnection type such as star, ring, bus etc.)
- ✓ Geographical considerations (such as office, business, campus environment etc.)
- ✓ Horizontal and vertical dispositions of computers being interconnected.

Normally, the horizontal cable runs are limited to about 300 ft and for vertical runs, the backbone cabling is installed for distances up to 300 ft with shielded twisted-pair (STP) and up to 2600 ft with unshielded twisted-pair (UTP). An outline on twisted pair cabling adopted in data communication networking is as follows: As stated before, the most commonly used cabling medium in telephone and data communication is the twisted-pair cabling. It consists of two insulated copper wires twisted approximately 20 turns per foot. The common wire gauge is 24. Two pairs are kept enclosed in a jacket.

Specific to data transmission wiring, 22 and 26 AWG wires are used. Again, the two versions of twisted copper-pair of wires are UTP and STP. The types of such twisted-pair cables are commonly categorized as *unshielded twisted-pair Type 3 (UTP-3)* and *Type 5 (UTP-5)*, *unshielded twisted Type 6 (UTP-6)* and the STP. Each of these cable types has associated transmission rate limitations. Twisted-pair is installed between end-systems and a wiring closet, usually located on a floor of a building. Connections to a workgroup are usually made at this point, along with links to the building or campus optical fiber backbone.

The various types of UTP cable available are categorized in the order of increasing quality and are specified as *Cat 3 (UTP-3)*, *Cat 4 (UTP-4)*, and *Cat 5 (UTP-5)*. The *Cat 6* is capable of 300 and 600 MHz response. It is intended to support higher bandwidth connections in comparison to Cat 5's 100 MHz transmission rate. (It should be noted that the transmission rate of the data, 25 Mbps, 100 Mbps, 155 Mbps, and so on, is not the coding rate at which the information is carried over the wire.) An important consideration in the transport of high data rates over twisted-pair is the resulting electromagnetic radiation. For example, in Europe, emission levels should not exceed the values specified within the specification EN55022. The quality of a given cable is dependent upon a number of factors, including impedance, attenuation, near-end crosstalk, and far-end crosstalk. The total quality of the cable can be specified by a single parameter known as the *attenuation crosstalk ratio* (ACR). It depicts a relation between the NEXT and attenuation.

STP is the 150 ohm cable used commonly in *token-ring* networks. It consists of Types 1 and 2, both of which are capable of up to 300 ft. runs. An additional STP, Type 6, is suitable for use as *patch cabling*. In installing UTP and STP, care should be taken not to exceed 270 ft between the patch panel and the wall connector. This will leave 30 ft for connections within the patch and between the wall plate and the user. Installed cable will consist of two-or four-pair 100 or 120 ohm in the case of UTP, or two-pair 150 ohm STP.

Sharing the cable system in data communication networks
In LAN implementation, the cabling does the following:

- Facilitating a physical connection between the DTEs
- Implementing an orderly access to the shared network cable system
- Providing electrical signaling.

Consistent with the dogma "networks are for sharing", the users of LAN share common facilities and resources (such as printers etc). Cabling is the physical medium that enables this sharing via interconnections, and there are standard protocols (developed over years by IEEE, EIA and CCITT) which specify the dictum or rules of agreement among the different parts of the network on how data are to be transferred, once interconnection is physically enabled by the cabling. (Industries develop their network products in conformity with these protocols.) Within the scope of such protocols, a LAN adapter enables each network node an orderly access to the cable to send or receive the data. This operational scheme is known as medium access control (MAC). Commonly, there are three MAC schemes adopted. They are, ARCnet, ethernet, and token-ring. For the purpose of understanding cabling principles vis-à-vis LAN environment, a brief note on the MACs is indicated below [3.10]:

Medium access control (MAC)

Ethernet: In this scheme, the link provides for a data rate of 10 Mbps. The stations can be separated up to 1.75 miles (or 2.8 km). Coaxial cabling is used for interconnecting the stations. The line code used is the Manchester encoded digital baseband.

In 1990, IEEE came up with ethernet specifications for transmissions via twisted-pair wiring. It is known as 802.3i 10BaseT standard. It indicates a rate of transmission of 10Mbps. It uses baseband signaling on twisted-pair wiring with star topology.

The baseband transmission involves no high frequency and the digital data is set by +15 V or −15 V to represent the logical binary states.

The architecture of IEEE 10BaseT consists of a central wiring hub and a separate run of cable to each node. A punchdown block can be used for better accessibility as illustrated in Fig. 3.29.

Permissible cable-segment length for ethernet is 185 meters (606 ft). Node-to-node separation should be at least 2 ft (0.5 meters) and 30 nodes can populate a single cable-segment. With star wiring, using a hub, the maximum length is restricted to 100 meters (328 ft).

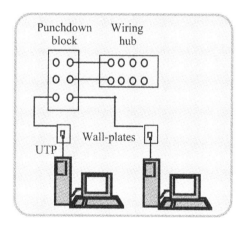

Fig. 3.29: IEEE 10 BaseT – based cabling architecture of DTEs

ARCnet: Attached resource computing architecture network (of Data Point Corporation) is a system which uses a broadcast architecture in which all the stations on the network can receive all the messages broadcast into the cable, approximately at the same time. (In contrast with the token-ring, there is a repeater activity of nodes, to be indicated later.) ARCnet works reliably at 2.5 Mbps speed and is typically compatible for office installations. A newer version, namely ARCnet plus allows even higher rates 20 Mbps. ARCnet uses traditionally RG/62 coaxial cable in star topology. The station-to-station separation is identical to that of the ethernet.

Token-ring: The relevant concept and its cabling scheme was developed by IBM in the early 1980s. It is standardized through IEEE 802.5 specifications. It is a very reliable and robust architecture. Token-ring has "designated to survive" cabling that reaches out to every node from a central wiring hub (called multistation access units, MAU). Though the architecture is star-configured, the data passed from node-to-node around an electrical ring.

There are a few cabling options for token-ring. STP can be used under 4 to 16 Mbps speeds and it is also compatible for high-speeds (100 Mbps). The use of UTP in token-ring refers to IEEE 802.3 10Base T/IBM Type 3 cable. It supports speeds like 4 Mbps.

If the legacy LAN environment is expanded to support high-speed ethernet, the relevant types of cable required vis-à-vis the transmission rates involved are indicated in Table 3.6.

Table 3.6: High-speed copper-based cables

Networking Type	Bandwidth	Media	Cable pairs	Distance
100Base-TX (802.3u)	100 Mbps	Cat 5 UTP	2 pair	
		Type 1 STP	2 pair	
Full duplex (802.3x)	200 Mbps	Cat 5 UTP	2 pair	
100Base-T4 (802.3u)	100 Mbps	Cat 3 UTP	4 pair	
		Cat 4 UTP	4 pair	330 ft
		Cat 5 UTP	4 pair	
100Base-T2 (802.3y)	100 Mbps	Cat 3 UTP	2 pair	
		Cat 4 UTP	2 pair	
		Cat 5 UTP	2 pair	

Note: 100BaseT: Ethernet standard 802.3-transmission on UTP with a maximum cable-section length of 330 ft (100 meters) at a rate of 10 Mbps to 100 Mbps using baseband line coding.

Interfacing...

This refers to how an entity/device such as a telephone, computer etc. is attached on to a transmission (network) facility.

An interface has the following four characteristics:

Mechanical interface refers to the type of mechanical connector (such as 25 pin connect EIA 232E etc) used in actual physical connection between DTE and DCE.

Electrical interface specifies signal levels between DTE and DCE in electrical values. For example

− 3 V with respect to ground: Binary 1
+ 3 V with respect to ground: Binary 0

Functional interface depicts the functional meaning assigned to each interchange circuit with respect to the signal's:

❑ *Data*
❑ *Control*
❑ *Timing*
❑ *Common ground potential.*

Procedural interface corresponds to the considerations on the sequence of events in transmitting data.

On asynchronous or statistical multiplexing...

In a statistical multiplexer (STATMUX) the idle slots are used to make the system more efficient. STATMUX allows the sum of the input speeds to exceed the output speed, because it is assumed that not all inputs are transmitting simultaneously. That is, with a STATMUX, it is possible to have link speeds less than the sum of the (input) terminal speeds.

Statistical multiplexing is based on the stochastical network dynamics wherein it is assumed that not all terminals would constantly send in messages simultaneously. Statistical multiplexers assign a time slot according to user channel activity, and in some cases, may compress the data using appropriate coding (based upon character frequency). Statistical TDM systems are "intelligent" in the sense that they use complex algorithm to allocate time slots to users.

Why twisted-pair?

The two wires in a twisted-pair are identically prone to the interference in a noisy environment (due to their proximity to each other). Therefore, the extent of noise induced as a result of this common mode interference in both wires is likely to be identical. As such, this common mode noise induced on both twisted-pair wires can be rejected by using an appropriate differential amplifier at the receiver. Thus, use of twisted-pair cabling can reduce RFI and/or cross-talk interference.

Characteristics of twisted-pair cables

Typical UTP and STP cables are illustrated in Fig. 3.30.

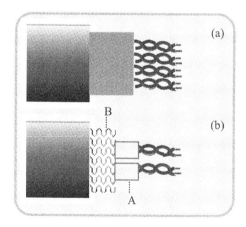

Fig. 3.30: Twisted pair copper cable: (a) 4-wire UTP and (b) 2-wire STP constructions
(A: Aluminum foil; B: Braided shield)

For LAN applications the UTP and STP characteristics are specified in the EIA/TIA 568A standard, the summary of which is presented in Table 3.6.

Table 3.6: UTP and STP cable characteristics (EIA/TIA 568A Standards)

Cable	Gauge (AWG)	Zo (ohm)	Remarks
UTP	22, 24, 26	85-110	▪ Minimum 2 twists/ft. ▪ No limit on number of pairs per cable ▪ Type and applications: 　Level 1: Voice and low speed data communication Level 2: Level 1 applications plus data communication up to 4 Mbps Category 3: LAN applications up to 10 Mbps plus Level 2 applications Category 4: Category 3 applications plus LAN applications up to 20 Mbps Category 5: High-speed LAN applications up to 100 Mbps
STP	22, 24, 26	150	
Token-ring UTP and STP	22, 24	150	IBM cabling system Types and applications: Type 1: Two pair/22AWG/STP/data grade support rates up to : 16 Mbps Type 2: Two pair/22AWG/STP/data grade Four pair/26AWG/voice grade Type 3: Four pair/22 or 24 AWG/category 3, 4 or 5 UTP supports rates up to 4 Mbps token-ring networks

Coaxial cables

A coaxial cable is made of a solid copper wire surrounded by a dielectric (such as PVC or Teflon[TM]). A braided conductor (copper) is woven on to the dielectric constituting an outer conducting shield. Normally, a protective covering made of protective vinyl material surrounds the coaxial structure as a sleeve.

Coaxial cable is used for *broadband* and *baseband* communications networks (and for cable television). Transmission through coaxial lines is free from external interference due to the inherent shielding offered by the outer conductor of the coaxial structure (see Fig. 3.31).

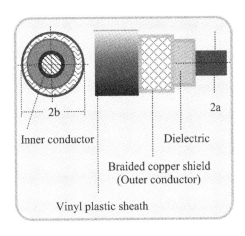

Fig. 3.31: A coaxial line

Coaxial lines offers a broader bandwidth, higher transmission rate, better noise immunity, and rugged operation superior to twisted-pair. But they are relatively bulky. A frequency response up to 500 MHz can be realized with coaxial lines. The various types of coaxial lines and their applications are summarized in Table 3.8.

Table 3.8: Coaxial lines

Coaxial line type	Application	Zo (ohms)	Attenuation dB/ft	Central conductor (AWG)	Outside diameter (in)
RG 58/A-AU	• IEEE 802.3 Thin-net	53	1.250	20	0.195
	10Base 2	50	1.000	20	0.195
RG 58 Foam	• Node-to-node daisy-chain wiring				
RG 8/A-AU	• IEEE 802.3 Thick-net 10Base 5	52	0.585	12	0.405
RG 8 Foam		50	0.572	12	0.405
RG 58/A-AU	• Cable TV wiring	73	1.130	20	0.242
RG 58 Foam		75	0.875	20	0.242
RG 62	• ARCnet	93	-	-	0.180
	• IBM mainframe installations				

Note: 10 Base 2: Transfer rate of 10 Mbps and maximum cable-segment of 10.5 ft (185 m)
10 Base 5: Transfer rate of 10 Mbps and maximum cable-segment of 1650 ft (500 m)

Baseband and broadband transmissions

The application of twisted-pair and coaxial cables are relevant to two types of transmission: The first type refers to the *baseband LAN transmission* in which transfer of digital data refers to d.c. to 100 MHz. Ethernet and token-ring protocols are typically baseband LANs in which appropriate line codes are used for baseband transmission.

In *broadband LANs*, frequency division multiplexing is adopted to segregate the available channel bandwidth into multiple channels for analog transmissions. Cable TV refers to broadband transmission. The splitting of the available channel bandwidth of a coaxial line for various subchannel transmissions is illustrated in Fig. 3.32.

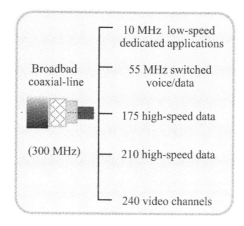

Fig. 3.32: Broadband coaxial line transmissions

Optical fibers

Fiber optic cable supports the transmission of information in the form of light [3.11 3.14]. An optical fiber consists of a core material surrounded by a cladding sheath. The refractive index of the clad is lower than that of the core so that, under appropriate conditions, the light energy launched into the core is retained within the core itself and travels along the fiber length. The mechanism that retains the light within the core is the *total internal reflection* facilitated by the condition on the refractive indices chosen (Fig. 3.33).

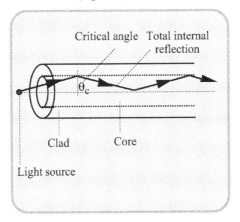

Fig. 3.33: Transmission of light along an optical fiber

The fiber optical transmission has enormous bandwidth capability (ranging into GHz). The attenuation can be as low as 0.03/kft. Further, the line is totally free from EMI. The transmission is completely secured. The fiber is light-weight and has size compatibility for gross telecommunication channels. Standard sizes for the core are 62.5 and 100 microns. The cladding is about 125 or 140 microns thick.

The optical fibers can be classified in terms of the refractive index profile across its cross-section as follows:

- Step-index fiber

- Graded-index fiber.

These are illustrated in Fig. 3.34

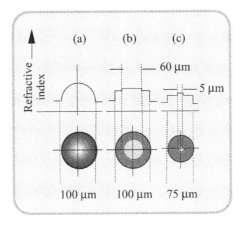

Fig. 3.34: Step and graded index fibers: (a) Graded index fiber, (b) multi-mode stepped index fiber, and (c) single-mode stepped index fiber

The fibers can also be grouped as *single mode* and *multimode fibers*. In the single mode transmission, the electromagnetic wave representing light energy is confined close to the center axis of the fiber throughout the length of transmission. Therefore, the emergent beam is seen as a single-spot centered at the axis. In terms of ray theory, the light essentially traverses the fiber length as a single ray. In multimode transmission, there could be several paths in which the light ray may travel. As a result the emergent beam can be seen as multiple spots around the axis. The refractive index profile together with the core-clad dimensions decide whether a fiber is of single-mode or multi-mode type.

Telephone cable versus other wire-line media

The performance of a telephone cable, in terms of its attenuation characteristics versus frequency (relative to other wire-line media) is presented in Fig. 3.35. The implications of the relevant performance profiles of various media will be addressed in the subsequent chapters while discussing the appropriate traffics the media would support in practice.

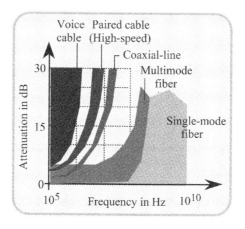

Fig. 3.35: Attenuation versus frequency for different media

3.3.3   Switching

As mentioned before, this is primarily the task of establishing (or releasing) a known type of connectivity between any end-entities by "bridging" them together (or snapping the bridge) across the network cloud. In the POTS, switching simply implies making or breaking an electric loop (or a "circuit") between the calling and the called subscribers [3.6]. In such circuit-switching, once a connection is established, the electric loop remains dedicated between the users. Therefore, conceptually it is a *message-switching* technique. That is, until the entire message (conversation) is transferred between the subscribers, the "line" or the "circuit" remains "hogged" and therefore, no other connection can use the resources tied up in this circuit.

In reference to the digital means of information transfer, the concept of switching took a new turn with the emergence of so-called *packet-switching*. In this technique, the digital bits of message originating from a source are chopped into segmented parts (or *packets*); by appropriately routing them, all the packets addressed to the same destination are reassembled back when received into the original message format.

In the following subsection, the relative performance consideration vis-à-vis circuit-switching and packet-switching are elaborated.

Circuit-switching

This technology establishes a dedicated path between any pair or group of users attempting to communicate, though originated in POTS, and is still applied in *narrow-band integrated services digital network* (N-ISDN). The associated considerations are:

- It is based on TDM (time-division multiplexing) principle in transferring the information from one end to another

- This technique is also synonymously referred to as *synchronous transfer mode* (STM)

- Though originally conceived in analog telephony, it has been adopted in the digital telecommunication systems as well

- In digital form, the bit transfer is done with a certain repetition frequency. For example,

  ✓ 8 bits in every 125 μs for 64 kbps channel rate of transmission
  ✓ 1000 bits every 125 μs for 8 Mbps channel rate of transmission.

The basic unit that represents a repetition along the time-scale is called a *"time-slot"* and the following can be attributed to the time-slot unit:

  ✓ Several time-slots are joined together or *"multiplexed"* for transmission over a link
  ✓ The association of a set of time-slots constitutes a *"frame"*, which can again be repeated with a certain frequency
  ✓ When a *"connection"* or *"circuit"* is established between the end-entities, the same time-slot in the frame is used by this connection over the complete duration of the session. In other words, each connection gets a dedicated time-slot during a session.

Thus, time-division multiplexing (TDM), also known as *synchronous time-division multiplexing* (STM), transmits specific patterns of digital signals (belonging to different channels), over one link allocating each channel a portion of the available time. Fig. 3.36 shows, for example, two channels carrying digital signals into a TDM multiplexer (MUX). Using its clock, the MUX transmits the data from the first input over every odd time-slot and transmits the data from the second input over every even time-slot. Hence, data from both inputs are transmitted on one link (as available at the MUX output). In this scheme, there are, however, idle time-slots which would be wasted.

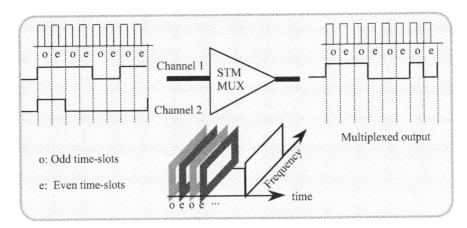

Fig. 3.36: Synchronous time-division multiplexing (*See PLATE 5*)

Example 3.6

Suppose a time-slot accommodates a specified number of bits (say, 8). The set of these bits is referred to as a "*word*" and the cardinality of this set is called the "*word size*". The frame-size is then defined as: The number of channels multiplexed × word size. Construct the frame by interleaving the words.

Solution

Each frame is constituted by word interleaving as illustrated in Fig. 3.37.

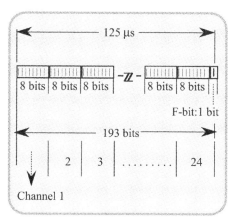

Fig. 3.37: Construction of a frame via word interleaving

Multiplexing

Relevant application of the multiplexing scheme can be seen in the so-called T-1 telephone voice communication system. The basis of TDM hierarchy is DS-1 (digital signal level 1) transmission format (Fig. 3.38), which multiplexes 24 channels of basic voice channels (DS-0) as illustrated.

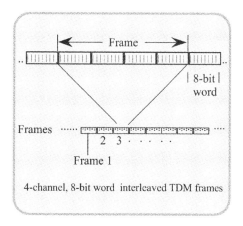

Fig. 3.38: T-1 framing

Each frame contains eight bits per channel plus a framing bit leading to (24 × 8) + 1 = 193 bits. Specific to voice transmission, it refers to the following: Each voice channel contains one word of digitized data. The original analog voice signal is digitized using pulse-code modulation (PCM) at a rate of 8000 samples per second. Therefore, each channel slot and hence each frame must repeat

8000 times per second. With a frame length of 193 bits, a data rate of $8000 \times 193 = 1.544$ Mbps is realized. For five of every six frames, 8-bit PCM samples are used. For every sixth frame, each channel contains a 7-bit PCM word plus a signaling bit. These eight bits constitute a stream (for each voice channel) that contains network control and routing information. (The control signals are used for call-setup or disconnect purposes as indicated before.)

The DS-1 format (developed originally for voice) is also used in digital data service. The rate of transmission is again 1.544 Mbps. Lower data rates are also supported on the DS-1 transmission by using a technique known as *subrate multiplexing*. Details on related multiplexing methods are furnished later.

When a frame is formed by means of *bit interleaving* or *word interleaving*, it is necessary to have a mechanism to synchronize the frame at the transmitting and receiving ends. That is, the start of the frame should be unambiguously identified at the receiving end. *Frame synchronization* can be done in a number of ways. In all the relevant methods, there are one or more framing bits with an identifiable data sequence. Frame bits may be added additionally to data bits. Alternatively, some of the data bits can be used as a frame bit set. For example, as indicated above in the Bell System T-1 channel frame, an additional bit is introduced for every frame, which alternates in value. That is, the T-1 channel structure multiplexes 24 channels and hence the frame has a length of 193 bits ($24 \times 8 + 1$). In one of the earlier TDM channel structures, the British Post Office used 7-bit PCM samples with one bit per octet for framing and signaling. Thus the data rate for the 24-channel TDM worked out as $24 \times 64 \times 10^3$ ($= 1.536$) Mbps including framing plus signaling bits.

Example 3.7

Higher level multiplexing: This refers to each higher level in the hierarchy being formed by multiplexing signals from the next lower level. Suppose the basic data rate is 1.544 Mbps and there are four sources interleaved 12 bits at a time. Determine the transmission rate.

Solution

The transmission rate: $1.544 \times 4 = 6.176$ Mbps

Example 3.8

A receiver collects 20 frames and checks for the occurrence of false framing. Suppose the tolerable extent of false framing refers to a probability of false framing equal to 0.00125. Determine the maximum number of frames, which fail out of 20 scanned by the receiver.

Solution

Assume the occurrence of false framing is due to equally-likely binary errors. Let the maximum number of false frames (out of 20) be X

$\therefore$ Probability of false framing = Probability of more than $(20 - X)$ agreements out of 20

$= 0.00125$

$$\equiv \binom{20}{20}\left(\frac{1}{2}\right)^{20} + \binom{20}{19}\left(\frac{1}{2}\right)^{20} + \binom{20}{18}\left(\frac{1}{2}\right)^{20} + \ldots \left(\frac{20}{20-(X+1)}\right)^{20}$$

Solving for, X

$$X = 16$$

At the receiver end, *frame synchronization* is achieved by searching for the frame synchronization pattern. For example, in the case of T-1 channels, an alternating bit pattern that occurs after every 193 bits is searched for. Framing is established by examining bit after bit successively until a sufficiently long framing pattern is detected. In such efforts, average *frame acquisition time* F_t is an important factor in designing frame synchronization schemes. Now, F_t for a scheme, which uses one framing bit per frame with alternating ones and zeros, can be deduced. In general, F_t has two components, F_{tA} and F_{tB}, defined as follows:

$F_{tA} \triangleq$ Average time required to scan the required number of bits before a framing bit is observed.

$F_{tB} \triangleq$ Average time required to test and ensure that the chosen bit is a frame bit

and,

$$F_t = F_{tA} \times F_{tB}. \tag{3.7}$$

Eqn.(3.7) implies that the bits are chosen one by one and tested one after another sequentially. (It is also possible to scan the bits in parallel. However, the present approach is confined to a sequential search only.) Suppose

p = the probability of finding a 1 in a bit position.
q = the probability of finding a 0 in a bit position.

Then, it follows that

$$p = (1 - q). \tag{3.8}$$

Assuming the first bit observed is 1, then the probability of a mismatch that occurs at the end of one frame is p, at the end of two frames is $q.q = q^2$, at the end of three frames is $q.p.p = qp^2$ and so on. Hence, the average time required to ensure that a chosen bit is not an information bit is given by the following in terms of frame times:

$$F_{tA} = (p + 2q^2 + 3qp^2 + 4pq^3 + 5p^3q^4 + 6p^2q^4) \times (\text{frame duration}). \tag{3.9}$$

With a simplified assumption that $p = q$, Eqn. (3.9) becomes

$$\begin{aligned} F_{tB} &= (p + 2p^2 + 3p^3 + 4p^4 + \ldots) \times (\text{frame duration}) \\ &= [p/(1-p)^2] \times (\text{frame duration}). \end{aligned} \tag{3.10}$$

Taking $p = 1/2$, $F_{tA} = 2 \times (\text{frame duration}) = 2 N$ bit times, where N is the number of bits per frame duration. If the search is started randomly, the average number of bits that must be tested before the framing bit is encountered is equal to N/2; that is, $F_{tB} = (N/2)$. Therefore, $F_t = (2N)(N/2) = N^2$ bit durations.

When extra bits are added to the data stream, they may interfere with the periodic sampling process by introducing aperiodicity, which in turn may result in sample time *jitter*. But the use of data bits for framing will not interfere with the periodic transmission rate. However, it will amount to trading off with data quality. More recent techniques use dedicated channels for framing which affect neither the periodicity of sampling nor the data quality. Such techniques are known as *added-channel framing*. Here, framing digits are added in a group such that an extra channel is formed. The scheme allows considerable performance and flexibility to the framing process. An example of added-channel framing is the CCITT multiplexing standard, which recommends 32 channels per frame with one channel carrying framing information, another channel for signaling information, and the remaining 30 channels supporting the data. The average frame acquisition time F_t for multibit frame code grouped as separate channel can be shown to be:

$$F_t = N^2/2(2^L-1) + N/2 \qquad \text{bits} \tag{3.11}$$

where N is the of the frame duration, and L is the length of the frame code.

(Again Eqn.(3.11) assumes that 1s and 0s occur with equal probability in the frame.)

3.3.4   *T-1 carrier system*

A T-1 carrier system refers to the North American system of multiplexing voice channels and transmission via twisted-pair copper wires [3.4, 3.5, 3.7, 3.15 - 3.17]. It carries 24 voice channels, which are encoded using the PCM technique. Some of the voice channels which are digitized, are pre-empted to send signaling bits. Hence, the 64 kbps used for one voice channel contains not only the voice bits but also the signaling bits. (Signaling is the control information that sends off-book conditions, dial tones, dialing digits, etc as mentioned before.)

The (24 channels × 64 kbps rate) gives a value equal to 1.536 Mbps. However, for synchronization purposes (namely, *framing*), another 8 kbps is added to give the total rate of 1.544 kbps as elaborated in the example shown below. The term "T-1" is typically meant for using twisted wire pairs to send this signal. However, this signal in general is called a DS-1 (digital signal level 1), which can be sent over any medium.

Example 3.9
Construct the digital signal at the first level (DS-1) in the U.S. digital hierarchy

Solution
Number of channels: 24
Number of samples of analog voice signal: (2 × 4000) samples/s
(Nyquist rate sampling)
Number of bits/sample: 8
∴ Bit rate = (8000 samples/s) × (8 bits/sample)
        = 64,000 bps
        ≅ DS-0 (Zeroth level digital signal)
DS-1 rate
        = (24 channels) × (64 kbps)
        = 1536 kbps
This corresponds to: (24 channels) × (8 bits/word)
                = 192 bits/frame
∴ Total bit/frame = 192 + 1 frame-bit (F-bit) =193

Transmission rate
including the F-bit: 193 bits/frame
        × 8000 frames/s
        = 1,544 kbps
The difference (1,544 – 1,536) kbps = 8 kbps results from the addition of framing bit.

Example 3.10
Calculate the duration of a frame bit and the duration of the 8 bit code word in DS hierarchy.

Solution
A frame in DS hierarchy corresponds to (1s/8000 samples), that is, 125 μs. The bit duration is, therefore, equal to : 125 μs/193 = 648 ns. Each pulse-amplitude sample of the voice is encoded into an 8-bit code. Therefore, duration of a code word is: 648 ns × 8 = 5.148 μs.

DS-1 frame
    The DS-1 frame consists of 24 8-bit binary code words as illustrated in Fig. 3.39. At the end of the 24th channel, a framing bit (*F-bit*) is attached to the frame. This bit is the 193rd bit of the frame as indicated above. This frame bit enables synchronization as well as other operations and maintenance (O & M) requirements.

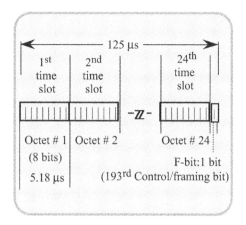

Fig. 3.39: DS-1 framing format

What is the line code scheme adopted in DS-1 transmissions?
The DS transmission employs the AMI line code as illustrated in Fig. 3.40.

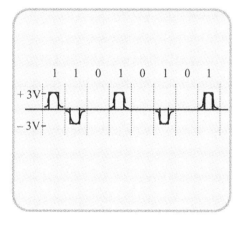

Fig. 3.40: An example of AMI encoded 8 bit DS word constituted by bipolar, raised cosine pulses

The AMI encoded pulses sent on the cable have a 50% duty cycle. That is, the width of the pulsed is one-half the time-slot allotted for each pulse (642/2 = 324 ns). The advantages of AMI encoding are as follows: (i) Most part of bipolar signal energy is concentrated at one-half of the pulse repetition frequency. Hence, if all the bits in a word are 1s, the maximum frequency of DS1 signals is 1.544/2 = 0.772 MHz. (ii) The reduced signal energy leads to a reduced cross-talk coupling. (iii) The bipolar transmission avoids the d.c. component, thereby permitting simple transformer coupling at field regeneration equipment. Lastly, (iv) the unique alternating pulse pattern enables error detection inasmuch as errors lead to violating an ongoing repetitive pattern. That is, AMI facilitates bipolar violation-based error detection as discussed in Chapter 2.

Asynchronous digital hierarchy
Multiplexing of basic DS0/DS-1 systems leads to higher levels of digital hierarchy. There are five levels of multiplexing in the North American asynchronous digital hierarchy. Two DS-1 signals can be combined to form a 3.152 Mbps signal containing 48 voice frequency (VF) channels. The multiplexer (MUX) which does this operation is known as *M1C MUX*. (1 refers to "first-level in" and C refers to "combined level out". The signal output at M1C MUX is called *DS1C*.

A combination of four DS-1s produces a 6.312 Mbps stream called digital signal at the second level, DS-2. It supports 96 VF channels. The MUX, which performs this combining of four DS-1s is called M12 (meaning first levels in and second level out).

Digital signal cross-connects (DSXs)

DSX represents the set of equipment frames with relevant jack panels, which serve as channel bank and MUX cross-connect interfaces at the telco office. The frames are designated as: DSX-0, DSX-1, DSX-1C, DSX-3, and DSX-4 for each of the DS rates. Each frame connects equipment, which operates at the respective DS-n rate. Fig. 3.41 illustrates the digital hierarchy.

Note: Channel bank is a hardware that combines analog VF signals into a TDM (or FDM) signal. In the TDM system, the channel bank performs quantization, coding, μ-law compression, and multiplexing. At the receiving end, the channel bank performs demultiplexing, decoding, μ-law expansion, and reconstruction of the voice signal.

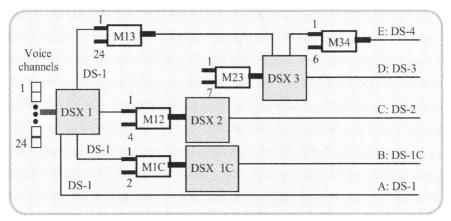

Fig. 3.41: The digital hierarchy: Development of higher levels of DS by multiplexing
[A: DS-1 – 1.544 Mbps (24 Voice frequency channels); B: DS-1C – 3.152 Mbps (48 voice frequency channels); C: DS-2 – 6.312 Mbps (96 voice frequency channels); D: DS-3 – 44.736 Mbps 672 voice frequency channels); E: DS-4 – 274.176 Mbps (4032 voice frequency channels)]

Equipment facilities of T-1 system

The Fig. 3.42 is illustrated the general equipment facilitation of the T-1 system across the network from the customer premises to the CO.

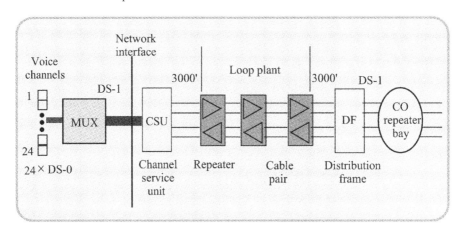

Fig. 3.42: Equipment facility of the T-1 system
Note: CSU: This provides impedance matching and loopback (for testing). It generates bipolar/AMI line coded signals. It also supports any special framing required.

Superframe (SF) and extended superframe (ESF) formats for "voice only" digital transmissions

In *voice only* transmission, the PCM enables required densities of 1s during talk-spurts, and during idle (silence) periods, the transmitter sends, for example, 00000010 (with a forced 1 bit in the 7th position). An unassigned channel is filled by the octet set, 11111111.

A sequence of frames (each of 193 bits) is used to construct a superframe (SF) format. It is called the D4 format, and it contains $12 \times DS-1$ frames with 2,316 bits.

Problem 3.3

Show that the superframe (D4 format) takes 1.5 ms when sent at DS-1 rate.

Shown in Fig. 3.43 is the structure of (12) 193 bit DS-1 frames constituting the D4 layout.

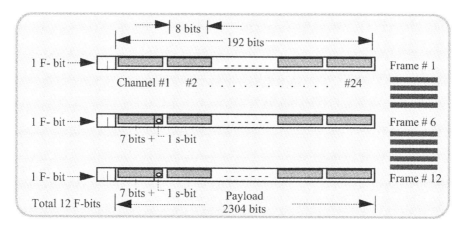

Fig. 3.43: D4 format (SF)

Note: s-bit is the voice channel signaling bit.

Extended superframe (ESF) layout

A set of 24 DS-1 frames constitutes an ESF, which contains 4,632 bits. It has 24 framing bits for the following purposes:

✓   Synchronization of terminal equipment (F-bits)
✓   Securing together multiframe alignment (F-bits)
✓   Providing a facility data link (FDL) at 4000 bps to carry application-dependent information
✓   Error checking (C-bits).

The construction of an ESF is illustrated in Fig. 3.44.

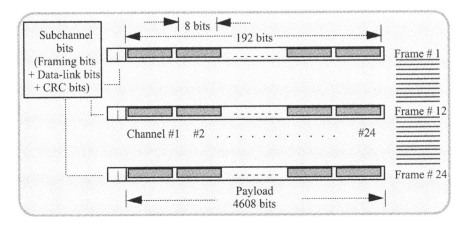

Fig. 3.44: Construction of an ESF

Problem 3.4
Determine the time line required to transmit an ESF at the DS-1 rate.

"Data only" transmissions

DDS offers a 56 kbps transport service for private lines. In reference to the 193 bit frame, 24 bits are used for signaling purposes and 1 bit is used for framing. Therefore, there are $193 - (24 + 1) = 168$ bits available for data transmission. It corresponds to a data rate of $(168 \times 1.544)/193 = 1.344$ Mbps. Since there are 24 channels, the data bit rate per channel is equal to 1.344/24 Mbps (= 56 kbps).

In essence 23 channels are used for data transmission. The 24th channel is reserved as a special sync byte, which allows faster and more reliable reframing in the event of a framing error. Within each channel, 7 bits per frame are used for data and the 8th bit indicates whether the channel for that frame has adopted data transmission. The data SF, which accomplishes this is shown in Fig. 3.45.

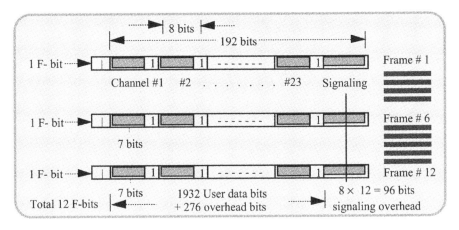

Fig. 3.45: Data SF structure

In the data only SF format (Fig. 3.45), the following characteristics can be observed:

- ✓ Per channel robbed bit signaling is eliminated and common channel signaling is used
- ✓ The 8th bit in all the 23 channels is always 1 meeting 1's density requirement
- ✓ The 24th time-slot provides for data synchronization, alarming of a failure in the outgoing direction, and surveillance of data transmission.

1's density requirement...

As indicated earlier, DS-1 employs AMI wherein 1s are signified by alternating positive and negative raised-cosine pulses, and 0s are represented by a zero level. Suppose there is a stretch of 0s. It would cause the repeaters to lose synchronization. Therefore, to maintain reliable working status, the signal stream of bits is designed to contain at least 1 in every 8 bits. The 12.5% 1s are maintained and no more than 15 consecutive 0s are allowed. This is called 1's density provision.

Example 3.11

Suppose the BER in a T-1 transmission is 10^{-8}. Data is sent on this line in blocks of 1000 binary words. Calculate (i) the probability of block error, and (ii) determine the average requests for retransmission of the blocks. Assume each binary word is 1 byte.

Solution

(i) Error occurrence of error in each bit refers to a statistically independent process. Therefore, the correct transmission probability (q) of 1000 words × 8 bits is given by:

$$q = (1 - BER) \times (1 - BER) \times \ldots 8000 \text{ times} = (1 - BER)^{8000}$$

$$\therefore \ p = \text{probability of block error} = (1 - q) = 1 - (1 - 10^{-8})^{8000} \cong 10^{-4}$$

(ii) Rate of retransmission request = 1 in 10^4 blocks transmitted.

Table 3.8: Multiplexers for digital hierarchy

MUX Type	Customer access	Signaling	Framing bit rate (bps)	Output
M 24	24 Voice channels (original T-1)	By robbing 8th bit in every 6th frame	8000	DS-1
M44	44 Voice channels $\equiv 44 \times 32$ kbps voice channels $\equiv 22 \times$ DS-0 channels + 2 × DS-0 (used for signaling)	2 DS-0 for common channel signaling	8000	DS-1
M48	48 Voice channels $\equiv 48 \times 32$ kbps voice channels $\equiv 24 \times$ DS-0 channels	Per channel signaling uses bit robbing	8000	DS-1

Continued …

M88	88 Voice channels $\equiv 88 \times 16$ kbps voice channels $\equiv 22 \times$ DS-0 channels $+ 2 \times$ DS-0 (used for signaling)	2 DS-0 for common channel signaling	8000	DS-1
M96	96 Voice channels $\equiv 96 \times 16$ kbps voice channels $\equiv 24 \times$ DS-0 channels	Per channel signaling uses bit robbing	8000	DS-1
M24/64	24×64 kbps data channels	1×64 kbps channel for data sync, alarms, and signaling	8000	DS-1 Clear channel capability (CCC) or clear 64 (customer uses the 64 kbps channel) on ESF format
M24/56	23×64 kbps data channels			DS-1 (See Fig. 3.45 on SF format)

Subrate data

The original T-1 system developed for voice communication was later conceived to include multiplexed digital transmissions of the following types:

- ✓ 64, 32, 16 kbps voice on DS-1 channel
- ✓ 54 kbps, $n \times 64$ kbps ($n = 1, 2, \ldots$) data on DS-1 channels
- ✓ 2.4, 4.8, 9.6, or 19.2 kbps sub-rate data on DS-0 channels.

The MUXs adopted in realizing the subrate data are indicated in Table 3.8.

The subrate multiplexing involves "robbing" an additional bit from each channel to indicate that subrate multiplexing is provided. This leaves a total capacity of $6 \times 8000 = 48$ kbps. This capacity is used to accommodate (5×9.6) kbps, (10×4.8) kbps, or (20×2.4) kbps channels.

For example, if channel 2 is used to provide 9.6 kbps service, then up to five data subchannels can share channel 2. The data for each subchannel will appear as 6 bits in channel 2 in every 5th frame.

Now considering the subrate data at 2.4, the set of 4.8, 9.6, and 19.2 kbps rates are multiplexed with other DS-0 channels on a DS-1 line. Subrate multiplexers are as follows:

- ✓ M-5: Customer access $\Rightarrow 5 \times 9.6$ kbps data channels; output $\Rightarrow$ DS-0B
- ✓ M-10: Customer access $\Rightarrow 10 \times 4.8$ kbps data channels; output $\Rightarrow$ DS-0B
- ✓ M-20: Customer access $\Rightarrow 20 \times 2.4$ kbps data channels; output $\Rightarrow$ DS-0B.

Subrate MUX is combined with n24/56 to produce DS-1 signals. For example, (23 M5 + M24/56).

Some definitions…

- ❑ *DS-0A: A DS-0 signal containing data from a single subrate station*
- ❑ *DS-0B: A DS-0 signal containing data from several subrate stations.*

Data and voice transmission

The DS-1 format can be used to carry a mixture of voice and data channels. In this case, all 24 channels are utilized and no sync byte is facilitated.

North American versus other digital hierarchies

In Table 3.9, the digital hierarchy adopted around the world are presented.

Table 3.9: Digital hierarchies around the world

Used in Australia, Canada, Japan, and United States				Used in CCITT countries and Europe			
Signal Level	Carrier system	Rate in Mbps	Number of channels	Signal level	Carrier system	Rate in Mbps	Number of channels
DS-O	-	000.064	1	CEPT0	-	000.-64	1
DS-1	T-1	001.544	24	CEPT1	E-1	002.048	30
DS-1C	T-1C	003.152	48	CEPT2	E-2	008.448	120
DS-2	T-2	006.312	96	CEPT3	E-3	034.368	480
DS-3	T-3	044.736*	672	CEPT4	E-4	139.264	1920
DS-4	T-4	274.176	4032	CEPT5	E-5	565.148	7680

* 32.064 Mbps in Japan

Channel multiplexing in the 24-channel North American frame (32-channel CCITT frame), is only the first level of multiplexing. Groups of channels can also be multiplexed so that high quality digital links can be constituted over long distances to carry data at much higher rates. Accordingly, a hierarchy of digital multiplexing is used by various agencies for carrying digital information. Different standards are indicated in Table 3.9. Each level of multiplexing requires additional digits to indicate framing, control, and signaling information. As a result, the data rate increases by a factor that includes additional added bits. The rates are chosen to be multiples of 64 kbps (DS-0), so that the standard octet structure (originally formed for voice transmission) is maintained throughout the hierarchy.

Basic voice service became highly automated since the 1960s. Any phone dialing for a connection to reach any other phone within seconds has become a reality. However, the early data connections had to be leased or dedicated lines, which had to be ordered weeks or months in advance.

In the past, PSTNs (designed for voice) relegated data to a secondary position, on separate overlay networks. For the most part, manual labor was required to install, change, or remove a data circuit. Large distribution frames contained thousands of line terminations at the central offices. For some types of high-speed circuits, only certain types of cables and certain wires within those cables could be used, leading to additional delays as workers searched for and set up suitable connection paths. Recordkeeping for these circuits was intense, and not always computerized. Data connections typically stayed in place far longer than a voice call, but the pace of data service expansion would have exceeded carrier capacity to provision and maintained the growing number of data lines using manual methods.

The steps which were undertaken initially to automate data connections were made in several ways:

✓ The first digital connection on demand refers to the so-called switched 56 service. Data transmission was treated as a separate network, though parts of it were carried on the transmission equipment and switches of the PSTN. However, carriers for a long time considered it a premium service and priced it higher than the market would accept

✓ Sprint was the first to extend the data call service in par with a voice call. Users with digital access to Sprint's network could dial up a connection and use it for voice or data at the same cost. However, the need for T-1 accesses at both ends severely hampered potential subscribers

✓ AT&T was first with T-1 switching for data transmission applications. AT&T opened its digital cross-connect network to some user control in the 1980s. The limitation was that only the customer's pre-subscribed access lines were eligible to be the end-points on these connections

✓ MCI offered a broader service, allowing any user of the network to connect its own access points or to any other user on the network. Switched T-1 was fairly expensive and not economical if used many hours per week.

In summary, the T-1 carrier system is characterized by the following:

✓ It caries the voice signal in PCM encoded form in each channel bank and merged with 23 other such voice channels
✓ Each channel supports a bit rate of 64 kbps (8000 samples/s × bits/sample)
✓ The 24 channels produce a frame format as illustrated in Fig. 3.38
✓ A single framing bit is added as the 193rd bit $(24 \times 8 + 1)$
✓ Each frame of duration 125 μs is repeated 8000 times/s for a total line rate of 1.544 Mbps.

Note : European digital carrier systems use the same 64 kbps, but multiplexed to 32 rather than 24 channels for 2.048 Mbps as indicated in Table 3.9.

Problem 3.5

Justify the rates of transmission presented in Table 3.9 for the various digital levels shown.

Plesiochronous multiplex

The European standard for PCM/TDM digital telephony multiplexing adopted internationally (except in North America, Japan, and Australia) is shown in Fig. 3.46. It has a basic access of 144 kbps and the multiplexing hierarchy yields (32 channels × 64 kbps) per channel equal to 2.048 Mbps. This TDM multiplex can be readily transmitted over 2 km sections of twisted pair cables (originally laid for analog voice transmissions).

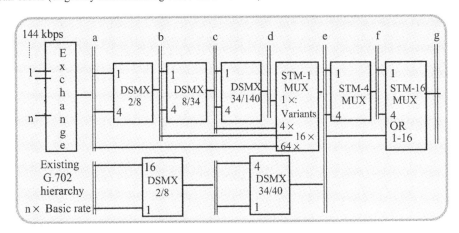

Fig. 3.46: European TDM hierarchy with SDH at upper levels (DSMX: Digital system multiplex)
Hierarchy levels and their bit rates: a: 2.048 Mbps, b: 8.448 Mbps, c: 34.368 Mbps, d: 139.264 Mbps, e: 155.520 Mbps, f: 622.080 Mbps, g: 2488.320 Mbps (New hierarchy G.707)
DSMX: Digital system multiplexer
STM: Synchronous transport module.

The ITU-T hierarchy provides for higher levels of multiplexing above 2.048 Mbps, combining four signals in the DSMX 2/8 and 8/34 (Fig. 3.46) so as to constitute the signal at higher multiplexing levels. At each level the bit rate increases by slightly more than a factor 4 since extra bits are added to provide for frame alignment and to facilitate satisfactory demultiplexing. The upper levels beyond 140 Mbps, from the synchronous digital hierarchy (SDH) will be described later.

The 2.048 Mbps multiplex level is also known as the *primary multiplex group*. The relevant frame is made of 32×8 bit slot-times. Time-slot zero is reserved for frame alignment and service bits. Time-slot is used for multiframe alignment, service bit, and signaling. The remaining 30 channels are used for information carrying, or payload capacity, each channel containing one 8-bit voice signal sample. The frame structure of PCM primary multiplex is illustrated in Fig. 3.47.

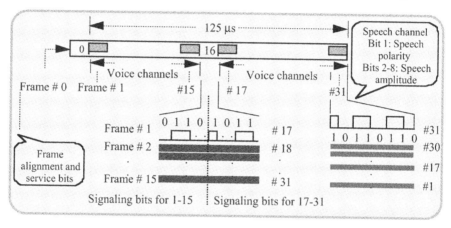

Fig. 3.47: The 2.048 Mbps multiplex level frame structure (PCM primary multiplex group)

The system for assembling the TDM telephony data stream assumes that the digital signal multiplexers (DSMX) in Fig. 3.47 are located at physically separated sites which implies separate free-running clocks at each stage in the multiplexing hierarchy. Hence, the name *plesiochronous multiplex* has been given to this system.

The clock oscillators of these multiplexers must run at speeds slightly higher than the incoming data so as to accommodate the local variations in the timings. This allows for small errors in the exact data rates in each of the input paths or tributaries but requires extra bits to be added (or stuffed) to justify and take into account the higher speed clock.

Elastic stores

These are used in a typical multiplexer to ensure sufficient bits are always available for transmission or reception. They are required because plesiochronous digital hierarchy (PDH) works by interleaving bytes or words from each 64 kbps tributory, rather than bit interleaving, to constitute the 2.048 Mbps multiplex. That is, whenever bit interleaving is deployed (at the 8 Mbps or higher levels of multiplexing) bits are accumulated for high-speed readout. Fig. 3.48 illustrates the plesiochronous frame structure and typical 2/8 multiplexer details which are self explanatory.

The down side of PDH is that is was originally designed for point-to-point applications. That is, the entire multiplex is required to be decoded at each end. It means full demultiplexing should be done at each level in order to recover the bit interleaved data and remove the stuffed bits. This is obviously an involved effort.

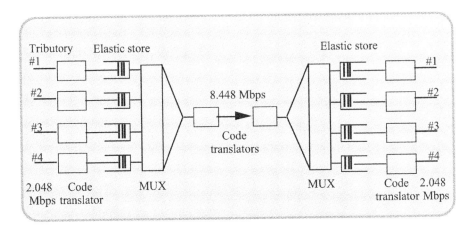

Fig. 3.48: A 2/8 multiplexer schematic

Example 3.12
Determine the maximum channel utilization capacity of the nth multiplex level of the PDH supporting standard PCM voice signals.

Solution
PDH level 1 of primary multiplex group carries 30 voice signals. Each of the subsequent levels combine four tributaries from the previous levels. Therefore, at the nth level, the number of voice signals (N) present can be deduced as follows:
$N = [4 \times 4 \ldots (n-1) \text{ times}] \times 30 = 30 \times 4^{n-1}$.
Further each signal level corresponds to 64 kbps. Denoting the bit rate of nth level as R_n bps, the channel utilization efficiency $\eta_c(n)$ can be obtained as follows:

$$\eta_c(n) = \left[\frac{\text{Number of voice channels at } n^{th} \text{ level}}{\text{Number of possible voice channels that can be accommodated at the } n^{th} \text{ level rate}} \right]$$

$$= \frac{30 \times 4^{n-1}}{(R_b / 64 \times 10^3)} \times 100\%.$$

Where and how is circuit-switching done?
Circuit-switching is done internally at a switching node. It is done by three methods, namely,

- ✓   Space switching
- ✓   Time switching
- ✓   Combination of space and time switching.

Space switching
Basically a switch is made up of two parts, namely, *switching fabric* (switching network) and *switching control*.

- *Switching fabric*: Here, the individual lines and trunks are connected to complete the communication path

- *Switching control*: This refers to the mechanism which signals the elements in the fabric when, and how to make the connections.

Examples of fabrics and control mechanisms are as follows:

Fabrics
In manual switching, the fabric is constituted by the following:

✓ Plugs
✓ Patch cards
✓ Jacks
✓ Lamps etc.

In automated switching, the fabric refers to:

✓ Strowger (rotary) switch
✓ Electronic switching.

Control mechanisms
In manual switching, the control mechanism is synonymous with the operator.
In automated switching, the control mechanism is facilitated by:

✓ Selective energization of the relays
✓ Electronic control circuits.

Considering a switching fabric associated with a set of inputs x, y, and z and corresponding outputs x', y', and z' (to which the inputs are connected respectively), a *space division switching* can be realized as shown in Fig. 3.49(a).

In Fig. 3.49, the left side switching performs *concentration* since each one of x, y, or z reduces several inputs to one path. That is, it selects one of the input switches. Switches on the right side (x', y', or z') perform *expansion*, since they select the proper outputs for the connections.

In space-division switching, the following can be observed:

✓ The switches are preset and do not change for the duration of the entire call
✓ Two simultaneous calls can be established over different physical paths
✓ Since physical paths are separated through spatial routings set by the switches, this method is known as space-division switching.

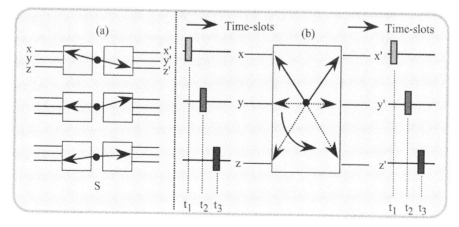

Fig. 3.49: Switching fabric: (a) Space-division switching and (b) time-division switching (*See PLATE 6*)

Time-switching
This is illustrated in Fig. 3.49(b). Here one switch is shown in three different instants of time-slots. During the first time interval, the connection for x is made. During the second interval the connection for y is made, and so on. This process is repeated and the information between the input

and output terminals is passed, each connection taking its turn. Since the three connections have to share a single physical path, they have to take turns sharing the path in a cyclic fashion.

The input end is termed as a *sampler* consistent with the fact that in each cycle, the information is sampled pertinent to one connection. The connections do not divide the physical space of the switches between themselves, but rather they divide the available time instead. Hence the name *time-division switching* (TDS) has been given. Time-division switching is similar to time-division multiplexing (TDM). The difference is as follows: In TDM, the connections between input terminals and output terminals are fixed; whereas in TDS, a given input can connect to any of the other inputs determined by the control mechanism in real time.

Time-space-time switching

Time-space-time switching is a technique used commonly in PBXs and telco switches. It allows many time-division switched buses to be interconnected with no blocking. In Fig. 3.50, there are four such buses shown existing in devices called modules A through D. These buses are connected to each other via a *matrix switch* providing a fully non-blocking capacity.

An intramodule call, (that is, a call between two phones, which are connected to the same module) does not get switched through the matrix. An intermodule call must be first switched in the time-domain on the caller's module, then be switched in the space-domain in the matrix and finally be switched again in the time-domain in the called party's module. Hence, the term *time-space-time* is used as an adjective for this type of a switch.

Fig. 3.50 shows four bits on the bus in A module, which are to be switched to the B, C, and D modules. First, the connection is made for C and then D. This way, the bits are properly routed to their destination. The clock of the matrix switch must be four times that of the clocks of the time buses, so that it can accommodate the switching four times that of the clocks of the time buses. This would allow the switching required for all four modules within one time-frame of the bus. Bits for an intramodule call are simply redirected to the originating module. It is evident that the control elements for such a switch are quite complex, because they must be able to direct various components of the switch in strict synchronization. In short, suppose two bits arriving from A should go to B. Between the times that these two bits are switched, the matrix switches a bit for each of the other modules, for example to D as shown.

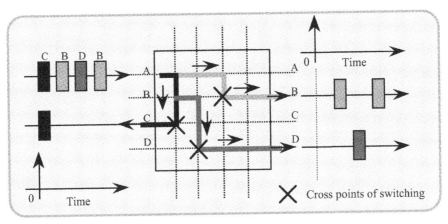

Fig. 3.50: A time-space-time switch (*See PLATE 6*)

Example 3.13
Does the matrix-switch (coordinate switch) illustrated in Fig. 3.51 use space-division or time-division switching?

Solution

Notice the switch has three inputs on the left and three outputs on top. To provide a full set of connections between them, there are a total of nine switches. Since each connection uses a different physical path, the switch is considered to use space-division switching, although it has more switches than the one illustrated in Fig. 3.49a.

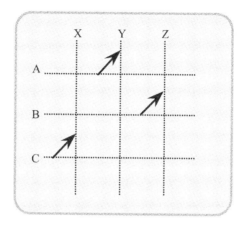

Fig. 3.51: A matrix switch

There are only 2-position switches in Fig. 3.49a while the one in Fig. 3.51 uses 3-position switches. (In modern hardware, switching in a matrix fabric is accomplished by using semiconductor technology such as VLSI, very large scale integration, which reduces several transistors on a single chip.)

Multiple switch network... a PSTN

The following terms are pertinent to the architecture of a PSTN:

- *Subscribers/entities: These correspond to end-entities attached to the network via a user-network interface (UNI)*
- *Local loop (subscriber loop): It is a link between the subscriber and the network*
- *Exchanges: Switching centers or central offices for the networks; they represent the subscriber end-offices*
- *Trunks: Branches between the exchanges*
- *FDM trunks: MCVF (multiple channel voice frequency) circuits*
- *TDM trunks: Synchronous TDM circuits.*

3.4 Packet-switching

As indicated earlier, the network architecture in digital systems may be based on either circuit-switching or packet-switching. In essence, for comparison, these two switching types can be briefly restated as follows:

Circuit-switching

A given pair of users (message sender and message receiver) has a dedicated connection for the length of their communication, with a guaranteed transmission rate. Time division multiplexing (TDM) is used to support low rate channels on to a single channel, such as a 64 kbps facility (using appropriate padding/de-padding techniques, if necessary).

Packet-switching

In packet-switching, messages from each user of the network are segmented into smaller units of packets of appropriate length. The packets traverse the network as individual units, which are subsequently reassembled at the receiver in proper order. Packets from different parts of the message, as well as packets from several users, are statistically multiplexed before transmitting them over digital links that operate at a standard, specific bit rate.

In circuit-switching a communication link is shared between the different sessions using that link on a fixed allocation basis. In packet-switching, on the other hand, the sharing is done on a demand basis, and therefore it has an advantage over circuit-switching in that when a link has traffic to send, the link may be more fully utilized.

3.4.1 Packet-switching networks

The principle adopted in networks supporting packet switching is known as the *store-and-forward* technique. Specifically in a packet-switched network, any message larger than a specified size is sub-divided (or "chopped") prior to transmission into segments. These segments as mentioned before are called the *packets*. The original message is reassembled at the destination on a packet-by-packet basis. The network may be viewed as a distributed pool of network resources (namely, channel bandwidth, buffers, and switching processors) whose capacity is shared dynamically by a community of competing user stations wishing to communicate. (In contrast, in a circuit-switched network, resources are dedicated to a pair of stations for the entire duration they are in session.) Accordingly, packet switching became highly compatible for the computer-communication environment in which "bursts" of data are exchanged between stations on an occasional basis. More so, the delay that may be encountered in the store-and-forward strategy is not of significance in data communication. This is because unlike voice communication, data information is not delay sensitive. Any delay encountered would not affect the quality of received text. The use of packet-switching, however, requires that a careful control be exercised on user demands; otherwise, the network may be seriously abused.

The concept of packetization evolved at the Rand Corporation (in the early 1960s) first as a *datagram*; it involved breaking up the message into small parts for the purpose of transmission between two data terminal equipment (DTE). Each datagram has a *header* added to the front of it, providing the destination address, source address, datagram number, and other such information. Datagrams are sent into the network via various links. The nodes, observing the header and their own prewritten routing tables, decide which link to use to forward a datagram towards its destination.

Packetized data refers to a fragmented message, which offers security of transmission. Further multiple routing feasibility at a node for a given source and destination enables *fault tolerancy* as well as immunity from node failure or sabotage. That is, if one route fails, an alternate route is available for transmission.

In summary, datagrams are packets constituted out of one message. These packets use different routes depending on which links are available on their passage towards this destination.

- Since datagram chooses a route on an ad hoc basis, the security of transmission is more robust

- Due to the possibilities of several routes, the duration of calls may vary. Hence, the tariff has to be specified accordingly.

In 1967, the Department of Defense (DoD) in the United States started the ARPANET (Advanced Research Projects Agency Network) connecting several government agencies and universities across the United States, using a datagram type of protocol. CCITT standardized the DTE-to-packet switched network interface and the associated protocol(s) in 1974 leading to the so-called *X.25 protocol*.

What is a protocol?

The essence of telecommunications is the correct and efficient routing of information from source to destination across a network. In order to accomplish this, the signaling between sources, switching centers, and destination must be understood by end-entities as well as intermediate nodes. Certain pre-established procedural considerations are therefore adopted in the information transfers. These are known as protocols.

X.25 protocol

X.25 is the ITU-TS recommended protocol for packet switching over public data networks. It addresses the interface between data terminal equipment (DTE) and data circuit terminating equipment (DCE) operating in the packet mode on public data networks. (More details of X.25 will be discussed in Section 3.6.)

X.25 uses a call-connect phase to establish a link with the distant DTE; that is, during the call-connect phase, on most X.25 implementations, a path is selected through the nodes and links, and all the data during the data transfer phase travels on this given route. This fixed route is called a *virtual circuit (VC).*

- Inasmuch as X.25 decides a designated route, the security is not assured

- Since the assigned (dedicated) path and the duration of the call are known, the billing for a connection can be similar to that of PSTN.

Packet-switched networks: A summary:

Network service should cater the following:

    ✓   Fast connection time
    ✓   Reliability of service
    ✓   Cost-effectiveness.

Additional objectives of the packet-switched networks are:

    ✓   Voice connection facilitating phone calls
    ✓   Data transmission facilitating interactive data connections
    ✓   High quality of service (QOS).

The similar functions of PSTN and packet networks are:

    ✓   Both telephone networks and packet networks use a common mesh topology. That is, they enable a terminal to access another terminal regardless of the type of terminal entity namely, a phone or a DTE
    ✓   Once a connection is established, the protocol in telephone as well as in packet networks allows voice communication or modulated data communication using codes such as ASCII, EBCDCIC, etc.

Dissimilar aspects of PSTN and packet networks:

    ✓   A telephone network is inherently a switched-circuit network. Once a connection is established between the end-points, there is one circuit dedicated to carry the conversation, and no other conversation can be transmitted on the same connection/circuit
    ✓   In a packet-network, the link between two end-points may be shared by other connections

✓ Over the PSTN, digitized voice is time-division multiplexed on its intermachine trunks

✓ In packet networks, the packets of 128 bytes statistically multiplexed are transmitted on each link making the connection

✓ PSTN primarily supports 4 kHz voice. But an X.25 network conveys units of information in *octets* or bytes

✓ A long-distance call in PSTN may take a connection time in the order of 10 seconds. However, in packet networks, it usually takes less than a second.

3.4.2   Construction of a packet

In contrast to circuit-switched networks, which facilitate a circuit between end-points for a dedicated use of two or more stations, packet-networks provide virtual circuits, which have many of the characteristics of a switched circuit except that the circuits are time-shared rather than dedicated to the connection.

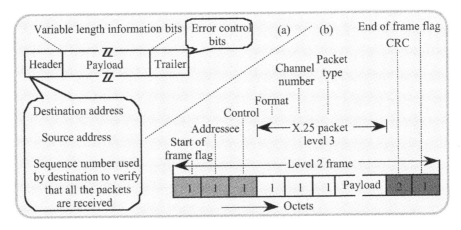

Fig. 3.52: X.25 protocol: (a) A typical packet and (b) level 2 frame enclosing X.25 packet

The packet used by CCITT's X.25 protocol is shown in Fig. 3.52. A packet network consists of multiple nodes that are accessed through dedicated or dialed connections from the end user, using one of the several possible access options. The switching nodes control the access to the network and route the packets to the destination over a backbone of high-speed circuits. These nodes consist of processors that are interconnected by backbone circuits. Essentially, information flows as packets through such networks. A packet consists of an information field sandwiched between the *header* and *trailer* records.

In Fig. 3.52, *octet* refers to the CCITT word for a *byte*. The X.25 packet is enclosed in a frame consisting of one octet flag having a distinctive pattern that is not replicated in the data field. The second octet is an *address code*, which permits up to 2,555 addresses on a data link. The third field is a *control octet* that sequences messages and sends supervisory commands. The X.25 packet consists of three format and control octets and an information field, which is filled by the user. The final three octets of the trailer are a 16 bit cyclic redundancy check field and an end-of-frame flag.

A message is chopped into packets at a *packet assembler/disassembler* (PAD) unit. The PAD communicates with the packet network using a packet network protocol X.25. It is a CCITT protocol recommendation that is used by most of the public and many private data networks. The packet network uses the address field of the packet to route it to the next node en route to its destination. Each node hands off the packet following the routing algorithm of the network until it reaches the final node. At the terminating node, the switch sequences the packets and passes them off to the PAD. The PAD removes the data from the headers and trailer records and reassembles it into the original completed message format.

Error checking is done on a link-by-link basis in the network. If a block is received with an error, the node rejects it and receives a replacement block. Due to such resending of block whenever an error occurs, and because blocks may take different paths to the destination, it is possible for the blocks to arrive at the destination out of sequence. The receiving node contains buffers to store the message so that blocks are presented to the PAD in the proper sequence.

The packet networks are similar to message-switching or store-and-forward networks with some difference. The underlying similar or dissimilar considerations can be ascertained from the following:

- Packet-switching networks are more effective and better suited for real-time operation. On the other hand, message switching networks store messages for later delivery. Though the time may be short, it could, however, be a substantial fraction of a minute under heavy traffic load conditions

- Message networks typically store a file copy of the message for a given retransmission period, if a node ahead makes a request. Packet networks, however, clear the messages from their buffers as soon as they are delivered

- Messages are transported as a collective unit in message networks; whereas in packet networks they are chopped into shorter slices. These sliced entities are reassembled into the message form at the receiving end of the circuit.

Packet size

Normally, in order to optimize the throughput of information, the packet size is established by the network designer appropriately. Since each packet consists of a fixed length header and trailer record, short packets would reduce throughout because of the additional time spent in transmitting overhead bits in the header and trailer. On the other hand, long packets reduce throughput for a different reason. It is because the switching node cannot forward the packet until all bits are received. This would increase buffer requirements at the node. Also, the time spent in retransmitting error packets is greater if the packet length is longer. In practice, the packet networks operate with a packet length of 128 to 256 bytes. Further, the packet size is variable in conventional packet-switching. (In contrast, it will be indicated later that all the packets are of constant length in the ATM transmission.)

What is meant by fast packet-switching?

Fast packet-switching refers to the technology that handles information in packetized form at high speed. The question is then "How much show high?" In the voice transmission point of view, a fast network would accommodate real-time voice connections with acceptable network quality.

The classical specifications, namely, X.25 and SNA (*System network architecture:* A IBM's data communication scheme) are *slow protocols*. They were conceived to carry large data, which may suffer long and/or variable delays. Such delays are, however, tolerated inasmuch as data is regarded as *delay-insensitive*. These slow protocols were not intended to support toll-quality voice.

The slow packets were conceived for transmission over the already available voice telephone lines (twisted copper-pair) via modem technology. Unfortunately such voice grade lines are susceptible to noise and impulses which are largely ignored in telephone conversations. However, these impairments would create bit errors in the data stream. The bit error rate (BER) could be even in the order of 10^{-3}. The loss of bits in the data transmission is not tolerable.

Hence, *error correction* became mandatory over every link in the X.25 architecture. The *error correction* (EC) procedure involves an error check (via CRC or other methods) in a frame at every receiver node. When error is detected, the node makes an automatic retransmission request (ARQ) with the originating node of the packet. Every node, therefore, keeps a copy of the packet transmitted until its safe delivery is confirmed.

The packets belonging to the same message and addressed to the same destination, may, however, be routed over different links when they arrive at a node. That is, the node forwards a packet to a link, which is free of congestion. As such, packets sent through different links would suffer different extent of delays and experience varying extent of error susceptibility.

Example 3.13

Suppose the bit error rate is BER. It denotes the probability of a bit arriving erroneously. Or, $(1 - BER)$ is the probability of a bit arriving without an error. Assuming the number of bits per packet as N and the number of links (that is, hops per path) as L, determine the probability (P) that a packet may arrive without error.

Solution

Given that,

BER = probability of bit error

$\therefore (1 - BER)$ = the probability of a bit in a packet arriving without error in a link

❑ There are N bits per packet and the corruption of a bit is independent of the impairment of the other bits

❑ There are L links and errors occurring in a link are independent of such occurrences in the other links

❑ Taking the statistical independence of the entities as above, the probability (P) that a packet arriving without error can be deduced as follows:

Suppose

$$Q = [(1 - BER)(1 - BER) \ldots N \text{ times}] = (1 - BER)^N$$

$$\therefore P = [Q \times Q \times \ldots L \text{ times}] = Q^L = [(1 - BER)^N]^L = (1 - BER)^{NL}.$$

Problem 3.6

Suppose BER $= 10^{-3}$, which represents the worst case while meeting telco specifications. Calculate the probability that a packet arrives without error over (i) 1 link, (ii) 3 links, and (iii) 5 links for a packet with a data field of 100 octets. Assume an overhead of 50 bits.

Example 3.14

On an average, the worst case BER in optical fibers is the order of 10^{-8}. Determine the error-free probability that a packet of 1 k data bytes and 250 bit overhead be transmitted over 15 links.

Solution

Given that:

Number of links (L) = 15

Packet size = 1 k byte (information field) + 256 bit overhead

= 1256 bits

= N

BER = 10^{-8}

$\therefore$ Error-free packet arrival probability:

$$P = (1 - BER)^{LN}$$

$$= 99.87 \%$$

Two error correction strategies ...

In packet transmission the EC procedure can be applied by two ways:

▪ Link-at-a-time strategy

▪ End-to-end strategy.

The link-at-a-time method refers to applying EC per link. That is, under high BER conditions, the error correction is facilitated at each link. However, if the line refers to a high quality channel (such as fiber), the BER is small and hence EC can be done on an end-to-end basis.

3.4.3 Virtual circuit (VCs)

Packet networks work on two versions of virtual circuits, namely *permanent* and *switched*. The permanent virtual circuits are the packet network counterparts of a dedicated voice circuit. In the permanent VCs, a path between end-entities is first established on which all the packets would be routed through the network. In a switched virtual circuit, the network path is established with each session.

Packets are of two types namely, *control packets* and *data packets*. Control packets contain information to indicate the status of the session. They are analogous to signaling in a circuit-switched network. For example, a call-setup packet is used to establish the initial connection to the destination host. The destination host then returns an answer packet. Control packets are also used to interrupt calls in progress, disconnect calls, and perform other supervisory signals on telephone networks. With switched virtual circuit operations, these control packets establish a session. In permanent virtual circuit operations, since the path is preestablished, no separate packets are needed to connect and disconnect the circuit.

Summary of the definitions vis-à-vis packet networking...

Packet: A unit of data information consisting of a header, user information, error detection, and trailer record.

Packet switching : A method of allocating network time by forming data into packets and relaying it to the destination under control of processors at each (major) node. The network determines packet routing during transport of the packet.

Packet assembler/dissembler (PAD): A device used on a packet switched network to assemble information into packets and to convert the received packets into a continuous data stream.

Virtual circuit: A circuit that is established between two terminals by assigning a logical path over which data can flow. VCs are of two types, namely,

- ❑ *Permanent VC: Here, terminals are assigned permanent path*
- ❑ *Switched VC: Here, a circuit is reestablished whenever a terminal has to send data.*

3.4.4 Transmission time: Messages and packets

In reference to the transmission of packets, the following can be observed:

- Normally, messages from different sources (users) will fluctuate in length as shown in Fig. 3.53

- Long messages may cause other users to experience significant delay while their messages wait in the queue of a high-speed line.

In such situations, long messages can be sliced into shorter segments (or packets). These packets as shown in Fig. 3.53 can then be individually interleaved on to the shared link known as the *bearer line*. At the receiving end of the link, the packets are sorted out and user messages are reconstructed.

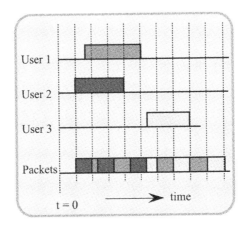

Fig. 3.53: Variations in message lengths of different users and slicing them into packets

Transmission times of a message with and without packetization
    Transmission time of a message without packetization: Suppose a message source C_0 is linked to C_k possible destinations via K transmission facilities. Let the propagation time (delay) over each link be T_{NP} and the total transmitted bits corresponding to a given message be equal to $(B + b)$, where B is the message length in bits and b refers to the overhead bits.
    On a time scale, the duration of each bit is, τ equal to $(1/f_b)$ where f_b is the bit rate in bps. Therefore, the total transmission time of $(B+b)$ bits will be, $(B+b)/f_b$. Since there are K links, the total time T_{Np} required for the transmission of the message without packetization is given by:

$$T_{NP} = [K(B+b)/f_b] + T_p$$
$$\approx (KB/f_b) + T_p \quad (3.12)$$

since ordinarily $b \ll B$.
    Transmission time of a message with packetization: Suppose the message is sliced into P packets, each with a number of bits/packet equal to B/p and the overhead bits needed is b. Therefore, in order to transmit this single packet over K, links will require a time T_{WP1} given by:

$$T_{WP1} = [K(\frac{B}{P}+b)/f_b] + T_P. \quad (3.13a)$$

    The next packet is processed immediately after the first and will arrive after the first one delayed by a time $(B/p + b)/f_b$. Since, after the first, there are only $(P-1)$ packets left, the total time T_{WP} required to transmit the entire message by packets is given by:

$$T_{WP} = \{[K(\frac{B}{P})+b)]/f_b + (P-1)[(\frac{B}{P})+b]/f_b\}. \quad (3.13b)$$

To find the value of P, which minimizes the packet transmission time, Eqn. (3.13) is differentiated with respect to P and set equal to zero. Hence, one can find minimum T_{WP} when $P^2 = (K-1)B/b$. Therefore, it follows that:

$$(T_{WP})_{min} = (\frac{b}{f_b})[\sqrt{\frac{B}{b}} + \sqrt{(K-1)}]^2 + T_P. \quad (3.14a)$$

If $(B/b) \gg K$, then

$$(T_{WP})_{min} \cong (B/f_b) + T_p \quad (3.14b)$$

and,

272

$$(T_{WP})_{min}/T_{NP} \cong [(B/f_b) + T_p]/[(KB/f_b) + T_p].$$ (3.14c)

3.5 Synchronous and Asynchronous Transmissions

In practice, there are two basic transmission techniques adopted in telecommunications to establish a time reference between the transmitter and the receiver. They are known as *asynchronous* and *synchronous* transmissions.

3.5.1 Asynchronous transmission

This refers to the transmission of data (bits) in a serialized fashion. That is, bits representing an encoded message, voltage level etc. are carried over the several channels one after another in a specific order. For example, the earliest form of electrical communication, namely, telegraphy, refers to a serial communication in which dots and dashes are sent one after another, as they are generated when encoding is done (via Morse code) on the alphanumeric characters. Another example of asynchronous transmission refers to sending a serial data of characters with start and stop bits as illustrated in Fig. 3.54.

When a serial data channel is idle (that is, when no data is sent), it rests in the ON (or the 1 state). When the transmission of a character begins, the idle state is removed and a *start bit* is sent. The start bit is the first bit preceding every data word (which represents the bit form of a character). The serialized data bits are usually eight bits. (Sometimes seven or nine bits are also used.)

The end of each set of data bits (that is, at the end of each character sent), a *stop bit* is added to indicate the end of the character. An asynchronous transmission format of a character is shown in Fig. 3.54.

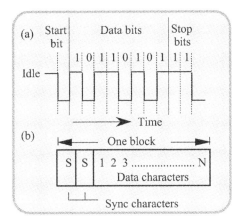

Fig. 3.54: Transmission format of data: (a) Asynchronous transmission format of a data byte and (b) synchronous transmission format

A parity bit is added immediately after the data bits for an even or odd parity check. In asynchronous transmission, two independent clocks, one at the transmitter and the other at the receiver are used. They are set to operate at an identical nominal frequency based on the bit rate of the transmission used. But the clocks are independent and are matched only in respect to the nominal frequency chosen (within a certain tolerance). Therefore, the instantaneous rates of these clocks are likely to vary. When a long stream of bits is transmitted, such differences in frequency and phase of the two clocks would cause erroneous reception. Suppose the clocks are synchronized at the beginning of a transmission. This will ensure a correct reception over a short stream of bits, before the clock pulses deviate and cause erroneous reception. This is the principle adopted in *asynchronous transmission*, which is illustrated in Fig. 3.52. The start bit at the beginning of the transmission synchronizes the phase of the receive clock with that of the transmit clock. The incoming data bits are then sensed correctly as long as the difference in the phase of transmitter and receiver clock frequencies remain within a pulse period. In essence, each character by carrying its own start/stop indications, bears inherent synchronization information.

In the asynchronous transmission described above, the number of bits b_n that can be transmitted reliably depends on the frequency tolerances of the clocks. A stop bit specifies the end of transmission, and normally, the extent of a stop bit may correspond to 1, 1.5, or 2 bits. Further, the number of stop bits used would depend on the type of receiver used. The stop bits provide the minimum buffer period required by the receiver equipment between two successive transmissions. In asynchronous transmission, which is also known as *start-stop transmission* for reasons specified above, the bit stream length is usually eight bits corresponding to a character. Hence, this transmission is often termed as *character mode transmission*. The use of two additional bits, one start and one stop bit for every eight bits represent a 25% transmission overhead. Therefore, it is not economical to transmit a large volume of data using the asynchronous mode.

The UART

The UART or universal asynchronous receiver transmitter is a device made of an array of logical units (in the IC form) which converts the parallel data from a computer's data bus into a serial format (or vice versa). It also creates the start bit, and selects and inserts the stop bits. It can determine the parity type and add parity bits as needed.

The asynchronous transmissions of alphanumeric and various control symbols, is based on U.S. standard 7-bit ASCII code (*American Standard Code Information Interchange*) corresponding to 7 data bits plus a parity check bit, a start bit (0) and a stop bit (1).

A set of rules to effect an orderly transfer of data from one locale to another using the common facilities and resources of the network denotes a protocol. Normally, at the transmitting and receiving ends exit a set of layers (as will be discussed later) which implement peer-to-peer basic functions such as accessing the medium (for data transmission), synchronization, and error-control. This is called a *data link layer* and the associated protocol is known as the *data link control (DLC)* protocol.

Data-link control (DLC)

The objectives of DLC are as follows:

- *Frame synchronization*: The beginning and the end of each frame are recognized by means of preamble and postamble bit patterns

- *Flow control:* Maintenance of sending frames at a rate not faster than the rate at which receiving station can handle them

- *Error Control*: Correcting the bit errors introduced by the transmission system

- *Addressing*: Identifying different stations under multipoint accessible situations

- *Control and data on the same link*: Distinguishing data and control information sent on the same link

- *Link management*: Facilitating coordination and cooperation towards call initiation, maintenance, and termination.

The asynchronous transmission corresponds to a start-stop DLC, using the start/stop format and the DLC facilitating a maximum rate of data transmission allowed. The relevant transmission is designated as isochronous transmission. It essentially follows the character synchronization.

3.5.2 Synchronous transmission

Synchronous transmission reduces the transmission overhead by using a single clock for both transmission and reception as illustrated in Fig. 3.53. Here, the transmitter clock reference is sent to the receiver, which uses the same clock to sample the incoming data. A separate channel can be used to send the clocking information to the receiver. Alternatively, synchronous receivers can be equipped with special tuned circuits that resonate at the desired clock frequency when excited by a received pulse. Special *phase locked-loop* (PLL) circuits that are capable of deriving the clocking information from the data itself can also be used at the receiver.

In asynchronous transmission, the transmission can take place at any instant of time. The start-stop bits indicate the presence of such transmissions. As such, asynchronous transmission is not "governed by a clock". On the contrary, synchronous transmission uses a constant rate clock, which determines the exact instant at which bits are sent and received. Hence, there is no need for the start-stop bit sequences to be associated with the data bits. But, the transmissions are "governed by a clock".

The robustness of extracting clock from the data relies on the extent of signal transitions that occur in the data. If the signal is a continuous stream of 1s or 0s for a sufficient time, the clock extraction process would deteriorate. As a result, the receiver and transmitter may go out of synchronism. Therefore, a number of techniques have been developed to ensure adequate density of signal transitions in the data transmitted in synchronous mode. They are as follows:

- ✓ Using a restricted format for the source code
- ✓ Facilitating a dedicated set of timing bits
- ✓ Incorporating bit stuffing in the data stream
- ✓ Forcing deliberate bit errors in the data stream
- ✓ Incorporation of data scrambling
- ✓ Line coding.

Restricted source code format

Since an n-bit binary code set has 2^n code words, one can restrict the valid code set to m words, where m is less than 2^n. The logic behind the choice of adopting a restricted code set (out of a redundant set) refers to eliminating those code words which do not have adequate signal transitions. This will ensure that the receiver clock remains in synchronism with the transmit clock. The main drawback of the technique is that new applications where the source does not exclude the unwanted data patterns cannot be supported on the network.

Use of dedicated timing bits

Similar to the start bit in asynchronous transmission, transition-bearing timing bits can be introduced at regular intervals in a synchronous transmission. The interval is so chosen as to ensure an adequate number of signal transitions per second required for reliable clock synchronization exist. It may, however, be noted that these added timing bits will enhance the transmission overhead.

Incorporation of bit stuffing

Bit stuffing is a technique where an extraneous bit is introduced in the data stream whenever no signal transition is observed in the data for a continuous period of time. For example, a signal transition bit may be introduced in the data whenever a series of seven 1s or 0s is observed on the data line. Bit stuffing breaks the continuous flow of data and delays that data stream by one bit, every time a bit is inserted. In isochronous, delay-sensitive services like voice, this may cause *jitter*, thereby affecting the quality of service.

Data scrambling

Data scramblers randomize the data patterns. This would prevent the continuous transmission of repetitive data patterns on the line. Data scramblers can produce adequate timing information, which can be usefully exploited for synchronization.

In summary, synchronous and asynchronous transmissions correspond to the following: Data transmitted over a connection such as RS-232-C between computers and peripherals (keyboard, tape driver, printer etc.) are sent and received in a character format. One character is sent at a time, and any two characters are separated by a minimum but arbitrary time interval. A character is represented by a set of seven or eight bits, depending on the code used. This type of transmission refers to *asynchronous transmission* because successive characters can be transmitted at arbitrary times, other than for the minimum separation.

Typical asynchronous connections include: RS-232-C EIA (Electronic industries Association) 1969. The connection supports transmission rates up to 38,400 bits per second over four to twelve wires for distances up to 50 ft. There are also other asynchronous connections, namely, RS-449, RS-422-A and RS423-A, which came into existence in the 1970s to support faster asynchronous transmissions.

Further, in asynchronous transmission, the transmission takes place at any instant of time and is not governed by a clock. Data bits of each character are proceeded and followed by special start and stop bit sequences to prompt and synchronize the receiving terminal.

3.5.3 Asynchronous and synchronous DLC protocols

Pertinent to asynchronous and synchronous transmissions are DLCs, which have been developed for peer-to-peer communication as indicated in Table 3.10.

Table 3.10: Asynchronous and synchronous transmission DLC

Transmission type	DLC Protocol	Versions
Asynchronous	Start-stop DLC	–
Synchronous	Character-oriented	▪ Bisync (IBM's BSC) ▪ ARPANET DLC
	Bit-oriented	▪ SDLC ▪ HDLC ▪ X.25
	Byte-count oriented	▪ DDCMP

Note:
SDLC: Synchronous data link control
HDLC: High-level data link control
X.25:   IEEE 802 medium access control
DDCMP: Digital data communication message protocol

Asynchronous start-stop DLC protocol

The data link control is done in the most rudimentary form. In reference to the synchronous transmission considerations outlined earlier, the following can be enumerated as the salient aspects of the asynchronous DLC protocol:

- *Transmission type*: This refers to character-by-character transmission. That is, the appearance of each character is asynchronous (and statistical) depending on how it originates from a user at a keyboard or other terminal. Each character on the line is identified individually by virtue of the start bit at the beginning and the start bit at the end of the data stream as illustrated earlier (in Fig. 3.54). The format of start/stop representations may vary. Thus, the functions of the asynchronous transmitter essentially includes structuring the data bits (representing a character) between start/stop bits plus adding a parity bit if needed and sending the characters one after another. Start/stop DLCs are implemented largely with UARTs

- *Reception of start/stop data*: At the receiving end, the UART looks for a 1-to-0 transition, which indicates the beginning of a start bit during the periods when the line is idle. The voltage level is then sensed at approximately the middle of each bit during the time of character transmission. The sensed voltage is translated into the associated start and stop bits (plus the parity bit if one is used), and 1s and 0s representing the character. The character is then retrieved by stripping the start/stop and parity bits. If parity check fails, appropriate error conditions are set as flags. The start-stop DLC protocol performs the basic character synchronization. That is, via appended start/stop bits, the information transmitted is recovered character-by-character

- *Error control in start/stop DLC:* This is essentially done via parity check. Either the even or odd parity check sum can be used to detect a single bit error. This however, does not provide error protection or correction. The detection of an error is simply sent to higher protocol layers which may use or ignore this error condition

- *DLC overhead:* The overhead in asynchronous start/stop DLC is due to the appended bits towards character synchronization and for parity check. The overhead is typically 20 to 44%. The lower value applies for eight bit character transmission with no parity and using a single stop bit. The higher side refers to five bit characters plus parity bit and the stop flag using two bits. These overheads do not include any overhead that arises as a result of retransmission requests made as necessary

- *Shortcomings of start/stop DLC protocols:* The asynchronous start/stop DLC protocols are simple and offer support limited to character synchronization and basic parity check. These protocols are useful for low-speed links (up to 1200 bps) with simple terminals. (For example, the Bell 103 modem.)

    Start/stop DLC protocols are not intended to perform more involved functions such as activation and termination of data link controls, sender/receiver identification, framing of data for transmission, flow control, link management and any abnormal condition recovery functions. Nor do they relegate such functions to higher layers.

    Hence, for higher speeds (> 1200 bps) of data transmission and to include more functions to the protocols, the synchronous transmission standards were developed as narrated in the following section.

Synchronous transmission: DLC protocols

As indicated in Table 3.10, there are three versions of DLC protocols for synchronous transmission, namely, character-oriented, bit-oriented, and byte-count oriented protocols.

Since the start/stop bits are not used in synchronous data transmission, it is necessary to distinguish one character from the other (as well from the parity bit). Synchronizing the receiver to the transmitter data clock before the message is transmitted completes this task. The differences in synchronous DLC protocols refer to how this framing-based transmitter-receiver synchronization is done in different ways. But all these protocols, in general, are based on the following procedure.

A transmitter sends an 8-bit sync character (called a *frame start flag*) that the receiver recognizes. Relevant clock recovery circuitry in the receiver uses this bit pattern to synchronize its clock. This process is called *bit-timing recovery*.

In order to understand synchronous DLC protocols, it is necessary to review and recall the basics of data networking:

A quick look-back at data networking…

A data network is made of nodes or communication stations and a mesh of transmission lines, which facilitate interconnection of the nodes shown in Fig. 3.55. There are two possible configurations. Fig. 3.55a depicts a centralized scheme with a single processing station which controls all traffic between point-to-point and/or multipoint connections. The distributed network illustrated in Fig. 3.55b refers to an interconnection of more than one centralized network.

Point-to-point connection: Refers to an individual connection on a dedicated basis between a pair of nodes done via the CPS. For example, the leased lines in a PSTN with dial-up connections.

Multipoint connection: At any given time a multipoint line is shared by two or more stations under the control of a CPS. The CPS manages the flow of data between the various tributary stations via polling and selection. That is, an "invitation to send" or polling is performed by the CPS to the tributary terminals in a predetermined sequence. Also, the CPS does selection or issues "an invitation to receive" notification to a specific tributary station. Many stations can be "concentrated" on a few multidrop lines (all under a single central control). This is done by using multiplexers and/or concentrators.

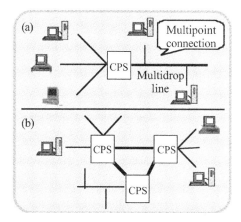

Fig. 3.55: Data networking: (a) Centralized networks and (b) point-to-point single-node based connections to the central processing station (CPS)

Switched networks for synchronous transmissions: Information can be switched through the network by (i) circuit switching, (ii) message switching, and (iii) packet switching.

Circuit-switching: Here a circuit path is established for the entire duration (session) of the call as done in the POTS connections.

Message-switching: A message in full is sent to a switching center, where it is stored until it is forwarded to a station closer to the destination. That is, a store-and-forward technique is pursued in message switching. Typically, telegrams and e-mail transmissions are message-switched across the network.

Packet-switching: In this technique, a message is divided into variable length units called packets, which are switched through the network with other packets asynchronously multiplexed, (that is, statistically concentrated). The statistical concentration involves a processor, which continuously samples input buffers and output lines. Any pause in data transmission sensed on a transmission line will let the processor redirect input data packets (waiting in the buffer) to that line on a first-in/first-out basis.

Each packet carries a destination address and other protocol information as its overhead. A packet received at a node is stored in a buffer for forwarding on queue. That is, the head of the line (HOL) packet is switched to the output buffer space, as available for transmission to the next node.

Character-oriented synchronous DLC protocols

IBM's BSC or Bisync protocol

These use special binary characters to segregate different segments of the transmitted information frame. These special characters are serial, binary-coded characters made of text information (message) and/or heading information (message on identification and destination). *Bisync* describes the transmission codes, data-link control and operations, error checking, and message format.

Essentially in this class of synchronous DLC, the characters follow each other in a synchronous fashion (that is, at clocked intervals). They are based on *control characters* in a character set. Most commonly, the ASCII character set constitutes the control characters. Other transmission code sets accommodated by Bisync are EBCD1C and six-bit Transcode. The control characters designate and control various portions of the information being transmitted. Bisync is designed to operate both on point-to-point and multipoint configurations. A typical *Bisync protocol* format is illustrated in Fig. 3.56.

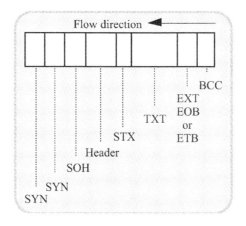

Fig. 3.56: Bisync protocol format

SYNC: *Synchronous idle*
> A character used to provide a signal from which synchronization may be achieved or retained. It may also be transmitted as a fill character in the absence of any other character.

BCC: *Block-check character*
> This is used for error protection.

ETX: *End of text*
> A character terminating sequence of characters started with STX and transmitted as an entity.

EOB: *End of block*
> All but the last message block are ended with EOB. The last block of the message ends with ETX (ETB: End of text block).

TXT: *Text*
> Message consisting of one or more blocks.

STX: *Start of text*
> A character that precedes a sequence of characters (text) to be treated as an entity and transmitted through to the destination. STX may also terminate a sequence of characters started by SOH.

Header: *Address etc.*
> This is for station control and priority.

SOH: *Start of header*
> A character used at the beginning of a set of characters and contains address/routing information. STX terminates the heading.

Bisync implementation

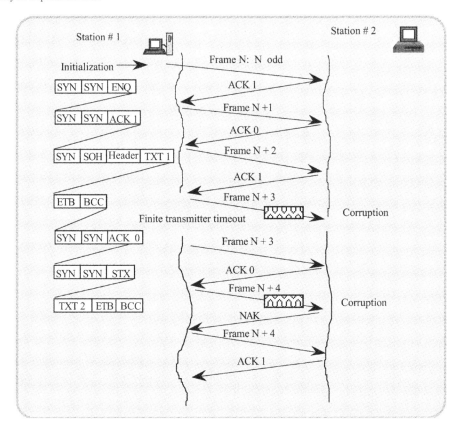

Fig. 3.57: Stop-and-wait automatic repeat request (ARQ) with bisync transmission

The Bisync protocol based data transmission in character-oriented fashion can be summarized as follows:

- Bisync operates only in half-duplex mode. That is, a station can transmit or receive, but it cannot do both at the same time. The station that sends in data is called a *master station* and the one that receives is known as the *slave*

- Data transmission is done via the frame format illustrated in Fig. 3.57. All frames begin with at least two SYN characters. Stations on the data link idle in a "sync search" mode await to identify a transmitted SYN character (such as 0110100 in ASCII). When two successive SYN characters are observed, the character availability flag is hoisted. With the SYNC received, the receiver clock is synchronized to receive the transmitted message with the remainder of the protocols facilitating the maintenance of the circuit during this process

- The phases of Bisync based transmissions are as follows:

  ✓ *Initialization phase*: In a point-to-point operation, the master station bids for the use of the line using a control character ENQ (enquiry). In ASCII, it is 1010000. (If two stations bid simultaneously, the station that persists gets the line, unless a predetermined priority scheme exists.)

  ✓ *Acknowledgement phase*: A character called ACK (acknowledgement) is transmitted back by a receiver as a positive response to a sender. There are two "go ahead" responses, ACK0 and ACK1, which are used alternately during the text block. (This would facilitate a running check whether each reply corresponds to the immediately preceding message block). In ASCII, the ACK0 and ACK1 are specified as follows:

$\Rightarrow$ ACK0: Acknowledges even-numbered data frames. It is transmitted as a two character pair namely, DLE0 $\Rightarrow$ DLE DLE. [DLE (data link escape) is a character which changes the meaning of a limited number of contiguously following characters. In ASCII, DLE = 0000100.]

$\Rightarrow$ ACK1: Acknowledge odd-numbered frames. ACK1 is DLE1 $\Rightarrow$ DLE.

Negative acknowledgement (NAK) is a character transmitted by a receiver as a negative acknowledgement response to the sender, and a wait or delay before ACK is accepted is a character designated as WACK.

Error-checking: Error control uses stop-and-wait ARQ. Each frame with user data must be acknowledged, either positively (with ACK1/ACK0) or negatively (NAK). An ACK is sent if the format of the frame is satisfactory and BCC indicates no error. Otherwise a NAK is returned. The ACK and NAK are sent as SYN SYN ACK(1, 0) and SYN SYN NAK. A typical stop-and-wait ARQ is illustrated in Fig. 3.57.

ARPANET DLC

It is a variation of Bisync DLC protocol. Essentially, it is a Bisync transparent text mode. But it uses a three-byte CRC (replacing the one-or two-byte CRC of Bisync). There are problems associated with this protocol. The frames may be received out of order since the protocol uses same logic channel-based transmissions. As such there is a burden placed on higher layers in reordering the frames.

Character-stuffing: Transparent text mode

During normal operation, characters transmitted could be a chosen character set such as ASCII characters. Alternatively, it is also possible that a transmission may correspond sending arbitrary bit patterns. This is known as the *transparent text mode*.

If the transmission is in transparent text mode, it is possible that the text may contain bit patterns identical to control characters such as DLE, ETX etc. For example, suppose the text contains the bit pattern of ETX or ETB. Then, the receiver will mistake it as a protocol control of the end of the text or a text block. It will lead to the rest of the actual frame being ignored and BCC will probably declare the frame as erroneous. Therefore, a technique known as *character stuffing* is done to protect the transparent text mode from such false controls.

Suppose the transparent text mode is done with the Bisync protocol. Then the transmission is initiated by the transmission of character sequence with DLE STX instead of STX. The data link escape (DLE) informs the protocol handler that the succeeding text is of the transparent mode. Likewise, the end of transparent text is terminated by a two-character sequence DLE ETX (or DLE ETB) instead of simple ETX (or ETB).

Further, if a DLE pattern is encountered in the text part of the frame, it is transmitted as DLE DLE so that if this repeated DLE is received, the receiver deletes the second DLE and considers the remaining bit pattern as that of the text.

Bit-oriented DLC protocols

These are more flexible and offer better speed than the character-oriented protocols. Descriptions on this family of protocols are as follows:

High-level data-link control (HDLC)

This refers to the ISO's DLC protocol. It is bit-oriented as illustrated in the frame format depicted in Fig. 3.59. It transmits bit streams regardless of the subdivision of these streams into characters.

A corresponding standard specified by ANSI refers to *advanced data communications control procedure* or ADCCP. A subset of ADCCP is the IBM version known as *synchronous data link control* or SDLC.

Essentially all bit-oriented DLC protocols are nearly compatible with each other except for certain variations in options. HDLC specifics link structures of three types as shown in Fig. 3.58.

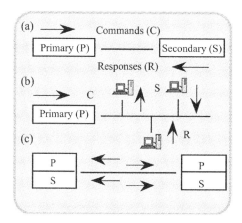

Fig. 3.58: Link structures of HDLC: (a) Point-to-point (unbalanced), (b) multipoint (unbalanced), and (c) point-to-point (balanced)

The relevant frame format used in HDLC is presented in Fig. 3.59.

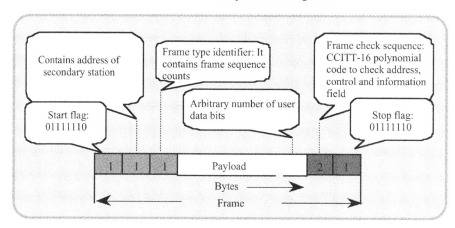

Fig. 3.59: An HDLC frame

---

16 bit CRC (CCITT V.41)

The generating polynomial of 16 bit CRC is: $G(x) = x^{16} + x^{12} + x^5 + 1$

Description of frame fields (Fig. 3.59)

- *Flag field*

  This marks the beginning or end of each frame. It has a special pattern 01111110. A station receiving this pattern recognizes that an HDLC frame would follow this pattern. Normally, there is no restriction on the pattern of the bit stream in the text field. Therefore, if the text pattern contains the special pattern of the flag namely, 01111110, then it may be interpreted as the flag. To avoid this, *bit stuffing* is done. The sending station monitors the bit stream between the flags before they are sent. If it detects five consecutive 1s, then it inserts (stuffs) an extra bit after the fifth 1. This splits the pattern potential to be mistaken as the flag. Upon reception, if the receiver finds five consecutive 1s followed by a zero, it will destuff or remove the zero from the stream.

- *Address field*

  The need for address field is self-evident. The primary station places the destination address of the secondary station in the frame. The address field is normally 1 byte or on an extended format, it could be 2 bytes. The response from the secondary station would contain the sender's identity. It is also possible that the field may contain broadcast address (all 1s) so that all secondary stations (of a predefined group) will be the destinations for the frame transmitted.

- *Data or information field*

  The field length here is variable. It carries the information bits. It may carry stuffed bits also as necessary.

- *Frame check sequence (FCS)*

  The block code character (BCC) is computed from the address, control, and information fields. Collectively, the FCS is performed for CRC error detection via 16 bit (standard) or 32 bit extended fields.

- *Control field*

  This is normally 1 byte long and in the extended form, it is 2 bytes. It is used to send status information or issue commands. Its contents depend on the type of the frame. There are three types of frames, namely, information frame, supervisory frame, and unnumbered frame.

- *Information frame*

  This governs the information (data) field transfer using the so-called go-back-n or the selective repeat sliding window protocols. In reference to Fig. 3.60, N(S) is a modulo eight (2^3) "send count" specifying the number of the frame transmitted and N(R) is a modulo eight "receive count". N(R) and N(S) correspond to frame acknowledgement and frame number indications respectively. N(R) is a *piggy-back acknowledgement* on all the frames up to N(R) – 1 received. (*Piggybacking*: When a station both sends and receives, it can avoid sending a separate acknowledgement by including the acknowledgement in the data frame itself.)

  The P/F bit in Fig. 3.60 refers to *poll/final* bit. It depicts whether the frame is sent by the polling station (primary station) or the secondary station. Essentially P/F specifies whether the frame is a command or a response. In command mode, it is a P bit and in the response mode, it is an F-bit. The F-bit (final bit) indicates the end of the sequence of frames.

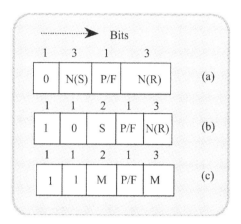

Fig. 3.60: (a) Information frame, (b) supervisory frame, and (c) unnumbered frame

- *Supervisory frame*
  Either station, to indicate its status or to send negative acknowledgement (NAK) for the frames received with error, uses this frame. In reference to Fig. 3.60, the N(R) and P/F bits perform the same functions as in the information frame. The two-bit S field is intended to specify the following:

  S field

  ✓ RR ⇒ *Receive ready* (00) is a general-purpose frame sent when a station wants to indicate its readiness to receive information. RR frame is also used to acknowledge all the received frames with N(S) numbers less than N(R) and to inform that it is looking for a valid frame containing N(S) = N(R)

  ✓ REJ ⇒ *Reject* (01) is for a negative acknowledgement (similar to NAK) for the frame and requests the other stations to resend all outstanding frames starting with the one whose number is N(R). This situation occurs whenever a frame arrives corrupted

  ✓ RNR ⇒ *Receive not ready* (10) frame is used to temporarily stop the flow of incoming frames, whenever the buffers at a station are full or an error is detected on the link associated with the station. This is similar to WACK

  ✓ SREJ ⇒ *Selective reject* (11) frame (similar to NAKs) is for selective repeat only on negatively acknowledged frames. It is based on the requests made to other stations to resend the frame whose number is specified by N(R).

- *Unnumbered frames*
  These establish how the protocol should proceed. These are used for housekeeping purposes, including startup and shutdown. They are identified by 11 as the first two bits of the control field.

  The five bits labeled M are modifier bits specifying the type of unnumbered frame. HDLC can use selective repeat or go-back-n, adopt different frame sizes, and communicate in one of three possible modes as indicated below. How the stations decide on when to do what, is governed by the unnumbered field of the control frame.

HDLC modes of communication...

Normal response mode (NRM): Here the primary station controls the communication. The secondary station has sent only capability and is instructed by the primary station. The configurations of the links refer to those indicated earlier in Fig. 3.58.

Asynchronous response mode (ARM): Here the secondary station is more independent and it can send data and control information to the primary station. It is not a total slave of the primary station. But it cannot send commands and the responsibility of establishing, maintaining and eventual termination of the connection, still resides with the primary station.

Asynchronous balanced mode (ABM): Here either station can send, receive, and control information and send commands. Typically in connections between computers with the X.25 standard, ABM is used.

Sliding window protocols: Go-back-n algorithm and selective repeat paradigm

Sliding window protocol specifies a window which slides down as frames are acknowledged and new ones are sent. It implies three sections, namely, (i) frames before the top of the window have been acknowledged; (ii) frames within the window are outstanding; and, (iii) no frame below the bottom of the window should be sent until the window slides down.

This strategy is needed if the number of frames sent are excessive. If a free-flow (unrestricted protocol) is used, it will "flood" the channel and overwhelm the receiver. Alternatively, a stop-and-wait protocol can be used in which the sending end waits until an acknowledgement is received before sending the next frame. Thus rather than pushing all the frames in rapid succession, the stop-and-wait protocol follows send-wait-receive acknowledgement-send next... types of transmissions of frames. However, it is obvious that this scheme is rather too slow. Therefore, the sliding window method indicated above is a compromise between unrestricted and stop-and-wait protocols. There are two versions of sliding window protocol:

- ❏ *Go-back-n protocol*
- ❏ *Selective repeat protocol.*

The above protocols are variants of automatic repeat request (ARQ) performed as per sliding window algorithm. By the time a negative acknowledgement is received at the originating (primary) transmitter, it may have already transmitted several frames following the one that faced an error while on transmission.

Go-back-n algorithm specifies that the erroneous frame and all the succeeding frames already sent should be retransmitted.

In selective repeat protocol, only the erroneous frame is retransmitted.

Other members of bit-oriented protocols

Table 3.10 provides a summary of other bit-oriented protocols and their relation to HDLC.

Table 3.10: Bit-oriented protocols (other than HDLC)

Protocols	Characteristics	Relation to HDLC
SDLC: Synchronous data link control (IBM/1970)	• Uses go-back-n • Part of SNA (IBM terminal-to-computer communication)	• HDLC was derived from SDLC and it is an ISO version of SDLC • DLC can have an information field length of any size. The SDLC information field must be an integral multiple of a byte • The SDLC control field option allows up to 7 frames to be transmitted before ACK is required. In HDLC, 127 frames is the limit.
ADCCP : Advanced data communications control procedure	• Similar to SDLC with several options	• ANSI version of SDLC
LAP: Link access protocol LAPB (B: Balanced) LAPD (D: Digital)	• Used for packet-switched networks • Used for ISDN	• ITU's adoption and modification of HDLC for use in its Z.25 network interface standard
LLC: Logic link control IEEE standard	• Used for LAN-to-LAN and LAN-to-WAN interconnections	• Similar to HDLC

Byte-count oriented protocols

Digital's data communications message protocol (DDCMP) developed by Digital Equipment Corporation is another DLC protocol, which can be used on synchronous or asynchronous data links. It is suitable for half- and full-duplex communications. Also, it is compatible for point-to-point or multipoint configurations. It is similar to Bisync.

Multi-access networks

In reference to data transmission, an essential consideration is that the transfer of data should be error-free. Hence, as discussed above *data link protocols* were developed which represent a set of procedures the establish point-to-point computer links with automatic error-control facilities. These protocols ensure the preciseness of synchronous transmissions.

The data link protocols were responsible for the emergence of indirect connection computers — or what is known as *store-and forward transmission* as briefed earlier. Suppose that computer B is connected to computers A and C by means of the two point-to-point links so that A, B, and C are collinear. Then, it is possible to send messages from A to C by routing them first from A to B and then from B to C. In this arrangement, however, a store-and-forward transmission from A to C via B would be more efficient. That is, the transmission from B to C can start prior to that from A to B is completed. This can be done by fragmenting the messages into small packets so that transmission from B to C would start as soon as B has received just one packet from A. The saving of time achieved by this packetized transmission increases with the number of intermediate nodes. For example, if there are N intermediate nodes, the transmission of the undivided message takes (N + 1) minutes while that of the 60 packets takes only one minute and N seconds. The transmission of messages in terms of packets is called *store-and-forward packet-switching*.

The store-and-forward networks in general can be wide area networks (WANs). They typically use telephone lines leased from telcos. The transmission rates can be a few tens of thousand of bits per second. Examples of multi-access networks are as follows:

   ✓   ALOHA
   ✓   Ethernet
   ✓   Token-ring.

ALOHA is a packet-radio network developed at the University of Hawaii in the early 1970s. A variation of the protocol used by ALOHA is used in the ethernet, developed in the middle of the 1970s at Xerox PARC [3.18 – 3.24].

Ethernet is distinct from ALOHA in that the transmitters interrupt their transmission as soon as they detect simultaneous transmissions. This detection may occur before the packets are transmitted completely (at least, when the cable is not too long). Consequently, ethernet is more efficient than ALOHA. The development of ethernet was concurrent to the evolution of personal computers. It is the most widely used local area network (LAN).

Token-ring is another LAN version of multiple access networks developed by IBM in the 1980s. It uses a token passing protocol to regulate access to the channel.

3.6   X.25 Protocol

As indicated earlier, in a packet-switching connection, a data terminal equipment (DTE) may connect with any of the other terminals in a network. That is, in practice, a packet-switched network can constitute a usable X.25 network with many links and switches connected in a mesh topology (to provide alternative routings) and greater reliability. That is, any terminal can connect itself with any other terminal on the network, as long as addresses are known (Fig. 3.61).

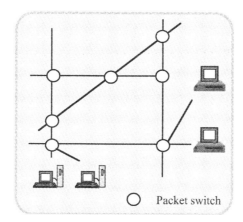

Fig. 3.61: Packet-switching based interconnection of DTEs

The sprouting of packet-switched networks across the world created a niche to produce a standard protocol compatible for international internetworking. Hence, the ITU-TS came up with the X.25 protocol for packet-switching over public data networks (PDNs). The X.25 was written in Geneva in 1976 and was amended subsequently in 1980. It addresses the interface between a DTE and DCE operating in the packet mode on PDNs.

X.25 has three layers of protocol designated as layer 1, 2, and 3 as shown in Fig. 3.62. The layered protocol structure of X.25 is in conformity with the so-called *Open Systems Interconnect* (OSI) reference model (to be detailed later) which offers a structural guideline for interoperation between various computers, terminals, and networks. The layered concept enables the functions of each layer to depend on the adjacent lower layer's functional interaction with the network. Hence, considering the three-layer architecture of X.25 protocol, the functional aspects of each layer are as follows:

   ✓   *Physical layer*: It is the lower most layer and deals with the physical interface between the attached station to the packet switching node

   ✓   *Link access layer*: It provides for the transfer of data as a sequence of frames. The relevant synchronous communication follows a data link control protocol, namely, *link access protocol/balance* (LAPB) indicated earlier as a bit-oriented DLC protocol

✓   *Packet layer*: This layer facilitates an external virtual circuit.

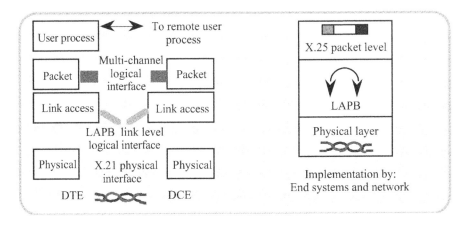

Fig. 3.62: X.25 interface protocol layers

The CCITT's X.25 and its related protocols can be summarized in terms of the network configuration and the associated equipment as depicted in Fig. 3.63. Essentially, the following are the cousins of X.25 protocol:

X.3: Services of an asynchronous PAD (packet assembler/dissembler)
X.28: Interaction between PAD and a non X.25 terminal
X.75: Protocol between signaling terminal equipment (STEs)
X.121: Protocol to address the interconnected networks properly. It specifies:

✓   Country code
✓   Network code
✓   Terminal number.

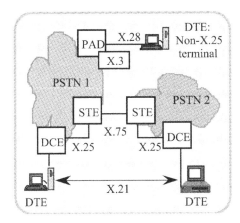

Fig. 3.63: Application of various protocols related to X.25
(CE: Data communication equipment; DTE: Data terminal equipment; PAD: Packet assembler/disassembler and STE: Signaling terminal equipment)

User classes within public data networks (PDNs)

Table 3.11: <u>PDN dependent user classes</u>

Type	Class #	Bit rate (in bps)
Start/stop mode	1	300
	2	50-200
Synchronous mode	3	600
	4	2,400
	5	4,800
	6	9,600
	7	48,000
	19	64,000
CCITT X.25	8	2,400
	9	4,800
	10	9,600
	11	48,000
	12	1,200
	13	64,000
CCITT X.28	20	50-300
	21	75-1,200
	22	1,200
	23	2,400
ISDN	30	64,000

Existing varieties of data transmissions are standardized as user classes as per ITU-R Recommendation X.1 as indicated in Table 3.11.

Functioning of X.25 interface and the associated demerits

The X.25-based transmission requires a considerable overhead. This can be observed in reference to the packet-switching arrangement. The salient considerations in reference to X.25 transmissions are as follows:

- The flow of data in the link frame format involves the transmission of a single data packet from source-end to destination-end and the return of an acknowledgement packet

- At each hop through the network, the data link control protocol requires the exchange of a data frame and an acknowledgement frame.

- At each intermediate mode, state-tables must be maintained for each VC to deal with the cell management and flow control/error control aspects of the X.25 protocol

- The entire overhead associated with the X.25-based transmissions are justifiable if significant error probability is anticipated. However with the advent of modern digital systems, where technology offers high quality, reliable transmission links (such as fiber optic links) with high data rate feasibility, overhead, as warranted in X.25, is rather unnecessary. It degrades the effective utilization of available data rates.

3.7   Frame-Relay Concept

The frame-relay concept is an advancement over the traditional X.25 and packet-switching techniques described above. It was seen that packet-switching in its traditional form makes use of

X.25, and X.25 determines the UNI internal architecture/design of the network. The essence of packet-switching / X.25 approach is as follows:

- Call control packets required to set up and release VCs are carried on the same channel and VC as the data/information packets. That is, *in-band signaling* is adopted

- Multiplexing of VCs takes place at layer 3 of the X.25 interface protocol illustrated in Fig. 3.64

- Layers 2 and 3 (Fig. 3.64) include flow control and error control mechanisms

- The physical layer deals with the physical interface between an attached station to the packet switching node. The link provides transfer of data as a sequence of frames. This link layer is referred to as a link access protocol/balanced (LAPB)

- The packet layer provides an external VC.

The frame-relay technique was therefore, developed to overcome the overhead of X.25 imposed on end-user systems and packet-switching networks. The differences between frame-relaying and the conventional X.25 packet-switching are as follows:

- Call control signaling is implemented in the frame-relay transmissions on a separate logical connection from user data. This is known as *out-of-band signaling*. Thus, intermediate nodes need not maintain state-tables or process messages relating to call control on an individual per-connection basis

- In the frame-relay operation, the multiplexing and switching aspects of logical connections take place at layer 2 (instead of layer 3 of the protocol stack as in X.25) thereby eliminating an entire layer of processing (Fig. 3.64)

- The hop-by-hop flow and error controls are removed in the frame-relay. Any end-to-end flow and error control, if they are employed at all, would be the responsibility of a higher layer. Thus, the intermediate nodes are relieved of the burdens on error and flow controls.

A single user data frame, for example, is sent from the source to the destination and an acknowledgement, generated at a higher layer is carried back in a frame as illustrated in Fig. 3.64.

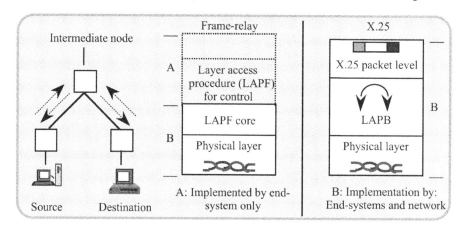

Fig. 3.64: Frame-relay network: Frame-relay protocol stack versus X.25 protocol stack

290

The frame-relay network has also its own demerits. They are enumerated below:

- Ability to do link-by-link error control is lost

- Hop-by-hop link control on error/reliability is lost. (However, advanced technology offers highly reliable switching and/or transmission facilities. Therefore, as mentioned earlier, a hop-by-hop check on reliability becomes rather unnecessary.)

The merits of a frame-relay network are as follows:

- Frame-relay is a streamlined communication process. That is, required protocol functionality and the UNI internal networking processing is reduced as well

- As a result of streamlined communication, lower delay and higher throughput are feasible in the frame-relay transmissions.

Frame-relay protocols (FRP)
FRP supporting frame mode service needs two separate planes of operation, namely:

- *Control plane* (CP): To establish and disconnect logical connections

- *User plane* (UP): To facilitate transfer of user data between subscribers.

Relevant to these two planes, FRP can be divided into:

- ✓ CP protocols (between a subscriber to a network)
- ✓ UP protocols, (which refers to end-to-end functionality).

3.8 Open Systems Interconnection Reference Model
In the 1970s, a number of companies developed computer networks – each company using its own version of network architecture. The commonness between them was, however, the layered architecture consisting of:

- ✓ A bottom layer transmitting the bits
- ✓ The next higher layer transmitting packets on one link between two nodes
- ✓ A third layer supervising the end-to-end transmission of packets on one link between two nodes
- ✓ A few layers (on the top of third layer) facilitating end-to-end transmission of packets to implement end-to-end communication services and making them available for user applications.

Notwithstanding the aforesaid common functional attributes, the evolution of data communication saw vagaries in the protocols, equipment, and architecture set by the whims and fancies of vendors and network providers. Therefore, it became crucial to formulate a protocol structure which would allow otherwise incompatible systems to interoperate and communicate regardless of their underlying architecture. Thus emerged the *open system*. Considering the implications of interoperability, the ISO addressed the problem of allowing many devices to communicate and developed its *open systems interconnect* (OSI) reference model in 1970. Its purpose was for compatible network designs and it was structured as a stated form as described below.

3.8.1 Layered architecture
This refers to a seven layered architecture. The stacking of the layers is depicted in Fig. 3.65. The functions of the seven OSI layers are as follows:

✓   Layer 1 (physical): Transmission of bits
✓   Layer 2 (data link): Transmission of packets on one given link
✓   Layer 3 (network): End-to-end transmission of packets
✓   Layer 4 (transport): End-to-end delivery of messages
✓   Layer 5 (session): Setup and management of end-to-end conversation
✓   Layer 6 (presentation): Formatting, encryption, and compression of data
✓   Layer 7 (application): Network services (such as e-mail and file transfer).

These layers collectivity function and coordinate so as to establish a reliable transmission of information between any two hosts, but each layer shares certain individual functions.

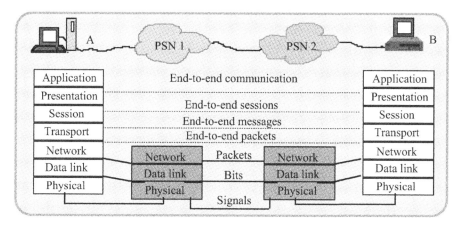

Fig. 3.65: OSI model

Each layer performs specific functions and communicates with the layers directly above it and below it. Higher layers deal more with user services, applications, and activities, and the lower layers are concerned more with the actual transmission of information. Layering the protocols separates out specific functions and makes implementing them transparent to other components. Layering allows the independent design and testing of each component. The details on the functions of each layer and the communication protocols between the layers can be summarized as indicated below:

Application layer

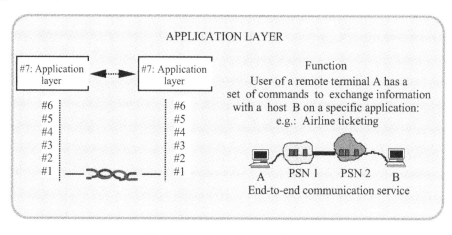

Fig. 3.66: Application layer function

The function of this layer is as follows: It provides the user of a remote terminal (Node A) with a simple set of commands to exchange messages with a remote entity (Node B) on the other end. The application layer program is partly located and runs at Node A as well as partly at Node B (Fig. 3.66).

Note:

- Actual mode of message transport is irrelevant in the design of application layer

- The application software always assumes that the two remote end application layers would facilitate a reliable communication between the ends by virtue of the service provided by the layer below.

Presentation layer

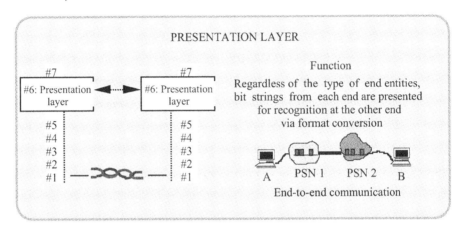

Fig. 3.67: Presentation layer function

The application layer program, for example, which runs at Node B, invariably may not know the specific type of terminal entity at Node A. Therefore, regardless of the type of bit string format generated at either end, applications should be converted as appropriate for recognition at the opposite end. This format conversion is required to accommodate different data representations and different terminal types at the Nodes A and B. Essentially, the presentation layer performs this format conversion. Relevant presentation layer programs run at both ends (Fig. 3.67).

Note:

- Application layer at Node A gives message to the presentation layer (at A) for conversion into a standard format for onward transmission to the presentation layer at Node B where this standard format will be recognized and converted into a necessary format expected by the application layer at B

- Sometimes, the presentation layer may also perform encryption or decryption on an ad hoc basis.

Session layer

The functions of this layer are: When the connection is first set up, Nodes A and B may agree on certain rules of dialogue such as establishing a full-duplex communication (namely, simultaneous two-way communication). The session layer performs this negotiation.

Note:

- Session layers also periodically monitor the synchronization points, and in the event of a failure, data sent since the last synchronization point is retransmitted.

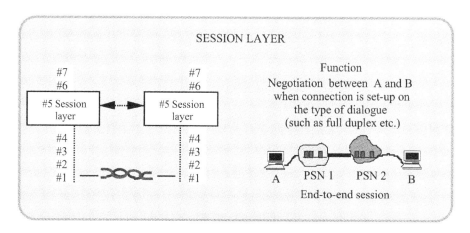

Fig. 3.68: Session layer function

Transport Layer

Functionally, the transport layer uses the end-to-end packet transmission facilitated by the network layer to perform the following:

- Supervision of the message transmission broken into packets which are numbered

- Implementation of an end-to-end message transmission service.

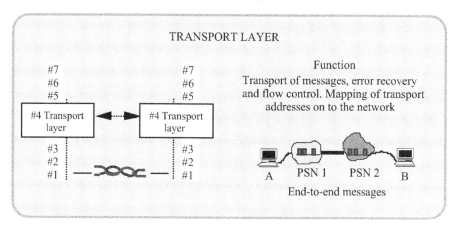

Fig. 3.69: Transport layer function

Network layer

The functions of this layer are as follows: Network layer uses the packet transmission service to carry packets from the source host to the destination host. This is done by transmitting packets on the successive links along the paths from the source to the destination. The network layer selects the paths. That is, the network layer implements an end-to-end packet transmission service between terminal nodes or host entities of the network.

Note:

- Network layer, for end-to-end packet transmission, may use :

  ✓ Datagram transport
  ✓ Virtual circuits.

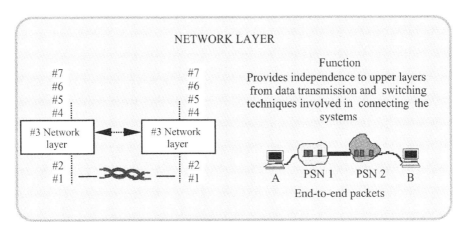

Fig. 3.70: Network layer function

Data link layer

Data link layer implements packet transmission service from one node to the other node which are directly connected by a link. Data link layer may request for retransmission of a packet, which is not received correctly.

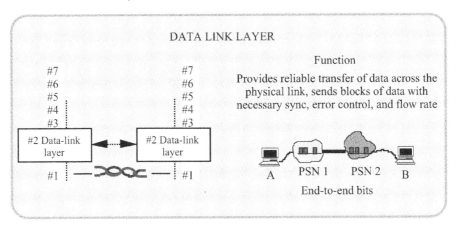

Fig. 3.71: Data link layer function

Physical layer

The physical layer facilitates a *bit pipe* available to the data link layer for the transmission of information as electrical or optical signal waveforms. That is, the data link layer distributed into two nodes can exchange bits by means of the physical layer.

Note:

- The physical layer pertinent to a node converts bits to electric/optical format and transmits these signals. The signals are converted back into bits by the physical layer at the receiver node

- For the data link layer, it does not matter whether the information was sent over a coaxial line or a fiber from the physical layer of the sending end to the physical layer of the receiving end

- Bit transmission service provided by the physical layer is used by the data layer to construct a packet transmission service.

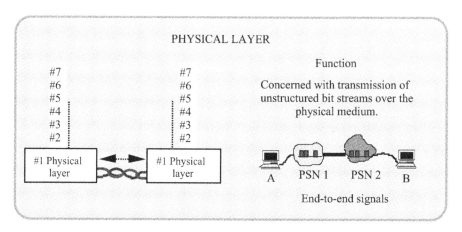

Fig. 3.72: Physical layer function

Datagram transport

Datagram packet switching refers to the transport of successive packets independent of one another. Therefore, it is possible for different packets from the same file to follow different paths in the network. This allows quick changes in paths in the event of path/link failure. In this transmission, it should be noted that:

✓ The packets may arrive in an order distinct from that of their original format

✓ In short, in datagram packet-switching, the bits are grouped as packets. Each packet is labeled with an address for destination, and the packets are routed independently of one another

✓ Datagram transport does not require any connection setup. It is ideal for short distance transmissions.

Virtual circuit packet-switching

The characteristics of VC-based packet-switching are as follows:

▪ The packets belonging to the same communication service are transported along the same path, called the virtual circuit

▪ The packets are streamed to follow one another as if they have been provided with a dedicated path (or circuit), even though they night have been interleaved with other packet streams

▪ Virtual circuits may include checking the correctness at each node of the packet it receives, and if an incorrect packet is received, a retransmit request can be made to the previous nodes

▪ Error verification is facilitated by the error correction code adopted

▪ Link error control is done by a positive acknowledgement (that is, a node acknowledging the reception being a correct one). If an uncorrectable error is detected, the link error control indicates a negative acknowledgement and requests for a retransmission.

Setting up virtual circuits

- Requires the selection of a path that will be used for the duration of the file transmission

- Packets of this transmission are labeled with a virtual circuit number which designates the path

- Routing designations are made when the VC is set up

- Each node stores these routing decisions in a routing table which indicates the path to be followed by a packet as per its labeled VC number

- Once a VC is set up, the node determines the path to be followed by a packet, by looking at the routing table. The node, therefore, makes no complex routine decisions. Further, since packets are sent in correct sequence, reassembly is simple

- These features permit fast and long duration communication.

OSI protocol: Summary

Layers	Functions
Application	A communication service of logical connectivity providing user services (such as e-mail, file transfer etc.) using application programs
Presentation	A communication service deliberated in an encoded format
Session	A session organizer for the logical connectivity
Transport	An end-to-end connectivity of message transportation across the network
Network	An end-to-end routing of packetized messages
Data link	A link level passage of packets between nodes preserving the semantic integrity
Physical	A bit pipe for the transmission of bits in the physical media

Functioning of a layered architecture

Referring to Fig. 3.73, layer N entities are known as *peer entities* or *protocol entities*. Suppose Nodes A and B jointly execute a specific protocol. In doing so, the peer entity at Node A would engage only in a virtual communication with the peer entity at Node B.

That is, it may appear as if a direct communication takes place between Nodes A and B. But, in reality, the actual exchange of information (between them) would involve several complicated steps facilitated by a chain of service primitives. Relevant considerations are discussed below.

The execution of a protocol refers to the following steps:

- ✓ A request is made from N^{th} layer to $(N–1)^{th}$ layer on the format of communication
- ✓ An indication on this request is forwarded at Node B to the protocol entity (at B)
- ✓ A response is sent back from Node B to Node A as a reply to the indication

✓ Based on the response sent from Node B, a confirmation at Node A is given to the protocol entity.

The details of PDU and SP considerations are as follows: Suppose protocol entities (at Nodes A and B), say at layers N, are to be engaged in exchanging messages (called PDUs) as indicated in Fig. 3.73.

To facilitate the exchange of PDUs as above, the necessary service primitives (SPs) are provided by layer (N–1) as shown in Fig. 3.73.

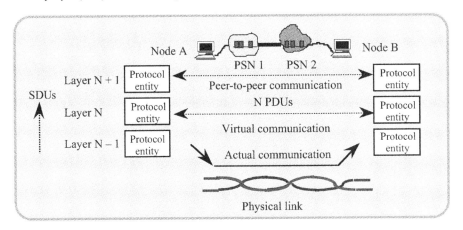

Fig. 3.73: Virtual communication between Nodes A and B

The services associated with SPs are as follows:

N-request: This service primitive refers to a request placed (say at Node A) by sending some *interface control information* (ICI) to layer N. The functional aspects of ICI are:

✓ It specifies the request type
✓ It carries the parameters such as addresses of the nodes
✓ It indicates the protocol entities inside the nodes
✓ It declares the desired quality of information transfer service
✓ It contains a pointer to the N PDU, that is, the location of that data in the memory of Node A.

(N – 1)-Indication: This (N–1) service sends an indication to the protocol entity in layer N at Node B.

(N – 1)-Response: The protocol entity at Node B (layer N) sends a response to the (N – 1)-indication.

(N – 1)-response: This contains parameters that describe the reply to the (N – 1)-request.

(N – 1)-confirm: This is the confirmation when the response eventually gets back to the protocol entity in layer N at Node A.

Thus the N-PDUs, (N – 1)-PDUs etc. constitute the virtual communication between peer entities, and (N – 1) service data units, N service data units etc. depict the actual communication between layer to layer. In summary, the PDU and SDU considerations are as follows:

Protocol data units (PDUs) and service primitives (SP)

In Fig. 3.73, layer-N performs N services available to N^{th} layer. These services are used by N^{th} layer protocol entities to exchange N *protocol data units* (PDUs). Layer N uses the (N – 1) services via *service primitives*.

Protocol data units (PDU)

- In a layered architecture, a service of layer N uses the service of layer (N–1)

- Typically, a service of layer N is executed by peer protocol entities in different nodes of a communication network.

- Messages exchanged by peer protocol entities of layer N are called Layer N *protocol data units* (N-PDUs).

Service data units (SDU)

- Messages exchanged by a service of layer N are known as: Layer N *service data units* (N-SDUs)

- Communication between layers proceeds by exchanges of the following *service primitives*:

  - ✓ Requests
  - ✓ Indications
  - ✓ Responses
  - ✓ Confirms.

Encapsulation

When the information flows from application layer to the physical layer, specific details can be added to it at each layer. That is, ICIs are added at successive layers. For example,

- Presentation layer may indicate which compression algorithm is appropriate to reduce the number of bits transmitted

- Session layer adds a session identification number

- Transport layer could specify packet sequence numbers

- Data link layer adds a link sequence number in order to keep track of the packets being acknowledged by the next node etc.

- Physical layer adds the error control bits.

As the information at the receiving end proceeds from physical layer to the application layer, the intermediate layers would remove these ICIs added by the corresponding layers at the sending end.

Architectures other than OSI

The IEEE 802 standards for LANs

As will be indicated later, the standards developed by the IEEE 802 working groups specify the physical layer and the data link layer of LANs. The data link layer is decomposed into the media-access-control (MAC) layer and the logical-link-control (LLC) layer.

Technical office protocols (TOPs) and manufacturing automation protocols (MAPs)

TOP is a set of protocols for office applications. MAP is a set of protocols for manufacturing automation and TOP is designed for ethernet and token-ring networks and

MAP is intended for token-bus networks. The main application layer programs in TOP are file transfer and management (FTAM), message-handling system (MHS) used for electronic mail, virtual terminal (VT) for remote terminal applications, directory service (DS), and job transfer and manipulation (JTM) used for remote execution of jobs. MAP provides a manufacturing message service (MMS) for communicating with control processors. It specifies primitives for reading and setting the registers.

Department of Defense (DoD) protocols

Internet networks use the DoD protocols. The application protocols include the file-transfer protocol (FTP), the simple mail-transfer protocol, (which refers to the electronic mail application) and the Telnet or the remote terminal program. These applications run on top of transport protocol namely, the connection-oriented transport-control protocol (TCP) or the connectionless user datagram protocol (UDP). These transport protocols rely on the Internet Protocol (IP), which refers to a datagram network layer. Internet is the name given to a collection of interconnected networks that use the TCP/IP protocols with its addressing scheme. These networks include APRANET, USENET, BITNET, NSFnet, and Cypress.

---

3. 9   LAN, MAN, WAN and GAN

3.9.1   *Local area networking*

Computers as they grew through generations became large, expensive, and manifested as complex mainframes that required a tremendous amount of space and specialized maintenance. A crew was required to steer the system administration and a computer expert was required to run the machine for various batch jobs. Such jobs that were submitted for execution would not come back immediately with the results.

With the advent of semiconductor technology, VLSI implementation, and supporting hardware, it has become economically feasible to own and interactively operate computers directly from desktops. However, in order to accommodate the computer-to-computer communication, a new form of networking came into existence to cope with the associated interoperation. It is called the *local area network* (LAN).

In the modern context, LAN is a communication network capable of providing an intrafacility internal exchange of voice, computer data, word-processing, facsimile, video conferencing, video broadcasting, telemetry, and other conceivable forms of electronic messaging. According to the IEEE 802 Committee, "A Local Area Network is a data communications system which allows a number of independent devices to communicate with each other."

A LAN can be further described as follows:

- LAN is intra-institutional, privately-owned and user administrated. It is not generally subjected to FCC regulation. The exclusions refer to common carrier facilities, including both public telephone systems and commercial cable television systems

- LAN represents a networking that is integrated through interconnection by a continuous structural medium. That is, multiple services may operate on a single set of cabling

- Variety in the designs of LAN configurations exists and offers a full connectivity profile

- In general, LANs can support low speed and high-speed data communications. Further, LANs are not subject to speed limitations imposed by traditional common carrier facilities. They can be designed to support devices ranging in speed as low as 75 bps to about 150 Mbps with fiber optic backbone

■ Today any LAN architecture is available off-the-shelf. The LAN market is, however, still volatile, notwithstanding the standards evolved and persistent vendor supports. The announced products in the market could still be in the beta-test stage.

LANs by definition link the hosts within a limited geographical area — a few miles from each other. Studies indicate that about 80% of the communications normally take place within the designated local environment and the remaining 20% may spill outside the local geographical area. A LAN can interconnect these resources in a manner that best suits one's needs to perform everyday communication tasks. Examples of these resources are as follows:

Data terminal equipment(DTE)

  ✓ Personal computers
  ✓ Main frame computers.

Peripheral devices

  ✓ Laser printers
  ✓ Graphics plotters etc.

Storage means

  ✓ Mass storage devices such as optical disks
  ✓ Magnetic storage media
  ✓ Servers and server farms.

Data communication equipment (DCE)

  ✓ High speed modem.

3.9.2 *LAN topology*

The topology of LAN refers to the geometrical outlay or configuration of intelligent devices and how they are linked together for communications. The intelligent devices on the network are called the *nodes*. Nodes in a network are addressable units. The interconnection between any two nodes is known as the *link*. Nodes are also referred to as *stations*. And the two terms are used interchangeably. The link constitutes the *communication channel*. In the design of a LAN, the topology is specified so as to be the best suited in meeting a particular environment. In other words, there are advantages and disadvantages pertinent to various topologies used. Factors like message size, traffic volume, costs, bandwidth, reliability, and simplicity are the governing issues. The most common LAN topologies are described below:

The star configuration

The star topology is the most common configuration used in telecommunication networking. It was first used in the telegraphic networks and later in the telephone. Subsequently, it was adopted into data communication networks. In recent years, it has become the most popular topology for all networks. The distinct feature of the star topology is that each node is radially linked to a *central node* as a point-to-point connection in a hub-and-spokes fashion. The central node of the star topology is referred to as the *central control, control switch,* or *hub*. The central node, typically a computer, controls the communications within the LAN. Any traffic between the outlying nodes must flow through the central node. Central control also offers a convenient base for troubleshooting and network maintenance.

The star configuration is very pragmatic in cases where most of the communication occurs between the central node and outlying nodes. When traffic is extensive between outlying nodes, the burden rests on the central node to control and ease the situations, which otherwise would cause a bottleneck and induce undue message delays. The star topology lacks reliability, performance, and robustness of the traffics stemming from all of the nodes connected to the LAN. This means that, if the central node fails, the network becomes paralyzed. However, if an outlying node fails, the remainder of the system can still continue to operate independently regardless of that node's failure. Fig. 3.74(a) illustrates the star topology.

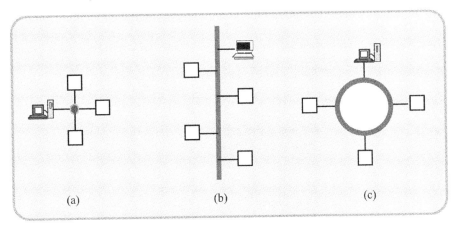

Fig. 3.74: LAN topologies: (a) Star, (b) bus, and (c) ring

The bus topology

The bus topology is essentially a multipoint or multidrop configuration of interconnecting nodes on a shared channel. The Ethernet LAN standard specifies the bus topology. In this configuration, control over the communication channel is not centralized at any particular node, contrary to the star topology. The distinguishing feature of a bus LAN is that the control of the bus is distributed among all of the nodes connected to the LAN. Whenever data are to be transmitted from one node on the bus to another, the transmitting station "listens" to the current activity on the bus. If no other nodes are transmitting data, that is, when the channel is clear, then it begins its transmission of data, and all stations connected to the bus become receivers of that transmitter. Then the function of each receiver is to decide whether the received data packet corresponds to its own address. If it does, the data packet is received; otherwise the receiving node discards the data. Fig. 3.74(b) depicts the bus topology.

Inasmuch as the control of bus LAN is not centralized, a node failure will not hinder the operation of the remaining part of the LAN.

The ring topology

As its name implies, the ring topology interconnects nodes point-to-point in a closed-loop configuration (Fig. 3.74(c)). The so-called *token-ring* configuration uses this topology. Normally, a message is transmitted in simplex mode (that is, in one direction only) from node-to-node around the ring until it is received by the original source node. Any node in the loop can act as a repeater, retransmitting the message to the next node. Each node also shares the responsibility of identifying if the circulating message is addressed to itself or another node. In either case, the message is received by the destination node and returned to the original source node. The source node then verifies that the message (token) circulated around the ring is identical to the originally transmitted message.

LAN standards

The type of protocol and access method used in a LAN system depends on the specific type of LAN standard that a vendor envisages. The IEEE 802 Standards Committee established the LAN standards. Briefly, these standards refer to the following:

802.1: This is the highest level interface standard. It is concerned with interfacing with other networks. Most of the specifications are yet to be defined.

802.2: *LLC protocol*: This is equivalent to the second layer of the OSI model. It provides a point-to-point link control between devices at the protocol level. Many applications for data communication on LAN use 802.2 so that they can be interfaced with the layers of the OSI model. The logical link control protocol can be used with any of the 802 configurations.

802.3 *CSMA/CD baseband bus*: This is a physical layer protocol, which addresses the *carrier sense multiple access with collision detection*. It can be used with a bus or star topology. When used with a bus configuration, such LANs as mentioned before depict the ethernet.

802.4 *Token passing bus*: This defines an alternative layer-1 protocol for a token-bus, again suitable for either a bus or star topology.

802.5 *Token passing ring*: This defines a layer-1 protocol for use on a token-ring topology.

802.6 *Metropolitan area networking protocol.*

The protocol standards applied to the different topologies may vary slightly from each other. The version of the physical layer protocol and the type of the LAN to be adopted depend largely on individual preference and the compatibility of the existing computers to be placed on the LAN. The description of various protocols and their relative merits are as follows:

CSMA/CD (IEEE 802.3, ISO 8802.3): Ethernet

The carrier sense multiple access with collision detection (CSMA/CD) is a contention protocol. It describes the techniques by which any device on a bus can transmit when the medium interface determines that no other device is already transmitting. This type of LAN uses twisted pair or a coaxial line as its physical medium. On a CSMA/CD LAN, the terminals do not request permission from a central controller before transmitting data on the transmission channel, they rather contend for its use. Before transmitting a packet of data, a sending terminal "listens" to check whether the path is already in use, and if so, it waits before transmitting its data. Even when it starts to send data, it needs to continue checking the path to make sure that no other nodes have commenced sending data at the same time. If the output of the sending terminal does not match with that of the one which is simultaneously monitored on the transmission path, it means that there has been a *collision*. In order to receive data, the medium access control (MAC) or layer-1 software in each terminal, monitors the transmission path and decodes the destination address of each packet passing through so as to ascertain whether that terminal is the intended destination. If it is, the data is read and decoded; otherwise, the data is ignored.

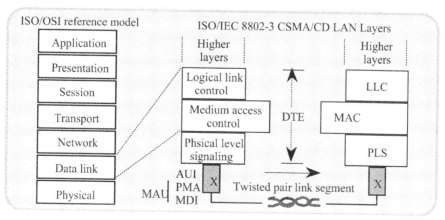

Fig. 3.75: Ethernet specification in relation to seven-layer ISO/OSI model
AUI: Attachment unit interface; PMA: Physical medium interface; MDI: Medium-dependent interface and MAU: Medium access unit

The most common network that employs CSMA/CD is the ethernet. Developed in 1980 by the combined efforts of the Xerox Corporation, Digital Equipment Corporation and Intel Corporation, Ethernet has been accepted under the IEEE 802.3 specification. The ethernet specification addresses the physical and data link layers of the ISO/OSI seven-layer model as discussed earlier.

In the existing market, ethernet LAN components are relatively inexpensive. Further, the bus topology is easy to realize and manage and is resilient to transmission line failures. Therefore, ethernet has become a popular type of LAN.

IEEE 802.4, ISO 8802.4: Token-bus

Here, a device on a bus topology transmits only when it receives a token. The token is passed in a user-predetermined sequence and guarantees network access to all users. That is, a token bus LAN controls the transmission of data onto the transmission path by the use of a single token. Only the terminal with the token may transmit packets onto the bus. The token can be made available to any terminal wishing to transmit data. When the terminal has the token it sends frames of data kept ready for transmission. Then it passes the token on to the next terminal. To verify whether its successor has received the token correctly, the terminal confirms when the successor commences transmitting the data.

Token-bus networks are most common in manufacturing premises. They are not commonly used in an office environment where ethernet and token-ring networks are more widely used.

IEEE 802.5: Token-ring

The token-ring standard is similar in operation to the token-bus in making use of the token to pass the "right to transmit data" around each terminal on the ring in turn. However, the sequence of token passing is different: That is, the token itself is used to carry the packet of data. The transmitting terminal sets the token's flag, by placing the destination address in the header to indicate that the token is full. The token is then passed around the ring from one terminal to the next. Each terminal checks whether the data is intended for it, and passes it on. Eventually, the token reaches the destination terminal where the data is read. Receipt of data is confirmed to the transmitter by changing a bit value in the token's flag. When the token gets back to the transmitting terminal, the terminal is obliged to empty the token and passes it to the next terminal in the ring.

Token-ring, like ethernet, is common in office environments, linking personal computers for the purpose of data file transfer, electronic messaging, mainframe computer interaction, or file sharing.

IEEE802.6: Metropolitan area network (MAN)

Metropolitan area networks (MAN) connect locations that are geographically located beyond 5 km up to 50 km. They support transmission of voice, data, and video. Coaxial cables, optical fibers, and wireless media are used as the physical media for transportation of signals.

Large enterprises are customers of MANs. Their specific needs to communicate at high-speeds within a metropolitan area are met by MAN topology. MAN providers offer tariff rates lower than telcos and facilitate faster installations over a diverse routing plus providing backup lines to meet emergency situations.

Wide area networking (WAN)

WAN is made of networks spread over a large geographical area. The WANs " cross public right of way" and use common telcos. They adopt a variety of communication media for interconnection, which includes switched and leased lines, private microwave circuits, optical fiber, and coaxial lines. WAN also supports voice, video, and data transmissions.

When taken across countries or regions (through satellite circuits), WANs explode into global area networks (GANs).

3.10   A Passage through PSTN to ISDN...

The principal function of PSTNs as they were originally conceived was to support telephone conversation service. However, with the advent of the need for data communication, as indicated earlier, telcos facilitated data service on telephone links (for example on T-1 service). But, still there are other service requirements mooted: What the carriers needed was a universal switched service that would allow the customer to set up a connection to any location for any type of information — on demand and at an attractive cost compared to leased lines. *Integrated switched data network* (ISDN) was therefore, created as a solution.

3.10.1   What is ISDN?

It refers to a global telecommunication service using the integration of transmission and switching functions in a circuit-switched telecommunication network in order to support voice and digital data communications [3.22, 3.23].

In the context of modern practice, the users want information transmission pertinent to voice, video, fax, image etc. In order to achieve such an integrated transmission, the originally implemented "voice only" system required additional resources be included so as to support the integrated digital communication. Further, the central office (CO) as well as the subscriber ends required the provision of compatible digital equipment. Thus, ISDN is an evolution of PSTN – from voice telephony to digital data transmission.

In essence, ISDN is a shared facility; that is, with ISDN all terminals, which plug into the same jack (Fig. 3.77), share the same wiring. On the contrary, under a non-ISDN environment, the PSTN supports voice and data transmissions as two independent transmissions. That is, voice and data terminals need separate wiring as illustrated in Fig. 3.76. In ISDN, however, one common outlet RJ-45 provides integrated access to a variety of services.

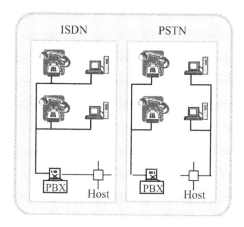

Fig. 3.76: PSTN versus ISDN

The merits of ISDN are as follows:

✓   Flexibility
✓   Portability of entities at a terminal
✓   Efficiency: A single line is shared by multiple terminals
✓   Devoid of limitations on transmission rates and quality of modems
✓   Residences can continue using the same 2-wire copper pair for the local loop connection.

3.10.2   Access interfaces of ISDN

The two methods of accessing ISDN are:
Basic rate interface (BRI): These are intended for:

▪   Residential customers

▪   Small business customers

▪   Individuals with a large business.

Primary rate interface (PRI): These are intended primarily for:

- Large businesses.

Bit rate interface (BRI)

This consists of four channels: Two *B-channels* (bearer channels), each supporting a rate at 64 kbps and two *D-channels* (data channels), which support a speed of 16 kbps. The functional aspects of these channels are as follows:

- One pair of metallic wires from the CO to the subscriber(s) allows voice communication on the B-channel with the information encoded as PCM

- Data connection is facilitated from one entity to the other at 64 kbps on the other B-channel

- If voice is *compressed* (see the indented note below) at 16 kbps at both ends, then four separate conversions are feasible on a B-channel using multiplexing and demultiplexing to switch these four channels as appropriate

- It is also possible to send fax, slow-motion video, or similar information on B-channel.

The D-channel is intended for the following:

- It facilitates connect-disconnect for the B-channels as required by control signaling

- It can also be used for low-speed packet-switching, telemetry etc.

Voice compression

As indicated in earlier chapters, the telephone quality speech is normally sampled at 8000 samples/s and quantizied at 8 bits/sample (corresponding to a rate of 8 x 8 = 64 kbps). This represents an uncompressed voice. However, simple compression algorithms like adaptive differential PCM (DPCM) can be utilized in which the correlation that prevails between adjacent samples can be used to reduce the number of bits by a factor of 2 to 4 with almost imperceptible distortion in the audio.

Much higher compression can be also obtained with algorithms like linear predictive coding (LDC), which models speech as an autoregressive (AR) process and sends only the parameters of the process as opposed to sending the speech itself. With LPC-based methods, it is possible to code speech at less than 4 kbps. (However, if very low bit rates are adopted, the compressed speech would sound artificial and synthetic.)

3.4.3   *ISDN services: A summary*

- Residential subscribers are provided with basic access consisting of two full-duplex 64 kbps channels (B-channels)
    Note: Half duplex – Stations not capable of simultaneous sending and receiving, regardless of communication facility.
    Full duplex – Stations capable of simultaneous sending and receiving

- One-full duplex, 16 kbps channel (called D-channel, or data channel). D-channel is used for telemetry and exchanging network control

- One B-channel is used for digitized voice, videotext etc.

- Primary access: This consists of 23 B-channels (64 kbps each) and one D-channel of 64 kbps made available to larger customers

- The basic ISDN access is equal to 2B + D. Each B channel is a full-duplex 64-kbps channel. The D channel is a full-duplex 16-kbps channel

- The envisioned services for ISDN include facsimile transmissions, access to databases, videotext, electronic mail, and alarms, in addition to the telephone service, as indicated in Table 3.10

- ISDN implementation requires upgrading of subscriber loops, making the 64-kbps channels available, and facilitating packet-switched channels.

Table 3.12 provides a summary of various services channel types and the associated channel rates.

Table 3.12: ISDN services based channels and their bit transmission rates

Services	Channel rate	Channel type
Voice		
Telephone	64 kbps	BC
Security and metering		
Alarms	100 bps	D
Utility metering	100 bps	D
Energy management	100 bps	D
Data communication		
Videotext	2.4-64 kbps	BP
Electronic mail	4.8-6.4 kbps	BP
Facsimile	4.8-64 kbps	BC
Slow scan TV	64 kbps	BC

3.11. Broadband Integrated Services Digital Network (B-ISDN)

Commensurate with the implementation of ISDN, (in the late 1980s) the telecommunications standards community considered a paradigm shift towards a second-generation ISDN, namely, the *broadband ISDN (B-ISDN)*. This network was conceived as an all-purpose digital network of the future to provide multiple capabilities to users that are not available in the "narrow-band" ISDN [3.22, 3.23].

The demands of new graphical computer software, the emergence of video telephony, and cable television, and the political pressure for the development of a super information highway jointly led to the development of B-ISDN.

Narrow-band ISDN (N-ISDN) in its original form falls incredibly short of the ultimate needs of a multi-service network because of the relatively restricted bit rate available on individual channels and due to the lack of service flexibility arising from the fact that it evolved out of classical telephone network principles. The maximum rate possible on original ISDN is n × 64 kbps, up to 2 Mbps. Such rates are not enough for very high speed file transfer between mainframe computers, or for interactive switching of broad bandwidth signals such as high definition television and video.

Increasing the unit bit rate (64 kbps) of N-ISDN would increase bit rates and make broadband and multimedia services possible, but it would be at the expense of gross inefficiency in network resources, particularly when carrying bursty data signals. So B-ISDN is not merely an extension of N-ISDN, but instead, it also capitalizes on the latest data switching techniques. Its protocols, architecture, transmissions and switching technology and platforms are different from ISDN.

B-ISDN is not a totally new concept. Many of the ideas are extracted and enhanced from N-ISDN and other telecommunication and data communication protocols. The reason for such extraction is that the fundamental objective of B-ISDN is to achieve complete integration of services, ranging from low-bit-rate bursty to high-bit-rate continuous real-time signals.

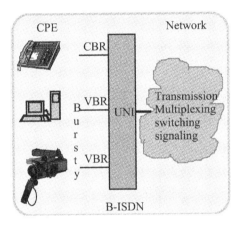

Fig. 3.77: The concept of B-ISDN

From the point of view of users, B-ISDN will offer a single interface that provides for all the required communication needs — voice, data, video-and will carry both signal and user information. The interface between user and the network will be identical for all users of B-ISDN.

Now, a network operator or service provider views B-ISDN at a different angle. The following are some of a network operator's aspirations for any network:

✓   Maximum utilization of the network, considering grade of service to maintain performance levels demanded by the users of the network
✓   Single-operation interface for all network elements and terminals attached to the network
✓   Ability of networks operators to rapidly identify, isolate and thus minimize the impact of faults occurring in the network
✓   Ability to manage the traffic transiting through the network in terms of automatic rerouting in case of congestion.

The services offered by B-ISDN networks can be largely split into two categories: *Interactive services* and *distribution services*. Interactive services are normal communications between just two parties. There are three subtypes, namely, *messaging, conversational,* and *retrieval services*.

An example of a conversational service is a telephone conversation or point-to-point data connections, where the two end-points of the communication are in real-time connection with one another and thus converse.

A message service is a telecommunication service that is often compared to the postal service. A message is submitted to the network. (This corresponds to a posted letter.) Subsequently, the message is delivered to the given address. Message services are usually not guaranteed. The network will not be able to check whether the recipient's address is valid, and may return no confirmation to the sender of receipt. Thus "no reply" may result either because the recipient never got the message or because the recipient remained silent.

A retrieval service is one in which a caller accesses a central server, database or storage archive, requesting the delivery of certain specified information.

Distribution services refer to the information from a single source distributed to many recipients at the same time. Distribution services are subdivided into those with or without individual user presentation control. An example of a distribution service without user presentation control is the broadcasting of a television program. All the viewers of the television receive the same program at the same time. An example of a distribution service with user presentation and control is video-on-demand. Here, only those subscribers who wish to pay would receive a given video.

In short, the objectives of B-ISDN are focused on defining a user interface and a network that will allow a wide range of telecommunication needs as described below:

- *Facilitation of interactive and distributive services*: Interactive services refer to a two-way exchange of information between subscribers or between a subscriber and a service provider. These services include conversational service, messaging services, and retrieval services. On the other hand, distributive services involve information transfer primarily in one direction, from a service provider to a subscriber. They correspond to broadcast services

- *Supporting continuous and bursty traffics*: Continuous traffics (for example, voice and interactive video) require guaranteed bandwidth so as to meet prescribed QOS requirements (such as low information loss and delay). On the other hand, a bursty data traffic characterized by short and sporadic transmissions can be handled more cost effectively with a lower grade of service, by using shared network resources on a statistical basis

- *Semantic and temporal transparency:* Certain traffics such as voice and video are isochronous in the sense that their transmissions should be effected almost on a real-time basis. This is because, the relevant information borne by these traffics is delay sensitive. Hence, transmission of isochronous traffics has to be provided for in temporal transparency. On the other hand, messages corresponding to data (such as file transfer) are not affected if a delay is perceived in their transmissions. However, such data traffics should be protected from bit errors or their semantic transparency will be impaired. B-ISDN is conceived to offer delay-free and/or error-free transmissions

- *Point-to-point and point-to-multipoint connections*: Some services such as voice would require the use of point-to-point connections, whereas other services such as multimedia communications benefit from parallel connections between end points.

The rate of transmission for broadband ISDN involves data rates in excess of 45 Mbps, with a vision to extend beyond 2.5 Gbps. In such cases, the optical fiber is the only option on the transmission medium that can support such high data rates. Therefore, the introduction of B-ISDN became a surrogate to the pace at which optical fiber links were built.

Another distinctive aspect of B-ISDN is the way in which the switching mechanism is implemented. It is done by a new user-network interface protocol known as the *asynchronous transfer mode (ATM)* proposed by the CCITT. The asynchronous transfer mode (ATM) briefly can be described as a high-bandwidth, low-delay, packet-like technique used for switching and multiplexing. It is independent of the physical means of transport. Its "low-delay" characteristics would support real-time services (such as voice) and its "high-bandwidth" feature is set to handle the video.

It is expected that the comprehensive deployment of B-ISDN would lead to the evolution of an "information society" enclaving an all-purpose communication system. That is, the associated services would then support all the conceivable types of communication media. Moreover, it would be possible to combine and integrate communication media for individual users and their application on an ad hoc basis in such broadband integrated systems.

Suppose users with access to larger computers in remote locations would like to bring the high-quality graphics and video capabilities of those computers to their desktop workstation. Today's desktop workstations are almost equitable to the past mainframe computers in respect to computing power and operating speed, thanks to the advances in microelectronics. With this colossal computing power at their disposal, the users of desktop workstations naturally look for more flexible interfaces with easy access to telco backbones.

Modern workstations at the desktop level are also playing new roles in interpersonal communications, enabling users to interact with each other from their homes or offices by exchanging information in different media. For example, computer-based desktop conference systems can support real-time conversations among the participants through a coordinate exchange of voice, video, and computer data (such as text, graphics and spreadsheets). Further, the conference participants can be either close to one another, exchanging information over local area networks, or, they can be geographically far apart, in which case the exchange of information takes place over WANs. In any case, the common objective of computer-based, desktop conference systems is to emulate realistic attributes of tête-à-tête conversations, sitting in the comfort of the users' home or at the office. And, B-ISDN is intended to offer a feasible technology to realize this scenario.

In a similar context, there is a general trend among business enterprises toward the development of high-capacity private communication networks interconnecting corporate centers in different geographic locations. These networks are required to meet a distinct business need so as to have access to and transport large volumes of data from one location to another. Further, such networks should include the relevant infrastructure to provide such services in a highly secured fashion. The construction of this kind of communication network is the outcome of a natural inclination on the part of business customers, regardless of the size of the business, to exercise control over their own telecommunication services with flexibility and ease in a cost effective manner.

The emergence of B-ISDN, together with rapid advances in storage technologies, make it feasible to provide multimedia-on-demand type of information services. The service offered by a multimedia server is analogous to that of a video rental store. The multimedia server digitally stores multimedia information such as entertainment movies, documentaries, and data bases on a large assortment of extremely high-capacity storage devices (like optical and magnetic disks). These storage devices are accessible with a short seek-time. They could be facilitated permanently online. Subscribers can select the multimedia information of their choice, and the multimedia server has the ability to satisfy their enquiry in an interactive manner, provided a broadcast transfer of information is available in the associated telecommunications.

3.11.1   ITU-T definition of B-ISDN

B-ISDN is a service requiring transmission channels capable of supporting rates greater than the *primary rates*. The B-ISDN access and typical B-ISDN services conceived thereof are as follows:

- ✓   Digital TV (about 100 Mbps, uncompressed)
- ✓   Digital HDTV (about 150 Mbps, compressed)
- ✓   Digital Hifi (about 2 Mbps, uncompressed)
- ✓   Multimedia terminals : Video-phone, graphics, data
- ✓   High speed retrieval of graphics and voice from databases
- ✓   Interconnection of LANs.

The specific aspects of B-ISDN can be listed as:

- ✓   The B-ISDN would make 150 Mbps channels available to users
- ✓   The transmission will be rendered on optical fibers
- ✓   The switching may use a combination of circuit-switches and *ATM switches*

✓ The shared buffer ATM switch is one of the many designs being explored.

As discussed earlier in this chapter, ISDNs were developed as telecommunication design options to provide an end-to-end digital connectivity that supports a wide range of services (including voice and non-voice services) to which users have access by a limited set of standard multipurpose user-network interfaces. The interface structures of ISDN as discussed before can be summarized as follows:

Basic access interface

✓ 2 × 64 kbps B-channels
✓ 1 × 16 kbps D-channel (for signaling).

Primary rate access

✓ Gross bit rate 1.5 or 2 Mbps constituting a high speed H-channel
✓ 64 Kbps signaling channel
✓ Mixture of H and B channels as indicated in Table 3.13.

Table 3.13: Various ISDN channels and interface structures

Channel/bit rate (kbps)	Interface
B/64	Basic access
H0/384	Primary rate access
H11/1536	Primary rate access
H12/1920	Primary rate access
D16/16	Basic access
D64/64	Primary rate access
(2B + D16)/192	Basic access
(23B + D64)/1544	Primary rate access
(3HO + D64)/1544	Primary rate access
(H11 etc)/1536	Primary rate access
(30B + D64)/2048	Primary rate access
(5HO + D64)/2048	Primary rate access
(H12 + D64 etc)/2048	Primary rate access

Goals of B-ISDN

Within the existing framework of ISDN (narrow band version), the goals of B-ISDN were focused to accomplish the following:

▪ Adding new high-speed channels to the existing channel spectrum of narrow band ISDN

▪ Defining a set of new broad-band user-network interfaces

▪ Relying on the 64 kbps ISDN protocols, which already exist and modifying (or enhancing) them only when absolutely needed.

Consistent with the goals as enumerated above, the natural queries, which arose towards the evolution of B-ISDN are:

▪ How many flexible options on B and H channels are possible?

- Can there be flexible interface options for channel combinations (other B and/or H)?

- If channel structures of different versions are to be adopted at the interface, do they have to be fixed at the time of subscription or can they be dynamically changed?
- How to handle the cohabited bursty and streamlined traffics?

Discussions and opinions on the above queries led to a concept of adopting ultimately an interface model. This model refers to breaking down the payload capacity into small pieces or packets of constant size, called *cells* each of which can serve or match any type of traffic involved, and each cell may be employed to carry information to any type of connection. Thus, answers to the above question sowed the land of information technology to reap eventually the crop of so-called *ATM telecommunications*.

Feasible B-ISDN services

Consistent with the objectives and scopes of the B-ISDN, the service categories intended to be supported are enumerated in the following tables (Table 3.14 to 3.17):

Table 3.14: Messaging B-ISDN service categories

Type of information	Broadband service applications
Video and sound	▪ Video mailbox service for the electronic transfer of moving pictures and accompanying sound
Document/text	▪ Electronic document mailbox service for the transfer of mixed documents, which contain text, graphics, still and moving picture information as well as voice annotations

Table 3.15: Retrieval B-ISDN service categories

Type of information	Broadband service applications
Multimedia: Text, data graphics, sound, still images, video retrievals	▪ Video-on-demand ▪ Remote education and training/virtual class-room ▪ Tele-shopping ▪ Tele-advertising ▪ News retrieval
Retrieval of video files	▪ Entertainment purposes ▪ Remote education and training/virtual classroom
Retrieval of high-resolution images	▪ Entertainment purposes/video-on demand ▪ Remote education and training/virtual classroom

...continued

312

- Professional image communications and medical imaging

Retrieval of data base of mixed document	• Mixed documents retrieval from library, information centers, archives etc.
Heavy picture data retrieval	• Telesoftware, special documents such as satellite images

Table 3.16: Conversational B-ISDN service categories

Type of information	Broadband service applications
Person-to-person video telephony	• The transfer of voice (sound), moving pictures and video-scanned still images and documents between two locations (person-to-person)
Point-to-multipoint video teleconference	• Multipoint communication for the transfer of voice (sound), moving pictures, and video-scanned still images and documents between two or more locations (person-to-group, group-to-group). • Tele-education/virtual classroom • Business conference • Tele-advertising
Multilocation video surveillance	• Building security • Traffic monitoring
Audio-visual systems	• TV signal transfer • Video/audio dialogue
Multiple sound mixed quality audio signal	• Multilingual commentary channels • Video plus high fidelity sound
High-speed large-volume data plus video/picture transmission	• High-speed transfer of data files on LAN and MAN • Transfer of text plus video information • Multi-site interactive computer aided design
High volume text above information	• Data file transfers with encryption if needed
Fast tele-control information	• Real-time controls • Telemetry of complex signal waveforms • Multisite alarms
Broadband high-speed fax transmission	• User-to-user transfer of text, basic images, drawings, etc.
High-resolution image communication service	• Professional images such as medical images, DNA structures etc. • High-quality video games

Table 3.17: <u>Distribution services of B-ISDN</u>

Type of information	Broadband service applications
Data distribution	▪ Distribution of unrestricted data at high-speeds
Distribution of text, graphics, and still images	▪ Electronic document distribution ▪ Electronic publishing ▪ Electronic newspaper broadcasting and announcements
Distribution of moving pictures	▪ Distribution of movies on broadcast basis and/or on demand
Existing quality TV distribution (NTSC, PAL, SECAM)	▪ TV program distribution
High quality TV	▪ Enhanced definition TV distribution service
Pay TV (pay-per-view, pay-per-channel)	▪ TV program distribution

3.11.2 Layered protocol structure of B-ISDN

The B-ISDN protocol architecture is a vertical layered architecture, which covers the transport, switching, signaling and control, user protocols, and applications and services. The architecture model is comprehensive and covers the complete set, including management. For an individual function such as switching or transmission, a subset of protocol model applies, along with its respective upper-layer functions, such as the management functions. The protocol architecture divides the functions of each layer so that appropriate functions are used to support a given application.

The B-ISDN *reference protocol model* (B-ISDN PRM), consists of three planes as shown in Fig. 3.78 They are:

- Control plane: Handles control information. It specifies how connections are made and released

- User plane: Concerns itself with user information. It addresses data flow control, error detection, and correction

- OAM plane: Performs operations, administration and maintenance through information packets that switches the exchange to keep the system running effectively.

The role played by each of these layers can be illustrated by considering a communication between two end-users. Suppose a user A wants to establish a connection (say, video, voice, or data) with another user B. The steps involved vis-à-vis the role of the protocol planes are as follows:

Step 1

When user A makes a call request to the network, appropriate switching would eventually establish the requested connection as a *virtual circuit* (VC). The relevant signaling, call set-up etc. is a control function managed by the control plane. Once the VCs are set up, the user information does not require any further routing decisions.

Step 2

Further, when the user information starts flowing along the virtual circuit, the user plane governs maintenance of bit transmission.

Step 3

Control information is transmitted by a store-and-forward transport strategy with error control provided at the link level. Any error in the transmission of control information should be corrected before it causes the switches to configure erroneously.

Management plane

Two types of functions exist in this plane, namely *layer management* function and *plane management* functions. All the management functions that relate to the whole system are located in the plane management. Its task is to provide coordination between all the different planes. No layered structure is used within this plane.

The layer management functions are in a layered structure. This structure performs the management functions relating to the resources and parameters residing in its protocol entities, such as signaling.

User plane

The function of the user plane is to transfer the user information from point A to point B in the network. All associated mechanisms, such as flow control, congestion control, or error recovery are included. A layer approach is used with the user plane in order to identify the different functional components involved in providing services to the user.

Control or signaling plane

Here again, layered structure is used. This control plane is responsible for call control and connection control functions related to setting up a connection and tearing down a connection of a call. These are all the signaling functions necessary to set up, supervise, and release a call or connection.

B-ISDN protocol layer

Implementation of B-ISDN also follows a layered architecture of stacked protocols as illustrated in Fig. 3.78. The associated layers and their functions are described below:

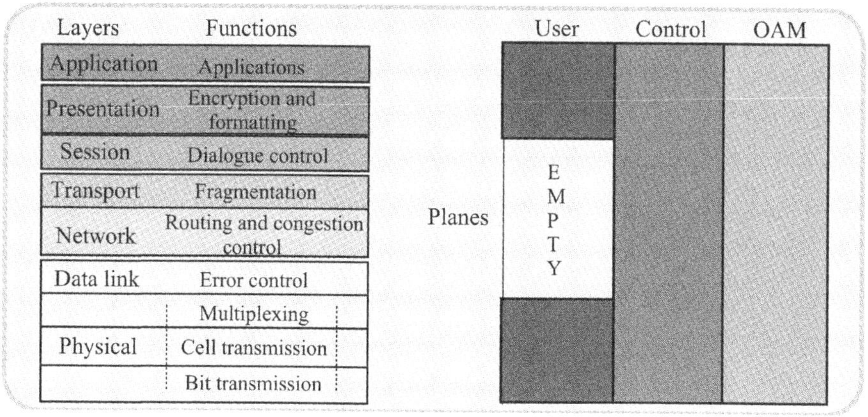

Fig. 3.78: B-ISDN protocol architecture

Physical layer

This essentially implements transmission and reception of bit stream via optical fiber (physical sublayer in Fig. 3.78). The other related functions are:

    ✓   Providing access to cells (physical sublayer 2)
    ✓   Execution of multiplexing and demultiplexing functions (physical sublayer 3)
    ✓   Bit transmission. (The relevant performance is similar to that in OSI architecture.)

3.12 ATM: An Overview of Its Evolution as a New UNI Protocol

3.12.1 ATM - a solution B-ISDN

As mentioned earlier, ISDN was the original idea, proposed in the 1970s to support integrated services pertinent to voice, video, and data on digital communication systems. Subsequently, ATM stemmed from a project to define and standardize a transmission mechanism for broadband ISDN (faster than DS-1). In 1990 CCITT selected ATM as the target architecture for basic transport of heterogeneous traffic (data, voice, and video) within the scope of B-ISDN [3.24, 3.25]. It prescribed the so-called *synchronous optics network* (SONET) and/or *synchronous digital hierarchy* (SDH) as the relevant backbone architectures [3.17]. In its designation as the target B-ISDN architecture, ATM was coupled with periodic framing of SONET/SDH, which set up the TDM channels for the cells to flow through. However, ATM cells can travel over any synchronous channel. (A brief outline on SONET/SDH as the platform supporting ATM telecommunication transmissions is presented later in this section and will be comprehensively addressed in Chapter 4.)

What are broadband systems vis-a-vis modern telecommunication?

Broadband systems in general, are capable of transporting high bit-rates. The transition profiles extending across narrow band to broadband telecommunications bearing different types of bit rate services are as follows:

- Narrow band services
  (Data rate below 1.544 Mbps)

- Wideband services
  (1.544 Mbps to 45 Mbps)

- Broadband services
  (Data rate more than 45 Mbps to 1.5 Gbps and beyond).

As indicated earlier, an integrated system of digital transmission refers to the integration of transmissions (and the associated switching functions) to support cohesively traffics pertinent to narrow band as well as broadband systems. *Why was such an integrated system warranted and what are the associated problems in implementing the integrated traffic?* The answers for these questions are buried in the following considerations:

1.   Evolutionary aspects of telecommunication traffic types, which include,

    ✓   Original type of traffic namely, voice or telephone conversation
    ✓   Emergence of data transmission requiring faster bit rates
    ✓   Use of telecommunication systems for high quality sound transmissions
    ✓   Implementation of high-speed document distribution such as text, graphics, and still images
    ✓   Realizing video telephony/video conference etc.
    ✓   Transmission of moving pictures, animations plus sound
    ✓   Opting for services like movie-on-demand, pay TV etc.
    ✓   Transmission of multimedia services.

2.   Evolutionary considerations in the type of switching or interconnections facilitating the traffic flow. These include,

- ✓ Original electromechanical telephone switching systems
- ✓ Emergence of electronic switching
- ✓ Replacement of analog communication by digital versions
- ✓ Network connections enabling voice and/or data transmissions
- ✓ Emergence of high speed data transmission networks (such as LANs etc)
- ✓ Changes in the modes of information transmission at the physical layer level : Electrical transmission to optical transmission
- ✓ Rapid growth in wirelines (cellular) telephony
- ✓ Feasible aspects of video transmission with compression etc.
- ✓ Customers inclination for ISDN facilities
- ✓ Emergence of multirate circuit-switching
- ✓ Packet switching.

The trend in modern telecommunications is set by customers' expectations to get teleservices of unbounded structures and applications. This trend naturally caused a search for compatible solutions. But relevant considerations in designing commensurate networks however, had to face the following shortcomings: Each network is conceived to support a specific type of traffic service for which it was originally developed. Sometimes, with additional equipment, (such as a modem), the traffic service can be expanded to a certain extent. Notwithstanding the emergence of such retrofit service supports, multiple services, if envisaged, would warrant multirate circuit connections. Hence the foreseeable obstacles in realizing such multiservice networking are as follows:

Inflexibility

Rapidly changing technologies and advancements in VLSI and signal processing methods enable new (voice) coding feasibility and compression of video etc. The existing networks may have difficulties in adopting such technology trends.

Inefficiency

Resources available in one network service may not prevail or be easily accessible in the other participating networks.

Nonstandardizations

Standards to include multirate multiple traffics across different types of existing networks may have not been totally specified.

The obvious question posed therefore, is as follows:

- How to evolve a single type of network for broadband ISDN traffics, which is service independent (meaning that is capable of transporting all or any type of service as required)

- How to enable that network to share all the resources between different network services

- How to include a design in such networks so as to be flexible and adapt itself to changing trends in communication and circuit technologies following a set of well-defined standardization specifications.

Ideally speaking, "it may not be possible" could be the straight answer. However, the same questions motivated the development of a system with the constraints stipulated as above. The result was the emergence of *ATM - asynchronous transfer node* - telecommunications.

ATM, therefore, was conceived as a system, which can be characterized and described as follows:

- ✓ A B-ISDN system
- ✓ Capable of supporting narrow band as well as broadband traffics

    ✓   Compatible with emerging optical and semiconductor technologies. That is, it includes the art of coding algorithm, use of fiber optic transmission and intricacies of VLSI technologies. Hence, the underlying telecommunication is directed to conserve bandwidths (via compression techniques), let the bits to flow at a Gbps rate, use VLSI based hardware switching and tailored to be adopted as a flexible future-safe strategy

    ✓   Capable of sharing the resources between services on an optimal statistical basis

    ✓   Being standardized in respect to designing, manufacturing, and maintaining/operation

    ✓   Being standardized towards global application involving all parties customers, operators and manufacturers following the concept of the universal net

    ✓   Being made cost effective and less expensive to design, manufacture, and operate

    ✓   Intended to meet residential subscriber and business subscriber expectations

    ✓   Designed to adopt message fragmentation and packetization procedures compatible for fast packet/cell switching

    ✓   Expected to perform information transportation with fixed size cells. Voice, data, and video information cells are asynchronously interleaved into a high speed transmission

    ✓   Compatible to support a high bandwidth, low delay, packet-like switching mode and multiplexing independent of physical means of transport

    ✓   Configured to exhibit a low delay feature which is needed to support real-time services (such as isochronous voice traffic)

    ✓   Conceived to offer a high bandwidth feature required to accommodate video traffics.

3.12.2   *Network service versus ATM transmissions*

The *network service* in general, refers to the functional attributes of the network cloud that intervenes between the customer premises equipment (CPEs). Essentially, the network service is defined as the user-network interface (UNI). It includes the following items:

■   *Transmission*
Core network technology, which transmits a digital form of information from one end to the other

■   *Multiplexing*
Core technology network, which allows multiple digital bit streams to be combined into a single bit stream

■   *Switching*
Core network technology, which interconnects multiple transmission links

■   *Signaling*
Core network technology, which permits a user to communicate with the network.

In reference to the above network services, an important attribute of ATM transmission refers to the fast switching functions of the network dealing with a heterogeneous set of traffics (data, voice, and video). It is done via *cell-switching*. The concept of cell-switching adopted in ATM transmissions can be tracked to the emergence of fast packet-switching, flow-charted in Fig. 3.79.

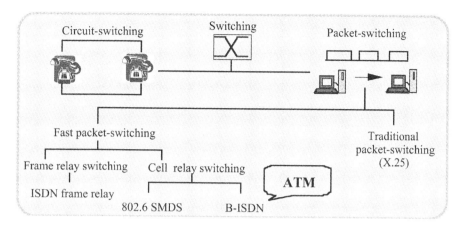

Fig. 3.79: Switching in telecommunications systems

The cell-switching aspect of ATM network service has the following distinct characteristics:

- Packets of fixed size (53 octets), known as *cells* are used in the information transfer

- The asynchronous (statistical) multiplexing is also constrained by priority considerations. That is, delay-sensitive cells can overtake nonisochronous traffic cells at the FIFO-based asynchronous multiplexer upon contentions

- Voice, data, and video cells are asynchronously interleaved or multiplexed

- Fixed-size cells enable switching at the hardware level with low-delay and high-speed transmission capabilities

- Information is transported transparently. Transparency implies the following:

    ✓ *Semantic transparency*: The function of the network guaranteeing the correct delivery of the bit at the destination
    ✓ *Time transparency*: The function of the network that guarantees the timely delivery of information at the receiver
    ✓ Broadcast packets are allowed.

In the classical STM and packet switches, only point-to point connections are available, because information has to switch from one logical inlet to another logical outlet. However, in the broadband network an additional requirement arises. As a matter of fact some services have a "copy" nature, and thus expect, in the ATM switches, the capability to provide multicast and broadcast functionality. (*Broadcast* can be defined as the provision of the information from one source to all destinations, whereas *multicast* provides the information from one source to many destinations).

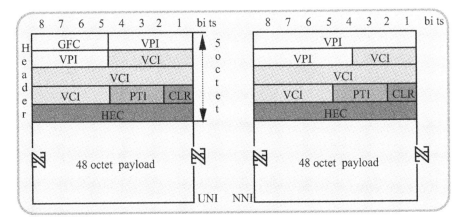

Fig. 3.80: An ATM cell

Why a cell size of 53 octets?

The packet-size adopted in ATM transmissions is fixed. This fixed size packet is called a *cell* and it contains 53 octets (bytes). This cell size (of 53 bytes) resulted from the following compromise which standardized ATM telecommunications.

- The network carriers of Europe wanted a very small cell, as short as 16 octets, to minimize latency (delay) and thus optimize the system for voice transmission. They wanted to simplify echo-cancellation. That is, most European intra-country distances are short enough and do not need echo cancellers, but the additional delay introduced if long cells are adopted would force them to use echo cancellers

- U.S. carriers on the other hand, wished to have a longer cell, over 200 octets, in order to reduce the number of headers that would have to be processed to complete a data file transfer. That is, shorter cells would warrant more cell transmissions per unit time, with corresponding increases in the number of headers. This would reduce the throughput thereof. Therefore, the argument of the U.S. carriers was in favor of larger cell size. Further, greater distances in the United States mean that all transmission equipment needed echo cancellers anyway

- However, objections concerning the use of long cells were raised from quarters of Europe. Such objections refer to the possible delay that could be introduced in a voice connection when a voice cell queues behind a long data frame. Since every switch and concentrator has a transmission queue, this unpredictable delay could be added many times as a cell crossed a network. Short cells on the other hand, eliminate the potential for adding a large and variable queueing delay

- Compromises among CCITT members, including European and North American standards organizations, finally resulted in a cell of 48 data octets plus a header of 5 octets, for a total of 53 octets – an example of "rather than choose either we'll do neither". Thus, a truce was signed with the magic number of 53 octets ending the battle of packets across the Atlantic Ocean.

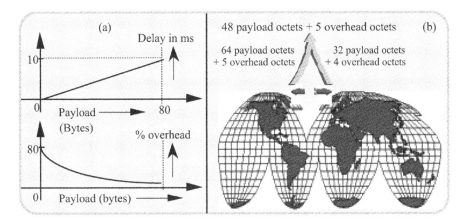

Fig. 3.81: Why 53 octets?: (a) Percentage overhead and packetization delay versus payload of 64 kbps voice signal; (b) A compromise was reached on an ATM cell size of 53 octets (ITU-TS Study Group XVIII in June 1989)

What is a transfer mode?

Transfer mode refers to a specific way of transmitting and switching information in a network.

In reference to the time-division multiplexing (TDM) technology, suppose a heterogeneous mix of packets (or cells) is derived at a multiplexer with respect to packets (or cells) arriving from different sources. There are three ways this multiplexing can be done. They are; (i) *synchronous multiplexing*; (ii) *asynchronous or statistical multiplexing*; and, (iii) *asynchronous multiplexing with priority considerations upon contending packets*. Earlier in this chapter, the synchronous multiplexing and asynchronous multiplexing were described.

In the context of understanding transfer modes specified as the synchronous transfer mode (STM), statistical packet-switching transfer mode, and asynchronous transfer mode (ATM) with designated priorities on the packets (cells) from different sources, the relevant multiplexing strategies are reconsidered and illustrated in Fig. 3.82.

Suppose there are three traffic sources: data, voice, and video (Fig. 3.82). Each source has distinct bit rates and bit rate fluctuation (if any) characteristics. Correspondingly, each source would send cells or load the line in a statistical manner. In short, the arrival pattern of cells at the multiplexer from each source would bear no deterministic relation to those of other sources. Now, the multiplexing of these cells can be done as mentioned earlier. The synchronous multiplexing refers to designating a time-slot for each category of cells (data, voice, and video) on the multiplexed (or bearer) line. In Fig. 3.82, the designated slots are 1, 2, and 3 for data, voice, and video respectively. As the cells arrive, each category of the cells will be placed only on the designated slots. If an incoming line is idle and carries no cell at an instant, its corresponding time-slot on the outgoing line will go empty or idle.

In the asynchronous packet-switching (such as X.25 and frame-relay), there are no designated time-slots. The cells from each source will be placed on the bearer line on a first-in/first-out (FIFO) basis. Each packet is identified in its respective category by virtue of the associated header information in the packets.

The third strategy is same as asynchronous (statistical) multiplexing indicated above except that the data, voice, and videos are prioritized. That is, taking into consideration that video and voice are isochronous (delay-sensitive) traffics, video is given the first priority, voice is next, and the data carries last priority. That is, in the event of contention at the multiplexer, the higher priority packet will be transmitted first. The lower priority traffics have to wait until higher priority traffics are sent out. However, under no contention instants, the usual FIFO policy is adopted.

In view of the above considerations, the synchronous and asynchronous transfer modes are specified as described below:

What is meant by synchronous transfer mode (STM)?

In reference to time-division multiplexing systems, successive time slots are allocated cyclically to the different channels as shown in Fig. 3.82 and as described earlier. That is, the transmission link is accessed by individual signals constituting independent channels by dividing the time into slots. In STM, a data unit associated with a specific position in the transmission frame is shown in Fig. 3.82. For example, voice packets appear at second positions in the transmission frame in Fig. 3.82.

What is meant by asynchronous transfer mode (ATM)?

Contrary to STM, in ATM (Fig. 3.82), the cell associated with a specific virtual channel may occur at any position on an FIFO basis (except under contention) as described earlier, above. In short, the ATM is a system in which information is transferred asynchronously with reference to its appearance at the input of the system and consistent with any priority specified. Further, asynchronous transfer facilitates the information to be buffered as it arrives and is inserted into an ATM cell when there is enough to fill the cell, and then the cell is transported across the network. The multiplexing is such that, a cell from a particular stream is transmitted as soon as there is an unused ATM cell is available to carry it. If there is no information available to be transmitted, an unassigned cell is transmitted as shown in Fig. 3.82.

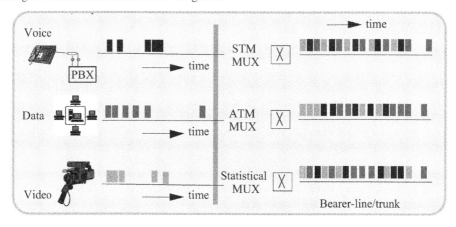

Fig. 3.82: ATM, STM, and conventional statistical multiplexing

Note: (1) Sequence used in STM: video-voice-data, (2) Priority schedule used in ATM: video-voice-data and (3) Statistical multiplexing policy FIFO.

Table 3.18: <u>ATM packet-switching versus conventional packet-switching</u>

ATM Network	Conventional packet-switching network
Cells are much shorter and are of fixed size (53 octets)	Variable size packets
Minimized overhead to realize high bit-rates	Larger overhead present: For example, X.25 — Error correction is enabled via necessary overhead
Cells are transported at regular intervals. No space between cells prevails and idle periods on the link carry an assigned cell	Packets of variable lengths are placed sequentially
Order in which cells arrive is guaranteed to be the same as the order in which they are transmitted. That is, cell sequence integrity is guaranteed	No cell sequence integrity guaranteed.

As could be evinced from the above narration, ATM essentially corresponds to packet-switching. However, there are differences as indicated in Table 3.18.

The asynchronous aspect of ATM stems from the fact that the cells are not produced with reference to a fixed synchronous cycle; rather, they are generated in accordance with the information sprouting from the application (source) at hand. ATM was adopted in preference to a synchronous transfer mode (STM) so as to overcome two limitations of the synchronous approach, namely

- ✓ STM does not provide a flexible interface to meet a wide variety of needs especially in dealing with heterogeneous streams of packet
- ✓ The switching system is more involved when multiple high data rates are used.

On the other hand, ATM switches allows considerable flexibility and opportunity for an efficient sharing of network resources (such as bandwidth, buffers, and processing horsepower) facilitating traffic flow for a variety of applications with extreme bit rate characteristics. The primary issue focused in developing ATM was therefore an efficient way in which network resources could be used. In addition, there is an issue of what is known as *quality of service* (QOS), which is measured in terms of two parameters:

- ▪ *Cell loss ratio* (CLR) defined as the ratio of the number of cells lost in transport across the network to the total number of cells pumped into the network

- ▪ *Cell transfer delay* (CTD) defined as the time taken for the cells of a particular service to propagate across the network.

As mentioned before, voice traffic is delay-sensitive but tolerant to cell-loss. The high-speed data traffic used for file transfers is loss sensitive but delay insensitive, whereas interactive video traffic can be both loss sensitive and delay sensitive. The design implications of ATM therefore, refer to the provision of an adequate quality of service for a wide range of traffic, while at the same time making efficient use of network resources. Only having an appropriate traffic management strategy can satisfy these requirements. That is, an effective mechanism for congestion control in an ATM-based transport network is needed.

What are variable bit rate (VBR) and constant bit rate (CBR) sources?

In certain signal sources bit rate may not be constant. For example, in a video, suppose a scene changes from say, a commentator to a juggler. The commentator's image (with occasional facial expression changes and portrayal of gestures) corresponds to almost a still picture with a low rate of information to be transmitted. The moving action of the juggler on the contrary depicts a high rate of bits to be transmitted. Hence, the question of VBR arises in supporting such heterogeneous information emitting sources.

When a stream of message such as voice or a simple text data has to be transmitted, the corresponding bit stream has almost a constant flow rate. Such transmissions are referred to as *constant bit rate* (CBR) transmissions.

Cell transfer delay (CTD) and cell delay variation (CDV)

Cell delay variation (CDV) could arise from the variable delays introduced in the network by the queues at switches and multiplexers and this would lead to a change from what would be expected in the gap between the cells of a given source multiplexed/switched asynchronous onto the outgoing line. This is illustrated for a constant bit-rate (CBR) source in Fig. 3.83. The problem of CDV is most acute with services where the difference in delay affects the quality of service perceived by the user, the so-called *delay-sensitive* services. A particular example of such a service is speech transmission, which is obviously an important service in B-ISDN.

A second problem is concerned with cell assembly delay and it arises because information from a source is buffered until there is sufficient space to fill a cell with 48 octets of information. The duration over which the information waits in the buffer obviously depends upon the rate at which it is arriving and it will be longer for low bit-rate sources. Because the time spent in the buffer represents a delay, it will have a significant impact on delay-sensitive services.

Voice telephony is again the most common service to be affected by the delay. At 64 kbps, it takes 6 ms to assemble 48 octets and this delay is significant when it comes to considering effects such as echo.

These two problems do not arise with the synchronous transfer mode used in the conventional narrow band ISDN (N-ISDN). Therefore, an apprehension is cast on the wisdom of choosing ATM and a query is put forward as to whether the flexibility and statistical multiplexing achievable with ATM is worth the extra complications in delay-sensitive services, particularly in voice telephony. The answer to this question is yes because CCITT agreed (in 1988) that ATM will be the target transfer mode for B-ISDN and the problems concerning delay-sensitive services were passed on a challenge to telecommunication engineers rather than being an insurmountable obstacle to the introduction of ATM itself.

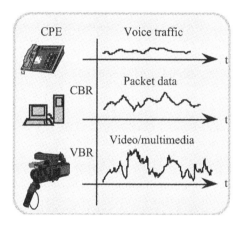

Fig. 3.83: CBR and VBR traffics

Statistical multiplexing gain

The ATM matches the transmission of the cells to the generation of information. Hence it does not waste capacity during the low-activity periods of the source by transmitting wasted cells; the cells that could be transmitted during this period are available for other sources inasmuch as ATM provides inherent statistical multiplexing on the link. If there is a reasonably large number of VBR sources generating information to be transported over a link, then a significant statistical multiplexing gain can be achieved via sharing the link capacity. This is illustrated in Fig. 3.84.

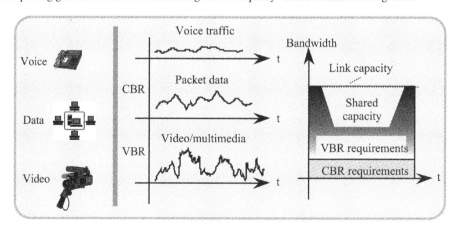

Fig. 3.84: Statistical multiplexing gain

ATM allows for the transport of digital information in the form of small, fixed-size packers called cells. This form of fast packet switching is made possible by exploiting the advances in microelectronics and developments in protocols, software, and switch architectures, which match the ATM systems. The use of cells makes it possible to provide "bandwidth on demand", in the sense that they are available on demand to users needing them. Each cell has a header with a label called the connection identifier, which explicitly associates the cell with a specified virtual channel on a physical link; a virtual cell, which denotes certain logical connection in ATM, is the basic unit of switching in B-ISDN. In today's packet-switching network, packets are combined with other traffic using byte-to-byte multiplexing at the physical layer of the OSI reference model.

Similarly one can observe that in ATM, flexible handling of dynamically varying loads is provided by means of cell-by-cell multiplexing with each cell having a labeled header used to identify which cells belong to which components of the aggregate traffic stream.

ATM transmission hierarchy

The stacked protocol of ATM transmission hierarchy confirms to B-ISDN-PRM. The lower layers of B-ISDN PRM are divided into three types, namely:

- ✓ Physical layer
- ✓ ATM layer
- ✓ ATM adaptation layer (AAL).

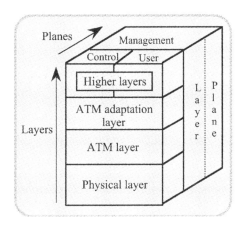

Fig. 3.85: The ATM protocol layers

The functions of these layers are as follows:

Physical layer function

The physical layer function is divided into two sub-layers known as:

- ✓ Physical medium (PM) sub-layer
- ✓ Transmission convergence (TC) sub-layer.

Physical medium (PM) sub-layer

The lowest layer of the B-ISDN protocol, it includes the functions that are only one physical-medium-dependent. The physical medium provides bit transmission capability, including bit alignment. Its functions include line coding and electrical to optical conversion (or vice versa). The physical layer itself depicts the media such as optical fiber, coaxial, or free-space.

Transmission convergence sub-layer

It is the second layer of physical layer. Its functions are:

- ✓ Cell rate decoupling

325

✓   HEC header sequence generation
✓   Cell delineation
✓   Transmission frame adaptation
✓   Transmission frame generation/recovery.

ATM layer functions

The next layer of the B-ISDN protocol is the ATM layer. It has characteristics independent of the physical medium. In simple words, the function of this layer is switching. The functions provided by this layer are categorized as follows:

✓   Generic flow control
✓   Cell header generation
✓   Cell virtual path identifier
✓   Cell multiplexing and demultiplexing.

ATM adaptation layer (AAL) functions

The basic function of AAL is the enhanced adaptation of the services provided by the ATM layer until the requirement of the higher layer's services are met. In this layer, the higher layer PDUs are mapped onto the information field of the ATM cell, which is 48 bytes long. The ATM layer adds a header with appropriate VPI and VCI values to this cell. To perform its functions, AAL is divided into two sub-layers: The convergence sub-layer (CS) and the segmentation and reassembly sub-layer (SAR), the details of which will be indicated in Chapter 4.

ATM switching technique

Much of the initial development work which led to B-ISDN started in the data networking field, as telecommunication engineers tried to develop fast packet-switching techniques suitable as a carriage for emerging video and other "image" applications. The switching technique developed thereof, constitutes the basis for B-ISDN and supports the ATM transmissions.

The "packets" conveyed by ATM as mentioned before are called cells. Like the packets of X.25, each cell of ATM consists of an information field and a header which tells the network where to deliver the packet, and provides a sequence number so that cells can be reassembled in the correct order at the receiving end. These cells are all of a fixed length (48 byte payload plus a 5 byte header). The fixed length gives the scope for overcoming the limitations of earlier packet networks.

3.13   Fiber Optics: A Physical Medium Support for B-ISDN/ATM Transmissions

With the introduction of low-loss fibers in the 1970s, optical fiber became the physical transmission medium of choice for high-speed wide-area communication systems. This development, along with research on fast-packet networks, made it possible to transmit digital data faster and more reliably (with fewer bit errors) than ever before. In 1984, an international effort began to standardize worldwide optical signal levels. The proposal that resulted from this effort suggested a hierarchical family of digital signal rates. This proposal eventually led to a draft standard of optical rates and formats. The basic optical rate and format chosen were a compromise between existing digital hierarchies in use throughout the world. The history of this standardization effort is quite interesting. First the international standards bodies had to develop a universal hierarchy of signal rates that would support the existing digital rates (for example, T-1 rates of 1.544 Mbps in the United States and 2.048 Mbps in Europe) as well as the higher optical rates in the future.

The internationalization of signal levels by the standards bodies of the International Telecommunications Union (ITU) resulted in a series of recommendations for the broadband integrated services digital network (B-ISDN) efforts. Driven by the emerging needs for high-speed communications and enabling technologies to support new services in an integrated fashion, the quest for large bandwidths that resulted could only be quenched with the transport capabilities of optical fibers. The optical data rates, synchronization, and framing format chosen for B-ISDN, are called the *synchronous digital optical network* (SONET) in North America. A corresponding strategy of CCITT is known as the *synchronous digital hierarchy* (SDH), which is followed in Europe and in most of the other parts of the world.

Once the transmission hierarchy for optical signal levels was established as a worldwide standard, work began on a universal multiplexing and switching mechanism to support integrated transport of multirate traffic. The major objective was to support the diverse requirements of multiple-bit-rate traffic sources and provide flexible transport and switching services in an efficient and cost-effective way. Hence as mentioned earlier, in 1998, ATM was chosen as the switching and multiplexing technique for B-ISDN. The ATM standard is designed to efficiently support high-speed digital voice and data communications. The expectation is that by the next decade, most of the voice and data traffic generated in the world will be transmitted by ATM technology.

3.13.1   SONET and SDH
Synchronous optical network (SONET)

Bellcore developed SONET. Subsequently, it was standardized by ANSI. SONET defines the rates and format for the optical transmission of digital information. Its specifications define a hierarchy of standardized data rates from 51.84 Mbps to 9.953 Gbps.

Corresponding to SONET, CCITT standardized a similar system called SDH (*synchronous digital hardware*). The term "optical" was omitted by CCITT because the system could do transmission not only via optical means, but also by other media such as digital microwave. (ITU-T recommendations: G.707-G.709)

SONET basically takes advantage of the broadband capabilities of optical fiber transmission, concurrently enabling high-speed digital transport of information.

Why SONET?

SONET has become popular as a broadband transport technology for the following reasons:

- ✓ Worldwide uniform standard
- ✓ Worldwide synchronous digital hierarchy
- ✓ Multivendor interpolation
- ✓ Standardized operations and management.

So, what is SONET?

In a broad sense the attributes of SONET can be enumerated as follows. SONET is:

- ▪ A new standard for optical transport in the North American context (ECSA/ANSI)

- ▪ A standard equivalent of that incorporated into the synchronous digital hierarchy (SDH) defined by CCITT

- ▪ A system based on 2 levels:

  - ✓ Optical carrier (OC) level
  - ✓ Electrical equivalent of OC designated as synchronous transport signal (STS).

SONET multiplexing

SONET multiplexing involves the following:

- ✓ Input acceptance of different types
- ✓ Byte interleaved multipelxing
- ✓ Conversion of all inputs to a base format: STS-1 (51-84 Mbps)
- ✓ Multiple STS-1 multiplexing into an STS-N electrical signal
- ✓ Visibility of low speed signals in the format
- ✓ Direct conversion from electrical to optical (OC-N) signal.

SONET: Definitions

In reference to SONET networking the following definitions are specified:

      ✓   Line: Termination equipment intended to terminate/or originate optical line (OC-N)

      ✓   Section: Transmission facility between two regenerators or between a *network element* (NE) and *a regenerator*

      ✓   Path: A logical connection between the point at which a standard format signal is assembled and the point at which the signal is disassembled.

The above units of SONET are illustrated in Fig. 3.86, where vendor X and Y meet at a point known as the *optical interface level* (OIL).

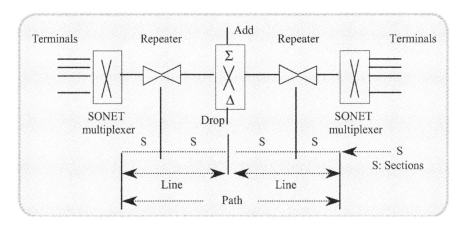

Fig. 3.86: Line, section, and path

Layered hierarchy in SONET

The SONET system hierarchy consists of four layers, as described here:

Photonic layer: This is the physical layer which includes specifications on types of optical fiber that may be used, required minimum laser powers, dispersion characteristics of the transmitting lasers, and required sensitivity of the receivers

Section layer: This second layer creates the basic SONET frame and converts electronic signals into photonic ones; it also has monitoring capabilities

Line layer: This third layer is responsible for synchronization and multiplexing of data into SONET frames; it also performs protection and maintenance functions and switching

Path layer: This final layer of SONET is responsible for end-to-end transport of data at the appropriate rate.

STS framing

The basic SONET building block is the STS-1 frame. It consists of 810 octets, which are transmitted once every 125 μs for an overall data rate of 51.84 Mbps. Construction of an STS frame stems from the implementation considerations of multiplexing standards beyond the existing DS-3 (44.736 Mbps) level.

Basically, a DS-3 is equal to 28 T-1s (= 28 × 24 = 672 channels). (The other digital signal levels studied earlier are reproduced in Table 3 for quick reference.)

For increased use of optical transmission systems, originally vendors introduced proprietary schemes combining anywhere from 2 to 12 DS-3s into an optical signal. However, SONET provides a standardized hierarchy of multiplexed digital transmission rates that accommodate existing North American and ITU-T (European) standards. Hence SONET received a distinct frame structure.

Table 3.19: Digital signal hierarchy

Digital signal type	Number of channels	Bit rates (Mbps)
DS-1	24	1.544
DS-1C	48	3.152
DS-2	96	6.312
DS-3	672	44.736
DS-4	4,032	274.176

SONET, essentially addresses the following specific issues:

1. Establishment of a standard multiplexing format using any number of 51.84 Mbps signals as building blocks. Because each building block can carry a DS3 signal, a standard rate is defined for any high-band width transmission system that might be developed

2. Establishment of an optical signal standard for interconnecting equipment from different suppliers

3. Establishment of extensive operations, administration, and maintenance (OAM) capabilities as part of the standard

4. Defining a synchronous multiplexing format to carry lower-level digital signals (DS-1, DS-2, and ITU-T standards). The synchronous structure greatly simplifies the interface to digital switches, digital cross-connect switches, and add-drop multiplexers

5. Establishment of a flexible architecture capable of a accommodating future applications such as broadband ISDN with a variety of transmission rates.

Thus, SONET has emerged to prepare telecommunications for future sophisticated service offerings, such as virtual private networking, time-of-day bandwidth allocation, and support for the broadband ISDN/ATM transmission techniques. To meet this requirement, a major increase in network management capabilities within the synchronous time-division signal is implicit in structuring SONET.

Signal hierarchy of SONET

The SONET specification defines a hierarchy of standardized digital data rates (Table 3.20). The lowest level, referred to as STS-1 (synchronous transport signal level 1), is 51.84 Mbps. This rate can be used to carry a single DS-3 signal or a group of flower-rate signals, such as DS-1, DS-2, plus the plesiochronous ITU-T rates (for example, 2.048 Mbps). Multiple STS-1 signals can be combined to form an STS-N signal. Interleaving bytes from NSTS-1 signals that are mutually synchronized creates the signal.

For the ITU-T synchronous digital hierarchy, the lowest rate is 155.52 Mbps, which is designated STM-1. This corresponds to SONET STS-3. The reason for the discrepancy is that STM-1 is the lowest-rate signal that can accommodate a ITU-T level 4 signal (139.264 Mbps).

Table 3.20: SONET/SDH signal hierarchy

SONET designation	ITU-T designation	Data rate (Mbps)	Payload rate
STS-1		51.84	50.112
STS-3	STM-1	155.52	150.336
STS-9	STM-3	466.56	451.008
STS-12	STM-4	622.08	601.344
STS-18	STM-6	933.12	902.016
STS-24	STM-6	1244.16	1202.688
STS-36	STM-12	1866.24	1804.032
STS-48	STM-16	2488.32	2405.376

System hierarchy of SONET

The SONET capabilities have been mapped into a four-layer hierarchy (Fig. 3.86) as indicated earlier, namely photonic, section, line, and path. The physical realization of the logical layers refers to the following:

Section: The basic physical block — a single run of optical cable between a pair of fiber optic transmitters and receivers (with repeaters for longer distances, if needed).

Line: This represents a sequence of one or more sections within which signal integrity is unchanged. End-points of a line correspond to switches/multiplexers that add/drop channels.

Path: This depicts an end-to-end circuit — data assembled at one-end is dissassembled only at the end of the path.

Optical cable and fiber optics transmission…

Limitations of telephone wires

Conventional POTS version of twisted pair telephone lines are highly bandwidth limited. The bandwidth of the line is decided by the product $L \times C$ where L and C are the distributed line parameters depicting,

$$L \quad = \quad \textit{Inductance/unit length: H/M}$$
$$C \quad = \quad \textit{Capacitance/unit length: F/M}$$

The channel bandwidth of the line is proportional to $1/(LC)^{1/2}$. Considering a twisted pair cable, it has a significant per unit length L and C values. Therefore, it poses a shrinking value of a bandwidth with increases in line length.

Optical fiber

An optical fiber is constructed of the following:

Core plus cladding: This part carries the light

Silicone coating: This prevents moisture and acts as a buffer

Kevlar sheath: It adds mechanical strengthening to the fiber

Polyurethane cover: This is the protective sheath

Principle of light transmission through an optical fiber

Light-beam enters via the core (with a refractive index, η_c) at an angle and obliquely incident on the internal surface of the core-clad interface [3.11 – 3.14]. Then, in order for total internal reflection to take place, the refractive index of the clad (η_{cl}) should be less than that of the core, namely, η_c: and the angle of incidence θ should be such that $\sin (\theta) \leq (\eta_d/\eta_c)$. This is known as the critical angle condition for total internal reflection. Further, using laws of reflection, it follows that, at the grazing angle of incidence $\theta_i \rightarrow \pi/2$, which specifies that $(\eta_d/\eta_c)^2 \leq 2$.

Optical transmission via fiber is characterized by

- Low fiber attenuation (allowing larger repeater distances)
- High transmission bandwidth (up to several hundred Mbps)
- Small diameter = facilities with low volume and weight
- High mechanical flexibility of the fiber
- Immunity against EM interface
- Low transmission error probability
- No crosstalk between fibers
- Tapping is very difficult: Secrecy of the message is retained.

Single mode and multimode fibers

Since the optical ray transmission through the fiber is dependent on the incident angle θ_i (for a given η_d/η_c ratio), larger core diameter fiber would accept rays with varying angles of θ_i. However, this would let a number of propagation modes through the fiber. On the contrary smaller diameter core would admit rays of restricted angles of θ_i, constituting a single-rate transmission (or also known as fundamental mode transmission).

The fiber can also be characterized in terms of the refractive index profiles across the core and the clad. If the refractive index changes abruptly, the fiber is known as the step-index fiber. When a graded transmission occurs, the fiber is termed as a graded index fiber.

Single mode versus multimode fibers

Single mode fibers are "the best" (though expensive). They introduce no intermodal interference. They can support bandwidth — to the order of 50 GHz. They are highly suitable for very long distance applications. Their shortcomings are:

- Small size
- Difficult to manufacture (precision considerations).

Grade multimode fibers are the "second best" and are moderately expensive. They can support bandwidth ~10 GHz. The multimode step-index fibers are well suited for general purpose applications and are relatively cheap. Their bandwidth is limited to 10 MHz for 1 km links. Their size and dimensions allow easier manufacture.

Description of an in-line fiber data bus coupler

An inline fiber data bus coupler is a passive coupler which can be used to remove a portion of the optical signal from the bus trunk-line or inject additional light signals onto the trunk. An in-line coupler is also known as a T-coupler. Being a passive coupler, optical signal is not regenerated at each terminal node. Insertion and output losses at each tap plus the fiber-losses between the taps, limit the network size to a small number (about 10 or less) in a symmetrical (tapout-taping symmetry) coupler.

Geometry of the coupler
It has four ports, namely,

- ✓ Two for connecting the device onto the bus: (A,B).
- ✓ One for receiving the tap-off data: (C)
- ✓ One for inserting the data on to the BUS: (D).

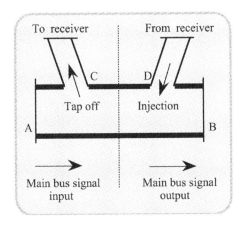

Fig. 3.87: Link coupler

In the receiving mode, the (tap-off) coupling is in the upstream (with respect to transmitter injection port) to avoid receiver overloading from the collocated transmitter-injection. Coupling is done via fusing the main bus fiber and the auxiliary fibers through which tap-off or injection is done.

Rigorous performance evaluation
(i)

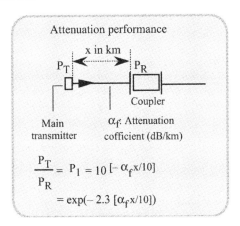

Fig. 3.88: Performance evaluation

(ii)

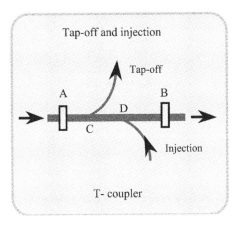

Fig. 3.89: Tap-off

In Fig. 3.89, A, B, C, and D are the coupling junctions and they correspond to the four ports indicated earlier. At each port, there is a connector, which facilitates a junction. Suppose each connector introduces a loss equal to a fractional power P_c. Therefore, power lost at each connector/junction:

$$P_{LC} = -10\log_{10}(1 - C_T) \text{ dB} \qquad (3.15)$$

(iii)   Power tap-off: $PT = -10\log_{10}(1 - C_T)$ dB where C_T is the tap-off factor

(iv)   Intrinsic loss: As the beam traverses a coupler, due to geometrical and optical imperfections, there will always be an intrinsic power loss, P_i.

$$P_i = -10\log_{10}(1 - F_i) \text{ dB} \qquad (3.16)$$

where F_i denotes the fractional intrinsic power loss factor.

(v)   Power coupling efficiency: η_c

This is the launching efficiency of the transmitting end fly-head into the main bus fiber.

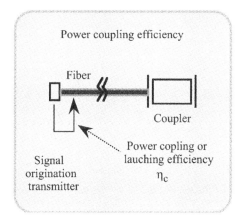

Fig. 3.90: Power coupling at the transmitter

(vi)   Calculation of the number of stations:

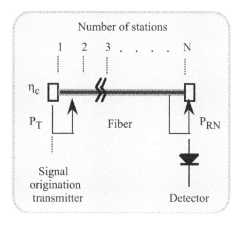

Fig. 3.91: Stations along the fiber

Let P_{RN} be the power tapped-off and detected at the Nth station.

(a) $\therefore$ Fiber attenuation-based loss resulting from $(N-1)$ links plus accounting for coupling efficiency at the fly-head (η_c):

$$P_{RN} = \left[P_i^{(N-1)} \right] \eta_c \tag{3.17}$$

(b)   Power connected into (via a connector) and out of (via a connector) each coupler: $(1-P_c)(1-P_c) = (1-P_c)^2$

$\therefore$ If N couplers are inserted, the corresponding total connector loss experienced is equal to $(1-P_c)^{2N}$.

(c)   At each coupler suppose tap-off exists.

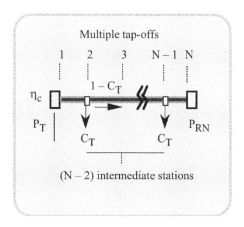

Fig. 3.92: Multiple tap-off situations

After letting off a fraction of $(1-C_T)$ power at each $(N-2)$ intermediate station, the Nth station will ultimately tap-off a fraction of C_T of the power (arriving at this N^{th} station).

$$\therefore \text{Total power tapped-off loss} = \left[(1-C_T)^{N-2} \times C_T \right] \tag{3.18}$$

(assuming power tap-off exists at each intermediate station)

(d)   Intrinsic loss can be assumed to exist at each intermediate station traversed by the beam.
       $\therefore$ Total intrinsic loss through $(N-2)$ couplers: $(1-F_i)^{N-2}$

Final result
The ratio between the transmitted power (PT) and the received power at the Nth station: P_T/P_{RN}
Taking the reciprocal,

$$P_{RN}/P_T = (a) \times (b) \times (c) \times (d)$$

$$= \eta_c [P_i^{N-1}] \times [(1-P_c)^{2N}] \times [(1-C_T)^{N-2} \times C_T] \times [(1-F_i)^{N-2}] \tag{3.19}$$

$[C_T] = [(1-C_T)^{N-2} \times C_T]$ if no tap-off exists in the intermediate stations. Sometimes, $(1-C_T)$ is substituted by $(1-2C_T)$, if (unwanted) tap-off exists at the injection port (D) of each coupler. If it exists, this unwanted (tap-off) power into such ports is rendered evanescent (or decayed off) so as not to interfere with any signal coming through for injection into the bus, for a transmitter.

Dynamic range (DR)

Worst case dynamic range refers to maximum optical power range to which any station detector (along the bus) must be able to respond. DR depends on:

- ✓ Smallest difference in transmitted and received power (which occurs between station 1 and 2)
- ✓ Largest difference in transmitted and received power (which occurs between station 1 and N_{max}).

It is given by:

$$\therefore DR = \left[(P_{RN} / P_T)_{N=2}\right] / \left[(P_{RN} / P_T)_{N \to N \max}\right]$$

$$= \frac{1}{\left[P_i(1 - P_c)^2(1 - C_T)(1 - F_i)\right]^{(N_{max}-2)}} \tag{3.20}$$

Problem 3.7

A planning stage concept design of an inline optical fiber data bus operating at 10 Mbps is based on the following considerations:

Separation between stations:	10 meters
Fiber loss:	3 dB/km
Optical source(s):	Laser diode 1 mw available at the fiber fly-head
Detector(s):	APD, – 58 dBm sensitivity
Couplers:	Power coupling efficiency: 10%
	Power tap-off factor: 5%
	Fractional intrinsic loss: 10%
Connectors:	Power loss: 20% (1 dB) per connector

Sketch the variation of PRN in dBm (say from 0 to –60 dBm) as a function of the number of stations N.

Determine the system operating margin for 8 stations.

Determine the worst-case dynamic range for the maximum allowable number of stations if a 6 dB power margin is implemented.

The conventional network topologies are illustrated in Fig. 3.93. One way of multichannel implementation is to have many channels multiplexed together, transmitted over some distance, and then demultiplexed at a common destination. This represents the simplest type of topology and is know as a point-to-point system. In addition, several other common network topologies exist for electrical networks, each of which can be modified to facilitate optical communications networks. These are the ring, bus, star, and tree configurations as indicated in Fig. 3.93. The names are descriptive of the topology, and each type is conceptually straightforward.

The ring structure has nodes (for example, users) that are periodically connected in a closed ring formation. The single ring is unidirectional, but a second inner ring can be added that would facilitate bidirectional communication, with each ring representing a different direction of propagation. Furthermore, the inner ring provides an alternative protection path in case of a single-link outage, enabling all nodes to still communicate with each other even if a circuitous route must be taken; such a protection path provides for a survival ring network.

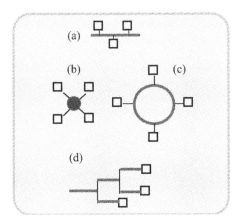

Fig. 3.93: Basic network topologies: (a) Bus, (b) ring, (c) star, and (d) tree

A similar topology to the ring is the bus network, which is simply a ring that has failed to close on itself and represents a situation in which the network nodes are connected to a common "backplane". Fig. 3.93 is a single bus. A dual bus can also be implemented, with one "rail" (for example, fiber) used for transmitting data to nodes upstream and the other rail used for transmitting data to nodes downstream. The bus is quite easy to build given almost any geographical configuration of the network, but a break in the bus will isolate a node or nodes from communicating with the rest of the network.

Historically, the ring and the bus have been highly favored for electrical TDM networks. One drawback in the implementation of an optical bus or ring is that a passive optical tap typically would be required at each node for single injection and/or recovery; therefore, the optical losses would become intolerable for a large number of users. The optical power available at the Mth user node in a bus, P_{bus}^M, for an M node network is given by

$$P_{bus}^M = P_T C[\beta(1-C)]^{M-1} \qquad (3.21)$$

where C is the ratio of the amount of optical power tapped off at each node. β is the excess loss at each tapping node, and γ is the excess loss in an M-by-M star.

Another network topology is the star, which has some advantages for an optical network. The star is configured such that a central device interconnects many nodes and each node can transmit to every other node through the central star. The star may involve active routing, but generally it is thought of as a passive element, such as a passive star coupler. This passive star coupler has N input fibers (connected to each node's transmitter) and N output fibers (connected to each node's receiver). The passive star splits the input light from any given input port equally among the N output ports. The optical power available at each of M output ports in a star P_{star}^M is given by

$$P_{star}^M = \frac{P_T \gamma}{M} \qquad (3.22)$$

The inherent optical loss in a star grows slower with the number of users, M, than does the loss in a bus or ring, and the difference can be significant in large networks (optical amplifiers can be used to compensate for some of these losses). Any node in a star can transmit to and access any other node, and a break in a link will disrupt communications only with that node, as opposed to a bus, for which a break could disrupt communications for a significant portion of the entire network. Since all transmissions must pass through the central star, the star has the disadvantages of having a higher propagation delay, requiring more fiber, and requiring the geography to accommodate a central device.

The final topological example is the tree, which is a favorite of broadcast and distribution, systems. At the base of the tree is the source transmitter, from which emanates the signal to be broadcast throughout the network. From that base, the tree splits many times into different branches, with each branch either having nodes connected to it or further dividing into subbranches. Whereas the other topologies are intended to support bidirectional communication among the nodes, this topology is useful for distributing information unidirectionally from a central point to a multitude of users. This is a straightforward topology and is in use in many electrical systems, most notably cable television (CATV).

The network users being referred to in this chapter can take many forms and require vastly different bit rates. The nodes may be video stations, supercomputers, library mainframes, personal computers, multimedia centers, personal communications devices, or gateways to other networks. Moreover, the selection as to which topology to implement depends on many factors, including the multiplexing scheme, geography, power budget, and cost. In fact, it is quite likely that a hybrid topology will be used in a large network. Normally, smaller rings, buses, and stars are interconnected to form larger networks.

The size of the larger network depends on the architecture of the network which, in turn, depends on the network's geographical extent. The three main architecture types are the local- metropolitan- and wide-area networks (LAN, MAN, and WAN, respectively). Although no rule exists, the generally accepted understanding is that a LAN interconnects a small number of users covering a few kilometers (for example, intra-and inter-building), a MAN interconnects users in a city and its outlying regions, and a WAN interconnects significant portions of a country (hundreds of kilometers). A WAN is composed of smaller MANs, and a MAN is composed of smaller LANs. Hybrid systems also exist, and typically a WAN may consist of smaller LANs, with mixing and matching between the most practical topologies for a given system. For example, stars and rings may be desirable for LANs, whereas buses may be the only practical solution for WANs. It is, at present, unclear which network topology and architecture will ultimately and most effectively take advantage of high-capacity optical carrier systems.

Problem 3.8

Assume there are $M = 50$ users (i.e., nodes). A bus topology when used leads to an excess loss of 0.5 dB and has a tapping ratio 105. If, on the other hand, a star topology is adopted, the excess loss is 3 dB. Determine the ratio between the available powers at the Mth user in star and bus topology.

3.14   Concluding Remarks

Networking in telecommunication is a blend of art and science in a technological framework as perceived in the realm of engineering. It involves hardware strategies and software methods. It is based on careful planning and artful management. The scope of telecommunication networking has widened ever since the concept of service integration and broadband implementations came into play as a result of consumer demands. ATM networking is still in its embryonic stage. However, its rapid deployment is foreseen in the new millennium.

Bibliography

[3.1]   E.H. Jolley: *Introduction to Telephony and Telegraphy* (Hart Publishing Co., Inc., New York, NY: 1970)

[3.2]   A.M Noll: *Introduction to Telephones and Telephone Systems* (Artech House, Inc., Norwood, MA: 1991)

[3.3]   W.D. Gayler: *Telephone Voice Transmission Standards and Measurements* (Prentice-Hall, Inc., Englewood Cliffs, NJ: 1989)

[3.4]   J. Bellamy: *Digital Telephony* (John Wiley & Sons, Inc., New York, NY: 1991)

[3.5]   B. Carne: *Telecommunications Primer* (Prentice Hall PTR, Upper Saddle River, NJ: 1995)

[3.6]   T. Viswanathan: *Telecommunication Switching Systems and Networks* (Prentice-Hall of India Pvt. Ltd., New Delhi, India: 1995)

[3.7]   R.G. Winch: *Telecommunication Transmission Systems* (McGraw-Hill Inc., New York, NY: 1993)

[3.8]   F.J. Derfler, Jr., and L. Freed: *Get a Grip on Network Cabling* (Ziff-Davis Press, Emeryville, CA: 1993)

[3.9]   C.N. Herrick and C.L. Mckim: *Telecommunication Wiring* (Prentice Hall PTR, Upper Saddle River, NJ: 1998)

[3.10]   H. Hegering and A. Läpple: *Ethernet* (Addison-Wesley Publishing Co., Workingham, England: 1995)

[3.11]   J.C. Palais: *Fiber Optic Communications* (Prentice-Hall, Inc., Upper Saddle River, NJ: 1998)

[3.12]   G. Keiser: *Optical Fiber Communications* (McGraw-Hill, Inc., New York, NY: 1991)

[3.13]   S.L.W. Meardon: *The Elements of Fiber Optics* (Regents/Prentice-Hall, Inc., Englewood Cliffs, NJ: 1993)

[3.14]   G. Ghatak and K. Thyagarajan: *Introduction to Fiber Optics* (Cambridge Press, Cambridge, England: 1998)

[3.15]   H. Hioki: *Telecommunications* (Prentice-Hall, Inc., Upper Saddle River, NJ: 1998)

[3.16]   M. Cole: *Telecommunications* (Prentice Hall PTR, Upper Saddle River, NJ: 1999)

[3.17]   U. Black and S. Waters: *SONET & T-1: Architectures for Digital Transport Networks* (Prentice Hall PTR, Upper Saddle River, NJ: 1997)

[3.18]   M. Khader and W.E. Barnes: *Telecommunications Systems and Technology* (Prentice-Hall Inc., Upper Saddle River, NJ: 2000)

[3.19]   J.D. Spragins, J.L. Hammond and K. Pawlikowski: *Telecommunications Protocols and Design* (Addison-Wesley Publishing Co., Reading, MA: 1991)

[3.20]   M. Schwartz: *Telecommunication Networks Protocols Modeling and Analysis* (Addison-Wesley Publishing Co., Reading, MA: 1998)

[3.21]   T. Ramteke: *Network* (Prentice Hall Education, Career and Technology, Englewood Cliffs, NJ: 1994)

[3.22]   W. Stallings: *ISDN and Broadband ISDN with Frame Relay and ATM* (Prentice-Hall, Inc., Englewood Cliffs, NJ: 1995)

[3.23]   W. Stallings: *Advances in ISDN and Broadband ISDN* (IEEE Computer Society Press, Los Alamitos, CA: 1992)

[3.24]   D.E. McDysan and D.L Spohn: *ATM Theory and Applications* (McGraw-Hill, Inc., New York, NY: 1998)

[3.25]   D.E McDysan and D.L. Spohn: *Hands-on ATM* (McGraw-Hill, Inc., New York, NY: 1998)

4

ATM NETWORK INTERFACES AND PROTOCOLS

Fenced out or thoroughfare?

"Interface ... a blessed barrier between doorsteps of users and gateways of networks. Without thee what does all the info mean?"

P.S. Neelakanta, 1999

4.1 Introduction

4.1.1 Standardization of B-ISDN by CCITT

It was indicated in the last chapter (Chapter 3) that the ATM (or asynchronous transfer mode) is the ground on which B-ISDN is being built. Or, in the words of ITU-T Recommendation I.121, "ATM is the transfer mode for implementing B-ISDN ..."

Transfer mode specifies the particular method of transmitting, multiplexing, and switching information across a network. In the context of B-ISDN, the associated transfer mode can be understood by revisiting the basics of ATM [4.1, 4.2]:

- The information transmitted is packetized into fixed size cells

- In each cell, the 48 octets information field is available for the user

- A five-octet header field attached to the cell carries details pertinent to a certain ATM layer's functionality. Inside the header, a label field is defined to recognize individual communications (or traffic types) supported by the cell

- Like conventional packet-switching, ATM provides communication with a bit rate which is set to match the actual need of the user. Such bit rates are not only tailored on an ad hoc basis, but also may change with time in a particular service depending on the burst of information emitted from the source

- ATM defines a new transfer mode in reference to the multiplexed transmission involved. It refers to the cells allocated to the same connection exhibiting an irregular (statistically irregular) pattern of occurrence. That is, "cells are filled in according to the actual demand" posed by the information being transmitted. As such, a cell belonging to a specific channel may occur at any position irregularly along the time scale. This "asynchronous" occurrence of cells, therefore, depicts a time-dependent stochastical pattern of arrival and/or departure at any point of observation.

ATM versus STM ...

As against ATM, in the so-called synchronous transfer mode (STM), a data unit associated with a given channel is identified by its designated position in the transmission (time) frame. The transmission frames have a rigid structure made of:

- *A user frame containing an information signal*
- *A framing signal, which is an indicator of the start position of the frame.*

Further, channel bit rates are fixed in STM as predefined values. This means, bit rates cannot be varied on an ad hoc basis, (once it is selected by a predefinition). Examples of preset bit rates are B Channel 64 kbps for basic access interface, H1 channel (32-34 Mbps) etc. Fixed channel bit rates are deliberately adopted in STM so as to avoid the need for using different types of interface units.

- Unlike STM, ATM-based networks follow multiplexing and switching of the cells independent of the application (that is, the service type). In other words, the same type of equipment will handle low bit rate connections or high bit rate connections no matter whether the bit flow pattern is "streamy" or "bursty" as it emerges from the source

- In order to accommodate the low bit rate (LBR) as well as the high bit (HBR) rate transmissions, a dynamic bandwidth allocation on demand (corresponding LBR or HBR conditions) is made feasible in ATM transmissions

- ATM may introduce impairments such as cell-loss, cell-transfer delay, and cell-delay variations (jitter) as a result of asynchronous multiplexing

- The tolerance of each specific service to these impairments should be made as an a priori connection request depicting a set of quality of service (QOS) parameters

- ATM is a combination of circuit-oriented and packet-oriented techniques. The connection-oriented aspects included in the ATM are as follows:

  - ✓ Low overhead and the associated processing
  - ✓ Once the connection is established, the low overhead considerations set the transfer delay of information carried as low.

- The packet-oriented consideration included in the ATM is as follows:

  - ✓ Flexibility prevails in terms of bit rate assigned to individual virtual connections.

ATM, is therefore, a circuit-oriented, hardware-controlled, low overhead concept of virtual channels (VCs) supporting heterogeneous types (voice, video, and data) of information. The VCs have no flow control on error recovery (in contrast to X.25 access), and the implementation of these virtual channels is done by (relatively) short, fixed-size cells, which provide for both switching and multiplexed transmission on an asynchronous (statistical) basis. Consistent with this defined profile, ATM bears the following attributes:

- Short cells and high transfer rates (B-ISDN rates) with low cell transfer delays (CTDs) and restricted cell-delay variations (CDVs)

- Compatible for delay-sensitive transmissions such as voice and video

- Cell level multiplexer and switch support flexible bit rate allocation, (which is characteristic of packet networks)

- ATM provides a link-by-link cell transfer capability common to all services

- Service specific adaptation functions of ATM are as follows:

✓   Mapping higher layer information into ATM cells on an end-to-end basis

✓   Packetization of continuous bit streams into ATM cells

✓   Segmentation of larger blocks of user information into ATM cells.

■   ATM networks have the feasibility of combining several virtual channels (VCs) into one *virtual path* (VP). VC is a concept used to describe unidirectional transport of ATM cells associated by a common, unique identifier value known as *virtual channel identifier* (VCI) located as a part of the cell header.

ATM: Networking considerations

ATM follows the general architecture of B-ISDN. The information transfer and signaling capabilities of B-ISDN include:

✓   Broadband capabilities on information transfer

✓   64 kbps ISDN capabilities

✓   Inter-exchange signaling

✓   User-to-user signaling

✓   ATM guarantees cell sequence integrity (when no faults are present). Cell sequence integrity implies that a cell identified as belonging to a particular VC connection will not overtake (ahead in the time scale) another cell belonging to the same VC connection sent earlier

✓   ATM is a connection-oriented technique

✓   ATM has packet (cell) transmission capability.

A connection within the ATM may consist of one or more links, each of which is assigned an identifier (located as 5-octet unit header in each cell). Over the entire duration of the connection between two end-entities, these identifiers will remain unchanged.

✓   ATM has two distinct capabilities, namely, information transfer and signaling

✓   Signaling information for a given connection is conveyed using a separate identifier: This is known as *out-of-band signaling* (Fig. 4.1).

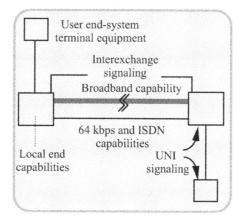

Fig. 4.1: Information transfer and signaling

4.1.2   Hierarchical levels in ATM performing transport functions

ATM architecture is based on ITU-T Recommendation I.321 for the layered structure of B-ISDN as shown in Fig. 4.2. The associated definitions vis-à-vis transport functions are indicated below. The layered ATM architecture organizes the various functions that must be performed in ATM network devices. Hence, the layers in the ATM architecture refer to *functional layers*.

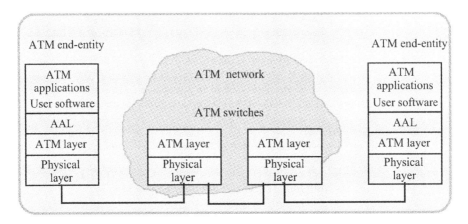

Fig. 4.2: Layered ATM architecture

The ATM architecture can be discussed in reference to its stacked layers of hierarchy, which working from the bottom up consist of (Fig. 4.2):

    ✓   The physical layer
    ✓   The ATM layer
    ✓   The ATM adaptation layer.

The functional attributes of these layers are illustrated in Fig. 4.3; in reference to the stacks and planes of B-ISDN specifically, the hierarchical levels, which constitute the ATM transport network are the physical layer and the ATM layer.

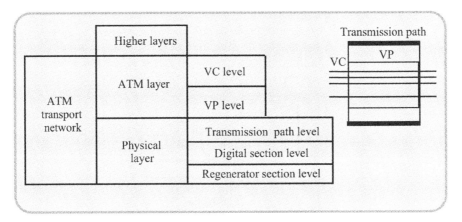

Fig. 4.3: ATM transport network

A summary on the functional characteristics of the aforesaid three layers are as follows:

▪   *The physical layer*: This sends and receives information either in electrical or in optical format over an appropriate physical medium. The physical communication process involved thereof may refer to the following:

    ✓   Converting cells in and making cells from a continuous bit stream or transmission frame format
    ✓   Applying various forms of encoding and decoding to the data contained in each field.

- *The ATM layer*: This layer is primarily responsible for cell-switching. An ATM layer entity in an ATM device accepts cells received over a transmission path. Then, it determines the outward transmission path over which each of those cells should be sent. Prior to such relaying, the ATM layer formats the cell-header

- *ATM adaptation layer* (AAL): This layer provides the interface between ATM user software (applications) and the ATM network. It functions only at the end-systems and deal thereof with end-applications. This layer is not needed in the ATM switches used in between the end-to-end connectivity. The AAL entity accepts the bits from the end-application software and structures them in the form of cells compatible for transport across the ATM network. A complementary function is performed by the AAL entity at the receiving end. That is, the receiving side AAL entity accepts the cells from the network, reconstructs the original bit flow structure and passes it to end-user application software.

4.1.3   *ATM functional layers versus OSI model functional layers*
The OSI model as indicated in the last chapter (Chapter 3) was designed to provide a common base for the coordinated usage of different standards in vogue. It is intended for interconnecting open systems. It is "open" to accommodate crisscross uses of applicable standards.
The seven layers of the OSI model and the three layers of ATM can be juxtapositioned as illustrated in Fig. 4.4 so that the correspondence between the layer functions can be observed.

4.1.4   *The physical layer*
The physical layer part of the ATM architecture is indicated in Fig. 4.4.

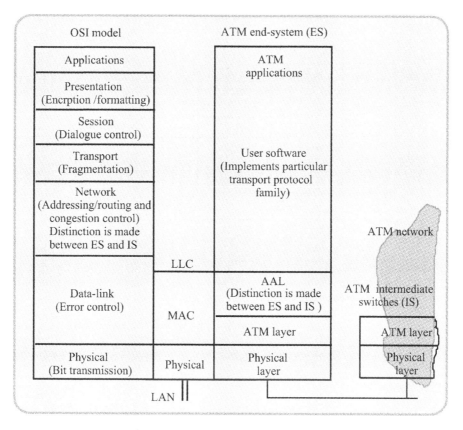

Fig. 4.4: ATM architecture: The physical layer

Associated with the physical layer are hardware parts comprised of ATM switches, end-entities, and transmission paths.

ATM switches

These perform routing and relaying of information playing the role of an *intermediate system* (IS) between end-users. The ATM switch, which is a part of the network of the telco service provider is known as the *public ATM switch*. It is called a *network node*. If an ATM switch is owned and maintained by a private enterprise, then it is known as a *customer premises node* (CPN).

ATM end-entities

These are devices at the end-points connected directly to public or private switches. These devices act as sources or destinations of user information. These are also known as *end-systems* (ES).

Transmission paths (TPs)

These represent physical communication lines used between ATM end-points and switches for physical interconnection purposes.

Functions of the physical layer

This is the lower most layer of the ATM hierarchy and it consists of two sublayer structures, namely:

      ✓   Physical medium sublayer (PM)
      ✓   Transmission convergence (TC) sublayer.

These sublayers are illustrated in Fig. 4.5

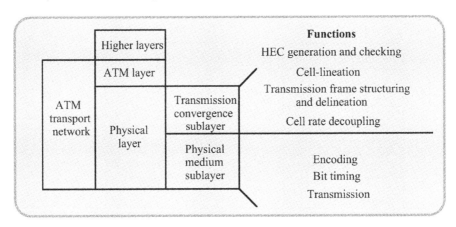

Fig. 4.5: The sublayers of ATM physical layer

Functions of PM sublayers

The *physical medium sublayer* is the lowest (sub)layer in the hierarchy and includes only the following PM-dependent functions:

      ✓   It offers bit transmission and bit alignment capabilities
      ✓   It enables electrical/optical conversions
      ✓   It accounts for those transmission functions, which are medium-specific.

Its other tasks include bit-timing functions, such as:

      ✓   Generation and reception of waveforms suitable for the transmission medium
      ✓   Insertion and extraction of bit-timing information
      ✓   Line coding, if required.

Functions of TC sublayer

The TC sublayer accepts cells from the ATM layer for outgoing transmissions. It combines cells to construct a data stream and passes the data stream to the PM sublayer. For incoming transmissions, the PM sublayer passes a data stream to the TC sublayer, which extracts the cells and passes them to the ATM layer.

The specific functions of TC sublayer are as follows:

   ✓   Generation and recovery of the transmission frame

   ✓   *Transmission frame adaptation*: This is responsible for all actions, which are necessary to adapt the cell-flow according to the payload structure in use in the sending direction. In the opposite direction it extracts the cell-flow from the transmission frame. Some physical layer interface protocols may send data on the link in the form of structured transmission frames. For these protocols, TC sublayer would structure the outgoing cells into appropriate frame formats and extracts cells from incoming frames

   ✓   *Cell delineation*: This mechanism facilitates the recovery of cell boundaries [4.3]. It also facilitates information scrambling and descrambling to guard against any malicious invasions. (See the indented note below)

Cell-lineation algorithm

    The PM sublayer processes the data as a continuous stream of bits. The TC sublayer when accepting the received bit stream is responsible for identifying the cell boundaries in the bit stream. The identification/recovery of cell boundaries is performed by a cell delineation algorithm.

    The cell-lineation algorithm is illustrated in Fig. 4.6. In the HUNT state, a cell delineation algorithm is performed bit-by-bit to determine if the HEC coding law is observed (that is, whether a match exists between received HEC and calculated HEC). Once a match is achieved, it is assumed that one header has been found, and the method enters the PRESYNC state.

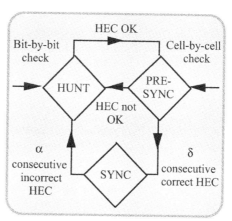

Fig. 4.6: Illustration of cell-lineation algorithm

    In the SYNC state, the HEC is used for error detection and correction. Cell-delineation is assumed to be lost if the HEC coding law is recognized as incorrect α times consecutively. In the PRESYNC state, a cell structure is now assumed. The cell-delineation algorithm is performed cell by cell until the encoding law has been confirmed consecutively

δ times. As stated earlier, cell-delineation is the process that allows identification of cell boundaries.

The values of α and δ in Fig. 4.6 are design parameters. Larger values of δ lead to longer delays in establishing synchronization. But they offer greater robustness against false delineation. Greater values of α result in longer delays in recognizing a misalignment but in greater robustness against false misalignment. The impact of random bit errors on cell delineation performance for a set of α and δ is such that an increase in BER decreases the in-sync time and increases the acquisition time. The in-sync time refers to the average amount of time that the receiver will maintain synchronization in the face of errors, with α as a parameter. The acquisition time is the average amount of time to acquire synchronization as a function of error rate, with δ as a parameter. In general,

$\alpha = 7$, $\delta = 8$:   for cell-based interfaces

$\alpha = 7$, $\delta = 6$:   for SDH based interfaces.

$\alpha = 7$:   for 155.52 Mbps ATM transmissions where the system will be in SYNC for more than a year with probability of error $\simeq 10^{-4}$

$\delta = 6$:   The above ATM with the same probability of error (10^{-4}) will need about 10 cells or 28 μs to reenter SYNC after loss of cell synchronization.

- *Header error control* (HEC) sequence generation:

  ✓ This is done in the transmit direction. The HEC is inserted appropriately in the field of the cell within the header
  ✓ At the receiver, HEC value is analyzed and compared with the received value. If possible, header errors are corrected; otherwise the cell is discarded.
  (More details on HEC will be furnished later.)

- *Cell-rate decoupling*: For most physical layer protocols, the PM sublayer requires a continuous stream of data without gaps. If there are no cells to send, the TC sublayer inserts idle (empty) cells in the outgoing transmission stream. That is, in the sending direction this mechanism inserts idle cells in order to adapt the rate of ATM cells to the payload capacity of the transmission medium. In the receiving direction, this suppresses all the idle cells. The idle cell has the following field settings in the cell-header: VCI = 0, VPI = 0 and CLP =1.

Transmission of ATM cells

B-ISDN specification on ATM cell transmission corresponds a basic transport within SONET and/or SDH. The relevant transmission structures have been defined (ITU-T Recommendation I 413) in respect of 155.52 Mbps. They are:

  ✓ Cell-based physical layer
  ✓ Synchronous digital hierarchy (SDH)-based physical layer.

Cell-based physical layer

In this structure, the cell transmission is characterized by the following considerations:

- No framing is imposed

- The interface structure consists of a continuous stream of 53 octet cells

- Since no external framing is imposed, some form of synchronization is provided

- This synchronization is achieved via the HEC field in the cell-header. That is, cell boundaries are identified by the cell-delineation.

Structure of cell-based interface

In the ATM transmission of the continuous stream of 53-octet cells, the maximum spacing between successive physical layer cells is 26 ATM layer cells.

What are ATM layer cells?

The ATM layer resides above the physical layer. Its features are independent of the physical medium. In the transmit direction, cells from individual VPs and VCs are multiplexed into a single resulting cell stream by the cell multiplexing function. The composite stream is normally a noncontinuous cell flow. At the receiving end, the cell demultiplexing function splits the arriving cell stream into individual cell flows appropriate to the VP or VC.

After 26 contiguous ATM layer cells, the insertion of a physical layer cell is enforced in order to adapt the transfer capability to the interface rate. The physical layer cells are inserted when no ATM layer cells are present.

SONET/SDH-based physical layer

As discussed in Chapter 3, ATM cells can be carried across a network at the physical layer level using SDH or SONET. This physical layer imposes a periodic framing, which sets up TDM channels for the cells to flow through. (It should, however, be noted that ATM cells can travel over any synchronous channel, not just on SONET/SDH.)

The frame formats of SONET and SDH

SONET: The basic frame format of the electrical signal in SONET is STS-1. This consists of 810 octets transmitted once in every 125 μs for an overall data rate of 51.84 Mbps. It is constructed via a matrix of 9 rows of 90 octets each. Its transmission corresponds to one row at a time from left to right and top to bottom. Further, its first 3 columns (30 octets × 9 rows = 270 octets) are used for overhead, and the remainder of the frame refers to the payload, which is provided by path layer. In SONET, the multiplexing hierarchy is defined in increment of 51.84 Mbps.

SONET is thus, essentially, a TDM framing on a transmission line that provides reference marks so that the receiver knows how to interpret the bit stream. The SONET framing imposes at least a 4.4% overhead. Other characteristics of SONET framing are:

- "Pointers" in the overhead section of the SONET frame indicate explicitly where the bit stream starts. This allows realizing rates of higher speed via direct multiplexing without resorting to intermediate multiplexing stages. That is, the pointer in the TDM overhead directly identifies the position of the payload. A synchronism is also maintained over the entire end-to-end transmission across the network by means of an accurate clock frequency source

- Since all cells are aligned on octet and flow together in the SONET payload, it is easier for the receiver to delineate the cells

- There is a bandwidth available in the SONET overhead bytes for operations and communication channels, separate from the payload so as to ease network management and control functions.

SDH: The ITU-T defined standard for the physical layer of high speed optical transmission is similar to SONET in terms of transmission speeds; however, differences in overhead functions exist. The basic building block of synchronous multiplex hierarchy in SDH is called synchronous transfer module-1 (STM-1) with a rate of 155.53, which exactly corresponds to the STS-3 rate of SONET.

A comprehensive description on SONET/SDH based physical layer in supporting the ATM cell transmission is presented in a later section.

Access to ATM backbone services

With the evolution from permanent virtual circuits to switched virtual connections (SVC), local exchange carriers (LEC) are expected to offer customers access (switched access) to (switched) long distance ATM services. Relevant effort involves the following:

✓ Conversion of any existing dedicated access to switched access
✓ Local loop technology and data rate will be retained as they are
✓ For the deployment of the conceived-switched access, tariff rates may change.

To incorporate the switched access indicated above, the initial attempts were as follows:

✓ Some carriers (LEC) announced an ATM service at DS-3 (namely, traditional T-3 transmission at 45 Mbps)
✓ Alternatively, OC-3 access on single-mode fiber with SONET framing at 155.52 Mbps. This is preferred (over the previous one) as it complies with international standards.

ATM Forum options on access facilitation

- Via multimode fiber at 100 Mbps and 155-52 Mbps

- Via twisted pair at 51 Mbps

- Via a serial data port using data exchange interface (DX1) protocol, compatible upto at least 50 Mbps.

Further considerations indicated by the ATM Forum are

- Close association of ATM and SONET in supporting transmission rate towards 4.8 Gbps

- Documentation on B-ISDN specifies deploying ATM and SONET together towards broadband, multiaccess transmissions.

It should be note however, that ATM is not synonymous with SONET. It can operate on other carriers as well, for example, T-3, E-3, E-1, T-1 fractional T-1/E-1; or over a serial interface even at 9.6 kbps. Similarly, SONET does not need ATM. For example, mapping the constant bit rate (CBR) streams into portions of a SONET frame (known as *virtual tributaries*) is feasible. The virtual tributaries refer to the synchronous format at sub-STS-1 levels. They are synchronous signals to transport low speed signals as tabulated in Table 4.1.

Table 4.1: Virtual tributaries

Type	Transport for	VT rate
VT1.5	1 DS 1 (1.544 Mbps)	1.728 Mbps
VT2	1 CEPT 1 (2.048 Mbps)	2.304 Mbps
VT3	1 DS1C (3.153 Mbps)	3.456 Mbps
VT6	1 DS2 (6.312 Mbps)	6.912 Mbps

4.1.5 *The ATM layer*

The ATM layer is implemented in ATM end-point devices and in ATM switches. It is at this layer that the task of pushing the information through the network really takes place.

As mentioned before, the transmissions in an ATM network are essentially connection-oriented. However, the ATM network can perform three versions of transmission services namely,

    ✓   Connection-oriented data transmission services as facilitated by a WAN data link

    ✓   Connectionless data transmission services akin to those in a legacy LAN data link

    ✓   Voice and video transmissions constituting isochronous delivery services.

The ATM layer overlays the physical layer in the ATM protocol reference models as shown in Fig. 4.7.

The physical interface associated with the functions of the ATM layer are the UNI and NNI. Each of these interfaces is specified by defining the format of ATM cells which traverse across these interfaces.

The ATM layer provides a set of services to the ATM adaptation layer above it when requested through a set of protocols operating in the ATM layer. The major protocol functions of the ATM layer performs are,

    ✓   Connection establishment
    ✓   Call switching
    ✓   Error detection
    ✓   Flow control
    ✓   Congestion control.

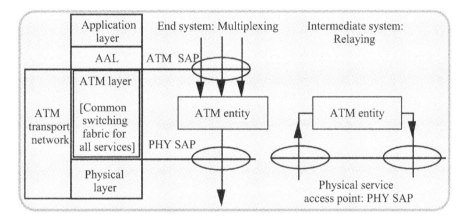

Fig. 4.7: The ATM layer in the stack of ATM architecture

Explicitly, these protocol functions can be summarized as follows:

- *Connection establishment*: When an ATM user requests a switched connection, the user's end-entity checks whether the network has sufficient capacity vis-à-vis resources available to support the type of connection requested. Accordingly, the network decides the route to be deployed for the connection and updates the logical routing tables in the ATM switches en route

- *Call switching*: This refers to the relaying task performed by each ATM switch in the network. That is, when an ATM switch accepts each cell arriving on an input TP with a specified set of VPI/VCI designations, then cells are transferred to an output TP with new VPI/VCI values. The ID swapping will be such that the cell is brought closer to its final destination. Thus cell switching implies a routing function by which ATM switches update their routing tables that they use to decide on appropriate logical TPs

- *Error detection*: The ATM layer functions under the assumption that the physical medium of transport (such as fiber optics) adopted in ATM transmissions is highly robust and reliable. Thus, the need for error check on the entire payload field may not be needed and, therefore, is not provided for in the ATM layer. If any error detection is warranted, it is left to the layer above the ATM layer. However, the ATM layer does single-error correction and double-error detection on a limited basis only on the cell header fields. This allows the detection of a corrupted cell header

- *Flow control function*: This attempts to ease the network traffic by preventing the network from overloading. A method adopted towards flow control is known as the *input rate control*. This places a limit on the rate at which each switch accepts cells arriving over input TPs

- *Congestion control*: This control mechanism is invoked if the network (or a part of it) becomes overloaded despite the flow control mechanism implemented. A strategy followed in ATM networks for congestion control is to discard cells on an ad hoc basis.

In summary, the ATM layer defines the following:

- ✓ Transmission of data in fixed-size cells
- ✓ Logical connections.

The associated functions are facilitated as follows:

- The ATM layer uses ATM connections to provide its service

- ATM connections are established and released using various signaling methods

- Cells of ATM connections are relayed by a series of intermediate ATM-entities located in intermediate systems (*Cell-relaying*)

- Intermediate ATM-entities map ATM cells from one physical connection to another

- ATM-entities multiplex ATM cells from several physical connections into one physical connection; alternate demultiplexing is also performed: Cell multiplexing/demultiplexing

- ATM-entities may provide copies of ATM cells from one physical connection to multiple physical connections: Point-to-point or point-to-multipoint connectivity

- Delay handling:

  - ✓ Connections are distinguished in terms of different delay requirements (for example, in terms of cell transfer delay and cell delay variation)
  - ✓ All cells of the same connection are subject to the same delay handling.

- *Cell-loss priority processing*: This refers to discarding cells during congestion according to network policies

- *Usage parameter control*: This control allows monitoring a connection to identify those cells that are not in compliance with any traffic descriptor negotiated through traffic contract

- *Explicit forward congestion notification* (EFCN): An IS in a congested state sets an EFCN indicator in the cell header so that an appropriate action may be taken at an end station

- Cell construction for UNI and NNI

- *Unassigned cell generation*: Insertion of unassigned cells into the flow of assigned cells to be passed to the physical connection for transmission

- *Unassigned cell extraction*: Extraction of unassigned cells from the flow of cells received from the physical connection

- *Cell header validation*: Verification of the first four octets of cell header to check whether they form an invalid pattern. Also this validation verifies that the VPI and VCI form an assigned value.

Links and connections

Transmission path: The transmission path extends between network elements that assemble and disassemble the payload of a transmission system. A payload represents a carriage of user information plus relevant overheads. It depicts the complete signal transmitted.

Digital section: This extends from the network element that assembles the bits into a stream to the other network element, which disassembles the continuous bit or byte streams.

Regenerator section: This section is a portion of the digital section extending between two adjacent regulators/repeaters.

Consistent with the illustrations of VC, VP, and transmission path in Fig. 4.8, the definitions thereof, are as follows:

✓   A transmission path consists of a bundle of virtual paths (VPs).
✓   A VP consists of a bundle of virtual channels (VCs).

Each ATM level has four architectural components as illustrated in Fig. 4.8.

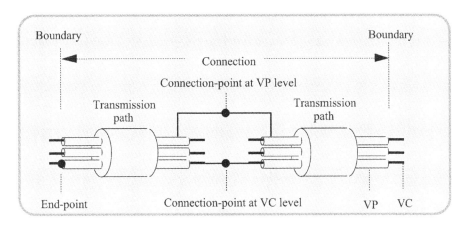

Fig. 4.8: Architectural components of ATM

Note:

VC: A concept used to describe unidirectional transport of ATM cells and associated with a unique *virtual channel identifier* (VCI) value, which remains unchanged between end-points at corresponding levels. The VCI is specified as a part of the cell header.

VP: A concept used to describe unidirectional transport of cells belonging to virtual channels that are associated with a common identifier value (known as *virtual path identifier*, VPI). VPI is also specified as a part of the cell header.

Definitions…

❑   *Connection end-point: This is located at the level boundary (for example, between VC level and VP level) and provides the connection termination function*

❑   *Connection point: This is located inside a connection where two adjacent links come together at a level where information is routed transparently*

❑   *Connection: This provides the relevant capability to transfer information between end points. It represents the association between end-points*

❑   *Link: This provides the capability to transfer information transparently. A link represents an association between two contiguous connecting points.*

❑   *Virtual channel link: This is a means of unidirectional transport of ATM cells between a point where a VCI is assigned and the point where that value is translated or removed.*

❑   *Virtual path link: It is a transport line terminated by the points where a VPI is assigned and translated or removed.*

Virtual channel connection (VCC) is a concatenation of VC links and a *virtual path connection* (VPC) is a concatenation of VP links. As mentioned before, an ATM cell contains a label in its header to identify explicitly the virtual channel (VC) to which the cell belongs: This label has two parts, namely,

     ✓   Virtual channel identifier (VCI)
     ✓   Virtual path Identifier (VPI).

Summary on VC and VP levels

-      A VCI identifies a particular VC link for a given virtual path connection (VPC). A specific value of VCI is assigned each time a VC is switched in the network

-      A VC link has unidirectional capability for the transport of ATM cells between two consecutive ATM entities where VCI value is translated. A VC link is originated or terminated by the assignment or removal of the VCI value

-      Routing functions of virtual channels are performed at a VC switch/cross-connect. It involves translating VCI values of the incoming VC links into VCI values of the outgoing VC links

- Virtual channel links are concatenated to constitute a *virtual channel connection* (VCC). A VCC is constituted between two VCC end-points, or in case of multipoints, between more than two VCC endpoints

- At the VCC endpoint, the cell information field is exchanged between the ATM layer and the user of the ATM layer service.

4.1.6   Switching of VCs and VPs: Logical ID swapping

The switching of VCs and/or VPs in ATM is a two-stage process. It involves two levels of logical identification (ID) pertinent to the swapping of VPI and VCI. In order to understand the underlying concept, the following salient considerations are recapitulated:

- ✓ VCIs and VPIs in general, have significance for one link
- ✓ In a VCC/VPC, the VCI/VPI value is translated at VC/VP switching entities
- ✓ "*Switch*" and "*cross-connect*" can be synonymously used while explaining VC/VP concepts: A switch is a "space-division switch". It is an electronic "patch panel"
- ✓ VP switches terminate VP links and therefore have to translate incoming VPIs to the corresponding outgoing VPIs according to the destination of the VP connection
- ✓ VC switches terminate both VC links, and VP links. Hence, both VPI and VCI translations are performed
- ✓ VC switching implies VP switching inclusive. (However, VC switching in principle may also perform only VP switching.)

The VP/VC switching considerations are consistent with the ATM network capability and offer a VP level service and/or VC level service. The execution of VP/VC switching refers to the ID swapping across an ATM switch, which in essence performs a *cell-switching function*. This function decides how to forward cells from one TP to another across the network. Thus, the switching function can be summarized as follows:

- Choosing of appropriate output TP for an incoming cell

- Placing an appropriate ID (VPI/VCI) tag in the field space provided in the header of that cell.

The switching function, hence corresponds to a *routing function*. To enable the input-to-output ID swapping, the switching makes use of routing table/routing databases. When a VC connection is established, the relevant switches in the network make an entry into their routing table databases concerning the details on the identifiers of appropriate TPs versus VPIs/VCIs. Normally, the switches across an ATM network all would run a distributed routing algorithm so as to create the routing tables that each switch may use for ID swapping on a given class of cell, which arrives at that switch.

As indicated above, the switching function may correspond to:

- ✓ Setting only VPI values (*VP switching*)
- ✓ Setting both VPI/VCI values (*VC switching*).

The procedure followed in each case is addressed below:

Virtual path switching

This is illustrated in Fig. 4.9. ATM switches B, D, and E in Fig. 4.9 are VP switches and do ID swapping across their input-to-output only on VPIs. That is, each of these switches assigns a unique VPI (number) to each physical transmission path attached to its input and output side.

Fig. 4.10 shows how an ATM switch X assigns the VPIs 10, 20, and 30 to the VPs at the transmission paths shown as XA, XB, and XC, respectively. Routing information is based on the details gathered when connections are established and it is shared among the switches. A switch would use this routing information to prepare its own VP routing table for each physical TP over which it may receive the cells. The routing table for the given input TP contains all the VPIs the switch can expect in the header of the cells received from that TP.

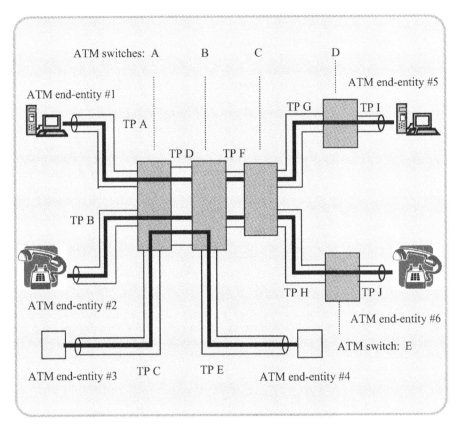

Fig. 4.9: Implementation of VP/VC switches across a typical ATM network

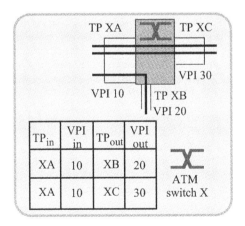

Fig. 4.10: Construction of VP routing table

VC switching

Here again, a VC switch exchanges routing information with other switches to maintain routing tables. The pertinent details enable the switch to choose an output transmission path for each cell it receives. The switch then assigns both VPI and VCI values in each header of the outgoing cell.

Thus, in essence, ATM switching implies logical IP swapping across its input to output ports.

Applications of VC and VP connections

VCCs and VPCs are employed between the following:

✓ User-to-user
✓ User-and-network
✓ Network-to-network.

All cells associated with an individual VCC/VPC are transported along the same route through the network by virtue of the VPI/VCI assigned to them. Further, cell sequence integrity is maintained throughout for all VCCs and VPCs. (That is, the first sent cell is received first as mentioned earlier.)

In reference to the types of VCCs and VPCs indicated above, the specific functions are as follows:

- *User-to-user VCCs*: These carry all user data and signaling information

- *User-to-network VCCs*: These are used to access all local-connection related functions (like user-network signaling)

- *Network-to-network VCC*: These perform the network traffic management and routing

- *VPC between users* is intended for the following:

  ✓ It provides them with a transmission "pipe"
  ✓ The VC organization within this pipe is left to the choice of the user (for example, organizing a LAN-LAN coupling).

- *VPC between user-to-network*: This is used to gather traffic from a user to a network element such as a local exchange or a specific server

- *Network-to-network VPC*: This is used to organize user traffic in accordance with a predefined routing scheme or to define a common path for the exchange of the routing of network management information.

In summary, an ATM network can provide a *virtual path level service*, a *virtual channel level service*, or both. In a network that provides a virtual path level service, a switch receives a cell with a specified virtual path and virtual channel identifiers. It then performs a table-lookup to determine how the virtual path identifier value should be remapped for forwarding to the next switch or end-system. The virtual channel identifier value is not remapped. This type of switching is called *virtual path switching* (VP switching).

Similarly, in a network that provides a *virtual channel level service*, again a switch receives a cell with specified virtual path and virtual channel identifiers. It then assigns a new set of virtual path and virtual channel identifier values to the cell before forwarding it to the next switching or end-system. This type of switching is called *virtual channel switching* (VC switching).

The following example illustrates the constructional procedure of a VC table.

Example 4.1

Illustrate, by an example, the VPI/VCI usage on link and end-to-end considerations.

Solution

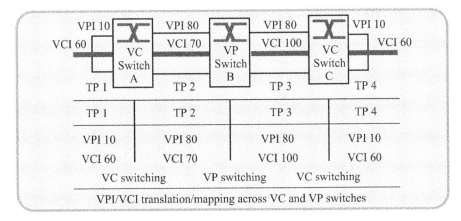

Fig. 4.11: VPI/VCI application on linked end-to-end connectivity

The VPIs and VCIs are both used to route cells through the network. It should be noted that VPI and VCI values should be unique on a specific transmission path (TP). Thus, each TP between two network entities (such as ATM switches) would use VPIs and VCIs independently. This is illustrated in Fig. 4.11. Each switch maps an incoming VPI/VCI to an outgoing VPI/VCI. The switches A and B have a single transmission path (TP) between them. Over this TP, there could, however, be multiple virtual paths (VPs). There are three switches indicated in Fig. 4.11. Switches A and C are VC switches, which perform both VPI and VCI translations. Switch B is a VP switch, which does only VPI translation.

At the ATM UNI, the input device to switch A provides, for example, a video channel over virtual path 10 (VPI 10) and virtual channel 60 (VCI 60). Switch A then assigns the VCI 60 to an outgoing VCI 70, and the incoming VPI 10 to outgoing VPI 80. Thus, on VPI 80, switch B specifically operates on virtual channel (VC) number 70 (VCI 70). This channel is then routed from switch B to switch C over a different path and channel (VPI 80 and VCI 100). Thus, VPIs and VCI are gathered onto each individual link across the network. This is akin to frame relay, where data link connection identifiers (DLCIs) address a virtual circuit at each end of a link. Finally, switch C translates VPI 90 into VPI 10, and VCI 100 to VCI 60 at the destination UNI. It should be noted that the destination VPI and VCI need not be the same as at the origin. The sequence of VPI/VCI translation across the switches can be viewed as a network address in an extrapolation of the OSI layer 3 model.

Example 4.2

How do VCs support multiple applications? Illustrate with an example.

Solution

Multiple traffics depicting different applications are facilitated in ATM transmissions via *cell segmentation* as illustrated in Fig. 4.12.

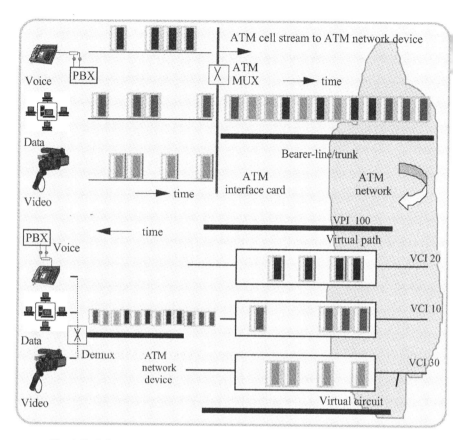

Fig. 4.12: Cell-segmentation in an ATM transmissions supporting multiple applications

In Fig. 4.12, the ATM switches receive a set of user information on data, voice, and video messages and segmentize each of them into fixed length cells. These cells are multiplexed asynchronously into a single bit stream, which is transmitted across a physical medium. As an example of the state-of-the-art multimedia applications, consider a military dispatch from a war-front campsite to a command headquarters detailing a tactical war-front scenario of a launched missile tracking an enemy target. Along with the commentary, the text message on the tactical data and a real-time video on the high-speed missile in its tracking mode would constitute the heterogeneous traffic involved. Thus, the networking should be facilitated to support these applications on individual VCs. Though a simple commentary may warrant a 64 kbps voice transmission, it is presumed here that a high fidelity audio transmission (in addition to voice) supporting certain high frequency audio recordings is transmitted for analysis purposes. Such transmissions can be interposed between talk-spurts. Thus, the audio transmission assumed refers to a high fidelity signal with an adequate bandwidth.

Video and voice are isochronous and hence, are very time-sensitive. The associated information cannot be delayed, and also any delay perceived cannot have significant variations. Disruptions in the video image of a missile and/or the target, that is, the distortions of the video display, would destroy the interactive, near real-life tactical aspects of this multimedia application under discussion. Data can be sent in either a connection-oriented or connectionless mode. In either case, the data is not nearly as delay-sensitive as voice or video traffic. But, data traffic is very sensitive to loss of bits. In the present example, the loss of information can lead to semantic impairments in the mission criticality involved. Therefore, ATM must offer transmission feasibilities, which are distinct in respect to voice, video, and data traffics, giving voice and video traffic priority and a guarantee-bounded delay. Simultaneously it should assure that data traffic has very low bit loss. The procedure to support such heterogeneous traffic is as follows:

❑ A VP is established between the workstation at the tactical site and the remote command station

❑ Over this VP, three VCs are defined for text data, voice/audio, and video

- VCI 10: Text data
- VCI 20: Voice/audio
- VCI 30: Video.

The ATM network devices (switches/cross-connects) may perform necessary logical ID swapping across the network between the end stations.

Example 4.3

In reference to Example 4.2, explain how user traffic is segmented into ATM cells, switched through a network, and processed by the receiving user-end.

Solution

Pertinent to the context of ATM transmission presented in Example 4.2, there are simultaneous transmissions of text, voice/audio, and video traffics from a workstation. The workstation contains an ATM interface card, where a segmentation procedure "slices and dices" each type of information into 48-octet data segments as shown in Fig. 4.12. In the next step, the relevant task refers to providing an ID (via an appropriate address) to each payload by prefixing it with a VPI, a VCI, and other necessary fields of the 5-octet header. The result is a stream of 53-octet ATM cells from each source: Voice, video, and text data. These cells are generated independently by each source, such that there may be contention for cell slot times on the interface connected to the workstation. The text, voice, and video are each assigned a VCC. For example, VCI = 10 for text data, VCI = 20 for voice/audio, and VCI = 30 for video, all on a VPI = 100. (This example has been grossly simplified, as there would normally be many more than just three active VCI values on a single VPI.)

Fig. 4.12 shows an example of how the war-front tactical site terminal sends the combined voice/audio, video, and text data. The terminal may shape the transmitted data in intervals of ten cells (about 100 μs at the DS-3 rate) allowing one voice cell, then six video cells, and finally what is left – three text data cells – to be transmitted. This corresponds to about 4 Mbps for voice/high-fidelity audio, 25 Mbps for video, and 12 Mbps for text data. All data sources (text, voice/audio, and video) would contend for the bandwidth in each shaping interval of ten cell times, with the voice/audio, video, and then text data being sent in the above proportion. Cells will wait in the buffer, in case all of the cell slot times were full in the shaping interval.

Example 4.4

Illustrate the functional aspects of an ATM switch using the multimedia scenario of Example 4.2.

Solution

See Fig. 4.13.

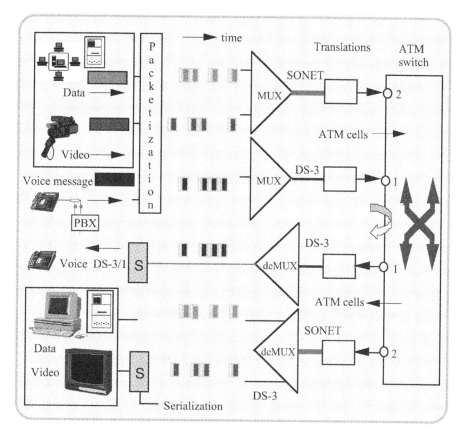

Fig. 4.13: Functional aspects of an ATM switch

In reference to the multimedia transmission considered in Example 4.2, an illustration of the functional attributes of the ATM switch is shown in Fig. 4.13. The continuous video source is indicated as the input to a *packetizing function*, with a logical destination VPI/VCI address. The continuous bit stream is sliced into fixed length cells comprised of a header and a payload field. The video transmission has a rate is greater than the continuous DS-3 bit stream.

The text data is directly packetized from the computer handling the missile tracking and is addressed to a corresponding data logging terminal.

These sources are time-division multiplexed over a transmission path, such as SONET or DS-3.

The primary function of the ATM switch shown in Fig. 4.13 is to translate the logical address to a physical outgoing switch port address and to an outgoing logical VPI/VCI address. This requires a header be prefixed to each input ATM cell. There are three point-to-point virtual connections in the Fig. 4.13. The address of DS-3 is translated into a corresponding address destined for physical port 1. The video source is translated into an address destined for port 2. The computer source is likewise translated to an address destined for port 1.

The ATM switch utilizes the physical destination address field to deliver the ATM cells to appropriate physical switch port and the associated transmission link. At the output of the ATM switch, the physical address is deleted by a *reduce function*. The ATM cells bearing logical addresses are then time-division multiplexed onto the outgoing transmission links. Next, these streams are demultiplexed to the appropriate devices. The video and the DS-3 connections will then have the logical address removed, and reclocked to the information sink by means of a *serialize function*. Devices, such as workstations, can receive the ATM cells directly. Again, the ID swapping performed at the ATM switch complies with the following considerations indicated earlier:

❑ VC switches terminate both VC links and necessarily VP links. Both VPI and VCI translations are performed

❑ VP switching implies VPI translation only

❑ VC switch may also perform only VP switching, in principle.

Problem 4.1

An ATM network is as shown in the diagram of Fig. 4.14. The elements of the network are as follows: End-stations, a set of VC switches, a VP switch, and ATM trunks with access interfaces. VCCs and VPCs are identified as indicated in the diagram.

Identify and assign VCIs and VPIs for all trunks and access interface ports for all the connections indicated.

Construct a typical switching table. (That is, tabulate all the incoming VPI/VCI values and outgoing VPI/VCI values.)

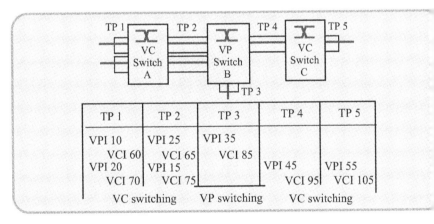

Fig. 4.14: VPI/VCI assignments across a cross-connect

Construction of VP/VC routing tables

Illustrated in Fig. 4.14 are step-by-step construction procedures to construct the VP/VC routing tables in reference to an ATM end-to-end connectivity with interposed VP and/or VC switches (shown in Fig. 4.14).

4.1.7   ATM cell structure

A core consideration in the asynchronous transfer mode of information is the unique version of packets adopted in carrying the bits. The following features characterize the ATM cell structure:

✓ Fixed size cell
✓ 5 octet header
✓ 48-octet information field.

The ITU-T Recommendation I.361 on B-ISDN/ATM layer specification defines the detailed bits and bytes of the ATM cell format, – their meaning and their use. Detailed cell structure, field coding, and relevant protocols are included in the mentioned recommendation. It also details the UNI/NNI cell formats.

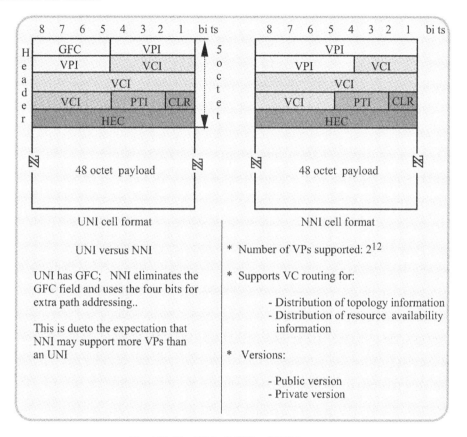

Fig. 4.15: The ATM cell: UNI and NNI formats

The small and fixed size attributes of the cells offer the following advantages:

- Queueing delay is reduced for a high priority cell, as it waits less if it arrives slightly behind a low priority cell that has (already) gained access to a resource

- Fixed size cells can be switched more efficiently which is important for very high data rates

- Switching mechanism (in hardware) can be implemented more easily with fixed size cells.

The numbering conventions (ITU-T Recommendation I.361) pertinent to an ATM cell contents are specified as follows:

- Octets are sent in increasing order starting with octet 1. Therefore, the cell header will be sent first, followed by the information field

- Bits within an octet are sent in decreasing order starting with bit 8

- For all fields, the first bit sent is the most significant bit (MSB).

The ATM cell ...

The term cell is essential for B-ISDN, and it is defined in the ITU-T Recommendation I.113 (Vocabulary of terms for broadband aspects of ISDN) as: "A cell is a block of fixed length. It is identified by a label at the ATM layer of the B-ISDN PRM". More detailed definitions for the different kinds of cell are presented in ITU-T Recommendation I.321 (B-ISDN protocol reference model and its applications).

Idle cell: A cell that is inserted/extracted by the physical layer in order to adapt the cell flow rate at the boundary between the ATM layer and the physical layer to the available payload capacity of the transmission system used.

Valid cell: A cell whose header has no errors or has been modified by the cell header error control (HEC) verification process.

Invalid cell: A cell whose header has errors and has not been modified by the cell HEC verification process. This cell is discarded at the physical layer.

Assigned cell: A cell that provides a service to an application using the ATM layer service.

Unassigned cell: An ATM layer cell which is not an assigned cell.

Only assigned and unassigned cells are passed to the ATM layer from the physical layer. The other cells carry no information concerning the ATM and higher layers and therefore will be only processed by the physical layer.

Example 4.5
How did the ATM Forum arrive at a payload size (number of octets per cell) as 53?

Solution
(See also the relevant discussions furnished in Chapter 3, Fig. 3.81)

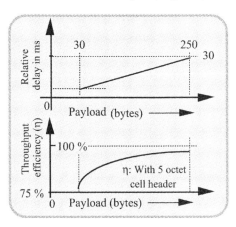

Fig. 4.16: Throughput efficiency and packetization delay versus cell-size

When a standard cell size was under discussion by the ATM Forum, there was a debate on using a 320 octet versus a 64-octet payload size. The decision on the 48-byte payload size was the compromise between these positions. The choice of the 5-octet header size was a separate consideration in choosing between a 3-octet header and an 8-octet header.

There is a basic tradeoff consideration between efficiency and packetization delay versus cell size as illustrated in Fig. 4.16 where efficiency is computed for a 5-octet cell header. Packetization delay is the amount of time required to fill the cell at a rate of 64 kbps, that is, the rate to fill the cell with digitized voice samples. Ideally, high efficiency and low delay are both desirable, but cannot be achieved simultaneously. As could be evinced from Fig. 4.16, better efficiency is realized at large cell sizes at the expense of increased packetization delay. In order to carry voice over ATM and interwork with two-wire analog telephone sets, echo cancellation must be used. Two TDM to ATM conversions are required in the round-trip echo path. Allowing 4 ms for propagation delay and two ATM conversions, a cell size of 32 octets would avoid the need for echo cancellation. Thus, the ITU-T adopted the fixed-length 48-octet cell payload as a compromise between long-cell sizes for time-insensitive traffic (64 octets) and smaller cell sizes for time-sensitive traffic (32 octets).

Cell header description

ATM cell-header is characterized with respect to two cell formats corresponding to two key physical interfaces adopted in practice. These physical interfaces are the user-network interface (UNI) and the network-network interface (NNI) as illustrated in Fig. 4.17.

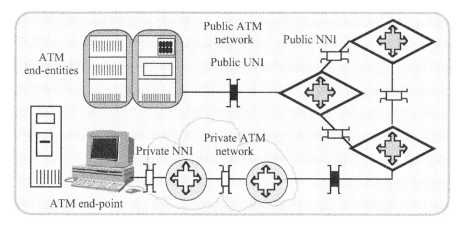

Fig. 4.17: ATM network interfaces

UNI

This is the interface that exists between an ATM end-point (end-system) and an ATM switch (intermediate system). Though ATM Forum distinguishes the public UNI and private UNI, currently, the ATM cell format for both of them are taken to be the same.

The public UNI refers to the interface between a service provider's public ATM network node (switch) and an ATM end-entity of a workgroup or the private ATM switch (customer premise node) belonging to a private network of an organization.

The private UNI corresponds to the interface between a private ATM switch and an ATM end-entity.

NNI

This is the interface between two ATM switches in a common carrier ATM network. As per ATM Forum, UNI is also of two versions namely, *public NNI* and *private NNI*. Currently, the same cell format is used in either case.

The public NNI exists between two ATM switches in the same ATM public network or between an ATM switch in one public network and an ATM switch in another public network.

The private NNI refers to the interface between two private switches (customer premise node). It can exist between two private ATM switches in the same ATM network of an organization or between private ATM switches belonging to two different organizations.

ATM cell formats of user-network and network-network interfaces

In reference to the cell formats of UNI and NNI shown in Fig. 4.15, it can be observed that the cell header structure in each case is distinct from the other. The following table (Table 4.2) summarizes the contents of the headers.

Generic flow control

Table 4.2a: Field contents of UNI/NNI cell headers

Field	UNI	NNI
Generic flow control (GFC)	4-bit field intended to define a simple multiplexing scheme. Currently, ATM Forum has indicated only the uncontrolled mode, where the 4-bits GFC is encoded always as zeros.	Not used in NNI cells

The concept of GFC as perceived through the standardization process and recommended in 1995 (ITU-T Recommendation I.361) governs the point-to-point configuration. It allows a multiplexer to control contention for a shared trunk resource through the use of traffic-type selective controls.

In reference to ATM UNI, GFC is a part of the cell header with 4-bit field. It forms about 1% of the available ATM cell payload rate.

When the GFC field is set with all 4 bits as zeros, it corresponds to a default coding of the GFC being null. It means that the interface is not under GFC, which is called the uncontrolled mode. (In a controlled mode, the terminal responds to commands from the multiplexer as it understands the GFC, not set as null.) The GFC carries distinct meaning depending on the direction of cell transmission. (Currently, GFC is set as null in ATM transmissions to allow interoperability.)

Table 4.2b: Virtual path

Field	UNI	NNI
VPI	8-bit field	12-bit field
	This is an identifier that groups virtual channels into path for the purpose of routing.	

Virtual path is a route through some part of the ATM network that begins at one ATM device and terminates at another. An ATM network has the capability to accommodate a large number of VPs. A given VP is implemented by the set of transmission paths that connects the two ends of the VP. (A transmission path refers to the physical link over which, the data is transferred from one ATM device to another).

A VP may involve a short portion of the whole network. Therefore, between end-to-end connectivity, there could be several VPs.

Table 4.2c: Virtual channels

Field	UNI	NNI
VCI	16-bits field	16-bits field
	This is an identifier of a particular virtual channel within a virtual path.	

VCs depict the second level in the hierarchy of logical connectivity and define how TPs are shared among multiple users. Like VPs, VCs also have two ends. A VC defines a route between two communicating entities over which cells are transported from one ATM device to another. The end-entity of a VC could be an ATM device or an ATM switch. The VCI specifies the unique identity of a VC within a VP.

Table 4.2d: Payload type (PT)

Field	UNI	NNI
Payload type (PT)	3-bit field	3-bit field
	This is an identifier of the type of information contained in the payload field.	

The meaning of 3-bit PT encoding (ITU-T Recommendation I.361) is as follows: The right most bit of PT encoding is an AAL indication bit. The middle bit denotes the upstream congestion and the first left discriminates between data and operations cells.

❑ *AAL indication – Specification on the AAL protocol in user information cells*

❑ *Congestion indication – Explicit forward congestion indication (EFCI) in user information cells*

❑ *Operations cells: OAM and resource management cells (These do not indicate AAL or congestion.)*

PT coding	PT meaning
000	*User information, EFCI = 0, AALI = 0*
001	*User information, EFCI = 0, AALI = 1*
010	*User information, EFCI = 1, AALI = 0*
011	*User information, EFCI = 1, AALI = 1*
100	*OAM cell (segment operation)*
101	*OAM cell (end-to-end operation)*
110	*Resource management cell*
111	*Reserved for future VC functions*

Note: AALI ⇒ (ATM layer) user - to - (ATM layer) user indication

Table 4.2e: Cell-loss priority (CLP)

Field	UNI	NNI
Cell-loss Priority (CLP)	1-bit field	1-bit field
	This field is used by an ATM equipment to determine whether the cell can be discarded in the event of congestion.	

The CLP is set as 0 or 1. The value 0 implies that the cell is of the highest priority and the network will not discard a CLP = 0 cell. If CLP = 1, the network may selectively discard the corresponding cell during congested periods so as to avoid a low loss rate for the high-priority cells. Such a CLP = 1 tag can be placed by the network as a result of policing. Thus, CLP bit plays the relevant role in the traffic and congestion control efforts.

Table 4.2f: Header error check (HEC)

Field	UNI	NNI
Header error check (HEC)	8-bit field	8-bit field
	The HEC field contains an error detection and correction code value to detect and sometimes correct errors in the header octets of the cell.	

The HEC provides for error checking of the header for use by the transmission convergence (TC) sublayer of the physical layer.

In the ATM header, the last octet denotes the HEC field. It is described in CCITT I.432. This octet is intended only for error detection and correction (ED & C) on the ATM header only. It is not provided for ED & C of the actual information field contained in the 48 octets. HEC addresses a robust issue in regard to any error happening in the header itself. Should such an error happen and it is not correctable, it is logical that the corresponding cell be discarded as it has lost its identity. That is, the VPI and/or VCI of this cell (specified as 8 and 16 octets in the header) has become invalid and, therefore, the cell will be misdelivered.

HEC tries to "fix" the error encountered by correcting single-bit error; in the event of 2-bit error occurring, the HEC simply depicts it and it does not have the capability to correct it [4.3, 4.4]. If a multiple bit (> 2) error occurs, the HEC has neither the capability to detect nor to correct it. The cell will be passed as an uncorrupted entity and will be eventually misdelivered. HEC uses an algorithm thereof, known as SECDED (*single-error correction, double-error detection*). This has been prescribed as an optimum strategy taking into account that the physical medium of the ATM is fiber optics and the associated BER is low. However, if the twisted copper-pair is used, the burst-errors would lead to multiple bit errors, which may be uncorrectable. Likewise, in wireless transmission BER could be high and hence the HEC strategy may not be an optimal solution.

Since the HEC algorithm is adapted for single-bit error correction, rather than merely to indicate errors and request retransmissions, (unlike in other error-correction schemes of the past), further details on HEC operation may be useful.

Fig. 4.18 illustrates the actions that a receiver undertakes to implement the HEC verification processing. The associated SECDED algorithm has two modes:

- ✓ Correction of single-bit errors
- ✓ Detection of double-bit errors.

Normally, HEC is in correction mode. If no HEC error is detected when the receiver's calculated HEC value on the received cell header is compared with the value received with the cell, no action is taken. Even after the receiver has corrected a single-bit error and recalculated the proper HEC, the receiver will transit to detect mode.

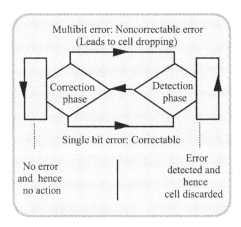

Fig. 4.18: States depicting HEC processes. (*Note*: Triple or more bit errors would go undetected)

No errors are corrected in the detection mode and all errored cells are dropped. (As mentioned above, such detected and dropped cells correspond to those which experienced a 2-bit error.) The dropping of a cell means that all sequences (consecutive runs of cells of any number) of cells containing errors are not allowed through the network. When a cell with a valid HEC is found in detection mode, the receiver will transition back to correction mode again. Thus the receiver will in practice correct only every other cell in a sequence of cells with invalid HEC values. The logic behind this is that on a fiber-based network, single-bit errors occurring twice within a span of 53 bytes (in consecutive cell header) may not happen with any specified regular pattern, and even if they did, the TC sublayer has more to do at the receiving side than to spend processing time correcting the bit errors in runs of consecutive cell headers.

In the above procedure, suppose two bits (in the total of $5 \times 8 = 40$ bits) at the header get corrupted. When these errors go undetected, it may lead to invalid VPI/VCI. As such, misdelivery of the corresponding ATM cells would occur. Thus, the primary source of misdelivered cell error or misinserted cells in ATM is due to such undetected bit errors.

Performance of HEC algorithm

The function of the HEC receiver is tailored for fiber transmission systems in which the following are valid considerations:

✓ Most errors are single-bit errors
✓ Sparse occurrence of bursty errors.

The consequent HEC functions have, therefore, been trimmed to the following:

✓ Correction of single-bit errors and detection of 2-bit errors
✓ Discarding of cells when bursty errors (more than 2 bits) are detected.

The HEC algorithmic procedure refers to on operation only on the 40 bits of the header and the HEC generation process leads to generating 8 bits only. This is rather simple when compared to the conventional CRC, which works on thousands of bytes and produces 16 and/or 32 bit-long frame check sequences.

The HEC method has been considered fairly "good" inasmuch as in its implementation, the probability of cell-loss is about 10^{-13} corresponding to an encountered bit error of 10^{-8}. Further, the chances of a cell with an undetected error passing through to the network is as small as 10^{-20}. These figures indicate that it may be rather unnecessary to waste process time and resources to perform more exquisite ED & C on ATM cells than the simple HEC procedure on the header indicated above. This optimistic view, is however, true when the transmission is supported by the robust physical medium, namely, the optical fiber.

A distinguishing feature of ATM protocol architecture is that the HEC check (ED&C) refers only to single-bit errors in the ATM header. The related error control has two unique considerations:

✓ HEC error-control check is applied only on the first 4 bytes of the ATM header. (The 5th byte refers to the HEC field itself.) No part of the ATM cell payload (48 bytes) is protected by the HEC
✓ Unlike most other data protocols, the HEC field of ATM is capable of facilitating the SECDED protection on the cell header contents. It does not simply perform the ARQ request for retransmission of cells when the received traffic has error contents.

As mentioned earlier, SECDED offers single-error detection and correction or it provides for double-error detection in the header. Since a physical medium such as optical fiber guarantees the semantic integrity of bits transmitted, most of the time the single-error detection and correction function offered by the HEC will suffice. This is also compatible with the situation when the ATM supports time-sensitive information where any delay (such as waiting delay arising, if an ARQ-based retransmission is adopted) cannot be tolerated.

The double-error detection (with no correction feasibility) enabled by HEC is much similar to the error control techniques of other protocols like X.25 or SDLC.

What happens if there are multiple errors accounting for more than 2 bits in the 40 bit cell header sequence? The HEC has no provision to detect such errors. As such, the corresponding cell may be allowed to pass through the network and probably will be misdelivered. The wisdom of the ATM Forum not considering such eventualities is quite obvious. With the robust physical medium dependent transmission supporting the ATM, encountering bit errors in excess of 2 bits in the header section is rather remote.

The HEC field is described in the ITU-T Recommendation I.432, Sec. 4.3.2/HEC sequence generation as follows: "The HEC field shall be an 8-bit sequence. It shall be the remainder of the division (modulo-2) by the generator polynomial $x^8 + x^2 + x + 1$ of the product x^8 multiplied by the content of the header excluding the HEC field."

In Chapter 2, the mathematical considerations on a generator polynomial were presented. For the ATM HEC sequence, the generator polynomial used to generate the HEC sequence corresponds to the bit string 100000111. This 9-bit-long bit string is used by the sender as a divisor of the 32 bits from the first 4 header bytes of the ATM cell. The division adopted is the modulo-2 type. The following example is used to illustrated the generation of 8 bit HEC check sequence.

Suppose the contents of an ATM cell are as follows: VPI = 7, VCI = 11, PTI = 0, and CLP = 0. In reference to the UNI cell header format depicted in Fig. 4.15, the header contents specified carry the following meanings:

VPI (= 7):   This is 8 bits in length and specifies the logical routing of the cell
0111 0000
VCI (= 11):   This is a 16-bit field that identifies a particular VC within the VP adopted
1011
PTI (= 0):   This is a 3-bit field that identifies the type of information contained in the
000   payload field

There is a generic flow control (GFC) field, which is 4 bits in length used in the cell defining the UNI. It is intended to specify a simple multiplexing scheme. The current ATM standard defines only a simple uncontrolled mode for which GFC = 0000.

Further, there is a CLP field of 1-bit length, which denotes the cell loss priority specifying whether the cells can be discarded when congestion occurs. CLP = 0 means that the cell is of the higher priority and the network is unlikely to discard the CLP = 0 cells under congestion. A value of CLP = 1 means that the network may selectively discard relevant cells during congested intervals. In the present example, CLP is taken as 1. The HEC procedure is as follows:

Step 1: The sender appends eight 0 bits to the header values as indicated below:

         0000 0000 0111 0000 0000 00001011 000 1 00000000
         GFC   VPI   VCI   PT CLP Appendix

         With the eight 0 bits appended to the 32 bits of the first 4 bytes of the ATM cell header, there are 40 bits in total constituting the "product x^8"

Step 2: The 40-bit string indicated above is divided by the generator polynomial string, namely 100000111 ($\Rightarrow x^8 + x^2 + x + 1$) using the XOR binary (modulo 2) division rule

Step 3: When the division is performed as above, the remainder is obtained as 01111001. *Note*: This remainder is 8 bits in length inasmuch as the divisor was 9 bits long. This remainder, namely 01111001 string constitutes the HEC sequence. It is appended to the 4-byte cell-header when sent across the link to the receiver. The division operator simply refers to the logical operation implemented as a series of shift registers with XOR gates

Step 4: The receiver, such as the ATM network node (switch) or an end-entity (CPE), recomputes the HEC check (called the *syndrome*) based on the received value of the first 4 header bytes, and this computed value is compared with the received value of the HEC field. This comparison is done via the XOR operation. If the compared entities are identical, the result will be zero. That is, the received header is accepted as valid and forwarded to the ATM layer for switching and/or for further processing. A non-zero result implies the presence of an error in the cell-header. As indicated earlier, a single-bit error can be detected (or located in the header content) and corrected. A 2-bit error can only be detected and cannot be corrected. The receiver then may discard the cell. If there are more than two corrupted bits due to bursty errors, this HEC procedure does not work. The cell will be passed on as a valid one leading to misinsertion at some destination

Step 5: When a single-bit error occurs, the question is how it is detected and corrected. Suppose in the example under discussion, a single-bit error has occurred in the PTI field, say the field reads as 010 (digital 2) instead of 000 (digital 0). Hence, the received sequence will read as:

0000 00000111 0000 0000 0000 011 010 1 00000000

Bit error (30th position)

Performing the division of the above sequence with the divisor 100000111, the remainder is obtained as 01100101.

Step 6: The original transmission sequence carries the HEC sequence 0111 1001 corresponding to PTI = 000. This sequence is compared with the computed sequence obtained on the corrupted transmission at the receiver. That is, 0110 0101 is XORed with 0111 1001:

$$\begin{array}{l} 0111\ 1001 \\ \otimes \\ \underline{0110\ 0101} \\ =\quad 0001\ 1100 \quad\longleftarrow \text{Index} \end{array}$$

The result is known as an *index*. It is used for in a table (Table 4.A1 in Appendix 4.1) to determine what position the single-bit error has occurred. Table 4.A1 indicates the single-bit error position is 30 in respect to the index where 00011100 was calculated. This confirms the fact that this 30th position is on the PTI field where the original 0 (binary 000) got changed to 2 (binary 010).

Table 4.1A is made available at the receiver end and it tabulates the results on a single-bit error (0 to 1) presumed to occur position-by-position along a 40-bits sequence set all as 0s, and the 100000111 is the generator polynomial used.

ATM information payload

A complete cell at the ATM level can be regarded as a *protocol data unit* (PDU). That is, it is sent (in its entirety) to a receiver in the same layer of a protocol stack in another piece of equipment. It contains both user information as well as the address where it should be delivered.

The ATM cell payload contains 48 octets (384 bits) of user data and/or additional control information. There are different formats used for the payload field corresponding to different versions of data delivery services that can be facilitated by an ATM network.

Within the cell, each 48-octet payload refers to a *service data unit* (SDU). That is, the SDU is a service received by the layer above in the protocol stack. The ATM function is to accept 48 - octet SDUs, scramble them to randomize bit patterns, apply an address (also received from the higher layer), then multiplex multiple connections on the physical medium that carries that electrical or optical signal.

Within the SDU, there can be any number of formats, as will be explained in later sections. The payload of an idle cell, one with an address of all 0s, is filled with a fixed bit pattern in each octet.

The 48-octet SDU seldom contains 48 bytes of user information. Rather, the information field usually holds a 48-octet segmentation unit which will itself contain a header (a sequence number, for example) and perhaps a trailer. These depend on the form of adaptation used, which in turn depends on the type of information being carried. Details on the process of breaking longer blocks into cells, and the overhead required, are follows:

The space for "real" data (including frame headers and trailers) is less than 44 octets out of 53 (83% of the bandwidth). In measuring the "true" information-carrying capacity, one must deduct the following from this 83%:

- Any additional LAN or packet headers, SONET formatting, or similar overhead that might be part of a higher protocol layer or physical transport format

- Stuffing or padding to fill out partially loaded cells (since they should be of constant length); a "remainder" segment from a higher protocol must occupy a full cell.

Details on the "inputs" to an ATM cell from higher layers of ATM protocol will be explained later.

Hence, for short LAN frames or packets with long headers, the overall efficiency can drop into the 60s. Fortunately, "bandwidth availability will keep up with the double whammy of more information flow and lower bandwidth efficiency (utilization)".

In summary, the ATM cell structure allows a "cell-relay" transmission in a connection-oriented networking. It facilitates the following operational characteristics of the ATM networking:

- ✓ Transfer of data in discrete chunks
- ✓ Multiple logical connections multiplexed over a single physical interface (similar to packet-switching and frame-relay)
- ✓ Information flow in each logical connection via organized fixed size packets (cells)
- ✓ Developed as a sequel to B-ISDN, ATM is considered for non-ISDN environments where high data rates are required
- ✓ In comparison to frame-relay, ATM can support data rates several orders of magnitude larger
- ✓ ATM protocols are streamlined for minimal error and flow controls (with reduced processing overhead)
- ✓ Reduces number of overhead bits in each cell enabling high data rates.

4.2 Parameters of ATM Transmissions

4.2.1   Temporal transparency of ATM

This refers to the transmission of information across the network with the end-entities not perceiving the end-to-end delay and/or its variations [4.3]. These are two ATM-specific temporal transparency parameters of interest. As indicated in Chapter 3, they refer to the *cell-transfer delay* (CTD) and *cell-delay variation* (CDV). A real-time transfer of cells (that is, CTD → 0) would be an ideal transmission objective, which can, however, be hardly achieved. A finite extent of CTD is unavoidable and so will be the associated stochastical variations.

In the so-called "delay-sensitive" transmissions (such as voice and video), it is crucial to maintain a tight temporal transparency with minimal CTD and/or CDV. The associated considerations are discussed below.

Cell-transfer delay

This can be defined as the elapsed time between a cell exit event at measurement point-1 (for example, at the source UNI), and the corresponding cell entry event at measurement point-2 (for example, the destination UNI) for a particular connection. Thus, the cell-transfer delay (CTD) depicts the sum of the total inter-ATM node transmission delay and the total ATM node processing delay between the two measurement points.

There are several contributing factors which ultimately decide the extent of CTD. Briefly they can be enumerated as follows:

- *Coding delay*: This refers to the time required to convert a nondigital signal to digital bit patterns. It may also include the data compression delay, if any, in addition to the delays arising from analog-to-digital conversion

- *Packetizing delay*: This delay occurs while accumulating the required number of bits to constitute an ATM cell and depends on the type of ATM adaptation used as well as on the source bit rate

- *Propagation delay*: This is due to the finite speed of signal transmission between the source and destination

- *Transmission delay*: This is dependent on the speed of the link and becomes negligible as the transmission speed increases

- *Switching delay*: This delay corresponds to the total delay incurred by a cell to traverse a switch. It depends on the interval switch speed and the amount of overhead added to the cell for routing within the switch

- *Queueing delay*: ATM switches may have buffers at the input ports, at the output ports, or a combination of input, internal and output buffers, depending on the type of switch fabric. The resulting traffic-handling at the buffers causes the queueing delays

- *Reassembly delay*: Depending on the application (service), several cells of a frame are collected at the receiver before they are passed to the application. For example, to provide a continuous 64 kbps constant rate service, a number of voice cells are collected before the frames are started to be played out.

Thus, the delays listed above are encountered as a part of the transmission profile of the information involved and the associated traffic considerations, which are largely stochastical in nature. And the traffic profile invariably would not follow any repetitive pattern either. Hence it is rather justifiable to assume the time interval between reception of data and the transmission of that data in a cell across a user network interface can be regarded as a stochastical variable.

Cell-delay variation

This is another traffic parameter used in defining the quality of service (QOS) associated with virtual channel connections in an ATM network. It refers to the upper bound of variability in temporal pattern of cell arrival observed at a single measurement point with reference to the peak cell rate of an ATM connection.

The end-to-end delay of the ith cell can be specified as $(D + W_i)$ where D is a constant, and W_i is a stochastical variable representing the cell-delay jitter. That is, W_i represents the random delay arising out of buffer overflows within the network due to asynchronous multiplexing as well as cell-loss arising from uncorrectable bit-errors. Writing in terms of interarrival times of the cells,

$$(D + W_{i+1}) - (D + W_i) = \delta_i \ . \tag{4.1}$$

This delay δ_i will be zero only when $W_i = W_{i+1}$, which corresponds to the interarrival times being equal to the interexit times of the cells. However, inasmuch as randomness persists due to multiplexing and cell-losses, it is inevitable that $W_i = W_{i+1}$. Therefore δ_i has a non-zero value and represents a random variable as decided by the stochasticity of the CDV. Suppose a jitter is perceived in the cell delay. This happens when the difference between the values of the transit delay in the cells of a connection varies randomly. In terms of probabilistic attributes, suppose

$$\text{Probability} (W_{i+1} - W_i) > w_1 \tag{4.2}$$

where w_i is a designated upper limit. Then, $(W_{i+1} - W_i)$ is regarded as a measure of the jitter portraying the cell-delay variation (CDV). Alternatively, the extent of delay variation from the mean can also be considered as a measure of CDV that is, considering,

$$\text{Probability} (W_{i+1} - E [W_i]) > w_2 \tag{4.3}$$

where w_2 is a specified upper limit and E[.] represents the expected value of the variables, $(W_{i+1} - W_i)$ is again a metric of CDV. Likewise, it is also possible to specify the variations of transmission delay in a connection in terms of the variance of W_i [4.3].

Upper bound on CTD: In reference to CTD and CDV, there are two-end-to-end delay parameter objectives, which are negotiated towards ATM specifications as illustrated in Fig. 4.19. They are:

    ✓   Maximum CTD
    ✓   Peak-to-peak CDV.

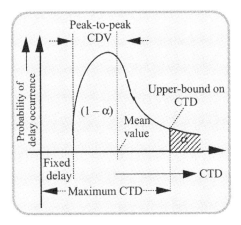

Fig. 4.19: Probability distribution of CDV

Maximum CTD pertinent to a connection refers to the $(1 - \alpha)$ quantile of CTD, as indicated in Fig. 4.19 where α is the probability of cell-delay variation as specified by the upper bound on CTD. In order to determine explicitly the upper bound value of the CTD and α, the extent of tolerable *cell-loss ratio* (CLR) requested at the connection request time is required. That is, the maximum CTD is the limit, and the cells incurring delays beyond this value would be dropped. The loss of such cells expressed as a fraction of total cells involved represents the CLR, and α is the probability of the occurrence of such extent of delay variations, which lets the cells be dropped.

CLR refers to the ratio of the number of cells lost to the total number of cells sent by a user within a specified time-interval. It is an ATM-specific metric. In general, cell-loss may occur due to two reasons namely, (i) buffer overflows and (ii) occurrence of two bit errors in the cell header, which are detected but cannot be corrected.

The first type arises due to the randomness of the traffic and limited buffer considerations. That is, a cell arriving at a switching node may find the buffer full and be lost with a nonzero probability. The second type is due to a change in the cell header bit pattern during the transmission. (As indicated earlier, the ATM cell header which is facilitated with an 8-bit header error control (HEC) field can provide a single-bit error correction and maintain a low probability of corrupted cell delivery. If a 2-bits error is detected in the header, it will not be corrected and the corresponding cell will be discarded).

It attempting to relate the CLR parameter with the maximum CTD, in terms of the upper bound of α, it is necessary to consider an associated peak-to-peak CDV. It is defined as the $(1 - \alpha)$ quantile of the CTD minus the fixed CTD that could be experienced by any delivered cell on a connection during the entire connection holding time as shown in Fig. 4.19. In Fig. 4.19, the fixed delay (F) includes the propagation and switching delays. Further the term "peak-to-peak" refers to the difference between the best and the worst cases of cell delays, where the best case is equal to the fixed delay (F), and the worst case is equal to a value likely to be exceeded with a probability no greater than α.

Typical cell-delay rules and corresponding jitter experience in practical ATM transmissions are indicated in Table 4.3.

Table 4.3:Cell-delay and cell-delay variations in typical application

Applications	Delay (ms)	Jitter (ms)
64-kbps video conference	300.00	130.00
5 Mbps MPEG NTSC video	5.0	6.5
20 Mbps HDTV video	0.8	1.0
16-kbps compressed voice	30.00	130.0
256-kbps MPEG voice	7.0	9.1

The cell-jitter due to multiplexing (buffering) can be controlled at the receiver and at the expense of larger buffers and delaying the cells appropriately. In particular, it is possible to store temporarily the arriving cells in a jitter removal buffer so that the departure of cells from the buffer are close to the interexit times of cells at the receiver.

In order to remove the jitter completely, adequate buffer size at the decoder can be facilitated to compensate for the maximum delay that a cell may incur in the network. However, it is also necessary to consider the delay requirements of specific applications in determining the buffer size. In particular, it may not always be possible to delay the cells to compensate for the maximum network delay, in which case, it is preferable, rather to drop those cells delayed by more than an acceptable value rather than attempting to control the cell-delay variation itself.

4.2.2   Semantic transparency of ATM

The *semantic transparency* refers to an error-free transmission of information. That is, between the end-entities it is expected that the network handles the teletransfer of information in tact. However, due to finite signal-to-noise ratio, there would be uncorrectable bit errors in the bit stream transmitted, which eventually would lead to cell-losses. Likewise, those cells, which are delayed beyond a maximum prescribed value are useless and will be dropped automatically. Again, the loss of such cells refers to impairment in the semantic transparency. The explicit value of the finite signal-to-noise ratio (SNR) that leads to the bit losses is measurable in terms of the bit-error rate (BER) parameter. The loss of cells is specified by the cell-loss ratio (CLR).

The other consideration, which masks the semantic transparency in ATM transmission, is that due to misdelivered cells. Again, the relevant aspects of information loss are self-explanatory. The corresponding parameter is known as the *cell-misinsertion ratio*. The presence of any errored cells expressed as a fraction of the total cells involved can be defined as *cell-error ratio* (CER).

4.2.3   Traffic and network parameters

During a connection establishment, a traffic contract is negotiated, which refers to the description of the traffic meeting the specifications of a QOS class.

Traffic, in general, is characterized by a set of parameters. That is, a *traffic parameter* is a specification of particular traffic type. For example, quantitative aspects of average connection holding time, other delays mentioned earlier etc. as well as qualitative considerations such as the description of the source type (for example, voice-telephone or video telephone) etc., may constitute typical traffic parameters. Hence, the generic set of such parameters, which can be used to capture the traffic characteristics of an ATM connection, constitutes an ATM (connection) *traffic descriptor*. It would include the source traffic descriptor with the tolerance applicable at the user network interface for an ATM connection. These parameters are user-specific and are declared by the user at the time of connection setup. They can be listed as follows:

- ✓   Peak cell rate (PCR)
- ✓   Sustainable cell rate (SCR)
- ✓   Available bit rate (ABR)
- ✓   Maximum burst size (MBS)
- ✓   Burst tolerance (BT)
- ✓   Source type.

The peak cell rate (PCR) is defined at the physical layer as the inverse of the inter-arrival time between two consecutive basic events. That is, PCR is the reciprocal of the peak emission interval of the connection. PCR characterizes the maximum rate at which a source can transmit cells as per a specified algorithm (such as the so-called *leaky-bucket algorithm*, to be discussed later) with a tolerance set by a parameter known as *cell-delay variation tolerance* (CDVT).

Sustainable cell rate is of interest in VBR video services. Although, PCR provides an upper bound of the cell rate of the connection, sustainable cell rate (SCR) is specified in order to allocate resources more efficiently. This optional parameter is a choice of the user, placing an upper bound on the realized average cell rate at an ATM connection to a value below the PCR (that is, SCR < PCR). SCR characterizes a bursty source. It defines the maximum allowable rate for a source in terms of the PCR and the *maximum burst size* (MBS). It is equal to the ratio of the MBS to the *minimum burst arrival time*.

Available bit rate refers to an ATM service category in which the network delivers limited loss, if the end user responds to any flow control feedback adopted. The ABR does not control CDV.

Maximum burst size is a traffic parameter that specifies the maximum number of cells that can be transmitted at the PCR in a connection such that the maximum rate averaged over many bursts is no more than the SCR.

Burst tolerance is a source parameter and denotes the "time-scale" over which cell-rate fluctuations are tolerated. It determines an upper bound on the length of a burst transmitted in compliance with a connection's PCR. BT is proportional to MBS. It measures the conformance checking of the SCR as a part of user parameter control function in the network.

A subset of ATM traffic parameters is known as the *source traffic descriptor*. The (traffic) parameters pertinent to the source traffic descriptors can vary from connection-to-connection. This set addresses the intrinsic traffic characteristics of the connection requested by a particular source.

As indicated earlier, CDV occurs due to the statistical multiplexing process and its tolerance is defined in reference to the peak cell rate. For a given virtual connection, it quantitatively measures the *cell-clumping* phenomenon due to the slotted nature of ATM, the physical layer overhead, and the ATM protocol considerations. As such, CDV is more of a network-dependent parameter rather than a source traffic parameter.

The source traffic descriptors together with the transmission attributes such as CDV and unambiguous specification of conforming cells constitute the *connection traffic descriptor* set. This is used by the cell admission control (CAC) to allocate the resources judiciously and monitor the user parameters. That is, the connection traffic descriptor provides for information required to perform conformance testing on cells in an ATM connection at the UNI.

Network parameters

These refer to the parameters of the network which sustain the maintenance of QOS negotiated. They are control parameters adopted for the purpose of bit error control. Bit errors in the cells caused by noisy channels can be corrected to a certain extent by error protection (control) methods applied to the cell field contents. Further, the lost and misinserted cells should also be detected, otherwise, severe performance impairment due to the loss of semantic transparency would occur. For example, suppose a constant bit rate (CBR) voice service is facing lost and misinserted cells. On real-time, this would cause an upset in the synchronism between sending and receiving ends. Lost and misinserted cells can be detected by monitoring a sequence number in the cell information field or any such equivalent mechanism.

4.2.4   Quality of service (QOS) considerations

The quality of service (QOS) parameters refer to the ATM layer performance and the higher layer performance. The ATM layer performance is decided by ATM cell-transfer characteristics.

What is meant by cell-transfer?

When cells belonging to a specified virtual connection are delivered from one point in the network to another, then *cell-transfer* is said to have occurred.

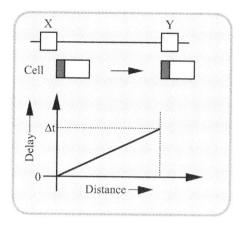

Fig. 4.20: Cell-transfer and CTD

In Fig. 4.20, X and Y are end-points of a VC depicting the network boundaries on an ATM connection, and $\Delta t > 0$ if a finite delay is experienced in the cell-transfer performance by the network. In defining the cell-transfer, the following definitions are usually observed as per ITU-T Recommendation I-356:

Cell-exit: This corresponds to event occurrence when the first bit of the ATM cell has completed transmission across X.

Cell-entry: This corresponds to the event occurrence when the last bit of the ATM cell has completed transmission across Y.

Quality of ATM cell-transfer *(ITU-T Rec. I.356)* is decided by the following outcome entities:

- ✓ Successfully transferred cell
- ✓ Errored cell
- ✓ Lost cell
- ✓ Misinserted cell
- ✓ Severely errored cell-block.

Note:

- ▪ Even if a cell arrives within an acceptable delay, if there are one or more bit errors in the received cell information field, which cannot be corrected, then the cell is declared an errored cell

- ▪ The occurrence of a lost cell corresponds to an outcome in which the cell arrives after a maximum specified delay or when it never arrives at Y (probably due to misrouting)

- ▪ A cell that had not originated from X, but arrives at Y is a misdelivered cell and it constitutes a misinserted cell at Y. Header errors not detected or errorneously corrected may produce such misinserted cells.

4.2.5   The traffic contract

ATM performance parameters are specified with respect to cells, in conformation to a traffic contract (ITU-T Recommendation I.371 ATM Forum UNI Specification: Version 3). The *traffic contract* is an agreement between a user and a network across a user-network interface (UNI) regarding the following interrelated aspects of any VPC or VCC based ATM cell flow:

- ✓ The quality of service (QOS) that a network is expected to provide
- ✓ The traffic parameters that specify the characteristic of the cell flow

    ✓   The conformance checking (policing) rule used to interpret the traffic parameters

    ✓   The network's definition of a *compliant connection*. (A compliant connection can identify some part of the cells to be non-conforming, but no more than the portion which the ideal conformance-checking rule would identify as nonconforming.)

The ATM performance parameters entered into the traffic contract are in reference to time and semantic transparency considerations discussed earlier. Explicitly, these parameters are as follows:

- Time transparency
  Delay parameters

      ✓   Cell transfer delay
      ✓   Cell-delay variation.

- Semantic Transparency
  Error parameters

      ✓   Cell loss ratio
      ✓   Cell misinsertion rate
      ✓   Cell error ratio
      ✓   Severely errored cell block ratio.

The traffic contract is a *consensus ad idem* or a memorandum of understanding between the subscriber and the network established at the time of a new connection. The contract is intended to stipulate the following:

- The network is responsible for supporting the negotiated traffic at a tolerant level on a particular connection

- The subscriber is responsible for complying with the traffic characteristics from the source not to exceed the performance limits that were originally negotiated.

A traffic control is therefore mandated so as to enforce the traffic contract. That is, to monitor whether the subscriber complies with the traffic requests made and control the network's performance of its consistency in providing the contracted service. Further, the network may accept a new connection, only if the available resources permit that connection without prejudicing the QOS of existing connections. The traffic contract is uniquely specified for every VPC/VCC. It governs the following interrelated issues:

    ✓   Traffic parameters characterizing the ATM cell flow
    ✓   QOS aspects anticipated from the network
    ✓   Conformance checking on traffic parameters
    ✓   Defining a compliant network connection for the user on the basis of resource availability.

The traffic control considerations can be summarized as follows:

- Type of traffic or traffic class

      ✓   CBR traffic
      ✓   Real time VBR (rt VBR)
      ✓   Non-real time VBR (nrt VBR)
      ✓   Available bit rate (ABR)
      ✓   Unspecified bit rate (UBR).

Definitions...

CBR traffic: This ATM service category supports applications like voice and video requiring a constant bit rate (CBR), constrained CDV, and low CLR connection. The PCR and CDVs traffic parameters define the characteristics of a CBR traffic/connection. The service category attributes and guarantees of CBR service can be summarized as follows:

- ❑ *Traffic descriptor: PCR*
- ❑ *Feedback control, if any: Nil*
- ❑ *Guarantees on: CLR, CDV and bandwidth*

- ❑ *Suitable applications*   –   *Ranking*

  - • *Telephony, video conferencing*   *I*
    video distribution, interactive
    multimedia, circuit emulation
  - • *Critical data*   *II*
  - • *LAN interconnect, WAN data*   *III*
    transport, compressed audio

VBR traffic: This refers to the traffic that supports information emitted by sources intermittently. The ATM forum divides VBR into a real-time and non-real time (rt VBR and nrt VBR) service categories in terms of the associated CDV and CTD considerations.

rt VBR
rt VBR is suitable for carrying packetized voice and audio. In general, it supports time-sensitive applications. But its time-varying transmission rate is limited to an average rate by SCR and MBS parameters. The service category attributes and guarantees of rt VBR traffic are as follows:

- ❑ *Traffic descriptors: PCR, SCR, MBS*
- ❑ *Feedback control, if any: Nil*
- ❑ *Guarantees on: CLR, CDV and bandwidth*

- ❑ *Suitable Applications*   –   *Ranking*

  - • *Compressed audio*
    and interactive multimedia   *I*
  - • *Circuit emulation and video distribution*   *II*
  - • *Critical data, LAN interconnect,*   *III*
    and WAN data transport

Note: Suitability of rt VBR videoconference is yet to be established.

nrt VBR
      The non-real time VBR is an ATM service category for traffics aptly suited for packet data transfers. It may experience significant CDV. But such lack of temporal transparency is not of concern in delay-insensitive data transmissions.
      Thus, nrt VBR supports applications that are not constrained by CTD and/or CDV. Bursty traffic considerations and variability in the transmission rate are freely permitted. Notable applications are: packet data transfer, terminal sessions, and file transfers. Networks can effectively perform asynchronous multiplexing of these VBR sources. The service category attributes and the associated guarantees are as follows:

- ❑ *Traffic descriptors: PCR, SCR, MBR*
- ❑ *Feedback control, if any: Nil*

❑   *Guarantees on: CLR and bandwidth (and not on CDV)*

❑   *Suitable applications   –   Ranking*

     •   *Critical data*   *I*
     •   *LAN interconnect, WAN data*
          transport, compressed audio and
          interface multimedia   *II*
     •   *Video distribution*   *III*
     •   *Circuit emulation*   *Not suitable*
     •   *Telephony and video conference*   *Suitability yet*
                                                     to be established

Available bit rate

     ABR is a service category in which the network delivers limited cell-loss if the end-user responds to flow control feedback. ABR does not however, control the CDV.

     Hence, ABR service category works in cooperation with sources that can change their transmission rate in response to the rate-based network feedback used in the context of closed-loop flow control. This technique allows a user dynamic access to bandwidth currently not used by other service categories. As a reciprocal incentive, the network provides a service with very low CLR. Hence, the attributes of this service category and the related guarantees are as follows:

❑   *Traffic descriptor:*   *PCR specifying the maximum transmit-rate bandwidth, maximum required bandwidth (known as minimum cell rate or MCR), and behavior parameters*

❑   *Feedback control, if any:*   *yes*

❑   *Guarantees on:*   *CLR and bandwidth. No guarantee on CDV*

❑   *Suitable applications   –   Ranking*

     •   *LAN interconnect, WAN data transport*   *I*
     •   *Compressed audio*   *II*
          and interactive multimedia
     •   *Critical data*   *III*
     •   *Telephony and video conference and video*   *Not suitable*
          distribution

UBR service

     The unspecified bit rate service category refers to a network making a "best effort" to deliver the traffic. It carries no QOS guarantees but has a single traffic parameter, namely PCR. Thus, UBR service is not tightly constrained by delay or delay variations. It is not intended to offer a guarantee on QOS and/or throughput. The traffic is set to flow on its own risk. Hence, the service category attributes and guarantees can be summarized as follows:

❑   *Traffic descriptor:*   *PCR only*
❑   *Feedback control, if any:*   *Nil*
❑   *Guarantees:*   *Nil*

❑   *Suitable applications   –   Ranking*

     •   *LAN interconnect and WAN data transport*   *II*
     •   *Compressed audio*   *III*
          and interactive multimedia

- *Critical data, circuit emulation, telephony*   *Not suitable*
  video conference, and video distribution

Note: In all the service categories indicated above, the suitable applications are ranked as I, II, or III in the order of corresponding qualitative descriptions of ranking being the best, very good, and good.

4.3   The ATM Adaptation Layer (AAL)

This is the top layer of the ATM architecture. It provides the required interface between ATM user network (application) software and the ATM network itself. That is, it performs the necessary mapping between ATM layer and the next higher layer. Further, the ATM adaptation layer may enhance the service provided by the ATM layer to the requirements of a specific service (I.362). The AAL is implemented only in ATM end-point devices; that is, at the edge of the ATM network. It is not used in the intermediate switches (Fig. 4.21).

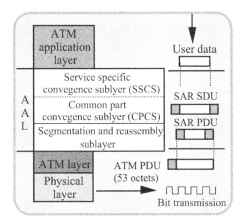

Fig. 4.21: The ATM adaptation layer in the ATM architectural hierarchy

4.3.1   Service classes

An ATM network can be regarded as semantically independent as well as time-independent. The semantic independency refers to the ATM network being independent of the telecommunication services it supports that is,

✓   User payload is carried transparently by the ATM network
✓   ATM network is devoid of processing the user payload
✓   The ATM network does not know the structure of the data unit.

Time-independency means the following: There is no timing relationship between the clock of the application and the clock of the network. That is, the network is expected to cope with any application bit rates launched onto it.

Thus, there is an implicit curtain between the telecommunication system and the ATM network implemented to support the services of that system. An ATM adaptation is therefore necessary to place the telecommunication services at the disposal of an ATM network. This adaptation is facilitated by the AAL. In order to facilitate the relevant adaptation, the telecommunication services are classified into four classes as described below.

It was indicated earlier that the ATM network is intended to support a variety of services comprised of voice, video, data etc. As such, the AAL must be capable of handling different types of information – voice telephony, packetized data, picture transmissions, video images, high quality audio, and multimedia. For the purpose of dealing with these vagaries in the information types, the standards set forth for the AAL define the following set of *service classes*:

- *Class A service*: This refers to *circuit emulation*. It is designed to have many characteristics of services supported on analog channels of the classical telecommunication network. Hence, this service is *connection* or *circuit-oriented*. It requires a connection be established between the end-entities prior to the transmission of user information. Further, the transmission is presumed to be of CBR type constituted by a continuous bit stream pattern

- *Class B service*: This is a variable bit rate service. It also assumes the transmission as a flow of continuous bit stream. It is an isochronous class of traffic with a clock synchronization provided to match the receiver bit rate with that of the transmitter. This service is distinct from Class A in respect to its flow rate being variable. This class of traffic can support voice and video (encoded and compressed), which exhibit a distinct flow pattern along the transmission link

- *Class C service*: This service corresponds to the type warranted in WAN links supporting computer network or in a data link used in packet-switched data networks, (such as those based on X.25 protocol). Class C service requires point-to-point connectivity be established prior to data transfer. Once the connection is established, data can flow in packets of different sizes. The ATM end-point user software ultimately receives these packets. With Class C service, the arrival rate of packets at the destination may or may not correspond to that of the transmitter

- *Class D service*: This refers to a connectionless data transmission service provided by complex networks, which have connectionless configurations. Such mode of operation can be observed in legacy LANs. Class D service can also be used in place of service provided by a connectionless WAN network (such as the Internet) or a private WAN that supports the TCP/IP protocol structure.

In summary, the service classifications of AAL protocols are based on:

- ✓ Timing relation between source and destination
- ✓ Bit rate considerations
- ✓ Connection-mode involved.

The different classes of service enumerated above are tied to a set of AAL types as indicated in Table 4. The mapping between classes of services and AAL types depicted in Fig. 4 is set forth by a series of protocols to be described later.

Table 4.4: Mapping between classes of services and AAL types

Class	Characteristics		
	Timing relation between source and destination	Bit-rate	Connection-mode
A: AAL-1	Required	Constant	Connection oriented
B: AAL-2	Required		Connection oriented
C: AAL-3/4	Not required	Variable	Connection oriented
D: AAL-5	Not required		Connectionless

Consistent with the classes of service indicated above, the AAL performs, in essence, the necessary mapping between the ATM layer and the next higher layer. This is done in the terminal equipment or at the terminal adapter. (That is, only at the edge of the ATM network.)

The AAL function corresponds to the task of directing the data flow sent by the user to the upper layers at the receiving end after duly taking into account the effects seen at the ATM layer:

- The data flow viewed at the ATM layer may show the errors in the transmission that occurred and the CDV suffered by the cells as a result of variable delay in buffers or due to congestion in the network

- Result?

     ✓   Loss of cells
     ✓   Misdelivery of cells.

The AAL is designed to cope with such encounters so as to offer a temporal and semantic transparency on the transmission to the end-users as required.

4.3.2   Sublayers of AAL

The AAL layer consists of two sublayers, namely:

     ✓   Convergence sublayer (CS)
     ✓   Segmentation and reassembly sublayer (SAR).

SAR:

     ✓   It does segmentation of higher layer information into a suitable size for the information field of an ATM cell prepared for transmission. In the receiving direction, SAR does the reassembly of the contents of ATM cell information field into a higher layer information format. This format is compatible for use application software of the (receiving) ATM end-user.

CS:

     ✓   Reconstitution of information sent as above is not sufficient. There are other functions to be performed such as processing of CDV, end-to-end synchronization, handling of loss, and misinserted cells. CS carries out such functions.

     ⇒   CS is a service-specific layer
     ⇒   There can be many convergence sublayers on the top of the same SAR, each serving a particular type of service.

Between the CS and SAR no *service access point* (SAP) has been defined. The SAP refers to the locale at which the services of a layer are provided. Normally, each layer defines an SAP at which the entities (the active elements within the layer) in the layer above request the services of that layer. Each SAP has an SAP address, by which the particular entity employing a layer service can be differentiated from all other entities that might also be concurrently used with that layer service. The need for SAPs between CS and SAR is yet to be defined.

The AAL structure and its sublayers are shown in Fig. 4.22.

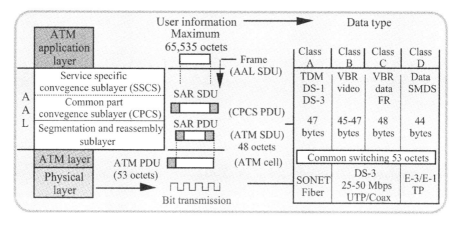

Fig. 4.22: The AAL structure and its sublayers

Convergence sublayer functions

In reference to the illustration in Fig. 4.22, it can be seen that the CS offers those functions needed for a particular type of service being provided. The *service specific convergence sublayer* (SCCS) functions are application-dependent. The *common part convergence sublayer* (CPCS) provides functions required for all users of a particular type of service. These functions may however, vary from one type of AAL service to another.

Segmentation and reassembly sublayer functions

The SAR at the source-side ATM end-point device does the packing of the 48-octet payload information field using the data bits arriving from the CS. The SAR then passes individual payload fields down to the ATM layer for inclusion in the cells for transmission.

The SAR at the destination ATM end-point device accepts the information payload from the ATM layer below and reconstructs the message using the payload field. The constructed message is passed upward to the CS.

Data units of AAL sublayer

The data units defined with respect to the AAL sublayer are based on the layered system:

Protocol data units (PDU): Data units between peer units
Service data units (SDUs): These refer to data units which pass across service access points (SAPs)

The making of an ATM cell through the AAL from the user information is elaborated in Fig. 4.22. Strictly speaking, the definitions of all the functions and protocols within the AAL are not yet totally defined. However, any definitions and functions associated with the AAL follow the usual layered system nomenclature. The functional aspects of AAL have been defined as a series of protocols, which specify the formats of the cell payload fields on the AAL protocol control information. This protocol control information constructed in the 5-octet cell-header is used to control the operation of the ATM layer and the physical layer, and the payload control information contained within the payload field of certain cells is used to control the functions required to provide four versions of AAL service.

AAL type 0

This has not been officially defined. It refers to the AAL with empty SAR and CS. That is, no AAL function ability is required and the content of the cell information field is directly and transparently transferred to the higher layer. In essence, AAL-0 defines a null service interface and the protocol essentially does nothing. In the context of AAL-0 protocol, the ATM user (application) software would break the user data into cells; further, it would also totally format the cell-header (5-octet) structure. Thus, the AAL would do no processing on the cells and it would simply pass them to the ATM layer below. The ATM-0 can be used by ATM users who like to communicate using their own private protocol.

AAL type 1

This protocol (AAL-1) refers to the payload field formats and protocol procedures required to provide ATM users with AAL Class A service. Such a service (as indicated earlier) depicts a connection-oriented service supporting a constant bit rate with a fixed-time relation between the sending and receiving ends. Hence, with the AAL-1, the type of input and output from the user-ends is a continuous stream of bits at a constant bit rate. Hence, CBR services (Class A) of the AAL-1 can be summarized as follows:

- It receivers and delivers SDUs with CBR to and from the upper layer

- Timing information is also transferred between source and destination

■ If necessary, information about data structure can also be transferred

■ Indication of errored information is conveyed to upper layer, if the error cannot be fixed within the AAL.

In the case of circuit emulation, the AAL-1 is concerned with the following:

■ Monitoring of end-to-end QOS at the CS

■ Calculation of the CRC for the information carried and transferring it to the receiver within the information field of a cell or in a special OAM cell.

Circuit emulation is an important feature of B-ISDN as it allows the circuit-based signals such as 1.5 (or 2 Mbps) to be transported meeting the specifications on delay, jitter, BER etc. for such signals. The user will not be even aware of the transfer mechanism involved.

The functions listed below may be performed in the AAL-1 in order to enhance the layer service provided by the ATM layer. The SAR-PDU format is given in Fig. 4.23.

✓ Segmentation and reassembly of user information
✓ Handling of cell delay variation
✓ Handling of cell payload assembly delay
✓ Handling of lost and misinserted cells
✓ Source clock frequency recovery at the receiver
✓ Recovery of the source data structure at the receiver
✓ Monitoring of AAL-PCI (*protocol control information*) for bit errors
✓ Handling of AAL-PCI bit errors
✓ Monitoring of user information field for bit errors and perform possible corrective action.

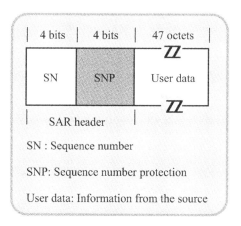

Fig. 4.23: SAR-PDU payload field format of AAL-1

As mentioned before, the SAR sublayers at the source-end packages the bits from the input bit stream into the 48-octet payload field for transport across the ATM network, and at the receiving end, the SAR sublayer constructs an output bit stream using the received cells. Fig. 4.23 illustrates the data format used for the 48-octet payload field within the 53-octet cell in reference to the ATM end-points using the AAL-1 protocol.

When all the 47 octets in the user data field (data being sent from the source to the destination) are filled with the bit received from the input bit stream, *sequence number* (SN) and *sequence number protection* (SNP) values are calculated and added as the 48th octet. At the SAR level, the sequence number protection (SNP) field, which provides 1-bit error correction and 2-bit error detection, is processed. If the result is right (no error detected or error detected and corrected), the sequence number (SN) field is sent to the CS level, which processes it depending on the application. Four CSs have been identified for the following applications:

✓ Circuit transport to support both asynchronous and synchronous circuits. Examples of asynchronous circuit transport are 1.544, 2.048, 6.312, 8.448, 32.064, 34.368 and 44.736 Mbps. Examples of synchronous circuit transport are signals at 64, 384, 1,536 and 1,920 Kbps
✓ Video signal transport for interactive and distributive services
✓ Voice-band signal transport
✓ High-quality audio signal transport.

In summary, the AAL-1 protocol processes the SAR-PDU, which has 48 octets. The first octet includes the protocol control information (PCI), the constitution of which is as follows:

4-bit sequence number (SN)

Consists of convergence sub layer indication (CSI) bit plus a 3-bit sequence count field

Sequence count value of SN makes it possible to detect the loss or misinserted cells

When CLR is very high, this method is not robust since the 3-bit sequence is rather short

The CSI bit can be used to transfer timing information and/or information about the data structure

4-bit sequence number of protection (SNP)

Consists of a 3-bit CRC which protects the SN field plus an even parity bit, which has to be calculated over the resulting 7-bit code word.

SNP provides error detection and error correction on single-bit errors and error correction on 2-bit errors

The calculations of SN and SNP refer to the following:

SN: This contains a 3-bit sequence number. It is used for sequence checking in reference to detect lost, misinserted, or damaged cells. The fourth bit (called the *CS indicator*), is used for different purposes depending on the type of data being transmitted.
SNP: This is a 4-bit CRC value. It is used for error detection and error correction on the SN field.

Convergence sublayer (CS)

The functions of CS can be summarized as follows:

▪ Tasks, which are dependent on service supported

▪ Handling of lost and miniserted cells

▪ Source clock frequency recovery

▪ Transfer of structure information between source and destination

▪ Forward error correction (FEC).

In reference to the AAL-1 protocol, the CS performs the following functions:

- ✓ Smoothing the cell delay variations
- ✓ Handling of lost, misinserted and corrupted cells
- ✓ Synchronizing the clocks
- ✓ Supporting framed data transport.

Smoothing of CDV

Jitter is smoothed in the AAL-1 with the CS compensating for the variations in cell transfer delay. This is done by placing the data as it is received in a buffer. When the CDV exceeds the ability of the buffer to send bits to the destination at CBR, the CS layer would facilitate two actions: (i) It would introduce extra bits if the buffer is empty; or (ii) it would drop excess bits, when the buffer overflows.

Handling of lost,
Misinserted, and
damaged cells

The task of cell handling refers to inserting dummy cells or discarding extra cells. In some cases, error-correction will be recommended to reconstruct the loss cells; in other cases, the lost cells will be replaced by cells with a specified (predefined) payload field pattern.

Synchronization of
the clocks

Since the AAL-1 is required to provide isochronous Class A service, the destination should receive the data at the same rate as that delivered at the source-end. This warrants a synchronism of the clocks at both ends controlling the processing. This synchronism can be achieved by two techniques, namely:

⇒ Synchronous residual time stamp (SRTS)
⇒ Adaptive clock (AC) mechanism.

SRTS: In this method the end-points use different clocks. There is a reference clock at the network level. The network (reference) clock is of lower resolution and cycles much lower than the end-point clocks. The source end keeps track of the number of its own clock cycles per one cycle of the reference clock. This count is periodically stamped and sent to the destination, and the destination end uses these time stamps and adjusts its own clock rate. Thus, a synchronism is achieved.

AC: In this technique, no reference clock is used. A buffer is used at the receiver end AAL-1 entity to store the bits received before passing them to the user. The content of the buffer will remain constant if its input rate is the same as its output rate, which will be the case under the synchronized state of both ends. If asynchronism occurs, the buffer content will fall below (or rise above) the constant value. Then, the clock will adapt running faster or (slower) so as to keep the buffer content constant. AAL-1 processing adaptively controls the setting of clock rate. The AC method is effective in controlling jitter.

Structured data transfer
(SDT)

The conventional methods of data transfer (such as PDH or SONET/SDH) refer to framed data transport. SDT is used in conjunction with framed data transport. Here, a single transmission frame might be carried by a sequence of ATM cells.

With SDT, the CSI bit in the cell-header indicates that the first octet of the user data filed contains a pointer. The pointer refers to the location within the user data field that begins a new physical transmission frame. Since not all cells contain the start frame, the pointer octet is used only when the cell contains the start of a new transmission frame.

Example 4.6
Suppose the signal (such as DS-1) has a service clock frequency f_s.
Specify an AAL/SRTS function to convey sufficient information so that the destination recovers f_s. Assume that the common clock frequency between the source and the destination, located in the network is f_N.

Solution
The AAL-1 SRTS operation utilizing a 4-bit residual stamp encoded in CSI bit for sequence count, 1, 3, 5, 7 is illustrated in Fig. 4.24. The objective is to pass enough information via the AAL so that the destination can accurately reproduce this clock frequency. The network clock frequency f_N is divided by X such that $1 \leq f_N /X/ f_s \leq 2$. The source clock frequency f_s divided by N, samples the 4-bit counter driven by the network clock (f_N /X) once every N $= 47 \times 8 \times 8 = 3008$ bits generated by the source. The SRTS-CS utilizes the CSI bit in the SAR-PDU to transmit this 4-bit RTS in an odd-numbered sequence (1, 3, 5, 7).

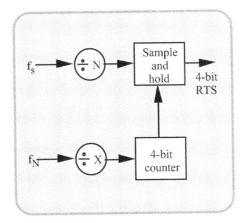

Fig. 4.24: AAL-1 SRTS arrangement

4.3.4   AAL type 2
At the end of the 1989-1992 CCITT study period, the AAL-2 had not been defined. However, it was indicated that possible requirements and the services provided by AAL-2 to the higher layer may include:

      ✓   Transfer of service data units with a variable source bit-rate
      ✓   Transfer of timing information between source and destination
      ✓   Indication of lost or corrupted information which is not recovered by AAL-2.

Hence, AAL-2 was proposed for VBR services with a specified timing relation between source and destination (Class B: VBR audio/video services). By September 1997, ITU-T approved the basic definition of the AAL-2 protocol in its Recommendation (I.363.2). Essentially, the recommended use of AAL-2 protocol refers to ways and means of minimizing delay and improve throughput efficiency for real-time voice and video. The relevant services provided by AAL-2 include the following:

      ✓   Identifying and multiplexing multiple users over a common ATM layer connection
      ✓   Transferring SDUs at VBR
      ✓   Indication of lost or erroneous information.

The AAL-2 protocol offers distinct advantages in transporting voice over ATM (VTOA). They include,

    ✓   Voice efficient bandwidth usage due to silent detection and suppression
    ✓   Idle voice channel detection.

AAL-2 protocol structure

In contrast with other AALs, the AAL-2 has no SAR. Instead, it uses a structure illustrated in Fig. 4.25:

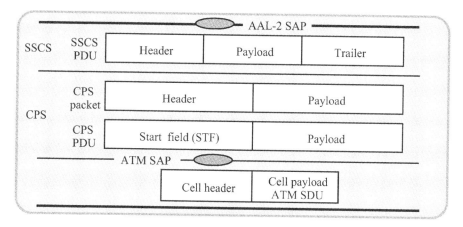

Fig. 4.25: The structure of AAL-2 protocol

The features of the AAL-2 protocol structure are:

    ✓   As in other AALs, an ATM-SAP is provided to interface with the ATM layer and an AAL-2-SAP is facilitated to interface with the higher layer (user applications)
    ✓   The common part sublayer (CPS) is divided into:

        ⇒   CPS-packet
        ⇒   CPS-PDU.

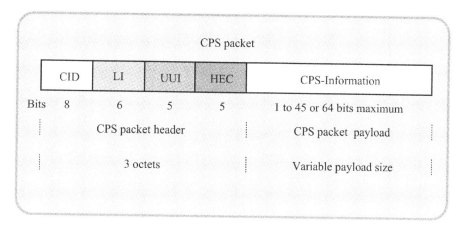

Fig. 4.26: CPS packet structure
(CID: channel identifier; LI: length indicator; UUI: User-to-user-indication; HEC: Header error control)

The CPS provides a way to identify AAL users multiplexed over a single ATM VCC. It manages the assembly/disassembly of VBR payloads for each user. It enables interface to the SSCS. Further, CPS facilitates an end-to-end service via concatenation of a sequence of bidirectional AAL-2 channel operating on an ATM VCC.

Each AAL-2 user generates CPS packets as illustrated in Fig. 4.26. The CPS packet has 3-octet header and a variable length payload.

CPS packet format

CPS-PH

CID: *8-bit channel ID used to multiplex multiple AAL-2 users on to a single VCC. The CID field supports up to 248 users/VCC plus eight CID values reserved for management functions and/or future applications*

LI: *6-bit length indicator field: This indicates the number of octets (minus one) in the variable payload structure, which can go up to 45 or 64 octets. With 45 octets, the CPS packet rigidly fits into the 48 byte ATM cell structure*

UUI: *This 5-bit field provides for identifying the particular SSCS layer together with source support for OAM functions*

HEC: *This error-control field (5-bit in length) enables ED&C for the CPS-PH. It is intended to protect the AAL-2-CPS packet header, the absence of which would affect more than one connection, if an error goes undetected.*

CPS-PDU

The CPS sublayer collects CPS packets from the users of AAL-2 who are multiplexed onto a single ATM VCC over a specified time interval. CPS-PDUs correspond to 48 octets of CPS packets. Fig. 4.27 depicts the CPS-PDU format.

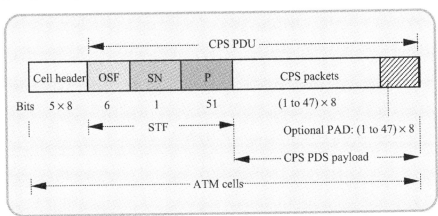

Fig. 4.27: CPS-PDU format

CPS-PDU format

STF: *This is a start field (8-bits). It is followed by 47-octet payload.*

OSF: *This 6-bit offset field within the start field identifies the starting point of the next CPS packet header within the cell.*

    Whenever more than one CPS packet is present in a cell, the length indicator (LI) in the CPS packet header comes into play to compute the boundary of the next packet.

The offset field allows CPS packets to span cells without any wasted payloads.

SN/P: The start field is critical to the reliable operation of AAL-2. Therefore, the single-bit sequence number (SN) and parity (P) are provided for error detection and recovery.

PA: AAL-2 provides for real-time delivery. Therefore, the protocol times out whenever no data is received. In such an event, a variable length padding (PAD) field is added to fill out the structure of 48-octed ATM cell payload.

Example 4.7

Construct a AAL-2 operation sequence leading to ATM cells assuming that AAL-2 multiplexes 3 real-time VBR sources identified as X, Y, and Z into a single ATM VCC. Also, assume that each source generates 32 octet samples.

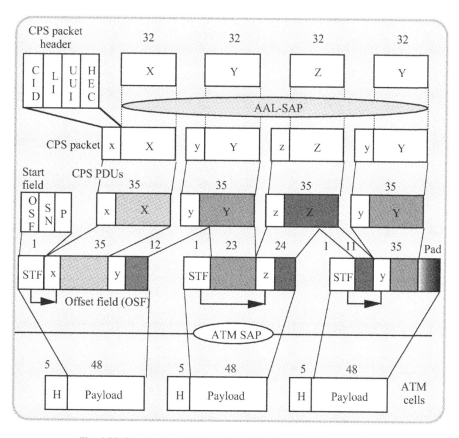

Fig. 4.28: Construction of ATM cells via AAL-2 protocol: An example

Solution

Referring to Fig. 4.28, the construction of ATM cells via the AAL-2 protocol commences at the top of the figure.

Step 1: Each rt VBR source generates 32-octet samples. These samples pass across the AAL-2 SAP to the common part sublayer (CPS)

Step 2: At the CPS, CPS-packets are constituted by adding (prefixing) a 3-octet CPS packet header (CPS—PH), with x, y, and z as channel Ids (CIDs), for each of the CPS-packet as shown

389

Step 3: The CPS sublayer collects CPS packets over a specified duration so as to construct CPS-PDUs totaling 48 octets worth of CPS packets using a 1-octet start field STF (see Fig. 4.28)

Step 4: In this case, each CPS-packet is made of 32 octets in the CPS-PDU. The STF offset will point to the immediately seen position of the CPS-PDU boundary as shown by arrows

Stet 5: In order to maintain real-time delivery, the AAL-2 times out when no data is received and fills out the CPS-PDU with a PAD of necessary octets (1-octet, in the present example, as shown) so as to make up for ATM payload structure of 48 octets

Step 6: The protocol maps the CPS-PDUs to ATM cell format across the ATM-SAP by adding the ATM cell header (5 octets)

Step 7: The ATM cells are ready for transmission over an ATM VCC

Remarks

(a) AAL-2 reduces packetization delay by multiplexing multiple sources together
(b) It controls CDV, by means of PAD, if the duration of source inactivity exceeds a specified timer threshold
(c) Minimal packetization delay is critical for VTOA due to echo control considerations
(d) Such CDV control is also essential and critical for video transmissions.

Problem 4.1

Multiplexing of four VBR sources (A, B, C, D) on a real-time basis is done via the AAL-2 protocol for transmission on a single ATM VCC. Each source generates 16-octet samples.

For the following sequence of 16-octet samples, develop an AAL-2 operation sequence leading to the connection of ATM cells for transmission of the single ATM VCC.

Source sequence: BAACDA…

(*Hint*: See Section 13.3.2 in [4.1])

4.3.5 AAL type(s) 3/4

Originally, standards defined two distinct protocols, namely,

✓ AAL type 3: Designed for Class C (connection-oriented data)
✓ AAL type 4: Designed for Class D (connectionless-oriented data).

Presently, these two have been merged as AAL Type 3/4, which is intended for both connectionless and connection-oriented variable rate service (ITU-T Recommendation I.363.3).

The AAL-3/4 protocol defines payload field formats and protocol procedures necessary to both Class C and Class D services used for the transport of computer network and data applications like connectionless broadband data service (CBDS) and switched multimegabit data service (SMDS).

For both Classes (C and D), the input to the network is in the form of discrete packets of information. Such packets may carry bits of variable rate. Further, in Class C and Class D services, it is not necessary to maintain a timing relation between the endpoints of an ATM transmission. AAL-3/4 conforms to the generic AAL model since it has SAR and CS.

The AAL-3/4 protocol is intended to operate either in *message mode* or in *streaming mode*.

Message mode

In this mode of transmission, the input blocks can be set as fixed or variable in length. The characteristics of the message mode are as follows:

▪ It can be used for framed data transfer with a high level data-link control frame

- It transports a single AAL-SDU in one (or optionally more than one) CS-PDU, which may construct one or more SAR-PDU

- This service provides the transport of fixed length or variable length AAL-SDUs.

Multicasting implementation

Connectionless computer networks use multicasting schemes that permit a network user to send a message to two or more destinations in a single operation. For example, on connectionless LAN data links, each frame sent out is received by every node on the LAN. The destination address value in the frame allows that particular node, for which the frame was intended, to accept the frame and process it.

Under a multicasting scenario, the destination address values are intended for a group of nodes, rather than for a single station. The nodes in this context would process the frames with group addresses as well as those frames with designated addresses.

The multicasting in an ATM network is done by converting the physical structure of the ATM network into a logical tree structure. Here, the source-end refers to the root of the tree and the destination ends correspond to the leaves of the tree.

This logical tree-based multicasting is implemented via a multipoint connection with a set of ATM switches as shown in Fig. 4.29.

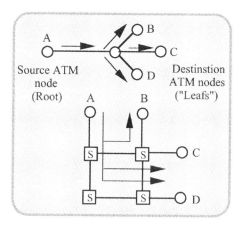

Fig. 4.29: Logical tree concept of multicasting

In the multipoint connection, an appropriate ATM switch makes copies of the cells in question and relays them to all the intended destinations via multiple transmission paths. The logical tree, on which the basis of such a multipoint connection is performed, is either predefined and/or defined dynamically on an ad hoc basis for the network administrator through a network management action.

Multipoint connection does not support a leaf-to-leaf transmission. A separate logical tree should be defined for any node playing the role of source station.

Multiplexing considerations

Suppose more than one source wishes to send frames to the same destination in one ATM network. It means a single ATM connection has to be shared by multiple ATM source-ends.

The ATM-3/4 protocol includes multiplexing feasibility so as to allow the cells associated with different ATM users to be distinguished from each other.

The MID value in the SAR-PDU header is used to identify all the cells carrying data from a particular frame. The cells belonging to a particular frame carry the same MID value. Thus, the use of distinct MID values for the cells of different frames allows cognizable multiplexing such that the cells can be assigned to appropriate frames when they are reassembled properly at the destination.

The example furnished below would explain the multiplexing scheme facilitated for services handled via the AAL-3/4 protocol. The implementation of multiplexing in a connectionless, LAN-like service supported on an ATM network is illustrated in Fig. 4.30.

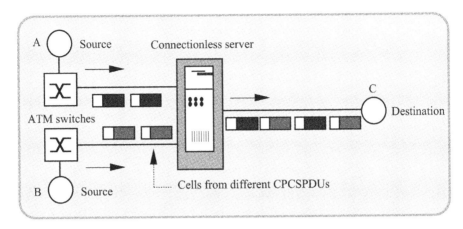

Fig. 4.30: Multiplexing via AAL-3/4 protocol

CPS-PDU and SAR-PDU

The AAL-3/4 CPCS-PDU/SAR-PDU format is illustrated in Fig. 4.32.

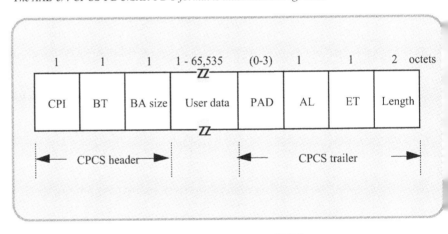

Fig. 4.31: AAL-3/4: CPS-PDU/SAR-SDU format

The associated definitions of Fig. 4.31 are as follows:

- *CPI:*   *This is the common part indication defining how other CPCS header and trailer fields are encoded*

- *BT/ET:*   *BT refers to the beginning tag, which contains a value that is incremented for each CPCS-PDU processed. The same value is placed in the ending tag (ET). On the receiving side, the two values are compared to check for errors*

- *BA size:*   *This specifies the buffer allocation size required when the CPCS-PDU is received and reassembled*

- *User data:*   *This is the data unit passed down from the ATM user software (applications) to the AAL*

- *PAD:*   *This is the padding added to keep the PDU length multipliers of 4 octets. This is to ensure that the CPCS trailer begins on a 32-bit boundary. This would make the processing more efficient on some computers*

❏   *AL:*   *This depicts the alignment field and is added to make the CPCS trailer 4 octet in length*

❏   *Length:*   *A binary value denoting the length of entire CPCS-PDU.*

Example 4.7

Illustrate the sequence of AAL-3/4 protocol operations involved in multiplexing two inputs at a data communication terminal, each of 75-byte packets arriving simultaneously and destined towards a single ATM output port

Solution

See Fig. 4.32.

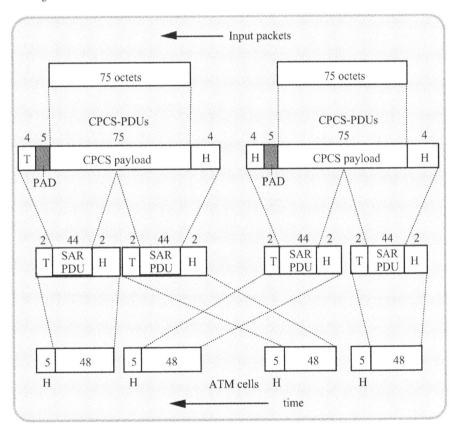

Fig. 4.32: AAL-3/4 protocol in multiplexing two inputs

Problem 4.2

Solve Example 4.7, assuming that the two inputs mixed at the DTE correspond to 96 byte and 69 byte packets. (*Hint:* See Section 13.4.4 in [4.1])

In message transport, each block of information is passed down to the CS, which transports the user data through the ATM network format of one or more CPCS-PDUs.

Streaming mode

Here, one or more fixed-size AAL-SDUs are transported in one CS-PDU. That is, input blocks are fixed in length, and multiple input blocks may be combined in one CPCS-PDU with a single CS header and trailer. Transfer of all these AAL interface data units may occur at different times.

The characteristics of streaming mode service are:

■   This service provides transport of variable length AAL-SDUs

■   It also includes an abort service by which the discarding of an AAL-SDU partially transferred across the AAL interface can be requested

■   AAL-SDU can be as small as one octet because this unit has to be recognized by the application.

Structure of AAL Types 3/4

Fig. 4.33 shows the general structure of this AAL. The CS is split into a common part (CPCS) and a service-specific part. The service-specific convergence sublayer (SSCS) is application dependent and may be null.

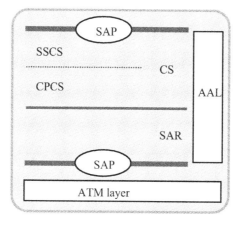

Fig. 4.33: Structure of AAL-3/4

SAR-PDU and ATM-PDU

In Fig. 4.33, the SAR sublayer divides the SAR-SDU into 44-octet segments. (If necessary, padding is done at the last segment to make up a total of 44 octets.) The SAR sublayer then adds the SAR header and trailer fields to each segment to constitute the SAR-PDU, which is passed down to the ATM layer as an ATM-SDU. Support of connectionless service is provided at the SSCS level, (The SAR-PDU encoding and protocol function and format are closely identical to the so-called L2_PDU of IEEE 802.6.) Fig. 4.33 illustrates the SAR-PDU/ATM-PDU.

ST: This defines the segment type. It has a 2-bit value and is used to indicate the cell that contains the beginning, middle, or end of a message:

❑   *Beginning of a message:*   *10 (BOM)*
❑   *Continuation of message:*   *00 (COM)*
❑   *End of message:*   *01 (EOM)*

❏   *Both beginning and end of*
         single segment message: *11 (SSM)*

The ST field is used for message segmentation and reassembly.

SN: *This is a 4-bit value sequence number used to perform sequence checking so as to detect any lost or misinserted cells*

MUX-ID: *This refers to the multiplexer ID. It has 10-bit value and used to multiplex multiple user connections over a single physical ATM connection*

CPCS segment: *This contains a 44-octet portion of a CPCS-PDU on control information and data sent from the source to the destination*

LI: *The number of 4 octets contained in the user data field is specified by a 6-bit length indicator*

CRC: *This cyclic redundancy check field is used for error-detection and correction of the CPCS segment field.*

The protocol aspects of AAL-3/4 can be summarized as follows:

❏   *Used in data applications*
❏   *Peer-to-peer service provided by both message and streaming services*
❏   *Assured operation: Retransmissions of missing or errored AAL SDUs: Hence, flow control is offered as a mandatory facility*
❏   *The assured operation is, however, restricted to point-to-point connections at the ATM layer*
❏   *Nonassured operation: In this mode, Loss or errored AAL-SDUs are not corrected by retransmission. (Delivery of corrupted AAL-SDUs can, however, be an optional feature for the user)*
❏   *Flow control, in general, is restricted to point-to-point ATM layer connections and not provided for point-to-multipoint ATM layer connections*
❏   *Segmentation and reassembly sublayer*
❏   *CS-PDU are of variable length*
❏   *When accepting such a PDU, the SAR sublayer generates SAR-PDUs containing up to 44 octets of CS-PDU data.*
❏   *CS-PDU is preserved by the SAR sublayer. This requires segmentation.*

Hence, the following segment type indication (ST-I) and a SAR payload-type indication (PT-1) result from the segmentation adopted:

ST-I: *Identifies an SAR-PDU as a beginning of a message (BOM), as a continuation of a message (COM) and as an end of a message (EOM) or as a single segment message (SSM)*

PT-1:
❏   *Represents number of octets of a CS-PDU contained in the SAR-PDU payload*
❏   *In the message-mode service, the SAR-PDU payload of all BOMs and COMs contains exactly 44 octets; whereas, the payload of EOMs and SSMs is of variable length*
❏   *In streaming-mode, the SAR-PDU payload of all segments depends on the AAL-SDUs.*

ST	*SN*	*MID*	*SAR-PDU Payload*	*LI*	*CRC*
2 bits	*4 bits*	*10 bits*	*44 octets*	*6 bits*	*10 bits*
SAR-PDU header				*SAR-PDU trailer*	
SAR-PDU					

Fig. 4.34: Structure of SAR-PDU

There are three major mechanisms which can be specified under AAL-3/4 protocols. They are:

❑   *Error handling*
      •   *Error detection mechanism*
      •   *Error recovery mechanism*
❑   *Providing connection-oriented service*
❑   *Providing connectionless service*
      •   *Use of connectionless services*
      •   *Multicasting mechanism*
      •   *Multiplexing mechanism.*

Error handling

ATM networks are conceived to provide a highly reliable service. Occasional errors, which will not cause undue impairment to the quality of the signal, when perceived by the ATM layer are simply discarded relevant (errored) cells during transmission. AAL-1 and AAL-2 protocols are designed to cope with such occasional errors which can be ignored in transmissions supporting voice and video applications.

AAL-3/4 handles Class C and Class D services designed to support data transmissions. These services are not delay-sensitive but are highly sensitive to loss of information arising from bit errors. In short, these services require a guaranteed semantic transparency and the data should, therefore, be transmitted through the network essentially, in an error-free manner. Therefore, AAL-3/4 protocol incorporates error handling procedures (which are not defined for AAL-1 and AAL-2 protocols). The functions of these procedures are as follows: Allowing the AAL to provide an error-free link by detecting and recovering missing information that results from cells the ATM layer discards whenever transmission errors occur.

Error detection

The mechanism of *error detection*, in reference to AAL-3/4 is as follows: The relevant protocol facilitates several versions of error checking to ensure that no uncorrected errors get through during transmission. The error-handling mechanism adopted on each cell sent by the SAR sublayer down to the ATM for transmission corresponds to:

✓   CRC check on cells, which might have been corrupted during transmission
✓   Insertion of the SN field to detect the lost or misinserted cell
✓   Use of MUX identifier (MID) and ST type fields to ensure that cells are properly reassembled into a CPCS-PDU
✓   Use of the HEC field to detect error in the cell-header
✓   Additional mechanism

⇒   As a part of a complete CPCS-PDU, beginning and end tag (BT/ET) fields are appended so as to ensure that all the cells making up a complete CPCS-PDU have been received and reassembled correctly.

Error recovery

In addition to error-detection in SAR-PDUs and/or CPCS-PDUs, the AAL-3/4 protocol provides for the following three distinct efforts towards *error recovery*:

✓   *Nonassured operation with discard*: In this mode, retransmission is not attempted and the user information contained in corrupted CPCS-PDUs is not passed to the user.
✓   *Nonassured operation with delivery*: Here again, the retransmission is not attempted, but the information contained in the corrupted CPCS-PDU is passed to the user

✓ In this mode, the AAL handles a retransmission of corrupted CPCS-PDUs. The ATM user software (application) perceives, thereof a complete error-free data link. The retransmission procedure is, however, a complex process. It may require acknowledgement, and retransmission of all those cells, which constitute a complete CPCS-PDU. Relevant protocol issues are in progress.

Facilitation of connection-oriented services

Class C connection-oriented services can be converged onto AAL-3/4. Inasmuch as the concept of ATM transmission itself is connection-oriented, the ATM switching provides for the connectivity required for Class C service users.

Facilitation of connectionless services

In a connectionless service (Class D), no connection can be established between the end users before data transmission take places. Then, the question is how the connections required during ATM network operations are made use of for the Class D, connectionless data delivery service.

The Class D connectionless service is comparable to the service realized in a shared-medium LAN data link or by a connectionless computer network such as the Internet.

Taking into consideration the fact that the ATM operates essentially on a connection-oriented basis, a method is required to realize the connectionless mode of operation on the connection-oriented base of ATM. The following strategies are, therefore, considered:

Use of connection servers

Here, the end-point workstation can include a network interface card (NIC) plus appropriate software to implement the ATM end-point functions. The workstations transmit frames in the usual manner as in conventional shared-medium LAN. The frames are then segmented into the cells by the AAL in the workstation and sent over an ATM VCC to the connectionless server. This server analyes the first cell of each frame, identifies its destination address, and routes the cell through the best ATM connection towards its destination. The connectionless server then forwards all the cells of the frame to their destination via the chosen ATM connection.

4.3.6 AAL type 5

This is defined as an alternate protocol to support Class C and D services. This protocol is however, simpler than AAL-3/4 and can be adopted when multicasting is required without any multiplexing.

In terms of protocol overhead, the AAL-3/4 and AAL-5 can be compared as in the following table (Table 4.5).

Table 4.5: <u>Overhead considerations in AAL-3/4 and AAL-5</u>

Overhead structure	
AAL-3/4	AAL-5
• 4 octets in each cell for header and trailer • 8 octets for CPCS header and trailer in each frame • 3 octets of padding	• CPCS adds an 8-octet trailer plus sufficient padding to the entire CPCS-PDU • Resulting CPCS-PDU is an even multiple of 48 octets. This allows CPCS-PDUs be evenly placed into cells • No protocol control information in individual cells. That is, all 48 payload octets are used to carry user data • For frames larger than 80 octets, AAL-5 is more efficient in terms of using the payload, (relative to AAL-3/4).

AAL type 5: The CPCS-PDU format

The CPCS-PDU format of AAL-5 is illustrated in Fig. 4.35.

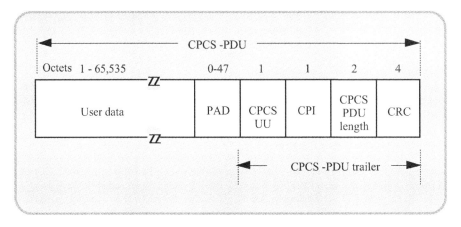

Fig. 4.35: The format structure of AAL-5/CPCS-PDU

The constituent parts of the formats indicated in Fig. 4.35 are as follows:

AAL-5/CPCS-PDU format

User data: *The information exchanged between the end-entities*

PAD: *Padding of octets done to render CPCS-PDU a multiple of 48 octets*

CPCS-UU: *CPCS user-to-user indication — the control information passed from source CPCS to destination CPCS*

CPI: *Common point indicator — this aligns the trailer to a 64-bit boundary*

Length: *This denotes the length of the user data field. Used by the receiving AAL entity to segregate the user data and padding*

CRC: *32-bit cyclic redundant check on entire CPCS-PDU. Normally, CPCS-PDU will discard corrupted PDU.*

The AAL-5 is regarded as a simple efficient adaptation layer (SEAL). It is a "light-weighted" protocol in comparison with AAL-1, AAL-2, and AAL-3/4.

The common part (CP) of AAL-5 supports VBR traffic, both connection-oriented and connectionless. The following example would illustrate the AAL-5 operation.

Example 4.8

Suppose an AAL SDU corresponds to 3297 bytes.

Perform necessary operation on these AAL SDUs via AAL-5 protocol to construct ATM cells.

Solution

The relevant solution is illustrated in Fig. 4.36.

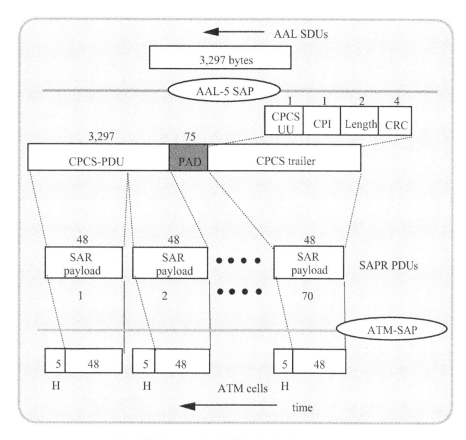

Fig. 4.36: Construction of ATM cells via AAL-5 protocol

Problem 4.3
Repeat Example 4.8 with AAL-5 SDU being 6000 bytes.

Suppose the multiplexing operation is done using AAL-5 protocol. The relevant difference when using AAL-3/4 versus AAL-5 can be understood by redoing the example with AAL-5 protocol.

Example 4.9
Repeat Example 4.7 assuming AAL-5 protocol in lieu of AAL-3/4

Solution
Relevant solution is illustrated in Fig. 4.37.

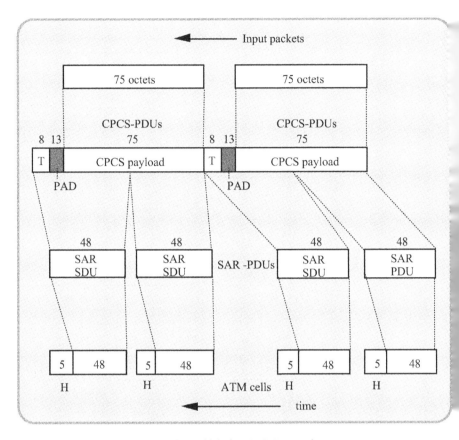

Fig. 4.37: Multiplexing: AAL-5 protocol

Note:

❑ In AAL-5, the entire packet need not be received before the protocol can begin the SAR function (in contrast to the procedure of AAL-3/4 with the insertion of correct buffer allocation size (BA size) field)

❑ The transmission is done in a serialized fashion (with the possibilities of larger CDV).

4.4 Synchronous Optical Network (SONET) and Synchronous Digital Hierarchy (SDH)

Physical layer interface standards and transmission hierarchy

Physical layer interface standards have been defined to specify the technological means intended to carry bit streams over the physical media. In reference to such standards there are two primary backbone hierarchies in practice known as [4.5]:

✓ Pleisiochronous or asynchronous digital hierarchy
✓ Synchronous digital hierarchy/SONET.

Both technologies refer to aggregating multiple lower bandwidth connections so as to create high-bandwidth trunks.

The interface standards of the aforesaid transmission technologies cohesively address the details on a combination of bit rate, physical media type, encoding scheme, and connector types. They correspond to public and private interfaces.

4.4.1   Pleisiochronous digital hierarchy (PDH)

This was developed by the telecommunications industry to facilitate a common method of supporting multiple voice channels on a single high-speed circuit. There are two versions of PDH: One type of hierarchy is intended to multiplex the North American T-system. The other refers to the multiplexing scheme for the European E-system. The constituent digital levels and the associated transmission rates of these hierarchies are indicated in Table 4.6.

Table 4.6: <u>PDH hierarchies</u>

North American and Japanese PDH						CEPT/Europe/(CCITT)PDH				
Level	DS-0	T-1/ DS-1/J-1	T-2/ DS-2/J-2	T-3/ DS-3	T-4/ DS-4	E-1	E-2	E-3	E-4	E-5
Speed in Mbps	0.064	1.544	6.312	44.736	274.176	2.048	8.448	34.368	139.264	564.992
Number of voice channels	1	24	96	672	4032	30	120	480	1920	7680

Note: Japanese PDH has J-3 rate equal to 32.064 Mbps

The pyramidal build of PDH transmission hierarchy pertinent to the both versions of Table 4.6 is presented in Fig. 4.38.

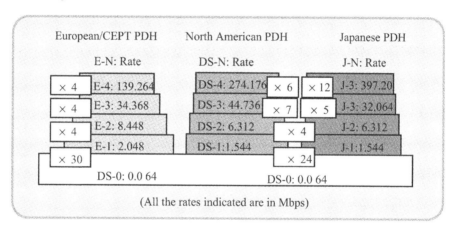

Fig. 4.38: PDH transmission hierarchy

The term "pleisiochronous" in PDH implies "almost or near-synchronous". That is, the PDH is concerned with such signals having nominally the same frequency but differing within a defined tolerance. This pleisiochronous attribute arises due to the following: Considering the PDH presented in Table 4 and illustrated in Fig. 4, a higher order multiplexer receives incoming (tributary) signals, which by themselves have originated at the lower order multiplexer. This lower level multiplexing is done at its own clock source. Hence, the clock frequency tolerance at each level would cumulatively lead to a pleisiochronous status of higher order signals. Multiplexing pleisiochronous signals obviously involves a synchronization problem. Hence, some "*justification*" (or *stuffing*) may be required in PDH based transmissions.

The question of falling out of perfect synchronism and the necessity to include justification/stuffing bits can be understood by considering a communication between the two autonomously timed digital switches depicted in Fig. 4.39.

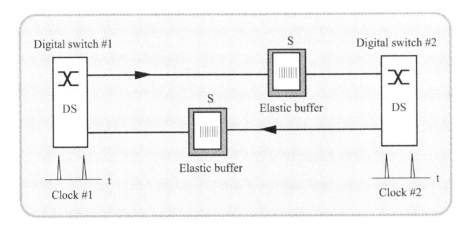

Fig. 4.39: Autonomously clocked, interconnected digital switches

In Fig. 4.39, suppose there is a shift in the frequency of one autonomous clock. The relative offset in the two clocks cannot be reconciled, no matter how small the offset may be. Any elastic buffer introduced in the interconnection path may store or absorb the variation to a certain extent, but not totally. The result is a *slip* in the synchronism between the switches.

This slip can be observed in terms of the difference between the input data rate and the output data rate at the elastic buffer. This difference eventually would lead to disruptions in the data stream and is referred to as "slips".

The need for clock synchronization is to prevent loss of data by way of slips. The procedure adopted to avoid slips (and hence achieve synchronism) refers to *pulse stuffing* (in North America) and "*justification*" in Europe. (*Note*: Pulse stuffing does not mean inserting pulse in the line code so as to enable required timing adjustment. It means rather adding bits on an ad hoc basis as described below.)

The basic concept of pulse stuffing involves the use of an output channel whose rate is purposely higher than the input rate. Thus, the output channel can carry the sum total of all input data plus a variable number of "*null bits*" or "*stuff bits*". That is, the "null bits" are not a part of the incoming data. They are inserted so as to *pad* the input data stream to match the higher output rate. Obviously, these extraneous stuff bits should be recognized at the receiver and destuffing should be performed to segregate and recover the original data stream.

The evolution and practice of pulse stuffing in the history of digital TDM hierarchy can be summarized as follows:

When it was required to combine lower rate tributaries (for example DS-1s) into higher levels (for example DS-2s), it became necessary to accommodate tributaries operating at slightly different rates. The generic terminology of combining such unsynchronized signals is known as *asynchronous multiplexing*. It refers to the context of multiplexing (slightly) unsynchronized tributaries into a higher level signal using pulse stuffing. (*Note*: The adjective "asynchronous" does not, however imply *asynchronous transmission*. The multiplexed higher level signal is always carried on a *synchronous transmission* link).

In constructing higher level framing formats via multiplexing using lower level data streams plus necessary stuffing, the following general principles are pursued:

- Use of fixed-length master frames accommodating the input channels with a provision for stuffing

- Imposing a distinct specification on the redundant stuffing introduced

- Distribution of noninformation bits (stuffing bits) across the master frame.

While the purpose of stuffing is to avoid loss of data when the digital interconnected transmission lines are unsynchronized with each other, it is possible that a stuffed bit may itself be erroneously interpreted as an information-bearing (data) bit. Therefore, redundant bits (*C-bits*) should be added to indicate whether the last bits carry data or stuff.

PDH implies the relevant multiplexing scheme from T-1 to T-3 and higher (or a corresponding scheme in E-hierarchy). It differs from the so-called *synchronous digital hierarchy* (SDH), where (unlike in PDH), the digital multiplexing scheme involves a plan in which all levels are synchronized to a single master clock.

The history of PDH dates back to the 1950s, when the classical analog voice signal transmissions were converted to digital signals in metropolitan areas for transmissions on already installed copper cabling. The conversion (envisaged by the Bell Labs in the United States) refers to assigning a level to each digital stream/signal (DS) as indicated in Table 4 (for the North American T-system).

In the PDH scheme that multiplexes several lower-numbered DS levels into higher-numbered DS levels, though there is certain specified frequency tolerance, there is no fixed relationship existing between the data belonging to the various levels of the hierarchy, except at the lowest level (that is, at DS-0 corresponding to 64 kbps rate).

As described in Chapter 3, the multiplexing scheme uses different multiplexers M1C, M12 etc. as illustrated in Fig. 4.30 (which is essentially same as Fig. 3.41).

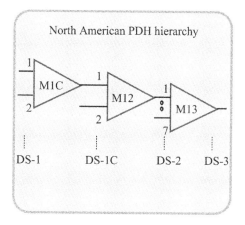

Fig. 4.40: North American PDH (more popularly known as, asynchronous digital hierarchy scheme)

The European PDH scheme (CCITT Recommendation G.703) is shown in Fig. 4.41.

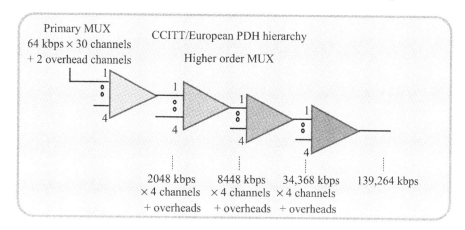

Fig. 4.41: CCITT/European version of PDH

The CCITT recommendations are intended to cover the telecommunication practice internationally. (However, regional authorities may have local standards. For example, in the United States, the regional standards are defined by ANSI.)

In Fig. 4.41, four incoming tributaries constitute a higher order multiplexing. Since each of these tributary signals has originated at a lower order multiplexer with a clock of its own, each tributary can be regarded to be of the same bit rate as the other, though with a specified tolerance. The result is a "near-synchronous" status of higher order digital levels. Hence, the designation of PDH is attributed to the multiplexing scheme. (As indicated before, the term pleisiochronous or near-synchronous is used more exclusively in CCITT context; in North American standards, the pleisiochronous signals are called *asynchronous signals*.)

Justification or stuffing (using justification/stuffing bits) is adopted to match the precise signal rates of the tributaries to the signal rate of the multiplexed signal. This is done while constructing a multiplexed frame structure. The following example can illustrate relevant considerations with respect to constructing a multiplex frame structure for (i) DS-1C and for (ii) an 8.448 Mbps signal as per CCITT Recommendation G.742.

Example 4.10
Describe a bit interleaved DS-1C frame format.

Solution
The DS-1C bit stream is organized into a block of 52 bits with 26 bits from each DS-1 input signal as illustrated in Fig. 4.42.

In Fig. 4.42, asynchronous multiplexing uses bit interleaving. A sequence of bits from port 01 (DS-1) and a bit from port 02 (another DS-1 signal) is repeated into a pattern sequence as shown.

A control bit precedes each block, designated as an M, C, or F bit. The 24 control bits are distributed in the M-frame and they constitute a *control sequence* or *word*. The resulting 1272 bit block is called an *M-frame*. The functional aspects of the control bits are as follows:

M-bits:
They reside at the 1st, 7th, 13th and 19th positions of the control sequence. Its sequence is 011X. Here, 011 identifies the M- frame format and the start of the four 318-bits subframes. The X bit is for the maintenance channel to send alarm conditions from the receiving end to the sending end. (X =1 means no alarm and X = 0 means that there is an alarm).

F-bits:
This sequence is made of alternate 0s and 1s forming a pattern 010101. It resides at every 3rd bit of the control sequence. The receiving terminal uses F-bits to identify the frame and control bit time-slots

C-bits:
This sequence identifies the presence or absence of stuff bits in the DS-1 parts of each subframe. The relevant rules are as follows:

- ❑ $111 \Rightarrow$ A stuff bit (pulse) should be inserted into the subframes
- ❑ $000 \Rightarrow$ No stuff will occur
- ❑   Stuffed time-slot: This refers to 3rd information bit following 3rd C bit in the subframe
- ❑   Stuffing for 1st DS-1 signal: 1st and 3rd subframes
- ❑   Stuffing for 2nd DS-1 signal: 2nd and 4th subframes
- ❑   Maximum stuffing rate: 4,956 bps for each DS-1 signal, and the nominal rate is 2,264 bps.

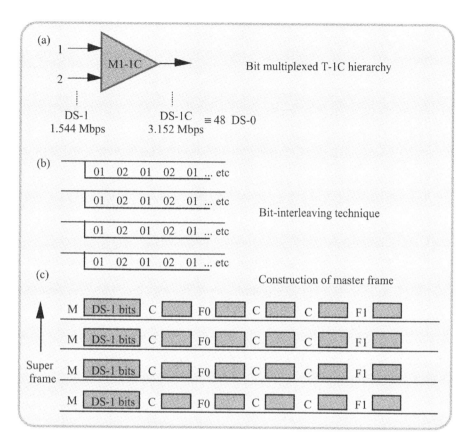

Fig. 4.42: Development of master frame of DS-1C

Example 4.11
Construct a multiplexed frame structure at 8.448 Mbps in PDH format.

Solution

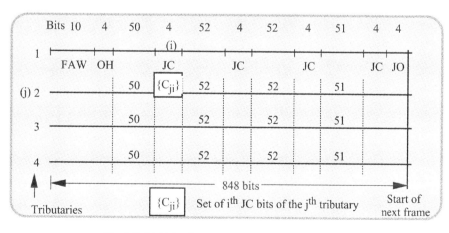

Fig. 4.43: Multiplex frame structure for 8.448 Mbps signal

Relevant CCITT Recommendation G.742 specifies a total of 848 bits in a frame as indicated below:

10 bits: Frame alignment word

2 bits: Signaling word (one for alarming and the other for "national use")

820 bits: Data from 4 tributaries. Each tributary data blocks consisting of $(50 + 52 + 52 + 51)$ bits $\times$ 4 tributaries $= 820$ bits

16 bits: For justification process (12 for *justification control* and 4 for *justification opportunity*)

Stuffing/justification:

The tolerance of each signal level in the bit-rates of E-system expressed in parts per million (ppm) is as follows:

Level 1:	2,048 kbps	± 50 ppm
Level 2:	8,448 kbps	± 30 ppm
Level 3:	34,368 kbps	± 20 ppm
Level 4:	139,264 kbps	± 15 ppm

(Levels 1-3 adopt HDB3 line coding; and, level 4 uses CMI)

Corresponding to 8.448 Mbps signal, the tolerance is ± 30 ppm. That is, the rate may swing in the range 8447747 bps to 8448253 bps. The maximum data rate per channel is, therefore, given by:

$$R_{max} = \frac{205}{848} \times 8448253 \text{ bps} = 2042325 \text{ bps}$$

The data rate arriving from the tributary is 2048 kbps ± 50 ppm. This covers the range 2047898 bps to 2048102 bps. Hence, the maximum arrival rate from a tributary is:

$$R_A = 2048102 \text{ bps} > R_{max}$$

That is, the data rate (R_{max}) carried by a tributary in the multiplexed stream is less than the data rate of that tributary arriving at the input of the multiplexer.

The difference is therefore, "justified" or "stuffed" with extra bits, designated as *justification opportunity* bits in Fig. 4.43. Without this justification or stuffing, the buffer at the multiplexer will experience an overflow since the incoming data is faster than the outgoing data.

Suppose for the tributary 1 under consideration, the data increases to 206 bits. The corresponding multiplexer output will then be larger than the input rate. In this case the justification opportunity (JO) bit need not be used. Thus, by using JO in some frames and not in others, the input-output rates at the multiplexer can be evenly matched. The presence of JO is specified for each tributary by setting the corresponding justification control bits as 0 or 1. For example, for tributary 1, if JO not used, $C_{11} = C_{21} = C_{31} = 0$; otherwise it is 1. From Fig. 4.43, it can be noted that justification is done independently on each tributary.

Problem 4.4

Construct the frame format for the digital signal used for 6.312 Mbps DS-2 signals in the North American digital hierarchy. (*Note*: A DS-2 signal is derived by bit interleaving four DS-1 signals at the M12 multiplexer and adding the appropriate overhead bits.)

Limitation of justification

PDH implies a rigid modular structure of transmission system. It was evolved mainly on the considerations of multiplexing. There are however, other functions such as OA&M, which warrant flexible and reconfigurable structures.

There is also another difficulty faced with PDH. This refers to the "*add-and-drop*" channels. That is, if a multiplexed channel exists, adding a primary rate channel (for example) on the way would require first demultiplexing the higher order level down to the primary level, adding the new channel, and then again multiplexing the channels up back into the higher level. This procedure is required because the original multiplexing had already experienced justification and adding a new tributary straight into this multiplexed stream is obviously not a preferable option. Hence, whenever add-and-drop is involved demultiplexing/multiplexing is necessitated. This would enhance the cross-connect complexity at the add-and-drop points. To overcome these shortcomings, a new set of standards was evolved to replace PDH. They are known as CCITT's *synchronous digital hierarchy* (SDH) and ANSI's *synchronous optical network* (SONET), both of which are described in the forthcoming sections. They are deployed in the wide area network at data rates from 155 Mbps to 9.6 Gbps. Both SONET and SDH have become support architectures to bear ATM transmissions. Before considering the relevant aspect of SONET and SDH and their association to ATM, the art of supporting B-ISDN/ATM via the traditional T- and/or E-system is described below.

ATM over PDH

With the emergence of B-ISDN, originally the carriers offered native ATM at DS-3 (the traditional T-3 transmission speed of about 45 Mbps). This called for an interoperation between the pleisiochronous digital hierarchy of DS-3 transmissions and the synchronous digital hierarchy of ATM. In other words, the ATM cells had to be "framed" so as to be compatible for transmission in the DS-3 architecture. That is, the ATM cells had to be converged on a pleisiochronous line as per a *physical layer convergence protocol* (PLCP)

To understand the underlying concept of this PLCP, first the DS-3 physical layer protocol structure at the UNI level can be revisited by considering the evolution of DS-3 via the associated multiplexing (Fig. 4.44):

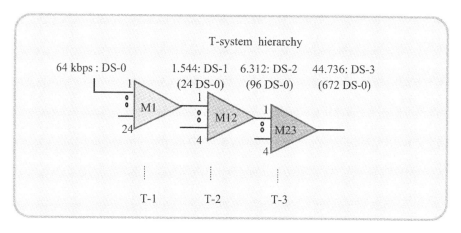

Fig. 4.44: T-System-multiplexing. (Evolution of telco's asynchronous digital hierarchy)

Traditional DS-3 frame format (M23 format)

The traditional DS-3 frame structure consists of 4760 bits made of (680 bits/subframe) × (7 subframes) as shown in Fig. 4.45.

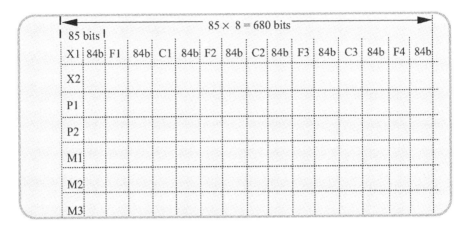

Fig. 4.45: Traditional DS-3 frame structure

The traditional DS-3 signal is mounted in frames called *M-frames*. Each M-frame consists of seven subframes as depicted in Fig. 4.45. The subframe is divided into eight blocks each of 85 bits, of which, 84 bits are *payload bits* and the first bit is a *control bit*. Now, what is this control bit?

The conversion of seven DS-2 channels into a DS-3 channel (Fig. 4.44), warrants a control bit added to 84 payload bits, and bits are stuffed in specific slots to equalize the speeds of the individual streams and achieve the speed of 44.736 Mbps.

In the DS-3 frame format of Fig. 4.45, the functions of the control bits are as follows:

X1, X2, P1, P2, M1, M2 and M3: $\Rightarrow$ Multiframe alignment bits; X1, X2, P1 and P2 are defined in ANSI T1.107

X $\Rightarrow$ Bits for alarm and status functions

P $\Rightarrow$ Parity value bits: P1 = P2 = 0 means that sum of all payload bits in the previous frame is 0; P1 = P2 = 1 indicates that the sum is 1

M $\Rightarrow$ Bits for frame synchronization M1 = 0, M2 = 1 and M3 = 0

F $\Rightarrow$ Bits for subframe synchronization F1 = 1, F2 = 0, F3 = 0 and F4 = 1

C $\Rightarrow$ Bits to identify the presence of stuffing bits in block 8. A single bit can be stuffed into one, some, or all of the last blocks of the seven subframes. The presence of such stuffing is indicated by setting three C bits to 1. This stuffing is necessitated due to the following reason: the DS-3 rate is not an integral multiple of the DS-1 rate (*Note*: 44.736/1.544 = 28.974, which is not an integer.) Hence, bits are stuffed ("justified") as needed, so that individual DS-1 frames *slide* relative to DS-3 frames. C1, C2, and C3 should be assigned as per IAW ANSI T1.404-1994 for C-parity.

With the stuffing done as above, it becomes necessary to locate those bits, which belong to a specific DS-1 input signal. This amounts to demultiplexing the entire stream.

In reference to the pleisiochronous digital hierarchy (that is, the asynchronous digital hierarchy) of the North American T-system, there is an another format of DS-3 frame known as *C-parity* as described below:

C-parity

In this DS-3 format layout, seven DS-2 signals are obtained via multiplexing exactly 6.306272 Mbps and a stuffing bit is inserted in every subframe. The result is a DS-3 signal exactly at 44.736 Mbps.

In this case, there is no need to indicate the presence or absence of stuffing bits (since, by default, a stuffing bit is always present in every subframe). Therefore, there are 21 C bits available for other purposes. For example, the three C bits in subframe 3 denote the parity status of the preceding frame at the origination point of the stream. At the receiver, a comparison of locally calculated (P1, P2) with this value indicates whether an error has incurred or not. Other bits can be used for data links, alarms, and status information purposes.

Problem 4.5

Show that the maximum and minimum input channel rates accommodated by an M12 multiplexer are 1.5458 Mbps and 1.5404 Mbps respectively.

PDH transmission framing with ATM cells

The B-ISDN considerations towards ATM transmission specify broadband requirements. Such considerations therefore define relevant operations over SONET/SDH data rates. However, many initial deployments of ATM adopted the existing PDH infrastructure. The reason was the absence of SONET/SDH at the local loop level. Also enterprises shied away from the expensive tariff on the SONET/SDH based 155 Mbps rate of transmission. More so, sometimes they did not even require such high bit rates. Their requirement was around 34 or 45 Mbps.

Hence, the ATM over PDH was evolved to operate at the E-3/DS-3 and E-1/DS-1 levels. There are two mapping techniques to place ATM cells into an E-3 frame:

> ✓   *Physical layer convergence protocol* (PLCP) as defined in G.851 for mapping the distributed queue dual bus (DQDB) into 34.368 Mbps
> ✓   HEC-based ATM mapping (as per G.804 and G.832) to DS-3 PLCP.

The first technique requires extra overhead. Therefore, the second method has become a preferred standard and has been adopted by equipment vendors. It is also more closely aligned with the mapping of cells over STM-1 (a frame structure of SDH to be described later). ATM cells are mapped into 530-octet data payload with overhead functions concatenated to the start of the 9×59 octet frame.

In the 45 Mbps/T-3 transmission, adding the PLCP in the transmission convergence sublayer facilitates the PDH frame for ATM access. The ATM to PLCP on a DS-3 pleisiochronous line is illustrated in Fig. 4.46.

The PDH transmission frame format (Fig. 4.46) for the ATM cell transport adopted with DS-3 interface consists of the following:

> ✓   12 ATM cells
> ✓   4-octets *protocol control information* (PCI) preceding each cell
> ✓   13 or 14 nibble trailer at the end of each complete frame.

Synchronization bits: These contain delimiters to identify the beginning of the frame. That is, the delimiter plus a constant 125 μs frame time maintain the synchronism between the sending and receiving end. Since the cells are located in the designated positions within the frame, they can be located directly without the need to perform time-consuming delineation calculations.

The physical layer convergence protocol for ATM cells on a DS-3 (45 Mbps) pleisiochronous line includes ample framing overhead and an adjustable stuff (13 or 14 nibbles) to accommodate differences in clock rates between synchronous digital hierarchy (facilitated by SONET) and the asynchronous digital hierarchy (PDH) of DS-3 transmission. Relevant aspects of PLCP are as follows:

> ▪   DS-3/PLCP: Defined by IEEE 802.6 specifications. PLCP is intended for T-3 service supporting C-bit parity, to remove variable stuffing of standard T-3 constituted of 28 T-1s with slightly unsynchronized clocks

> ▪   ATM cells are enclosed in 125 μs frame defined within the DS-3 main frame

> ▪   PLCP mapping transfers 8 kHz timing across the DS-3 interface. This corresponds to a throughput of (53 octets $\times$ 8 bits/octet $\times$ 12)/125 μs = 40.707 Mbps maximum (and a payload of 36.864 Mbps).

✓ This is somewhat inefficient. That is, 40.704 Mbps refers to about 92% of $(7 \times 8 \times 84$ bits$)/(7 \times 8 \times 85$ bits/44.736 bps$) = 44.21$ Mbps, which is the payload rate of direct mapping DS-3. This inefficiency is due to the overhead incorporated.

PLCP framing		POI	POH	PLCP payload	
A1	A2	P11	Z6	ATM cell 1	
A1	A2	P10	Z5	ATM cell 2	
A1	A2	P 9	Z4	ATM cell 3	
A1	A2	P 8	Z3	ATM cell 4	A1: 11110110
					A2: 00101000
A1	A2	P 7	Z2	ATM cell 5	P0-P11: POI
					Z1-Z6: Growth octets
A1	A2	P 6	Z1	ATM cell 6	- 00000000
					F1: User channel
A1	A2	P 5	F1	ATM cell 7	B1-: BIP-8
					G1: PLCP path status
A1	A2	P 4	B1	ATM cell 8	C1: Cyclic stuff counter
A1	A2	P 3	G1	ATM cell 9	
A1	A2	P 2	M2	ATM cell 10	
A1	A2	P1	M1	ATM cell 11	
A1	A2	P0	C1	ATM cell 12	Trailer (125 µs)
1	1	1	1	53 octets	13/14 nibbles
octets			BIP-8 calculation		

Fig. 4.46: ATM to DS-3 PLCP mapping

The details on DS-3 PLCP frame format (shown in Fig. 4.46) are as follows: In the ATM mapping, the PLCP does the rate adaptation by optionally stuffing an additional nibble (that is, stuffing of 4 bits) at the end of every third PLCP frame. The 4 synchronization, signaling, and control bytes prefixed to each ATM cell in Fig. 4.46 carry the following functional attributes:

■ Framing
A1: 11110110
A2: 00101000

✓ Recovery of cells at the receiver relies on locking onto the PLCP frame. The A1 and A2 bytes facilitate a comfortable way to recover synchronization.

■ POI: Path overhead identifier

✓ There are 12 P types (P11 to P0), each of which has a different binary value to identify the field in the path overhead (POH) of the PLCP that follows immediately.

- POH: Path overhead

  - ✓   Z1-Z6:   Growth octets = 00000000
  - ✓   X: Unasssigned
  - ✓   B1:   PLCP bit interleaved parity-8 (BIP-8)
  - ✓   G1:   PLCP path status
  - ✓   C1:   Cycle stuff counter.

- The BIP check included (in the B1 byte) has a field which reflects an " even parity" on all the bits in the same position (1-8) of the bytes in the previous PCLP frame excluding the A1, A2, and P bytes. A BIP-8 refers to eight independent checks longitudinally on the data.

- G1 byte refers to two alarm fields:

  - ✓   *Far-end block errors* (FEBE) in the first 4 bits. This counts the number of bits (0000 to 1000 binary) in the previous BIP-8 that did not match the even parity check. A FEBE set as 1111 indicates that the BIP/FEBE is inactive. Other values are not valid
  - ✓   *Remote alarm indication* (RAI): This is the 5th bit in G1. It is returned on a UNI when receive failure takes place.

- C1 byte: This is used to alert the receiver should a stuffing take place (14 nibbles instead of 13 in the trailer). See Table 4.7 wherein the stuffing control in DS-3 PLCP is indicated.

Table 4.7: <u>DS-3 PLCP: Stuffing</u>

PLCP value of subframe	C1 byte in the trailer	Nibbles
1	1111 1111	13
2	0000 0000	14
3	0110 0110	13 (No stuffing)
	1001 1001	14 (Stuffed)

Summarizing, in order to support ATM cells on a DS-3 asynchronous digital hierarchy, the UNI adopted refers to the following: A PLCP is added in the transmission convergence sublayer.

This protocol is intended for the T-3 service that supports *C-bit parity* to remove the variable "stuffing" of standard T-3 whose rate adapts 28 T-1s at the same time allowing slightly unsynchronized clocks (at the multiplexer).

Direct mapping

As indicated above, under PLCP, the available bit rate for the transport of ATM cells is 40.707 Mbps (and a payload of 36.864 Mbps). Hence, a preferred trend is towards a direct mapping using the traditional DS-3 mapping of Fig. 4, where the 84 bits correspond to ATM cells (data).

Example 4.12
Describe the PDH UNI for ATM over T-1 and E-1 lines.

Solution
The extension of the lower limit of recognized ATM speeds down to T-1 came about in mid-1993. Certain vendors announced the corresponding T-1 interface. Bellcore defined a scheme to support ATM cells in a T-1, which provided byte alignment as well as significant overhead for synchronization. Also, the associated HEC cell delineation is more bandwidth efficient. Fig. 4.47 illustrates the T-1 PLCP, which accommodates ATM cells. Bellcore originally defined this format for SMDS access loops.

PLCP framing		POI	POH	PLCP payload	
A1	A2	P 9	Z4	ATM cell 1	
A1	A2	P 8	Z3	ATM cell 2	
A1	A2	P 7	Z2	ATM cell 3	A1: 11110110
A1	A2	P 6	Z1	ATM cell 4	A2: 00101000 P0-P11: POI Z1-Z6: Growth octets
A1	A2	P 5	F1	ATM cell 5	- 00000000
A1	A2	P 4	B1	ATM cell 6	F1: User channel B1-: BIP-8
A1	A2	P 3	G1	ATM cell 7	G1: Path status M2, M1: Management
A1	A2	P 2	M2	ATM cell 8	information C1: Cycle/stuff counter
A1	A2	P1	M1	ATM cell 9	
A1	A2	P0	C1	ATM cell 10	Trailer 3 ms
1	1	1	1	53	6 octets
			BIP-8 coverage		

Fig. 4.47: T-1 PLCP

The receiver assumes, initially, the location where the cell starts on a random basis, and then calculates the HEC. A comparison of that value with the bits found in the position, where the HEC field output to be, should offer a match, if the assumption made is correct. Otherwise, the receiver shifts the assumed starting point of the cells to the next bit position and recalculates the HEC. This procedure is continued until the correct match is realized. This HEC calculation is done in hardware. The checking can be done in less than a half second even on the slowest standard ATM interface. Thus HEC allows a receiver to delineate cells very swiftly without any reference to the physical medium framing. T-1 access to ATM is attractive in making the last leg of ATM to the desk-top level economical.

Other versions of ATM on PDH transmissions

Though DS-3/E-3 and DS-1/E-1 have been adopted for ATM on PDH, an interface at 6.312 Mbps (known as J-2) in Japan and another at 8.448 (E-2 in Europe) have also been defined for ATM cell mapping by the ITU-T.

Access to ATM backbone services

With the transition from using permanent virtual circuits to switched virtual connections (SVC), local exchange carriers (LEC) are expected to offer customers access (switched access) to (switched) long distance ATM services. Such services involve the following:

✓ Conversion of any existing dedicated access to switched access. (Time table …?)

✓   Retention of local loop technology and data rate as they exist in the current practice

✓   Determination of new tariff rates for the deployment of the conceived switched access.

To incorporate the switched access indicated above, the initial attempts have been as follows:

✓   As indicated earlier, some carriers (LEC) came up with an ATM service at DS-3 namely, the traditional T-3 transmission at 45 Mbps

✓   Subsequently, the OC-3 access over a single-mode fiber with SONET framing at 155.52 Mbps became the choice of the LECs. This has the preference (over the previous one) as it complies with international standards.

ATM Forum options on access facilitation are as follows:

✓   Via multimode fiber at 100 Mbps and 155.52 Mbps
✓   Via twisted pair at 51 Mbps
✓   Via a serial data port using *data exchange interface* (DX1) protocol, which is good up to at least 50 Mbps.

Multimode fiber and shielded twisted pair copper wire transmissions in ATM access strategies

▪   Multimode fiber:

In facilitating ATM access via multimode fibers, the following considerations are envisaged:

✓   Bit stream is encoded as per fiber channel (FC) standards
✓   Transmission rate refers to 10 baud/8 bits (8B/10B) at 194.4 Mbaud on 63.5/125 micron graded indexed multimode fiber
✓   Source: 1300 nm
✓   Encoding: This is done similar to 4B/5B coding of FDDI. This conversion of data characters to layer symbols creates additional symbols which can be used for framing, synchronization, alarm, and control

⇒   5 special characters replace the header of the OA&M cell to provide positive synchronization of the 27 cell frame and to delineate the cells

⇒   The 1 to 26 ratio of OA&M cell to data (or idle) cells again produces a payload that exactly fits in an OC-3c

⇒   The OA&M cell is the 1st byte which provides for alarm indication far-end receiver failure, and errored frame indication (a bit set to return an alarm upstream when an invalid 8B/10B is received).

▪   Shield twisted pair

✓   For short distances (up to 100 meters), the same electrical signal (194 Mbps) that drives the optical transmitter on multimode fiber is transmitted over 150 Ω shielded twisted pair (EIA/TIA 568). In such applications, Type 1 or 2 cable usage is anticipated.

✓   Physical Connector: DB-9 connector is used for physical interface.

✓   Electrical signal: Square pulses on the balanced pair or with differential voltage 1 to 1.6 V

✓   Line coding: 8B/10B. (same as for FC standard).

"Slower" ATM interface

Since 1993 there has been a growing demand to provide ATM access (transmission and switching) to every individual at desk level. That is, to facilitate an UNI adoption at a speed in SDH below 155 Mbps. An approach towards such facilitation has been conceived and implemented as follows:

- Physical medium: Use of twisted pair copper wires

- Speed: Use of slow bit rates compatible for the copper wire transmission from ATM transmission to desk-tops. The associated considerations are:

  ✓ *Cell format*: Head-to-tail sequencing of cells without gaps. This facilitates cell location easily and efficiently. (1991 standardization by vendors)
  ✓ Cell sequencing is done in the presence of an HEC field
  ✓ No assumption on physical layer framing
  ✓ Any serial or parallel stream may carry the cells
  ✓ HEC: The receiver makes an initial random assumption about where the cell starts. Then, it calculates the HEC value. This calculated value is matched against the value that should be corresponding to the guessed position in the HEC field (known a priori)
  ✓ If the guessed position is wrong, the search is iterated until a matching is obtained. Since there are 121 bits in a cell, the probability of a correct guess is only 1/421
  ✓ This brute-force strategy/HEC calculation and iterated search is done on hardware.

The demarcation point of ATM, SDH, and B-ISDN at customer premises depicts a user network interface (UNI). The UNI is not based on SONET equipment. However, the two technologies are not separable.

The close relationship between UNI and SONET is maintained by:

  ✓ Using the same cell throughput as an STS-3c payload
  ✓ 149.76 Mbps, rate on a serial interface over copper or fiber. (*Note*: 149.76 = 155.52 x 26 / 27)

The nominal speed of 155 Mbps includes the SONET overhead, which is replaced by overhead cells on a serial interface. A serial link is adjusted to the payload of an STS-3c by inserting one OA & M cell after every 26 data cells.

The OA & M cells maintain the same ratio of overhead to payload as in the STS-3 frame. This allows a UNI to be fed directly into a SONET network and exactly fill an OC-3. Between the OA & M cells, all cells are either data or idle.

4.4.2 SONET/SDH: Defining a new TDM format

In contrast with PDH, SONET/SDH was developed as new form of TDM framing on a transmission system that provides reference marks explicitly so that the receiver would know how to interpret the bit stream. This digital hierarchy is *synchronous* and is indicated exclusively on an optical physical medium in the North American telecommunications. As indicated earlier, it is referred to as SONET or *synchronous optical network*. A corresponding standard of CCITT is known simply as *synchronous digital hierarchy* (SDH).

Both SONET and SDH framing impose overheads (at least 4.4 %). However, they offer the following advantages:

  ✓ Explicit indication where an octet starts via pointers provided in the overhead position of the SONET/SDH frame

✓   Ease of network management and control using the designated bandwidth provided via overhead bytes for operations and communication channels separate from the payload.

In contrast to PDH, as mentioned earlier, a strategy that is based on digital multiplexing of signals with all levels of the signals synchronized to the same master clock is known as *synchronous digital hierarchy* (SDH). This scheme was adopted as a standard for optical transport by ECSA/ANSI and has been designated as the *synchronous optical network* or *SONET.*

SDH is a variation of SONET and has been adopted by CCITT. The term "optical" was omitted in the CCITT standard, inasmuch as that the standard has been made applicable to other nonoptical transmission media, like microwave (ITU-T Recommendation G.707-G.709).

SONET

In 1985 Bellcore developed the SONET and ANSI standardized it. SONET defines the rates and format for the optical transmission of digital information. Its specifications define a hierarchy of standardization with data rates from 51.84 Mbps to 9.953 Gbps.

ANSI developed the transmission standard for SONET, which allows a fiber system to transport many digital signals that have differing bandwidth requirements without wasting any of the capacity of the fiber. This is done by transmitting digital signals in frames, consisting of headers and error-correcting bytes along with a payload of data. Further, SONET eliminates several layers of multiplexing and provides higher speed transmission with the ability to add, drop, or insert send/receive stations along the transmission line.

Thus, SONET defines a technology to carry several signals of different capacities through a synchronous, flexible, optical hierarchy. This is accomplished by means of a *byte-interleaved multiplexing* scheme. Byte interleaving simplifies multiplexing and offers tractable end-to-end management.

The specific issues pertinent to SONET are as follows:

▪   The SONET standard is established for a multiplexing format of any number of 51.84 Mbps signals. The building block for this strategy is the DS-3 rate

▪   It specifies an optical signal standard for interconnecting equipment manufactured by different vendors

▪   SONET standards comprehensively address the enhanced OA&M capabilities

▪   SONET defines synchronous multiplexing of lower level digital signals (*tributaries*) with simplified interfaces to digital switches, cross-connect, and add-drop multiplexers

▪   It allows flexible architectures to embrace future B-ISDN rates, as well.

The driving force behind the development of SONET is comprised of the following:

✓   Need to extend the multiplexing feasibility beyond DS-3 rate in a simplified manner
✓   To realize the telecommunication of a smaller extent of traffic facilitated within the bulk payload of optical signal in an economical way
✓   To achieve the capability to enhance sophisticated types of services and futuristic needs with almost unlimited bandwidth availability.

The primary task in the SONET multiplexing process involves the generation of the lowest level or *base signal*. In SONET, this base signal is referred to as *synchronous transport signal level-1*, or simply STS-1. It operates at 51.84 Mb/s. Higher level signals are integer multiples of STS-1. That is, by interleaving bytes from N × (STS-1) signals, which are mutually synchronized, it generates the family of STS-N signals as indicated in Table 4.8. Thus, an STS-N signal is composed of N byte-interleaved STS-1 signals. Table 4.8 also includes the optical counter-part for each STS-N signal, designated as OC-N (*optical carrier level-N*). These OC-N signals are obtained by scrambling the STS-N signal (except framing bytes) and converting electrical signal to optical signal.

Table 4.8: ANSI/SONET hierarchy

SONET level (electrical/optical carrier level)	Bit rate Mbps	Capacity in terms of DS/T-levels
STS-1/OC-1	51.840	$28 \times DS\text{-}1$ or $1 \times DS\text{-}3$
STS-3/OC-3	155.520	$84 \times DS\text{-}1$ or $3 \times DS\text{-}3$
STS-12/OC-12	622.080	$336 \times DS\text{-}1$ or $12 \times DS\text{-}3$
STS-24/OC-24	1244.24	$672 \times DS\text{-}1$ or $24 \times DS\text{-}3$
STS-48/OC-48	2488.320	$1344 \times DS\text{-}1$ or $48 \times DS\text{-}3$
STS-192/OC-192	9953.280	$5376 \times DS\text{-}1$ or $192 \times DS\text{-}3$
STS-N/OC-N	$N \times 51.84$	$N \times 28 \times DS\text{-}1$ or $N \times DS\text{-}3$

The asynchronous digital hierarchy used towards construction of SONET hierarchy is reproduced in Table 4.9 for quick reference.

Table 4.9: Asynchronous digital hierarchy: ANSI rates

Signal	Bit Rate	Channels
DS-0	64 Kb/s	1 DS-0
DS-1	1.544 Mb/s	24 DS-0s
DS-2	6.312 Mb/s	96 DS-0s
DS-3	44.736 Mb/s	28 DS-1s

The transmission hierarchy of SONET is shown in Fig. 4.48.

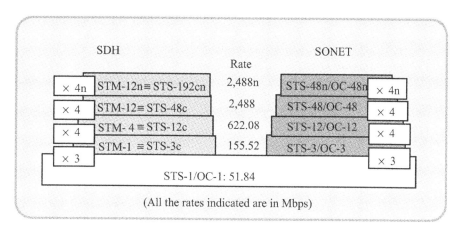

Fig. 4.48: Transmission hierarchy of SONET and SDH

STS: The art of making a new superframe...

The frame format of STS-1 is shown in Fig. 4. The frame is divided into two parts: *Transport overhead* and the *synchronous payload envelope* (SPE). The SPE is again divided into two parts: *STS path overhead* and the *payload*. The payload corresponds to the revenue-producing traffic being transported and routed over the SONET network. Once the payload is multiplexed into the synchronous payload envelope, it can be transported and switched through SONET without having to be examined at intermediate nodes. Transport overhead is composed of *section overhead* and *line overhead*. The STS-1 path overhead is part of the synchronous payload envelope.

STS-1 frame structure

STS-1 has a specific sequence of 810 bytes (6480 bits), which includes various overhead bytes and an envelope capacity for transporting payloads. It can be described as a 90 column × 9 row structure as shown in Fig. 4.49.

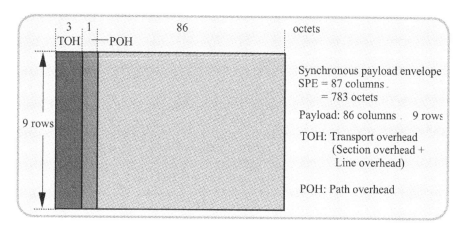

Fig. 4.49: STS-1 Frame structure

In view of a frame length being 125 µs (or 8000 frames per second), STS-1 has a bit rate of 51.840 Mbps as indicated below. The transmission of bytes is done row-by-row from top to bottom, left to right. As illustrated in Fig. 4.49, the first three columns are for the *transport overhead*; each column contains nine bytes. Of these, nine bytes are overhead for the section layer, and 18 bytes are overhead for the line layer. The remaining 87 columns depict STS-1 envelope capacity (payload and path overhead). The signal rate, is therefore calculated as: 9×90 bytes/frame $\times 8$ bits/byte $\times 8000$ frames/s $= 51,840,000$ bps $= 51.840$ Mbps. This is known as the *STS-1 signal rate*. The optical equivalent of STS-1 is known as *optical carrier level-1* (OC-1), which is deployed in the transmission across the fiber.

The STS framing is very much analogous to T-1 framing. The difference is SONET inserts "a block of octets" periodically whereas a bit is inserted in T-1. Fig. 4.50 demonstrates this the resemblance consideration by depicting the STS frame in a linear fashion.

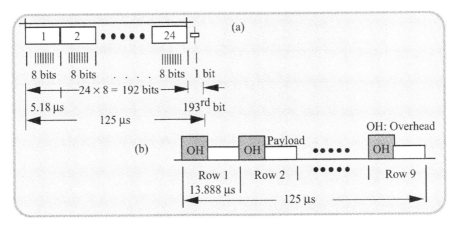

Fig. 4.50: (a) T-1 frame and (b) a linear layout of STS frame

The SONET frame (like T-1 frame) is transmitted every 125 μs. That is, the frame rate is 8000 kHz. Therefore, the STS rate is equal to (overhead rate + information envelope rate) $= (9 \times 3 \times 8 \times 8000) + (9 \times 87 \times 8000) = 51.84$ Mbps. That is, 810 bytes are transmitted every 125 μs.

Problem 4.6
In reference to the STS-1 channel structure show that: (i) Information payload $= 50.112$ Mbps; (ii) transport overhead (TOH) $= 1.728$ Mbps; (iii) path overhead (POH) $= 576$ kbps; and (iv) line overhead $= 1152$ kbps.

Example 4.13
Calculate the data rate of the STS-3 frame.

Solution
The frame structure of STS-3 is depicted in Fig. 4.51. It represents $3 \times$ STS-1 framing.

❑   STS-3 payload capacity $= 258$ bytes
❑   STS-3 SPE $= (258 + 3) = 261$ bytes
❑   Transmission rate $= 270 \times 9 = 2430$ bytes transmitted every 125 μs
                                   $= 155.52$ Mbps

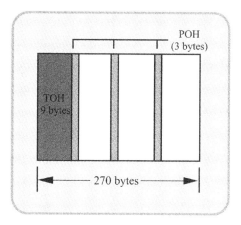

Fig. 4.51: STS-3 frame

Problem 4.7

Determine the payload rate of transmission in the concatenated STS-3c shown in Fig. 4.52.

The concatenated STS-3c is identical to STS-3 except that the POH is not replicated three times. If not, there is a wastage of bandwidth due to redundant replication.

- • TOH: 9 bytes
- • Payload: 260 bytes
- • SPE: $(260 + 1) = 261$ bytes.

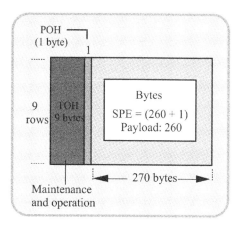

Fig. 4.52: STS-3c frame

Example 4.14

How many ATM cells can be accommodated in the STS-3c frame?

Solution

The transmission rate of STS-3c frame corresponds to (270×9) bytes in every 125 μs. That is, each STS-3c frame provides 155.52 Mbps. Out of the total 2430 bytes, 90 bytes are used by the TOH and POH leaving 2340 bytes (96.3 %) of the SONET frame for the payload. That is, 96.3 % of 155.53 Mbps ≈ 150 Mbps is available to accommodate the payload. Considering the 53-byte ATM cells, the number of such cells per frame, which can be carried by the payload of STS-3c is given by:

$$\frac{2430 \text{ bytes/frame} - 90 \text{ bytes/frame}}{53 \text{ bytes/ATM cell}} \approx 44 \text{ ATM cells/frame}.$$

Example 4.15

Illustrate ATM-cell mapping on a STS-3c frame.

Solution

Presented in Fig. 4.53.

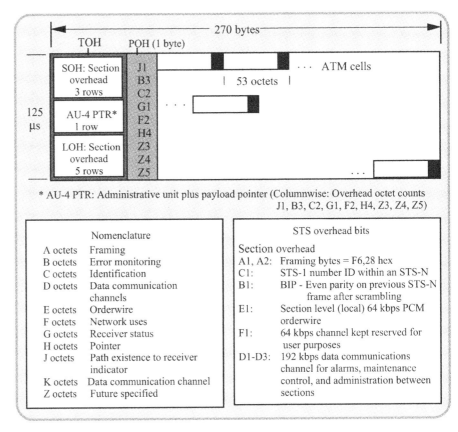

Fig. 4.53: ATM-cell mapping on a STS-3c frame

Payload = $9 \times 2608 \times 8/125$ μs = 149.76 Mbps

Note: STS-3c also corresponds to STM-11 (synchronous transport mode) of SDH

Problem 4.7

Considering the (5 + 48) octet ATM cell, what is the bandwidth available for user information (with the exclusion of any AAL processing needed)?

SONET physical hierarchy

The SONET physical hierarchy consists of path, lines, and sections as shown in Fig. 4.54

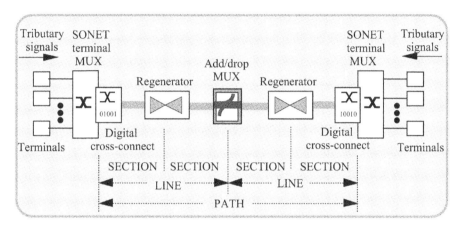

Fig. 4.54: Physical hierarchy of SONET

Sections, lines, and path ...

Section:	*Transmission medium and associated equipment between (i) a network element and a regenerator or (ii) two regenerators. Associated equipment includes optical interface and SONET processing equipment at the originating or at the terminating end. Section spans are not individually protected. Section span also allows network performance to be maintained between line regenerators or between a line regenerator and a SONET network element allowing fault localization. Regenerators rejuvenate the attenuated signals and revive the stream of digital waveform for onward transmission.*
Line:	*This span consists of a transmission medium plus the associated equipment that provides the means of transporting information between two consecutive network elements. A line span is protected with respect to equipment failure and performance deterioration. Line span allows network performance to be maintained between transport nodes and provides the majority of network management reporting.*
Path:	*This span deals with the transport of services between path termination equipment (PTE). It refers to the logical connection between the point at which a standard format signal is assembled and the point at which the signal is disassembled. Path span allows network performance to be maintained from a customer service end-to-end perspective.*

System hierarchy of SONET

The system hierarchy of SONET consists of four layers, namely:

- *Photonic layer*

  This is the physical layer, which includes specifications on types of optical fiber that may be used, minimum laser power required, dispersion characteristics of lasers, and required sensitivity of receivers. It provides optical transmission at a very high bit rate. Electrical/optical equipment communicates at this level. That is, the conversion of electrical signal to optical signal takes place at this layer.

- *Section layer*

  This creates basic SONET frames and converts electric signals into photonic ones. It also has monitoring capabilities. Other functions include framing and scrambling.

- *Line layer*

  This is responsible for synchronization and multiplexing of data into SONET frames; it also performs protection and maintenance functions and switching.

- *Path layer*

  This is the final layer of SONET responsible for end-to-end transport of data at the appropriate rate.

The four level hierarchy of SONET is presented in Fig. 4.54. Each level in Fig. 4.54 communicates horizontally to peer equipment in that level, and each level also processes certain information and delivers it to the next level.

Viewing top-down, the network services (DS-1s etc.) are inputs to the path layer. The path layer transmits horizontally to its peer on the other end. The path layer maps the services and POH into SPEs, which it hands over vertically to the line layer. The line layer transmits to its peer entities horizontally and also the line layer overhead. It maps the SPEs and line overhead into STS-N signals and hands them over to the section layer. The section layer again transmits to the peer entities horizontally. It transmits the STS-N signals and section layer overhead. It maps STS-N and the section overhead into pulses that are handed over (vertically) to the optical transceiver. It should be noted that not all pieces of equipment support all layers.

STS-N frame structure

An STS-N is a specific sequence of N × 810 bytes. Byte-interleaving STS-1 modules form the STS-N, Fig. 4.55.

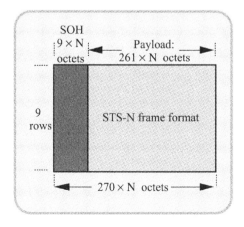

Fig. 4.55: STS-N Frame format

SONET overheads

SONET provides substantial overhead information, allowing simpler multiplexing and greatly expanded *operations, administration, maintenance,* and *provisioning* (OAM&P) capabilities. There are several overhead layers:

- *Path-level overhead* is carried from end-to-end; it is added to DS-1 signals when they are mapped into virtual tributaries and for STS-1 payloads that travel end-to-end

- *Line overhead* is for the STS-N signal between STS-N multiplexers

- *Section overhead* is used for communications between adjacent network elements, such as regenerators.

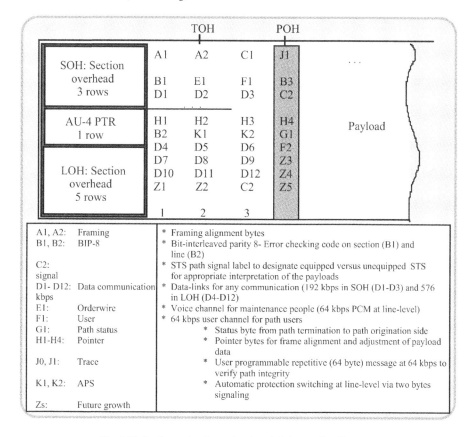

Fig. 4.56: Section and path overheads and their associated bytes

The section overhead contains nine bytes of the *transport overhead*. This overhead supports functions such as:

- ✓ Performance monitoring (STS-N signal): Parity check, STS-1 identification
- ✓ Data communication channels to carry information for OAM&P, voice communication channel (orderwire), and user channel
- ✓ Framing: Frame alignment pattern.

The bytes of section overhead are presented in Fig. 4.56.

Table 4.10 describes the functions of individual bytes of Fig. 4.56.

Table 4.10: <u>Section overhead bytes: Functions</u>

Bytes	Functions
A1-A2	These two bytes (11110110 00101000) provide a frame alignment pattern. These exist in all STS-1s with an STS-N. These two bytes identify the beginning of the SONET STS-1 frame.
C1	This byte is given a binary number as per its order of appearance on the byte-interleaved STS-N frame. It can be used for the framing and deinterleaving process so as to determine the position of other signals. This is also provided in all STS-1s within an STS-N with the first STS-1 set as # 1 (0000 0001).
B1	This byte enables error monitoring via the *bit-interleaved parity* (BIP-8) code using even parity. In an STS-N, the section BIP-8 is calculated over all bytes of the previous STS-N frame after scrambling, and the computed value is placed in the B1 of the STS-1 before scrambling.
E1	This byte enables a local orderwire voice communication channel between regenerators and network elements.
F1	This is reserved for user's purpose and is terminated at all section-level equipment.
D1-D3	These 3 bytes facilitate data communication channels for messages of OAM&P, monitoring, alarm, and other communication needs at 192 kbps rate between section termination equipment.

Line overhead contains 18 bytes and supports the following functions:

- ✓ Communication between the line-terminating equipment (LTE)
- ✓ Locating the SPE in the frame
- ✓ Multiplexing or concatenating signals
- ✓ Performance monitoring/Error detection
- ✓ Automatic protection switching
- ✓ Line maintenance
- ✓ Synchronization between LTE.

The line overhead consists of 18 bytes of the STS-1 and the functional aspects of these bytes are indicated in Fig. 4.57.

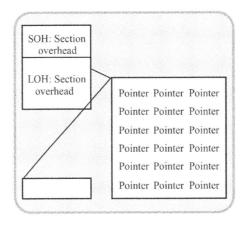

Fig. 4.57: Functional aspects of LOH

Table 4.11: Constituent (18) bytes of STS-1 line overhead

Bytes	Functions
H1-H2	These three bytes evaluate the operation of the STS-1 *payload pointer*. These are provided for all STS-Ns in an STS-N
B2	This byte facilitates BIP-8 line error monitoring. The line BIP-8 is calculated over the entire bits of the line overhead and payload envelope capacity of the previous STS-1 frame before scrambling is done. The calculated value is placed in the B2 byte. This byte is designated for STS-1 in an STS-N signal.
K1-K2	These two bytes enable *automatic protection switching* (APS) signaling between line equipment. These are defined only for STS-1 in an STS-N signal.
D4-D12	These line bytes facilitate a data communication channel at 576 kbps for messages on administrations, monitoring, maintenance, alarms, and other conditions needs between line termination equipment. D4-D12 are defined only for STS-11 of an STS-N signal.
Z1-Z2	Functions are yet to be defined for these two bytes
E2	This is for express orderwire channel for voice communications between line terminating equipment. This byte is defined only for STS-1 of an STS-N signal.

Path overhead (STS POH) provides for communication between the point of creation of an STS SPE and its point of disassembly. This overhead supports the following:

✓ Performance monitoring of the STS SPE
✓ Signal label (the content of the STS SPE)
✓ Path status
✓ Path trace.

The path overhead in STS-1 frame is illustrated in Fig. 4.56: The functions provided by the POH include end-to-end transport of services, sequencing of cells, path-terminating element status, continuity, error detection, and user-defined functions.

Table 4.12: 9 bytes of the POH in the SONET frames

Bytes	Functions
J1	This is used for repetitively transmitting 64-byte information. This is a fixed-length string. Hence, a continued connection to the source of the path signal can be verified at any receiving terminal along the path.
B3	This is intended for BIP-8 based path error monitoring. The path BIP-8 is computed over all bits of the previous SPE, and the calculated value is placed in the B3 byte before scrambling.
C2	This byte provides for constructing the STS-SPE by means of a label value assigned from a list of 256 possible values (8 bits).
G1	This is adopted to convey back to the originating STS-PTE the path termination status and performance. This facilitates two way path monitoring at any point or at the ends of the path.
F2	User's purpose byte between the PTEs.
H4	For multiframe phase indication for VT payload.
Z3-Z5	Reserved for future use.

Pointers

SONET uses a concept called *pointers* to compensate for frequency and phase variations. The use of pointers avoids the delays and the loss of the data associated with the use of large (125 μs frame) slip buffers for synchronization. Pointers permit dropping, inserting, and cross-connecting of the payloads in the network in the simple way. Transmission signal wander and jitter can also be readily minimized with the pointers.

Fig. 4.58 shows an STS-1 pointer (H1 and H2 bytes) which allows the SPE to be separated from the transport overhead. The pointer is simply an offset value that points to the byte where the SPE begins. If there are any frequency or phase variations between the STS-1 frame and its SPE, the pointer value will be increased or decreased accordingly to maintain synchronization.

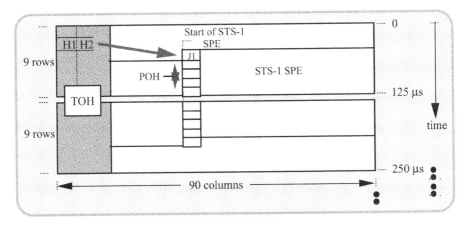

Fig. 4.58: SPE position in the STS-1 Frame

Concept of pointers

❑ *Pointer points to the byte where synchronous payload envelope (SPE) begins*
❑ *Pointers dynamically phase align STS and VT payloads permitting ease of dropping, inserting, and cross-connecting*
❑ *Phase variation between STS-1 frame and its SPE is adjusted by increasing and decreasing the pointer.*

Need for pointers/pointer adjustment can be seen by considering conventional circuit-switched networks and SONET in their respective way of locating different channels (multiplexed) within a payload.
Pointer functions in circuit-switched and in SONET networks

Circuit switched networks

❑ *Suppose a signal received (from A) (=> DS-0 single channel 64 kbps) is addressed to B and the rest is passed on to C. In order to accomplish this, the single DS-0 should be isolated at B. That is, B must demux every bit of 1.544 Mbps, remove the required data, remux every bit and pass it on to C.*

SONET

❑ *To isolate a specific channel data from a payload, SONET uses pointers.*
❑ *A set of pointers locate channels with in a payload (and also locate different payloads within a frame*

426

❑　*Pointer information is contained in the POH (path overhead) and indicates channel locations in a payload. (Pointer information in LOH (line overhead) indicates payload location in a frame.)*

How is SPE maintained across the electrical STS transmission and SONET (optical transmission)?

❑　*Payload is allowed to slip through an STS-1 frame increasing or decreasing thereof the pointer value at intervals by one byte position*

❑　*If payload rate is greater than local STS frame rate then the pointer is decreased by one octet position so that next payload will begin one octet sooner than the earlier payload*

⇒　*To avoid the loss of an octet on the payload that is thus squeezed the H2 octet is used to hold the extra octet for that frame*

❑　*Similarly, if the payload rate lags behind the frame rate, the insertion of the next payload is delayed by one octet. In this case, the octet in the SPE that follows the H3 octet is left empty to allow for the movement of the payload.*

Concatenated payloads

For comprehensive applications, the STS-1 may not have enough capacity to carry some services. Therefore, SONET offers the flexibility of concatenating STS-1s to provide the necessary bandwidth. STS-1s can be concatenated up to STS-3s. Beyond STS-3, concatenation is done in multiples of STS-3s.

Virtual tributaries (VT)

In addition to the STS-1 base format, SONET also defines synchronous formats at sub-STS-1 levels. The STS-1 payload may be subdivided into *virtual tributaries*, which are synchronous signals used to transport lower-speed transmissions. Table 4.13 indicates the size of the VT.

Table 4.13 Virtual tributaries

VT type	Bit rate	Size of VT
VT 1.5	1.728 Mb/s	9 rows, 3 columns
VT 2	2.304 Mb/s	9 rows, 4 columns
VT 3	3.456 Mb/s	9 rows, 6 columns
VT 6	6.912 Mb/s	9 rows, 12 columns

In order to accommodate mixes of different VT types within an STS-1 SPE, the VTs are grouped together. An STS-1 SPE that is carrying VTs is divided into seven VT groups, with each VT group using 12 columns of the STS-1 SPE.

SONET multiplexing

The multiplexing considerations in SONET refer to the following:

▪　*Mapping* – This refers to a process used when tributaries are adapted into virtual tributaries by adding justification bits and path overhead (POH) information

▪　*Aligning* – This process occurs when a pointer is included in the STS path or VT path overhead, to allow the first byte of the VT's to be located

- *Multiplexing* – This is used when multiple lower-order path layer signals are adapted into higher-order path signals that are adapted into the line overhead

- *Stuffing* – SONET has the ability to handle various input tributary rates from asynchronous signals. As the tributary signals are multiplexed and aligned, some spare capacity has been designed in the SONET frame to provide enough space for all these various tributary rates

- Therefore, at certain points in the multiplexing hierarchy this space capacity is filled with "fixed stuffing" bits that carry no information, but are required to fill up the particular frame

- For the purpose of transporting and switching payloads smaller than STS-1, the STS SPE can be subdivided into smaller structures known as *virtual tributaries*. All services below the DS-3 rate are transported in the VT structure.

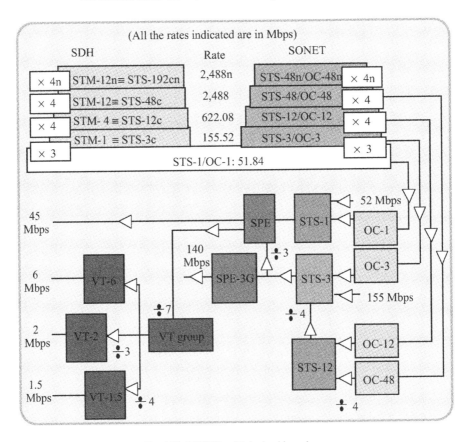

Fig. 4.59: SONET multiplexing hierarchy

Various types of service adapters can accept different types of services, starting from voice and ending with high-speed data and video. The adapter maps the signal into the payload envelope of the STS-1 or virtual tributary (VT).

Except for concatenated signals, all inputs are eventually converted to a base format of a synchronous STS-1 signal (51.84 Mb/s or higher). Lower inputs (such as DS-1s) correspond to the first bit (or the byte) multiplexed into VTs. Several STS-1s are then multiplexed together to form an STS-N signal.

SONET network elements
The essential network elements of SONET are as follows [4.6 – 4.10]:

- Terminal multiplexer

- Regenerator (Due to the long distance that normally prevails between multiplexers, the signal level may become low. Therefore, a regenerator is placed in such long hauls. It clocks itself off the received signal and replaces the section overhead bytes before retransmission)

- Add /drop multiplexer
  (It provides interfaces between the different network signals and the SONET.)

Add-drop multiplexer(ADM)

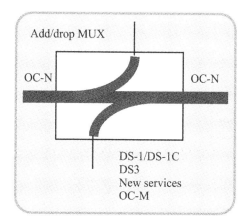

Fig. 4.60: Add-drop multiplexer

The add-drop multiplexer (ADM) is adopted at intermediate site locations. Normally, it corresponds to a single stage where signals needed to be accessed are dropped or inserted, while other traffic continues straight through. ADM replaces the conventional back-to-back devices in DS-1 cross-connects. It is actually a synchronous multiplexer, which is used to add or drop DS-1 signals onto the SONET ring. It can be reconfigued to allow for continuous operations in the event of a ring failure. The ADM terminates (accommodates) both OC-Ns and electrical signals. It operates bi-directionally. Hence, it uses both E/O and O/E interfaces. (E: Electrical; O: optical).

Cross-connect
There are two versions of cross-connects used as SONET network elements. They are,

    ✓   Broadband cross-connect
    ✓   Wideband cross-connect.

Broadband cross-connect
This performs the switching at STS level. It accepts various optical carrier states. It is used at SONET hubs for grooming (aggregating or segregating) STS-1s. The architecture of broadband cross-connect is similar to a DS-3 cross-connect.

Wideband cross-connect
This is used to perform switching at the VT level. It accepts DS-1, DS-3, and optical carrier signals. It is ideal for DS-1 *grooming* applications.

SONET network configurations

The configuration adopted in SONET networks are as follows:

- *Point-to-point network*: This involves two terminal multiplexers linked by fiber with or without a regenerator in the link (Fig. 4.61)

- *Point-to-multipoint network*: This configuration facilitates adding and dropping tributary channels at intermediate points in the network

- *Hub network*: This is a star configuration (Fig. 4.62), which accommodates unexpected growth and change more easily than other configurations. It concentrates traffic at a central site and allows easy reprovisioning of the circuits

- *Ring network*: This configuration allows uninterrupted service. That is, if a fiber is cut, the multiplexers send the services affected via an alternate path through the ring without interruption (Fig. 4.63).

Linear add/drop strings

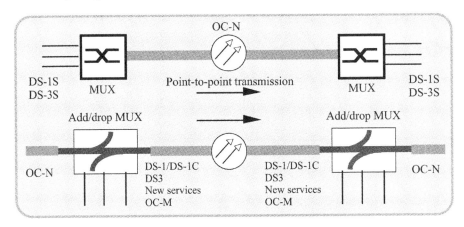

Fig. 4.61: Linear point-to-point SONET network

Point-to-multipoint tree configuration

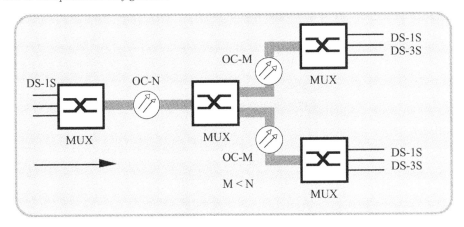

Fig. 4.62: Star SONET network

Ring configuration/diverse routing

Ring topology exists in different forms. The arrangement of *counter rotating rings* shown in Fig. 4.63, is known as *unidirectional self-healing ring* (USHR). In reference to the two fibers shown (in Fig. 4.63), one is called a *working fiber* and the other is a *protection fiber*. Suppose there is a failure on a fiber or at an interface to a node (such as the ADM in Fig. 4.64); the ring will take corrective measures (or self-heal) and cut out the problem area. The reconfigured arrangement is depicted in Fig. 4.63b.

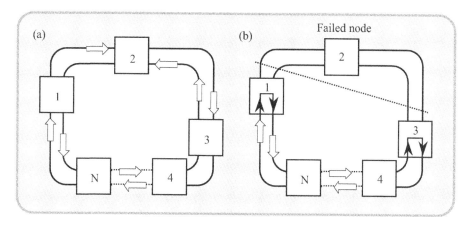

Fig. 4.63: Ring topology
(a) Counter rotating rings and (b) reconfigured network after a node experiences a failure

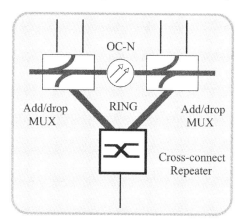

Fig. 4.64: Ring SONET network

Another possible topology refers to using four fibers and operates with a second arrangement. It is called a *bidirectional SHR* (BSHR). Here, the traffic shares the pairs of working and protection fibers. The traffic is sent over the shortest path between nodes.

A dual ring architecture plus a digital cross-connect (DCS) system is shown in Fig. 4.64: Here, the DCSs are used to cross-connect the VTs. One of their principal functions is to process the contents of the TOP and POH signals and map various VTs to others. DCS (Fig. 4.65) is effectively a switch that provides a central point for the grooming and consolidation of user payload between two ring systems. It is also used for diagnostic purposes. It responds to alarms and failure notifications. It performs switching at the VT level. It can segregate large bandwidth and low bandwidth signals and send them to appropriate ports.

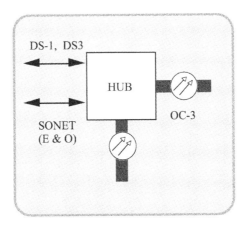

Fig. 4.65: A digital cross-connect system

Typical SONET equipment has a terminal multiplexer, add-drop multiplex (ADM), and a built-in *integrated timing supply* (BITS). The terminal multiplex performs packing of incoming T-1, E-1 and other signals into STS payloads for network use. The terminal multiplexer has architecture made of a controller, which is software driven. There is also a transreceiver that is used to get access for lower speed channels. A *time-slot interchange* (TSI) is used in SONET equipment, which feeds signals into higher speed interfaces.

SONET equipment uses synchronous clocks for timing and BITS to implement it. Timing is distributed to the network elements with BITS, which is used at these elements to synchronize the output onto these lines.

4.4.3   ATM cell processing into SONET frame
The processing of ATM cells into the SONET frame is done as follows (Fig. 4.66):

■   The received cells are stored in (read into) a transmit-side FIFO buffer

■   An option is made available to calculate the ATM cell HEC. That is, the fifth byte in the ATM header is calculated and the 5-byte ATM header is completed

■   Using a self-synchronizing scrambler polynomial in order to improve the efficiency of the cell delineation process, it then scrambles the 48-byte ATM cell payload. ATM cell header is not, however, scrambled

■   The cells are multiplexed into SONET payload, once a complete cell has arrived in the FIFO buffer

■   If a cell is incomplete in the FIFO, idle cells are placed into the SONET payload. That is, if the FIFO does not contain 53 bytes of information at the start of a cell insertion cycle, an idle cell is inserted

■   To ensure that a SONET frame is always full, the transmit circuit inserts idle ATM cells when no ATM cells are received from the SAR

■   The transmitter generates the SONET header, places the multiple ATM cells into the SONET payload, and scrambles the SONET frame. Finally, the SONET frame is transmitted every 125 µs

- On the receive side, the receive clock is generated from the incoming serial data stream; the receiver phase-lock loop (PLL) can produce an accurate clock within 5 ms. Once the clock is recovered, ATM cells are recovered from the SONET STS-3c SPE.

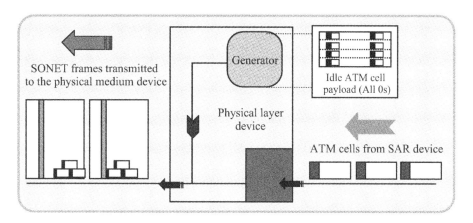

Fig. 4.66: ATM cell processing into SONET frame

As indicated earlier, each SONET STS-3c frame contains just over 44 ATM cells. It corresponds to 350,000 ATM cells being transmitted every second. If an ATM cell is not ready to be transmitted, as mentioned before, an idle cell is sent. This idle cell holds a place in the SONET frame. It consists of a payload set to all 0s and a 5-byte header set to all 0s with only the CLP bit set to 1 so that when congestion occurs, the idle cell has low priority and will be dropped. On the receiver side the idle cell is discarded. The ATM cell processing at the receiver is as follows:

Over every 125 μs, a SONET frame is received. Typically, the serial data stream is converted into byte wide data; next the SONET frame is descrambled, checked for parity, and the ATM cells are extracted. Finally, the ATM cells are descrambled and the byte wide information is passed to the ATM layer.

A typical ATM/SONET biCMOS receiver-transmitter is TDC 1500 SABRE ^{TM} (of TI). It provides the TC sublayer functions. It includes two analog PLLs, which facilitate receive clock recovery and transmit clock generation. They offer necessary digital logic to insert and extract ATM cells into/out of a SONET frame. The associated SAR device provides for the AAL and ATM layer functions. It includes a host interface to allow the use of host memory for segmentation and reassembly of the CS-PDU, thereby reducing the amount of local memory needed for an interface card. The SAR device provides full hardware support for AAL-5 and partial support for AAL-3/4 and AAL-1. It meets the ATM Forum UNI specifications.

4.4.4 SONET versus SDH

In 1989, ITU-T, formerly CCITT, published the *synchronous digital hierarchy* [4.5]. The SONET can be regarded as a subset of SDH. But, the SDH as defined by the ITU " is an outgrowth of the North American SONET standard developed by Bellcore in the United States". The time division multiplexing (TDM) of ANSI combines 24 × 64-kbps channels (DS-0s) into one 1.544 Mb/s DS1 signal. On the other hand, the ITU-T TDM multiplexes 32 × 64-kbps channels (E-0s) into one 2.048 Mb/s E-1 signal. The differences are given in the Table 4.14.

Table 4.14: SONET/SDH hierarchies

SONET signal	Bit rate Mbps	SDH signal	SONET capacity	SDH capacity
STS-1/ OC-1	51.840	STM-0	28 DS-1s or 1 DS-3	21 E-1s
STS-3/ OC-3	155.520	STM-1	84 DS-1s or 3 DS-3	63 E-1s or 1 E-4
STS-12/ OC-12	622.080	STM-4	336 DS-1s or 12 DS-3	252 E-1s or 4 E-4s
STS-48/ OC-48	2488.320	STM-16	1344 DS-1s or 48 DS-3	1008 E-1s or 16 E-4s
STS-192/ OC-192	9953.280	STM-64	5376 DS-1s or 192 DS-3	4032 E-1s or 64 E-4s

Note:
STM = Synchronous transport module (ITU-T)
STS= Synchronous transfer signal (ANSI)
OC = Optical carrier (ANSI)

Table 4.15: Asynchronous hierarchies-ITU-T rate

Signal	Bit rate	Channels
64-kbps	64-kbps	164-kbps
E1	2.048Mb/s	1 E-1
E2	8.45Mb/s	4 E-1
E3	34Mb/s	16 E-1
E4	144Mb/s	64 E-1

Although SDH STM-1 has the same bit rate as the SONET STS-3, the two signals contain different frame structures. SDH recommendations define methods of subdividing the payload area of an STM-1 frame in various ways so that it can carry combinations of synchronous and asynchronous tributaries. Using this method, synchronous transmission systems can accommodate signals generated by equipment operating from various levels of the asynchronous hierarchy.

The differences in the frame structure of SONET STS-3c frame and the corresponding STM-1 frame of SDH are as follows:

Table 4.16: Differences in STS-3c and STM-1 frames

Byte	SONET	HEX	SDH	HEX
1st C1 byte	0000 0001	01	0000 0001	01
2nd C1 byte	0000 0010	02	0000 0000	00
3rd C1 byte	0000 0011	03	0000 0000	00
1st H1 pointer byte	0110 0010		0110 1010	
2nd H1 pointer byte	1001 0011		1001 1011	
3rd H1 pointer byte	1001 0011		1001 1011	

Further, the SONET frame may contain place holding cells, namely, an idle cell (with header = 00 00 00 01 52) and an unassigned cell (with header = 00 00 00 00 55).

SDH is specified in G.707, G.708, and G.709 Recommendation of ITU-T and SONET is specified in ANSI T1-105 and Bellcore GR-253.

The similarity between SDH/SONET permits vendors to develop both the North American and European variants of equipment such as cross-connects and multiplexers.

Example 4.16

In reference to the multiplexers indicated in Fig. 4.67, identify the DS/STS types at the output of each multiplexer and construct a table comparing the each output in respect of (a) frame time, (b) frame per second, (c) bps, and (d) percentage overhead.

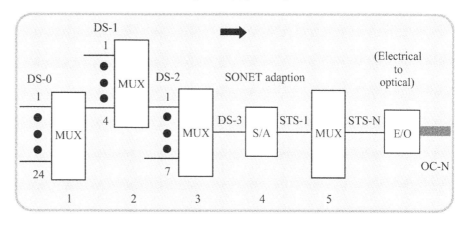

Fig. 4.67: DS/STS multiplexing

Solution

MUX 1: DS-1

Frame-time	= 125 µs
Frame/s	= 8000 frame/s
Bit rate	= 1.544 Mbps
Framing bit	= 1 bit = 8000 bps (overhead)

MUX 2: DS-2

Frame-time	= 125 µs
Frame/s	= 8000 frame/s
Bit rate	= 6.312 Mbps

Framing + stuff bit = $6.312 - 4 \times (1.544) = 136$ kbps

MUX 3: DS-3

Frame-time	= 125 µs
Frame/s	= 8000 frame/s
Bit rate	= 44.736 Mbps

Framing + stuff bit = $44.736 - 7 \times (6.312) = 552$ kbps

MUX 4: STS-1

Frame-time	= 125 µs
Frame/s	= 8000 frame/s
Bit rate	= 51.84 Mbps
Overhead	= 27(TOH) + 9(POH) octets = 2.304 kbps

MUX 5: STS	STS-3	STS-3c
Frame-time	= 125 µs	125 µs
Frame/s	= 8000 frame/s	8000 frame/s
Bit rate	= 155.52 Mbps	155.52 Mbps
Overhead	= 81(TOH) + 3 × 9(POH)	81(TOH) + 1 × 9 (POH)

Table 4.17:<u>SONET signal hierarchy</u>

OC level	STS level	Line rate (Mbps)
OC-1*	STS-1	51.84
OC-3*	STS-3	155.52
OC-9	STS-9	466.56
OC-12*	STS-12	662.08
OC-18	STS-18	933.12
OC-24	STS-24	1244.16
OC-36	STS-36	1866.23
OC-48*	STS-48	2488.32
OC-96	STS-96	4876.64
OC-192	STS-192	9953.28

* The more popular implementations of the present technology

General formula for STS-n
The STS-n frame is illustrated in Fig. 4.66. Its percentage overhead can be determined as follows:
Bit rate = $51.84 \times n$ Mbps
% overhead = 4n/90n = 4/90 = 0.444 = 4.44 %

Similarly, the general formula for the percentage overhead can be obtained as follows:
Bit rate = $51.84 \times n$ Mbps
$$\% \text{ overhead} = \left(\frac{3n+1}{90n} \right) \times 100\%$$

Table 4.18: <u>Summary of results</u>

	Frame Time (μs)	Frame/s	Bit rate	Percentage overhead (Excluding overhead in lower level DS frame in payload)
DS-1	125	8000	DS-1 = 24 DS-0 DS-1 frame = (24×8) + 1 overhead (OH) bit = 193 bits Data rate = 193 bits/frame $\times$ 8000 frame/s = 1.544 Mbps	Overhead size = 1 bit Overall frame size = 193 bits Overhead ÷ 0.518%
DS-2	125	8000	DS-2 = 4 DS-1s DS-2 frame = (7×193) + 17 OH bits = 789 bits Data rate = 789 bits/frame $\times$ 8000 frame/s = 6.312 Mbps	Overhead size = 17 bit Overall frame size = 789 bits Overhead ÷ 2.155%
DS-3	125	8000	DS-3 = 7 DS-2s DS-2 frame = (7×789) + 188 OH bits = 5592 bits Data rate = 5592 bits/frame $\times$ 8000 frame/s = 44.736 Mbps	Overhead size = 188 bit Overall frame size = 5592 bits Overhead ÷ 3.36%
STS-1	125	8000	See Note A as described below	
STS-n	125	8000	See Note B as described below	
STS-nc	125	8000	See Note C as described below	

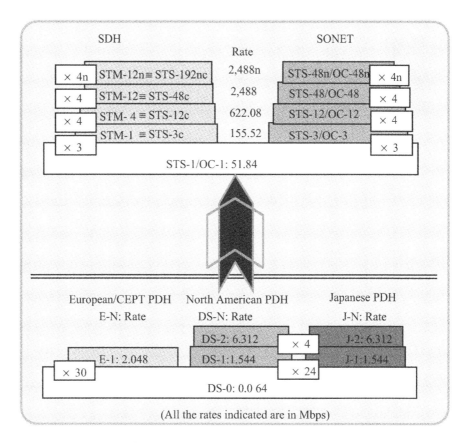

Fig. 4.68: SONET support for current technology

In the early 1980s, the T-1 committee developed the SONET standard by proposing 50.688 Mbps as the initial standard. Between 1984 and 1986, they settled on the STS-1 rate as a base standard; however, ITU-T disagreed for STS-1 but preferred STM-1 as a ITU base standard. Then, the T-1 committee tried to adapt STS-1 for converging with Europe standard (STM-1). A concatenated rate by 3, 51.84 Mbps is accepted to be STS-1 for converging to 155.52 STM-1 (multiplexing STS-1 for 3 times).

In reference to Table 4.18, the notes indicated (Note A, Note B, and Note C) are as follows:
Note: A
(a) DS-3 (44.736 Mbps) maps to STS-1 (51.840 Mbps) as illustrated in Fig. 4.67.
Assume that the premise in a channel end has a data rate equal to 44.736 Mbps (DS-3 level premise). Then the corresponding frame format is as follows:

> 699 DS-3 bytes
> 57 synchronizing DS-3 bytes
> 18 OH multiplex VTs bytes
> 9 POH bytes
> 27 TOH bytes
> Total: 810 bytes in the STS-1 frame of duration 125 μs

Therefore, the percentage overhead = $(810 - 699) \times 100/810 = 13.70\%$.

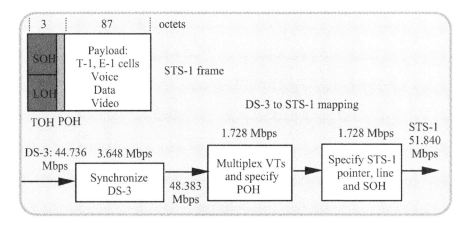

Fig. 4.69: Frame format for STS-1 and DS-3 mapping into STS-1

(b) 28 DS-1 are mapped into STS-1 as shown in Fig. 4.70.

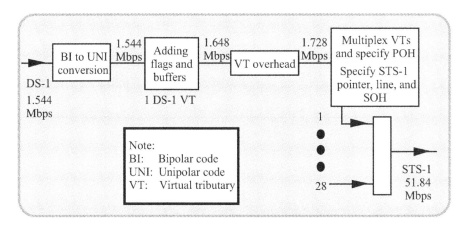

Fig. 4.70: T-1 to STS-1 mapping

Assume that the premise in each channel has a data rate equal to 1.544 Mbps (DS-1 level premise). The resulting frame format is as follows:

$$
\begin{aligned}
(24.125 \times 28) &= 675.5 \text{ DS-1 bytes} \\
(1.625 \times 28) &= 45.5 \text{ flag and buffers bytes in DS-1} \\
(1.25 \times 28) &= 35 \text{ VT overhead} \\
& \quad 18 \text{ OH multiplex VTs bytes} \\
& \quad 9 \text{ POH bytes} \\
& \quad 27 \text{ TOH bytes} \\
\text{Total:} & \quad 810 \text{ bytes in the STS-1 frame of duration } 125 \text{ μs}
\end{aligned}
$$

Hence, the percentage overhead = $(810 - 675.5) \times 100/810 = 16.60\%$.

(c) 21 CEPT1s (2.048 Mbps in Europe standard) are mapped into STS-1 as depicted in Fig. 4.71.

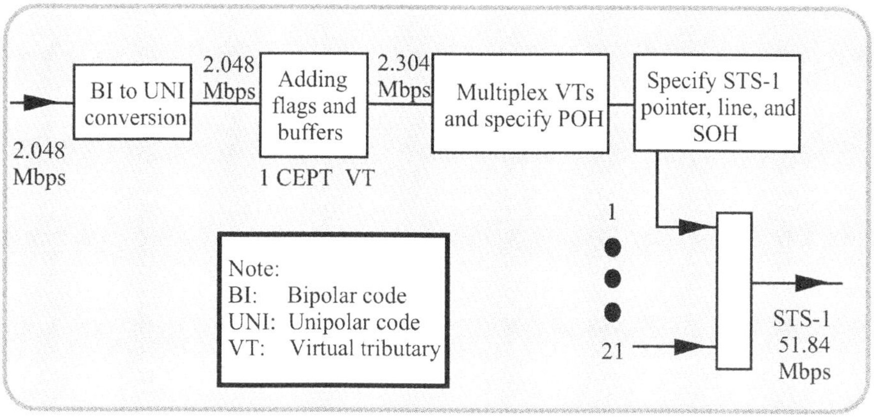

Fig. 4.71: E-1 to STS-1 mapping

Assume that the premise in each channel end has a data rate equal to 2.048 Mbps (CEPT1 level premise). The resulting frame format is:

(32×21)	= 672 CEPT1 bytes
(8×21)	= 84 flag, buffers, and VT overhead bytes in CEPT1
	18 OH multiplex VTs bytes
	9 POH bytes
	27 TOH bytes
Total:	810 bytes in the STS-1 frame of duration 125 µs

Hence, the percentage overhead = $(810 - 672) \times 100/810 = 17.04\%$.

Note: B

In reference to STS-n or STM-n, the bit rate is the n multiplier of STS-1 for STS-n or STM-1 (155.52 Mbps) for STM-n such as, STS-3 = 51.84×3 = 155.520 Mbps.

The percentage of overhead is the same as STS-1 since in STS-n both overhead and data load section is multiplied by n. (13.70%, 16.04%, and 17.04% for DS-3, DS-1, and CEPT1 respectively).

Note: C

Considering STS-nc, the bit is the same as in STS-n = STS-1 $\times$ n. But payload section has more space than STS-n because only 1 POH is used for n > 1. This means that the data load space increases by $(n - 1) \times 9$ bytes.

For n DS-3 mapped to STS-nc
$$\text{Overhead size} = (57 + 18 + 27)n + 9 = (102n + 9) \text{ bytes}$$
$$\% \text{ overhead} = \frac{100 \times (102n + 9)}{810n} \%$$

For 28n DS-1 mapped to STS-nc
$$\text{Overhead size} = (45.5 + 35 + 18 + 27)n + 9 = (125.5n + 9) \text{ bytes}$$
$$\% \text{ overhead} = \frac{100 \times (125.5n + 9)}{810n} \%$$

For 21n CEPT1 mapped to STS-nc

Overhead size $= (84 + 18 + 27)n + 9 = (129n + 9)$ bytes

$$\% \text{ overhead} = \frac{100 \times (129n + 9)}{810n} \%$$

Table 4.9: <u>Overhead percentage of STS category</u>

#C	Percentage of overhead					
	DS-3_STS-n	DS-3_STS-nc	DS-1_STS-n	DS-1_STS-nc	CEPT1_STS-n	CEPT1_STS-nc
1	13.7	13.704	16.6	16.604	17.04	17.037
3	13.7	12.963	16.6	15.864	17.04	16.296
9	13.7	12.716	16.6	15.617	17.04	16.049
12	13.7	12.685	16.6	15.586	17.04	16.019
18	13.7	12.654	16.6	15.556	17.04	15.988
24	13.7	12.639	16.6	15.540	17.04	15.972
36	13.7	12.623	16.6	15.525	17.04	15.957
48	13.7	12.616	16.6	15.516	17.04	15.949
96	13.7	12.604	16.6	15.505	17.04	15.938
192	13.7	12059837963	16.6	15.499	17.04	19.932

Note: #C: Number of channels.

Example 4.17

Determine how many ATM cells can be accommodated when packed head-to-tail in the payload sections of the following transport frames: STS-1, STS-3, STS-3c, DS-3/IEEE 802.6 PLCP, ESF and superframe.

Solution

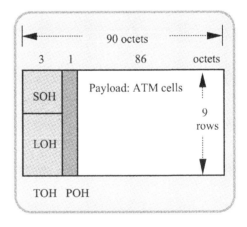

Fig. 4.72: STS-1 frame

1 ATM cell = 53 octets

Hence, STS-1 frame can carry = 774/53 = 14.6 cells

$\Rightarrow$ 14 cells/frame + 32 free-space octets/frame

STS-3 = STS-1 × 3 $\Rightarrow$ (section overhead + line overhead) × 3 + (path overhead × 3 + payload × 3)

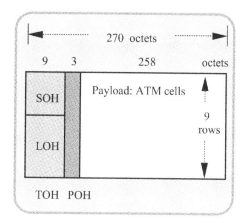

Fig. 4.73: STS-3 frame

STS-3 payload space = $(261 - 31) \times 9 = 258 \times 9 = 2322$ octets
Hence, STS-3 frame can carry = 2322/53 = 43.81 cells
$\Rightarrow$ 43 cells/frame + 43 free-space octets/frame

STS-3c = STS-1 × 3 $\Rightarrow$ (section overhead + line overhead) × 3 + (path overhead × 1)
+ payload (3 STS-1 payload + 2 free space POH)

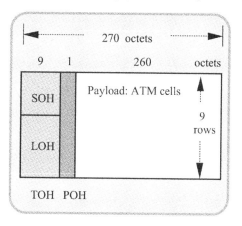

Fig. 4.74: STS-3c frame

STS-3c payload space = $(261 - 1) \times 9 = 260 \times 9 = 2340$ octets
Hence, STS-3c frame can carry = 2340/53 = 44.15 cells
$\Rightarrow$ 44 cells/frame + 8 free-space octets/frame

DS-3/IEEE 802.6 PLCP

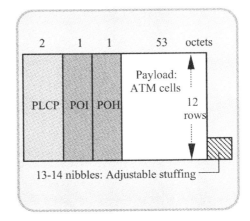

Fig. 4.75: DS-3/IEEE 802.6 PLCP frame

DS-3 PLCP payload space = 53 × 12 = 636 octets
Hence, DS-3/IEEE 802.6 PLCP frame can carry = 636/53 = 12 cells

ESF superframe

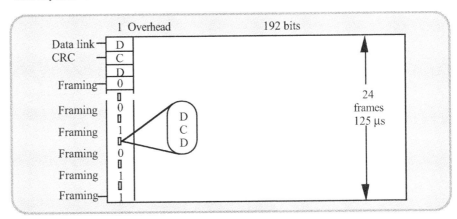

Fig. 4.76: ESF frame

ESF payload space = 192 × 9 = 4608 bits = 576 octets
Hence, ESF frame can carry = 576/53 = 10.87 cells
$$\Rightarrow 10 \text{ cells/frame} + 46 \text{ free-space octets/frame}$$

4.4.5   Convergence aspects of ATM, video, and optical protocols (SONET/SDH)

Convergence is a trend towards the delivery of audio, data, images, and video through diverse transmission and switching systems, which supply high-speed transportation over any medium to any location. With its modular, service-independent architecture, SONET provides vast capabilities in terms of service flexibility. Because of the its bandwidth capacity, SONET is a logical carrier for ATM.

Existing experimental switching systems use CMOS technology as well as emitter-coupled logic techniques to realize systems that are capable of operating up to the Gbps range. As a first step, optical switching matrices using space or *wavelength division multiplexing* (WDM) systems are being considered for use in optical cross-connects.

WDM allows carriers to divide and condense standards fiber optic transmissions into separate wavelengths. Each wavelength carries different content of teleservices. This permits carriers to expand their nearly saturated fiber interstructures, and helps them to get the most for their deployment dollars. Signal propagation over more than 50 km, without the use of electrical regeneration, can be made possible through optical amplification.

Dense wavelength division multiplexing (DWDM) segmentizes a standard OC-48 into as many as 8, 16, or 32 channels at as much as 2.5 Gbps per channel. This technology allows transmission of 48 non-compressed video channels and 298 CD-quality audio channels up to a distance of 40 km. It is ideal for high volume point-to-point or backbone links with minimal switching and routing requirements.

As discussed earlier, ATM uses time division multiplexing and information is transported in cells. This requires the cell header to be processed in each switching element so that it can be forwarded to its proper destination. In optical networking, the signal at switching points must be converted from optical-to-electrical form so that the header can be analyzed. The signal then must be converted to the optical domain for correct routing. The optical-to-electrical conversion process reduces the efficiency of the link as a result of added noise and the bandwidth restrictions imposed by the electronics. Emerging silicon waveguide technology provides an all-optical switching alternative that is potentially reliable and economical enough to facilitate deployment in local and metropolitan area networks DWDM.

The optical network configuration can be a single-hop and multihop network. Single-hop network assumes that the information is transmitted from the sender to the receiver as light, without being converted back to electrical form. The implementation for this kind of network, however, requires significant effort.

Multihop can be used to reduce the implementation complexity. In these networks, the signal from the source first passes through an intermediate node, where the optical signal is converted into electrical and analyzes it. This node decides whether the signal is addressed to it or if it has to be transmitted to the next node. Also, the intermediate nodes introduce additional delays. In this kind of network, the electrical information may be from SDH-frames or ATM-cells or any other structure. The advantages of these optical cross-connects and ATM networks are:

- ✓ Almost unlimited bit rates
- ✓ Universal transport network carrying signals with different formats and bit rates
- ✓ Future-proof system because new applications can be integrated very simply by using an unoccupied wavelength
- ✓ Cost-effectiveness by avoiding unnecessary optical-electrical conversion and by reducing the efforts for the electronic processing.

4.4.6   SONET/SDH: A summary

The SONET and SDH represent a part of modern telecommunications depicting the deployment of a high-bandwidth set of technologies. These technologies in the present time, correspond to the national and international standards of transmission systems. SONET/SDH standards depict interfaces, signal rates, and formats, which have become the key to the evolution of an interoperable, efficient, and economical fiber transport infrastructure for public networks. Even though there are other ways to deploy fiber for similar gains in bandwidth, a network that follows the SONET standards can interconnect equipment from a variety of vendors without converting back and forth between optical and electrical signals as may be required without standardization.

SONET is structured to encapsulate the signals generated under existing standards for digital hierarchies in networks based on electrical interfaces. This implicitly means the following: The end-users, as well as service providers, who have equipment that handle signals according to certain hierarchies can continue to use the same equipment while they take advantage of SONET features, such as virtually error-free transport.

What remains transparent to them is how SONET ramps the bandwidth up from, say, the 1.5 Mbps of a DS-1 signal, the basic building block of existing digital networks, to several megabits and even gigabits per second. SONET is conceived to start at about 52 Mbps, which encapsulates the DS-3 rate of 45 Mbps, and extends above 2 Gbps.

Originally, SONET facilities were deployed within public networks, with no direct customer access. Subsequently, rings that connect customers have been used to provide improved reliability for DS-1 and DS-3 rate data services. This has been improvised via new serving arrangements that include pairs of access paths and service options for more than one central office. Further, the associated redundancies assure that any particular customer's service may survive if there is a breakdown in one piece of the network.

The services, which provide "fast packet" switching can ride on SONET. Such facilitation meets the customers needs for LAN interconnection and other applications. These services include SMDS, and frame-relay. Most service providers – local-exchange, inter-exchange, and value-added-network offer both services, because of customers' demands. As such, riding these services on SONET has become a popular consideration in modern telecommunications. Equipment vendors, such as those that manufacture routing equipment, also support both services.

Considering packet-switching, it is in general, suited for "bursty" traffic, with short spells of a lot of information and longer gaps. Bursts of such information can be loaded into trains of smaller packages for transport on a packet-network. As indicated in Chapter 3, both frame-relay and SMDS transmit data over public networks more quickly and economically than earlier services, but each is especially well suited to certain complementary kinds of data needs.

ATM plus SONET jointly represent a technology that would contribute significantly to the realization of broadband networking. ATM is a flexible and convenient technique for switching and multiplexing "cells" (of 53 bytes each). ATM technology supports the transport of traffic with a fairly continuous bit rate, such as video, and variable-bit-rate traffic such as electronic messaging or file transfer. It also supports both connection-oriented and connectionless services.

This flexibility comes from the rigid size of the cells that are switched on a hardware basis. Each cell includes a header that identifies which set of cells – which streams of information – it belongs to and, as in a connectionless packet system, where it is going. A stream of cells with the same header constitutes a virtual circuit. The cells from different virtual circuits are interleaved, or multiplexed, by the ATM switching system so that the overall speed and burstiness of each reflect the speed and burstiness of each original data stream that came into the network through a customer-access interface. A continuous-bit-rate video circuit will get its cell into the overall bit stream in a regular pattern, separated by regular numbers of other cells. A bursty circuit for file transfer might be allocated a number of cells in a row, but there would be longer stretches in which none of the file-transfer cells appeared.

Broadband applications generally include a gamut of services: Interactive video, video telephony, advanced television, multimedia conversational messaging and information-retrieval services, high-speed image transfer for medical imaging, and burstiness graphics. Further, the broadband networking may include customer networking, host-to-host, or interconnecting of local-area networks.

4.5   Concluding Remarks

Envisioning the future, there is an avenue to incorporate many emerging high-speed, public-networking standards, including SONET and ATM switching. The applications for networks operating at such speeds are primarily computer to computer or visual, or they use combinations of media. This is because customers do not absorb other forms of data at those rates – the rough equivalent of thirty thousand pages of single-spaced text each second. But the applications are real, and customers will be looking for them soon enough.

The progress in telecommunication relies on how public carriers build the future highways for information networking. It should, however, be noted that they cannot build them alone. They need support from and communication with all of the people and corporations that have interests in the smooth operation of those highways and the ramps that lead onto them. Such a cooperative venture may not be without difficulties, but collective solutions can serve all of those who stand to benefit from the final products. Specifically, equipment manufacturers, applications, and service providers, many kinds of carriers, and the customers who are already waiting to use capable, reliable, smoothly inteoperable, high-speed public data networks will be the eventual beneficiaries of the conceived telecommunication via SONET/SDH plus ATM.

Bibliography

[4.1]   D.E. McDysan and D.L. Spohn: *ATM Theory and Applications* (McGraw-Hill, Inc., New York, NY: 1998)

[4.2]   D. Ginsberg: *ATM – Solutions for Enterprise Networking* (McGraw-Hill, Inc., Boston, MA: 1995)

[4.3]   R.O. Onvural: *Asynchronous Transfer Mode: Performance Issues* (Artech House, Inc., Boston, MA: 1995)

[4.4]   W.J. Golarski: *Introduction to ATM Networking* (McGraw-Hill, Inc., New York, NY: 1996)

[4.5]   U.D. Black and S. Waters: *SONET & T1: Architectures for Digital Transport Networks* (Prentice Hall PTR, Upper Saddle River, NJ: 1997)

[4.6]   P.K. Cheo: *Fiber Optic – Devices and Systems* (Prentice-Hall, Inc., Englewood Cliffs, NJ: 1985)

[4.7]   J. Powers: *An Introduction to Fiber Optic Systems* (R.D. Irwin, Inc., Co., Chicago, IL: 1997)

[4.8]   D.C. Agarwal: *Fiber Optic Communication* (Wheeler Publishing Co., New Delhi, India: 1996)

[4.9]   J. Franz and V.K. Jain: *Optical Communication Systems* (Narosa Publishing House, New Delhi, India: 1996)

[4.10]   G. Keiser: *Optical Fiber Communications* (McGraw-Hill, Inc., New York, NY: 1999)

Appendix 4.1

Table 4.A1: <u>Bit configurations and corresponding hex values adopted to correct single-bit errors in header bit position (from 1 to 40)</u>

Position	Bit configurations		Hex values
1	0011	0001	31
2	1001	1011	9B
3	1100	1110	CE
4	0110	0111	67
5	1011	0000	B0
6	0101	1000	58
7	0010	1100	2C
8	0001	0110	16
9	0000	1011	0B
10	1000	0110	86
11	0100	0011	43
12	1010	0010	C2
13	0101	0001	51
14	1010	1011	CD
15	1101	0110	D6
16	0110	1011	6B
17	1011	0110	B6
18	0101	1011	5B
19	1010	1110	AE
20	0101	0111	57
21	1010	1000	A8
22	0101	0111	54
23	0010	1010	2A
24	0001	0101	15
25	1000	1001	89
26	1100	0111	C3
27	1110	0000	E0
28	0111	0000	70
29	0011	1000	38
30	0001	1100	1C
31	0000	1110	0E
32	0000	0111	07
33	1000	0000	80
34	0100	0000	40
35	0010	0000	20
36	0001	0000	10
37	0000	1000	08
38	0000	0100	04
39	0000	0010	02
40	0000	0001	01

Note: Bit positions 33 through 40 in the HEC field itself are protected from single-bit errors.

5

ATM SIGNALING AND TRAFFIC CONTROL

How to get connected

"Let's make a call ... to reach out and touch everyone"

C. Gadecki and C. Heckart
ATM for Dummies, 1997

5.1   Introduction

5.1.1   *What is signaling?*

Signaling is a means by which communication is established, released, monitored, controlled, and maintained between end-entities via the intervening networks in a telecommunication system. It is one of the two major functions of a telecommunication network. The primary function is the transportation of a message transmission across the network depicting an information transfer from a sending end to one or more destinations. An adjunct effort is essentially what is known as the *signaling function*. That is, in order to facilitate an information transfer, a communication path should be established between the sending and receiving ends; further, this path should be monitored while it is in use and it should be disconnected when the transfer of information is completed.

This signaling consideration essentially depicts a set of control functions associated with the telecommunication networks and it is broadly known as the *network signaling*. The signaling or control tasks can be itemized as follows:

- *Connection establishment*: This refers to establishing a communication path from a source (from which information originates) to a sink where the information is destined

- *Supervision*: This effort may depict one of the following:

   ✓   To monitor the status of a line or a circuit so as to determine whether it is busy, idle, or requesting a connection be established
   ✓   To supervise and maintain a reliable progress of a communication during the entire period of a communication session.

- *Alerting*: This is an indication function deliberated at the destination end to inform the user of the arrival of an incoming message

- *Addressing*: This is the process of routing the information across the network from the point of origination to the point of destination

- *Connection release*: This is the last phase of the control function done in the event of the sender or the receiver terminating the communication session. This function leads to the release of the connection established.

5.1.2   *Signaling in POTS*

The essence of signaling in a telecommunication network can be understood by considering the control functions in the plain old telephone system (POTS). The sequence of communication tasks involved in telephony can be enumerated as follows [5.1-5.4]:

Local call signal

A call requires a communication circuit (connection) between two subscribers. In reference to local calls, the sequence of signaling functions can be listed as follows:

- *Call establishment phase*

  The calling subscriber who wishes to make a call, lifts the telephone set off the hook. This state (of telephone set being off the hook) is *supervised* in the local-loop circuit. If it is a manual telephone exchange system, this supervision/monitoring is done at the switch-board by the human operator, who would then communicate with the calling subscriber and obtain the called subscriber's number. At the next phase, the operator would monitor the *idle* state of the called subscriber's line. If it is free, the operator would send a ringing tone to alert the called subscriber of an incoming call. If the called subscriber responds (by lifting the telephone set off the hook), the corresponding state is monitored by the supervising operator, who would then patch up the calling and called subscriber, thus establishing a connection.

  In an automated system, after lifting the telephone off the hook, the calling subscriber hears a dial-tone, which indicates the readiness of the exchange to receive the dial pulses. Then the calling subscriber dials the called subscriber number. The automatic switch (either rotary, crossbar, or electronic) hunts for the called subscriber line. In the event of this line being free, the ringing tone is applied to the called subscriber line. When this alert is acknowledged (that is, if the called subscriber lifts the telephone set off the hook), the connection is established via the appropriate line. Suppose the called subscriber line is engaged, a busy tone is returned to the sender, who may hang up and retry later.

- *Progression (session) supervision phase*

  During the session of the call, the network performs certain control functions such as maintenance of the established route, fault indication, recording of session duration for tariff/billing purposes etc.

- *Call release phase*

  This phase in telephony refers to one of the party (called or calling subscriber) placing the telephone set back on the hook. It is an indication to the network that the established connection be released.

Illustrated in Fig. 5.1 is a flow chart of basic steps involved in a local telephone communication depicting the associated information transfer and control functions involved.

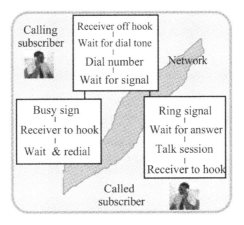

Fig. 5.1: A flow chart of basic steps in a local telephone communication

A connection set up and release diagram can be drawn to illustrate the steps of Fig. 5.1. It is presented in Fig. 5.2.

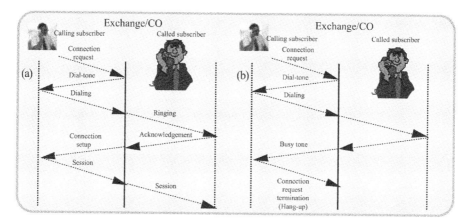

Fig. 5.2: Connection setup/release diagram for a local telephone connection (a) when called subscriber line is free and (b) when called subscriber line is busy

Trunk call signaling

The telephone network consisting of exchanges, trunks, and subscriber lines is essentially circuit-switched and the associated signaling functions are carried out with signaling messages moving through the network.

There are two types of calls: Intra-exchange call refers to a call involving subscribers attached to the same exchange. Inter-exchange calls depict those between subscribers attached to other local exchanges.

The switched connection over the telecommunication network largely involves analog signaling. Even if the bulk of the facility is digital, the subscriber loop usually remains analog. An overview of the signaling system across a long-distance trunk depicted in Fig. 5.3 is as follows:

    ✓   In the idle state, the subscriber loops have a battery on the ring side of the line and an open circuit on the trip. No loop current flows in this state

    ✓   Signaling equipment is placed between the local COs and the toll office providing a single frequency (SF) of 2600 Hz signaling tone indicating the idle circuit status. The associated switch continuously monitors (scans) the subscriber line and trunk lines to find any change in their busy or idle state

    ✓   Suppose subscriber A lifts the telephone off hook. Then the current that flows in the local loop signals the local CO of an incoming call from subscriber A

    ✓   The CO on A's side responds by making the calling line busy and returns a dial tone to the calling subscriber (A). Dial tone is one of the call progress signals between the telephone network equipment and the calling party. It is an indication that the CO is ready to receive the addressing signals

    ✓   Next, subscriber A sends digits in the form of dial pulses or DTMF signals to the CO

    ✓   The equipment at this CO translates the digits into the address of the destination, namely subscriber C

    ✓   Address translation also accompanies a look up of routing table procedures to determine the path of connectivity to subscriber C

    ✓   Hence, the equipment at the CO (on A's side) decides the routing of the incoming call (from subscriber A) to subscriber C via toll office B

    ✓   The CO at A checks for the idle/busy status of the trunks to B and seizes the idle trunk. If no idle trunk is available, the CO at A reorders or sends a busy call progress tone to subscriber A

- ✓ If an idle trunk is seized, automatic message accounting (AMA) equipment at the calling end CO does an initial entry identifying the calling and the called subscriber addresses. Further, the trunk seizure would remove the single frequency (SF) 2600 Hz tone to indicate its seized state. The equipment at toll office B responds to the CO at A with a momentary interruption of the signaling tone (called wink) indicating its readiness to receive the digits

- ✓ When the wink is detected, the CO at A sends addressing pulse either by interrupted SF tones or by multifrequency (MF) dial pulses

- ✓ The toll office B would continue to send on-hook tone towards subscriber A until subscriber C answers. At this point, the CO at A has completed all its call origination functions and awaits the completion of the call

- ✓ Toll office B translates the received dial pulses to determine the address of the destination subscriber, selects an idle trunk leading to the CO at C, and seizes it. The CO at C detecting the seizure sends a start signal to B and prepares to receive the digits. Subsequently, toll office B sends the dial digits forward

- ✓ The CO at C checks the called subscriber (A) line for its busy/idle status. If busy, it returns a busy tone over the voice channel. Then the calling subscriber recognizes the busy tone and hangs up and the connection is "torn down" or released. AMA makes no tariff entries

- ✓ If the called subscriber line (C) is idle, the CO at C sends a 20 Hz alerting signal (ringing tone) to the called party. (This audible ring tone is another call progress tone.) This tone is also returned to the calling party. The ringing continues until subscriber C answers or subscriber A hangs up or the equipment times out

- ✓ If the called subscriber C answers (by lifting the telephone off hook), the line current flows in the local loop. This current trips the ringing tone. Correspondingly, the CO at C would change its status to an off-hook condition by interrupting the SF tone and informing B and subscriber A thereof. The conversation session sets in and AMA starts recording the tariff

- ✓ When either of the party hangs up, the change in the line current indicates a status change to the corresponding CO. The CO forwards the status change and restores SF tone conditions releasing all the equipment and AMA from the seized condition. Now, the system is ready to handle signaling for another incoming call.

A connection setup/release diagram of a trunk call environment is presented in Fig. 5.3

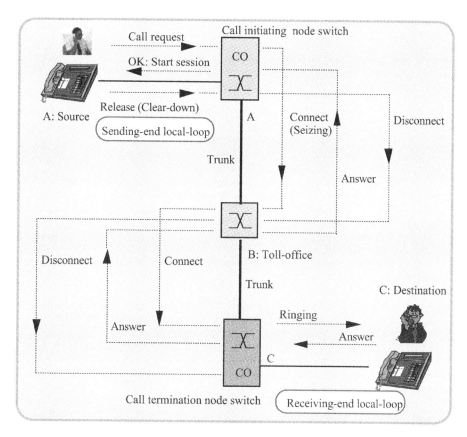

Fig. 5.3: Connection setup/release diagram: Trunk call between subscriber A and subscriber C via toll office B

Table 5.1: Operations at an exchange in respect of signaling functions

No.	Operations	Description	Associated software routines
1	Scanning inputs at the periphery of the system for supervisory transmission and digits	• Inputs can be lines, trunks or both • Signal transmission: • On the lines: On-hook and off-hook • On the trunks: Change of state denoting the arrival of a seizure (connect) signal sent to next station	• Apply dial-tone and wait for digits • Read digits and collect them from the periphery • Delete dial-tone on detection of the first digit
2	Applying output signals to the periphery	• These signals include the following: • Dial-tone (call origination detected) ringing, interoffice trunk signaling (the connect or seizure signal), and call-progress tones to the originator (such as audible ringing and busy signals)	• Set up talking path through switch and create terminal process to control terminating phone or trunk • Determine whether busy or on-hook at termination
3	Routing calls through the switch	• Allocating and establishing "talking paths"	• Apply ringing tone if on-hook • Detect off-hook at termination side; remove ringing tone at termination and at origin
4	Allocating the paths and tearing down the paths	• Call clearance phase	

Typically a local/tandem exchange (one that provides both local and transit functions) in a network should be capable of generating and interpreting the sequence of the signaling functions indicated above. There are four basic operations at an exchange associated with the signaling functions as shown in Table 5.1.

In-band signaling and common channel signaling

The traditional and subsequent versions of the signaling mode, namely, the methods of sending various signaling messages, are as follows:

✓   Classical/traditional version: In-band signaling or channel associated signaling
✓   Subsequent version: Common channel signaling.

In-band signaling

This refers to sending signaling messages over the same subscriber lines and/or trunks intended for the transmission of information-bearing signals.

Common-channel signaling (CCS)

In this method, the messages intended to perform the control functions are carried over separate signaling channels

The common channel signaling was evolved (by AT&T) due to the following drawbacks of in-band signaling:

✓   In-band signaling involves the occupancy of trunks and the associated blockage of information transfer to a substantial extent
✓   Vulnerability of the circuits to fraud.

CCS is a reliable and rapid method of establishing connections between the equipment. CCS in North America is based on AT&T's *common channel interoffice signaling* (CCIS). The international standard refers to CCITT *signaling system # 7* (SS7).

The concept and scope of CCS was further expanded (with necessary revisions in 1980 through 1988), and has become known as *SS7*. It is an open-ended CCS standard that can be used over a variety of digital circuit switched-networks. In SS7, while the network being controlled is circuit-switched, the control signaling itself uses packet-switching. Therefore, the functions in SS7 are defined assuming a packet-switched operation, but the actual implementation can be in circuit-switched nodes as additional functions.

With the advent of ISDN, specific considerations have been looked into while designing the SS7. That is, the internal control and network intelligence, which are essential in ISDN, are provided by SS7. The SS7 has been optimized to work with digital exchanges utilizing 64 kbps channels as described below.

5.1.3   Signaling in ISDN

An ISDN serves conventional (analog) subscribers and ISDN users. ISDN subscribers can communicate with each other in two modes: In a circuit-switched mode, the network sets up a dedicated connection for the call. The other mode refers to packet-switched mode.

A digital subscriber line (DSL) connects the user's terminal equipment to the local exchange. The constituents of an ISDN network are illustrated in Fig. 5.4.

ISDN network...

❑   *Circuit-switched network: This refers to 64 kbps digital trunks, and exchanges with digital switches. (On some transition networks, a mixture of analog (FDM) and digital trunks with exchanges supporting analog and digital switches may coexist).*
❑   *Packet-switched network: This is a data communication network adopted with an interface to ISDN local exchanges, or a new network*
❑   *SS7: Signaling system # 7.*

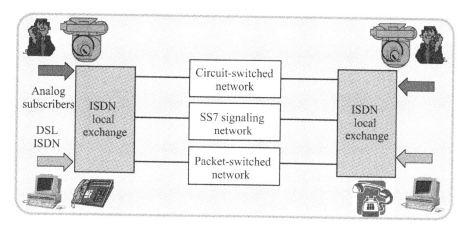

Fig. 5.4: ISDN network

The ISDN has two objectives: It is an integrated services network that provides circuit-switched (speech and data) and packet-switched (data) communications for its users. The second objective is user-to-user digital connectivity. This means that all components (lines, trunks, and exchanges) in an ISDN connection transfer 64 kbps of digital data.

In order to understand the CCS aspects of ISDN, the network terminology associated with the data communication network and the signaling network considerations can be specified distinctly as follows:

Table 5.2: <u>Data communication and signaling network terminology</u>

Network part/concept	Data communication network terminology	Signaling network terminology
• Node	• Node	• Signal transfer point (STP)
• Link	• Data link	• Signaling data link
• Data unit	• Packet	• Signal unit
• User	• Customer premises equipment (CPE) or data terminal equipment (DTE)	• Signaling end point

Circuit-switched mode ISDN services

There are three groups of ISDN serviced for the circuit-mode communication. They are as follows:

   ✓   Bearer service: This includes 3.1 kHz audio (voiceband modem data), or 64 kbps digital data
   ✓   Teleservice: This refers to 3.1 kHz audio and 64 kbps calls for data services such as facsimile, telex, teletext etc.
   ✓   Supplementary services: These are specific services, which may vary from country to country. Examples: malicious call handling, call line identification etc.

ISDN network layer

The two (expected) categories of ISDN network layer functions are as follows:

   ✓   Controlling of connection: establishment/release
   ✓   Message transport such as rerouting of signaling messages in the event of D-channel failure, multiplexing, message segmenting and blocking.

In reference to the signaling or control functions, the relevant messages for circuit-mode connection control are as follows [5.4-5.8]:

- Call establishment messages
  Alerting
  Call proceeding
  Connect
  Connect acknowledgement
  Progress
  Setup
  Setup acknowledgement

- Call information phase messages
  Resume
  Resume acknowledgement
  Resume reject
  Suspend
  Suspend acknowledgement
  Suspend reject
  User information

- Call clearing messages
  Disconnect
  Release
  Release complete

- Miscellaneous messages
  Congestion control
  Facility
  Information
  Notify
  Status
  Status enquiry.

Digital subscriber signaling system # 1 (DSS1)

This is used for signaling between ISDN subscriber terminal equipment (TE) and the local exchange to which the user DSL is attached. The DSS1 signaling messages are carried in the D-channel of the DSL, which is the common signaling channel for the TEs on a DSL.

DSS1 and SS7 are specified separately by CCITT. But there are similarities between them. The functional aspects of DSS1 and SS7 are depicted in Fig. 5.5.

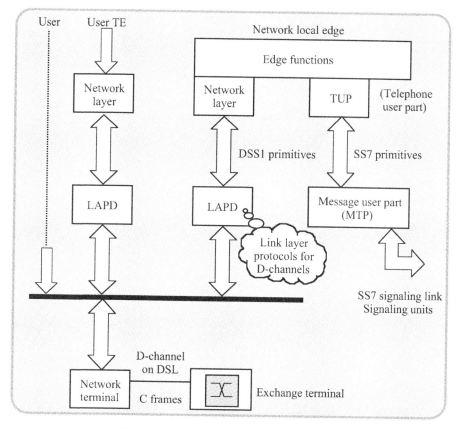

Fig. 5.5: DSS1 and SS7: Functional considerations

In Fig. 5.5, SS7 is organized as a hierarchy of protocols. The *message transfer part* (MTP) serves a number of SS7 user parts, such as the *telephone user part* (TUP) and ISDN user part (ISUP). Likewise, DSS1 is divided into the *link layer protocols for D-channels* (LAPD) and the network layer. The functions of the LAPD are comparable to those of the MTP. The network layer includes protocols comparable to those of ISUP. (These protocols are known as *Q.931 protocols*.) The LAPD and network layer communicate by passing the primitives.

Example 5.1
Describe a typical signaling sequence for an intra exchange call from user X to user Y as per Q.931 messages.

Solution
Illustrated in Fig. 5.6 is the required signaling sequence between TE-X to TE-Y via a local exchange.

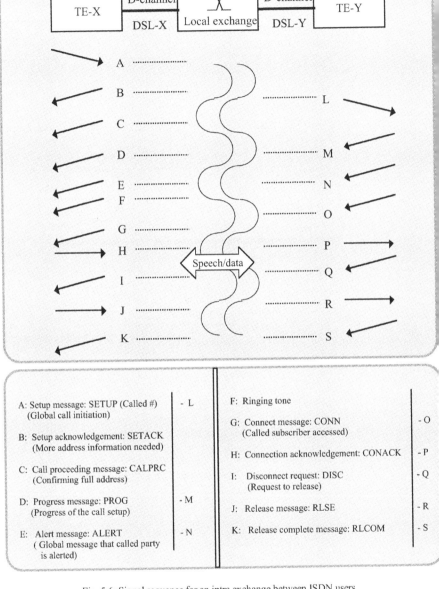

Fig. 5.6: Signal sequence for an intra exchange between ISDN users

A: Setup message: SETUP (Called #) - L
   (Global call initiation)

B: Setup acknowledgement: SETACK
   (More address information needed)

C: Call proceeding message: CALPRC
   (Confirming full address)

D: Progress message: PROG - M
   (Progress of the call setup)

E: Alert message: ALERT - N
   (Global message that called party
   is alerted)

F: Ringing tone

G: Connect message: CONN - O
   (Called subscriber accessed)

H: Connection acknowledgement: CONACK - P

I: Disconnect request: DISC - Q
   (Request to release)

J: Release message: RLSE - R

K: Release complete message: RLCOM - S

Example 5.2
What are call reference values (CRVs)? How do the exchange and the TEs inform each other about the assignment of CRVs and B-channels at the commencement of a call and the release of these items at the termination of the call?

Solution

Call reference value (CRV): An interface that identifies the call to which the message relates. Q.931 messages are call-related instead of trunk-related because the Q.931 protocol covers both circuit-mode and packet-mode calls. Fig. 5.7 illustrates the assignment/release procedures on CRVs and B-channels.

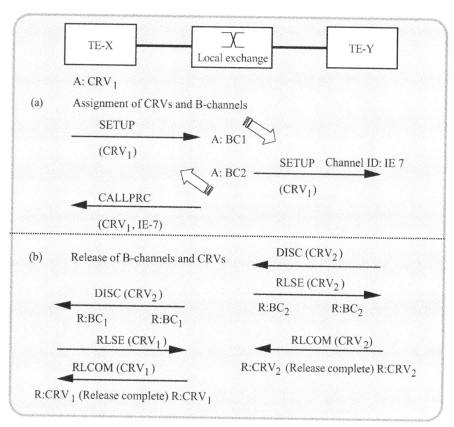

Fig. 5.7: Assignment/Release of CVS and B-channels

Note: *Information elements*: (IEs): These refer to three fields:

(1)   *IE identifier* (1 octet) such as bearer capabilities, called party number, calling party number etc. (IE.1 through IE.14: See Problem 5.)

(2)   IE length (1 octet) indicating the length (number of octets) of the content field

(3)   IE value: A variable length field that holds the actual information content of the IE.

Problem 5.1

Listed below are the IEs with their corresponding reference numbers.

Insert appropriate IEs into the signaling sequence for an intra exchange between the ISDN users described in Example 5.2.

Reference number		Information element (8bits)
IE.1	⇒	Bearer capability
IE.2	⇒	Called party number
IE.3	⇒	Calling party number
IE.4	⇒	Called party subaddress
IE.5	⇒	Calling party subaddress
IE.6	⇒	Cause
IE.7	⇒	Cleared ID
IE.8	⇒	High-layer compatibility
IE.9	⇒	Keypad
IE.10	⇒	Low-layer compatibility
IE.11	⇒	Process indicator
IE.12	⇒	Signal
IE.13	⇒	Transit network selection
IE.14	⇒	User-User information

There are provisions to handle setup failure in ISDN via a set of designated descriptions (F1, F2, ..., F6) and appropriate treatment procedures.

SS7 signaling

   ISDN user part: This refers to messages and procedures for the control of inter exchange calls between two (analog) subscribers, two ISDN subscribers, and between an ISDN user and a subscriber.

   ISUP call-control signaling is primarily link-to-link and it also has the provision for end-to-end signaling.

   The telcos with TUP signaling in their networks are connecting to ISUP signaling.

   CCITT/ITU-T has defined ISUP in Recommendation Q.761-Q.764. The ISUP locals are indicated in Fig. 5.8.

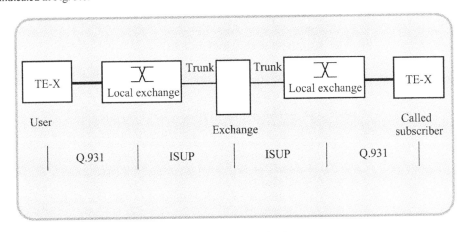

Fig. 5.8: ISUP signaling locales

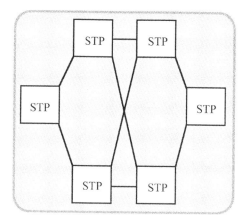

Fig. 5.9: SS7 signaling network topology

SS7: Signaling network

The signaling system # 7 of ISDN has a public network service architecture depicted in Fig. 5.9. It includes a quad structure consisting of signaling end points (SEPs) and signal transfer points (STPs). The STPs route the signaling messages from one SEP to another. Each SEP is connected to the SS7 network by redundant STSs to ensure reliability of operation.
SEPs can be one of the following:

   ✓ *Switching point* (SP): This refers to an end-office switch at which the customer gains access to the network. Using CCS, the switch provides normal call functions, namely call establishment, call maintenance, and call release

   ✓ *Service switching point* (SSP): This is embedded in the SP. It processes calls that require remote database translations. The SSP recognizes special call types, communicates with the *service control point* (SCP) and deals with the calls according to SCP instructions

   ✓ *Service node* (SN): This supports the provision of ISDN bearer services and supplementary services

   ✓ *Operator services system* (OSS): This provides operator assistance including directory assistance.

The service transfer point (STP) is of two types:

   ✓ *Integrated* STP: This is a SP which has the STP capability as well as, if it provides OSI transfer layer functions

   ✓ *Stand-alone* STP: This refers to a SP with only STP capability or STP and signaling connection control part (SCCP) capabilities.

The protocol structure of SS7

The SS7 has a 4 layer protocol architecture as shown in Fig. 5.10. The lower three levels are known as the *message transfer part* (MTP). This facilitates a reliable service for routing messages through the SS7 network. The lowest layer is called *signaling data link* (SDL). This is concerned with the physical and electrical characteristics of the signaling links between *signaling transfer points* (STPs) as well as between STPs and signaling points (SPs).

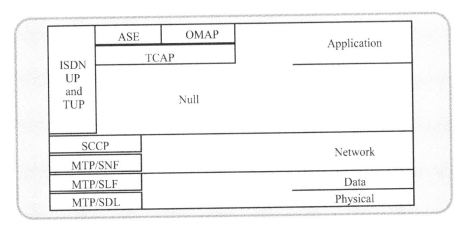

Fig. 5.10: SS7 Protocol architecture

The MTP in general provides reliable transport of signaling information across the signaling network. It is capable of responding to network/system failures. The three levels of MTP and their functions are as follows:

Signaling data unit (SDL)

In this physical layer, the signaling data link has two similar data channels operating in opposite directions. That is, a full-duplex signaling at rates from 4.8 kbps to 56/64 kbps is facilitated. (65 kbps: North American; 64 kbps: In other countries).

Signaling link functions (SLF)

This is a data link layer, which specifies a signaling link for transfer of signaling messages between two SPs in variable length messages. These messages correspond to signal-to-signal units (SUs) constituted by *message signal units* (MSUs), *link status signal units* (LSSUs), and *fill-in signal units* (FISUs). Error detection in these signal unit formats is achieved via 16-bit CRC.

MSU: This SU carries the signaling information field. Its format has necessary flags and error control fields.

LSSU: This is intended for flow control. The congested receiver notifies the sender of its condition with an LSSU denoting busy status. Its format also has necessary flags and error control fields.

FISU: When there is no message traffic FISUs are sent so as to sustain the error monitoring task and keep up to date with link performance.

Signaling network functions (SNF)

This level performs signal unit handling and management including recovery of routing under failures.

Signaling connection control part (SCCP)

This is the upper layer on MTP, which enhances the services of the MTP to provide full OSI network layer capabilities. Specifically, the addressing and routing capabilities are enhanced. It provides connectionless and connection-oriented classes of services.

Transaction capabilities part (TCAP)

This part of SS7 architecture enables a set of tools in a connectionless environment that can be used by an application at one node to invoke execution of a procedure at another node and exchange the results. The sublayers of TCAP are as follows:

Component sublayer: This is concerned with request for action at the remote end
Transaction sublayer: This is concerned with the exchange of messages that contain these components

Operations, maintenance, and administration part (OMAP)

This provides application protocols (*application service elements*: ASE) and procedures required to monitor, coordinate, and control all the network resources.

ISDN user part/telephone user part (UP/TUP)

This denotes the location of model layers invoked when the network employs the transport capabilities of the MTP and SCCP to provide call related services to the user.

SS7: ISDN call procedures

The sequences involved in establishing a circuit-switched ISDN call are illustrated in Fig. 5.11.

Simple circuit-switched calls

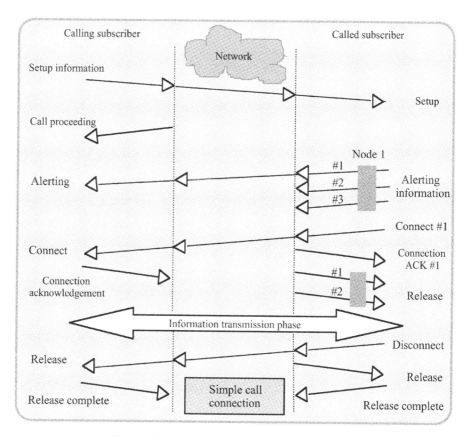

Fig. 5.11: Circuit-switched ISDN call setup/release diagram

The step-by-step procedure of ISDN call setup/release sequences is illustrated in Fig. 5.11 and it refers to the following:

- *Setup:* Customer requests a call. The calling terminal creates a message for the call request

  ⇒ This is a layer 3 process. The constructed message is transmitted via layer 2 to layer 3 of the digital switch

  ⇒ Using the call request message, the digital switch makes the routing to the called user. (Suppose a signaling message across a metropolitan area is to be sent. The setup message maintains a route through the switch. The switch then contacts the destination switch using the protocol of SS7.)

- *Alert:*   Upon receiving the setup message at the called terminal an alert message is returned to the caller as a setup acknowledgement

- *Connect:*   If the called party is available, the call acceptance message (connect) is sent to the calling terminal

- *Call proceeding*:   When the connection is completed across the network, the dialog session (call proceeding) takes place involving transfer of information (voice or data)

- *Clear down*:   This is the end of the communication phase. The disconnect can be initiated by either party at any time.

Call setup and release by ISUP of SS7

The basic call control procedures via ISUP of SS7 are illustrated in Fig. 5.12.

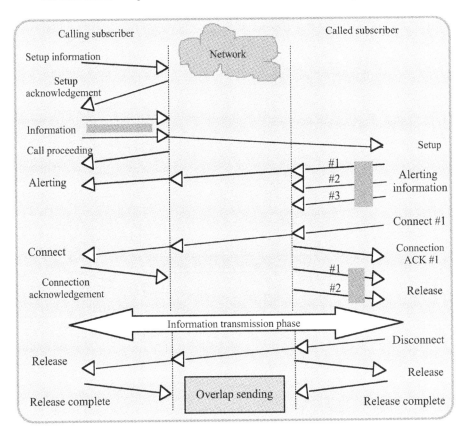

Fig. 5.12: Sequences of call setup/release procedure by ISUP of SS7

5.2   Broadband Signaling

It was indicated in earlier chapters that by adopting ATM cell switching and SONET transmission technologies, the broadband ISDN has been conceived to offer a rich set of service types and a wide variety of features and functions. But there is a key question:

How does a user invoke these services and functions?

A user relies on the capability of signaling so as to communicate with the network in order to invoke various types of services involved in the B-ISDN transmissions. The associated *signaling objectives* are as follows:

- Defining and developing capabilities and protocols which support the management of broadband services including the following:

  - ✓ Connection establishment and release functions
  - ✓ Management of various interface parameters such as the ATM parameters.

- Making relevant control functions consistent with narrowband ISDN for interoperatibility and cost-effective implementation. That is, the B-ISDN signaling objective explicitly specifies making use of the signaling technology associated with N-ISDN.

Thus the scope of conceived signaling in the framework of B-ISDN/ATM includes developing a comprehensive set of capabilities and protocols in a phased manner [5.8]. Further, such efforts should be compatible with narrowband ISDN for interoperability and economic provision of telco services.

Meeting the above objectives, the approach pursued in B-ISDN signaling refers to making use of the prevailing signaling technology associated with N-ISDN. Hence, relevant signaling capabilities of B-ISDN/ATM transmission are:

- 64 kbps ISDN-applications related signaling support. This support follows the ITU-T recommendation Q.931

- ATM-specific signaling capabilities, namely:

  - ✓ Capability to control ATM virtual channel connections and virtual path connections for information transfer. These capabilities refer to:

    - ⇒ Establishing, maintaining, and releasing ATM VCCs, and VPCs (on-demand or on a permanent basis)
    - ⇒ Supporting point-to-point, multipoint, as well as broadcast configurations
    - ⇒ Negotiating/renegotiating traffic characteristics for a given connection.

  - ✓ Capability to support multiparty and multiconnection calls. This would involve the following:

    - ⇒ Supporting symmetric and asymmetric (low bandwidth in one direction and high bandwidth in other direction) connections
    - ⇒ Simultaneous establishment/removal of multiple connections associated with a call
    - ⇒ Adding/removal of a party to/and from a multiparty call.

✓ *Multiconnection call*: This refers to several connections established to build up a "composite" call, comprising of, for example, voice, image, and data

✓ Multiconnection call should bear the possibility of adding new connections to the existing setup

✓ A telecommunication network, in general, would require a means of correlating all the connections pertinent to a call and it must be possible to release the call as a whole

✓ Such correlation efforts must be done at the call origination and destination end-switches so that transit nodes are not burdened with such tasks

✓ *Multi-party call*: This refers to *conferencing* wherein several connections between more than two end-points. The associated signaling should then correspond to:

⇒ Establishment and release of conference calls
⇒ Adding and removing a party or parties participating the conference.

▪ B-ISDN signaling also implies interworking between (B-ISDN) and non-(B-ISDN) services

5.2.1 *Signaling virtual channels(SVCs)*

In reference to signaling at VCC levels, there are three possible signaling VCs. They are:

▪ *Point-to-point signaling virtual channel*: Bidirectional signaling VC

Here, one virtual channel connection in each direction is allocated to each signaling entity and a standardized VCI value used

▪ *General broadcast signaling virtual channel (GBSVC)*: Unidirectional signaling VC

A specific GBSVC is reserved per VP at a user-to-network interface (UNI)

▪ *Selective broadcast signaling virtual channel* (SBSVC): Unidirectional signaling VC

Instead of GBSVC, a virtual channel connection for *selective broadcast signaling* (SBS) is used for call offering, in cases where a specific service profile is adopted. Examples are: SBS for voice calls and SBS for video calls.

▪ *Meta-signaling virtual channel (MSVC)*: Birectional signaling VC

In order to establish, check and release the point-to-point and broadcast signaling virtual channel connections, meta-signaling procedures are facilitated. For each direction, meta-signaling is carried in a permanent virtual channel connection having a standardized VCI value. This channel is called the *meta-signaling virtual channel*. The meta-signaling function is required to perform the following functions:

⇒ Managing the allocation of capacity to signaling channels
⇒ Establishment, release, and checking the status of signaling channels

⇒ Provision of a means to associate a signaling end-point with a service profile, if such profiles are supported.

⇒ A meta-signaling virtual channel is intended to manage signaling virtual channels only within its own VP pair with VPI = 0. That is, by designating a VPI value equal to zero, implies the presence of a meta-signaling channel with a standardized VCI value

⇒ Meta-signaling VC may be activated at the instant of VP establishment

⇒ A specific VCI value for meta-signaling is reserved per VP at UNI. For example, considering a VP associated with point-to-multipoint signaling configuration, a specified VC within this VP will be activated for meta-signaling.

More on meta-signaling VC (MSVC)...

❏ *There may be one MSVC per interface*
❏ *This channel is bidirectional and permanent*
❏ *MSVC is a sort of interface management channel used to establish, check, and release point-to-point and selective broadcast SCVs*
❏ *Point-to-point signaling access configurations on a pre-established SVC. In contrast, in a point-to-multipoint signaling access configuration, meta-signaling is required for managing the signaling virtual channels.*
❏ *Meta-signaling is not used for network-to-network signaling.*

Point-to-point MSVC

❏ *Point-to-point signaling channels are bidirectional and permanent*
❏ *Though MSVC is permanent, a point-to-point signaling channel is allocated to signaling end-point only while it is active*
❏ *They are used to establish, control, and release VCCs to carry user data.*

Point-to-multipoint MSVC

❏ *Broadcast SVC (BSVC) is permanently present and reaches all signaling end-points. Selective BSVC may also be provided as a network option. That is, BSVCs can be used to send signaling messages either to all signaling end-points in a customer's network or to a selected category of signaling end-points*
❏ *Broadcast SVCs are always unidirectional (network-to-user only).*

To illustrate the SVC concept of B-ISDN, different possibilities of carrying signaling information from the customer to the network and vice versa are described below:

❏ *A signaling VP link can transport all signaling information to be exchanged between a user and the local exchange, including meta-signaling*
❏ *The signaling VP can be distinct from a VP that carries only the user data with the corresponding VCs switched at the local exchange*
❏ *When signaling has to be facilitated to a point other than the local exchange, such signaling can be done on an extra virtual path connection (VPC), which may carry signaling as well as user data. This VPC goes through the local exchange but terminates at a point of interest*
❏ *A VP may carry only the user data, but this VP as a whole would pass transparently through the local exchange.*

Multicasting and broadcasting

ATM networks are expected to support broadcast or multicast connections. These are arranged so that a "root" station can reach many "leaf" stations by addressing a cell with one VPI/VCI. The network would recognize the address as a multicast, from the nature of original setup configuration, and replicates the cell as many times as needed for delivery to all associated "leafs".

Generally, a broadcast connection is one-way from root-to-leaf. Return messages are sent on point-to-point virtual connections or via another multicast connection configured around the formed leaf as the root station. A station can be both leaf and root on several connections.

Broadcast service is designed to aid routers in updating their routing tables and exchanging information.

5.2.2   Phased implementation of signaling functions in ATM-based networks

Extension of signaling concepts to the B-ISDN/ATM-based networking can be viewed in terms of the following considerations:

- ✓   Phased introduction of signaling into B-ISDN/ATM networking
- ✓   Protocol architectures at the UNI and NNI specific to signaling
- ✓   Signaling transfer mechanisms across broadband networks
- ✓   Signaling applications meeting B-ISDN/ATM functional requirements.

In the development of ATM-based networking pertinent to various service requirements, an integrated approach has been pursued [5.9, 5.10]. It refers to a phased in program set by ITU-T as Release 1, 2, and 3 with respect to the B-ISDN/ATM signaling time-table. The signaling approach and its evolution thereof are as follows:

- ■   *Short-term approach*

  This is concerned with access signaling and network signaling which can be described as follows:

  - ✓   *Access signaling*: Extension to ISDN access signaling Q.931 (layer 3) protocols to introduce an initial set of broadband call control capabilities at UNI
  - ✓   *Network signaling*: Extension to common channel signaling SS7 ISDN user part (ISUP) to support interoffice signaling capabilities at NNI.

- ■   *Long-term approach*

  This refers to the following:

  - ✓   Defining and developing new call models to support multiparty and multimedia services
  - ✓   Developing signaling capabilities to support advanced call control strategies.

5.2.3   Signaling standards: B-ISDN/ATM

The signaling standards spelled out towards the phased evolution of B-ISDN/ATM signaling are:

- ■   Phase 1 UNI Signaling, ATM Forum UNI Specification, Version, 3.0, October 1993

- CCITT Document TD-XVIII/10 "AAL Type 5, Draft Recommendation Text for Section 6 of I.363", January 1993

- ITU - TS Draft Recommendation Q.93B "B-ISDN User-Network Interface Layer 3 Specification for Basic Call/Bearer Control", May 1993

- ITU Document DT/11/3-28 (Q.SALL1) "Service Specific Connection Oriented Protocol (SSCOP) Specification", May 17, 1993

- ITU Document DT/11/3-XX (Q.SALL2) "Service Specific Connection Oriented Protocol (SSCOP) Specification", May 17, 1993.

ATM Forum UNI Version 3.0 ⇔ ITU-T Recommendation Q.2931
Signaling functions: Initial phase

This recommendation indicates the major functions of signaling with respect to ATM networks. Relevant functions refer to:

- Point-to-point connection setup and release
- VPI/VCI selection and assignment
- Quality of service (QOS) class request
- Identification of calling party
- Basic error handling
- Communication of specific information in setup request
- Subaddress support
- Specification of peak cell rate (PCR) traffic parameters.
- Transit network selection.

In reference to Q.2931, the ATM Forum UNI version 3.0 specification does not, however, require the following capabilities:

- No alerting message sent to called party
- No VPI/VCI selection or negotiation
- No overlap sending
- No interworking with N-ISDN
- No subaddress support
- Only a single transit network may be selected.

The ATM Forum UNI version 3.0 specification also defines the following capabilities in addition to Q.2931:

- Support for a call originator setup of a point-to-multipoint call
- Extensions to support symmetric operation
- Addition of sustainable cell rate and maximum burst size traffic parameters
- Additional information elements for point-to-multipoint endpoints
- Additional NSAP address structures.

Signaling functions: Follow up phase

- Specification of a call model where each call may have multiple connections, for example in multimedia
- Support for a distributed point-to-multipoint call setup protocol
- Renegotiation of traffic parameters during the course of a connection
- Support for multipoint and multipoint-to-point calls
- Specification of meta-signaling, which establishes additional connections for signaling.

Table 5.2: Signaling timetable: Release 1, 2, 3: ITU-T

Release 1
- This refers to signaling applicable to simple switched services with constant bit rate and interworking with the existing 64 kbps ISDN. The specific versions of connections and the associated signaling functions are as follows:

  ✓ Connection-oriented CBR service with end-to-end timing
  ✓ Point-to-point connections (uni and bidirectional, symmetric and asymmetric)
  ✓ Single simultaneous connection
  ✓ Peak bandwidth indication
  ✓ Peak rate allocation
  ✓ Interworking with 64 kbps ISDN
  ✓ Two signaling access configurations
    ⇒ Point-to-point or point-to-multipoint signaling access
    ⇒ Meta-signaling: Limited set of supplementary services.

Release 2
- This refers to more sophisticated VBR traffic signaling under the statistical multiplexing ambient. The associated connections and their signaling functions are as follows:

  ✓ Connection-oriented VBR services
  ✓ Quality of service indication by the user
  ✓ Point-to-multipoint connections
  ✓ Multi-connection, delayed establishment
  ✓ Use of cell-loss priority
  ✓ Negotiation and renegotiation of bandwidth
  ✓ Supplementary services
  ✓ With Release 2, call and control functions are separated. That is, connections can be established or released during a call.

Release 3
- This addresses signaling considerations to support a full range of services including multimedia and distributive services. The constituent connections and their signaling functions are as follows:

  ✓ Multimedia and distributive services
  ✓ Quality of service negotiation
  ✓ Broadcast connections

ATM Forum UNI signaling
    This is based on extensions to Recommendation Q.931, namely, Q.93B. It essentially reflects Phase 1 (Table 5.2) Release supports. In summary, the relevant functions and connections are:
  ✓ Demand (switched) connections
  ✓ Point-to-point and point-to-multipoint switched connections
  ✓ Connections with symmetric or asymmetric bandwidth requirements
  ✓ Single connection calls
  ✓ Basic signaling functions via protocol messages, information elements and procedures
  ✓ A single statically defined out-of-band channel for all signaling messages
  ✓ Public UNI and private UNI addressing formats for unique identification of ATM endpoints
  ✓ Multicast service addresses.

The point-to-multipoint connection refers to a collection of associated ATM, VC, or VP links and the endpoint nodes, with the following characteristics:

- ✓ One ATM link, called *root link*, serves as a *root* in a simple tree topology
- ✓ When the root node sends information, all remaining nodes (called *leaf nodes*) on the connection receive copies of information
- ✓ For Phase 1, only zero return bandwidth ("leafs"-to-root) is supported
- ✓ Leaf nodes cannot communicate directly to each other.

5.2.4 *Signaling messages of B-ISDN/ATM transmissions*

ATM Forum UNI signaling uses a message-oriented protocol. The signaling messages and their uses are as follows:

Messages for ATM call/connection control

- ▪ Call establishment messages

    - ✓ Call proceeding
    - ✓ Connect
    - ✓ Connect acknowledge
    - ✓ Setup.

- ▪ Call clearing messages

    - ✓ Release
    - ✓ Release complete.

- ▪ Miscellaneous messages

    - ✓ Status
    - ✓ Status enquiry.

Messages for ATM multipoint call/connection control

- ✓ Add party
- ✓ Add party acknowledge
- ✓ Add party reject
- ✓ Drop party
- ✓ Drop party acknowledge.

Message format

The signal messages are formatted in a specific manner and the characteristics of such formatting are:

- ✓ Message formatting for signaling is independent of the message type. That is, each message consists of the same structure
- ✓ Fields, which are not used by a message are coded with a "null" value.

The message format formats are of two types:

- A: General signaling message format
- B: Format for meta-signaling messages

General signaling message format

8	7	6	5	4	3	2	1	Octets
Protocol discriminator								1
0	0	0	0	Length of call reference value (octets)				2
Flag	Call reference value							3
Call reference value (continued)								4
Call reference value (continued)								5
Message type								6
Message type (continued)								7
Message length								8
Message length (continued)								9
Variable length information coded elements (as required)								10

etc.

Format used for meta-signaling (MS) messages

8	7	6	5	4	3	2	1	Octet Number
Protocol discriminator								1
Protocol version								2
Message type								3
Reference identifier								4...5
Signaling virtual channel identifier A								6...7
Signaling virtual channel identifier B								8...9
Point-to-Point cell rate								10
Cause								11
Service profile identifier								12...22
Null-fill								23...46
Cyclic redundancy check								47-48

The fields indicated in the two message formats considered above refer to:

Protocol discriminator (PD)

- PD distinguishes messages for user-network call control (Q-93B) from other messages

- In MS protocols, PD identifies messages on the MS channel as MS messages. If not, the message is identified to another protocol.

Call reference (CR)

In reference to messages related to a particular call attempt, each message contains a common mandatory information element known as the *call reference*. This is unique on a signaling interface. The attributes of call reference are:

- CR is used to identify a call at the local user-network interface to which a particular message applies (CR has end-to-end significance)

- Call reference values are assigned by the originating side of the interface for a call. These values are unique to the originating side only within a particular signaling virtual channel

- Call reference value is assigned at the beginning of a call and remains fixed for the duration of the call

- Call flag identifies which end of a signaling virtual channel originated a call reference. Originator always sets the flag to "0", and destination sets the flag to "1".

								Octets
0	0	0	0	0	0	1	1	1
Flag		Call reference value						2
Call reference value (Continued)								3
Call reference value (Continued)								4

Protocol version

- ✓ The protocol version differentiates between individual versions of MS protocol
- ✓ It identifies the general message format used.

Message type (MT)

Names of the messages are identified by the MT field

- ✓ It determines exact function and a detailed format of each message.

Constituents of signaling message

The signaling message consists of *information elements* (IE) made of two constituent parts, namely, *mandatory elements* (M) and *optional elements* (O).

(a) Mandatory information elements

These elements refer to the following signaling messages:

- ✓ ATM user cell rate requested
- ✓ Called party number
- ✓ Connection identifier (that is, the assigned VPI/VCI values)
- ✓ QOS class requested.

- *Optional information elements*

These are optional signaling messages adopted on an ad hoc basis. The following are relevant elements:

✓   Broadband bearer capability requested
✓   Broadband lower and higher-layer information
✓   AAL parameters
✓   Called party subaddress
✓   Calling party number and subaddress
✓   Transit network selection
✓   Cause code
✓   End-point reference identifier and end-point state number.

(All messages should also contain an IE for their type, length and protocol discriminator, that is, the set from which these messages are taken).

Table 5.3: <u>Variable length IE</u>

Information element identifier					
Extension	Coding Standard	Flag	Reserved	Spare	Message action indicator
Length of information element					
Length of information element (continued)					
Contents of information element					

Note:
Extension bit: "0" indicates more octet groups to follow; "1" indicates last octet. Coding standard: "00" indicates ITU-TS (CCITT) standardized.

IE instruction fields (bits 5-1 of octet 2) are interpreted only in the case of unrecognized information element identifier or unrecognized information element contents.

Flag, reserved, spare, and message action indicator bits are currently set to "0" in the present ATM UNI implementation agreement.

Message type (2 octets)
This is intended to identify the function of the message being sent and to allow the sender of a message to indicate explicitly the way the receiver should handle unrecognized messages.

Message type						
1	0	0	Flag	0	0	Message action indicator
	Spare			Spare	Spare	

Flag	Meaning
0	Message instruction field not significant (regular error handling)
1	Follow explicit instruction (supercedes regular error handling)

Action indicator	Meaning
00	Clear call
01	Discard and ignore
10	Discard and report status

11   Reserved

Message type (First octet)

000-----   Call establishment messages
00010   Call proceeding
00111   Connect
01111   Connect acknowledge
00101   Setup

010-----   Call clearing messages
01111   Release
11010   Release complete
0110   Restart
01110   Restart acknowledge

011-----   Miscellaneous messages
11101   Status
10101   Status enquiry

100-----   Point-to-multipoint messages
00000   Add party
00001   Add party acknowledge
00010   Add party reject
00011   Drop party
00100   Drop party acknowledge

Message length (2 octets)

This identifies the length of the contents of a message. (It does not include protocol discriminator, call reference, message type, and message length octets.) Message length is a 16-bit binary coded value.

Table 5.4: <u>An example: Call party number</u>

0	1	1	1	0	0	0	0	
1	Coding standard		Flag	Reserved	Spare	Message action indicator		
Length of information element								
Length of information element (continued)								
1	Type of number			Address/numbering plan identification				
0	Address/number digits (IA5 characters)							
NSAP address octets								

Note: Coding standard "00" ITU-TS (CCITT): Standardized flag, reserved, spare, and message action indicator all set to "0"

Type of number   Meaning

000   Unknown (to be used for OSI NSAP address)
001   International number (to be used for E.164)

Address/numbering plan	Meaning
0001	ISDN/telephony numbering plan (Recommendation E.164)
0010	OSINSAP

Client registration mechanism

- An end-system and a switch exchange an identifier and address information across a UNI

  - ✓ Network administrator manually configures ATM network address information into a switch port
  - ✓ A terminal attached to the switch port exchanges address information whenever the terminal is initialized, re-initialized or reset. (The network administrator need not configure the terminal address.)

5.2.5 *Private and public network addressing*

Private network addressing

- An ATM private network address uniquely identifies an ATM endpoint

- ATM private network address is modeled after the format of an OSI network service access point (ISO 8348 and CCITTX.213).

  - ✓ Structure of the low-order part (ESI and SEL) of the *domain specific part* (DSP) is as specified in ISO 10589
  - ✓ Three *initial domain identifier* (IDI) formats specified in phase 1 ATM Forum signaling: DCC, ICD, and E.164.

Address fields

- *Authority and format identifier* (AFI):This 1 octet field refers to the authority allocating the data country code, international code designator, or E.164 number

- *Data country code* (DCC): This 2-octet field signifies the country in which the address is registered. (Specified in ISO3166, BCD coded)

- *International code designator* (ICD): This 2-octet field identifies an international organization, ICD, maintained by the British Standards Institute

- *E.164* (8 octets): Identifies an ISDN number up to 15 digits long

- *Domain specific part format identifier* (DFI): This 1-octet field specifies the structure, semantics, and administrative requirements for the remainder of the address

- *Administrative authority* (AA): This 3-octet field depicts the authority for allocation of addresses in the remainder of the DSP (for example, in the United States, it is done by the ANSI-administered U.S. Registration Authority for OSI organization names for DCC IDI format.)

- *Reserved* (RSRVD): A 2-octet field reserved for future use

- *Routing domain* (RD): This 2-octet field identifies a domain that is unique within one of the following: E.164, DCC/DFI/AA, or ICD/DFI/AA

- *Area* (Area): This 2-octet field identifies an area within a routing domain

- *End system identifier* (ESI): This 6-octet field identifies an end system within an area. This identifier must be unique within an area

- *Selector* (SEL): This 1-octet field is not used for ATM routing but may be used by an end system.

Public network addressing

Public UNI is intended to support one of the following

- E.164 address structure

  ✓ Type of number field ⇒ International
  ✓ Numbering plan indication ⇒ E.164

- Private ATM address structure

  ✓ Type of number ⇒ Unknown
  ✓ Numbering plan indication ⇒ ISO NSAP

- Both

  ✓ E.164 numbers are:

    ⇒ Those defined by CCITT Recommendation E.164
    ⇒ Those administered by public networks

Multicast service address

- *End-system-identifier* (ESI) field (6 octets) satisfies the following multicasting requirements:

  ✓ Addressing schemes allow for multicast service addresses to be distinguished. For example, when an IEEE 48-bit MAC address used as an ESI, a multicast address is distinguished by a "1" in the multicast bit of the address
  ✓ An ATM endpoint can have multiple multicast service addresses. For example, when multicast is supported on top of the network by a multicast server, the server can have a separate multicast service address for each multicast address it supports, in addition to its own non-multicast address.

End-to-end compatibility parameter identification

- On a per connection basis, the following end-to-end parameters can be specified:

  ✓ AAL type (for example, Type 1, 3/4, or 5)
  ✓ The method of protocol multiplexing (for example, LLC versus VC)
  ✓ For VC based multiplexing, the protocol, which is encapsulated (for example, any of the list of known routed protocols or bridged protocols)
  ✓ Protocols above the network layer.

5.2.6 *Signaling protocols*
What does the signaling protocol specify? It specifies the following considerations:

  ✓ Sequence of messages to be exchanged

✓   Rules to verify the consistency of the parameters

✓   Actions needed to establish and release ATM layer connections.

How are signaling protocols specified?

✓   Via narrative text

✓   Via state machines

✓   Via a semigraphical specification definition language (SDL).

The ATM Forum UNI Specification version 3.0 uses a narrative method.

5.2.7   Protocol architecture for B-ISDN/ATM

As indicated in Table 5.2, ITU-T distinguishes two signaling access configurations at the UNI (Release 1):

✓   Point-to-point signaling access configuration

✓   Point-to-multipoint signaling access configuration.

■   Single signaling end-point on user-node

■   Signal terminal or an intelligent node (PBX) depending on customer network configuration

■   Only a single permanently established point-to-point SVC is required for this signaling access configuration. This channel is used for:

> Call offering
> Call establishment
> Call release

ITU-T Draft Recommendation Q.2120: Meta-signaling protocol

■   Meta-signaling protocol is used for establishing, maintaining, and removing the user-network signaling connection at the UNI.

■   Meta-signaling protocol operates only over the MSVC as indicated earlier.

Meta-signaling... Revisited

MSVC

❑   *A meta-signaling virtual channel manages only those signaling virtual channels within its own VP pair. A VPI = 0 depicts the presence of meta-signaling channel with a standardized VCI value*

❑   *Meta-signaling VC may be activated at the time of VP establishment*

❑   *A specific VCI value for meta-signaling is reserved per VP at UNI.*

MSVC definition

•   *MSVC is identified by VCI = 1 in every VP*

•   *Its default peak rate equal to 42 cells*

•   *In principle, MS protocol can operate on each active VP, but it is sufficient to use only one in the VP between the user and the local exchange so as to reduce implementation problems.*

MS protocol

- *It is a part of ATM layer*
- *It is located within the layer management plane*
- *It is under the control of plane management.*

Meta-signaling functions at the user access

- *Management of the allocation of capacity to signaling channel*
- *Establishment, release, and checking the status of signaling channels*
- *Providing a means to associate a signaling end-point with a service profile if profiles are supported.*

Scope and application of meta-signaling protocol

Meta-signaling protocol facilitates procedures for the following:

- *Assignment and removal of point-to-point SVCs (PSVCs) and their associated (BSVCs)*
- *Checking the status of these two channel types.*

Consistent with these protocols, the following are made possible

- *Associating a signaling end-point with a PSVC and a BSVC*
- *Allocating a cell-rate to SVC*
- *Resolving possible contention problems for SVC.*

In MS protocol, the following six messages are used towards MS protocol procedure

- *Assign request*
- *Assigned*
- *Denied*
- *Check request*
- *Check response*
- *Removal.*

The first three in the messages listed above refer to assignment procedure.

- ❏ *Assignment procedure is invoked by the user-side sending an "assign req" to the network asking for a PSVCI and BSVCI*
- ❏ *Depending on its condition, the network will either send an "assigned" message indicating the PSVCI/BSVCI pair*
- ❏ *It may send a "denied" message with appropriate reason.*

The next two messages correspond to the check procedure supervised by a timer at the user-end to cope with loss of messages. This procedure is initiated by the network sending a check request message and it waits for a check response. Simplest check refers to a single PSVCI/BSVCI pair. (However, more generally it may refer to all signaling channels.) The sixth message is intended for removal procedure.

- ❏ *The removal procedure can be initiated by the network side or by the user side*
- ❏ *In contrast to the other two procedures, here, no handshake procedure is adopted*
- ❏ *The initiating side sends only a "removed" message, followed after a random interval by a second "removed" message*
- ❏ *This double message tactic prevents error, if one message is lost*
- ❏ *User-side message is delayed for some random time before delivery to network to avoid certain network overloading.*

Assignment procedure, checking procedure, and removal procedure are independent of each other. Necessary coworking is accomplished via management plane. Further, exists a network option on the dynamic indication of whether the signaling access configuration is point-to-point or point-to-multipoint.

5.2.8   ATM connection setup

As indicated earlier ATM supports both permanent virtual circuits and switched virtual circuits. The former are always present and can be used at will, like leased lines. The latter have to be established each time they are used, like making phone calls. In this section how switched virtual circuits are established is discussed.

The connection setup is not part of the ATM layer but is done by the control plane using a sophisticated ITU protocol called Q.2931. Nevertheless, the logical place to handle setting up a network layer connection is in the network layer. That is, network layer protocols do connection setups as discussed here.

Several methods are available for setting up a connection. The most common type is to first acquire a virtual circuit for signaling and use it. To establish such a circuit, cells containing a request are sent on virtual path 0 (VPI = 0) virtual circuit 5 (VCI = 5). If successful, a new virtual circuit is opened on which connection setup request and replies can be sent and received.

The underlying basis for this two-step setup procedure is that this way the bandwidth reserved for VCI = 5, (which is barely used at all) can be kept extremely low. Further, an alternative way is facilitated to setup virtual circuits. Some carriers may allow users to have permanent virtual paths between predefined destinations or allow them to set these up dynamically. Once a host has a virtual path to some other host, it can allocate virtual circuits on it itself, without the switches being involved.

VC establishment uses the six message types listed in Table 5.5. Each message occupies one or more cells and contains the message type, length, and parameters. The messages can be sent by a host to the network or can be sent by the network (usually in response to a message from another host) to a host. There are also other status and error reporting messages, which are not, however, indicated here.

Table 5.5: Connection establishment and release messages

Message	Meaning when sent by host	Meaning when sent by network
SETUP	Establish a circuit	There is an incoming call
CALL PROCEEDING	The incoming call is noticed	Call request will be attempted
CONNECT	The incoming call is accepted	Call request was accepted
CONNECT ACK	Call acceptance noted and acknowledged	Acknowledgement of connection
RELEASE	Terminate the call	The other side has no more information to send
RELEASE COMPLETE	Acknowledgement for release	Acknowledgement for release

The common procedure for establishing a call is for a host to send a SETUP message on a special virtual circuit. The network then responds with CALL PROCEEDING to acknowledge receipt of the request. As the SETUP message travels toward the destination, it is acknowledged at each hop by CALL PROCEEDING.

As the SETUP message finally arrives, the destination host can respond with CONNECT to accept the call. The network then sends a CONNECT ACK message to indicate that it has received the CONNECT message. As the CONNECT message propagates back toward the originator, each switch receiving it acknowledges it with a CONNECT ACK message. This sequence of events is shown in Fig. 5.13.

The sequence for terminating a virtual circuit is rather simple. The host wishing to end the session would simply send a RELEASE message, which travels to the other end and causes the circuit to be released. At each hop along the way, the message is acknowledged, as shown in Fig. 5.13.

ATM networks are also designated to facilitate setting up multicast channels. A multicast channel has one sender and more than one receiver. These are constructed by setting up a connection to one of the destinations in the usual way. Then the ADD PARTY message is sent to attach a second destination to the virtual circuit returned by the previous call. Additional ADD PARTY messages can be sent afterward to increase the size of the multicast group.

In order to setup a connection to a destination, it is necessary to specify which destination, by including its address in the SETUP message. ATM addresses are specified by three versions: The first type is 20 bytes long and is based on ISO addresses. Its first byte indicates which of three formats the address is in; and, the bytes 2 and 3 specify a country. The byte 4 gives the format of the rest of the address, which contains a 3-byte authority, a 2-byte domain, a 2-byte area, and a 6-byte address, plus some other items. In the second address format, bytes 2 and 3 designate an international organization instead of a country. The rest of the address is in the same format as the first version. There is also an older form of addressing known as (CCITT E.164), which uses a 15-digit decimal ISDN telephone number. (More details are furnished in subsection 5.3.1.)

Example 5.3
Describe a signaling procedure in reference to a point-to-point call setup and release as per Q.2931 based signaling protocol for the ATM UNI.

Solution
Point-to-point call establishment procedure

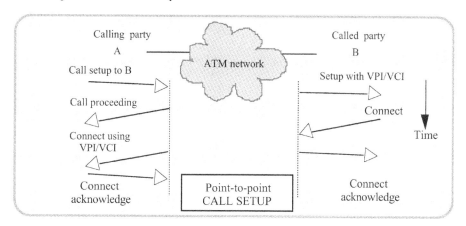

Fig. 5.13(a): Point-to-point call setup diagram

❑ Calling party A initiates the call attempt using a "setup message" indicating B as the called party

❑ The network routes the call to the physical interface on which B is connected That is, the network outputs a *setup message* indicating that the VPI/VCI that should be used if the call is accepted. (Optionally, the setup message may also communicate the identify of the calling party)

❑ The called party (B) accepts the call attempt by returning the *connect message*

• This connect message is propagated back to the originator of the call (namely party A) with minimal delay, so that call setup time is low.

❑ The *connect acknowledgement message* is used bilaterally by both parties,

• From the network to the called party
• From the calling party to the network
• (This is the final stage of a three way hand-shake confirming the connection is alive.)

Point-to-point *call release* procedure

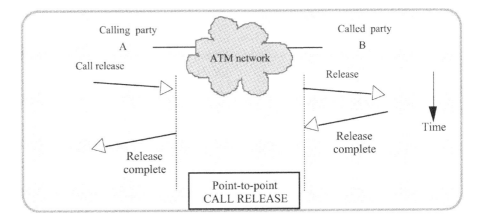

Fig. 5.13(b): Point-to-point call release diagram

- ❑ Release process can be requested by either party
- ❑ In this example, the calling party is assumed as the one who initiates disconnect process, that is, party A sends a *release message* across the network to party B
- ❑ Party B acknowledges the release request. (This is done by returning a *release complete message*)
- ❑ The release complete message is propagated back to the calling party across the network. (Release originator)
- ❑ Thus, a three-way hand shake completes the call-release process.

Example 5.4
Describe a point-to-multipoint call-setup and release procedure as per the signaling protocol for the ATM UNI.

Solution
The solution is illustrated in Fig 5.14. The calling party and called party are A and B respectively. Accordingly,

- ❑ Originator of the call is party A (Node A)

  - • Root node

- ❑ The call is intended for two "leafs" B and C connected to a local ATM switch on a single ATM UNI
- ❑ D is a third leaf node connected to a separate ATM UNI

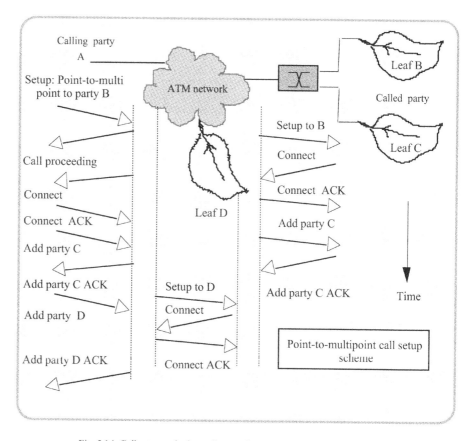

Fig. 5.14: Call setup and release diagram for a point-to-multipoint signaling

❑   Node A begins the single point to multipoint call by sending a set of messages to the network requesting setup of a point-to-multipoint call identifying leaf node B's ATM address

•   In the present example, Node A requests a call setup to Node B; and to network responds with a "call proceeding" message (similar to that in a point-to-point call)

❑   The network switches the call attempt to the intended destination and issues a setup message to Node B with assigned VPI/VCI

❑   The first node leaf then indicates its intention to join the call by returning a connect message

•   The network, in turn, acknowledges with a connect ACK message

Point-to-multipoint call establishment

❑   The network informs the calling root node (A) of a successful addition of party B through a connect and connect ACK hand-shake as shown in Fig.5.14

❑   Root node (A) now requests that party C be added through the Add party message, which the network relays to the same ATM UNI as party B through the Add party message to inform the local switch of the requested addition

❑   Part C responds with an Add party ACK message

•   This is propagated by the network back to A

❏   Now, the root node A requests another C party D be added through an Add party message

❏   The network routes this to the UNI connected to party D and issues a setup message, since this is the first party on the new ATM UNI

❏   Node D responds with a connect message to which the network responds with a connect ACK message

❏   The inclusion of leaf party D in the point-to-multipoint call is communicated to the root node A through the Add party ACK message.

Point-to-multipoint call release

❏   The leaves of the point-to-multipoint call can be removed from the call by the *drop party message*, if one or more parties would remain on the call on the same UNI.

•   Or, by the release message, if the party is the last leaf present on the same UNI.

❏   The root node should drop each leaf in turn and then release the entire connection.

Meta-signaling: Implementation

Since an ATM network is made of means to support several kinds transmission (voice, video, and data), the signaling needed must satisfy all the diverse connection speeds, delays, and types users require on the ATM network. To cope with theses issues, ATM has introduced the concept of meta-signaling indicated earlier.

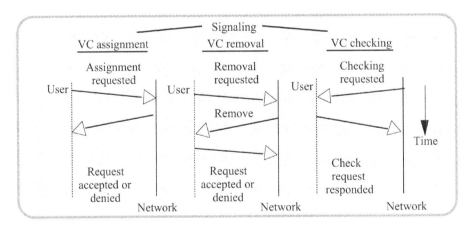

Fig. 5.15: Meta-signaling messages

This is an important function of the ATM layer involves the so-called concept of *meta* (beyond) signaling. This signaling is *beyond* the normal signaling of connections in majority network types inasmuch as it involves the setting up of the signaling channels themselves. ATM switched virtual connection (SVC) networks are setup via a call-setup and disconnect protocol, exactly as in other networks. But ATM networks are required to work not only with point-to-point connections for data but also with broadcast connections for video/audio and multipoint-to-multipoint connections for conferencing.

Meta-signaling at the ATM layer, in essence, is a simple protocol to establish and tear down signaling channels themselves. All information in meta-signaling is done with a one-cell message. Only three functions are required for ATM meta-signaling as illustrated in Fig. 5.15.

First, the meta-signaling channel may establish a new signaling channel with the VC assignment operation, via a two-phase exchange of cells between user and network. Second, the meta-signaling channel may disconnect an existing signaling channel by means of the VC removal operation, by means of a three-phase exchange of cells. Finally, the network can also check on the status of signaling channels, (which may exist for long periods without being used) using a VC checking operation. The VC checking messages performs a "hearbeat polling" function similar to the status enquiry requests on frame-relay networks. The role of meta-signaling cells could be more comprehensive than the three functions indicated above. However, it is not yet finalized.

5.3   ATM Signaling: Implementation

Apart from meta-signaling, ATM networks should implement the signaling used in other connection-oriented networks (such as voice) so as to establish and release connections. This protocol is a formal user request for network resources. It is essentially a user-to-network negotiation process. In this process, parameters such as connection type (PVC versus SVC, point versus multipoint), call endpoints (ATM network addresses), traffic contract (QOS parameters, bandwidth), service parameters (which ATM adaptation layer to use), and VPI and VCI number allocation are handled.

As specified earlier (in subsection 5.2.3), there are three standards proposed: The first is Q.93B from ITU-T, an adaptation of Q.931, standardized as Q.2931. The second is just known as ATM Forum signaling, which is also called "skinny Q.93B" and is a subset of Q.93B. The ATM Forum is committed to eventually making its own signaling standard compliant with Q.2931. The last is B-ISUP which is essentially the SS7 (signaling system 7, the international standard signaling protocol for voice networks) extended for ATM networks. B-ISUP stands for *broadband-interim signaling user protocol*. This has been designed by ITU-T.

5.3.1   Details on ATM Forum signaling

The signaling protocol provides a standard way of expressing the requested connection to a local ATM network node across a UNI. The signaling protocol is another variable-length frame following its own set of protocol rules.

There are some concepts that are specifically important in ATM signaling. The connection that the signaling protocol is setting up must use some form of network address. This is the same as in the telephone networks, where the network address is the telephone number.

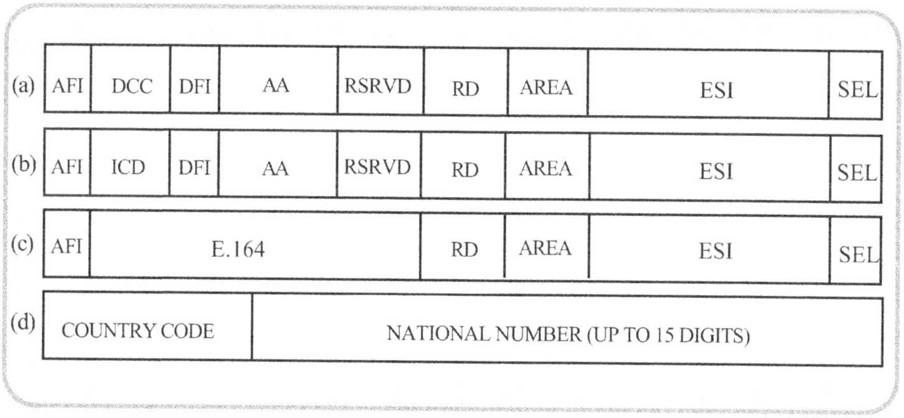

(a) AFI	DCC	DFI	AA	RSRVD	RD	AREA	ESI	SEL
(b) AFI	ICD	DFI	AA	RSRVD	RD	AREA	ESI	SEL
(c) AFI	E.164				RD	AREA	ESI	SEL
(d) COUNTRY CODE	NATIONAL NUMBER (UP TO 15 DIGITS)							

Fig. 5.16: The address formats of ATM

(a) DCC ATM format for private use, (b) ICD ATM format for private use, (c) E.164 ATM format for private use, and (d) E.164 ATM format for public use.

(b) *The data country code (DCC) ATM format*: In this format, the first field is the authority and format identifier (AFI) field, which has a value of 39 for this format. The DCC field itself (2 bytes) specifies the country where the address is registered. The *domain-specific part identifier* (DFI) field (1 byte) specifies the structure of the remaining fields. The *administrative authority* (AA) field (3 bytes) indicates the authority responsible for the rest of the address. Then specified is a reserved (RSRVD) field (2 bytes) intended for future use. The *routing domain* (RD) field (2 bytes) is used to specify a unique routing domain (that is, when the address is unique). The *area field* (2 bytes) identifies a unique area within an RD. The *end system identifier* (ESI) field (6 bytes) identifies an end-system within the area, and the *selector* (SEL) field (1 byte) is used by the end-system for "selecting" an endpoint.

(b) *The international code designator (ICD) ATM format*: Here, the first field has an AFI value of 47. The next field is the ICD field itself (2 bytes), which identifies an international organization. The codings are administered by the British Standards Institute, and the rest of the fields are identical to the DCC ATM format.

(c) *The E.164 ATM private format*: In this format, the fields are specified by the ITU-T in Recommendation E.164 originally planned for ISDN networks. The first field has an AFI value of 45. The next field is the E.164 address itself (8 bytes), which may contain up to 15 binary-code decimal (that is, the digits 0 to 9) values. The initial 4-bit nibble is set to 0000, and any trailing nibbles, if the address is less than 15 digits, are set to 1111.

(d) *The E.164 ATM public format*: This format is identical to the E.164 ATM private format. However, it is administered and assigned publicly conforming to the E.164 international format.

As mentioned earlier, the present ATM signaling systems are based on Q.931 signaling for ISDN and Q.93B signaling. The standardized version, namely, the Q.2931 protocol involves exchanging variable-length messages. Each protocol has a message type and message length field and a number of information elements (IEs). Each IE has parameter values for the circuit attributes being negotiated. (Some IEs are mandatory in some message, and some are optional as detailed in subsection 5.2.4.)

The Q.2931 messages are similar to those of ISDN. They correspond to ALERTING DISCONNECT, CALL PROCEEDING, etc. The initial standard refers to point-to-point and "root to leaf" point-to-multipoint. Multipoint-to-multipoint and video connection signaling are in the next stage. There are also some IE add-ons for ATM-specific functions such as ATM cell user rate and VPI/VPI identifier.

The ATM Forum specifies a subset of the full Q.93B/Q.2931, namely, the "skinny Q.93B", as a stable subset that allows rapid deployment and acceptance of equipment. Q.93B uses no meta-signaling at all.

VPI/VCI support for signaling: More details ...

The Phase 1 signaling specification supports *virtual path connection identifiers* (VPCI) to identify the virtual path across the UNI, with the restriction that a one-to-one mapping exists between VPCI and VPI, and hence values beyond 8 bits are restricted.

The following list describes the Phase 1 signaling capabilities in reference to VPIs, VPCIs, and VCIs:

    ✓   Phase 1 signaling enables the identification of virtual paths (using VPCIs) and virtual connections within virtual paths (using VCIs)

    ✓   It includes no negotiators of VPCIs or VCIs. However, it does not preclude negotiations

    ✓   It has no provisions to negotiate or modify allowed ranges for VPCIs or VCIs within virtual paths. But it does not preclude such provisions in future releases.

Support of a single signaling virtual channel

For single point-to-point signaling virtual channels, a VCI of 5 and VPCI of 0 are used for all signaling in Phase 1. The association between signaling entities should be permanently established. Meta-signaling is not supported in Phase 1. Broadcast signaling using a virtual channel is also not supported.

The ATM Forum uses the signaling default VPI = 0, VCI = 5. It is expected to be used and accepted extensively. It sheds off most of the "status" and "information" messages from Q.2931. But its scope concentrates on the most common set of features for all vendor equipment.

The ATM Forum signaling is actually an *extended* Q.2931. Due to the importance of conferencing to ATM network implementation, it was decided to add point-to-multipoint signaling capabilities. This refers to an unidirectional "root to leaf" connection. That is, the originator (root) may add parties ("leafs") or drop them, but the "leafs" may not generate signals at all. Additional messages such as ADD PARTY, ADD PARTY ACK, ADD PARTY NACK, DROP PARTY, DROP PARTY ACK, and DROP PARTY NACK are added for this purpose. The signaling message flow pattern across an ATM network UNI and NNI to establish a connection is same as that indicated in Fig. 5.14.

Point-to-point ATM signaling

Essentially, in reference to the signaling in Phase 1, the most important aspects are point-to-point and point-to-multipoint connections. The point-to-point signaling architecture is illustrated in Fig. 5.17.

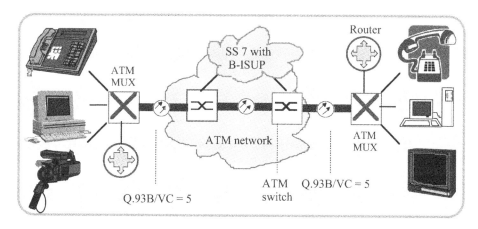

Fig. 5.17: ATM signaling: Point-to-point connectivity

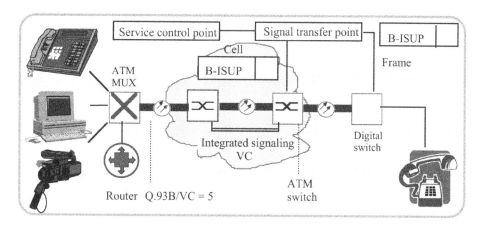

Fig. 5.18: ATM signaling across a SS7 network architecture

The ATM signaling based on the existing SS7 protocol is used in norrowband ISDN. Since this signaling is based on an existing signaling protocol, no new messages are required or created. In this protocol, however, the B-ISUP (Q.2761) (similar to ISUP for ISDN) messages are carried as cells in the signaling VC between the ATM switches. The same message is carried as frames between the STPs and non-ATM switches as shown in Fig. 5.18.

The point-to-point signaling is carried on VC = 5 and VP = 0. The protocol used for access and network signaling is based on Q.2931.

Point-to-multipoint signaling

In this connection, each virtual path has a separate signaling channel (VC = 1). The signaling is used to establish, monitor/maintain, and tear down signaling VCs. This signaling channel is also used to carry the service profile information between the end-points as illustrated in Fig. 5.19

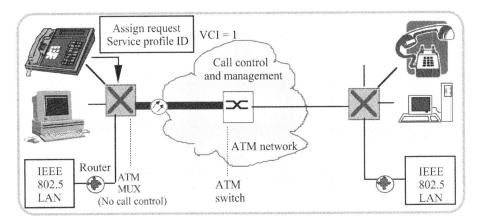

Fig. 5.19: ATM signaling: Point-to-multipoint connectivity (connection request phase)

Note: Meta-signaling is used to assign or delete signaling VCs. ATM equipment requests signaling VC via the meta-signaling channel (VCI = 1).

In the connection request phase, the ATM CPE equipment requests the signaling VC signaling channel VC =1 and VP = 0. Information such as the bandwidth requirement and *service profile identifier* (SPID) is carried within the cell to the ATM switch via the ATM MUX, where the actual call control or connection setup occurs. Once the cell with the information is received by the ATM switch, the switch performs a call control operation and forwards the cell to the appropriate "child node". Upon receiving the cell, the child node sends an acknowledgement with information on the channels to be used. This process is repeated for every "child node" of parent-list.

As the CPE equipment receives the appropriate channel for the transmission of information, the setup acknowledgement is transmitted on the appropriate channel (VC number). This is followed next by the actual information transfer.

Example 5.5

This problem refers to what may happen in regard to relevant ATM signaling when a user initiates a multimedia call. The following points may be noted:

❑   Multimedia combines voice and data sessions on a signal PC or workstation

❑   In order for the PC/workstation to have real-time voice and/or video capabilities, it should be facilitated with a separate network connection and communications card. (If not, then the voice and video must be downloaded over the data network in their entirety to the hard disk on the PC and then displayed/listened to.)

❑   In contrast, ATM handles this all through a single high-speed network interface (such as SONET); that is, a separate connection with just the right parameters is established for each activity, namely, video and data.

Assume User A has an ATM-equipped, multimedia PC with a camera, microphone, a set of speakers and the usual keyboard/mouse/windowing setup.

a.   User A places a call to User B who has a similar setup. Call initiation is done by a simple mouse-click on an icon: This amounts to an off-hook condition of the telephone. Assume this initial call setup refers only to a voice connection

b.   Next, during the progress of voice session, User B decides to activate the video camera on the PC/workstation so as to establish a video-link with User A. This leads to a video session (in addition to the already existing voice session)

c.   Suppose now, User A and B decide to view a video clip stored on the ATM network (at a video server site). This refers to a point-to-multipoint connection

d.   At the next stage, the Users A and B also decide to exchange data between each other. This connection should permit the two users to view the data, store it, and have it updated during the session

e.   During the progress of the above sessions, any one of the Users (A or B, but not both) may request the network to establishment a connection to a database stored on the ATM network (at a database server site) to download a text file

f.   Finally, the connection-release (for all sessions) is requested by, say, User B.

For the above scenario:

Sketch VCC connection setup/progress/release diagrams.

Indicate VCI and VPI values for signaling connections and for the various application-session connections involved.

Indicate the appropriate AAL # accessed for the different application sessions.

Solution

A) User A places a voice call to User B:

Voice transmission: Timing preserved, VBR, AAL-2, connection-oriented

Signaling VCI = 5, VPI = 0

❑   User A initiates the call attempting to use a SETUP message indicating User B as the called party

❑   The network routes the call to the physical interface on which it is connected

❑   The network outputs a SETUP message indicating the VPI/VCI that should be used (any available VPI/VCI that guarantees the QOS negotiated) if the call is accepted; if the call is accepted (optionally) the SETUP message may also communicate the identity of the calling party

❑   User B accepts the call attempt by returning the CONNECT message

❑   The connect message is propagated back to User A

❑   The CONNECT ACKNOWLEDGEMENT is used bilaterally to both parties, from the Network to User B and from User A to the Network.

B) Next, User B activates a video session with User A:

Video: Timing preserved, VBR, AAL-2, connection-oriented

Signaling VCI = 5, VPI = 0

❑   User B initiates call attempt using a SETUP message indicating User A as the called party

❑   The network routes the call to the physical interface on which A is connected

❑   The network outputs a SETUP message indicating the VPI/VCI that should be used (any available VPI/VCI that guarantees the QOS negotiated) if the call is accepted; if the call is accepted (optionally) the SETUP message may also communicate the identity of the calling party)

❑   User A accepts the call attempt by returning the CONNECT message

❑   The connect message is propagated back to User B

❑   The CONNECT ACKNOWLEDGEMENT is used bilaterally to both parties, from the network to User A and from User B to the network.

C) Users A and B connect to a video ATM server:

Video: Timing preserved, VBR, AAL-2, connection oriented

Signaling VCI = 1, VPI = 0

Note: This is a point-to-multipoint connection. According to ATM Forum Q.2931, only the originator may add parties (or drop them). This is known as "root-to-leaf" signaling.

Assume this point-to-multipoint video session is a new session and that User A is the originator. Thus the sequence is as follows:

- ❏ User A initiates a POINT TO MULTIPOINT call attempt using a SETUP message indicating User B as the called party
- ❏ The network routes the call to the physical interface on which User B is connected
- ❏ The network outputs a SETUP message indicating the VPI/VCI that should be used. (Any available VPI/VCI that guarantees the QOS negotiated can be used, if the call is accepted. (Optionally, the SETUP message may also communicate the identity of the calling party.)
- ❏ User B accepts the call attempt by returning the CONNECT message
- ❏ The connect message is propagated back to User A
- ❏ The CONNECT ACKNOWELDGEMENT is used bilaterally to both parties, from the network to User B and from User A to the network
- ❏ User A now requests that party C (ATM video server) be added via the ADD PARTY message
- ❏ The network routes the call to the physical interface on which C is connected
- ❏ The network outputs a SETUP message indicating the VPI/VCI that should be used. (Any available VPI/VCI that guarantees the QOS negotiated can be used, if the call is accepted. (Optionally, the SETUP message may also communicate the identity of the calling party.)
- ❏ Video server C accepts the call attempt by returning the CONNECT message
- ❏ The connect message is propagated back to User A
- ❏ The CONNECT ACKNOWELDGEMENT is used bilaterally to both parties, from the network to server C and from User A to the network.

D) Users A and B establish a data session
Variable delay acceptable, VBR, AAL 5, connection oriented
Signaling VCI = 5, VPI = 0

Note: Assume User A initiates the data session

- ❏ User A initiates the call attempt using a SETUP message indicating User B as the called party
- ❏ The network routes the call to the physical interface on which User B is connected
- ❏ The network outputs a SETUP message indicating the VPI/VCI that should be used. (Any available VPI/VCI that guarantees the QOS negotiated can be used, if the call is accepted. (Optionally, the SETUP message may also communicate the identity of the calling party.)
- ❏ User B accepts the call attempt by returning the CONNECT message
- ❏ The connect message is propagated back to User A
- ❏ The CONNECT ACKNOLEDGEMENT is used bilaterally to both parties, from the network to User B and from User A to the network.

E) One User (A or B) requests a data session with Database ATM server D:
Variable delay acceptable, VBR, AAL-5, connection oriented
Signaling VCI = 5, VPI = 0

- ❏ User initiates the call attempt using a SETUP message indicating server D as the called party
- ❏ The network routes the call to the physical interface on which database server D is connected
- ❏ The network outputs a SETUP message indicating the VPI/VCI that should be used. (Any available VPI/VCI that guarantees the QOS negotiated can be used, if the call is accepted.) (Optionally, the SETUP message may also communicate the identity of the calling party.)
- ❏ ATM database server D accepts the call attempt by returning the CONNECT

message
- ❏ The connect message is propagated back to User A
- ❏ The CONNECT ACKNOWLEDGEMENT is used bilaterally to both parties, from the network to Server D and from User A to the network.

F) User B requests release of all connections
   i) Voice session

- ❏ User B sends a RELEASE message across the network to User A
- ❏ User A acknowledges the release request by returning a RELEASE COMPLETE message
- ❏ This RELEASE COMPLETE message is propagated back to the calling party across the network.

   ii) Video session

- ❏ User B sends a RELEASE message across the network to User A
- ❏ User A acknowledges the release request by returning a RELEASE COMPLETE message
- ❏ This RELEASE COMPLETE message is propagated back to the calling party across the network.

   iii) Multipoint ATM video session
       According to ATM Forum Q.2931, only the originator (root 0) may add parties or drop them, and the leaves may not generate signaling at all. Thus User B cannot terminate this session because User A initiated the point-to-multipoint link. In later phases of ATM implementation, this may, however, be possible.

   iv) Data session

- ❏ User B sends a RELEASE message across the network to User A
- ❏ User A acknowledges the release request by returning a RELEASE COMPLETE message
- ❏ This RELEASE COMPLETE message is propagated back to the calling party across the network.

   v) Database session
       Database session is between User A and database server D; User B is not involved, therefore User B cannot drop the session.

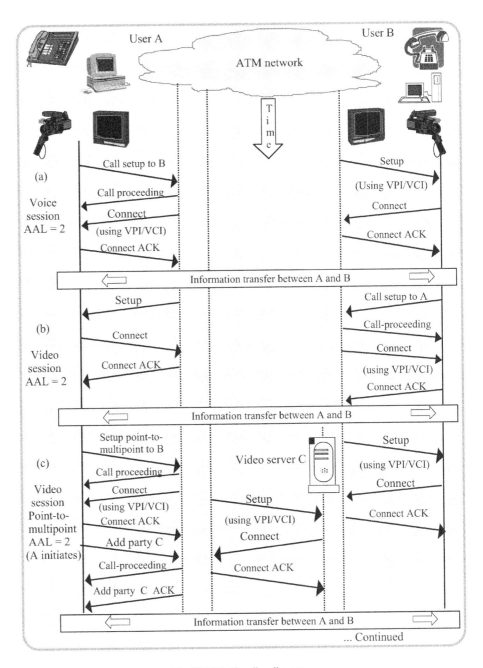

Fig. 5.20 (A): Signaling diagram

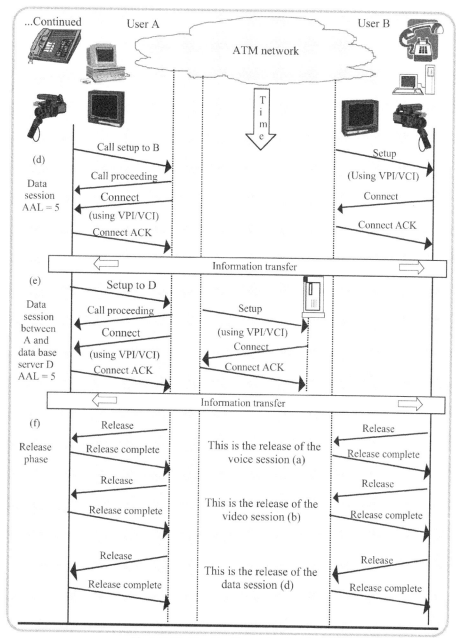

Fig. 5.20 (B): Signaling diagram (continued)

Example 5.6

This problem is concerned with ATM signaling pertinent to a conference call, which may involve multimedia access. The relevant sequences are described below:

a)   User A wants to establish a conference call with User B, C, and D who are at different locales

b)   Assume User A and B work with a PC/workstation having multimedia capabilities (multimedia can combine voice, video, and data sessions on a single PC/workstation)

491

c)   Users C and D participate in the conference only via telephones (voice connections)

d)   Assume initial call setup originating from User A is only a voice connection and this connection is multicast to Users B, C, and D

e)   During the progress of voice session, User B activates a video transmission for User A (in addition to the existing voice session)

f)   Next, User C excuses himself/herself from the conference and drops out of the session

g)   Now, Users A and B decide to view a video clip stored on the ATM network (at a video server site)

h)   A gives a running commentary (voice) on the video clip to User D

i)   User D, now invokes a low speed data transmission from his PC terminal, parallel to the voice communication already in progress with A and B

j)   User B ends the conference call.

Sketch VCC connection setup, progress, release diagrams.
Indicate VCI and VPI values for signaling connection(s) and for the various application-session connections involved.
Indicate the appropriate AAL 1 # accessed for the different application sessions.

Solution

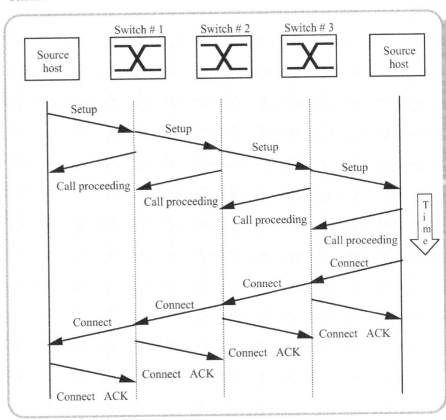

Fig. 5.21 (a): Signaling messages

For the scenario indicated in the problem, it is assumed that: (i) To facilitate the PC/workstation to have real-time voice and/or video capabilities, a separate network connection and communication card are needed. (Otherwise, the voice and video must be downloaded over the hard disk on the PC and then displayed/listened to); and (ii) multimedia transmission may require high-speed network interface. Hence the flow of signaling messages is indicated in Fig. 5.21(a). And, the signaling phases are depicted in Fig. 5.21(b).

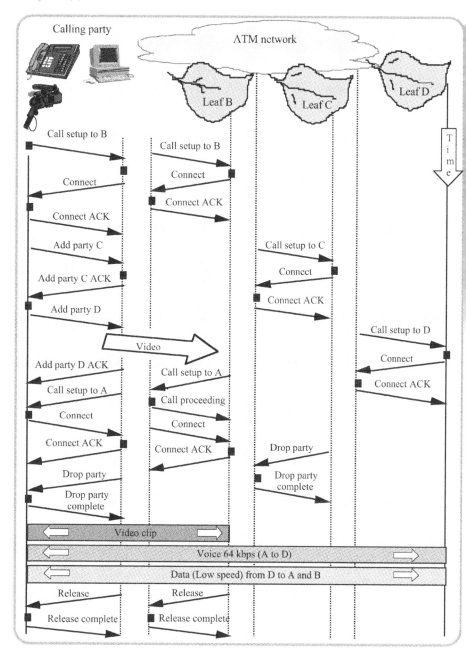

Fig. 5.21 (b): Signaling phases

In reference to the details furnished in Fig. 5.21 the following can be observed:

a) Number of messages sent to establish the circuit = (4 × setup + 3 × call proceeding + 4 × connect + 4 × connect ACK)

b) The signaling for point-to-multipoint connections is based on setting up multiple point-to-point connections among A, B, C, and D. In this connection, each virtual path (A to B, A to C, and A to D) has a separate signaling channel, thus VC = 1 and VP = 0 is used.

c) Since there is transfer or voice between A and D and video transmission between A and B, the AAL used is AAL-2. This specifies a connection oriented and variable bit rate (high) for audio and video. For the data transmission from D to A and B, AAL-1 is used to support the regular data transfer. This is a connection-oriented transmission and the bit rate is constant.

The various signaling sequences are indicated in Fig. 5.21 (b)

Problem 5.2

An ATM signaling scenario is described below.

Sketch the VCC connection setup/progress/release diagrams.

Specify VPI/VCI as appropriate for the signaling connections and for the various application session connections involved.

Indicate the appropriate AAL # accessed for the different application sessions.

Assume the following:

Multimedia combines voice, video, and data sessions on a single PC or workstation. ATM handles all these communications with guaranteed QOS parameters with the support architecture such as SONET.

User X: ATM-equipped multimedia workstation

User Y: Similar end entity as User X

User Z: Non-multimedia workstation with only a voice (telephone) and a simple PC

Session commenced by user X refers to a conference session involving Users X, Y, and Z.

- Phase 1: This session involves simple telephone conversation between users
- Phase 2: User X sends a text to both Users Y and Z
- Phase 3: Per some instructions received in phase 2, User Z gets released from the conference session and invokes a request call with a remote server and downloads some text
- Phase 4: Users X and Y are now on point-to-point basis exchange video information
- Phase 5: Users X and Y release themselves from the multimedia session but engage in a voice and text transfer session
- Phase 6: To inform Users X and Y of the data downloaded from the server, User Z joins the conference again
- Phase 7: Conference call is terminated by User Y.

Problem 5.3

This problem refers to a single point-to-point connection establishment/release procedure as per ITU-T Q.2931 Recommendation.

Suppose a point-to-point connectivity is required to be established (and released) between two end-entities via an ATM network. The type of information to be transferred is voice and large text with intermittent burstiness of data bits. Assume release is initiated by either one of the parties.

a)   Draw a connection establishment and release diagram to illustrate the connectivity indicated

b)   Specify appropriate ATM adaptation layer for the convergence of the traffic specified

c)   Indicate signaling VPI and VCI

d)   What are various information demands (IEs) involved in the signaling phases?

e)   Is it necessary that the user side respond to a SETUP with the CALL PROCEEDING or ALERTING message?

f)   Indicate the two-way handshake required to complete the call release phase.

(*Hint*: See [5.9], Section 15.2)

Problem 5.4
Describe the type of signaling procedure adopted in ATM telecommunication for the purpose of broadcasting audio, video, and data applications.
(*Hint*: See [5.9], Section 15.2)

5.3.2   *NNI signaling*
This refers to signaling between network interfaces and includes two possible domains:

✓   Private network domain
✓   Public network domain.

The associated protocols are briefly addressed below:

Private network domain
   Interim interswitch signal protocol (IISP) is the ATM Forum-specified protocol, which is useful in building multi-vendor ATM-switched VC networks. This is a simple strategy applicable when minimum levels of interoperability for multi-vendor ATM private networks are considered.

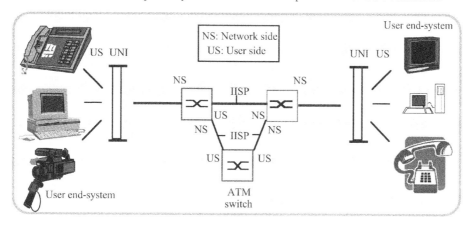

Fig. 5.22: IISP scenario

IISP protocol has been conceived as a pilot to be used prior to a more elaborate procedure called *private network-network interface* (PNNI). IISP is based on the UNI 3.0/3.1 protocol. It is illustrated in Fig. 5.22. There is set of ATM switches between the end-systems.

IISP provides only static routing and the associated switch-to-switch signaling is based on a slight variant of the UNI 3 signaling specification. IISP allows interoperalibity of vendor switches through standard routing functionality. But its capabilities are limited. It cannot dynamically share routing information. Its routing depends on static routing tables. IISP also cannot guarantee QOS for circuits because the associated switches do not share real-time information.

Private network-node interface (PNNI)

This is the ATM Forum's routing protocol. It reflects two possible uses: Connecting nodes within a network and interconnecting networks. That is, the protocol can be applied to interconnect both switches and networks. Hence, PNNI is a trunking, routing, and signaling protocol.

The essence of NNI signaling is that upon receiving a signaling request the network needs some way to establish the new connection. PNNI defines two methodologies thereof:

✓ Sharing routing information
✓ Establishing a connection with the requested QOS.

ATM routing is identical to the routing used by routers in the Internet and corporate networks with the added commitments to support multiple service categories.

The two protocols of PNNI, which enable call management in ATM networks are:

✓ *Routing protocol*: This is defined for distributing network topology information between switches and/or groups of switches
✓ *Signaling protocol*: This enables communication between the network switches to propagate UNI and other signaling messages throughout the network.

Distributed routing scheme of PNNI

This is based on the IETF's open shortest path first (OSPF) protocol. It is a "link state" routing technology. It uses a distributed maps database model. In this technique, each switch describes its local environment and propagates this information throughout the network. This information in ATM context includes the following:

✓ *Reachability* information: This basically consists of a list of addresses accessible through a switch
✓ *QOS metrics*: These specify the bandwidth, guaranteed CTD/CDV performance and cell-loss ratio.

Each switch builds a local database describing the above and updates it.

PNNI hierarchy

Hierarchical network is rather implicit with PNNI so as to allow the hierarchical routing necessary in large networks of complexity. The bottom of the hierarchy consists of small groups of switches called peer groups. All the members of this peer group have total knowledge on each other's *reachability* and *QOS metrics*. This bottom group represented by a single peer leader is then organized as groups at the next hierarchical level. This process is continued until only one group exists at the top of the hierarchy.

Best path for requested connection

To determine the "best path" for a requested connection, the connection request must pass to the lowest level hierarchy that contains both the source and the destination address. The objective of this process is to find out the best sequence of switches to submit the call request for establishing the requested connection.

The result of this PNNI processing is a *source route* that lists the combination of ATM switches and links through which a particular connection can reach its destination. The originating switch inserts this information into a signaling request and forwards it through the network to the destination.

PNNI also includes the signaling messages to exchange call management information among the network switches.

Suppose the "best path" chosen poses a locale where the resources are insufficient to let the call pass by. This may happen when the routing information gathered is not up-to-date. The *crank back* mechanism would send the message to the preceding node to try an alternate "best path" for rerouting.

The result of ATM call setup is an end-to-end connection following the switches and links identified by the routing process. All subsequent communication takes place over this connection as long as source and destination have traffic to share (without requiring any additional routing). This process differs from IP routing in the sense that in IP routing, each individual packet must bear the burden of finding its own route in the network.

PNNI: Addressing and routing

The PNNI relies on the *network service access point* (NSAP) address summarization to provide the routing hierarchy. That is, all members of a given peer group share that peer group's address prefix, enabling effective summarization. The addresses which allow summarization are "native", and those that inhibit summarization are "foreign". This is illustrated in Fig. 5.23

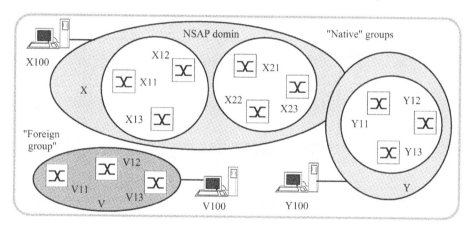

Fig. 5.23: PNNI routing: Native and foreign summarization

QOS support: PNNI

The path over which it is decided to route a message should pose resources to support the QOS guaranteed. That is, the PNNI link metrics and attributes should conform to different classes of transmission specified as CBR, rt-VBR, nrt-VBR, ABR, and UBR.

Generic connection admission control (GCAC)

The PNNI protocol tries to minimize the probability of failure in determining the first source route by defining the generic CAC algorithm. This algorithm allows the source node to estimate the expected CAC behavior of nodes along candidate paths based upon additive link metrics advertised in the *PNNI topology state packet* (PTSP: The PDU and flooding procedure used by PNNI to reliably distribute topology information throughput a hierarchical network) and the requested QOS of the new connection request. GCAC provides a good prediction of a typical node-specific CAC algorithm.

Designated transit lists (DTLS)

The source node encodes the selected route to the destination in a *designated transit list* (DTL), which describes the complete hierarchical route to the destination. The source then inserts the DTL into the SETUP signaling message, which it then forwards to the first node identified in the DTL.

Broadband intercarrier interface (B-ICI)

B-ICI supports PVCs as well as SVCs. The B-ICI version 1.0 of the ATM Forum defines the PVC towards service interconnection between carriers as shown in Fig. 5.24.

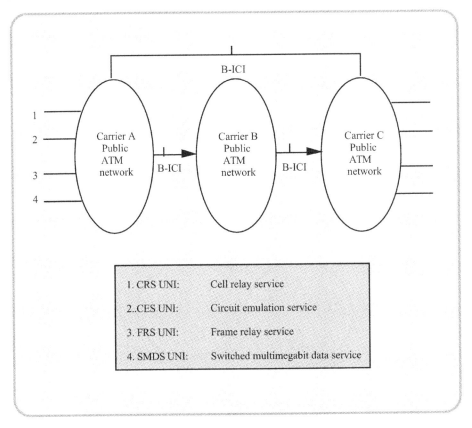

Fig. 5.24: B-ICI

B-ICI version 2.0 specifies support for UNI 3.1 SVCs across interfaces that connect carrier networks. It includes support capabilities such as bidirectional, symmetric/asymmetric connections, specified QOS classes etc.

Connections across the B-ICI may use either VPCs or VCCs depending upon the switch capabilities and services desired.

Addressing ...

Traditional phone sets use E.164 (ISDN) addresses and computer systems use network layer addresses such as IP and IPX. ATM-connected devices use ATM layer end-system addresses in addition to a possible network layer address.

For private networks, the ATM Forum has defined the use of network service access point (NSAP) format addresses for end-systems. The addresses refer to the following:

-     *ATM end-system (AES): NSAP format*
-     *E.164*
-     *Broadcast*
-     *Multicast*
-     *Anycast.*

5.4   ATM Traffic Control

In telecommunication systems, an attempt is made to offer and maintain the best possible quality of service in order to reduce any subscribers' concerns as well as to boost the image of service providers. Thus an objective of telecommunication systems is to implement such efforts by which the traffic in the system becomes well organized. In general, organized *telecommunication traffic* refers to the entire aggregate of message transfers over a group of circuits or trunks, the duration of such message transfers, as well as their number.

In reference to ATM telecommunications, organizing the incoming heterogeneous set of information efficiently requires an analysis of the *throughput* performance of the cell traffic under congestion. Throughput of a channel, in general, is the expression of how many data are put through. In other words, throughput is an expression of channel efficiency. It is defined quantitatively as the net useful bits put through per unit time (bps.). The throughput, however, can be hampered by congestion in the traffic. Further, the statistical multiplexing (as in ATM) would normally warrant a decision whether a connection should be accepted or not and a negative decision may cause additional degradation of throughput. Hence, appropriate congestion avoidance measures should be facilitated as necessary. Such measures involve the specific traffic control strategies as described below.

The *traffic control* in a telecommunication network refers to those measures undertaken to protect the network so as not to fall below the required performance objectives. In such efforts, the traffic and congestion control mechanisms may not rely on higher layer protocols (which are either application-dependent or service-specific. However, protocols may make use of the information in the ATM layer to increase their efficiency).

B-ISDN, as indicated in Chapter 3 is an expansion of ISDN. It is a general-purpose digital network that employs a hybrid of circuit-switching and packet-switching techniques, in which all data are packetized and virtual circuit-switched. Since B-ISDN/ATM is conceived to support a heterogeneous set of communication services, a strategy to share the resources is necessarily adopted. Therefore, as in any resource-sharing network, ATM would also encounter network congestion problems. These problems should be properly addressed via traffic control strategies [5.9, 5.10].

There are three levels of traffic control involved with ATM transmissions:

- ✓   Cell-transfer control
- ✓   Call control
- ✓   Network control.

Cell-transfer control

This level of control provides those functions which regulate the flow of cells. It includes *priority control* at the output buffers as well as *policing function* to keep the number of input cells within the declared traffic parameters.

Call control

This refers to the level that facilitates functions, which control the flow of calls across a network. Relevant functions correspond to *call admission control* and *path selection* for call setup requests.

Network control

This consists of network-wide controls like *routing control, link capacity assignment* and *call congestion control*. Within the ATM connections, there is a set of quality of service (QOS) parameters, which is specified at the connection setup. They are guaranteed by the network and the network must be able to obtain enough information from the user about the connection so as to ensure that no other connection that shares the resources degrades the QOS requirement.

The hierarchy of the aforesaid controls is illustrated in Fig. 5.25.

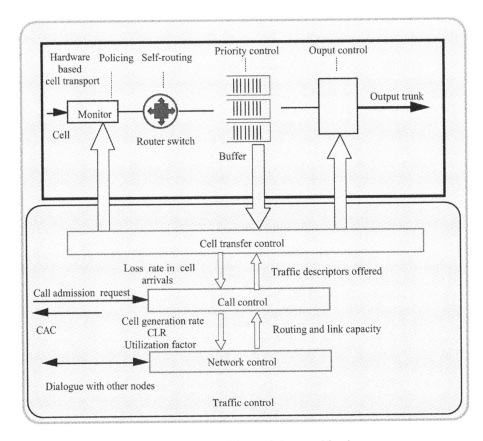

Fig. 5.25: Hierarchy of a ATM transmission control functions

In general, a VC setup (over a high-speed access interface), is seldom entitled to the full capacity of the interface. Therefore, it becomes necessary for the network to enforce a limit on individual VCs so as to prevent any single user sole use of the resources.

But how are these limits enforced? It is done at the time of connection establishment by prescribing certain constraints on the management traffic characteristics opted by the user, and these constraining factors enforced on the traffic management are known as *traffic parameters*.

Further, the way a network should respond in the event of excess traffic observed on a given connection is specified by the *QOS parameters* associated with the connections.

The above considerations constitute the so-called *traffic contract*, which imposes a responsibility on the network to support contracted traffic with the specified QOS parameters; and it prohibits the users from exceeding the performance limits requested at the time of establishing the traffic.

5.4.1   Traffic control functions

The locations of various traffic control functions in an ATM network are indicated in Fig. 5.26.

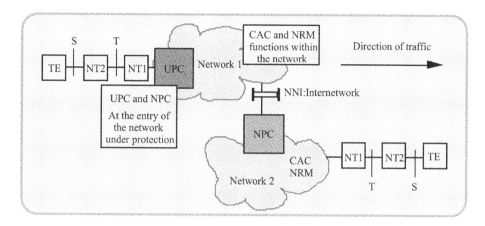

Fig. 5.26: The locales of traffic control functions in an ATM network

Feedback can be used to manage certain traffics (such as multimedia traffic) over ATM-based networks. In such efforts, a sampler obtains information regarding congestion level etc. in the network. This information acquired by the sampler is used by a control algorithm to calculate the control parameters for the next step.

In the various tasks of traffic management, the rate of management is shown in Fig. 5.27

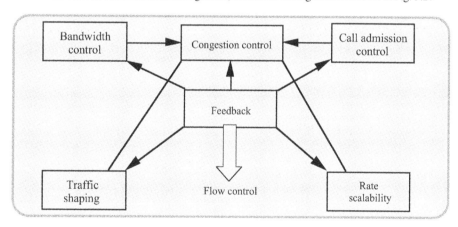

Fig. 5.27: Traffic management components

The state-of-the-art of traffic control procedures and their impact on resource management in ATM telecommunication can be enumerated as follows:

- Traffic control procedures for ATM are not yet totally standardized (within the scope of ITU-T recommendations)

- Any such procedures are being specified only as requirements

- There is reluctance on the part of network operators who would rather be inclined to facilitate a flexible/controllable set of network traffic tools matching customers' needs. It means that no single traffic control method would satisfy all the service providers and any "standardization" becomes rather questionable. (For example, peak bit rate conditions may render network efficiency low, if an average bit rate is specified as a call admission control parameter at the time of connection. (Network providers may not relish this.)

Hence, the traffic control goal often meets two conflicting situations, namely,

    ✓   Achieving a good ATM network efficiency
    ✓   Meeting the users' QOS requirements.

As a result, traffic control measures and/or resource management schemes have become complex procedures in practice. Nevertheless, some procedures have been accepted and adopted. The general considerations vis-à-vis control functions addressed on ATM networks are as follows: ATM-centric networks are expected to offer broadband performance in compliance with traffic parameter specifications as indicated in Chapter 4. To accomplish this, ATM-based networks are required to have a set of traffic control capabilities discussed above. ITU-T Recommendation I.371 identifies the following as the traffic control capabilities of B-ISDN/ATM networking:

    ✓   *Traffic control functions*

        ⇒   Network resource management (NRM)
        ⇒   Connection admission control (CAC)
        ⇒   Usage parameter control (UPC) and network parameter control (NPC)
        ⇒   Priority control (PC)
        ⇒   Traffic shaping (TS)
        ⇒   Fast resource management (FRM)

    ✓   *Traffic congestion control*

        ⇒   Selective cell discarding (SCD)
        ⇒   Explicit forward congestion indication (EFCI).

A summary of the above considerations is furnished below:

5.4.2   *Network resource management*

The basic tool for NRM is the *virtual path technique* adopted in ATM. The associated efforts can be listed as follows:

- Grouping all VCs into a VP: Whether such a grouped VC/VP connection can be accepted ("admitted") or not, is decided at the call set-up phase. That is, a call *admission control* (CAC) is facilitated via the VP technique as a part of NRM

- By aggregating the entire traffic in one VP, policing of user and network parameters can be done in a simplified manner: That is, *user parameter control* (UPC) and *network parameter control* (NPC) are done collectively for the entire traffic at the VP level in the NRM

- Traffics of different types may have different priorities on time-transparency parameters (such as CTD and CDV) as well as on semantic-transparency parameters (such as cell-losses). As an NRM requirement, such priorities are segregated/grouped (in respect of different qualities of service) through the VP technique and are handled distinctly

- The NRM conveys the messages on traffic congestion conveniently to each virtual circuit bundled in a VP

- The VP technique associated with the NRM also controls the segregation of statistically multiplexed traffics from one another so that those traffics with guaranteed bit rate would remain unaffected.

5.4.3   Connection admission control (CAC)

ATM transmissions require end-to-end connections to be established before the traffic can be transmitted. As indicated in earlier chapters, these connections are identified by the *connection identifiers*, namely, the VCI and VPI. A specific field in the cell header is assigned to designate the VPI and VCI. The VPIs are used to route packets between two nodes that originate, remove, or terminate the VPs, whereas the VCIs are used at the VP end-points to distinguish between different connections. In general, VCI values are unique only in a particular VPI value and VPI values are unique only in a particular physical link. A VP is a collection of VCs between two nodes in an ATM network. A VP link is defined between the point where the VPI is assigned and the point where it is removed or translated. A VC link consists of one or more physical links between the point where the VCI is assigned and the point where it is switched.

Call admission control refers to a set of actions taken by a network during the call setup phase (or during the call renegotiations phase) to decide whether a VC/VP connection can be accepted (admitted) or not. The general conditions for connection admission are:

- Call (or connection) admission is facilitated only when sufficient network resources are available to establish the connection end-to-end with the required QOS

- Admission of such a connection (that is, the new connection) should not affect the QOS of any existing connections.

Connection admission control is supported via the following parameters:

- A set of traffic parameters describing the inherent *source characteristics* specified by:

  ✓ Peak cell rate (PCR)
  ✓ Allowed cell rate (ACR)
  ✓ Burstiness ratio
  ✓ Maximum burst size (MBS)
  ✓ Burst tolerance (BT)
  ✓ Peak duration (or, equivalent set of parameters)
  ✓ Sustainable cell rate (SCR)

- A set of parameters, which identify the required QOS class and specified as *connection traffic descriptors*, namely:

  ✓ CTD
  ✓ CDV/jitter
  ✓ CDV tolerance (CDVT)
  ✓ CLR/BER

Currently, there are no specific standard sets to define the CAC, but a rule-based effort, however, does exist. It specifies a rule that a call should not be accepted if it requires more than 50% of the available resources. An example can illustrate this rule: Let the resources deployed be 100 units. Suppose a node has currently 10% of its resources occupied. This would leave 90% available for other connections. Consider a call (Call A) requested at this node, which requires 40 units of resource allocation. Then, the CAC decision is stipulated as follows:

$$\text{Resource requirement of Call A, } 40 < (90/2) \qquad \Rightarrow \qquad \therefore \text{ Call A will be accepted}$$

This results in 50% (40% due to Call A and 10% due to previous calls) commitment of available resources. Now, let Call B request a connection at this same node. Suppose Call B also requires 40 units of resource allocation. Then, the CAC decision will be specified as follows:

Resource requirement of Call B, $40 > (50/2)$ $\Rightarrow$ $\therefore$ Call B will be rejected

The available resource now still remains at 50%. Now, assume a third call, Call C, requires routing through this same node. Suppose Call C requires only 15 units of resource allocation. Then, CAC is decided as per the following:

Resource requirement of Call C, $15 < (50/2)$ $\Rightarrow$ $\therefore$ Call C will be accepted

The available resource left over at this node is now 35%.

The above rule-based strategy has two benefits: First, the resource utilization at the nodes or links never reaches 100%. Second, there is a good mixture of heavy and low utilization calls. This technique thus prohibits heavy utilization calls from monopolizing all of the resources.

Thus, CAC in one form or the other is defined as a procedure for decision whether a new connection request is accepted or rejected. A call is accepted if the network resources are available to satisfy the QOS requirements of the connection request and at the same time, the QOS of existing connections are not affected when multiplexed with this new connection. Therefore, the extent of bandwidth required by a connection for the network to provide the required QOS needs to be ascertained.

In order to facilitate, the lingering questions are:

✓ *What is the extent of bandwidth required by a new connection? How is it ascertained?*
✓ *By accepting a new connection, how can the network ensure that other (existing) connection requirements are unaffected?*

Bandwidth allocation considerations

Bandwidth allocation refers to the procedure to decide the extent of bandwidth required by a connection so as to support the prescribed QOS.

There are two approaches for *bandwidth allocations* known as *deterministic multiplexing* and *statistical multiplexing*. In deterministic multiplexing, each connection is prudently allocated its peak bandwidth. This is known as *peak-rate reservation*. By virtue of the highest rate allocated, the deterministic multiplexing can almost totally eliminate the congestion; although there is still a small amount of probability that buffers may overflow and cells will be lost. Also, by implementing deterministic multiplexing, there could be a large wastage of bandwidth especially when the burstiness ratio is significant. Further, deterministic multiplexing does not take advantage of the multiplexing capability of ATM and, therefore, restricts the utilization of network resources.

On the contrary, in statistical multiplexing, the amount of bandwidth of a VBR (variable bit rate) source is less than its value at peak rate, but it is greater than its average rate. Hence, with statistical multiplexing and the associated bandwidth allocation, the sum of the peak rates of multiplexed sources can be greater than the link bandwidth as long as the sum of the statistical bandwidths can be less than or equal to the provisioned link bandwidth. This means, the bandwidth efficiency of statistical multiplexing can be higher when the connection bandwidths get close to their average bit rate, and lower when all the connections approach their peak bit rates. Thus, the bandwidth of a connection depends on both its own stochastical characteristics as well as on those of existing connections in the network.

Call admission control (CAC) algorithms

Call admission control algorithms are intended to answer the queries posed above. That is, their purpose for the following:

✓ To determine the extent of bandwidth required by a new connection to support the type of traffic impressed on it
✓ To verify whether the service levels required by the existing connections would be affected when multiplexing together with the new connection.

Consistent with the above-said functions, CAC aims at the maximization (optimization) of utilizing the available network resources on a real-time basis. The relevant efforts will involve the following considerations:

- A set of parameters should be specified, describing the source activity adequately for accurate prediction of the performance metrics of interest in the network. Such specifications are, however, still open issues

- The prevailing strategies are, therefore, tailored on the basis of certain assumptions and approximations. The salient aspects of such strategies are listed below:

   ✓ *Superposition of arrival streams*: In ATM networks, *superposition* implies a collection of arrival streams buffered and mapped on to a single, down-stream transmission link. The buffer (of finite size or time-slots) stores the incoming cells temporarily, if the cell arrival rate is greater than the rate at which the cells will depart. A "server" assigned to the queueing system renders the superposition of the traffics

   ✓ The arrival process is assumed to be *Markovian*. Specifically, the *Markov modulated Poisson process (MMPP)* and the *Markov modulated Bernoulli process (MMBP)* are assumed. (Details on MMPP and MMBP structures are discussed in Section 5.)

There are a few algorithms which have been proposed in the literature [5.10] towards CAC implementation in the ATM environment. A summary of these paradigms is presented below.

Gaussian approximation algorithm

Given n multiplexed connections, this technique formulates an algorithm to deduce the total bandwidth C_o, required to support the superposed traffics from these n connections. This method does not take into consideration the involvement of buffer size. It decides the set $\{C_o, n\}$ so that the probability that the instantaneous aggregate bit rate exceeding C_o is less than a given value $\in$, known as the *desired overflow probability*. In the computations involved, each connection is characterized by the average value (λ_m) and standard deviation, σ of the bit rate associated with the aggregated (superposed) traffic. In terms of λ_m and σ^2 of the aggregated traffic, C_o is determined as follows:

$$C_o \approx \lambda_m{}' + \alpha\sigma' \qquad (5.1)$$

where α is the inverse of the Gaussian distribution, namely, $\alpha = \sqrt{2\ln(1/\varepsilon) - \ln(2\pi)}$. Further, $\lambda_m{}'$ is the new mean value given by $\lambda_m{}' = \lambda_m + \lambda_{m\,(n+1)}$, and σ' is the new standard deviation, $\sigma' = \sqrt{\sigma^2 + \sigma_{n+1}^2}$. Here, $\lambda_{m(n+1)}$ and σ_{n+1} represent the parameters posed as a result of a new connection request received.

Now, the CAC criterion is stated as follows: If $C_o \leq C$, namely, the provisioned link bandwidth, then accept the connection; otherwise, reject it.

The demerits of the Gaussian approximation technique are:

- The Gaussian approximation on the connection statistics is valid only when the connections multiplexed are large in number and have similar statistical characteristics of the traffics they support

- It is assumed that all connections have identical cell loss requirements. This may not be totally true

- Since buffer size and its presence are not taken into account, any statistical gain that may result from the multiplexing is not included in the algorithm.

Fast buffer reservation algorithm

To decide whether a new connection can be multiplexed with the existing connections on the link, a *fast buffer reservation* can be done as follows: First, the *excess demand probability*, namely, the probability of requiring more buffer slots than available, is calculated. In case the excess demand probability is greater than a predefined value, then the new connection is rejected; otherwise, it is accepted. This algorithm is based on assuming a two-state Markov structure in characterizing the statistics of a bursty source. The underlying principle is as follows: A predefined number of buffer-slots in the link buffer is reserved to meet the demand posed by the active period of the source. At the end of the active period, the reservation is removed. This process is repeated throughout the duration of the connection. The relevant algorithmic considerations can be specified as follows:

Given that,

λ_i = Peak bit rate of ith connection

m_i = Average bit rate of ith connection

s_i = Random variable representing the number of buffer-slots needed by the ith connection

q_i = Excess demand probability (P_r) given by:

$$q_i = \begin{cases} P_r\{s_i = B_i\} = (1 - m_i / \lambda_i) \\ P_r\{s_i = 0\} = m_i / \lambda_i \end{cases} \tag{5.2}$$

where $s_i = 0$ denotes zero buffer slots needed in the silent state and $s_i = B_i$ represents an active state (i) with B_i being the number of buffer slots required by that active state. The value of B_i is determined in terms of peak-to-link rate ratio, namely, $B_i = \lceil L\lambda_i/R \rceil$ where L is the total number of buffer-slots and R is the link rate.

If a link carries n connections, the total buffer demand B is the sum of n random variables, s_i. That is,

$$B = \sum_{i=1}^{n} S_i \tag{5.3}$$

In reference to the above procedure, the CAC criterion is as follows: If the excess demand probability q_i is greater than a predefined value, the new connection is rejected; otherwise it is accepted.

Instead of using the excess demand probability, the connection probability (p_i) of the ith connection can be used for CAC decisions. The *contention probability* is defined as the probability that the number of buffer slots requested is not available at the instant when the ith source transmits its burst. For large values of q_i, $q_i \rightarrow p_i$. For small values of q_i, the excess demand and contention probabilities differ significantly.

Equivalent capacity algorithm

This model assumes two states, one corresponding to the peak-rate bit flow duration (active period) and the other representing no-bit flow duration (silent period). It is essentially a "*flow model*". Its algorithmic considerations are as follows:

Let m_{n+1} and R_{n+1} be the parameters of a new $(n + 1)^{th}$ connection denoting the mean and peak bit rates of the connection respectively. Suppose b_{n+1} represents the average duration of the active period. Assuming the current values of the aggregated traffic statistics are m and σ depicting the mean and standard deviations of the link-speed (bit rate) respectively, the link-speed, C is then determined from the following relation:

$$C = R \frac{y - X + \sqrt{(y-X)^2 + 4X\rho y}}{2y} \tag{5.4}$$

with $y = \alpha b(1 - \rho)R$ and $\alpha = \ln(1/\varepsilon)$. Further, the other notations are as follows:

R = Peak bit rate of the connection
b = Average duration of the active period
X = Buffer size
λ = Transition rate out of the silent state ($\lambda = \rho/\{b(1 - \rho)\}$)
μ= Transition rate out of the active state ($\mu = 1/b$)
ρ= Source utilization.

The new parameters of the aggregated traffic statistics are determined as follows:

$$m' = m + m_{n+1} \tag{5.5a}$$

$$(\sigma')^2 = \sigma^2 + m_{n+1}(R_{n+1} - m_{n+1}) \tag{5.5b}$$

Next, the new value of link-speed C' is calculated from the following relation:

$$C' = \min(m + \alpha'\sigma, \sum_{i=1}^{n} C_i) \tag{5.6}$$

where ΣC_i is the sum of the equivalent capacities (C_i) of individual connections. Now, the CAC criterion can be stated as: If C' is less than the link bandwidth, then accept the connection; otherwise, reject it. The entity C' is the total bandwidth of n multiplexed connections and it takes into account the interaction between individual connections. It is done by capturing the effect of multiplexing via blending the Gaussian approximation with equivalent capacities.

Flow approximation to cell-loss rate algorithm

This technique uses an upper bound on the cell-loss probability in making a call admission decision. This procedure does not use the buffer size in decision-making. Hence, there is a restriction posed by the amount of statistical gain involved. Further, this method does not distinguish between cell-loss requirements of individual connections. It calculates a single cell-loss probability and compares it with the relevant QOS requirement of any connection in order to provide an accept or reject decision.

Heavy traffic approximation

This technique is based on the asymptotic behavior of the tail-end of queue-length distribution in an infinite capacity queue with constant service times. Further, a Markovian cell-arrival process governed by a probability matrix P(z) is presumed. Then, the condition P{z: queue-length > i}= p is used to make call admission decisions.

Relevant considerations on the tail-end behavior of the steady-state queue-length distribution of such a queueing system leads to the determination of P{queue-length > i} = p. Then, for an asymptotically large extent of i, p = P{z: queue-length > i} = $\alpha(1/z^*)^i$ where,

$$\alpha = \sum_{i=1}^{N} \rho_i \tag{5.7}$$

$$z^* = 1 + \frac{(1 - \alpha)}{\sum_{i=1}^{N} m_i(1 - \rho_i)^2 b_i} \tag{5.8}$$

with $\rho_i = m_i/R_I$, and in reference to the ith connectivity of N multiplexed connections,

$$R_i = \text{Peak rate}$$
$$B_i = \text{Average burst length}$$
$$M_i = \text{Average rate.}$$

If the probability P(queue length > i) = p with the new connection included is less than or equal to a desired loss ratio, then the new connection is accepted; otherwise it is rejected.

Nonparametric approach

The nonparametric approach method does not require any knowledge on the statistical distribution of the arrival process. It is based on the peak and average cell rates of connections. The call admission procedure is decided by an estimate of the blocking probability when there are n multiplexed connections on the link. In particular, suppose U(n,τ) is this estimate where τ is the observation period. If U(n, τ) is greater than or equal to the cell-loss ratio (CLR), then the call admission procedure is done by calculating U(n+1,τ). This is the CLR after the new connection request is superposed. If U(n+1,τ) is greater than the desired loss ratio, then the new connection is accepted; otherwise it is rejected.

Among the algorithms considered above, the equivalent capacity method achieves the highest level of statistical multiplexing while guaranteeing the desired CLR application at the link. The heavy traffic approximation becomes effective as the ratio of the buffer size to the burst length increases. If the buffer size is small, then it coincides with the deterministic allocation namely, the peak bandwidth allocation. The nonparametric approach is, in general, less sensitive to buffer size. In particular, increasing the buffer size may not result in a significant increase in the amount of statistical multiplexing that can be achieved with the nonparametric approach.

5.4.4   *User (or usage) parameter control (UPC) and network parameter control (NPC)*

User specific parameters refer to the user's choice of traffic parameters and *network specific parameters* depict the specifications imposed by the network. User parameter control is done at the user-network interface (UNI) and network parameter control is performed at network-node interface (NNI). *So, what are these UPCs and NPCs?*

UPC and NPC are the set of actions taken by the network to monitor and control traffic in terms of traffic offered and validity of the ATM connection at the user access and the network access respectively. Their main purpose is to protect network resources from malicious as well as unintentional misbehavior, which otherwise may affect the quality of service of already established connections. Detecting violations of negotiated parameters and taking appropriate actions thereof correspond to the parameter controls envisaged. Connection monitoring includes all connections crossing the user-network/network-node interface. UPC/NPC apply to both user VCCs/VPCs and signaling virtual channels as well.

The user parameter monitoring takes the following into consideration:

- Checking the validity of VPI/VCI values

- Ensuring that agreed parameters are not exceeded in the entire volume of traffic entering the network

- Overseeing the total volume dynamics of accepted traffic on the access link.

The action taken by UPC/NPC refers to the following:

- Discarding those cells which violate the negotiated traffic parameters

- Releasing a connection that is found guilty of violating the agreed traffic contract

- Instead of discarding those cells that overburden the traffic parameters, they may also be allowed to be transferred, as long as they do not upset/cause serious harm to the network traffic. This action is known as "*tagging of violating cells*".

Quality of service, usage parameters and service categories

As discussed in earlier chapters, ATM networks are intended to support a wide range of services – each with a distinct set of traffic characteristics. The ATM network handles each service with a stream of 53-byte cells traveling through a virtual circuit. However, the way in which each data flow is handled by the network depends upon the characteristics of the traffic.

Prior to a connection being established, certain parameters are negotiated between the user and the network. Essentially, there are three parameters that define the quality of service (QOS) of a connection and quantify the end-to-end network performance. These QOS parameters are measured at the network exit point and were defined in Chapter 3. For a quick reference, those definitions are reiterated below:

QOS and user parameters... a revisit

A definition of QOS is given in ITU-T Recommendation I.350, 1988: "Quality of service is defined as the collective effect of service performances that determine the degree of satisfaction of a user of the specific service."

❑ *Cell transfer delay (CTD): This is the amount of delay experienced as the cell traverses the network. It is the time-difference between transmission of the first bit in the cell to reception of the last bit in that same cell. It is usually specified as maximum CTD or mean CTD*

❑ *Peak-to-peak cell delay variation (CDV): This depicts the maximum allowable difference between the maximum and minimum CTD experienced during the connection*

❑ *Cell loss ratio (CLR): It is the ratio between lost cells to total transmitted cells on a connection.*

Five additional parameters (as defined in Chapter 3 and listed below) are also negotiated when the connection is being established. These parameters discipline the behavior of the user and are called usage parameters as indicated earlier. The network only guarantees to provide the QOS for those cells adhering to these usage parameters. The network at the entry point (UNI) enforces usage (or user) parameters and they specifically refer to the following (which were elaborated in Chapter 3):

❑ *Peak cell rate (PCR): This is the maximum instantaneous rate at which the user may transmit*

❑ *Sustained cell rate (SCR): This refers to the average cell rate when measured over a long interval*

❑ *Minimum cell rate (MCR): The minimum desirable cell rate specified by the user*

❑ *Maximum burst size (MBS): The maximum number of cells sent continuously to the network at the peak cell rate*

❑ *Burst tolerance (BT): This is the amount of time allocated for sending cells into the network at the peak rate.*

The BT and MBS are related to each other by the following algorithm:

$$BT = (MBS - 1)[(1/SCR) - (1/PCR)] \qquad (5.9)$$

Although BT and MBS appear to be similar in their definitions, they in fact describe two distinct parameters: *MBS* defines the number of cells that the user can present to the network assuming the peak transmission rate. However, the user must limit the duration of sending this burst of data (specified by BT). If the user transmits clumps of cells equal to the MBS, then a sufficient idle period is necessary to ensure that the overall rate does not exceed the SCR. Usually messages transmitted specify only the MBS; BT is determined by the above formula (Eqn. 5.9).

There are five *service categories* defined in ATM as indicated in Chapter 3 and summarized below:

Service categories

❑   *Constant bit rate (CBR) service: This service emulates circuit-switching and requires a fixed data rate that is continuously available. This service is delay- sensitive and requires tightly constrained values of CTD and CDV. Examples of CBR services are telephony, television, and video conferencing.*

❑   *Real-time variable bit rate (rt-VBR) service: This service is also delay-sensitive with CTD and CDV tightly constrained. However, the cell rate varies over time. An example of this service is compressed video.*

❑   *Non-real time variable bit rate (nrt-VBR) service: This service can be used for data transfers with critical response-times. The cell rate varies over time with tight constraints on CTD only. Examples of this service are airline reservations and banking transactions.*

❑   *Unspecified bit rate (UBR) service: This service would utilize the remaining capacity of the network. Additional capacity may be available due to the following considerations:*

-      *The sum of the resources used by VBR and CBR services may be less than the total capacity*
-      *The bursty nature of VBR traffic leaves open spaces.*

Further, no commitments are made to the source nor any feedback is adopted. This service is sometimes referred to as best-effort service. An example of UBR service is E-mail.

❑   *Available bit rate (ABR) service: This service also uses the leftover capacity of the network, but uses some form of feedback to control the rate of transmission. Examples of this service are critical data transfer and distributed file transfer.*

Fig. 5.28 shows the usage of the line versus time for each of the service categories.

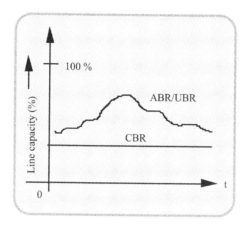

Fig. 5.28: Line capacity versus time

Constant bit rate (CBR) and variable bit rate (VBR) services: A summary ...

As elaborated in the earlier chapters, ATM networks are conceived to support CBR and VBR traffics. CBR traffic is delay-sensitive and connection-oriented with a fixed bit rate. That means, a channel with a fixed and constant transmission rate is assigned to connections during their duration. On the other hand, VBR transmission depicts a delay-sensitive and connection-oriented traffic with a variable bit rate. That is, connections are not assigned a fixed, dedicated bandwidth.

Architecturally, the ATM adaptation layer AAL1 is used for CBR services that require a timing relationship between the end points of connections; whereas AAL2, AAL3/4, and AAL5 are provided for VBR.

UPC mechanisms

The network must ensure that sources stay within their traffic contract provisions on connection parameters negotiated at the connection setup phase. This function as stated earlier is referred to usage parameter control (UPC). It resides at the access point to the network (that is, at the UNI). It is a *traffic policing function* and it should detect a nonconforming source as quickly as possible and take appropriate actions to minimize the potentially harmful effect of excess traffic. Further, this should be achieved transparently to conforming users, in that the traffic generated by such sources should not be artificially delayed at the interface.

Specific user parameter control algorithm has not been fully standardized. However, a number of desirable features of such control algorithms have been identified as follows:

 ✓ Capacity of detecting any illegal traffic situation (traffic policing)
 ✓ Selectivity over the range of checked parameters; (that is, the algorithm could determine whether the user behavior is within an acceptance region)
 ✓ Rapid response time to mitigate the parameter violations
 ✓ Simplicity of implementation.

The call admission control discussed earlier is not sufficient to prevent congestion. The main reason is that users may not stay within the connection parameters negotiated at the call setup phase. This is because:

 ✓ Users may not be aware of and/or they might have inadvertently underestimated the connection requirements

 ✓ User equipment may be malfunctioning
 ✓ User may be deliberately underestimating their bandwidth requirements in order, for example, to pay less tariff
 ✓ Users may be purposely trying to crash the network.

UPC algorithms refer to actions that a policing function can take when a source is detected to be nonconforming. Such algorithms enable the following:

 ✓ Dropping the violating cells altogether
 ✓ Delaying the violating cells in a queue until the departure from the queue conforms to the contract
 ✓ Marking violating cells distinct from those that stay within the negotiated parameters and transmitting them so that the network can treat them differently when congestion arises
 ✓ Adaptively controlling the traffic by informing the source whenever it starts to violate its contract.

The set of parameters to be monitored (policed) and controlled is that used to characterize a source. Since the latter remains an open issue, the former has been sparsely resolved. Nevertheless, various policing mechanisms have been proposed in the literature. In most of these schemes, the controlled source parameters largely refer to the peak and average bit rates.

Since traffic control is necessary to protect a network so that it can achieve the required performance objectives, UPC enforces a contract between the user and the network about the nature of the call. This prevents any user from causing excessive traffic and degrading the quality of service. As a basic contractual requirement, the UPC prescribes a simple format that refers to restricting the peak rate of the traffic from a source. But the impending challenge with peak-rate control is the impact of cell-delay variation (CDV), introduced by the access network. One of the conclusions that has been reached within the *European Commission RACE Programme* is that, because of this CDV, the peak rate cannot be controlled adequately by simple mechanisms (particularly those based on the so-called *window method*). Another parameter, namely, the maximum allowed CDV, has been, therefore necessitated into the traffic contract, which in effect, introduces a tolerance to the peak rate control. However, the consequence of introducing such tolerances may render traffic with quite different characteristics in conforming with the traffic contract. This may seriously affect the cell-loss performance in the network, because the traffic facilitated with a tolerance to the peak-rate would require significantly more bandwidth to be allocated than that required for the CBR connection. This can be illustrated by considering a *leaky-bucket scheme* of UPC.

A counter, normally dubbed as a leaky-bucket takes input traffic and shapes it into a desired cell stream according to a predefined criterion. The underlying principle behind this approach is that a cell, before entering the network, must obtain a token from a token pool. That is an arriving cell will consume a token and depart from the leaky-bucket. The token pool is preset to receive tokens generated at a constant rate. There is, however, an upper bound on the number of tokens that can be waiting in the pool and any cell arriving at a time when the token pool is full is discarded. Thus, the size of the token pool implicitly imposes an upper bound on the burst length and determines the number of cells that can be transmitted back-to-back, controlling the burst length. The maximum number of cells that leave the leaky-bucket is always greater than the pool size. This is because, while cells arrive and consume tokens, new tokens are generated and placed at the pool.

If an arriving cell enters when there is no token waiting, it will be dropped. Alternatively, instead of being dropped, the arriving cells may be placed in a buffer if they arrive when there is no token available at the pool. When a token arrives, the first cell waiting in the queue at the buffer will immediately consume the arriving token and depart the leaky-bucket. Therefore, there cannot be concurrently any tokens in the token pool as well as cells waiting in the buffer. The buffer is normally of finite size and as indicated before, the cells arriving at a time when the buffer is full, will be discarded. The operation of a leaky-bucket incorporated with a buffer is not transparent to the user as a result of the delays introduced by buffering. The size of the token pool can be set to a large value to reduce such buffering delays, but this would cause large bursts to enter the network, limiting the effectiveness of the method. Further, it may also cause the cell-loss probabilities to increase within the network.

One solution to this problem is the *spacer-controller* (a combination of using a *spacing function* and a *leaky-bucket*), which enforces a time minimum between cells, corresponding to a particular maximum cell-rate. Spacing is performed only on those cells that conform to the traffic contract, and hence this prevents clustered cells (cell-bunching) of the worst case traffic, or that caused by variations in cell delay from entering the network. *Note*: More details on the leaky-bucket scheme are furnished later.

Selective discarding

To compensate for the uncertainties due to the statistical nature of user traffic and to allow higher resource utilization, cells of nonconforming users may be accepted into the network after they are appropriately marked (tagged). When the congestion occurs, these cells can be dropped within the network if necessary. Moreover, users may choose to prioritize their cells before they are transmitted to the network. For example, a voice source may place the most significant bits of its frames into one cell and the least significant bits in another. Since the effect of losing cells of the latter type is less significant than the former type, different loss priorities can be assigned appropriately to the two types of cells. Similar framework can be employed in other types of applications, such as video, data etc.

There are two versions of discarding mechanisms: *Push-out* and *threshold*.

Push-out mechanism of discarding a cell

This mechanism prescribes priorities by replacing high-priority cells in lieu of low-priority ones under congestion. Both low-priority and high-priority cells are accepted into the network as long as there is a space available at the intermediate switch buffer. Whenever the buffer is full, if a low-priority cell arrives at that time, it will be discarded. But as far as a high-priority cell is concerned, it will be discarded only if there is no high-priority cell waiting in the queue; otherwise, it will be taken into the buffer replacing a low-priority cell. The only exception to such replacement is when the low-priority cell has already been committed for transmission.

Threshold technique of discarding a cell

In this method, a *threshold value* is set less than the buffer capacity in order to regulate the buffer occupancy between the high-priority and low-priority cells. Both types of cells are accepted into the queue as long as the total number of cells waiting in the queue is less than or equal to the threshold. Whenever the number of cells in the queue exceeds the threshold, all low-priority cells are discarded until the queue size falls below the threshold value. Again, as long as there is a space available in the queue, high priority cells will continue to enter the queue.

The major concern in this method refers to determining the value of the threshold itself. If it is set to a very low value, low-priority cells at the buffer may be unnecessarily discarded. On the contrary, if the threshold is set to a large value, then the performance of high-priority cells may degrade, inasmuch as there may not be adequate space left to accommodate them. The threshold value depends on the characteristics of both high and low priority cells and their QOS requirements. The threshold mechanism can be implemented on an FIFO basis.

UPC algorithms
The UPC algorithms are as follows:

- ✓ Leaky-bucket algorithm
- ✓ Jumping window algorithm
- ✓ Sliding window algorithm
- ✓ Exponentially-weighted moving average algorithm.

UPC, as defined earlier is a set of policing actions taken by the network to monitor and control traffic in terms of conformity with the agreed traffic contract and cell routing validity at the user access. UPC and NPC are identical except that UPC is performed at UNI and NPC is performed at NNI. The purpose of UPC/NPC is to protect the network resources from malicious, as well as unintentional, misbehavior that could affect the QOS of other established connections.

How is UPC carried out?

UPC is done by detecting violations of negotiated parameters and taking appropriate actions. These actions refer to discarding cells, tagging cells or dropping the connection, as indicated earlier. The location of UPC functions at the end of VP/VC link is indicated in Fig. 5.29.

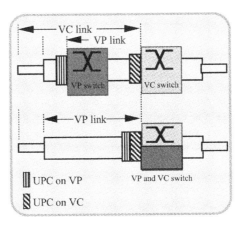

Fig. 5.29: Locales of UPC functions

In reference to the locations shown in Fig. 5.29, the specific functions carried out are:

- Checking the validity of VPI/VCI values; that is, checking whether proper VPI/VCI values have been assigned

- Monitoring the traffic entry from each VCC/VPC in order to check whether agreed parameters have been violated.

Generic cell rate algorithm (GCRA)

GCRA classifies each arriving cell as *conformant* or *nonconformant*. Once a connection is established, the node must have a mechanism to monitor the traffic on that connection so as to ensure that the traffic contract is met. This is referred to as the *generic cell rate algorithm* (GCRA), which determines whether a cell adheres to the provisions of the traffic contract. Two versions exist for GCRA, namely, *virtual scheduling* and *leaky (token) bucket algorithm*.

Virtual scheduling algorithm

The *virtual scheduling algorithm* can be represented in a flow-chart as depicted in Fig. 5.30. It is initialized with the arrival of the first cell at time $t_A(1)$. The algorithm updates the *theoretical arrival time* (T_{AT}), which is the target arrival time of the next cell. A *tolerance* (L) is also associated with this algorithm. The tolerance specifies the amount of time prior to T_{AT} that a cell may arrive at the UNI and still conform to the traffic contract. When the next cell arrives at the user interface, three possible scenarios exist:

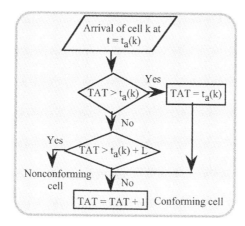

Fig. 5.30: Virtual schedule algorithm

First, the cell can arrive at a time greater than T_{AT}. This cell will be conforming to the contract and will be allowed to enter the network. The next expected arrival time is updated by adding an increment I to the cell's arrival time. The cell can also arrive before T_{AT}, but after the predefined tolerance level $(T_{AT} - L)$. In that case, again this cell is designated to be conforming to the traffic contract. However, the next arrival time is calculated by adding the increment I to the T_{AT} and not the cell's arrival time. Finally, the cell may arrive before $(T_{AT} - L)$. This cell does not comply with the traffic contract and is not allowed to enter the network. T_{AT} remains the same so as not to penalize the user more than once. With T_{AT} remaining the same, the probability increases that the next cell will conform to the traffic contract.

The main problem with the virtual scheduling algorithm is its inflexibility to data bursts. For example, suppose that a long idle period is followed by a transmission of closely spaced cells. Most of these cells will be discarded and may never enter the network. At the same time, this algorithm has no way of crediting the user for the long idle period in which no cells were transmitted. This can severely affect the throughput capacity of VBR services. In order to solve the problems present in VBR services, the *leaky-bucket* (sometimes referred to as the *token bucket*) algorithm can be used as described below.

5.4.5   Leaky-bucket algorithm

Leaky-bucket is a colloquial adjective used to qualify an algorithm adopted for conformance checking of cell flows against a set of traffic parameters. The *"leak rate"* of the *"bucket"* specifies a particular rate while the *"bucket depth"* determines the tolerance for accepting bursts of cells.

ATM Forum defines a reference model vis-à-vis leaky-bucket information pertinent to cell rate conformance in terms of the following parameters and tolerances: PCR, CDV tolerance < SCR, BT and maximum burst size (MBS).

In the leaky-bucket algorithm, the cell must pass a "gate" before entering the network. This gate is referred to as the *leaky-bucket*, which in reality, is just a *counter*. In order for a cell to enter the network, the counter must be non-zero. The counter is incremented (usually by one) n times per second. If the counter is non-zero, the cell will enter the network, and the counter will be decremented. When the counter becomes zero, the cell will not be allowed to enter the network. The leaky-bucket has also a defined maximum value (that is, the counter will not be incremented above a certain limit). For a cell to enter the network, it must remove a token from the bucket. If no tokens are available, then the cell does not enter the network. The bucket is replenished with tokens at some constant rate. When the bucket becomes full, any extra tokens generated are discarded.

In order to make this algorithm more dynamic, a common token pool can exist for multiple logical connections. Fig. 5.31 illustrates the relevant concept. All the logical connections have their own token pool. If an overflow occurs on anyone of these smaller token pools, the excess tokens are stored within a common, larger token pool. If a connection experiences a sudden burst of data, then it can access the extra tokens from the common pool. Priorities may exist on which connection has access to the common token pool.

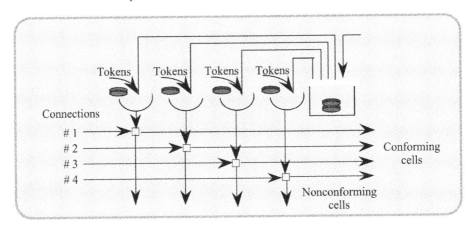

Fig. 5.31: Common token pool

Double leaky-bucket algorithm

In practice, the GCRA can be applied several times to each logical stream. One instance would check for conformance of cells arriving with CLP = 0. This function could tag non-conforming cells and pass them to a second GCRA that applies different criteria (traffic parameters) to the sum of CLP = 0 and CLP = 1 cells. This arrangement is called the *double leaky-bucket algorithm*, which is illustrated in Fig. 5.32.

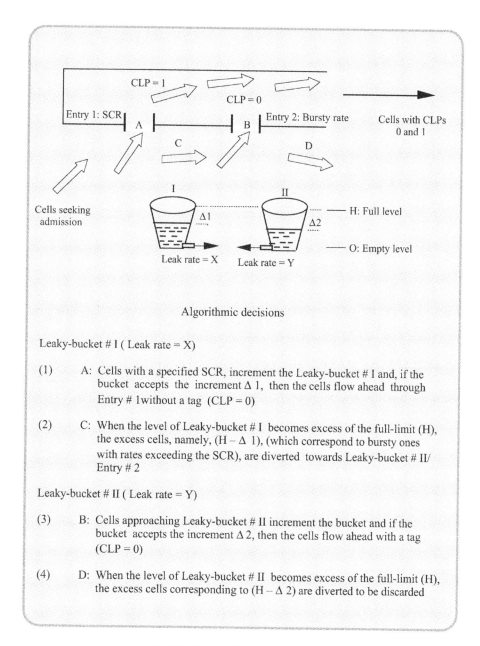

Fig. 5.32: Double leaky-bucket algorithm

Algorithmic decisions

Leaky-bucket # I (Leak rate = X)

(1)   A: Cells with a specified SCR, increment the Leaky-bucket # I and, if the bucket accepts the increment Δ 1, then the cells flow ahead through Entry # 1without a tag (CLP = 0)

(2)   C: When the level of Leaky-bucket # I becomes excess of the full-limit (H), the excess cells, namely, (H – Δ 1), (which correspond to bursty ones with rates exceeding the SCR), are diverted towards Leaky-bucket # II/ Entry # 2

Leaky-bucket # II (Leak rate = Y)

(3)   B: Cells approaching Leaky-bucket # II increment the bucket and if the bucket accepts the increment Δ 2, then the cells flow ahead with a tag (CLP = 0)

(4)   D: When the level of Leaky-bucket # II becomes excess of the full-limit (H), the excess cells corresponding to (H – Δ 2) are diverted to be discarded

More complex traffic descriptors may require more than two GCRA implementations.

GCRA implementation
Peak-cell rate conformance
        The peak-cell rate (PCR) traffic parameter for a connection is defined at the physical layer service access point (PHY-SAP) within an equivalent terminal representing the VPC/VCC in a reference model (Fig. 5.33). Denoting a basic event as a request to send an ATM-PDU (*protocol data unit*) in the equivalent terminal, the peak cell rate (R_p) of the ATM connection is defined as the inverse of the minimum interarrival time T_p between two consecutive basic events. T_p ($= 1/R_p$) is denoted as the peak emission interval of the ATM connection.

517

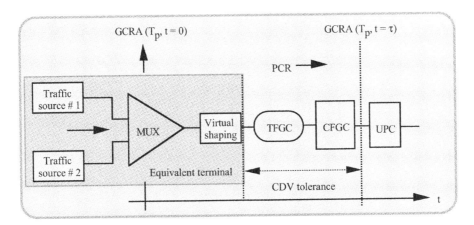

Fig. 5.33: Reference model (GCRA: Generic cell rate algorithm, CFGC: CPE functions generating CDV, and TFGC: TE Functions generating CDV)

The PCR is a mandatory source traffic parameter and it applies to ATM connections supporting both CBR and VBR services. Due to a possible multiplexing process (MUX) before the PHY-SAP, a virtual shaping function has been defined in the equivalent terminal (Fig. 5.33) to allow an unambiguous definition of the peak cell rate at the PHY-SAP.

Conformance of sustainable cell rate and burst tolerance

When ATM connections support VBR services, the peak rate provides an upper-bound of the connection cell rate. However, in order to allow the network to allocate resources more efficiently, an additional traffic parameter set, such as the sustainable cell rate (SCR) and burst tolerance τ_s, is necessary.

The (SCR, R_s) is an optional source traffic parameter set. A user/terminal may choose to place an upper-bound on the realized average cell rate of an ATM connection to a value below the peak cell rate. In order to be useful to the network provider, the SCR value must be less than the PCR value. (For CBR connections, the user would not declare the SCR and would only negotiate the PCR.)

The sustainable cell rate traffic parameter for a VPC/VCC supporting VBR service, is defined at the PHY-SAP within the equivalent terminal of the reference model. If a request to send an ATM-PDU in the equivalent terminal refers to a basic event, then the sustainable cell rate (R_s) of the ATM connection is the inverse of a average interarrival time T_s, between two basic events. Consequently $R_s = 1/T_s$.

The burst tolerance τ_s is a source traffic parameter and reflects the "timescale" during which cell rate fluctuations are tolerated. It is defined in relation to the sustainable cell rate according to the algorithm GCRA (T_s, τ_s) and determines an upper bound on the length of a burst transmitted in compliance with the connection's peak cell rate.

Conformance of cell delay variation tolerance

As indicated before, ATM layer functions may alter the traffic characteristics of ATM connections by introducing cell delay variation (CDV). When cells from two or more ATM connections are multiplexed, cells of a given ATM connection may be delayed while cells of another ATM connection are being inserted at the output of this multiplexer. Also customer equipment (CE) may introduce CDV. Further PCR and SCR would implicitly influence the CDV in a VBR service.

The other causes for CDV are as follows: Some cells may be delayed when the physical layer overhead or OAM&P cells are inserted. Therefore, some randomness affects the interarrival time between consecutive VPC/VCC cells as monitored by the GCRA at the UNI/NNI. This implies that UPC/NPC functions cannot solely rely on the PCR traffic parameter. In addition, the degree of the interarrival time distortion has to be defined and is called *CDV tolerance*, (τ).

The CDV tolerance, τ, is defined in relation to the peak cell rate according to the algorithm GCRA (T_p, τ), where T_p is the inverse of R_p (the peak cell rate). The CDV tolerance allocated to a particular VPC/VCC at the UNI/NNI represents, at this interface, a quantitative measure of the VPC/VCC cell clumping phenomenon due to the slotted nature of the ATM, the physical layer overhead, and the ATM layer functions. Consequently, τ is not a source traffic parameter. A user may explicitly or implicitly select a value for the CDV tolerance at the UNI/NNI for an ATM connection from amongst a set of values, which could be supported by the network.

To summarize, the connection traffic descriptor specifies the source traffic descriptors (peak cell rate, sustainable cell rate, and the burst tolerance) defined at the PHY-SAP within an equivalent terminal and the CDV tolerance (τ) specified at the UNI/NNI. The burst tolerance, τ_s accounts for the tolerance in relation to the SCR at the PHY-SAP while the CDV tolerance specified at the UNI/NNI is defined in relation to the peak cell rate by the GCRA (T, τ) rule. In Fig. 5.33, the burst tolerance τ_s^* specified at the UNI/NNI takes the CDV tolerance τ into account and is also defined in relation to the sustainable cell rate by the GCRA (T_s, τ_s^*) rule.

Granularity of ATM traffic parameters

UPC/NPC functions cannot be requested to handle any specific parameter value. Due to hardware limitations, only a finite and discrete set of PCR, CDV tolerance τ, SCR, and burst tolerance τ_s can actually be handled at the ATM layer.

The set of PCR values is referred to as the ATM peak cell rate granularity in ITU-T Recommendation I.371 and has been standardized to allow for simple interworking. In general, PCR and SCR have identical granularity.

Window-based algorithms
Jumping window (JW) technique

The *jumping window* (JW) places an upper bound m on the number of cells accepted from a source over a fixed time interval T. This fixed time interval is known as a *window*. It is assumed that a new interval starts immediately after the end of the preceding interval. That is, a new window 'jumps in" at the end of the previous one. Now, the traffic flow is regulated as follows: Once m cells are received and transported, all consecutively arriving cells are dropped and hence prevented to enter the network. Similar to the leaky-bucket scheme, these excessive cells can also be marked to have a low priority using the CLP bit in the cell and these cells can be accepted (instead of being dropped), if such an implementation is provisioned. The rate at which cells are offered to the network (λ_p) is equal to m/T where T ($= N$) is the *token generation time*. Suppose m = 6, and T = 12, $\lambda_p = 0.5$; or, with m = 12, T = 24, again $\lambda_p = 0.5$. That is, the rate at which cells are offered to the network is decided by the ratio m/T. *Then what is the difference when distinct rates of m are used?*
The difference has the implication to the extent of the actual number of cells submitted to the network during a window. Relevant considerations are:

- If N = T is large, so as to accommodate large m, then it takes a larger time to find out that the controlled rate is exceeded

- When N = T is small, then the JW scheme becomes nontransparent to users in reminding them to stay within the negotiated parameters.

Example
Determine the probability that the JW scheme may drop a cell originating from a nonviolating source.

Solution
The probability (P_v) of dropping a cell from a nonviolating source can be defined as follows:

$$P_V = \left(\frac{\text{Number of cells dropped}}{\text{Total number of cells generated}} \right)_{\text{Nonviolating source}}$$

$$= \left(\frac{\text{Total number of cells generated} - \text{Number of cells permitted to pass}}{\text{Total number of cells generated}} \right)_{\text{Nonviolating sour}}$$

$$= \frac{\lambda T - \left[N + \sum_{i=o}^{m-1} (N-i)X_i(T) \right]}{\lambda T}$$

where $X_i(T)$ is a random variable depicting the number of cells generated by a source during the interval, T and λ is the mean cell generation rate of the source.

Triggered jumping window mechanism

In the JW scheme discussed above, the window is not synchronized with the source activity. In the *triggered jumping window* (TJW) mechanism on the other hand, the window is triggered by the first arriving cell (of the busy period). As a result, consecutive windows in TJW are not necessarily sequential, but are triggered as per the first arriving cell.

Thus, the major difference between JW and TJW is that in a JW, the beginning of a window does not correspond to an arrival, whereas in a TJW the window starts with an arrival. This difference implicitly refers to the nature of the probability distribution of the number of arrivals during a period of time.

Sliding (moving) window (MW)

Similar to JW mechanism, in the MW technique, the maximum number of cells allowed during a predefined interval of time is constant. The difference between JW and MW schemes is that, in MW, each cell is remembered exactly over one window (that is, T time units). Therefore, the MW scheme can be interpreted as a window that is steadily "sliding" (moving) along the time axis.

To deduce the violation probability, the MW scheme can be modeled as an m-server queue with the service time equal to the duration of the window. A cell arriving when there are m cells in the system is assumed to be lost. The cell loss probability is then equal to the violation probability.

Exponentially weighted moving average

The exponentially weighted moving-average (EWMA) mechanism is similar to the JW scheme. Here again, the window size T is constant and a new window is triggered immediately after the preceding one ends. The difference between the two is that, in EWMA, the number of cells X_i, accepted during the ith window, varies from one window to the next. The maximum number of cells permitted within a fixed time, i^{th} window is a function of the mean number of cells (M) per window. This condition leads to an exponentially-weighted sum of the cells (m_i) accepted in the successive windows as given by:

$$m_i = \frac{\left[(M - (1 - \Gamma)(\Gamma X_{i-1} + ... + \Gamma^{i-1} X_i) \right] - \Gamma^{i+1} S_0}{(1 - \Gamma)} \qquad (5.10)$$

where S_0 is the initial value of the EWMA metric. The weight factor Γ decides the number of relevant preceding windows, which influence the number of cells permitted in the current window. A nonzero value of Γ permits more burstiness. For a value of $\Gamma = 0.8$, up to $5 \times M$ number of cells can occur in the first window. Thus, a large value of Γ increases the reaction time and it has been shown that the dynamic behavior of EWMA is the worst. Moreover, the implementation complexity of this scheme is higher than the leaky-bucket and other window-based schemes. The EWMA metric corresponding to an ith window, namely S_i can be specified by $S_i = (1 - \Delta) X_i + S_{i-1}$, where Δ is a factor that controls the flexibility of the mechanism with respect to the burstiness of the traffic. If $\Delta = 0$, m_i is constant and EWMA $\to$ JW.

Window-based schemes: A relative appraisal

It is generally agreed that the leaky-bucket performs better than window-based schemes, and is also recommended by ITU-T and the ATM Forum as the policing mechanism of choice for B-ISDN. In reality, neither type satisfactorily solves the problem of policing the source behavior, in which the metrics of interest to be included are the peak and average bit rates, burst duration, and the distributions of the active and silent periods. In general, all of these parameters are needed to characterize the traffic in the network accurately so as to monitor and control them effectively. But some of the problems associated thereof include the following:

- The average cell rate is often estimated from a small number of samples. This may lead to large errors due to poor statistics and hence incorrect policing decisions

- The time involved in reacting and taking appropriate corrective actions increases with the window size. As the sampling interval increases, the excessive cells might have already been submitted to the network by the time the source is detected to be violating its negotiated contract

- Inasmuch as there is some uncertainty in the characterization of source parameters themselves, the parameters of policing mechanisms are not accurately specified. This limits the effectiveness of the implementations involved.

Nevertheless, the leaky-bucket and the EWMA are regarded as effective in coping with short-term fluctuations in the cell streams.

Priority control

ATM cells have an explicit cell-loss priority bit in the header. Hence, at least two different ATM priority classes can be distinguished as indicated below:

- *Cell-loss based priority classes*
  Different traffic services are distinctly prioritized with respect to cell-loss that can be tolerated by each service. For example, considering a multimedia traffic environment, as shown in Fig. 5.74, the cell-loss probability which is acceptable for each type of traffic ranges from 10^{-4} for file transfer to 10^{-10} for interactive compressed video.

- *Cell-delay based priority classes*
  These refer to those classes of traffic that can be distinguished on the basis of the acceptable extent of CDV each traffic may suffer as indicated in Fig. 5.34.

These refer to service-specified priorities, namely, priority on semantic transparency, (cell-losses) and on time transparency (CTD, CDV).

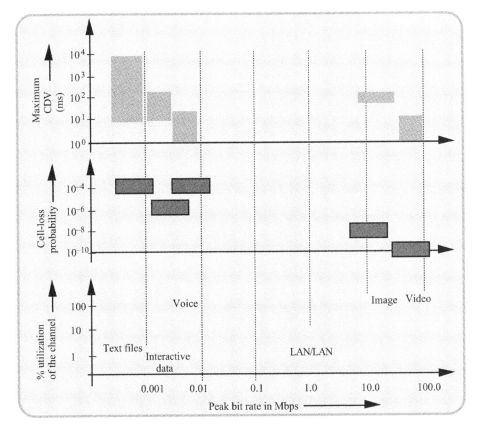

Fig. 5.34: Cell-loss and cell-delay-based traffic classes

How to implement priority control?

Priority control is done (with respect to cell-loss priority) by means of *"buffering"*. There are three versions of buffers, namely,

- *Common buffer with push-out mechanism* (for two cell-loss priorities)

    ✓ Cells of both priorities share a common buffer
    ✓ If the buffer is full and a high priority cell arrives, a cell with a low priority (already residing in the buffer) will be pushed out and lost
    ✓ Buffer management, however, would maintain the cell sequence integrity (But this is a complex mechanism)

- *Partial buffer-sharing mechanism*

    ✓ Access to low priority cells is denied, if the buffer-filling is more than a specified threshold level, (S), and S is less than the total buffer capacity
    ✓ High priority cells can access the buffer unconditionally
    ✓ The threshold level, S can be adjusted for optimal load conditions adaptively

- *Buffer separation mechanism*

  ✓ Here, two different buffers are used for two different priorities. But cell sequence integrity can be maintained only if a single priority is assigned to each connection.

Advantages of cell-loss priority (CLP) control

  ✓ Significant improvement in admissible traffic load
  ✓ CLP control allows smaller buffer sizes to be chosen, thereby reducing overall implementation complexity.

How is cell loss priority set?

Cell-loss priority is set via the payload type indication (PTI) in the ATM cell-header. The CLP field is 1 bit long. Due to the statistical multiplexing of connections, cell losses are inevitable in ATM networks. In such events, depending on the CLP set in any cell, the network may discard a cell during congestion. In particular, cells that arrive at a time when the transmission link buffer is full may be dropped at a switch.

5.4.6   Traffic shaping

VBR services and their delay bound attributes

The traffic generated by networking applications generally either alternates between the active and silent periods or has a varying bit rate generated continuously. Further, the peak-to-average bit rate of a VBR source is often much greater than 1. Presenting VBR traffic to the networks as CBR traffic by means of buffering or, instead, by artificially controlling its bit generation rate has the drawbacks of network resources under utilization and QOS degradation. Although CBR service simplifies the network management and control tasks, it is more natural to provide VBR services to VBR sources and thereby extend a better service as well as a framework to achieve higher resource utilization. For example, a voice source alternates between active and silent periods. Similarly, a video source (generally depending on the coding scheme used) generates a continuous bit stream at varying rates. Taking advantage of such source activities, VBR services can be facilitated with multiplexing gains. For example, studies indicate that about 2.5:1 multiplexing gains in bit rate compared to carrying a typical set of multiplexed videos in CBR mode can be achieved via facilitating multiplexed VBR service.

Bounds on CBV of VBR traffic

When tolerances are introduced in the UPC/NPC functions, distinct traffic patterns can be considered as compliant to the traffic contract and they can pass transparently through the UPC/NPC. Since the network is committed to achieving the requested QOS required by those connections, which comply to their traffic contracts, a conservative network allocation scheme is normally considered based on the most resource-demanding traffic pattern among all the compliant ones. This particular traffic pattern is often referred to as the *worst case traffic pattern* of the ON-OFF type. In this NPC-specified pattern, the maximum number of cells in the ON period is a function of T (T_s) and τ (τ_{s*}). Considering UPC monitoring, the maximum number of cells in the ON period is a function of peak and sustainable cell rates. The ON period for the PCR case is characterized by a clump of back-to-back cells at full link rate (see Fig. 5.35) while, for the SCR case, it is described by a burst of consecutive cells emitted at a peak cell rate.

Each source is characterized by a recurrent cell interarrival time pattern, decided by the following parameters at the UNI:

T_{min}:   Minimum cell interarrival time.
T_{max}:   Maximum cell interarrival time.
T_{rec}:   Period of the recurrent interarrival time pattern.

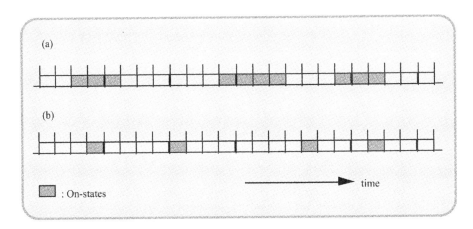

Fig. 5.35: Worst-case ON-OFF patterns: (a) Back-to-back cells are not allowed; burstiness, B = 1 and (b) Back-to-back cells allowed; B > 0

The identified type of deterministic ON-OFF sources is offered to a single stage multiplexed FIFO output queue of infinite capacity. A random phasing of the sources with respect to each other, uniformly distributed over T_{rec}, is assumed. In this way representative results for the buffer occupancy distribution can be derived from the aggregate superposition process.

Thus traffic in ATM transmissions can be dynamically shaped; that is, traffic characteristics can be altered on an ad hoc basis on a VPC/VCC in order to reduce the peak cell rate, limit of burst length, or reduce CDV. Such alteration can be done by spacing the cells on a time scale, while maintaining the cell sequence integrity. This *"traffic shaping"* and UPC can be combined so that the negotiated parameters can be rescheduled to achieve an improved network performance.

Traffic shaping refers to altering the traffic characteristics (that is, the inter-arrival times of the cells) of a VCC/VPC in order to achieve a desired modification of the traffic characteristics (such as reducing peak cell rate, limiting burst length, or reducing CDV). In its simplest form, traffic shaping does cell-spacing so as to smooth out the peaks in a cell rate at the expense of adding in more delay. However, this will not upset the cell sequence integrity. The salient features of traffic shaping are as follows:

- Cell-spacing compensates CDV and its consequent impact on peak cell rate by emitting cells at a declared maximum cell rate (that is, an inverse of minimum interarrival time). This avoids cell bunching due to CDV and leads to considerable performance improvement (*Note*: Cell bunching reduces the admissible traffic load within the network.)

- Another mechanism of traffic shaping is as follows: Phase relationship between different periodic cell streams with identical period lengths is adjusted by means of a suitable *shaping function*. This leads to an optimum scheduling of cells of different connections originating within the same access network

- A third traffic shaping method refers to adapting the peak cell-rate at which the next burst will be sent into the network so as to enforce an effective cell-rate used by CAC (This scheme refers to a *linear CAC scheme.*)

- Traffic shaping is an option at the user level as well as at the network operator's level

- Within a customer network, it should be in conformance to the traffic contract of the traffic across UNI

■   In the network-operators point of view, traffic shaping enables sizing the network cost effectively.

Control methods: Algorithms for traffic shaping

The two major algorithms considered for traffic shaping are:

✓   The window method indicated earlier, which limits the number of cells in a time-window
✓   The *leaky-bucket procedure* (LBP), which increments a counter for each cell arrival and decrements this counter periodically as explained before.

Traffic shaping via LBP algorithm

A general framework explaining the performance and underlying principle of a *traffic shaper* is as follows: The source is characterized by a peak rate λ_p, an average rate λ_a, and mean ON duration T_{ON}. It is assumed that the network access link at the output of the traffic shaper has a capacity equal to the peak rate of the source stream. Hence, any burst arrival is serviced fastest at the peak rate. A traffic shaper, which closely fits the model above is the *LBP with a peak prescribed rate policy*. In the following paragraphs, the characteristics of a LBP based output traffic is described.

In a generalized model of the leaky-bucket shown in Fig. 5.36, tokens are generated at a fixed rate as long as the token buffer of size b is not full, and $\lambda_a < \lambda_{T0} < \lambda_p$.

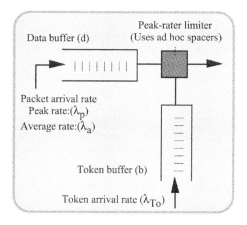

Fig. 5.36: A generalized LBP frame work

When a packet arrives from the source, it is released into the network only if there is at least one token in the token buffer. This scheme enforces the token arrival rate λ_{T0} on the input stream. Clearly, λ_{T0} should be greater than the average (packet) arrival rate λ_a (for stability) and should be less than the peak arrival rate λ_p so as to achieve bandwidth utilization. An input data buffer of size d permits statistical variations. An arriving packet finding the input buffer full is said to be a violating packet and can be dropped or tagged for preferential treatment at the switching nodes.

It is presumed that a peak-rate limiting spacer is an integral part of the leaky-bucket mechanism. When a burst of data arrives at the input, even if enough tokens are present, the packets are not instantaneously released into the network. Successive packets are delayed by the transmission time at the negotiated peak rate λ_p, where $\tau = 1/\lambda_p$. The LBP may be used to designate the leaky-bucket with a peak rate policer.

For the leaky-bucket parameters defined above, maximum burst size at the output is $b' = b / (1 - \lambda_{T0}/\lambda_p)$. This includes the new tokens that arrive during the transmission of the first b packets. The output of the leaky-bucket is characterized as follows:

- *Maximum burst size*

  For the LBP, the maximum burst size at the output is $b' = b/(1 - \lambda_{T0}/\lambda_p)$. This is obtained as follows. It is assumed that the largest burst starts at t_1. This would be possible only if the source generated an input burst after a prolonged OFF period of b/λ_{T0} where b is the token buffer size. Since the burst service is not instantaneous due to the peak rate policer, more tokens may arrive during the consumption of the existing tokens. Since tokens are removed at λ_p and arrive at λ_{T0}, the instantaneous token count in TB will be $b(t) = b + (\lambda_{T0} - \lambda_p).t$ and hence the token bucket empties at time $b/(\lambda_p - \lambda_{T0})$. The maximum burst size b' then becomes $b/(1 - \lambda_{T0} - \lambda_p)$.

- *Long-term output smoothness*

  Over a large time duration T, the number of packets sent out by the leaky-bucket $n(T)$ is $\leq (\lambda_{T0} \times T) = n_{T0}$. This relationship is also true for any time duration T starting from zero or any epoch when token buffer becomes empty. It is assumed here that the token buffer is empty at $t = 0$.

- *Short-term burstiness*

  Over the duration smaller than T mentioned in the previous item and exceeding the maximum burst size, the leaky-bucket output can be modeled as a *linear bounded arrival process* (LBAP) with parameters (α, ρ). Here, α represents the maximum burst size b' and ρ represents the token rate λ_{T0}. In terms of the smoothness that for any T starting from 0 (or from any epoch when token buffer is empty), LBP output is (n_{T0}, T) smooth.

What does traffic shaping do?

During the active period, the cells are buffered before they enter the network and form a queue. The departure rate from this buffered (queued) state is set less than the peak rate, (arrival rate) λ_p. This is done at the source equipment or at the network access point.

Example 5.8
Illustrate the effectiveness of traffic shaping on ATM traffic.

Solution

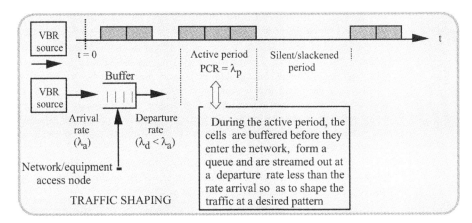

Fig. 5.37: Bursty and slackened bit transmission duration in a VBR traffic

Suppose the amount of bandwidth reserved for a connection is BW and the average bit rate (λ_a) < BW < peak bit rate (λ_p). Here, λ_a and λ_p are the lower bound and the upper bound of BW respectively. Suppose a VBR source is considered. The corresponding shaped traffic profile is illustrated in Fig. 5.37.

Variation of leaky-bucket schemes: A summary

The basic concept behind LBP, (as indicated earlier) is that a cell, before entering the network, should obtain a token from a token pool. An arriving cell will consume one token and immediately depart from the leaky-bucket if there is at least one available in the token pool. Tokens are generated at a constant rate and placed in a token pool. There is an upper bound on the number of tokens that can be waiting in the pool and tokens arriving at a time when the token pool is full are discarded. The size of the token pool imposes an upper bound on the number of tokens that can be waiting in the pool and tokens arriving at a time the token pool is full are discarded. The size of the token pool imposes an upper bound on the burst length and determines the number of cells that can be transmitted back-to-back, controlling the burst length. The maximum number of cells that can exit the leaky-bucket is greater than the pool size, since while cells arrive and consume tokens, new tokens are generated and placed at the pool. The following are typical variations of leaky-bucket based enforcement schemes (Fig. 5.38).

Scheme type 1

An arriving cell is dropped if it arrives when there is no token waiting in the token pool. Since tokens are generated at a constant rate, this scheme can be used to control either the peak or the average cell transmission rate (but not both).

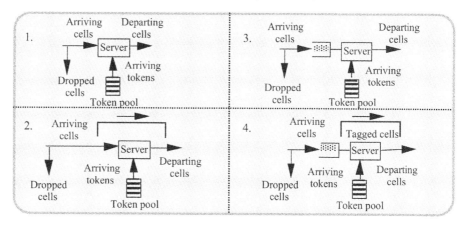

Fig. 5.38: LBP schemes: (a) Type 1 (b) type 2 (c) type 3 and (d) type 4

Instead of being dropped, the arriving cells in types 2 and 4 are placed in a buffer if they arrive when there is no token available at the pool. If there is a cell waiting in the queue when a token arrives, then the first cell in the queue will immediately consume the arriving token and exit the leaky-bucket. Hence, there cannot be any tokens in the token pool and cells in the buffer simultaneously. The buffer size is finite and cells arriving at a time when the buffer is full are discarded. In the buffered version, the operation of the leaky-bucket is no longer transparent to the user due to delays introduced by buffering. The token pool size can be set to a large value to reduce the buffering delays, but this would cause large bursts to enter the network, limiting the effectiveness of the method, and would cause the cell loss probabilities to increase within the network. Type 3 is identical to Type 1 except the tagging of cells is facilitated.

An alternative to discarding cells before they enter the network is to allow them to enter and discard them within the network at the congested nodes. The main reasoning behind doing this is to increase the resource utilization in the network. Furthermore, due to the statistical nature of user traffic, this approach provides a safeguard mechanism that penalizes sources for not knowing their traffic characteristics more accurately or for transmitting excess traffic in a short period of time. Types 1 and 3, or 2 and 4, are essentially the same, except that types 2 and 4 allow cells of nonconforming sources to enter the network. It is necessary to guarantee that the traffic of conforming sources is not negatively affected due to this excess traffic. The ATM cell header includes a reserved CLP bit that can be used for this purpose. As long as the source is observed to be staying within its connection parameters, cells are transmitted to the network with their CLP bits unset. As soon as the policing function detects that a source is nonconforming, the CLP bit is set before cells are transmitted. This is repeated for each cell until the source is observed to be conforming again. This framework is sufficient for the network to treat these cells differently. In particular, a network node discards the cells with CLP bits set before they discard cells with CLP bits unset when congestion occurs.

The selective discarding of cells indicated above refers to the following:

□   *Policing schemes are designed to protect the network from the excess traffic generated as well as to protect the conforming users from unconforming users*

□   *However, sometimes unconforming users are allowed/admitted because:*

     •   *The link may not be fully utilized all the time due to the statistical nature of the traffic*

     •   *Higher utilizations of resources may be necessary.*

Such unconforming user admission is done on the assumption that cell-dropping is requested on an ad hoc basis, only when the congestion occurs. Selective discarding of cells is done via push-out or threshold schemes indicated earlier. Apart from the leak-bucket technique, the other UPC mechanism available refers to the various window techniques described earlier namely, sliding window, jumping window, and exponentially moving average window. Among these, the leaky-bucket and exponentially moving average window techniques are proven to be flexible and less complex in implementations.

Fast resource management (FRM)

This is concerned with bursty traffics. In response to a user request to send a burst, the network may allocate appropriate capacity (such as bandwidth and buffer size). This should, however, be done with minimal delay. Corresponding resource management undertaken refers to *FRM*. That is, for services with stringent delay requirements, fast resource management schemes should be implemented in hardware so as to limit the resource negotiation phase to the propagation delay. FRM introduces a new QOS parameter, namely, *burst blocking probability*, which should be controlled via adequate traffic control mechanism.

5.5   Traffic Congestion Control

ATM is a kind of packet-switched network. Two types of traffic control, in general, exist and are known as *flow control* and *congestion control*. The flow control as explained in the previous section, is concerned with the regulation of the rate at which the sender transmits packets of data to match the rate at the receiver obtains so that it is not overwhelmed. It is done with various types of leaky-bucket and/or window techniques. Congestion control tries to reduce the possibility of congestion within the network. Each node in the network can regulate the traffic flow on its input links by forcing them to slow down their transmissions as the possibility of congestion increases. Congestion control mechanisms in ATM networks are classified into two categories: *Preventive* and *reactive* controls.

Preventive control

Preventive congestion control is the technique that attempts to prevent occurrence of congestion by taking appropriate actions. Preventive control is used to ensure that network traffic will never reach an unacceptable congestion level; however, It is not sufficient to eliminate congestion problems in ATM networks.

Reactive congestion

Because of the statistical nature of ATM traffic, it is not possible to totally eliminate the short periods of cell-losses in the network. The preventive techniques may reduce the buffer overflow probabilities, but only to a certain extent. Cell-loss, if it persists, would require retransmissions by the source nodes, thereby increasing network traffic and leading to the momentary buffer overflows and sustained periods of cell losses. This in turn may cause a crash, reducing the effective network throughput eventually to zero. Also, most preventive techniques allocate resources more than necessary, thereby restricting the efficient use of network resources. Finally, preventive schemes often require accurate source characterization. Therefore, in addition to their preventive counterparts, reactive control mechanisms are necessary to monitor the congestion level in the network, notify sources when congestion is detected, and take appropriate action based on the congestion information. Thus, the main objective of a reactive scheme is to prevent "momentary periods of overload from turning into sustained periods of cell-losses". Reactive control mechanisms have been satisfactorily used in the low-speed packet-switch network. However, the time to react to congestion increases proportionally to the (propagation delay × bandwidth) product, decreasing its effectiveness. As the link speeds increase, most cells in the network are in transit and are not stagnated at switch buffers. For example, the number of cells in transit on a 5,000-km OC-3 link could be more than 9,000, assuming a 5-μs/km propagation delay. In reference to applications specified by delay considerations, the buffer sizes required at the intermediate nodes to store such large numbers of cells in transit momentarily during a congestion at the downstream nodes may be prohibitively expensive or may not even be feasible at all. Cell buffers at the intermediate nodes are expected to be in the order of a few hundred cells, mainly to limit the end-to-end delays and the cell delay variations. More so, by the time the sources are informed of the congestion in the network, it may rather be too late to react effectively. Further, by the time sources are ready to react, there may no longer be a congestion to react to at all.

In short, reactive congestion control is a scheme provided for reacting when congestion occurs. It is a technique adopted to manage the traffic when real-time congestion occurs.

Since reducing network traffic upon congestion may cause degradation in throughput, still another alternate approach to reactive congestion control is feasible. It is based on a technique that creates the resources dynamically.

There are three network considerations that have generally been regarded as crucial in the context of network congestion: *Buffer space*, *link bandwidth*, and *processing speed*. The design of conventional communication protocols aims to maximize the link bandwidth utilization and minimize the use of buffers. Due to current technological advancements, fiber optic links can now offer almost error-free transmissions and relatively unlimited bandwidth.

In B-ISDN, *congestion* is defined as a state of network elements (such as switches, concentrators and transmission links) in which the network may not able to meet the negotiated quality of service objectives for the connections already established or for any new connection requests. This is because of traffic overload or control-resource overload conditions that may prevail. Congestion can be caused by unpredictable fluctuations of traffic flows or by fault conditions within the network. Congestion is to be distinguished from *queue saturation*, which may happen while still remaining within the negotiated quality of service.

Congestion control in a network means minimizing congestion effects and preventing congestion from proliferating. It can employ CAC and/or, UPC and NPC procedures to avoid overloading situations. For example, congestion control can reduce the peak bit rate available to a user and perform policing so that the new value is not exceeded. This is a prudent approach that can be categorized as preventive control.

Congestion in any telecommunication network occurs when the input rate of data exceeds the output link capacity. That is, when the sum of the input rates is greater than the available link capacity. A specific question then arises, "*What causes congestion?*"

The goal of ATM is to provide better bandwidth and resource utilization throughout the network. In order to accomplish this goal, the network allocates resources to users only when the user has something to transmit. Hence, *resource sharing* is invariably employed. In this endeavor, network resources are shared both along the links (trunks) as well as within the nodes (routers, hubs, switches). However, whenever resources are shared, there is a possibility that multiple users may try to gain access to the same resource (and this is called *contention*). In order to resolve the contention, usually a queue is associated with that resource at its entry point. The presence of a queue obviously implies both delay (that is, waiting for service) and possible variations in that delay (jitter) as well. Fig. 5.39 shows the effects of queue length as the data arriving to a node is increasing.

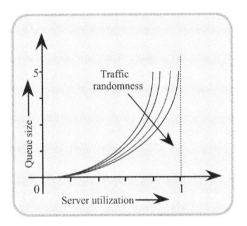

Fig. 5.39: Utilization versus queue length

Utilization indicated in Fig. 5.39, can be defined as the ratio between the average arrival rate to the average outgoing rate at a node. It is usually expressed as the percentage of time the server is busy. As depicted in Fig. 5, when the average rate of arrivals approaches the average rate of transaction processing (that is when utilization $\rightarrow 1$), the queue length would tend to infinity. This means that more data will be waiting to be processed at a particular node resulting in congestion at that node.

With the possibility and potentials of increasing queue lengths with the types of sources handled in ATM-centric networks, a set of network protocols are warranted to combat the associated problem. However, not all protocols may decrease the network congestion. For example, Fig. 5.40 shows how typically the throughput varies within a network as the load increases for different types of protocols.

In order to understand the significance of Fig. 5.40, it is necessary to understand the concept of throughput. *Throughput* is defined as the amount of data *successfully* being delivered to the end node. The curve A in Fig. 5.40 is typical of how the ethernet, IP, and ATM support certain types of traffic considerations. It can be observed that, as the load is increased, initially the network can handle the incoming data. However, a point is reached when the throughput of the network actually begins to decrease with increases in offered load. This effect is due to collisions within the network. An example pertinent to ATM can illustrate this scenario: Suppose congestion is detected within the network, and to resolve the situation a node discards say, 500 cells. It is very unlikely that all the cells at the node came from a single logical connection. Since ATM has no way of determining which cells were discarded for a particular logical connection, the user may have to re-transmit the entire packet of data. Assume that all 500 discarded cells were from different logical connections, and each connection was trying to transmit a stretch of some fifty cells. This means that discarding 500 cells would require a re-transmission of 25,000 cells. This happens concurrently to the situation that the network is already congested leading to a crash of the network. As a remedy, one might suggest implementing traditional techniques adopted in the packet networks (X.25) such as automatic repeat request (ARQ). With the use of ARQ, the sender has to retransmit only certain portions of the loss/errored data. However, these techniques are not suited for the high-speed environment in B-ISDN/ATM. They require complex software processing at the network nodes which slows down the routing of cells. This would be devastating to the delay-sensitive traffics such as voice and video handled in B-ISDN/ATM networking.

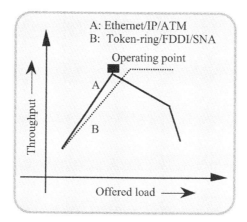

Fig. 5.40: Load versus throughput for various protocols

The curve B in Fig. 5.40 depicts the effects of a well-controlled network. As the load increases, the network continues to operate at full throughput capacity. This is possible if the queues in this scenario are located at the end-user devices (rather than in the intermediate nodes) so as to prevent overflow of the network queues. Obviously, this results in developing complex user devices.

Misconceptions about congestion control

A formal solution to congestion control in any telecommunication system is rather a complex issue. It depends on traffic parameters and services rendered. As evident from the previous section, the solution becomes even more complicated in ATM networks due to traffic characteristics and QOS guarantees. Further, there are a few misunderstandings considering the causes and solution techniques adopted towards congestion control.

First, one might assume that congestion is simply caused by the shortage of buffers. Thus, adding more buffers within the network may alleviate the congestion, but enhancing buffer size is by no means a cure-all strategy. Larger buffers are useful for only short-term congestion and will introduce undesirably long delays within the network. In essence, large buffers may simply postpone the discarding of cells, but they do not prevent it. Besides, long queue delays accommodated by large buffers are unacceptable for certain types of services.

Since congestion is caused by slow links, it is also a notion that by increasing the speed of the links one can simply solve the congestion problem. This is not always the case, and in some instances this solution might actually aggravate the congestion.

In the topology of Fig. 5.41, if both sources (S_1 and S_2) increase their peak rate to the destinations (D_1 and D_2), then congestion will invariably occur at the switch. That is, higher speed links will cause congestion to get worse at the switch.

A third conceivable proposal on congestion avoidance could be a strategy to simply increase the processing speed of the switch. Again, referring to Fig. 5.41, one can see the falsity of this solution. Faster processes will transmit more data over a specified time. If several nodes begin to transmit at their peak rate to a single switch simultaneously, the switch will be, however, overwhelmed quickly.

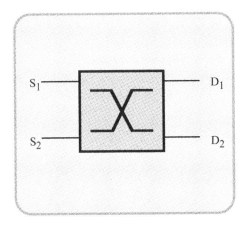

Fig. 5.41: A simple network topology

5.5.1   Congestion control techniques

Congestion control mechanisms are responsible for equalizing the input rates to the available link capacities. There are several mechanisms available to accomplish this, and the selection of which control scheme to implement is dependent upon the severity and duration of the congestion. Fig. 5.41 shows the relevant congestion techniques appropriate for specified congestion duration periods. As shown in Fig. 5.42, networks experiencing constant congestion require higher speed links and a redesign of the network topology to meet user demands. For sporadic congestion, calls can be routed based upon the load level existing on the links and rejection of any new connections can be implemented if all links are fully loaded. This is referred to as the *connection admission control* (CAC) as discussed before. For shorter congestion periods, an end-to-end or link-by-link feedback control scheme can be used. If the congestion occurs in spikes, then additional buffers may be used to resolve the problem. Note that solutions for short-term congestion are not the same as for long-term congestion, and vice versa. Thus, a combination of various techniques can be used inasmuch as all networks experience different lengths of congestion at different times.

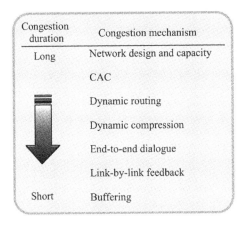

Fig. 5.42: Congestion duration and congestion control strategies

Link capacity

A question that arises in telecommunications is how to distribute fairly the connection speed of a particular trunk among various multiple users. An answer to this question should be known before other parameters are considered towards control strategies. If the link capacity (cell/bit rate) for a particular user is unknown, then connection admission control rather becomes a moot point of interest.

In a shared network, the throughput of any particular link depends on the demands stemming from all the users. The so-called "*max-min allocation*" is the most commonly used criterion for the correct share of bandwidth among all the sources. It provides the maximum possible bandwidth to the source, which is limited by other connection constraints.

Mathematically, the max-min algorithm can be defined as follows: Suppose there are n contending sources with the ith source receiving x_i amount of bandwidth. The allocation vector can then be represented as $\{x_1, x_2, \dots x_n\}$ and is feasible if the sum of these link load levels is less than or equal to 100%. The "*unhappiest source*" is defined as the source receiving the least allocation of the resource. In using the algorithm, first the vector that gives the maximum possible bandwidth to the unhappiest source is found. Next this "unhappiest source" is taken out, thereby reducing the problem to $(n-1)$ size. This procedure is continued for all the remaining sources.

Example 5.9

Suppose that a network has a topology as shown in Fig. 5.43.

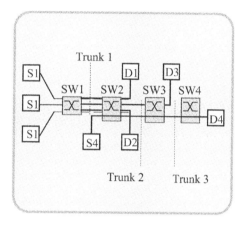

Fig. 5.43: Max-min allocation

There are four switches connected via three trunks. It is assumed that each trunk can support 150 Mbps of capacity. Fig. 5.43 shows the logical connections (VC) between the sources (S_i) and their destinations (D_i). The first trunk is shared by sources S1, S2, and S3. The second trunk is shared by sources S3 and S4. The third trunk is used only by source S4. Discuss the max-min algorithm procedure for resource allocation.

Solution

First, let the bandwidth be divided equally among each source. On trunk 1, sources S1, S2, and S3 each receive 50 Mbps capacity. On trunk 2, S3 and S4 are allowed 75 Mbps of capacity. Finally source S4 gets all 150 Mbps on trunk 3. Note that source S3 is allowed 75 Mbps on trunk 2 but only 50 Mbps on trunk 1. Thus, source S3 should be allocated 50 Mbps on all trunks. Now, source S3 with its link capacity (50 Mbps) that is removed. This means that the link capacities for trunk 1 and trunk 2 are now 100 Mbps each. Again, the sources S1 and S2 are reassigned 50 Mbps on trunk 1. Now, only source S4 is left on trunk 2 and it receives the remaining 100 Mbps. Finally, source S4 receives all 150 Mbps on trunk 3. Another discrepancy exists at source S4. It is allowed 150 Mbps on trunk 3, but only 100 Mbps on trunk 2. Thus 100 Mbps is assigned to source S4 on all trunks. This means that 50 Mbps is not utilized on trunk 3. The final results of the max-min allocation are as follows:

	Speed in Mbps		
Trunk # Source	1	2	3
S1	50	N/A	N/A
S2	50	N/A	N/A
S3	50	50	N/A
S4	N/A	100	100

In high-speed networks, waiting for congestion to occur before taking corrective measures may not serve the purpose. By the time the network informs a source to slow down its transmission rate, thousands of additional cells may have already been sent out. Also, there is no congestion control needed for CBR, VBR, and UBR traffic. Therefore, preventing congestion prudently from ever occurring becomes crucially important. For this purpose, the source must define the type of transmission that it requires for a particular traffic (that is, the usage parameters). The network will then try to route the connection (via a specified VC and VP) in order to meet these demands. If no route/resources are available to meet these needs, then no connection is made.

Resource reservation

After a connection is accepted, one method to prevent congestion is to allocate a certain amount of bandwidth for its use. The network provides a cumulative amount of capacity and performance on a VP, which is then shared by the VCs. If a VCC (virtual channel connection) passes through multiple VPCs (virtual path connections), then the overall performance of that VCC depends upon the performance of each of the VPCs and the performance of any VC-switch along the route. There are basically two ways that VCCs can be grouped within a VPC. First, all the VCCs within a particular VPC can be assumed to have generally the same traffic characteristics. In this case, all the VCCs should experience the same throughput and QOS. As a second case, the VCCs may have varying traffic requirements. In this scenario, the traffic contract for the VPC should be determined by the most demanding VCC.

Once the VCCs are grouped, two options exist for allocating capacity on the VPC. *Aggregate peak demand* is based on providing a data rate to the VPC equal to the sum of the peak cell rates of all the VCCs within that VPC. This will prevent any congestion in the network, but leads to the under-utilization of network resources. Another alternative is to use *statistical multiplexing*. Here, the data rate of the VPC should be slightly greater than the sum of the average cell rates of all the VCCs but less than the aggregate peak demand. This leads to a more efficient use of the network resources; however, a problem can still occur if all logical connections require maximum throughput simultaneously. For example, a connection has an SCR of 20 cells per second and a PCR of 100 cells per second. This can be accomplished by allocating one 20-cells per second connection and, when a sudden burst occurs, five such connections can be multiplexed together. The problem occurs when multiple connections require the five multiplexed circuits simultaneously. Then the throughput on some connections will be severely affected. In the worst case scenario, throughput may even dwindle to zero.

Credit-base rate control

Although the ATM Forum has elected rate-based control in preference to *credit-based flow control*, LAN environments are still using and implementing the *credit-based flow control*. It consists of link-by-link, per-VC window flow control, and uses buffers and credits to manage the flow of data. Each node maintains a separate queue for each VC. The receiver then monitors the queue lengths of each VC and determines the number of cells that the sender can transmit on that VC. The number of allowable cells is called *"credit"*. The sender can only send cells if credits have been allocated. There are two phases associated with controlling the flow of cells on a VC. First the amount of buffer space must be allocated to the VC. This involves determining the round trip delay time involved with the connection. On any connection, multiple cells may be in transit between the source and destination at any particular time. The receiver must allocate a certain amount of buffer space in order to cope with the arrival of these cells. In addition, the credits issued by the receiver will also take some time to reach the source. During this period, more cells can be transmitted by the source. Therefore, the total round trip delay must be taken into account when allocating buffer space.

Example 5.10

Assume that two nodes are separated by 200 kilometers and are connected via 155 Mbps link. Assuming that only propagation delays (approximately 5 µs/km) prevails; determine the memory space required for a credit-base rate control.

Solution

Memory space required is given by:
$(400 \text{ km} \times 5 \text{ µs/km}) \times (155 \text{ Mbps}) = 310 \text{ kbytes}$

Secondly, in the credit-based control, the credits must be controlled. This deals with determining the amount of available buffer space remaining for that VC, and then issuing credits to the sender. This type of credit-based flow control is referred to as *flow-control virtual channels* (FCVC). Fig. 5.44 shows the flow of credit cells and data cells in a network implementing credit-based flow control.

There are two problems with this version of FCVC. First, the credits could be lost or dropped during transmission. The sender will not be notified of new credits and may remain idle needlessly. The throughput severely affected since the source will remain idle for an unknown amount of time. To combat this problem, *credit resynchronization algorithms* have been developed. These consist of both source and destination maintaining a count of the credits issued. These credit inventories are then exchanged. If there are any discrepancies, new credits are issued by the receiver.

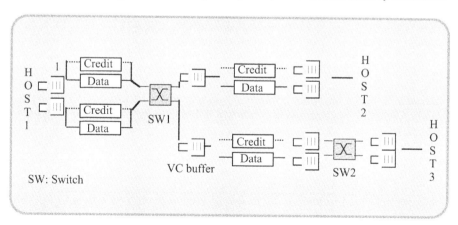

Fig. 5.44: Credit-based flow control

The second problem with FCVC deals with the amount of buffer allocated for a particular connection. Each VC needs to reserve enough buffer space for the entire round trip although several VCs may share the same link. In addition, the transmission rate of each VC may differ, which results is a "waste" of buffer allocation. To resolve this issue, an *adaptive FCVC algorithm* is used to allocate only a fraction of the buffer space needed for the entire round trip. This fraction will depend upon the rate at which the VC consumes the credit. Although the adaptive FCVC algorithm reduces the buffer requirements, a *ramp-up time* may be warranted. That is, it may take some time before a particular VC utilizes the full capacity even if there are no other users on that link. Although the maximum capacity is reached, the final throughput may be lower in some cases due to this ramp-up time. Another consideration is that the credit-based flow control will mainly be used in the LAN environment where the round trip delays are relatively low. Thus, buffer space allocation will also be relatively low.

Rate-based flow control

As stated earlier, the ATM Forum has chosen *rate-based flow control* for use in ATM networks. This congestion control mechanism dynamically adjusts the maximum sending rate by using feedback from the network. Rate-based flow control cannot be used with CBR and VBR traffic due to the real-time nature of the information. The network cannot ask these isochronous traffic types to slow down their data transfer. Further, with UBR traffic, there are no concerns about the throughput. Therefore, rate-based flow control is mainly used by ABR traffic. In rate-based flow control, the following parameters characterize the transmission from the source:

a) PCR $\Rightarrow$ Peak cell rate
b) MCR $\Rightarrow$ Minimum cell rate
c) ACR $\Rightarrow$ Allowed cell rate. This is the current transmission rate of the source
d) ICR $\Rightarrow$ Initial cell rate. This is the initial value of ACR.

The source begins by substituting the cells to the network setting ACR equal to ICR. It then adjusts the ACR based upon feedback received from the network. Periodic feedback is facilitated by using *resource management* (RM) cells. Usually, RM-cells are transmitted after 31 data cells or if more than 100 ms has elapsed since the last RM-cell was sent. Each RM-cell contains the following fields:

1) CI $\Rightarrow$ *Congestion indication.* Setting of this field indicates the presence of congestion in the network.
2) NI $\Rightarrow$ *No increase.* This field is set to inform the sources not to increase their transmission rates.
3) ER $\Rightarrow$ *Explicit rate.* This field is used to convey the desired transmission rate. $ICR \leq ER \leq PCR$.
4) CCR $\Rightarrow$ *Current cell rate.*

Most RM-cells are transmitted by the source and are termed as *forward RM* cells. Any ATM switch along the route of the VC or the destination may change the CI, NI, or ER fields. Once the forward RM-cell reaches the destination, the end system transmits a *backward RM* cell in the reverse direction. When the source receives the backward RM-cell, it adjusts the ACR as per the rules given in Table 5.6.

Table 5.6: Rules for rate-based flow control

NI	CI	Action towards the adjustment of ACR
0	0	$ACR = \max\{MCR, \min[ER, PCR, ACR + (RIF) \times (PCR)]\}$
0	1	$ACR = \max\{MCR, \min[ER, (ACR) \times (1 - RDF)]\}$
1	0	$ACR = \max\{MCR, \min[ER, ACR]\}$
1	1	$ACR = \max\{MCR, \min[ER, (ACR) \times (1 - RDF)]\}$

In reference to the feedback mentioned earlier, the ATM switch has numerous ways of providing feedback to the source.

1) EFCI $\Rightarrow$ *Explicit forward congestion indication.* Setting the appropriate bits within the header of a data cell (payload type field) can perform this. If the destination receives this indication, it will set the CI bit in the RM-cell.
2) CI or NI setting $\Rightarrow$ The switch may directly set the CI or NI bits on a forward or backward RM-cell.

3) ER marking $\Rightarrow$ The switch may reduce the value of the ER field in a forward or backward RM-cell.

The question now is how to determine the appropriate value of ER at a switch. Three algorithms are currently posed and used for this purpose. All of these algorithms try to allocate a "*fairshare*" of resources to the connections.

Enhanced proportional rate control algorithm (EPRCA)

In this algorithm, the switch keeps track of the average load from each connection, which is termed *mean allowed cell rate* (MACR). The relevant procedure is given by:

$$MACR_{new} = (1-\alpha)(MACR_{old}) + \alpha(CCR)$$ (5.11a)

$$Fairshare = DPF \times MACR$$ (5.11b)

where $\alpha = 1/16$; and,

$$DPF = Down\ pressure\ factor = 7/8$$ (5.11c)

Note that the previous value of MACR ($MACR_{old}$) is assigned more weight than the current cell rate in determining the new MACR. If the queue length at the switch exceeds a predetermined threshold, then the ER field is updated to:

$$ER_{new} = min[ER_{old}, Fairshare]$$ (5.11d)

There are two main problems associated with EPRCA. First, it reacts to congestion rather than avoiding it. Second, EPRCA determines the presence of congestion based upon queue length. This has proven to result in unfairness among resource allocation to the VCs. Sources that start transmission at later times than other sources will get lower throughput. This occurs because queue lengths may already be quite long. Therefore, these "late" sources will not be allocated high throughputs even though they did not contribute to the congestion.

The next two algorithms (ERICA and CAPC) indicated below try to avoid congestion. They both make adjustments to the throughputs based on a load factor (LF) defined as:

$$LF = (Input\ rate)/(Target\ rate)$$ (5.12)

where, input rate is equal to the total input data rate of the link measured over an interval of time, and target rate is equal to 85% to 90% of the link's bandwidth. When LF > 1, congestion will occur and the VCs should reduce their data rates. If LF < 1, the network is not congested and no rate reduction is required.

Explicit rate indication for congestion avoidance (ERICA)

In this algorithm, the following two parameters are calculated:

- Fairshare = (Target rate) / (Number of connections) (5.13a)

- VCshare = (CCR) / (Input rate)
  = (CCR/Input rate) × Target rate (5.13b)

A closer look at the VCshare can indicate that the first term on the right-hand side (in Eqn.5.13b) gives the amount of load a particular VC is presenting to the switch. Thus, by multiplying this expression by the target rate, the switch can elucidate the fraction of the target rate to assign a VC. ER is calculated as follows:

$$ER_{new} = min[ER_{old}, max(Fairshare, VC\ share)]$$ (5.13c)

where ER_{old} is the value assigned to the ER field in the incoming RM-cell. *Note:* ATM switches may only *reduce* the ER filed and not increase it.

Behavior of ERICA under specific circumstances

First, assume congestion is beginning to occur at a switch. This means that the input rate is exceeding the target rate (LF > 1). This will cause the throughput of some VCs to decrease while others will experience an increase in throughput. Now, what is required is a way to reduce the load factor (LF → 1) and provide an equal share of the resources among all VCs. That is, the VCs monopolizing the resources will have their rates reduced. These resources will then be shared among those VCs, which have data rates lower than the *fairshare*. Now, assume low loads (LF < 1) at a switch. Then, the throughput on all VCs should increase. However, this increase in throughput will not be equally shared among the VCs. Those VCs whose data rates are less than the *fairshare* will receive a greater increase in throughput. Thus, ERICA provides a "fairer" way to allocate resources than EPRCA. It also culminates in a final operating point almost 10 to 20 times faster than EPRCA.

Congestion avoidance using proportional control (CAPC)

CAPC is similar to ERICA except in its strategy for updating the *fairshare* parameter. The relevant algorithm is as follows:

$$\text{If LF} < 1, \text{Fairshare}_{new} = \text{Fairshare}_{old} \times \min[\text{ ERU}, 1 + (1 - \text{LF}) \times R_{up}] \qquad (5.14a)$$
$$\text{If LF} > 1, \text{Fairshare}_{new} = \text{Fairshare}_{old} \times \max[\text{ ERF}, 1 - (\text{LF} - 1) \times R_{dn}] \qquad (5.14b)$$

where, ERU is the maximum allowable increase (≈ 1.5), and

R_{up} = Slope parameter = 0.025 to 0.1
ERF = Minimum allowable decrease ≈ 0.5
R_{dn} = Slope parameter = 0.2 to 0.8.

CAPC has two distinct advantages over ERICA. First, the algorithm to implement CAPC is simpler than ERICA. Second, CAPC has an oscillation-free steady-state performance. Once the best resource allocation scheme has been found, CAPC will no longer adjust the VC's data rates.

Due to the inherent high-speed considerations and resource-sharing attributes of ATM networks, congestion control has become essential for maximizing the throughput of all connections. However, the varying traffic types carried by the ATM network may prevent the use of any single congestion control technique. Deciding which technique is to be implemented depends upon the severity and duration of the congestion involved.

Priori to any congestion control technique being implemented, the ATM switch must know the link capacity for all logical connections. The "max-min allocation" technique is used to determine the appropriate share of throughput among the sources. Once the connection is established, the network must monitor the user to ensure that the traffic contract is met. The generic cell rate algorithm (GCRA) accomplishes this task. Resource reservation can be used for all traffic types, but it is mainly used for CBR and VBR service categories. It reserves a certain throughput for a VC. Aggregate peak demand and statistical multiplexing are the reservation techniques utilized. For ABR traffic, either credit-based or rate-based flow control can be implemented. Credit-based flow control is usually implemented in LAN environments while rate-based flow control is used in WAN environments. Fig. 5.45 shows the typical implementation of credit-based and rate-based flow control in an integrated network topology.

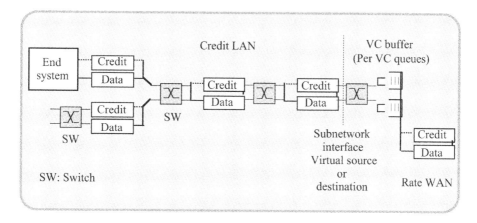

Fig. 5.45: Credit-based and rate-based flow control integration

End-node notification techniques

When the congestion is detected at the node, the end-node should be able to react. There are three techniques proposed for congestion notification in ATM networks:

- *Estimation by the end nodes*

  Transmitting a probe cell to measure the delay time between the source and destination is used in this technique. At the end node, the probe cell will be used to estimate the one-way delay. The congestion is thus a specified based on this estimation so that the source can adjust its appropriate rate.

- *Explicit backward congestion notification*

  This technique is to send back the congestion information from the congested node to the sources, which can react to congestion along their paths effectively. A special cell is therefore established and the congestion information is placed in it before transmitting to the source. However, this technique is not appropriate for ATM. This is because the ATM is a high-speed transmission link, which limits the effectiveness of this method, and more importantly, the use of special cells would impose a considerable processing burden on intermediate nodes. Furthermore, if the same VCI value is not used in both directions, then a congested intermediate node has to establish a connection to each source before the congestion notification cell can be transmitted.

- *Explicit forward congestion notification (EFCN)*

  This is an optional congestion control mechanism. A network element in a congested state may set an explicit forward congestion indication in the cell-header. At the receiving end, the user equipment may use this indication to implement protocols, which adaptively lower the cell rate of the connection during congestion.

  When a congestion occurs at the trunk, as observed by monitoring the queue occupancies reaching a predefined threshold value, all the cells passing through that trunk will be marked until the congestion period ceases. That means, the EFCN bit of a cell is set at a node and it will be sent to the receiver along the path without being modified at any node. Then, the receiver sends a notification back to the source for adaptive rate control. In an adaptive rate control, the rate at which traffic is submitter to the network is varied by a source depending on the congestion status information. In an adaptive rate control mechanism based on the EFCN scheme proposed, the source traffic passes through a variable-rate server. The rate of the server is controlled by the feedback information from the destination node.

The congestion control in ATM networks is not totally complete. It is still open for resources to find better solutions in order to satisfy new advanced applications.

Payload type indicator

Within the ATM cell there are a number of bits available for congestion and priority setting. These include the payload type indicator (PTI) and the cell loss priority bit (CLP). The PTI field is 3 bits long. The most significant bit is the left bit, bit 3, that is used to specify whether the cell carries user or operations, administration, and maintenance (OAM) data. Bit 2 is used to indicate if the cell has passed through one or more congested switches. Bit 1 with user data is currently used only by AAL-5 to distinguish the last cell of a user frame from the others. In addition to the specification of two types of OAM flows, PTI coding includes the definition of a resource management (RM) cell.

Fig. 5: Resource management cell Format

Table 1: *Payload Type Indicator*

PTI Coding	Meaning
000	*User data cell, congestion not experienced, SDU type = 0*
001	*User data cell, congestion not experienced, SDU type = 1*
010	*User data cell, congestion experienced, SDU type = 0*
011	*User data cell, congestion experienced, SDU type = 1*
100	*Segment OAM flow-related cell*
101	*End-to-end OAM flow-related cell*
110	*RM cell*
111	*Reserved*

(SDU = Service data unit)

The RM cell is used for congestion control of connectionless traffic. An ATM block is group of cells represented by two RM cells, one before and another after the last cell of the block. The recently defined ATM block transfer (ABT) service is used to deliver blocks of ATM cells by assigning the network resource needed for the transfer of an ATM block dynamically on an ATM block basis. Two ABT capabilities are defined: ABT with delayed transmission (ABT/DT) and ABT with immediate transmission (ABT/IT). In ABT/DT, the leading RM cell will be sent into the network and will wait for a response from the RM cell in network before transmitting the remainder of the ATM block. On the other hand, in ABT/IT the ATM block is transmitted immediately after sending the leading RM cell without waiting for the response from the network.

5.6 Scheduling Disciplines

An important issue of concern, which is addressed in ATM networks, is how to support real-time applications. Inasmuch as an ATM network should provide guarantees on bandwidth, delay, jitter, and packet loss rate, they must employ *traffic scheduling algorithms* within the network. Such algorithms are intended to determine the rate and order in which packets from different connections are serviced. Hence, they permit control of the bandwidth, delay, and packet loss rate.

Several scheduling algorithms have been proposed in the literature. Ideally, a relevant scheme should enable the network to treat users differently in accordance with their desired quality of service. At the same time, it must ensure isolation for real-time flows, and fairness for best-effort applications. That is, a few users should not be allowed to degrade the service rendered to other users to the extent that the traffic contract is violated.

The desirable features of a scheduling algorithm are as follows. Apart from the isolation and fairness properties mentioned above, the scheduling algorithm must be simple to implement and it should be able to support a large number of connections.

5.6.1 Traffic scheduling algorithms

There are two classes of traffic scheduling algorithm: *Work-conserving* and *non-work-conserving*. A work-conserving schedule is never idle when there is a packet in its queue. On the contrary, a non-work-conserving scheduler may stay idle even if there is a packet being transmitted. Further, the non-work-conserving schedules are usually adopted to control delay jitter in the network and may result in higher average delays than work-conserving ones.

The schedules are also categorized based on their structure. Accordingly there are two types of schedules namely, the sorted-priority version and frame-based version. Sorted-priority schedulers make use of a global variable, usually known as virtual time, for each output port (or link). This variable gets updated every time a cell arrives or departs from the switch. Also, every packet in the system has a timestamp indicating that it is extracted from the virtual time. Packets are transmitted in increasing order of their timestamps. Examples of sorted-priority schedulers are *weighted fair queueing, virtual clock, self-clocked fair queueing* (SCFQ), and *frame-based fair queueing*.

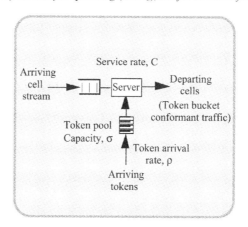

Fig. 5.46: Token bucket shaper

Frame-based schedulers bifurcate time into "*frames*". A frame could be of fixed or variable length. It is a time interval in which the various traffics sources make their bandwidth reservations. Typical examples of such schedules include *stop-and-go-queueing, weighted round-robin*, and *deficit round-robin*.

To determine the delay-bound of a session, the burstiness of the source traffic should be constrained by a *shaping model*. Typically, token-bucket constrained traffic model has been proposed in the literature for the worst-case analysis of delay. In reference to the leaky-bucket algorithms discussed earlier, the output of the token bucket is a (σ, ρ, C) conformant traffic stream where σ is the number of token bucket in the token bucket, ρ is the rate of the incoming tokens, and C is the peak rate of the server at the token bucket. If $A(t_1, t_2)$ is the amount of traffic that leaves the token bucket during the interval (t_1, t_2), then the constraint that holds it is: $A(t_1, t_2) \leq \min\{C(t_1, t_2), \sigma + \rho(t_1, t_2)\}$.

There are two important properties of a scheduler. They refer to the worst-case end-to-end delay bound and the fairness the scheduler provides. Normally, fairness is less important than the end-to-end delay for real-time connections. However, it is quite important for best-effort flows. Several measures have been used in the literature to characterize the fairness of scheduling algorithms. For example, an approach is to use the maximum differences between the normalized service received by two backlogged connections over an interval in which both are backlogged. Thus, the *service-fairness index* (SFI) has been specified as follows:

$$SFI = \left| \frac{W_i(t_1, t_2)}{\rho_i} + \frac{W_j(t_1, t_2)}{\rho_j} \right| \qquad (5.15)$$

where $W_i(t_1, t_2)$ and $W_j(t_1, t_2)$ represent the service received by connections i and j, respectively, during an interval (t_1, t_2) over which both are backlogged, and ρ_i, ρ_j are the allocated rates of the two connections.

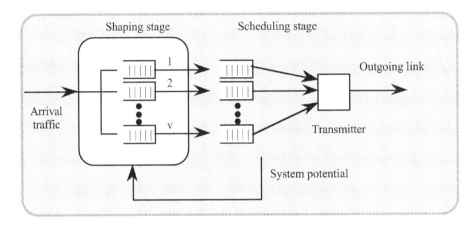

Fig. 5.47: A scheduler in the SRPS class

A *generalized processor sharing* (GPS) is considered as an ideal scheduler based on end-to-end delay bounds and fairness properties. However, since it is based on a fluid model, it cannot be implementable in practice. *Weighted fair queueing* (WFQ) is the packetized version of GPs. This again poses difficulties to implement because of the need to simulate a fluid-model GPS as its background. There are other schedulers that attempt to approximate WFQ at a reduced implementation complexity. The so-called *virtual clock* has the same delay bounds, but it is an unfair algorithm because a session may be denied service for an arbitrary time period because of excess bandwidth received in the past. The self-clock fair queueing (SCFQ) on the other hand avoids the fairness problem of virtual clock. However its worst-case delay bound is of the order of the number of active connections. Another strategy known as the *frame-based fair queueing* has both the delay bound aspects of WFQ and bounded fairness. This algorithm is similar to virtual clock but uses a distinct mechanism to recalibrate the system periodically so that the scheduler remains fair.

The schedulers, which address the problem of reshaping the traffic at each switch so as to minimize the burstiness of the sources and the buffering needed inside the network, are known as *shaped rate-proportional servers* (SRPS). They are based on combining a shaping mechanism with a scheduler from the class of *rate-proportional servers*. The concept is shown in Fig. 5.47. The arriving packets enter a shaper and are admitted into the scheduler only when they become eligible. The criterion to make a specific packet eligible is that the present value of the system potential in the scheduler is equal to or greater than the finishing potential of the previous packet admitted from the connection. The *system potential* is a function that characterizes the overall progress of work in a scheduler. Depending on the system potential function used, an SRPS scheduler may or may not be work-conserving. The simplest scheduler that belongs to the SPRS class is known as the *shaped virtual clock*. In *virtual clock*, the system potential function is actually the real-time itself. The shaped virtual clock scheduler is, therefore, non-work-conserving. It spaces the various packets from the same connection according to connection's bandwidth reservation. Hence, the burstiness at the output of the scheduler is optimal for the case of networks handling fixed size packets. This scheduler, thereby provides both low delay and jitter bounds.

5.7   Teletraffic Modeling of ATM Telecommunications

Teletraffic modeling is a branch of the telecommunication endeavor engineered to characterize and elucidate the traffic information that flows in the associated network. The classical concept of teletraffic refers to the telephone system. Relevant considerations were developed by H.K Erlang (circa 1915). The objective was to evaluate quantitatively the performance of (then existing) telephone systems. Specifically, determination of the expected number of call arrivals per hour, the statistics of the calls, determination of peak-hour traffic etc. were the efforts addressed by Erlang. The results of Erlang have been useful and could be applied in the engineering design of exchanges/switches, routing considerations etc.

The basis of the model developed for telephone traffic (due to Erlang) is essentially statistical. That is, Erlang's model was based on the random events of occurrence in a telephone service such as the number of subscribers placing calls, call arrival rates, and call duration. Subsequently, the teletraffic model based design of exchanges/switches and the related resources proved to be effective in providing subscribers with a telephone service with minimal *"busy tones"* (*blocking probability*) and sufficient grade of voice quality with low delay characteristics.

Thus the efforts of Erlang became the seed for modern teletraffic considerations and relevant extensions have become subsequently compatible for the B-ISDN and ATM era. Like in telephony, the major consideration in teletraffic modeling of other telecommunications is also the statistical attribute of the variables involved. That is, whether be it a voice traffic of the POTS or the data communication posed by computers or the integrated digital transmission of heterogeneous messages (constituted by voice, data, video etc.), the essence of the associated traffic is invariably stochastical in nature. Hence, the models developed thereof are stochastical algorithms.

This section is devoted to presenting highlights on the stochastical description and modeling considerations specific to B-ISDN/ATM telecommunications.

5.7.1   Queueing-theoretic models

The statistical models of telecommunication systems (as they were conceived at a primitive level for the telephone system by Erlang) are essentially based on the "queueing" considerations applied to the flow of information. It has been an inevitable choice to introduce the concepts of *queueing theory* in modeling and analyzing telephone systems as well as other telecommunication systems [5.11, 5.12]. Over the years, coping with evolutions in telecommunication engineering, a host of queueing-theoretic models of teletraffics have emerged to describe the traffic patterns, evaluate the information transfer performance, and design the intricate network web pertinent to various telecommunication systems indicated in Chapter 1.

In short, the telecommunication system models that have sprouted (and are still emerging) largely use the queueing-theoretic consideration built upon the stochastical aspects of the teletraffic involved. Such models aim at providing a system model, the performance of which mimics (as closely as possible) the real-life telecommunication systems they represent. Thereby, they further facilitate assessing the performance of real-systems implicitly to a reliable extent.

What is queueing theory?

The queueing theory refers to the stochastic theory applied to *"waiting lines"*. It is a study on a class of models in which customers arrive in some random manner at a service facility. This is typical of a telecommunication system, say, for example, in telephony – the sequence of random events associated in the processing of telephone calls often faces the "waiting lines" before the available resources could handle them.

As indicated above, queueing-theoretic considerations have been aptly and amply applied to evaluate the performance of various telecommunication systems, ever since the earliest attempts by Erlang to analyze telephone traffics. In such efforts, the following are definitions and fundamental quantities commonly used and are necessary in understanding queueing-theoretic models of telecommunication traffics and specifically the teletraffic engineering applied to B-ISDN/ATM.

Queueing-theoretic entities and definitions

❑   *Queue*:   *A waiting line of units demanding service at a service facility/counter*
❑   *Customer*:   *The unit demanding the service*
❑   *Server*:   *The entity/person who provides the service*

Notations
L: *Average number of customers in the system*
L_Q: *Average number of customers waiting in the queue*
W: *Mean time that a customer spends in the system*
W_Q: *Mean time that a customer spends in waiting in the queue*

The general queueing system is illustrated in Fig. 5.48

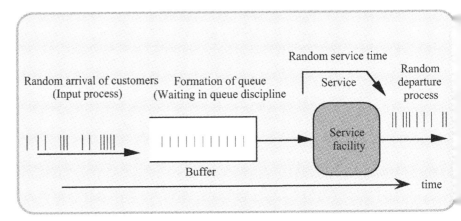

Fig. 5.48: Queueing system

λ: *Mean arrival rate (Average number of customers arriving at the input process per unit time at the service point)*
μ: *Mean departure rate (Average number of customers departing per unit time at the service point)*

Little's formula
     Average number of customers in the system is equal to: (Arrival rate × Average length of time staying in the system). That is,

$$L = \lambda W \hspace{3cm} (5.16)$$

Kendall's notation
     Considering the statistical aspects of arrival and the service process involved in a queueing system, a notation has been prescribed. It is of the form A/B/C/M/Z, where

     A ⇒ *Arrival distribution*
     B ⇒ *Service distribution*
     C ⇒ *Number of servers*
     K ⇒ *Maximum capacity of the queue (Default value = ∞)*
     M ⇒ *Population of customers (Default value = ∞)*
     Z ⇒ *Service discipline*

Probability distributions representing A and B are:

D: *Constant $\Rightarrow$ Deterministic law*
M: *Markov or exponential law*
G: *General law*
GI: *General independent law*
E_k: *Erlang's law of order k*
H_k: *Hyperexponential law of order k*

The Kendall's notation indicated above specifies the conventional categorization of queueing systems. Two examples of queueing systems, namely, the M/D/1 and M/M/1 are popular in describing telecommunication queueing systems. In terms of Kendall's notation, each of these queueing systems has Markovian arrivals (negative exponential or memoryless arrivals) at a rate of λ arrivals per second. The M/M/1 system has random length arriving entity with a negative exponential distribution (Markov), while the M/D/1 system has constant length of arrival entity. The parameter μ^{-1} defines how may seconds (on average) are required for the transmission link to send each arrival entity. For the M/M/1 system, this is an exponentially distributed random number with this average length, while in the M/D/1 system, this is a constant (or fixed length) for every arrival. Both systems have a single server (namely, the physical transmission link) and an infinite population (number of potential arrivals) as well as infinite waiting room (buffer space). The units of the buffer in the M/D/1 model are arrival entities, while in the M/M/1 case the units of the buffer are a clustered set of arrival entities.

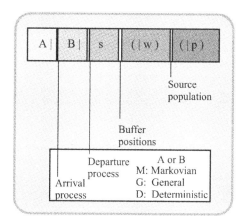

Fig. 5.49: Queueing system notation

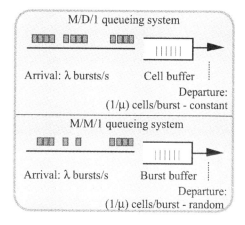

Fig. 5.50: Application of M/D/1 and M/M/1 queueing systems relevant to the ATM environment

Fig. 5.49 illustrates these physical queueing systems and their specific relationship to ATM. (Arrival entities: These may refer to bits, packets, cells, messages etc. in the context of modern telecommunication systems.)

Considering the tradeoffs encountered in the relevant modeling, the M/D/1 system in the ATM traffic environment represents the fact that the buffers in the switch are in units of cells; however, the bursts in the associated traffic are all assumed to be of fixed length. The M/M/1 system does not model the switch buffers accurately since it is in units of bursts and not in cells; however, the modeling of random burst lengths is more appropriate with multiple traffic sources. The M/M/1 model is also simple to analyze, and therefore it has been adopted extensively to illustrate specific tendencies in ATM systems. In general, if the traffic is more deterministic than the M/M/1 model (for example, more like the M/D/1 model), then the M/M/1 model will be pessimistic. In other words, there will be actually less queueing and less delay in the modeled network. If the traffic is more bursty than the M/M/1 model, then the M/M/1 results will be optimistic, indicating that there will actually be more queueing and more delay in the modeled network.

Link utilization/traffic intensity: $\rho = \lambda/\mu$ $\qquad\qquad$ *(5.17)*

Service priority disciplines:

$\qquad$ *FCFS: First-come/first served*
$\qquad$ *FIFO: First-in/first-out*
$\qquad$ n^{th} *arrival/first-sent (n > 1): LCFS – Last come/first sent*
$\qquad$ *FIRO: First-in/random out*

5.7.2 Arrival process

In queueing theory, this refers to the statistics of customer arrival process. In telecommunications, such an arrival process is concerned with the arrival of telephone calls, packetized message, ATM cells etc. Most often, the arrival process is described by the Poissonian statistics. That is, the arrivals are characterized via the *Poisson process*. In reference to Fig. 5.51, the following attributes of a Poisson process can be observed [5.6].

- ✓ Probability of an arrival event in the time interval $\Delta t \to 0$ is equal to $\lambda \Delta t$ < 1, where λ is a constant and it depicts the arrival rate
- ✓ Probability of no arrival in Δt is: $(1 - \lambda \Delta t)$
- ✓ The arrivals are *memoryless* corresponding to an arrival event in a given time interval of length Δt is independent of events in previous or future events. That is, random arrival processes can be described as the Poisson (or Markov) process as illustrated in Fig. 5.51. Poisson arrivals occur such that for each increment of time (T), the probability of arrivals is independent of any previous history. That is, a memoryless condition is assumed. The arrival events refer to either individual cells, a set of bursts of cells, service completions, or a set of cells (or packets) in telecommunication practice. The probability that the interarrival time between events t, as shown in Fig. 5.51, has a certain probabilistic value specified by an interarrival time probability density.

As stated above, the Poisson process is a memoryless process, because the probability that the interarrival time will be x seconds, which is independent of the memory of how much time has already elapsed. This fact greatly simplifies the analysis of random processes since no past history, or memory, is prevalent. This type of process is commonly known as a class of Markov processes.

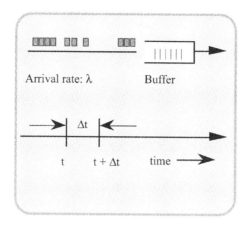

Fig. 5.51: Time interval used in the Poisson process

For the large, finite interval T, the probability p(k) of k arrivals in T is given by:

$$p(k) = (\lambda T)^k [\exp(-\lambda T)] / k! \qquad k = 0, 1, 2, \ldots \qquad (5.18a)$$

This is known as the *Poisson distribution*. Its mean (*expected value*) is given by:

$$E(k) = \sum_{k=0}^{\infty} k \, p(k) = \lambda T \qquad (5.18b)$$

and the *variance* $\sigma_k^2 = E[k^2] - \{E[k]\}^2$ is given by:

$$\sigma_k^2 \equiv E(k) = \lambda T \qquad (5.1c)$$

Note: E[·] depicts the expected value.

Rate parameter

The entity λ is defined as, $\lambda = E[k]/T$ and, as indicated earlier, λ denotes the average rate of Poisson arrival.

Interarrival distribution

Considering a variable τ depicting the random variable of time between arrivals, (that is, the interarrival time) in a Poisson process, the probability density function (pdf) of τ is given by

$$f_\tau(\tau) = \lambda e^{-\lambda \tau}, \qquad \tau > 0 \qquad (5.19)$$

This is an *exponential distribution* with a mean value $E[\tau] = 1/\lambda$ and variance $\sigma_k^2 = 1/\lambda^2$.

Statistics of calls in telephony

The earliest efforts in teletraffic engineering as mentioned before refer to telephone calls. The call attempts in telephony follow the Markov process. Suppose the call arrival rate is λ. Without any blocking, the average number of calls in progress during the busy hour is equal to λT where T is the call holding time (namely, the *call duration*). Thus, the average offered traffic load (a) is given by:

$$a = \lambda T \qquad \text{Erlangs} \qquad (5.20)$$

In reality, blocking of calls would occur with a probability P_B. Then, the telephone systems carries a load of only $(1 - P_B) \times a$ Erlangs.

Given an average offered load of a Erlangs and n trunks, the probability that the system may block a call attempt is given by (*Erlang-B formula*):

$$P_B(n,a) = \left[a^n / n! \right] / \sum_{k=0}^{n} a^k / k! \tag{5.21a}$$

with $P_B(0,a) = 1$.

Given an average offered load of ρ Erlangs, served by n operators, the probability that a system places a call in a queue to wait is given by (Erlang-C formula):

$$P_C(n,a) = W / \sum_{k=0}^{n-1} [(na)^k / k! + W] \tag{5.21b}$$

where $W = \left[(na)^n / n! \right] / \left[1/(1-\rho) \right].$

5.7.3   Combined stochastic processes: Bursty traffic considerations

Markov and *Poisson stochastic processes* can be combined to realize the so-called *Markov modulated Poisson process* (MMPP). It consists of two basic processes, the discrete set of events (which corresponds to ATM cells) and the set of continuous events, which corresponds to bursts of cells [5.10].

The labels on the arrows of Fig. 5.52 show the probability that the transitions at the source between active and inactive bursting states. In other words, during each cell time, the source makes a state-transition, either to the other state, or back to itself, with a probability of the corresponding action indicted by the arrow(s) in the diagram.

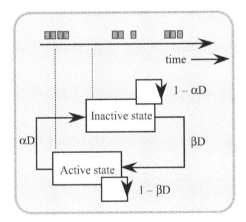

Fig. 5.52: Discrete time Markov process: Transition states
(αd: Probability that a burst begins in a particular cell time; βd: Probability that a burst ends in a particular cell time)

The *burstiness*, or *peak-to-average ratio* of the discrete source model is given by the following formula:

$$B = (\alpha + \beta)/\beta \tag{5.22}$$

where α is the average number of bursts arriving per second and β is the average rate of burst completion. Given that the *cell quantization time* (in seconds/cell) is d, the *average burst duration* (in seconds) is given by:

$$BD|_{mean} = 1/\beta d. \qquad (5.23)$$

If the cell quantization time is not considered, the state transition rate corresponds to a *continuous time Markov process*, which models the statistics of the time duration of bursts instead of modeling the individual cells. In the absence of d, $B = (\alpha + \beta)/\beta$ and $BD = 1/\beta$.

Another distribution that is occasionally used to model extremely bursty traffic is that of the *hyperexponential distribution*, which is, effectively the weighted sum of a number of negative exponential arrivals. This turns out to be a more pessimistic model than Poisson traffic because bursts and burst arrivals are more closely clumped together.

Research based on actual LAN traffic measurements indicates that these traditional traffic models may, however, be overly optimistic. Relevant results show that the LAN traffic has similar properties regardless of the time-scale on which it is observed. This is in sharp contrast to the Poisson and Markovian models, where the traffic tends to become smoother, and more predictable when time averages are computed over a larger period of time.

5.7.4   Delay performance statistics

In assessing the system delay and loss performance, it is necessary to define a parameter to specify the statistics of offered load. This is the *link utilization* or *traffic intensity*, namely $\rho = \lambda/\mu$ as indicated earlier, where λ is the average number of bursts per second, and that μ^{-1} is the average time in seconds per burst. Thus, the offered load has the interpretation of the average fraction of the resource capacity that is in use, under the bursty environment.

The *service rate* μ_B is computed as follows (for a burst of B bytes at a line rate of R bits per second), μ_B is given by

$$\mu_B = 8B/R \qquad (5.24)$$

The probability that there are n bursts waiting in the M/M/1 queue is given by the following formula:

$$\text{Probability}[\text{n bursts in M/M/1 queue}] = \rho^n(1-\rho) \qquad (5.25)$$

The average queueing delay (namely, waiting time) in the M/M/1 system is given:

$$\langle M/M/1 \text{ queuing delay} \rangle = \frac{\rho/\mu}{(1-\rho)} \qquad (5.26)$$

Generally, M/D/1 queueing predicts better performance than M/M/1. Indeed the average delay of M/D/1 queueing is exactly one-half of the M/M/1 delay. The probability for the number of cells in the M/D/1 queue is much more complicated, which is the reason why the M/M/1 model is used more often than M/D/1 model.

Gaussian approximation of Bernoulli processes

A *Bernoulli process* is essentially the result of N independent coin flips (or Bernoulli trials) of an "unfair coin." An *unfair coin* is one where the probabilities of heads and tails are unequal, with p being the probability that "heads" occurs as the result of a coin flip and $(1 - p)$ being the probability that "tails" occurs. The probability that k heads occur as a result of N repeated Bernoulli trials (coin flips) is called the *binomial distribution*, given by:

$$\text{Probability}[k \text{ "heads" in N "flips"}] = \binom{N}{k} p^k (1-p)^{N-k} \qquad (5.27)$$

where $\binom{N}{k} = N!/(N-k)!k!$.

The *Gaussian* or *normal distribution* is a continuous approximation to the binomial distribution when $N \times p$ is a large number. Gaussian and binomial distributions have basically the same shape, and for large values of $N \times p$, in the $N \times p \times (1 - p)$ region about $N \times p$, the Gaussian distribution is a reasonable approximation to the binomial distribution. The tail of the binomial distribution can be approximated by the *cumulative distribution of the normal density*, $Q(\alpha)$.

$$P[k > x] \approx Q\left(\frac{x - \mu}{\sigma}\right) = Q(\alpha) \approx \frac{1}{2}e^{-\alpha^2/2} \tag{5.28}$$

where $Q(\alpha) \equiv \dfrac{1}{\sqrt{2\pi}}\displaystyle\int_{\alpha}^{\infty} e^{-x^2/2}dx$.

Statistics of arrival processes of multiple traffic streams and multiplexed (superimposed) streams

The statistics of arrival of bits (or cells) pertinent to a given type of source depends upon the source-emission characteristics as mentioned earlier. Such emission would correspond to (as discussed before):

- ✓   Constant bit rate (CBR) emissions
- ✓   Variable bit rate (VBR) emissions.

Modeling of the above source outputs constituting the traffic stream of bits or cells is known as *source modeling*. For presentation purposes, let N arrival processes be considered, each characterized by a Poisson process with rate $\lambda(i)$, $i = 1,..., N$. Then the probability distribution of the number of cells in the queue can be exactly obtained with a single Poisson arrival stream with rate:

$$\lambda = \sum_{i=1}^{N} \lambda(i) \tag{5.29}$$

However, this formula is not applicable to queueing models for the ATM networks in representing the multiplexed traffic stream on the bearer-line. In particular, the multiplexing refers to the superposition of N independent *renewal processes* and such a multiplexed traffic is not a Poisson process if one or more of the components processed are not Poisson. Various source models of B-ISDN applications, such as *Markov modulated Poisson process* (MMPP) and *Markov modulated Bernoulli Process* (MMBP), are not even renewal processes and cannot be modeled with Poisson processes, since the variations squared coefficients of the interarrival times of cells are greater than one and possess also correlation. One may, however, assume the superposed process as Poisson for modeling purposes and analyze the queue accordingly, particularly when the number of individual components increases. But, the discrepancy between the performance metrics of the original queue with non-Poisson arrival streams and the queue with a Poisson arrival stream would increase to an unacceptable levels as queue utilization increases.

Various techniques, mostly approximate, have been proposed to obtain the superposition of arrival streams of B-ISDN applications. In most cases, the superposed traffic is approximately represented as a two-state MMPP or MMBP. A brief outline on relevant statistics is presented below along with considerations of multiplexing (superposition).

Markov-modulated Poisson process

The MMPP has been extensively used to model various B-ISDN sources, such as voice and video, as well as characterizing the superposed traffic. It has the property of capturing both the time-varying arrival rates and the correlation between the interarrival times. In addition to characterizing the desired properties of B-ISDN applications, these models are analytically tractable and produce fairly accurate results.

An MMPP is a *doubly stochastic Poisson process*. The arrivals occur in a Poisson manner with a rate that varies according to a k-state Markov chain, which is independent of the arrival process. Accordingly, the transition rate matrix of its underlying Markov chain and arrival rates characterize an MMPP. If the multiplexed corresponds to superposition of MMPP processes, the multiplex traffic is also a MMPP process.

Equivalently, the state of the superposition of N MMPPs is $(i_1, i_2, ..., i_n)$, where i_j denotes the state of the jth component MMPP. The total arrival rate at state $(i_1, i_2, ..., i_n)$ is the sum of the arrival rates of component processes, which depends on state i_j; that is,

$$\lambda(i_1, i_2, ..., i_n) = \sum_{j=1}^{N} \lambda_{i_j} . \qquad (5.30)$$

The transition rates from state $(i_1, i_2, ... i_j = k, ..., i_n)$ to state $(i_1, i_2, ... i_j = m, ..., i_n)$ is given by the rate of going from state $i_j = k$ to state $i_j = m$ in the jth component MMPP.

As the number of component process increases, the number of states of the superposed process increases exponentially. To reduce the complexity of solving queues with a large number of arrival streams, a simpler process that captures important characteristics of the original process as closely as possible may approximate the superposed process. The simplest model that has the potential to approximate an MMPP accurately with a large number of phases is the *two-phase MMPP* defined by four parameters namely, the arrival rates λ_1 and λ_2, and the variances σ_1 and σ_2. Then the problem is reduced to choosing the parameters of the two-state MMPP using the four metrics of the superposed process.

Markov-modulated Bernoulli process (MMBP)

Time in MMBP is discretized into fixed-length slots. The probability that a slot contains a cell is a Bernoulli process with a parameter that varies according to an r-state Markov process, which is independent of the arrival process. At the end of each slot, the Markov process moves from state i to state j with a probability p_{ij} or stays at state i with a probability p_{ii} such that:

$$\sum_{j=1}^{r} p_{ij} = 1 \qquad \text{For all } i = 1, ..., r. \qquad (5.31)$$

In state i, the probability that a slot contains a cell is α_i and no cell is $(1 - \alpha_j)$. The arrival probabilities of cells and the underlying Markov process are assumed to be independent of each other. Hence, an MMBP is characterized by the transition probability matrix P[ ] and the diagonal matrix $\Lambda_{[]}$ of arrival probabilities:

$$\lambda = \begin{bmatrix} \alpha_1 & 0 & 0 \\ ... & ... & ... \\ 0 & 0 & \alpha_r \end{bmatrix}; \qquad p = \begin{bmatrix} p_{11} & \cdots & p_{1r} \\ ... & ... & ... \\ p_{r1} & \cdots & \alpha_{rr} \end{bmatrix}. \qquad (5.32)$$

The superposition of MMBPs is a *switched-batch Bernoulli process* (SBBP). Time in an SBBP is divided into slots of equal length. The arrivals during a slot occur as a batch process with the batch size distribution varying according to a k-state Markov chain.

5.7.5 Analysis of multiplexed/buffered multiple traffic streams

This subsection summarizes some simple models of switch delay and loss performance characteristics as impacted by various aspects of switch buffer architectures. (Details on switch buffers are presented in Chapter 6.) For simplicity, Poisson arrivals and negative exponential service times are assumed. Output queueing delay performance then behaves as a classical M/M/1 system. Input queueing incurs a problem known as *head of line* (HOL) *blocking* (or *contention*). HOL blocking occurs when the cell at the head of the input queue cannot enter the switch matrix because the cell at the head of another queue is traversing the matrix.

As discussed in the literature, for uniformly distributed traffic with random message lengths, the maximum supportable offered-load for input queueing is limited to 50%, while the fixed message lengths increase the supportable offered-load to only about 58%. On the other hand, output queueing is not limited by utilization as in input queueing. Fig. 5.53 illustrates relevant considerations in terms of average delay *versus* throughput for input and output queueing.

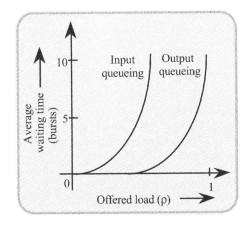

Fig. 5.53: Delay *versus* load performance for input and output queueing

Consequences of switch delay force all ATM switches to have some form of output buffering as will be indicated in Chapter 6. If input buffering is used on a switch it is to make sure that some means to address HOL blocking is implemented. Examples of methods to address HOL blocking are: A switch fabric that operates much faster than the cumulative input port rates, schemes where an HOL blocked cell can be bypassed by other cells, or by the use of priority queueing on the input.

When the burst size of a average higher layer protocol data unit (PDU) is P cells, the approximate *buffer overflow probability* is given by the formula:

$$\Pr obability[Overflow] \approx \rho^{B_C / P + 1} \qquad (5.33)$$

where B_C is the buffer capacity (in cells), and $\rho = nT$ is the offered load.

The buffer overflow objective as specified by the above probability offers a fair guideline to allocate *buffer size* in an ATM device. The buffer size B_C required to achieve an objective cell-loss ratio (CLR) is given by:

$$B_C = P \times \left[\frac{\log(CLR)}{\log(\rho)} \right]. \qquad (5.34)$$

Delay and cell-loss performance of CBR traffic

Fig. 5.54 illustrates the basic traffic source model. N identical sources emitting a cell once every T seconds, each beginning transmission at some random phase in the interval (0, T), define the traffic source model.

The cell-loss rate for such a randomly phased CBR traffic input is well approximated by the following formula:

$$P_L = \text{Probability [Cell loss]} \approx \exp[-2B_C^2 / n - 2B_C(1 - \rho)] \qquad (5.35)$$

where n is the number of CBR connections.

A closed form solution can also be derived for the number of buffers required to achieve a certain loss probability ($P_L = CLR$) from the above formula. It is given by:

$$B = \frac{\sqrt{[n(1-\rho)]^2 - 2n\ln(P_L)} - n(1-\rho)}{2}. \tag{5.36}$$

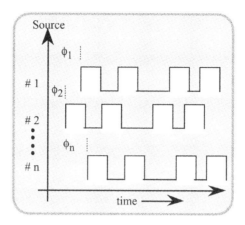

Fig. 5.54: Illustration of randomly phased CBR sources

The CDV (in µs) for each switch for a CBR buffer capacity B_C selected to achieve a specified CLR is given by:

$$CDV(B_C, R) = \frac{384}{R}\left\{B_C - \frac{\sqrt{n}}{2}\frac{1 - \Phi[(1-\rho)\sqrt{n}]}{\varphi[(1-\rho)\sqrt{n}]}\right\} \tag{5.37}$$

where φ and Φ are the standard normal density and distribution functions. Further, R is the link rate (in cells per second) and n is the number of CBR connections.

Problem 5.5

Plot the family of curves depicting the required CBR cell buffers (BC) versus number of CBR connections ($0 < n < 5000$) for different extents of the offered load ($0 < \rho < 1$). Assume a CLR value equal to 10^{-8}.

Priority queueing

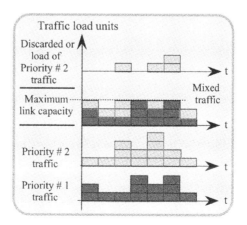

Fig. 5.55: Illustration of priority queueing with contentions

When the switch implements priority queueing, the performance measure indicated here can be applied independently of other performance measures. Fig. 5.55 illustrates cell-traffic originating from two virtual connections multiplexed onto the same transmission path, the higher priority traffic being numbered 1, and the lower priority traffic being numbered 2. Priority 1 cells are serviced first, and priority 2 cells are serviced only when they are not contending with priority cells for link resources. In the event that more cells arrive in a servicing interval than can be accommodated, the priority 2 cells are either delayed or discarded as illustrated at the top of Fig. 5.55.

It is assumed that priority 1 and 2 cells have Poisson arrivals and negative exponential service, priority 2 cells are delayed in an infinite buffer when priority 1 cells are being served (such that no loss can occur), and that 25% of the load is priority 1. The key to priority queueing is that the priority 1 traffic will observe a delay as if the priority 2 traffic did not even exist. On the other hand, the priority 2 traffic sees delay as if the transmission capacity were reduced by the average utilization taken by the priority 1 traffic. The formulas for the average priority 1 and priority 2 queueing delays are as follows [5.6, 5.12]:

$$\text{Average[Queueing delay for priority 1]} = \frac{\rho/\mu}{(1-\rho_1)} \qquad (5.38a)$$

$$\text{Average[Queueing delay for priority 2]} = \frac{\rho/\mu}{(1-\rho)(1-\rho_1)} \qquad (5.38b)$$

where μ is the service rate and (ρ_1, ρ_2) are the offered loads for priority 1 and 2 traffic, respectively, and $\rho = (\rho_1 + \rho_2)$ is the total offered load.

Fig. 5.56 illustrates the effect of priority queueing in which the average delay that would be seen by a single priority system (according to the M/M/1 model), corresponds to the priority 1 cell-delay, and the priority 2 cell-delay. It can be observed that the priority 1 performance is markedly better than the single priority system, while priority 2 performance degrades only slightly.

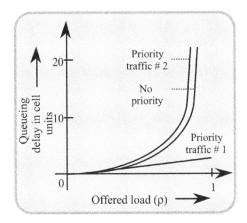

Fig. 5.56: Priority queueing performance example

Modeling ATM traffic: General considerations
Thus, the teletraffic models originally began their journey as a humble version of the Erlang model depicting the canonical telephone traffic. The associated queueing more or less has a Poissonian structure and memoryless attributes.

With the advent of data communication, a new insight into the teletraffic of digital information became imminent. Time-sharing resources, random arrival of bits (or packets), waiting at buffers, contentions, bandwidth allocations etc. have become vital considerations in teletraffic engineering. The relevant queueing models hence have become more than simple depictions via the Poisson or related processes.

Added consideration came into being with the deployment of integrated services (ISDN) where a set of heterogeneous traffics exist, which are to be modeled cohesively in the composite traffic structure.

The subsequent strategy, inculcated into broadening the bandwidths of the services associated with the integrated system, added variable bandwidth considerations into modeling strategies adopted for the teletraffics of B-ISDN/ATM.

The multiplexing of different services (or *statistical multiplexing gain*), cell-transfer delay considerations and the associated cell-delay and its variations pose stochasticity of their own making queueing-theoretic models far from being a humble Poissonian type.

Further, the bit rates representing the constant or variable levels commonly encountered in B-ISDN/ATM warrant unique considerations in evolving relevant queueing models. Specific implications also arise with the bursty bit rates of multimedia transmissions supported by B-ISDN/ATM traffics, which reflect intense Markovian attributes.

Teletraffic engineering models, many times, could be insufficient in the context of switching network where CAC is performed. Specific details on the bounds on the QOS parameters may be required to perform CAC more realistically and reliably. Inclusion of relevant considerations leads to the so-called *bounded traffic models*.

There are three common methods pursued towards analytical and/or numerical solutions of queueing models. They are as follows:

- *Matrix method*: This is adopted to find the steady-state probabilities for a finite buffer via the matrix formulation

- *Probability queueing function method*: This refers to the extraction of probabilities via a functional equation in a complex plane

- *Fluid-flow approximation method*: Here the discrete cell arrival process is depicted by a continuous "*information fluid*" in terms of a set of linear differential equations.

Level of ATM traffic models

In ATM telecommunication, the progress of cells along the virtual connection established is statistical. The relevant reasons are as follows: (i) The cells are guaranteed by different sources (such as voice, data, and video) and the bit emission characteristics of these sources are random in nature. (ii) The cells generated by different sources are also multiplexed statistically. Hence, the models established to describe the ATM traffic are all stochastical in nature. Such traffic models refer to the following levels of traffic:

- *Cell level traffic model:* This refers to the time-dependent random fluctuations of cell-flow and decides buffer dimensioning, namely, the size of ATM switches and multiplexers

- *Burst level traffic model:* This model describes the statistics involved in control measures undertaken and decides the cell admission control strategies.

ATM traffic has been modeled in a variety of mathematical formats. They are based on the following considerations and/or parameters of the traffic:

- ✓ Statistical aspects of information (bits) generated by a heterogeneous set of sources
- ✓ Burstiness (peak bit rate to average bit rate ratio) of bits occurring in variable rates
- ✓ Geometrically distributed burst lengths
- ✓ Switched Poisson process: On-off occurrence of bits. (For example, voice spurts and silent periods)
- ✓ Markov-modulated Poisson process (MMPP)
- ✓ Markov-modulated Bernoulli process (MMBP)

✓ Arbitrarily modulated deterministic process.

Other performance parameters included to model and analyze the ATM traffic are QOS parameters such as CTD, CDV, and CLR. Further, the types of ATM traffics subjected to modeling are:

✓ Constant bit rate (CBR) traffics
✓ Variable bit rate (VBR) traffics.

As indicated in earlier chapters, the classification of a traffic as CBR or VBR type is decided by the differences in sources emission characteristics, that is, by cell generation speeds. Typically, an ATM traffic has an *active period* (of transmission) which refers to the duration of a connection of the period over which a source generates a traffic; the other period is known as the *silent or idle period*. This refers to the time between consecutive active periods during which no traffic is generated.

Statistics of QOS parameters

QOS parameters are user-specific parameters. That is, these parameters denote the aspirations of the users in obtaining the quality of service from their traffic contracts with network providers. Since QOS is a users' view of service, absolute definitions of QOS parameters are rather difficult due to the following reasons:

✓ Existence of different types of users
✓ Variety in the types of services
✓ The subjective dependence on the user's view of the service.

CBR traffic stream

In modeling a CBR traffic, the parameter of interest is the maximum bit rate. Digital data generated by a source is presented to the network as a constant bit rate stream either by the use of smoothing buffers or by controlling the rate at which bits are generated. For example, telephone conversation (where silent periods may also exist) represents a typical CBR traffic stream.

VBR traffic stream

In this case the parameter(s) of interest are as follows:

✓ Maximum bit rate
✓ Average bit rate.

Here, the traffic generated by the source alternates between active and silent periods, or a continuous bit stream is generated at varying rates. Further, it exhibits a peak cell rate namely, the maximum bit rate at which a source generates traffic during its active period.

ITU-T definitions

Instantaneous peak cell rate: The reciprocal of the minimal interarrival time of cells during the active period.

Integrated peak cell rate: This refers to the number of cells generated by a source measured over a predefined short interval ΔT divided by ΔT.

True average cell rate: This indicates the number of cells generated (measured) during the connection duration divided by the length of the duration.

Average cell rate: This corresponds to the number of cells generated by a source over a long interval of time T divided by T.

VBR (bursty) source parameters

❑   *Burstiness-ratio of peak bit rate to average bit rate*
❑   *Average burst length: Mean active period during which a source generates a traffic at its peak rate*
❑ C^2 *parameter: Squared coefficient of the variation of interarrival times of cells*

$$= \frac{s^2 \, (\textit{interarrival time})}{\{E[\textit{interarrival time}]\}^2}$$

Notations:

$1/T$	=	R_p	: *Peak cell generation rate*
a^{-1}			: *Average duration of active periods*
S^{-1}			: *Average duration of silent periods*
(a^{-1}/T)	=	N	: *Average number of cells in an active period*
$(a^{-1} + S^{-1})$	=	T_i	: *Average time from the start of an active period to the start of the next active period*
$(a^{-1})/(a^{-1}+S^{-1})=$		m	: *Average cell rate*
(R_p/m)	=	β	: *Traffic burstiness*

Example 5.11

Describe a typical CBR video implementation.

Solution

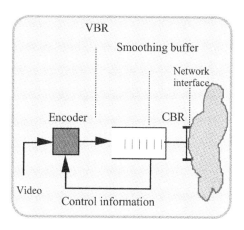

Fig. 5.57: CBR video implementation

This is used to control the bit generation rate at the encoder, when the smoothing buffer reaches a threshold. This avoids overflow problems.

CBR traffic versus VBR traffic

CBR traffic	VBR Traffic
• CBR traffic is easy to manage. A constant bit rate is reserved for each CBR connection throughout its duration, regardless of whether the source is actively transmitting or in a silent state	• VBR traffic management is rather complex, since resource and connection managements are to be tailored on need-based traffic congestion etc.
• Inefficient use of transmission capacity	• Transmission capacity is used more efficiently saving bandwidth
• Designation of CBR-1 and CBR-2 (as independent rates) leads to poor utilization of a network's transmission capacity.	• Since the amount of information generated by most applications varies over time, it is possible to reserve less capacity in the network than the peak rate of the applications. This allows more connections to be multiplexed and permits an increase in the resource utilizations
	• High utilization of transmission links is warranted in WAN's applications such as VCR quality video transmission (via optical links) ($\Rightarrow$ 20 Mbps).

Statistical multiplexing gain

The types of sources involved in ATM, as discussed earlier, are normally on-off and/or bursty in nature. These sources exhibit different cell rates. The multiplexed traffic is an aggregate of such transmissions and refers to composite statistics. In reference to such multiplexing, the associated (*statistical*) *multiplexing gain* (G) can be defined as follows:

$$G = \frac{\text{Number of sources supported}}{\text{Required number of channels}} \qquad (5.39)$$

Suppose there are N identical sources and B is the burstiness ratio. Then the required number of channels (N_C) required to achieve an objective CLR of $Q(\alpha)$ is given by:

$$N_C = (N/B) + \alpha\sqrt{N(B-1)}/B \qquad (5.40)$$

where α is a parameter determined by equating the CLR to the tail of the normal distributing, namely, $Q(\alpha)$ = CLR. Correspondingly, the statistical multiplexing gain (G) can be expressed n terms of the peak rate-to-link rate ratio (η) as follows:

$$G \approx \frac{\eta(\sqrt{\alpha^2(b-1)+4b/\eta} - \alpha\sqrt{b-1})^2}{4} \qquad (5.41)$$

5.7.6   *Principle of CAC modeling*

As discussed earlier, these are intended to determine the amount of bandwidth required by a new connection. That is, to decide whether the service levels required by existing connections would be affected when multiplexed together with a new connection.

The above said functions (in real time) require maximization of the utilization of network resources. They involve the following considerations:

■   A set of parameters required to describe the source activity adequately for accurate prediction of the performance metrics in the network. This is still an open issue. However, some basic modeling considerations can be presented at this stage. Relevant considerations are as follows:

■   Superposition of arrival streams

✓ ATM network can be regarded as a collection of queues connected in a manner determined by network topology

✓ Store the incoming cells temporarily if cell arrival rate is greater than the rate at which they can be transmitted.

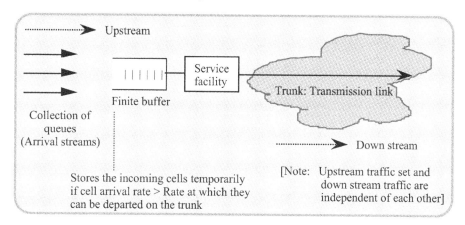

Fig. 5.58: Superposition of heterogeneous traffic streams

▪ Markovian model of the arrival process

Suppose there are N arrivals. Each is characterized by a Poisson process with rate $\lambda(i)$, i = 1, 2.., N. That is, the probability distribution of the number of cells in the queue can be distributed and obtained with a single Poisson arrival stream with a rate assumed as follows:

$$\lambda = \sum_{i=1}^{N} \lambda(i). \qquad (5.42)$$

This superposition of the Poisson process is not, however, applicable to ATM traffic/B-ISDN traffics. The more realistic models (at least is an approximate sense) as indicated before are:

✓ MMPP (Markov-modulated Poisson process)
✓ MMBP (Markov-modulated Bernoulli process).

The MMPP in the ATM context captures the following considerations in describing the traffic streams of ATM:

✓ Time-varying properties of arrival rates
✓ Correlation between interarrival times
✓ MMPP is a doubly stochastic Poisson process.

The doubly stochastic Poissonian characteristics can be understood by referring to Fig. 5.59.

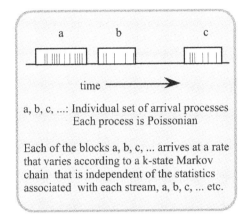

Fig. 5.59: Doubly stochastic Poisson process

In Fig. 5.59, a, b, c denote individual arrival processes and each process is Poissonian. Each of the blocks a, b, c... arrive at a rate which varies according to a k-state Markov chain which is independent of the statistics associated with each stream, a, b, c...etc.

Let i be the state of the Markov chain, $i \in \{1, 2,....,k\}$. Suppose σ_{ij}: The transition rate from state i to state j. $(i \# j)$, λ_i: Arrival rate when the Markov chain is in i^{th} state $\lambda_i > 0$,

$$\sigma_i = \sum_{j=1; i \neq j}^{k} \lambda(i), \text{ and } X_m \text{ is the time between } (m-1)^{st} \text{ and } m^{th} \text{ arrivals. A two-state Markov process,}$$

that is, $k = 1$ indicates that there are two types of streams. (For example, data and voice), which can be modeled as above under constant bit rate (CBR) conditions. MMPP, in summary refers to a stochastical process in which the concepts of the Poisson process and the Markovian structure of memoryless considerations are attributed to the arrival process. As indicated before, there are two versions of MMPP: The first type refers to *discrete MMPP*, which applies to ATM cells, and the second type is known as *continuous MMPP* corresponding to higher-layer PDUs that generate bursts of cells.

The MMBP model is appropriate for ATM links supporting variable bit rate traffic. Voice, video, and data signals serviced by an ATM network are digitized pulses grouped and transmitted as a stream of packetized cells of constant size. A major quality requirement of signal transmission, especially for voice (which is isochronous), is that the delay introduced on such signals across the network be short. Delay requirements are even more stringent with VBR video traffics with a high degree of burstiness. ATM systems are designed to minimize the processing and transmission overheads internal to the network so that very fast cell-switching and routing are made feasible; yet, cell-delays due to asynchronous multiplexing is inevitable. Further, such delays may also "jitter" or vary with time, randomly constituting cell delay variations (CDV).

Burstiness of VBR video traffic adds additional constraint on the ATM in determining the amount of resources required. That is, in order to guarantee a quality of service parameters (vis-à-vis the throughput and the delay), the network should support a resource reservation scheme to compromise the delay-sensitiveness.

Resource allocation based on average traffic rates is not, however, compatible when VBR video sources are present due to the extensive burstiness of the traffic. As a result, in order to avoid unacceptable delays, an upper bound should be prescribed taking into consideration the VBR and burstiness aspects of the relevant sources. That is, a bound-delay test has to be evolved to stipulate the call admission control for the ATM traffic that includes delay-sensitive VBR video with significant burstiness.

5.7.7 *Modeling of video traffic*

In the existing literature, there are two major models indicated to include VBR video traffic into network characterization. The first method is based on characterizing the VBR video traffic by a stochastic process [5.13, 5.14]. *Markov modulation, autoregressive considerations* and *self-similarity of burstiness* are some of the strategies adopted to specify the stochasticity of VBR video traffic. While the stochastic models are flexible in accommodating the statistical features of the sources and yield bounds to achieve higher network utilization, they do not exhaustively capture all the burstiness aspects and time-correlation features of VBR video. Further, their implementation is too involved and their statistical guarantees do not offer deterministic bounds on cell-loss ratio (CLR) and/or cell-transfer delay (CTD).

Delay-bound tests pertinent to VBR video lead to obtain deterministic guarantees on the cell admission control (CAC) via resource allocations decided on the basis of *"empirical envelope"* models of the bursty sources. These empirical models are specified to represent the most accurate, time-invariant, deterministic characterizations of such sources.

As indicated earlier, the asynchronous transfer mode of cell transmission itself, in general, causes an accumulation of cell-transfer delay (CTD) due to the statistical multiplexing involved leading to an eventual dropping of cells. The extent of such cell-dropping can be linked to the quality of service (QOS) and a cell-loss ratio (CLR) parameter is specified as the quantitative measure of QOS. It places an upper bound α on the probability ascertained from the statistical distribution of cell-delay variations (CDV) such that α specifies the maximum CTD before cell-dropping occurs.

Specific to ATM transmissions, which support VBR video, a strong correlation exists among cells originating from the VBR video source giving rise to a bursty traffic. A burst source generates cells at a peak rate or a near-peak rate for every short duration and remains almost inactive in between; that is, video codes, in general, produce a variable output wherein the code word corresponding to one frame does not always have that same size. Bursty traffic has alternate busy (or active) duration and off (or silent) duration. The duration of the active and idle states are *geometrically distributed*.

In an homogenous traffic or VBR video, all the cells in a burst have the same destination and the active and idle state of the burst are assumed to represent a two state Markov process.

Suppose a specific coding scheme is adopted in the VBR video traffic. Say, for example, the coding may correspond to an algorithm based on intrafield/intraframe differential PCM with or without motion compensation. The proposed modeling is in reference to a set of cells per frame associated with the encoded VBR video.

Let Y_n be the number of cells in the nth frame. The following are indicated as the characteristics of Y_n.

    ✓   The number of cells per frame is a stationary Markov chain
    ✓   The marginal distribution of Y_n is negative-binomial
    ✓   The correlation between Y_n and Y_{a+k} has the form p^k.

In the homogeneous model presumed, the arbitrary Markovian traffic sources are identical and finite-dimensioned. These sources (pertinent to a VBR video) are also time-reversible. Further, relevant to the two-state transitions of the Markov chain Y_n, the transitions matrix is given by:

$$\widehat{\rho}_{ij} = \frac{\text{Number of transitions i to j}}{\text{Number of transitions out of i}} \tag{5.43}$$

Let p represent the probability of the active state (or i^{th} state), which is assumed to last over τ_k time-slots in a frame, and q be the corresponding of the idle state (or j^{th} state) occupying τ_r time-slots in the same frame.

The *negative binomial distribution* of the Markov process concerning VBR video traffic being considered, in essence, represents the statistics of realizing k active states before r idle states occur in the associated bursts. In such a bursty environment, the active states are the ones which pose an enhanced demand on the traffic flow. That is, the higher the probability p, the larger the "active", bursty durations encountered over the transmission are envisaged. As a result, the corresponding excess demand on the traffic would cause buffer overflow conditions, leading to cell/frame losses.

The probability density function (pdf) of the negative binomial distribution under discussion is given by:

$$f(r;k,p) = \binom{k+r-1}{k} p^k q^r \qquad (k+r) = k, k+1, k+2, \ldots \qquad (5.44)$$

That is, by the independence of the trails (events), the probability for a given number of active events k to occur over $(k + r)$ Bernoulli events is specified by f_k. It is the same as the probability $\{k - 1\}$ active states in $(k + r - 1)$ total events of active plus idle events $\times$ probability {active state at $(k + r)^{th}$ event}. The mean and variance of the above distribution are follows:

Mean (m)	$=$	$k(1-p) / p$	(5.45a)
Variance (v)	$=$	$k(1-p) / p^2$	(5.45b)

with $0 < p < 1$, $q = (1 - p)$ and $r > 0$.

The negative binomial distribution offers an attribute to the presumed process that it refers to, the chance of attaining $(k + r)^{th}$ active plus idle states. This means that among the first $(r + k - 1)$ states, there are exactly r idle states and following, that is, the $(k + r)^{th}$ state refers to an active state. It is identical to the chance of winning k times in $(k + r)$ trials where the win probability in each trial is p. Therefore, f_k can be assumed to represent the probability density function (pdf) of a discrete waiting-time random variable. This is justified further in the following elaboration.

If a sequence of $(k + r)$ Bernoulli trials is performed at the rate of one per second, then the number of failures preceding the first success decides the waiting-time. Suppose in a particular sequence no success has occurred in the first v trails, so that $r > v$. The waiting-time from this trail to the next success is independent of the number of preceding failures. Hence the probability that the waiting-time will be prolonged by an additional time (say, T seconds) is independent of v and equals the initial probability of the total length exceeding T seconds. (It is to be noted that this property is not shared by waiting-times encountered in phenomena such as waiting lines before counters and lifetimes of machines.)

The probability of attaining a population of active plus idle states equal to $(k + r)$ after a stretch of time-slots (in a frame) specified by t and starting with an active state population of $(k + 1)$ at $t = 0$, is the same as the probability of attaining k active states in $(k + r)$ Bernoulli events. The negative binomial distribution (also known as *Pascal distribution*) is the k-fold convolution of the geometric distribution and can be specified by: $f(r; k, p) = \{q^r p\}^k$, $k \geq 0$; and, $q = (1 - p)$.

Problem 5.6

Develop a computer code to simulate the multiplexed traffic stream corresponding to the following parameters of the ATM multiplexer and the connection types:

Multiplexer

Output capacity	$= 600$ Mbps
Buffer space	$= 0.5$ Mb

Sources: VBR types (mean bit rate $= m$; peak bit rate $= h$)

1. On-off type

$P_{on} = m/h$ ($m = 10$ Mbps; $h = 40$ Mbps)

$P_{off} = 1 - (m/h)$

2. Binomial source

$$P_i = \binom{R}{i} \left(\frac{m}{h}\right)^i \left(1 - \frac{m}{h}\right)^{R-i}, \quad i = 0, 1, 2, \ldots, h$$

$m = 5$ Mbps; $h = 40$ Mbps

3. Binomial source

$m = 5$ Mbps; $h = 80$ Mbps

5.8   Statistical Performance Considerations in ATM

HEC Performance

The CRC encoding adopted for HEC in ATM enables error detection and correction feasibilities. It is of interest to calculate the following probabilities, in the event of random errors occurring in the cell header.

- ✓   P(F/D): Probability of falsely sending out an errored header under the condition that the node performs HEC error-detection
- ✓   P(Ds/D): Probability of discarding a cell by a node when the received cell header does not match the valid cell header under HEC error detection
- ✓   P(F/C): Probability of encountering invalid cells while performing HEC and implementing header error correction. (The invalidity refers to a false match occurring as observed in a codeword, which appears to have only one bit error and is then inadvertently corrected into a valid code word)
- ✓   P(Ds/C): Probability of discard given that the header correction is done.

The above probabilities can be obtained by considering ε errors in the 40-bit cell header. It corresponds to a binomial distribution b(40, ε) namely, $\begin{pmatrix} 40 \\ \varepsilon \end{pmatrix} p^{\varepsilon} (1-p)^{40-\varepsilon}$ where p is equal to the probability of bit error. Hence, with header error detection, the P(F/D) corresponds to the probability of randomly matching a valid codeword (1/256), if three or more bit errors occur. Hence, the following results are obtained:

$$P(F/D) = \frac{1 - b(40,0) - b(40,1) - b(40,2)}{256} \qquad (5.46a)$$

$$P(Ds/D) = \{1 - b(40,0)\}\{1 - p(F/D)\} \qquad (5.46b)$$

$$P(F/C) = \frac{41\{1 - b(40,0) - b(40,1) - b(40,2)\}}{256} \qquad (5.46c)$$

$$P(Ds/C) = \{1 - b(40,0) - b(40,1)\}\{1 - p(F/C)\} \qquad (5.46d)$$

Problem 5.7
Derive the various probability expressions of Eqns.(5.46a) through (5.46d) specified toward ATM HEC performance.

5.8.1   *HDLC performance*

High level data link control (HDLC) is a popular ISO and ITU-T standardized link layer protocol standard for point-to-point and multipoint communications. Though CRC is implemented, the bit-stuffing efforts (discussed in Chapter 2) in HDLC is susceptible for errors, both of random and bursty type. The bit-stuffing is done to eliminate the occurrence of a flag sequence within an HDLC frame. Suppose one or two bit errors appear in wrong places within a received frame. It would falsely indicate a valid flag field, thereby terminating a frame. Likewise, the trailing flag filed can be corrupted by bit errors. The relevant probability of undeleted error for HDLC is given by the following approximate formula:

$$\text{Probability(undetected error/HDLC)} \cong [1.36Kb + \left(\frac{m}{k}\right)p^4] \times 2^{-16} \qquad (5.47)$$

where k is the number of bytes in the HDLC frame, p is the BER, and m is the average HDLC frame length (in bits) after bit-stuffing; M is given by:

$$m = 8\left[\frac{64}{63}k + 2\right]$$ (5.48)

5.8.2 AAL-5 PDU performance

The AAL-5 uses a 32-bit CRC generator polynomial, which has 15 nonzero coefficients. That is, its Hamming distance is 15. Should there be any undetected error with respect to AAL-5 PDU, the following situations must arise:

    ✓   A particular pattern of exactly 15 random bit errors must occur
    ✓   The probability of such bit errors should be 2^{-32}.

Therefore, the undetected frame level error performance of AAL-5 for an information field of k octets is given by:

$$\text{Probability(Undetected error/AAL-5)} = b(X < Y, Z)$$ (5.49)

where b(..) is the binomial distribution, $X = 8\left(\frac{k+8}{48}\right)^{*}$, Y = 15 and Z = p = bit error probability. The asterisk in (x)* denotes the smallest integer greater than x.

Problem 5.8
Derive the expression specified in Eqn.5.49 depicting the probability of undetected error in reference to AAL-5 PDU.

5.9   Information-Theoretic Models

5.9.1 Entropy of cell losses

The information-theoretics refer to the subject of studying a phenomenology in the perspectives of information theory. It is concerned with the entropy of the statistics associated with the phenomenology and hence depends on the views and concepts of information-content (in Shannon's sense). The teletraffic considerations have been addressed in the information-theoretics domain largely in terms of the capacity of discrete-time queues representing such traffics. Studies combining queueing-theory with information theory have also been done pertinent to multiaccess systems [5.15, 5.16]. Information-loss considerations in packet voice systems have been addressed. Specific to ATM traffic, relevant entropy considerations have been studied. Fuzzy concepts have also been introduced into the queueing dynamics of ATM cell streams in the information-theoretic domain by the author [5.17]. Further, the information-theoretic characterization of multiplexed ATM traffic has been studied in [5.15, 5.18].

An information-theoretics based analysis of the heterogeneous teletraffic (such as ATM) can be done via complexity metric [5.18] considerations, which represent the cohesive aspects of CDV and CLR of the traffic involved. The following paragraphs are centered on relevant considerations.

5.9.2 CDV based cell-losses

Suppose the statistics of the CDV (δ_η) of a traffic (identified as the traffic from source # η) is assumed to be Gaussian. Then the probability density function of δ_η, namely $p(\delta_\eta)$ can be written as:

$$p(\delta_\eta) = \frac{1}{\sqrt{2\pi}\sigma_{\delta\eta}}\exp[-(\delta_\eta - \mu_{\delta_\eta})^2/2\sigma_{\delta_\eta}^2]$$ (5.50)

where μ_{δ_η} and σ_{δ_η} denote the mean and standard deviations of δ_η respectively. The maximum entropy vis-à-vis the aforesaid Gaussian statistics is given by:

$$(H_{max})_\eta = \frac{1}{2}\ln[2\pi e(\sigma_{\delta\eta}^*)^2] \tag{5.51}$$

where $\sigma_{\delta_\eta}^*$ is σ_{δ_η} normalized with respect to a time parameter such as the total time of transmission of M cells, T sec.

As deduced in Appendix 5.1, a critical transition from "*simple-to-complex*" state mediated by M cells arriving at a rate λ_η cells/s occurs when the *complexity parameter* (defined in Appendix 1), s $\to 1$. Using Eqn.(5A1.16), the corresponding maximum entropy associated with M cells flowing at a rate λ_η can be specified as:

$$(H_{max})_\eta = \ln\left[\frac{M+1}{\lambda_\eta T}\right]. \tag{5.52}$$

Hence, combining Eqns.(5.51) and (5.52), it follows that $\ln[(M+1)/\lambda_\eta T] = \ln[2\pi e(\sigma_{\delta\eta}^*)_{max}^2]^{1/2}$

Or, $(M+1)/\lambda_\eta = \sqrt{2\pi e}(\sigma_{\delta_\eta})_{max}$.

In other words, for a set of M cells for which end-to-end delay is assessed, the maximum permissible limit of δ_η is decided by the standard deviation given by:

$$(\sigma_{\delta_\eta})_{max} = \frac{(M+1)}{\sqrt{2\pi e}\lambda_\eta}. \tag{5.53}$$

The corresponding variance parameter then decides the upper limit of permissible cell-delay jitter beyond which the network is led to discard the cells.

5.9.3   Cell-loss ratio due to CDV

As indicated in Fig. 5.60 the probability distribution of end-to-end delay for a typical ATM connection has a tail-end probability of α, which can be used to depict the upper bound on delay due to jittering.

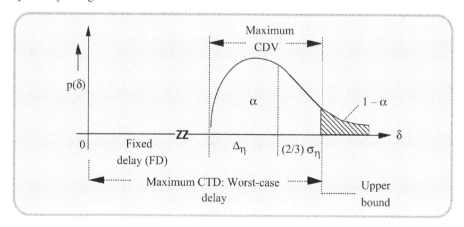

Fig. 5.60: Pdf of CDV in an ATM transmission

Now, considering the cell-losses, a probabilistic attribute α that places an upper bound on net CLR can be bifurcated in to α_1 and α_2 pertinent to cell-losses stemming from queueing congestion that induced multiplexing delay and that is due to buffer over-flow. That is, α_1 refers to the part linked with the cell-losses caused by jittered CTD (resulting from queueing congestion) exceeding a specified upper bound (Fig. 5.60). If the corresponding cell-loss ratio is designated as CLR1, then a relevant relation has to be formulated between α_1 and CLR1. Similarly, α_2 refers to the overflow aspects of the buffer of finite size and the corresponding CLR can be specified as CLR2. In summary, the required functional relations to be ascertained are:

$$\alpha_1 \leftrightarrow \text{CLR1} \Rightarrow \text{CTD1} \tag{5.54a}$$
$$\alpha_2 \leftrightarrow \text{CLR2} \Rightarrow \text{CTD2} \tag{5.54b}$$

where CTD1 and CTD2 are parameters which decide jointly a net upper bound on CTD leading to a corresponding net upper bound on CLR indicated in Fig. 5.60.

The net cell-losses thus manifest, is a parameter α constituted by α_1 and α_2. Therefore, the stochastical aspects of multiplexing and buffer overflow, which decide the probability α_1 and α_2 respectively specify implicitly the net CLR and the associated Shannon's information (loss) content pertinent to an ensemble of cells constituting the traffic of an ATM service.

The problem of linking α_1 and CLR1 is same as correlating the variance of CDV and the CLR1. That is, under worst case congestion, the CTD1 exceeding a maximum value equal to fixed delay (F) plus CDV with a probability no greater than α_1, specifies the maximum limit of cell transfer delay. This bound is required to be correlated to CLR1. In a similar fashion, the cell-losses due to buffer over-flow would lead to α_2 and hence, CLR2.

Relevant to a source η, suppose a cell-loss probability $P_{x\eta}$ is specified as an impairment parameter (of end-to-end traffic performance) due to congestion-induced CDV and $P_{B\eta}$ denotes the probability leading to relevant cell-dropping caused by finite buffer size. The strategy to implement the required inter-relations specified above in Eqn. (5.54) is based on the following heuristics:

CLR1 versus $P_{x\eta}$

The CLR1 parameter pertaining to cell-losses arising from CDV caused by congestion in ATM links can be specified by a corresponding cell-erasure probability $P_{x\eta}$ that leads to a loss of average information-content per unit bandwidth associated with M cells constituting the ATM traffic. The nonzero value of $P_{x\eta}$ can be regarded as an *equivalent probability of error* induced by an *impairment factor*, $C_{x\eta}$. That is, the cell-loss due to excessive CDV resulting from the congestion mechanism can be dubbed equivalently, as if such a cell-loss is a result of some impairing entity (analogous to noise) being present in the traffic flow. Therefore, from the considerations of the digital communication theory as applied to binary digit errors introduced (due to noise), $P_{x\eta}$ can be specified by an exponent relation. In reference to an ηth source ($\eta = 1, 2, \ldots, N$),

$$P_{x\eta} = \frac{1}{2}\exp(-C_{x\eta}) \tag{5.55}$$

In the above equation, the impairment factor $C_{x\eta}$ depicts an *erasure exponent* that sets a limit on CDV exceeding an upper bound.

Considering the sources representing a heterogeneous traffic environment, each source is facilitated with a random extent of asynchronous multiplexing of its cells at the MUX. This multiplexing process refers to a statistically independent source of the others on an FIFO basis, as long as there is no contention at the MUX. Otherwise, (in the event of contention), a higher priority traffic cell is served first under the conditional probability that there exist contending cells from lower priority traffics.

The MUX is a multiaccess shared-processor where cells from different sources compete for the processor time. The more sources are active at a given time, the less the rate of service each receives, since there is more contention. The total service rate rendered by the MUX depends on the state of the queue through the number of cells competing for multiplexing service.

Suppose m_η cells from the ηth source are processed at an ATM MUX (Fig.5.61). The arrival rate of these cells is λ_η cells/s. The MUX (a queue-server) facilitates asynchronous multiplexing at a rate $\psi(m_\eta) > 0$ per unit time. Also, it divides this multiplexing service rate equally among all cells emanating from a given source. That is, whenever there are $m_\eta > 0$ cells in the traffic stream from η^{th} source, each cell receives a service at a rate of $\psi(m_\eta)/m_\eta$ per unit time. And the cell departs from the MUX (once it has received the service) on the multiplexed line at a rate μ cells/s. If a priority schedule is imposed, a *contention priority coefficient* Υ_η, ($\Upsilon_\eta < 1$ and $\sum_{\eta=1}^{N} \Upsilon_\eta = 1$) can be used to weight the service rate rendered to the cells of the ηth source. In other words, the service rate for each cell can be written as $[\Upsilon_\eta \, \psi(m_\eta)/m_\eta]$ per unit time. The coefficient Υ_η, in essence, is a conditional probability that a cell of ηth source is served or not served depending upon (or conditioned by) the contention with cells from other sources.

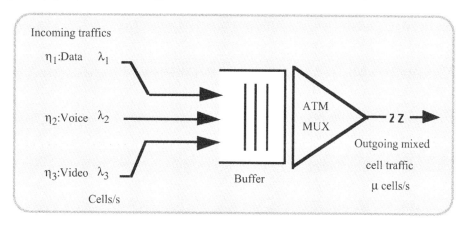

Fig. 5.61: ATM multiplexing

Corresponding to the service rate $\psi(\cdot)$ provided by the MUX, an information-impairment activity parameter can be specified to depict the cell-losses arising from the buffer over-flow at the MUX. The factor $P_{x\eta}$ indicated in Eqn.(5.55) denotes relevant cell-dropping statistics. Typically, the number of cells which do not drop corresponds to a geometrically distributed random variable with a mean equal to $1/(1 - P_{x\eta})$.

Denoting the cell-loss based rate of information-impairment by $\Phi(m_\eta)$ resolved per unit time, the rate of information-loss shared equally by m_η cells can be written as $(1 - \Upsilon_\eta) \Phi(m_\eta)/ m_\eta$. The factor $(1 - \Upsilon_\eta)$ implies that lower priority traffic suffers a higher rate of information loss. Explicitly $\Phi(m_\eta)$ is given by:

$$\Phi(m_\eta) = (m_\eta) \ln\left\{1 + \frac{1}{(m_\eta - 1) + \dfrac{1}{C_{x\eta}}}\right\}. \tag{5.56}$$

The steady-state distribution of information-impairment suffered by m_η cell at the MUX due to finite P_x is given by:

$$P(m_\eta) = [1/K\Phi_\pi(m_\eta)] (\ell_\eta)^{m_\eta} \tag{5.57a}$$

where $\ell_\eta < 1$ is the average fractional loading by the cells of ηth source on the multiplexed line via MUX. Suppose each source continuously emits cells and loads them on the multiplexed line. The fraction of such loading by each source can be denoted as ℓ_η^* and $\sum_\eta \ell_\eta^* = 1$. The relation between ℓ_η and ℓ_η^* is derived in the Appendix 5.2. Further, $\Phi_\pi(\cdot)$ in Eqn.(a) refers to: $\Phi(m_\eta) =$ $\prod_{n=1}^{m_\eta} \Phi(n)$.

In terms of information arrival expressed in nat per unit bandwidth and resolved per unit time, the following relation can be validly specified:

$$\ln(M_T \ell_\eta) = (\lambda_\eta/\mu) \, E[H_\eta]/(1 - P_{x\eta}) \tag{5.57b}$$

where M_T is the total number of cells from all the N active sources resolved per unit time. That is, $M_T = \sum_{\eta=1}^{N} m_\eta$.

The entity $E[H_\eta]$ in Eqn.(5.57b) is the ensemble mean of information-content resolved per unit time (expressed in nats per bandwidth) and loaded by the ηth source at a rate λ_η into the MUX. Suppose μ is the available bandwidth of transmission on the trunk line. The ratio (λ_η/μ) depicts the *traffic intensity* or the *utilization of* the ηth source. The extent of $E[H_\eta]$ is decided by the geometrically distributed cells that are not lost, and therefore, is weighted by $1/(1 - P_{x\eta})$. Further, $E[H_\eta]$ is dependent on the work-load or cell emission activity of the ηth source. Thus, ℓ_η can be regarded as a fraction that globally takes the traffic intensity as well as the work-load characteristic of the ηth source.

The factor K in Eqn.(a) is given by:

$$K = 1 + \sum_{m_\eta=1}^{\infty} (\ell_\eta)^{m_\eta} / \Phi_\Pi(m_\eta). \tag{5.57c}$$

In terms of $P(\cdot)$ of Eqn.(5.57a), the expected value of fractional cells lost causing information-impairment per bandwidth of trunk transmission and resolved in unit time can be written as:

$$L_\eta(m_\eta) = E[\text{lost cells}] = \sum_{\theta=1}^{m_\eta} \theta P(\theta). \tag{5.58}$$

The corresponding *average information-loss*, $H_\eta(L_\eta)$ resolved per unit time can be specified as:

$$H_\eta(L_\eta) = \ln [M_T L_\eta(m_\eta) + 1] \qquad \text{nats/bandwidth} \tag{5.59}$$

via Eqn.(5.57b). Further, the mean delay time $(\Delta_{1\eta})$ equivalence (representing the cell-loss) can be obtained from the well-known Little's formula [5.12]. That is, when $M_T L_\eta(m_\eta) \gg 1$,

$$\Delta_{1\eta} \approx M_T L_\eta(m_\eta)/\lambda_\eta \qquad \text{second} \tag{5.60}$$

where the subscript 1 corresponds to traffic-impairment specified by CLR1.

The expected loss of information caused by the cell-drops pertinent to η^{th} source as given by Eqn.(5.59), is identically equal to the entropy associated with the CDV. Hence,

$$\ln[M_T L_\eta(m_\eta)/\lambda_\eta] \equiv \ln [2\pi e \, \sigma_{1\eta}^2]^{1/2} \tag{5.61a}$$

or,

$$\sigma_{1\eta} = \Delta_{1\eta}/(2\pi e)^{1/2} \tag{5.61b}$$

which represents the standard deviation of the statistics describing the CDV equivalence of the cell-loss pertinent to the ηth source. It should be noted that both the mean value ($\Delta_{1\eta}$) and the standard derivation ($\sigma_{1\eta}$) are specified per unit bandwidth of trunk transmission and resolved per unit time.

CLR2 versus ($P_{B\eta}$ and α_2)

Additional delay $\Delta_{2\eta}$ is induced as a result finite buffer size at the MUX. It can be specified explicitly in terms of an erasure probability $P_{B\eta}$, which can be written as:

$$P_{B\eta} = \frac{1}{2}\exp(-B) \tag{5.62}$$

where B is a parameter depicting the size of the buffer facilitated at the MUX. That is, given $P_{B\eta}$ or its exponent B, the pertinent cell-loss ratio, namely, CLR2 linked to the CTD will give rise to a corresponding average delay $\Delta_{2\eta}$ with a variance $\sigma_{2\eta}$. The parameters $\Delta_{2\eta}$ and $\sigma_{2\eta}$ can be determined by the similar procedure indicated above replacing $P_{x\eta}$ by $P_{B\eta}$ and $C_{x\eta}$ by B. Again, the subscript 2, in Δ and σ refers to the conditions dictated by CLR2.

Thus, the net CTD specification pertinent to an ATM system can be deduced in reference to an ηth source in terms of the following:

1. Fixed-delay (F) due to propagation, delay induced by switching system and/or processes and due to fixed components.

2. Delay introduced by the congestion in the incoming traffic from different sources (of different bandwidths and bit rates). Its mean value and upper bound are specified by ($\Delta_{1\eta}$ and $\sigma_{1\eta}$).

3. Delay introduced by cell-drops due to finite buffer size prevailing at the multiplexer of the ATM link. Its mean value and upper bound are specified by ($\Delta_{2\eta}$ and $\sigma_{2\eta}$).

The combined effects of congestion plus multiplexing with a finite-sized buffer can be specified by a single Gaussian statistic of CDV with a mean value of $\Delta_\eta = (\Delta_{1\eta} + \Delta_{2\eta})$ and a standard deviation by: $(\sigma_\eta) = [(\sigma_{1\eta})^2 + (\sigma_{2\eta})^2]^{1/2}$. Hence, a corresponding effective CLR bound can be stipulated. The probability α in Fig. 5.60 is decided by a resultant upper bound set by the quartile value (equal to $2/3 \, \sigma_\eta$).

Thus, for a given set of specifications on ($L_{X\eta}$, $L_{B\eta}$), ($P_{X\eta}$, $P_{B\eta}$), λ_η and m_η of the ηth traffic, a Gaussian pdf curve can be constructed for δ_η when a total of M cells (from all the sources) are impressed on the input to the MUX. It is given by:

$$p(\delta_\eta) = (1/\sqrt{2\pi}\rho_\eta)\exp\left\{-\left[\frac{(\delta_\eta - \Delta_\eta)^2}{2\rho_\eta^2}\right]\right\}. \tag{5.63}$$

5.10 Concluding Remarks

The signaling framework of telecommunications facilitates procedures to establish, maintain, and terminate network connections. It provides a means to exchange information between the end-entities and the network as well as between networks/switches. Its basic infrastructure from the time of Graham Bell until today remains more or less the same. But the art of signaling has grown to support additional capabilities matching the emerging applications. In the information superhighway, signaling has a lane of its own, that lets the end-user traffic be not only established, maintained and disconnected but also makes it robust and reliable. The broadband features of ATM networking have added a unique technology profile to signaling.

Bibliography

[5.1]   E.H. Jolley: *Introduction to Telephony and Telegraphy* (Hart Publishing Co., Inc., New York, NY: 1970)

[5.2]   A.M. Noll: *Introduction to Telephones and Telephone Systems* (Artech House, Inc., Norwood, MA: 1991)

[5.3]   J.G. van Bosse: *Signaling in Telecommunication Networks* (John Wiley & Sons, Inc., New York, NY: 1998)

[5.4]   T. Viswanathan: *Telecommunication Switching Systems and Networks* (Prentice-Hall of India Pvt., Ltd., New Delhi, India: 1995)

[5.5]   J.D. Spragins, J.L. Hammond and K. Pawlikowski: *Telecommunications Protocols and Design* (Addison-Wesley Publishing Co., Reading, MA: 1991)

[5.6]   M. Schwartz: *Telecommunication Networks – Protocols Modeling and Analysis* (Addison-Wesley Publishing Co., Reading, MA: 1988)

[5.7]   D. Minoli and G. Dobrowski: *Principles of Signaling for Cell Relay and Frame Relay* (Artech House, Inc., Norwood, MA: 1998)

[5.8]   R.O. Onvural and R. Cherukuri: *Signaling in ATM Networks* (Artech House, Inc., Boston, MA: 1997)

[5.9]   D.L. McDysan and D.L. Spohn: *ATM Theory and Applications* (McGraw-Hill, Inc., New York, NY: 1998)

[5.10]   R.O. Onvural: *Asynchronous Transfer Mode Network – Performance Issues* (Artech House, Inc., Boston, MA: 1995)

[5.11]   W. Stallings: *High-Speed Networking – TCP/IP and ATM Design Principle* (Prentice-Hall, Inc., Upper Saddle River, NJ: 1998)

[5.12]   D. Bertsekas and R. Gallager: *Data Networks* (Prentice-Hall, Inc., Englewood Cliffs, NJ: 1992)

[5.13]   S. Jagannathan: *Estimation of Information-Theoretics Based Delay Bounds of MPEG Traffic Over ATM Networks*, M.S. (Computer Science) Thesis, Florida Atlantic University (Boca Raton, Florida 33431, USA), 1998

[5.14]   L. Wei: *Estimation of Information-Theoretics Based Delay-Bounds in ATM Networks*, M.S. (Computer Science) Thesis, Florida Atlantic University (Boca Raton, Florida 33431, USA), 1997

[5.15]   S. Hsu, P.S. Neelakanta and S. Abeygunawardana: Maximum cell transfer delay versus cell-loss ratio in ATM transmissions *Proceedings of IEEE ATM'96 Workshop*, (Aug 25 – 27, San Francisco, CA, USA), Paper # 4 – Session TP 2B-2

[5.16]   S.M. Abeygunawardana: *Studies on Traffic Characteristics in Asynchronous Transfer Mode Telecommunication*, M.S.E. (Electrical Engineering) Thesis, Florida Atlantic University, Boca Raton, Florida 33431, USA, 1996

[5.17]   P.S. Neelakanta and W. Deecharoenkul: Fuzzy aspects of queueing dynamics of ATM cell streams in information-theoretic domain *Proceeding of the 33rd Annual Conference on Information Sciences and System*, Vol. II, (March 17 – 19, 1999, Baltimore, MD, USA), 1999, 826 – 831

[5.18]   P.S. Neelakanta and M. Palaniappa: Information-theoretic characterization of multiplexed ATM traffic, *Proceeding of the 33rd Annual Conference on Information Sciences and System*, Vol. II, (March 17 – 19, 1999, Baltimore, MD, USA), 1999, 832 – 837

[5.19]   P.S. Neelakanta and W. Deecharoenkul: A complex system characterization of telecommunication services, *Complex Systems* (in press)

[5.20]   A.E. Ferdinand: A theory of system complexity, *International Journal of General Systems*, 1, 1994, 19-23

Appendix 5.1

A Complexity Metric for the Acceptable Threshold of CLR

Suppose $0 \le p(i) \le 1$ denotes the probability of occurrence of the ith cell. Let $i = 1, 2, ..., M$ represent the cells for which the end-to-end performance is assessed in ATM link. Then, the axiomatic probability requirement is that $\sum_{i=1}^{M} p(i) = 1$ and the mean value $\sum_{i=1}^{M} ip(i) \le \beta_0$. Here $\beta_0 > 0$ depicts the constraining value on the ensemble mean as decided by the limit of acceptable traffic performance. Given a set of VC parameters attributed to a VC traffic, the entropy (Shannon information) parameter of the epochs of cell occurrence is given by [5.19, 5.20]:

$$H(p) = - \sum_{i=1}^{M} p(i) \ln [p(i)] \qquad \text{nats.} \qquad (5A1.1)$$

Using the Lagrange multipliers a and b, one can define,

$$h = - \sum_{i=0}^{M} p(i) \ln p(i) - a \sum_{i=0}^{M} p(i) - b \sum_{i=0}^{M} ip(i). \qquad (5A1.2)$$

Differentiating Eqn.(5A1.2) with respect to $p(i)$ and setting the resulting derivatives identically equal to zero, the following set of equations are realized:

$$1 + a + \ln p(i) + bi = 0 \qquad \text{for all i.} \qquad (5A1.3)$$

The Lagrange multipliers can be determined from the conditions $\sum_{i}^{M} p(i) = 1$ and $\sum_{i}^{M} ip(i) \le \beta_0$. The first condition yields,

$$\exp[-(a + 1)] \sum_{i=0}^{M} \exp(-bi) = 1 \qquad (5A1.4a)$$

and the second condition leads to:

$$\beta_0 = \exp[-(a + 1)] \sum_{i=0}^{M} i \exp(-bi) = \frac{\sum_{i=0}^{M} i \exp(-bi)}{\sum_{i=0}^{M} \exp(-bi)}. \qquad (5A1.4b)$$

Let the following new parameter be defined for convenience [5.20]:

$$s = \exp(-b). \qquad (5A1.5)$$

Hence it follows that,

$$s \frac{d}{ds} \ln[S(s, M)] = \beta_0 > 0 \qquad (5A1.6a)$$

where $S(s,M) = \sum_{i}^{M} s^i$, and further differentiation leads to:

$$\frac{d^2}{db^2} \ln[S(e^{-b}, M)] > 0 \qquad (5A1.6b)$$

Therefore, the possible solution of Eqn.(5A1.6a) is unique and positive. Also,

$$\frac{\partial^2 H(p)}{\partial p(i)\partial p(j)} = 0 \qquad \text{if} \qquad i \neq j \qquad\qquad (5A1.7a)$$

$$= \frac{-1}{p(i)} \qquad \text{if} \qquad i = j. \qquad\qquad (5A1.7b)$$

Eqn.(5A1.7) stipulates that the matrix of the second derivatives of H(p) is negative definite and H(p) is, therefore, maximized.

By performing the summation of Eqn.(5A1.4a), one can obtain:

$$\exp(a+1) = (1 - s^{M+1})/(1-s) \qquad s < 1 \qquad\qquad (5A1.8a)$$

$$= (s^{M+1} - 1)/(s-1) \qquad s \geq 1. \qquad\qquad (5A1.8b)$$

Also from the set of Eqn.(5A1.3) it follows that,

$$p(i) = \exp[-(1+a)]\exp[-bi] \qquad\qquad (5A1.9a)$$

$$= s^i(1-s)/(1-s^{M+1}) \qquad s < 1 \qquad\qquad (5A1.9b)$$

$$= s^i(s-1)/(s^{M+1} - 1) \qquad s \geq 1. \qquad\qquad (5A1.9c)$$

The above relations (Eqn.(5A1.9)) indicate that p(i) maximizes the entropy functional and describe the probability of having i disorderly (performance-wise) subsets in the cell-space of total size, M. That is, p(i) depicts the probability distribution which maximizes the entropy H(p) of disorderliness associated with epochal occurrence of cell events and implicitly depicts the associated performance impairment such as jitter and/or bit-errors.

The introduction of a constraint in the Lagrange multiplier operation via β_0 is equivalent to the energy constraint in the energy functional optimization problems. The correspondence between β_0 and the new variables in the above derivations is similar to the introduction of the activity variable in the statistical mechanics.

Calculation of the mean value of β_0 is as follows: β_0 is the expected value of i, namely,

$$\beta_0 = E[s,M] = s\frac{d}{ds}(\ln[S(s,M)]) \qquad\qquad (5A1.10a)$$

$$= \frac{s}{(1-s)} - \frac{(M+1)s^{M+1}}{(1-s^{M+1})}, \qquad s < 1 \qquad\qquad (5A1.10b)$$

$$= -\frac{\rho}{1-\rho}\left[\frac{1}{\rho}\right] + \frac{(M+1)\rho^{M+1}}{1-\rho^{M+1}}\left[\frac{1}{\rho^{M+1}}\right] \qquad s \geq 1 \qquad\qquad (5A1.10c)$$

where $\rho = 1/s$.

Denoting β_0 by E[(s,M)], the entropy H(s,M) of the subsets can be specified as follows:

$$H(s,M) = \ln S(s,M) - E[(s,M)]\ln(s). \qquad\qquad (5A1.11)$$

With relevant substitutions, it follows that,

$$H(s,M) = \ln\left[\frac{(1-s^{M+1})}{(1-s)}\right] - \left[\frac{s}{(1-s)} - \frac{(M+1)s^{M+1}}{1-s^{M+1}}\right]\ln(s) \qquad \text{for } s < 1 \qquad (5A1.12a)$$

$$= \ln\left[\frac{(1-\rho^{M+1})}{1-\rho}\left(\frac{1}{\rho^M}\right)\right] - \left[\frac{\rho}{(1-v)}\left(\frac{1}{\rho}\right) + \frac{(M+1)\rho^{M+1}}{1-\rho^{M+1}}\left(\frac{1}{\rho^{M+1}}\right)\right]\ln(\rho), \text{ for } \rho \leq 1. \qquad (5A1.12b)$$

The function E[s,M] is strictly monotonic with respect to s, and for s > 0, it is also a strictly monotonic function of M. The entropy of the disordered subsets is a positive function increasing monotonically with respect to M for all values of s ≥ 0. It also increases monotonically with s for s < 1 but decreases monotonically with s for s > 1.

The coefficient s can be regarded as a *measure of complexity* associated with the ATM transmission experiencing cell-losses in an end-to-end connection. For a given stretch of cells, it depicts implicitly the extent of CLR expected. As such, when s = 0, the cells flowing between an end-to-end connection constitute a *"simple"* subsystem with an expected value of E[0,M] equal to zero. The other extreme situation refers to s → ∞ (in which case, the system is *totally complex* with E[∞,M] = M meaning that the cell-losses and the associated loss of information are too excessive to be tolerated).

When the number of disordered subsets of cell-loss in a cell population M → ∞, the complexity associated refers to the entire universe of the jittered and/or bit-errored cells. The corresponding expected value E[s,∞] for a unit size of the assembly can be deduced using the following relations.
With s ≥ 1 and in the limit M → ∞, the functions associated become nonanalytic at s = 1. Explicitly, it implies that

$$E[s,M]\,|_{M\to\infty} = \begin{cases} \dfrac{s}{(1-s)} & s < 1 \\[2mm] \dfrac{M}{2} & s = 1 \\[2mm] \dfrac{\rho}{(1-\rho)} & s = 1/\rho \geq 1 \end{cases} \qquad (5A1.13a)$$

and
$E[s,M]\,|_{M\to\infty} = E[s,M]$ for small s. Correspondingly,

$$H[<s,M>]_{M\to\infty} = -\ln(1-s) - \frac{s\ln(s)}{1-s} \qquad 0 \leq s < 1$$

$$= -\ln(1-\rho) - \frac{\rho\ln(\rho)}{1-\rho} \qquad s = 1/\rho \geq 1 \qquad (5A1.14)$$

and $H(s,M) \cong H(s,\infty)$ for small s. Relevant inferences pertinent to the above algorithmic derivations are:

- For very small extents of the cell-losses with s ≪ 1, the expected extent of cell-loss is almost independent of the number (M) of cells involved

- For very large extents of CLR with s ≫ 1, the expected extent of cell-loss is characterized by the number of cells (M) involved

- The characteristic value of s = 1 bifurcates the system as *"simple"* or *"complex"* with respect to the extent of information-loss perceived due to dropping of cells caused by queueing congestion, and multiplexing process involving a finite buffer size.

The "simple", or small extents of CLR when grown to a larger level would make the overall system performance be designated as "complex". That is, small values of cell-losses can be considered as *quasi-autonomous* (simple) subsets, but when grown to a large extent would render the system complex in terms of its performance assessed via CDV parameters. To model this consideration, the complexity coefficient s can be written as a function of M. Specifically, around $s = 1$, let $s = (1 - \Omega)$ where $\Omega = (A/M) \to 0$ as $M \to \infty$ and the constant A remaining invariant. Using Taylor's expansion, at $s = 1$, one has:

$$E[s, M] \approx \frac{M}{2}\left(1 - \frac{A}{3}\right) \bigg/ \left(1 - \frac{A}{2} + \frac{A^2}{6} + ...\right) \tag{5A1.15}$$

and

$$H(s,M) \approx \ln(M+1). \tag{5A1.16}$$

The above results correspond to defining a coefficient of cell-loss complexity in a functional form of the type:

$$s = \exp(-A/M) \tag{5A1.17}$$

with $s < 1$ and $A > 0$, that is, when the system is considered simple. With the exponential form of s given by Eqn.(5A1.17), the following results can be deduced with $A > 0$:

$$s(e^{-A/M}, M) = M/A[1 - e^{-A}] + \frac{1}{2}[1 - e^{-A}] + \vartheta(A/M) \tag{5A1.18}$$

$$E[e^{-A/M}, M] = MF(A) - G(A) + \vartheta(A/M) \tag{5A1.19}$$

$$H[e^{-A/M}, M] = \ln M + U(A) + \vartheta(A/M) \tag{5A1.20}$$

where

$$F(A) = \frac{1}{A} - \frac{1}{(e^A - 1)}$$

$$G(A) = \frac{1}{2} + \frac{1}{(e^A - 1) - \dfrac{Ae^A}{(e^A - 1)^2}}$$

$$U(A) = AF(A) + \ln\frac{1}{A}(1 - e^{-A}),$$

and, $\vartheta(\cdot)$ represents the "order of $(\cdot)$".

At the critical point of $s = 1$ and in its neighborhood, the mean value of system performance depicts an *extensive property* with respect to the possible number of impaired cells (M) and the propensity of impairment is directly proportional to $\ln(M)$, namely, the message content of M cells.

When $s \geq 1$, the exponential law can be modified as $s = \exp(A/M)$ in which case,

$$E[e^{A/M}, M] = \{1 - F(A)\}M - G(A) + \vartheta(A/M) \tag{5A1.21}$$

$$H[e^{A/M}, M] = \ln(M) + U(A) + \vartheta(A/M). \tag{5A1.22}$$

The aforesaid algorithmic considerations can be adopted appropriately to specify a *complexity parameter*, s as a measure of the extent of a quality of service parameter (such as cell losses) in an end-to-end connection.

Appendix 5.2

Fractional Loading Factor of Sources Emitting Cells in an Interrupted Fashion

In reference to $\eta = 1, 2, \ldots N$ sources, when each of these sources give out cells continuously (without interruption) but randomly, the corresponding fractional traffic load of ηth source is taken as, ℓ_η^*; and, $\sum\limits_{\eta=1}^{N} \ell_\eta^* = 1$.

In practice, the sources being of different types (such as video, voice, and data) would exhibit distinct time-varying (interrupted) characteristics in regard to their rates of bits emitted specifying the idle/active duration of such bit emissions. As such, the actual loading factor of each source has only an average value, ℓ_η. The relation between ℓ_η^* and ℓ_η can be evaluated with respect to data ($\eta = 1$), voice ($\eta = 2$), and video ($\eta = 3$) traffics as follows. (See Table 5 for the notations used):

$$\ell_1 = \ell_1^*(\lambda_{1h}/\lambda_{1\ell})[(P_{LH}\lambda_{1\ell} + P_{HL}\lambda_{1h})/(\lambda_{1h} + \lambda_{1\ell})] \times (\lambda_{1h}/\lambda_H) \qquad (5A2.1)$$

where λ_H is the largest rate of the traffics contending to be multiplexed.

$$\ell_2 = (\ell_2^*)(\lambda_2/\lambda_H)P_{on} \qquad (5A2.2)$$

and

$$\ell_3 = \ell_2^*(\lambda_3/\lambda_H)[P_{NB}\lambda_3 + P_{BN}(BP \times \lambda_3)]/(\lambda_3 + BP \times \lambda_3). \qquad (5A2.3)$$

Further, as stated before, $(\ell_1^* + \ell_2^* + \ell_3^*) = 1$.

Table 5: Notations on the statistics of traffic parameters

Service categories	Parameters
η_1: Data traffic (CBR: Interrupted Bernoulli process)	High bit rate = Λ_{1h} Low bit rate = $\Lambda_{1\ell}$ Transition probabilities From high-to-low rate = P_{HL} Low-to-high rate = P_{LH} Loading factor: ℓ_1^*
η_2: Voice traffic (CBR: On-off Bernoulli process)	On-time bit rate = Λ_2 On-state probability = P_{on} Off-state probability = P_{off} Loading factor: ℓ_2^*
η_3: Video traffic (VBR: Markov-modulated Bernoulli process)	Sustained (non-bursty) bit rate = Λ_3 Burstiness ratio = BP Peak-rate = $\Lambda_3 \times BP$ Transition probabilities: From bursty-to-nonbursty state = P_{BN} From nonbursty-to-bursty state = P_{NB} Loading factor: ℓ_3^* Bursty duration: BD (% of total time) Non-bursty duration: NBD (% of total time)

Note: $\ell_1^* + \ell_2^* + \ell_3^* = 1$; Cell-rate ($\lambda_\eta$) = Bit-rate ($\Lambda_\eta$ in bps)/(53×8) cells/s

Appendix 5.3

ARMA Process Model of Video Sequences

An ARMA process can be assumed to describe the video sequences so that the model effectively characterizes the key property of the video scenes, namely, the associated autocorrelation. The corresponding queueing-theoretic sequences of ATM cells can be specified by an ARMA model of orders p and q denoted as ARMA(p,q). The ARMA sequence can be defined as [5.13]:

$$Z(k) = \Phi_0 + \Phi_1 Z(k-1) + \Phi_2 Z(k-2) + \ldots + \Phi_p Z(k-p)$$
$$+ e(k) - \theta_1 e(k-1) - \theta_2 e(k-2) - \ldots - \theta_q e(k-q) \tag{5A3.1}$$

where $Z(k)$ is the event size generated in the kth frame of a sequence and $\Phi_0, \Phi_1, \Phi_2, \ldots, \Phi_p, \theta_1, \theta_2, \ldots \theta_q$ are the weighting coefficients of the ARMA process. Eqn.(5A3.1) depicts a recursive procedure, which generates a series of values. This current value in the series is computed as a linear combination of the sum of the p and most recent values of itself and a linear combination of the q most recent values of series e. The series e is a Gaussian random process. The sets ($\Phi_0, \Phi_1, \Phi_2, \ldots, \Phi_p$) and ($\theta_1, \theta_2, \ldots \theta_q$) are constant coefficients. They are derived empirically for a given statistical (teletraffic) process such as the video transmission.

Essentially Eqn. (5A3.1) has three time-frame locations, namely, at $(t-1)$, $(t-2)$ and $(t-3)$ located in reference to the time-instant t. Stochastically, the details (information) at these time-frames are correlated depicting a Markovian structure. The coefficients associated denote characteristic attributes to the time-dependent data sequence.

6

ATM SWITCHING AND NETWORK PRODUCTS

To switch or not to switch?

"ATM switching is among the simplest forms of virtual circuits defined by a routing table"

W.A. Flanagan
ATM User's Guide, 1994

6.1 Introduction

6.1.1 What is switching?

The efficacy of telecommunication systems can be specified in terms of the efficient and reliable way with which a teletraffic is "switched through" the network on an end-to-end basis. So, *what is meant by "switching" in the context of such teletransfers?*

Switching is a function carried out by a device (or a set of devices) in a telecommunication network and it refers to taking information from a multiplexed set of inputs and delivering this information to other outputs [6.1]. The switching function is the responsibility of a *switching element*, which is a basic unit of a *switch fabric*. At the input port of a switch, the routing information of an incoming message is analyzed and accordingly the message is directed to a correct output port. The switch fabric represents the central function of a switch, which buffers and routes the incoming primitive data units (PDUs) to the appropriate output ports.

Switching, in short, is an intermediate functionality between end-entities in a telecommunication network intended to transmit the electrical communication information – whether it is in the form of telegraphic dashes and dots, telephone conversations, or ATM cells carrying the bits from a heterogeneous set of applications.

Constituent parts of a switching element

Basically, a switch takes information from a particular physical link at a specific multiplexed locale and connects it to another output physical link. This is enabled by the following constituents of the switching element illustrated in Fig. 6.1:

- ✓ Interconnection network
- ✓ Input controller (IC) for each incoming line
- ✓ Output controller (OC) for each outgoing line.

In addition to the above constituents, *buffers* are provided as a part of the switching element.

Reasons?

When two or more information units compete for the same output simultaneously, the switch would experience an internal collision, which may lead to an information loss. Hence, in order to avoid such losses, the incoming messages are buffered at the switching element until the *contention* is resolved.

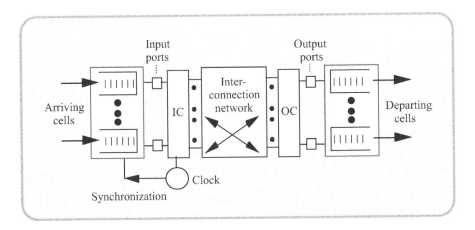

Fig. 6.1: A basic switching element

The operation of a switch, in essence, refers to the functional aspects of ICs and OCs of Fig. 6.1. These functions in summary depict the following:

- ✓ IC synchronizes the arriving cells to an internal clock
- ✓ OC transports cells, which are received from the interconnection network towards the destination
- ✓ ICs and OCs are coupled by the interconnection network
- ✓ Buffers are provided to store the cells so as to prevent collisions while the cells are contending for the same ports.

6.1.2 *Switching versus multiplexing*

The functional aspects of *switching* and *multiplexing* need to be distinguished at this stage: Multiplexing refers to the technique by which multiple streams of information (possibly from different applications or sources) are *converged* to share a common physical transmission medium.

In contrast, a switch acquires the information from a multiplexed information stream at its input and directs this information to an appropriate output. This input-to-output translation of information at a switch corresponds to relocating the information from a specified *multiplexing position* or *locale* at the input side to another multiplexing position on the output side.

Switching therefore, is basically an interconnected network of multiplexers.

Multiplexing positions

The switching of an information unit across the input-side to output-side of a switching element may correspond to the translation of one of the following five multiplexing positions:

- ✓ Space
- ✓ Time
- ✓ Frequency
- ✓ Address
- ✓ Code.

The considerations of multiplexing positions indicated above are illustrated in Fig. 6.2.

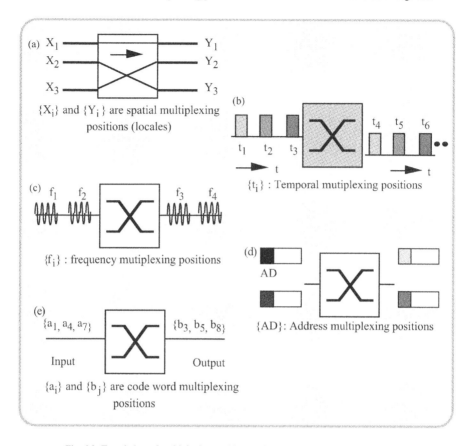

Fig. 6.2: Translation of multiplexing positions: Direct switching functional formats
a) Spatial, b) temporal, c) frequency, d) address (label), and e) code

6.1.3 Point-to-point and point-to-multipoint switching

In reference to switching translations of multiplexing positions illustrated in Fig. 6.2, each information unit with a specified multiplexing position on the input side is translated into a corresponding information unit at the output side, but with a different multiplexing position. Hence, each point on the input side refers to a corresponding point on the output point. That is, the switching functions of Fig. 6.2 refer to *point-to-point switching*.

On the other hand, a single information unit on the input side can also be switched on to multiple points on the output side of a switching element. This type of translation may represent either a *broadcast* or a *multicast* function. In general, it is known as *point-to-multipoint switching*, as illustrated in Fig. 6.3.

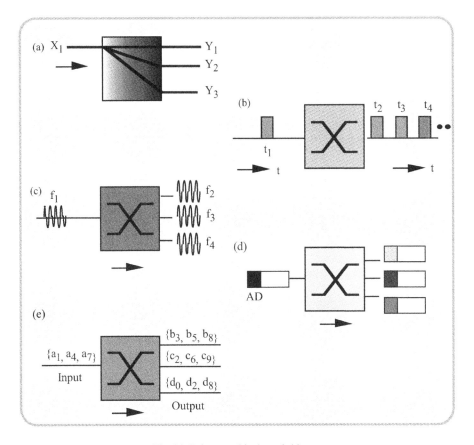

Fig. 6.3: Point-to-multipoint switching

6.1.4   Switching techniques

Pertinent to the types of multiplexing positional translation involved across a switching element, a host of switches has been conceived in practice. The following are the switching techniques evolved to realize the translation of multiplexing positions, namely, space, time, frequency, address, and code.

Space-division switching

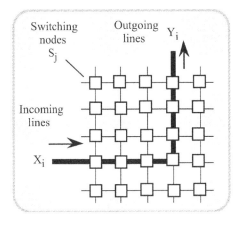

Fig. 6.4: A cross-bar matrix

Typically, a *cross-bar switch* can be regarded as an example of a *space-division switch*. The switching elements are located at the cross-point nodes of a set of input lines and a set of output lines as shown in Fig. 6.4.

The crossbar network is a *matrix fabric*. An input (spatial) multiplexing position can be translated into (that is, switched to) a desired output (spatial) multiplexing position via a cross-point node. In Fig. 6.4, for example, $X_i \Rightarrow Y_i$ switching is done by the S_i cross-point nodal function as illustrated.

In practice, the classical space-division switch fabrics were built with electromechanical and discrete electronic devices which are, in general, low speed devices. With the advent of modern integrated electronics, the relevant switches conceived allow cross-bar functions to operate up to several Gbps speeds. Trends realizing optical cross-point elements could even allow the capacities in the order of Tbps in the future.

The classical cross-bar switches were a part of POTS and PSTN. The extension of space-division switching in the modern era refers to *matrix switches*, *high-performance parallel interface* (HIPPI) switches, and 3/3 *digital crossconnects* (DCC/DXC).

Time-division switching (TDS)

Consistent with the fact that a switch is basically a set of interconnected multiplexers, TDS has been conceived by using time-division multiplexing (TDM) as the interface protocol. It is illustrated in Fig. 6.5.

On the input side, in Fig. 6.5, a TDM frame structure has an information unit at the kth time-slot. The incoming frames are mapped onto the input data memory of the switch. The nodal function of this switch is to remap this input entry onto an output address memory. Thus, the input frame in question (with the information unit residing at the kth slot) may become a frame in the output memory location with the information unit of interest, now residing at ℓ^{th} time-slot; that is, the information unit at the kth time-slot on the input-side gets translated to the ℓth time-slot locale (depicting a new temporal multiplexing position).

The TDS function is specified in terms of an *execution rate*, with I instructions per second. Suppose a switch handles a frame size of M time-slots, then the TDS function should occur in a time less than τ second such that,

$$(\tau \times I) \geq M. \tag{6.1}$$

Single-stage TDS may have a size M ranging from 12,500 to 125,000 and the corresponding I spans from 100 to 1000.

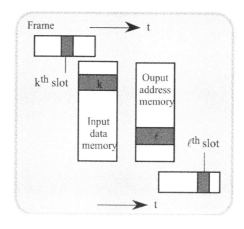

Fig. 6.5: Time-division switching

Frequency-division switching (FDS)

In the modern context, FDS essentially governs switching at optical levels. Therefore, it is more appropriate to designate FDS as *wavelength-division switching* (WDS). It is based on *wavelength-division multiplexing* (WDM). WDS uses the *shared-medium concept* applied to all-photonic network interconnecting a number of optical nodes or "end-systems" as depicted in Fig. 6.6.

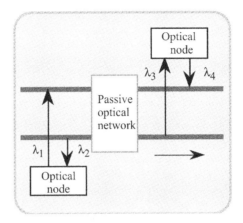

Fig. 6.6: Wavelength-division switching

The underlying principle of WDS is that the information identified on a multiplexing position (specified in terms of a wavelength, λ_i) is switched to another multiplexing position (specified in terms of a different wavelength, λ_j).

Address switching

This is applicable to data streams in which the data is packetized and each packet has a distinct information field (payload) and an address field (header). That is, the header contains address information. The switching node performs multiplexing position translation in respect of the address on each packet. Such switching decisions at each node allow the packet to progress towards the destination on a hop-by-hop basis. The address determines which physical output the packet is directed to, along with any translation of the header addresses. *Address switching* permits point-to-point, point-to-multipoint, and multipoint-to-multipoint architectures.

In the current practice, address switching operates at electrical link speeds up to 2.4 Gbps on ATM and packet-switching systems. The concept of address switching is the crux of ATM telecommunications.

6.2   ATM Switching

What is inside an ATM switch?

From an end-entity facilitated with an ATM port card the cells are directed across a switch matrix, which represents a set of *memory blocks* [6.2 – 6.10]. As a cell enters the matrix, its contents are written into these memory blocks. Each memory block is designed to hold the contents of several cells. The VPI/VCI values located in the header of each cell is checked. If the cell does not contain a VPI/VCI, it is discarded. If a VPI/VCI is found, it is checked against a routing table to determine the new required VPI/VCI. If the switch contains a 16 × 16 matrix, then memories of the cell's contents are erased from the other 15 memories associated with the matrix. The remaining cell, now with an updated VPI/VCI must compete with other cells in its path ahead. Whether a cell is read en route by the switch depends on the priority it has been assigned in addition to several other factors. The horizontal input, the vertical passage across the matrix, and horizontal exit of a cell in a switch architecture are illustrated in Fig. 6.7.

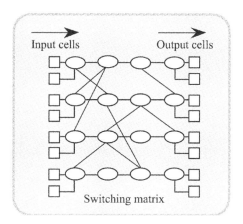

Fig. 6.7: An 8 × 8 ATM switching matrix for reading and writing out cells

6.2.1 Characteristics of ATM switches

In reference to ATM telecommunications, a switch as described above is essentially connected to a number of physical communication circuits each of which implements a *transmission path* (TP). The switching function then refers to receiving cells on incoming TPs and transmits them on outgoing TPs. The functional requirements of ATM switches can, therefore, be summarized as follows:

  ✓ Receiving a cell over any incoming TPs
  ✓ Deciding the appropriate outgoing TP on which an incoming cell be sent
  ✓ Placing appropriate values in the VPI and VCI fields of the cell header
  ✓ Sending the cell on the selected outgoing TP.

Typically, the physical communication circuits, which implement ATM switching, are full-duplex circuits supporting an *inbound TP* and an *outbound TP* as shown in Fig. 6.7. The ATM switch, however, views each transmission path as logically separated from all others.

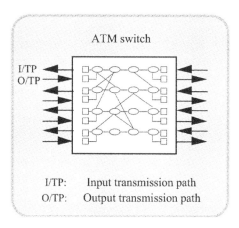

Fig. 6.8: Concept of an ATM switch

ATM switch: Design basics

The parameters and concepts, which are used in conceiving ATM switching architectures are as follows:

- ❑ *Throughput: It is a measure of the rate at which data passes through the switch. It highlights the following aspects of switching:*

  - • *Internal processing involved*
  - • *Transmission rate across the switch*
  - • *Degree of parallelism with the switch.*

- ❑ *Scalability: This refers to performance considerations of a switch vis-à-vis the scalable extents of TPs implemented*

- ❑ *Contention resolution: The switching function often has to cope with situations where contention for resources occurs. Buffers are used to hold cells until the contention is resolved. Size and placement of buffers are crucial in ATM switch designs in reference to contention resolving performance, congestion avoidance, and delay (waiting) considerations.*

- ❑ *Cost: The complexity of ATM switching fabrics decides the overall cost of the fabrics*

The major functions or tasks of ATM switching, in short, can be enumerated as follows:

- ✓ An ATM switching element is essentially a cross-connect node
- ✓ A switching node performs VPI/VCI translation
- ✓ Cells are transported, as they arrive at the input of a switch, to their dedicated outputs.

In reference to ATM switches, *switch fabric* or *switch architecture* has the following functional attributes:

- ✓ ATM switch fabric establishes a connection between the arbitrary pair of inputs and outputs within a switching node
- ✓ In principle, a switch fabric can be implemented by a single switch element
- ✓ Such a single-element fabric, is however, not sufficiently large for ATM switching node operations in practice.

Normally, larger switch architectures are built with a number of switching elements in the fabric. Existing technology aims at the following features in realizing ATM switch fabrics:

- ✓ Throughput of a switching node approaching Gbps regimes
- ✓ Low cross-node delay
- ✓ Low cross-node cell-loss.

The abovesaid objectives, however, forbid the use of a centralized control on cross-node functions. Therefore, switch fabrics are made of highly *parallel architectures*.

Blocking and nonblocking considerations

A *nonblocking switching network* is one in which there are enough switching paths so that an incoming information unit will always be connected to a desired outgoing line. That is, the switch will not block the input-to-output translation for want of switching resources.

A *blocking switching network* may deny an information unit at a specified multiplexing position on its input side to a multiplexing position on the desired outbound line. This can occur whenever the switching architecture is burdened with excessive traffic. The resulting blocking is then characterized in terms of a *blocking probability*.

The blocking and nonblocking attributes of switching fabrics were original paradigms observed with respect to telephone switches. Relevant concepts have been extended to specify a *figure of merit* in the context of ATM switching.

In circuit-switches, if an inlet channel can be connected to any unoccupied output channel, up to the point where all inlets are occupied, then the switch is said to be *strictly nonblocking*. A typical assumption associated with this definition is that the statistical distribution of inlet channels needing connection to specific output channels corresponds to uniform distribution of a random process.

Further, circuit-switches defined as above are specified as *virtually nonblocking*. That is, a small blocking probability occurs as long as no more than a certain fraction of inlet channels are in use.

An ATM switch uses a different paradigm than a circuit-switch. That is, in a circuit-switch, bandwidth is reserved on a dedicated basis. This bandwidth dedication is independent of other connections. But this is not true for ATM switches, in which:

    ✓   There are virtual connections (VCCs and VPCs) that arrive at input ports destined for potentially different output ports

    ✓   Cell-loss can occur depending on the statistical aspects of the VC traffic, which is handled by the cell switching and buffering strategies

    ✓   Invariably, cells are buffered at the switching element.

Further, the blocking performance of an ATM switch can be characterized as follows:

    ✓   It is sensitive to the switch architecture

    ✓   It is also dependent on source traffic characteristics.

In essence, source traffic versus type of switch fabric decides the cost of the switch for a specified blocking performance.

6.2.2 ATM switching architectures

Matrix/cross-point switch

The matrix or cross-point switch technology as indicated earlier was originally developed to support telephony and it refers to space-division switching. The cross-point switches use parallel data paths to connect input and output TPs. The parallel paths help to reduce congestion and increase throughput by allowing multiple cells to flow through the switch in parallel.

Blocking can occur in a matrix switch if two input paths need to transmit data at the same time over the same output transmission path.

Parallel data path makes cross-point switches efficient, and its relatively simple structure enables it cost-effective. However, the complexity of the architecture of the switch grows with the number of TPs attached becoming very large.

The matrix switch and its performance can be summarized as follows:

    ✓   The switch is internally nonblocking

    ✓   Interconnection network is realized via cross-points

    ✓   By this matrix, it is always possible to connect any idle input/output pair

    ✓   Whether a cross-connect point connects an input to an output depends on the routing information of the cell as well as the occurrence of collisions.

Switch buffers

The occurrence of switch blocking/cell-loss depends on the type of buffering strategy deployed at the switch. Technically, there are four feasible methods of buffering as shown in Fig. 6.9. The type of buffering depends on the type of queueing adopted. The associated parameters are the number of ATM ports and number of effective buffer positions per port.

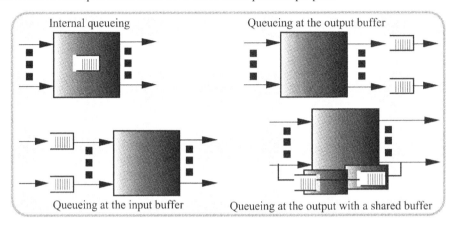

Internal queueing   Queueing at the output buffer

Queueing at the input buffer   Queueing at the output with a shared buffer

Fig. 6.9: Buffering schemes

In reference to Fig. 6.9, the associated queueing types are as follows:

Internal queueing

The characteristics of internal queueing are:

✓ Switch fabrics built with internal queueing have the potential to scale to large sizes
✓ However, it is difficult to implement other functions with the internal queueing facility, such as:

⇒ Priority queueing
⇒ Implementing large buffers
⇒ Facilitating multicasting

✓ Internal queueing is:

⇒ Simple to implement
⇒ But, its major disadvantage is as follows:

* If the cell at the *head of the line* (HOL) is not switched through the fabric, all the cells behind this are delayed accordingly.

Input queueing

The input queueing is based on the following considerations:

✓ Assumed Poisson arrival process of incoming cells
✓ Negative exponential service time at the switch

As specified in the last chapter (Chapter 5), the probability that the interarrival time between Poisson arrival events, designated by τ has a value x is given by:

$$\text{Probability } (\tau = x) = \lambda \exp(-\lambda x) \qquad (6.2)$$

where λ is the average arrival rate. (That is, there are λ events on an average per second.)

Suppose there are a number of cells waiting in a queue to be served. Focusing the attention on the output of the queue, a time can be marked at which a cell has received a service. Let the random interval between such service completions be denoted as r (Fig. 6.10). Further, this must also be the service time if the next cell is served as soon as the one in the service departs the system. When r is assumed to be exponentially distributed in time, with an average value E[r] = 1/μ, then, the probability density function of r, is:

$$f_r(r) = \mu e^{-\mu r} \qquad r \geq 0. \qquad (6.3)$$

That is, the service-time distribution is a negative exponential.

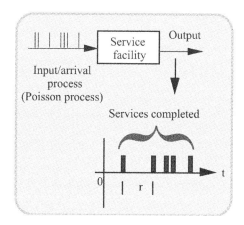

Fig. 6.10: Random internal service completions

An alternative way of depicting the input queueing refers to the *M/M/1 process modeling.* Relevant considerations are as follows:

- Input queueing occurs when HOL is blocked. That is, the cell at the head of the input queue cannot enter the switch matrix because the cell at the head of another queue is (already) traversing the matrix

- For uniformly distributed traffic with random message lengths, (that is, when messages of different, random sizes are equally likely to occur in the traffic stream), the maximum supportable offered load for input queueing is 50%

- For messages of fixed lengths, this percentage increases to 58%

- Offered load is defined as the average fraction of the resource capacity used and it is equal to: (Average number of arrivals) × (average value of time involved in servicing the arrivals). It is given by : λ × (1/μ)

- Input queueing limits switch throughput to only 50 to 60% of the port speed.

Output queueing

This queueing refers to the scheme of allowing the cells to wait in the line at the output side of the switch. It is theoretically optimal. The use of output queueing is consistent with the following considerations:

- Input queueing is not adequate for many applications. It is therefore, used in conjunction with other queueing methods, such as output queueing

- Hence, ATM switches have some form of output buffering (since it is optimally suitable taking throughput into consideration)

- A method of achieving most preferable queueing is to combine input as well as output queueing schemes. This is known as *shared output queueing*. The *shared output queueing scheme* is characterized by:

  ✓ Maximum throughput
  ✓ Fewest cell buffer positions.

Type of buffers
Input buffers

These facilitate the input queueing system indicated above and are located at the input controller. They correspond to FIFO buffers. Under the first-in first-out strategy of service, however, collision may occur, if two or more HOL cells compete simultaneously for the same output. Then, all but one cell are blocked. Even those cells destined for another available output, which are behind the blocked HOL cell will also be blocked. *How is this blocking problem overcome?*

FIFO buffers can be replaced by *random access memory* (RAM). If the first cell in the buffer is blocked, the next cell destined for an idle output is selected for transmission. This operation is, however, complex because the buffer control must seek and find such a cell being destined for an idle output. Further, total buffer capacity is logically subdivided in a load-dependent manner into single FIFOs, that is, one FIFO for one output.

An improved method refers to facilitating a buffer with multiple outputs. That is, by incorporating a method to transfer more than one cell simultaneously from the buffer to different outputs.

Output buffers

Here, *collisions*, namely, several cells hunting for the same output simultaneously, occurs only if the matrix operates at the same speed as the incoming lines. Collision can be compensated by:

  ✓ Reducing the buffer access time
  ✓ Speeding up the switching matrix.

Such provisions, however, are limited by technological considerations and by the size of the switching element.

A switching element with output buffers could be nonblocking, only if the *speed up factor* of the matrix is equal to b, namely, the number of cells hunting for the same output simultaneously, to be switched. In all other cases, additional buffers are needed at the input in order to avoid cell-losses caused by internal blocking.

Cross-point buffers

These apply to a *butterfly switching element*. Here, buffers are located at the individual cross-points of the matrix. Such locations of buffers prevent cells, which hunt for different outputs, from affecting each other. Further, if the cells destined for the same output are located in different buffers, a control logic is needed to choose which buffer should be served first. The demerit of cross-point buffers is that a control-logic based buffer location for priority service is required, which limits the buffer size to be small at each location; further, buffer-sharing is not feasible. Therefore, efficiency of this switching element is not as good as those with output buffers.

Type of switching elements
Central memory switching element

In this switching arrangement, all input and output controllers are directly attached to a central memory. The memory is written by all the inputs/input-controllers and read by all the output controllers. Further, the common memory can be organized to offer logical input and output buffers. Since all the buffers share one common memory, a significant total memory requirement is achieved in comparison with the systems with physically separated buffers. The disadvantage of central memory switching is that a high degree of *internal parallelism* should be maintained to enable the accessibility of the memory by the controllers in reasonable time (without excessive delays).

An example of a central-memory switch is a *Sigma switch*, which has logical output buffers. Consistent with ATM switching node functions (namely VPI/VCI translation, transport of cells from the input to a destined output), such functions in the Sigma switch are done through what is known as the *self-routing principle*. The relevant functional attributes are as follows:

       ✓   VPI/VCI translation is done only at the input

       ✓   After translation, the cell is *extended* by the switching network internal header. That is, another header is added preceding the original cell header

       ✓   This extension of the cell-header may call for an increase in the internal network speed.

The self-routing principle is depicted in Fig. 6.11a. It depicts a network with K stages wherein the internal header is divided into K subfields. The ith subfield contains the destination output field number in the ith stage.

In Fig. 6.11b, the generic realization of self-routing plus the central memory with logical output queues is shown. Here, switching is controlled by the routing information included in the internal cell header. Wide parallel memory interfacing with serial-to-parallel conversions at the input and parallel-to-serial conversion at the output are used in the central memory section. These facilitate the required buffer access speed tangible with the technological considerations.

The central memory-based control with address assigned to each outlet takes care of correct delivery of the cells. A self-routing method using the single-bus concept is as illustrated in Fig. 6.11a. It is a multiple bus switch fabric, which provides a broadcast bus for each input port. This eliminates the need for *bus arbitration*, (to be discussed later). It shifts the burden of controlling the blocking to the outputs. Since each input is broadcasting to every output, this switching is inherently a *multicast type*.

A nonblocking, self-routing switching element is shown in the Fig. 6.11b. Here the nonblocking feature is realized with a sophisticated set of internal elements. The augmented features on these fabrics include:

       ✓   Increased internal matrix speed
       ✓   Multiple connections between switching elements.

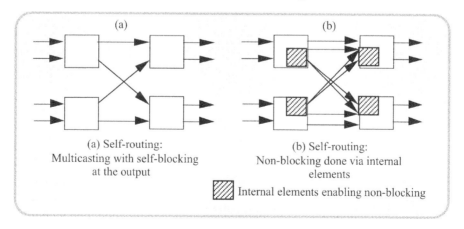

Fig. 6.11: Self-routing schemes: (a) Blocking type and (b) Nonblocking type

Bus-type switching element

       Single/multiple bus type switch elements indicated above, for example, are compatible for high speed TDM bus based interconnections. Here, conflict-free transmission can be assured only if the total capacity of the bus is at least equal to the sum of the capacities of all the inputs. Typically, 16/32 bit parallel data transmission could offer this kind of high capacity transmission.

       In the bus-type switching element, the control algorithm allocates a bus to individual input controllers at constant intervals. The input-controller is designed to transfer the arrival cell to the destined output within a short time. That is, the associated service time is such that the transfer is completed before the next cell arrives. Inasmuch as several cells may arrive at an output simultaneously, buffering is done at the output, to let the cell out, one after another. The relevant performance features correspond to those of matrix type switching elements with output buffers.

Ring-type switching element

Here, all input and output controllers are interconnected via a *ring network*, which is operated in a slotted-fashion to reduce overhead. A fixed time-slot allocation is used with ring capacity equal to the sum of the capacities of all the inputs. (Otherwise, a flexible ring capacity, if used, will require additional overhead.)

The advantage of a ring structure over a bus structure is as follows: With a time-slot allotted, within one rotation, the allotted time-slot can be used several times. This is true under the assumption that the output controller empties a received time-slot. This technique is useful when *destination release*, namely, the output controller emptying a received time-slot, is possible. Then, a utilization of 100% is realized.

The demerit of this system is that there is an additional overhead involvement for flexible ring capacity and for destination release facilitation.

6.2.3 Arbitration

Arbitration refers to determining the winning cell, when several cells compete simultaneously for the same output. Upon arbitration, the winning cell makes a throughput across the switch and all other (defeated) cells remain in the queue, that is, they will be delayed in the transfer to their destined outputs. The arbitration is concerned with:

      ✓   Minimization of blocking/cell-loss
      ✓   Minimization of CDV.

Strategies of arbitration

Random method: The line to be served first is chosen randomly from all the lines, which compete for the same destined output. The advantage of this method is that the implementation overhead is small.

Cyclic method: Here, buffers are served in a cyclic fashion. This method again needs only a small overhead.

State-dependent method: Here, the choice to be served is based on choosing the first cell from the longest queue. The relevant algorithm compares the lengths of the buffers hunting for the same output.

Delay-dependent method: This is a global FIFO method. It takes into account all the buffers that feed one output. Here, some overhead is warranted to maintain a relative order of arrival of the competing cells.

For a typical Bernoulli arrival process with a traffic load of 50%, the relative performance of different arbitration strategies studied vis-à-vis the delay involved lead to the following inferences:

- Random strategy causes highest CDV

- Cyclic strategy has a significantly improved performance

- Optimum strategies are:

      ✓   For minimum cell loss: State-dependent algorithm
      ✓   For optimal CDV: Delay-dependent algorithm.

6.2.4 *Performance of switching elements*

The essential performance parameters of switching elements are as follows:

- ✓ Mean cell-delay versus offered traffic load: This refers to the switch element throughput.
- ✓ Technological factors depicting the following considerations:

  ⇒ Sizes of:

    * Central memory
    * Input buffer
    * Output buffer

  ⇒ CMOS or BiCMOS implementation, which decides the chip area (memory part and random logic part, such as serial to parallel conversion).

The chip size and power dissipation rating considerations studied indicate that the central memory (CM) switching element offers:

- ✓ Minimal chip size
- ✓ Maximal power dissipation rating.

Switching elements of the output buffer category provides the least favorable choice on chip size and power dissipation rating. The reasons are:

- ✓ Memory section (MOS type) consumes least power
- ✓ High speed random logic consumes relatively higher power.

Relative proportion of the memory presence and the random logics in a switching element decides the chip size versus dissipation characteristics of the element.

6.3 Switching Networks

Switching networks are networks constituted by a set of switching elements. They are classified as *single-stage networks* and *multistage networks*.

6.3.1 *Single-stage network*

It consists of switching elements that are connected to the inputs and outputs of a switching network. The types of single-stage networks are *extended switching matrix, funnel-type network,* and *shuffle exchange network.* The details on each type are as follows:

Extended switching matrix

This is formed by a matrix of (b × b) switches with no restriction on the matrix size, and a switching element can be augmented by adding b inputs and b outputs in it as extensions. Input signals are relayed to the next column of the matrix via the additional outputs. The additional inputs are connected to the normal outputs of the switching element in the same column, but in the row above. The advantages of an extended switching matrix network are:

- ✓ Small cross-delay since cells will only be buffeted once when crossing the network
- ✓ This cross-delay depends on the location of the input.

The limitation of this network is its size (being limited to 64 × 64 and 128 × 128).

Funnel type network

This is an (N × N) nonblocking switching network. Here, switching elements are interconnected in a *funnel-like structure*. All switching elements consist of 2b inputs and b outputs. Each funnel corresponds to an N × b matrix and N/b are in parallel. The available technology corresponds to a 32 × 16 switching elements status.

Shuffle exchange network

This is based on a perfect shuffle permutation, which is connected to a stage of switching elements. In reference to Fig. 6.12, the dashed lines represent feedback. The feedback is necessary to reach an arbitrary output from a given input. In this structure, a cell may pass through the network several times before it reaches its destination output. Hence, this network is known as a *recirculating network*. At each switching element output, it has to be decided whether the cell can leave the network or it has to be fed back. The merit is that it requires only a small number of switching elements, but the performance is poor. That is, cross-delay depends on the recirculations involved.

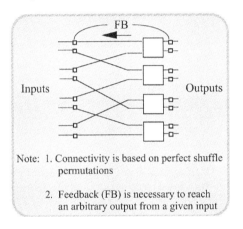

Fig. 6.12: Shuffle exchange network

6.3.2   *Multistage networks*

These are typically classified as:

    ✓   Single-path networks (Banyan networks)
    ✓   Multipath networks.

Single-path networks

Here, there is only one path to destination from a given input. While cross-point or cross-bar switches are based on a *matrix topology*, the single-path networks (also known as *Banyan networks*) are based on *tree topology*.

The tree topology refers to the following: Here, each input port is the root of a tree that branches over a number of intermediate switching elements, with its output ports as its leaves. A collection of trees (N) that share all the links and switching elements except the roots constitutes an (N × N) Banyan switching network. This is a multistage configuration, built on several stages, which are interconnected by a certain link pattern (tree structure).

The path available to reach a destination output from a given input is just one. Therefore, it is a simple structure considering the routing strategy (via a single-path) involved. However, there is a chance of internal blocking, which can occur due to the fact that an internal link (tree branch) can be used simultaneously by different inputs.

An example of (1 × 4) single-path multistage Banyan tree is illustrated in Fig. 6.13. Four delta-2 networks can be used to form a *Delta-S network*. A Delta-S network forms the topology of a baseline network. In Fig. 6.14, the thick-line indicates the path from input 1 to output 14 and it traces the binary destination address 1010.

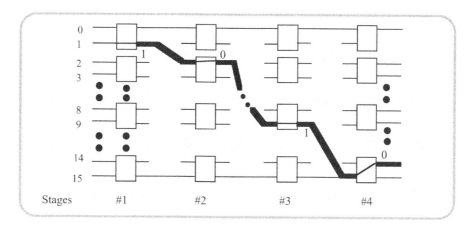

Fig. 6.13: Delta-2 network

The subgroups of Banyan networks are as follows:

Level Banyan network:
    This is characterized as follows:

        ✓   Switching elements of adjacent stages are alone interconnected
        ✓   Each path through the network passes through exactly L stages
        ✓   L-level Banyan network can be classified into: *Regular* and *irregular* types:

        ⇒   *Regular Banyans* are constructed of identical switching elements
        ⇒   *Irregular Banyans* are constructed of nonidentical switching elements, for example generalized delta network.

Regular Banyans denote the set {SW Banyans, delta networks}. SW Banyans are constructed recursively from (identical) basic switching elements with F input links and S output links. L-level SW Banyan is obtained by connecting several (L − 1) level SW Banyans with an additional stage of (F × S) switching elements.

Delta networks
    The delta network is a special implementation of SW Banyans. They correspond to L-level Banyan networks, which are constructed of (F × S) switching elements and have S^L outputs. Their characteristics are as follows:

        ✓   Each output is identified by a unique destination address, which is a number of base S with L digits. Each digit specifies the destination output of the switching element in a specific stage. This allows simple routing of cells through the delta network, which is called *self-routing* as indicated earlier
        ✓   In rectangular delta networks, the number of inputs is equal to the number of outputs (S = L).

The special classes of delta networks are those that remain as delta networks even if inputs and outputs are interchanged.

Multipath networks

In multipath networks, the internal path is determined during the connection setup phase. Relevant features are as follows:

✓   All cells on the connection will use the same internal path
✓   By maintaining the FIFO strategy, the cell sequence integrity can be assumed warranting no resequencing.

Depending on the different ways of forming tree structures, there are also different versions of multipath Banyan network architecture, as summarized below.

Three-stage folded switching network (Fig. 6.14)

✓   Here alternate, multiple paths exist between a given input and a destined output
✓   Internal blocking can, therefore, be reduced or even avoided.
✓   Each link represents the physical lines for both directions.

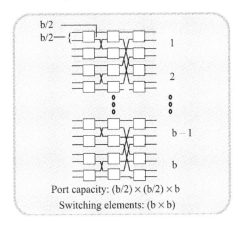

Fig. 6.14: A three-stage folded switching network

Folded versus unfolded structures

In *folded structures* all the inputs and outputs are located on the same side of the network and the network's internal links are operated bidirectionally. The advantage of folded networks is that they use paths of short length. For example, if the input line and the output line are connected to the same switching element (via *folding*), cells can be reflected at the same switching element and need not be passed to the last stage. In general, the number of switching elements to be traversed by the cells of a connection depends on the physical location of the input and the destined output. The port capacity (of a three-stage folded network as shown in Fig. 6.14) built with (b × b) switching elements, is equal to (b/2) (b/2) (b).

In modern technology, typical sizes of Banyan network architecture are given by:

$$(b \times b) \Rightarrow 16 \times 16 \Rightarrow 1024, \text{ ports 3-stage network}$$
$$\Rightarrow 32 \times 32 \Rightarrow 8192, \text{ ports 3-stage network}$$

Unfolded networks

Here, inputs and outputs are located on the opposite sides of the network. They are characterized as follows:

✓   Internal links are unidirectional
✓   All cells have to pass through the same number of switching elements.

Multipath unfolded networks
These are based on single-path unfolded networks using (b × b) switching elements.

6.3.3   Characteristics of Banyan networks

In Banyan networks, the number of switching elements is specified by (n × n). That is, each switching element is an (n × n) cross-bar switch. Suppose a switch consists of $\log_n(N)$ stages. Then, the total number of cross-points is equal to: $n^2(\dfrac{N}{n})\log_n(N) < N^2$, (where N^2 = Total number of cross-points in an (N × N) matrix or cross-bar switch).

Banyan networks, in essence, are *self-routing switches*. That is, a path can be uniquely established from an input to a destined output. Further, the proliferation of the path from the input can be done as follows: Considering the routing vector $(r_1, r_2, \ldots, r_k)$, with r_j depicting the jth port number, the cell is routed to the switching element in the jth stage of k stages.

Banyan networks are *internally blocking* exhibiting the following characteristics:

- ✓ The a blocking occurs when more than one cell attempts to use the same link between any two stages
- ✓ The blocking degrades the throughput performance
- ✓ It is proportional to the number of ports in the network
- ✓ Three methods adopted to reduce the internal blocking are:

  - ⇒ Internal speed up
  - ⇒ Internal/input technique/output buffering
  - ⇒ Using sorter networks preceding the switch fabric:

    - * *Internal speedup*
      This is done by speeding up the internal link passage of cells by N times faster than the fastest of the incoming link. (Here, N denotes the number of cells contending for the same output.) For example, suppose the link corresponds to OC-3 155 Mbps and the switch size is 16 × 16. Then, the internal link speed should be larger than 600 Mbps (and, it should be more than 1.2 Gbps in a 64 × 64 switch). In practice, the speed-up factors in ATM switches are limited 2 to 4 as constrained by the available technology.

Input, internal, and output buffering schemes
Relevant considerations and their relative aspects are presented in Table 6.1

Table 6.1: <u>A comparison of input, internal, and output buffering schemes</u>

Input buffering	Internal buffering	Output buffering
▪ Based on FIFO strategy ▪ Leads to HOL blocking ▪ It is not work conservative. That is, even non HOL cells are made to wait.	▪ Reduced probability of blocking ▪ Large buffers ▪ High chip complexity ▪ Random delays in the switch fabric may leading to CDV	▪ Better throughput ▪ Managing the connection of all outputs from the switching elements of the Banyan network to all the outputs by the buffer is rather difficult

... continued

- Internal buffering is most effective when the traffic is uniformly distributed at the output and the cell stream in an incoming link is not correlated to the cellstreams in other links. (Both are not true in ATM traffic. Here different frames are interleaved.)

- The above consideration is of concern, especially, when switching speed is high. (For example, OC-3 links need 1.2 Gbps bandwidth. Therefore, scalability is limited. That is, scaling up the switch to a layer size is difficult.)

Solution to internal blocking problems in Banyan networks.

Internal blocking in Banyan structures is counteracted by using a *sorter* preceding the switch. The underlying principle is as follows:

- In Banyan networks, there are various permutations of realizing N concurrent input/output (I/O) port connections

- The problem is then to *sort* the incoming cells in such a way that when they access the input ports of the switch, no two cells share an internal link, thereby making the network internally nonblocking

- To accomplish this, relevant sorting is done by *space-division-based sorting networks*.

Examples of such sorters are *Batcher sorters*, which pose least complex algorithms. There are two versions of Batcher sorters, namely,

✓ Up-sorter: (2×2) cross-point that sorts 2 numbers in descending order
✓ Down-sorter: (2×2) cross-point that sorts 2 numbers in ascending order

⇒ Here, the "number" corresponds to output port addresses.

Sorting procedure

✓ Two pairs of numbers are first sorted by using a (2×2) sorter
✓ The sorted list of 2 numbers are then sorted by using a (4×4) sorter
✓ Next, the sorted list of 4 numbers are sorted by using an (8×8) sorter, and so on.

In an (8×8) Batcher sorter connected to a Banyan network, for example, k stages of sorters are needed to sort 2^k elements with the ith stage sorter made of i stages of (2×2) sorters. Then the total number of stages involved is equal to $k(k+1)/2$. With N/2 sorters in each stage, the total number of (2×2) sorters used in a Batcher sorter is equal to $0.25 \times N \times \log_2(N)[1 + \log_2(N)]$.

Sorting/trap Banyan network (Fig. 6.15)

This is another version of an unfolded multipath network developed to reduce the internal blocking. It is a combination of a sorting network and a *trap network* in front of a Banyan network. In this architecture, the sorting network arranges the arriving cells in a monotonous sequence depending on the network's internal destination address. Cells with identical addresses are detected by the trap network and all but one of these cells are fed back to the input of the sorting network. The cells, which have to pass through the sorting network will again be assigned a higher priority in order to maintain cell sequence integrity. Cells entering the Banyan network thus are sent to their destination without any internal blocking.

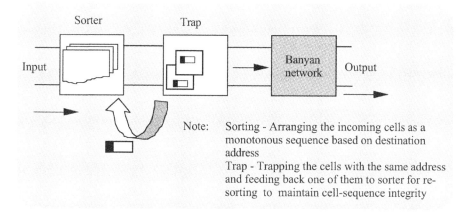

Fig. 6.15: Banyan network with a preceding set of sorting/trap arrangements

Distribution/Banyan network [Fig. 6.16]
 Here, a distribution strategy is adopted to reduce internal blocking. The salient features are:

  ✓ The distribution network preceding the Banyan network, distributes the cells as evenly as possible over all the inputs of the Banyan network
  ✓ This method reduces the probability of internal blocking
  ✓ But, cell sequencing may become upset. Therefore, resequencing is needed at the output to maintain cell-sequence integrity.

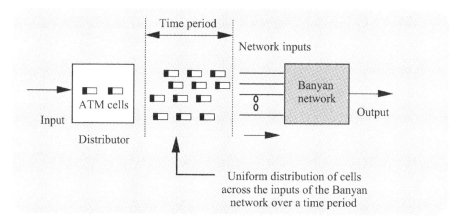

Fig. 6.16: Banyan network with a preceding distribution network

Scaling up of multipath networks
 Scaling can be done by using several planes of Banyan networks in parallel. It corresponds to a *vertical stacking*. Here, all cells belonging to the same connection pass through the same plane as determined at the connection setup phase. Alternatively, scaling can be done by adding a number of stages to a given Banyan network. This is called *horizontal stacking*. Here, by adding a baseline network with a mirrored topology to an existing baseline network, a multipath interconnection scaling topology is realized. Another network known as the *Benes network* is similar to the above except that the last stage of the baseline network coincides with the first stage of the mirrored baseline network.

Cell header processing in switch fabrics

As indicated earlier, the major tasks of ATM switching nodes are:

- ✓ VPI/VCI translation
- ✓ Transport of cells from the input to the destined output.

These tasks are accomplished by:

- ✓ Self-routing principle
- ✓ Table-controlled principle
- ✓ Self-routing via switching elements.

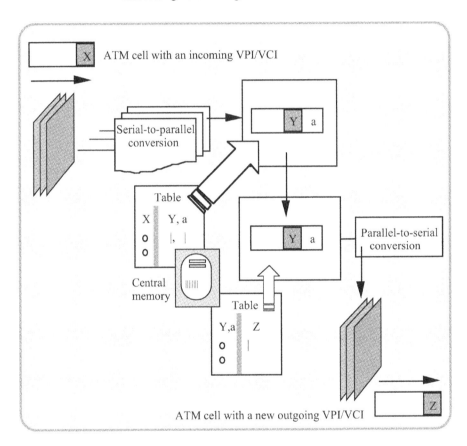

Fig. 6.17: Principle of self-routing and inclusion of a central memory and routing table

6.3.4 Principle of self-routing in ATM switches

This is illustrated in Fig. 6.17. The associated features are:

- ✓ VPI/VCI translation is done at the input
- ✓ Upon translation, the cell is extended by the switching element by including an internal header, which precedes the cell header
- ✓ The added internal header facilitates internal speeding up (so as to avoid blocking)
- ✓ A central memory is included in the generic realizations of the self-routing strategy. The reasons are:

⇒ The central memory has logical output queues controlled by logical information enabled in the internal cell header

⇒ Central memory is accessed via the parallel interface to achieve high buffer access speed

⇒ For parallel interfacing, necessary serial-to-parallel (at the input) and parallel-to-serial (at the output) conversions are facilitated

⇒ The central control monitors the correct delivery of cells.

Table-controlled switching (Fig. 6.17)

✓ Here, the VPI/VCI of the cell-header are assigned new values in each switching element and this method does not require adding an internal cell

✓ The working principle refers to the contents of the tables being updated at the time of establishing connection

✓ Each table-entry consists of:

⇒ A new set of VPI/VCI

⇒ A number designation of the appropriate output.

In reference to the cell-header processing methods indicated above, the self-routing is preferred due to the following reasons:

✓ Less control complexity

✓ More reliable and robust

✓ Burden of increasing the bit rate internally due to added cell length is not critical under the state-of-the art technology.

6.3.5 *Multicast functionality: Copy networks*

A drawback of Banyan networks is their inability to support *multicasting*. That is, Banyan networks are not able to multicast an incoming cell to a number of outgoing links. This led to the development of ATM switches that have the ability to "copy" the incoming cell and send it to several outlets consistent with the following considerations:

▪ If the self-routing strategy is extended for multicast purposes, it would need a header translation such that all the "copies" made should possess the same VPI/VCI values

▪ The above restriction is avoided by using a concentrator as described below :

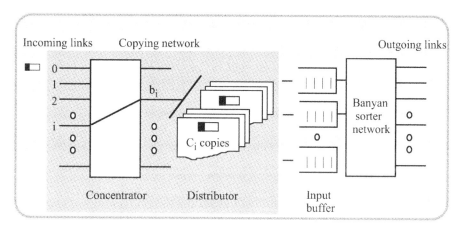

Fig. 6.18: Banyan network with multicasting

A copy distribution network and a compact concentrator precede the sorter Banyan network as shown in Fig. 6.18. In Fig. 6.18, all links connecting different modules are numbered from top to bottom, starting with 0. Suppose,

$$C_i \quad = \quad \text{Number of copies to be made out of the cell at the incoming } i^{th} \text{ link}$$

$$a_i \quad = \quad \text{An indicator (flag) bit for the } i^{th} \text{ link.}$$

The associated logic is $a_i = \begin{cases} 0 & \text{if } C_i = 0 \\ 1 & \text{if } C_i \geq 1. \end{cases}$ \hfill (6.4)

Now assuming that b_i is the output link number of the concentrator to which the incoming i^{th} link is routed, a *concentrator function* can be defined as follows: This moves the cell at the incoming i^{th} link to its output port b_i with $b_i = \sum_{k=0}^{i-1} a_k$. The copy distribution network takes this cell from its input b_i (from the concentrator) and makes C_i copies while transporting these cells to its output links m_i to M_i, where $m_i = \sum_{k=0}^{i-1} C_k$ and $M_i = \sum_{k=0}^{i} C_k - 1$.

In the *copying network*, after the copies are made by the copy distribution network, point-to-point addresses are needed to switch cells from the incoming links of the sorter/Banyan network to its output ports. Two identifiers are needed for this purpose:

✓ One, to uniquely identify the original cell
✓ Another, to identify the copies.

Two values are used for this purpose: *Broadcast channel number* (BCN) and *copy Index* (CI). This CI ($\bullet_i$) is computed as follows: $\bullet_i = (y_i - m_i)$ where y_i is the output port number at the last stage of the copy distribution network, and $m_i = \sum_{k=0}^{i-1} C_k$, which represents the original lower bound. Given BCN and m_i, the point-to-point self-routing address of the cells at the sorting/Banyan network can be obtained from an address translation table residing at the switching node. The packet header format of a copy network is depicted in Fig. 6.19.

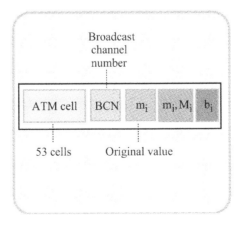

Fig. 6.19: Packet header of a copy network

6.3.6   *ATM subsystem development*

This development effort refers to deploying either *switches* (controlled by switching functions) and/or *cross-connects* (controlled by network management functions). In using switches and cross-connects, the hardware could be identical, but the software may differ.

Further, in ATM subsystem development, the following considerations are essential:

    ✓   Self-routing strategy is employed because of its simplicity and popularity
    ✓   Connection-related information is made accessible from peripheral units
    ✓   Non-blocking switching is preferred and adopted
    ✓   A unique cell-rate is specified internal to the switch.

The above are generic system characteristics, which can be specified by a set of *system modules* listed below:

    ✓   *Subscriber line module broadband* (SLMB): A subscriber is connected to the switching network or MUX via an SLMB. The bit rate supported is 155.52 Mbps or 622.08 Mbps
    ✓   *Trunk module broadband* (TMB): Connection to other switches/cross-connects is done by TMB. TMB supports up to 2.4 Gbps in pleisiochronous and SDH systems
    ✓   *MUX*: Used for local concentration of subscriber traffic within the switcher
    ✓   *Switching network*: Connects the interface module, the MUX and the control processor. It also provides internal dialogue between the node subsystems
    ✓   *Control processor*: System control via signaling or network management.

The aforesaid modules are illustrated in Fig. 6.20.

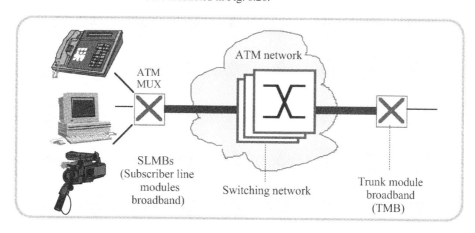

Fig. 6.20: Switch/cross-connect architecture

The functions of SLMB/TMB at the input of a node can be listed as follows:

    ✓   Cell extraction, such as from the SDH frame
    ✓   Adaptation of the node-internal speed
    ✓   Cell delineation and cell header error detection/correction
    ✓   VPI/VCI translation
    ✓   Generation of the new HEC value
    ✓   Usage parameter control
    ✓   Traffic measurement towards accounting, administration, and traffic engineering

    ✓   Generation of the system internal cell format
    ✓   Transmission of cells to both switching planes.

At the output of the node, the corresponding functions are:

    ✓   Selection of correct cells from the redundant switching plane
    ✓   Converting of the internal cell format to a standardized format
    ✓   Adapting the cell rate to the outgoing transmission speed by means of inserting idle cells
    ✓   Filling of cells in the transmission system as necessary (for example, SDH frame).

6.3.7   *Characteristics of ATM switches*

The following aspects of ATM switches define the expected attributes of such switches.

- *Modularity*:   This is defined as the incremental number of ports that can be added to a switch.

- *Maintainability*:   This measures the isolation extent of a disruption on the remainder of the switch.

- *Availability*:   This refers to continuity of operation in the presence of single or multiple faults.

- *Complexity*:   This is often measured in terms of logic gate counts, chip pin-out, and card pin-out in comparing different switch implementations.

- *Flexibility*:   This refers to the maximum number of switch ports supportable by the architecture.

Switching architectures

    Though the benefits of ATM have been the talk of the town, the hows and whys of various switching methods have not been explicitly portrayed in terms of design considerations even within the scope of technical writings. Notwithstanding the lack of such efforts, it can be visualized that a study on the practical aspects and desideratum of ATM switch fabrics is of prime importance to networking engineers who have to match a vendor's implementation to actual applications.

    In general, telecommunication product implementation issues are largely trade-offs between cost, performance, size, and flexibility. Among the various fabrics described above, the three versions of switches of practical interest are *shared-bus*, *Banyan matrix*, and *broadcast matrix* architectures. (The cross-point matrix is no longer considered as a viable hardware, due to its performance shortcomings, in the current generation of ATM switches). The following is a summary of these three switches, which provides a revisit to grasp relevant characteristics for their subsequent applications in ATM networking presented later.

Shared-bus switch

    The principle of a shared-bus (and shared memory) is well known. It is commonly used in almost in all computer and communications equipment. For example, PCs, routers, and T-1 multiplexers use shared-bus techniques. Basically, the shared-bus consists of a backplane and some number of I/O cards. In each of these cards, the bus interface components are facilitated with a selected clock rate. The faster the clock, the more expensive the associated components. This price versus performance curve is, however, nonlinear. That is, once the particular technology limits are reached, the price may go up exponentially even for a small increase in speed. Modern silicon technology cost-effectively delivers bus speeds from about half a gigabit to a gigabit per second. Higher bus speeds may pose cost penalties that may not be consistent with the realizable outcomes on a prorated basis.

Further to economic issues, the shared-bus implementations are not scalable. That is, once the bus components have been designed-in, the card adopted is limited to that bus-speed. To increase the bus-speed, one would have to swap-out all the I/O cards. This is not, obviously, a cost-effective solution. Hence, an alternative choice is to replace the switch with a better one (in terms of size and speed) and shifting the older node to a smaller site. This could be a prudent plan, especially when the budget in the interim period is tight.

Buffered-Banyan network

In the existing literature on ATM switching, the *Banyan*, *Batcher-Banyan,* and buffered-*Banyan matrix* have been indicated as promising technology. In almost all ATM switches adopted today, the Banyan matrix uses a self-routing header. As indicated earlier, this extra header is used to inform the internal switch elements how to move the cell from a given input through the fabric, until it exists the targeted output port.

Regardless of vendor implementations, as a cell arrives at an input card, the address is first examined and compared to an on-card table that has been previously loaded by the control card of the node. This table contains a specific *self-routing header* (SRH) that corresponds to every expected arriving cell address. This self-routing header is pre-pended to the cell and then moved into the fabric. The sequence of 1s and 0s in the SRH is arranged so that each switching element is ordered to switch the cell either up or down as the cell negotiates its way through the fabric and then out the other side. The actual switching in this procedure is done in hardware without passing through a centralized von Neumann bottleneck (similar to that in CPUs). Hardware-based switching makes ATM switches fast (and hence cost-effective). Software-based switching is usually limited to low rates (below one Gbps regime) in comparison with hardware-based switches, which allow speeds ranging from tens to hundreds of Gbps.

Banyan switches, however, have several deficiencies: Inasmuch as the number of switching elements and stages are finite, whenever multiple inputs are active, it is possible that cell collisions may take place.

Suppose the arriving cells correspond to a broadcast or multicast transmission, (which is state-of-the-art in client-server and multimedia applications). In such implementations, the cell is copied at the input and re-sent through the fabric with a unique self-routing header. This is referred to as the *cell gain*. When this process is combined with multicast on multiple input ports, significant cell-loss probabilities could be foreseen.

Buffered Banyan matrix switching adds memory buffers to every switch element so that whenever two cells arrive at a given element, one can be held while the other is switched through. In such efforts, a school of thinking exists that considers that the resultant implications pertinent to longer transit delays, variability in delay, and added complexity are not consistent with the goals of application-independent ATM switching being mooted. It is not known at this stage whether any production switches use this technique.

Another method that has been advocated extensively is the Batcher-Banyan architecture. It refers to having a separate stage (namely, the *Batcher sorter*), added in front of the Banyan switching fabric as discussed earlier. The Batcher stage arranges cells in an optimized order before actually entering the Banyan fabric. This process minimizes the probability of congestion. Though theoretically feasible, this method is rather complex, unstable, and too unpredictable to be used in production switches.

Table 6.2: <u>ATM switches: A comparison</u>

Characteristics	Single bus type	Multiple bus type	Self-routing type	Augmental self-routing type
Level of complexity	Low	Medium	Significant	Significant
Operational maximum speed	1-20 Gbps	1-20 Gbps	Up to 200 Gbps	Up to 200 Gbps
Scalability	Low feasibility	Medium feasibility	High feasibility	Excellent feasibility
Point -to-multipoint connection	Feasible with ease	Feasible with ease	Difficult to implement	Difficult to implement
Level of blocking	Not significant	Low/medium	Medium	Not significant
Other specific attributes	Low cost	Low cost	VLSI implementation feasibility	VLSI implementation feasibility

Broadcast matrix (Fig. 6.21)

A type of fabric that is acquiring popularity among the vendors of ATM switches is the *broadcast matrix*. It is a non-blocking, direct-wired, contentionlessly connected, output-buffeted, self-routing ATM fabric. This fabric has a dedicated contention-free path from each input to a dedicated buffer at each output. When a cell arrives, the input card pre-pends a self-routing header and shifts them both in. The cell, due to the multiple dedicated paths, arrives at all output ports simultaneously. Each port then simply checks the SRH to see whether the cell belongs to it. If not, the cell is cleared. Otherwise, the cell is shifted out. The entire process is collision-free.

In a multicast mode, the pre-pended header simply addresses multiple output ports. Outputs 2 and 3 in Fig. 6.21 clear their buffers of their cells while ports 1 and 4 shift theirs out. The process is collision-free and requires extra steps. The chances of mid-plane collisions in a broadcast matrix is almost nil. It has been regarded as the simplest and most predictable method for ATM switching.

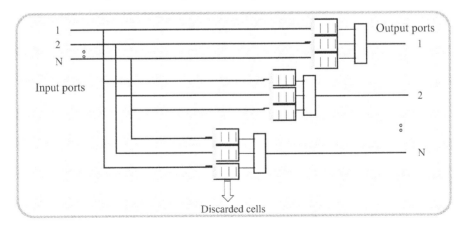

Fig. 6.21: Broadcast matrix switching architecture

The only possibility of cell-loss that can occur refers to a focused load condition, namely, the degenerative case specified for all switch architectures. This can be observed when each input wants to talk to a single output. This refers to a condition that no switch can tolerate over a period of time. This is where the need for output buffering comes in.

Ideally, the buffer depth should be in the neighborhood of one to two hundred cells deep. Greater depths to an extent of thousands of cells have been shown to add too much delay for diminishing gains in respect of collision avoidance.

For perfect delivery, message switching should be used with disc-drive buffers (and resultant hourglass delays). ATM architectures strive for the lowest absolute delay and lowest delay variability. ATM benefits can, therefore, be fully reaped only when all the associated hardware as well as software are optimized for maximum application independence.

Considering the broadcast matrix being near-perfect, the impending query is why not all the vendors opt for it. As with any technique, there are always diminishing returns. Using cost-effective silicon, the broadcast matrix is probably limited to a 16 × 16 or may be a 32 × 32 single stage matrix with a total fabric throughput of around 20 Gbps and I/O limits of around a Gbps each. All of the "dedicated paths and dedicated buffers chew up extensive real estate to remain bullet-proof and predictable". Hence, it would probably not be so cost effective to use this technique on a 200 Gbps CO switch.

Shared-bus techniques can be conceived effectively in small (1 to 2 Gbps) switches. In such cases, the long-term scalability is not a design issue. It is considered as a useful technique to expand the sub-gigabit port count on higher speed switch fabrics.

Above 20 Gbps speed regimes, capacities can be addressed best using Banyan techniques or in some cases by multistaging a broadcast matrix fabric (also known as a *Clos network*).

Which technique is appropriate for a network engineer to choose and deploy in the design of ATM networking? The answer is not simple since there has not been a unique way, which is best for any given application constrained by budget bounds. The bottom line is that the end-users should have a dialogue with the vendors and compare both short and long-term network life-cycle costs as they relate to the growing ambient of the company. Matching the switching technology to the application is the first step towards a successful program in switching implementation.

6.4   ATM Switching Products

6.4.1   ATM environment

The ATM technology warrants hardware specific to three environments, namely:

- ✓   Central office (CO)
- ✓   Customer premises equipment (CPE)
- ✓   Campus networking.

The internetworking devices needed thereof are as follows:

- ✓   Switches/cross-connect
- ✓   Routers
- ✓   Bridges
- ✓   Hubs
- ✓   Multiplexers
- ✓   Gateways
- ✓   Backbones
- ✓   ATM network interface cards for high-end work-stations.

Internetworking devices: Definitions

CPE: *This refers to customer premises equipment such as terminals, telephones, and modems installed at the customer sites and connected to a PSTN.*

Switch: *It is a fabric-controlled device deployed to interconnect lines to complete communication paths. In comparison with bridges (described below) switches are significantly faster since they are essentially hardware-*

based. As a network device, a switch may filter, forward, and flood frames based on the destination address of each frame. It operates at the data link layer of the OSI protocol. (Flooding refers to a traffic-passing technique used by switches and bridges in which traffic received on an interface is sent out of all the interfaces of that device, except the interface in which the information was originally received.)

Router: A device (hardware or software) operating at layers 1 (physical), 2 (data link), and 3 (network) of the OSI model. It handles the integration of networks that were disjointly fabricated with different network protocols and facilitates information transfer across different media. Routing is the process of moving information across an internetwork from a source to a destination. A router, in essence connects devices on LANs to devices on other LANs, usually via WANs. Their functions include:

- Serving segment LANs and WANs
- Deciding the best way to route the data to its destinations
- Communication with peer routers
- Deciding to route data whether by LAN or by WAN.

Bridge: To improve performance, a LAN architecture is normally subdivided into smaller ones and is interconnected either by a LAN switch or by a bridge. A bridge connects and passes packets between two network segments that uses the same communication protocol. It operates at the data link layer (layer 2) of the OSI reference model. In general, a bridge filters, forwards, or floods an incoming frame based on the MAC address of that frame. A bridge or a switch is somewhat "smart". A device connected to a bridge or a switch will not hear any of the information meant just for devices on the ports of the switch. A bridge, in essence, does (hardware or software) functions interconnecting LANs at the OSI data link layer, filtering and forwarding frames according to MAC addresses. That is, a bridge operates at the data link layer and connects individual LAN segments together to form one large logical network.

Hub: This is a device adopted as a center in a star-configured networking. An individual terminal is connected with the hubs via a single physical line, thus providing the full (physically) possible transfer capacity to each terminal. It may be a hardware or a software device that contains multiple independent but connected modules of network and internetwork equipment. It can be active (where they repeat signals sent through them) or passive (where they do not repeat, but merely split signals sent through them). It is typically used in older ethernet and token-ring networks. A device connected to a hub receives all the transmissions of all other devices connected to that hub. That is, a hub is used to connect devices together so that they are all on one LAN. While the cables used for ethernet with RJ-45 connectors facilitate connecting only two devices, a hub allows more than two devices on one LAN. Hubs are now replaced in many cases by LAN switches. A hub is not a smart device, it simply sends all the data from a device on one port to all the other ports. It means every device on the hub listens to everything transported, whether it is meant for it or not.

Multiplexer: A device that combines multiple streams of information to share a common physical medium.

Gateway: In OSI terminology, a gateway is a device performing protocol translations for all seven layers. In IP context, routers are synonymously called gateways wherein protocol conversion is restricted to the lower three layers of the OSI model. In essence, a gateway represents a combination of hardware and software that interconnects otherwise

	incompatible networks or networking devices. The term gateway is sometimes used to indicate a device (though not so commonly), which translates between disparate protocol stacks.
Backbone:	*It represents a LAN or a WAN connectivity between subnets across a high-speed network. It is a part of a network that acts as a primary path for traffic, which is most often sourced from, and destined for, other networks. A collapsed backbone is a nondistributed backbone in which all network segments are interconnected by way of an internetworking device. A collapsed backbone might be a virtual network existing in a device such as a hub, a router or a switch. In ATM context, for example, a backbone refers to high-speed connectivity such as an ATM OC-12 that interconnects lower speed networks such as ATM OC-3 via fiber optic cable.*
ATM network interface card:	*This is used with a device (such as a server or high performance workstation), which needs direct access to an ATM switch or network of switches. It does bus interfacing, ATM engine (SAR) functions, and physical I/O requirements.*

CO-based ATM hardware

CO-based switches are intended to handle large number of ports and bear heavy industrial strengths. It requires significant power supply accessories. Its throughput is greater than 5 Gbps.

CO ATM switches form the backbone architecture to ATM networks. They contain ATM native UNI. The CO environment, which accommodates ATM switches, provides for the required power supply and "rack and stack" physical arrangement. It concurrently enables scaling of processing port capacity as well.

CO ATM switches set up calls for CPE switches similar to the CO voice switches, which set up calls for CPE PBXs.

Typically, CO ATM switches are available as broadband switching modules with switching capacity scaling and mounted on a single shelf. They are capable of supporting DS-1, DS-3, OC-3, and OC-12 interfaces for ATM, frame-relay, and SMDS services. The CO-switch can also switch the video/voice traffic to the voice switches and other network as needed.

Campus-network-based ATM switches

These can support up to 5 Gbps. These are smaller than CO counterparts. Further, unlike the CO version, here many interfaces are made available. These interfaces refer to:

- ✓ Native LAN (ethernet and token-ring)
- ✓ MAN (FDDI and DQDB)
- ✓ SNA
- ✓ X.25
- ✓ Voice.

The other associated features are protocol conversion, LAN emulation (to be described later), and virtual networking.

These switches can run on regular AC power supply. They bear smaller port capacity and possess less processing power than CO types.

Typically, ATM switches designed for campus networking are virtually nonblocking types and can support DS-3/OC-3 and optical fiber at 100 Mbps (4B/5B) and 155 Mbps (8B/10B) carrying ATM cells. The switch control could be proprietary.

CPE-based units

These are smaller capacity units with low scalability and processing procedures. The CPE-based hardware requires very low power supply needs. The CPE ATM-ready router or hub allows clients and servers to communicate in a virtual network. Workstations and servers can also be directly connected to the local ATM switch constituting a high-end work group.

The deployment of CO-based, campus-networking-based, and CPE-based units is illustrated in Fig. 6.22.

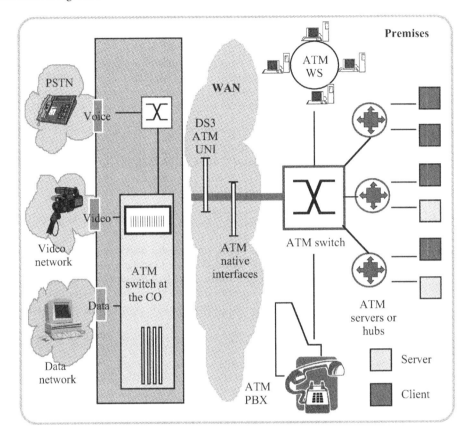

Fig. 6.22: CO-based, campus-networking-based, and CPE-based switches

ATM network interface card (NIC)

This is used when a device such as a server or a high-performance workstation requires direct access to an ATM switch or network of switches. That is, NIC is an adapter card for workstations and PCs connected via an ATM LAN. Its *bus interface function* prepares information in the workstation's bus for transmission over a communications network. That is, standard functions such as bus interfacing and parallel-to-serial conversion are performed here. In some NICs, appropriate traffic shaping and flow control as required by the ATM UNI are also carried out. Appropriate application programming interface software allows a structured environment for interfacing with the workstation's operating system (or network operating system) for extracting the packets and transmitting them over an ATM network.

The other functional area of NIC is the *ATM engine*, which enables segmentation and reassembly and signaling functions as specified by the chosen AAL in concurrence with the UNI. These functions are carried out via function-specific chips. There are two ATM engines: *Client engines* and *server engines*. The client engines are simple since the number of addresses required to be processed simultaneously are low. Correspondingly the complexity (in terms of memory etc.) is also small.

Server engines on the other hand require significant sessions of simultaneous SAR functions. The larger the number of logical connectivity, the larger the complexity requiring high-speed RAM.

The third functional aspect of NIC refers to physical I/O interface. Consistent with the emerging trends in ATM networks, users can choose low (25/5 Mbps) to high (622 Mbps) data rates, and use simple-/multimode fibers and/or category 3/5 UTP/STP.

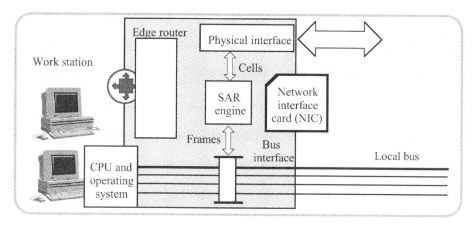

Fig. 6.23: ATM NIC

Legacy-to-ATM devices

These refer to smaller devices supporting hubbing or switching of legacy LANs (such as ethernet, token-ring, and FDDI). They provide an ATM uplink to the campus backbone.

LAN traffic is generally collected by a hub. As illustrated in Fig. 6.24, by resorting to switching hubs, large bandwidth to many users can be provided by means of a high-speed access line to the backbone.

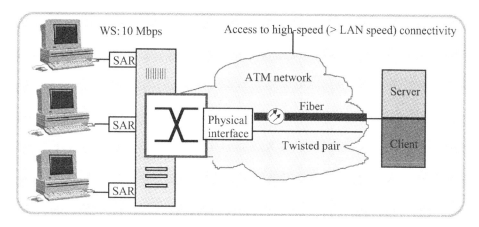

Fig. 6.24: ATM adapting hubs

ATM access to concentrators, adapting switches, and MUXs

Commensurate with the networking environment indicated above, the ATM hardware and software products have been conceived to drive "ATM out towards the edge of the network without forcing changes on those devices and applications simply don't need it. This is done with a variety of ATM adaptation devices".

In reference to private networking (such as in the campus environment), ATM adaptation devices can be identified as follows:

  ✓ ATM network interface cards (NICs)
  ✓ Legacy-to-ATM devices (hubs)
  ✓ ATM access switches, concentrators, and MUXs
  ✓ ATM work group switches
  ✓ ATM backbone switches.

The public networking equipment generally include the following:

- ✓ Adaptation devices
- ✓ ATM edge switches
- ✓ ATM core switches.

These switches may differ widely in their functionality and not all the features available in this class of equipment may be either useful for a specified ATM networking application.

The *concentrators* are essentially cell assembler-disassemblers. These are also known as *application-specific adaptation devices* (ASADS) or *network access devices* (NADS). They perform as a multiport digital (data) service unit (that converts RS-232 or other terminal interface to line coding for local loop transmission) and/or as a channel service unit (that interfaces the T-1 line terminating on the local loop).

Concentrators also do protocol-conversion pertinent to packet assembler/disassembler wherein conversion between packets (X.25, etc.) and sync or async data takes place. Concentrators find use in branch office applications. For example, a concentrator can provide a protocol specific adaptation on a port-by-port basis in respect of end-entities such as PCs and document images, so that an ATM feeder network (public or private) can be properly utilized.

In conjunction with an NIC, hubs and concentrators are largely function-specific in a network. On other hand, an *adapting switch* may carry out multiple functions at a time. Its role is to work as a *network consolidator*. For example, adapting ATM switches can emulate a traditional T-1 line by applying the SAR to the CBR stream (as described by AAL-1) as described below.

Example 6.1
Describe an adapting ATM switch as a network consolidating product that provides CBR support via T-1 emulation

Solution
As a network consolidating product, an adapting ATM switch has to support a variety of interfaces. For example,

- ❑ A 10Base T hubbing for a LAN
- ❑ CBR interface
- ❑ VBR interface.

An example of CBR support refers to T-1 circuit emulation. Suppose T-1 trunks emanating from a PBX are adapted onto a new backbone. That is, all of the site's traffic is consolidated onto a cost-effective T-3 facility at an interexchange carrier's (1XC's) point-of-presence (POP).

This has to be done without disrupting the existing system (that uses PBX) operating on dedicated T-1 lines and plugging the PBX outlets into the adapting ATM switch as illustrated in Fig. 6.25.

The T-1 emulation card renders a "look-like" T-1 transmission facility. It covers jitter minimization, constant CTD facilitation, and sourcing/sinking of network timing. The relevant function includes merging of CBR streams pertinent to voice from the PBX with other VBR/CBR applications on the site.

On the CO side, the switching is a mirror-reflection of emulation functions.

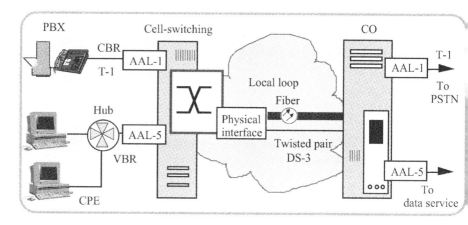

Fig. 6.25: T-1 emulation

Problem6.1
Describe integral video adaptation switching in ATM-based high resolution image networking.

ATM adaptation devices

LAN level adaptation
ATM access to workgroup switches
Workgroup switches refer to ATM LAN switches for adaptation to workgroups with less than 24 users who require high networking performance. These switches may provide connectivity to legacy LANs as well. The key features of workgroup switches are port types and configurations, cost, and performance levels.

ATM access to backbone switches
These refer to switches intended for building corporate LAN backbones on a campus. Such a backbone architecture warrants support for a large number of virtual connections and fast SVC call setups, minimal delay, and maximized throughput.

Public carrier level adaptation
These are small devices that provide adaptation of non-ATM traffic into ATM transmission. Such devices, in modern practice, are being shifted to CPE level. Typical adaptation devices for public carrier equipment are those that convert frame-relay, SMDS, and IP into ATM. Augmented versions include ISDN and dial-up data. Adaptation devices may also support native LAN services.

ATM edge devices
These are medium-sized switches for delivering ATM services. They can be used as trunking adaptation devices (depending on carrier architecture). The ATM services delivered correspond to fully-featured, cost-effective high-performance efforts.

ATM core devices
These refer to large machines operable at 20 Gbps and above. The core switches enable carriers to concentrate traffic into a few, high traffic density backbone trunks for very long distance transportation. High quality and reliability considerations, however, render these switches very expensive.

In summary, adaptation allows legacy equipment to migrate to ATM. Network topology considerations pertinent to customer premises, campus and/or CO levels, allow combining segments with a wide variety of concentration and adaptation devices being marketed.

Use of routers in ATM environment

The classical art of routing in internetworking has been studied to realize that in the context of network expansion, the complexity of routing grows enormously with a corresponding price tag on such expansions.

Usually, making additions and changes in a network is difficult because of the subnets, which are tied to physical ports. Changes, thereof, require complex paradigm shifts and addressing schemes.

Normally, the *collapsed backbone routing* may ease the situation to some extent. But, any network in its expansion profile will pose extreme demands on the router's backplane requirements. It means the box of the backbone will soon run out of space and the backplane will not accommodate the increase in bandwidth.

In the ATM environment, the above problem can be met by using a *routing server* that represents a consolidated set of interface cards over a shared bus. Fig. 6.26 illustrates a classical routing scheme where data transfer within a router occurs between interface cards over a shared bus. The control card runs routing protocols, but handles only a small traffic with I/O cards to update its routing table.

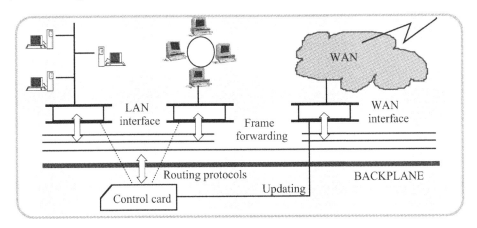

Fig. 6.26: Backplane based classical routing

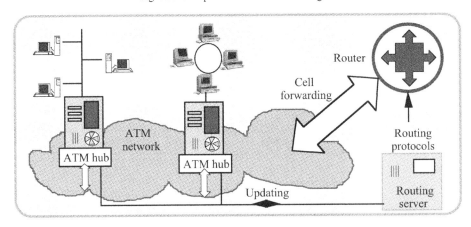

Fig. 6.27: Routing using a network of ATM switches

In a consolidated plan, the control cards are consolidated into a routing server as shown in Fig. 6.27. The routing interface cards are replaced by ATM hubs. The ATM backbone network handles data transfer among hubs as directed by the routing server.

6.4.2 Architecture features of ATM products

In reference to ATM products vis-à-vis equipment vendor architecture, the specific features to be looked into in product selection for ATM networking are as follows:

- ✓ Port types and numbers
- ✓ Size and type of switching fabric
- ✓ Buffer sizes and location
- ✓ Type of queue serving
- ✓ Management of congestion
- ✓ Virtual circuit support
- ✓ Signaling/SVC call handling
- ✓ Switch footprint:

  - ⇒ Power supply
  - ⇒ Certifications
  - ⇒ Clocking options.

- ✓ Feature supports
- ✓ Virtual LAN and WAN emulation
- ✓ Product performance:

  - ⇒ Switching delay
  - ⇒ Variations in cell arrival
  - ⇒ Video/voice/data traffic performance
  - ⇒ Congestion handling
  - ⇒ Congestion isolation
  - ⇒ Call setup times
  - ⇒ Call per second
  - ⇒ Throughput (packets/s)
  - ⇒ Broadcast and unknown server-broadcast speed.

- ✓ Interoperability options
- ✓ Upgradability and scalability options
- ✓ Network management
- ✓ Customer service
- ✓ Overall costs.

Example 6.2

Consider an enterprise network which employs ethernet and token-ring on individual segments (mostly on the same floor) and uses 100 Mbps FDDI as its backbone technology as shown in Fig. 6.28.

Suppose there is growth in the size and bandwidth demands of network applications in this enterprise. Traffic is found to be increasing exponentially. The processing of the packets at the multiple routers in the data path is observed to pose a high latency (delay). As a result, transferring of large data files is seen to face low router throughput.

Discuss a method of upgrading the network by incorporating switches to meet the traffic demands while keeping the original backbone topology in place.

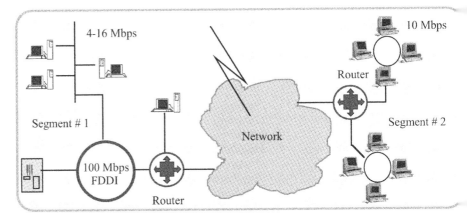

Fig. 6.28: Existing backbone architecture

Solution

A possible upgrading solution is illustrated in Fig.6.29. The solution refers to replacing the existing routers with switches and reallocate those routers to other areas of the network such as Internet gateways and WAN gateways.

Now data would not have to pass through successive routers facing high latency. Instead, it is zipped through the network. The switch performs necessary translation so that the 16 Mbps token-ring can transmit data to ethernet segments or FDDI rings swiftly and effectively. The switch ports act as hubs and extra ports could be still available for future expansion.

The switch solution allows the original backbone based on the 100 Mbps FDDI ring to remain in tact and seamlessly integrate the existing segments.

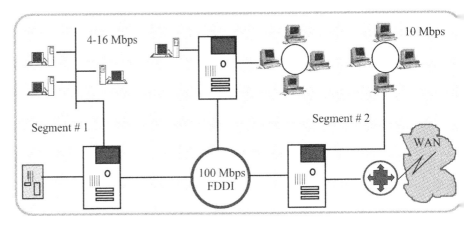

Fig. 6.29: Upgraded architecture

Problem 6.2

Suppose an existing network is made of multiple independent networks serving several departments. Assume some departments exist within the organization without any form of network connection. Suppose a server is added to cope with the company's growth so as to service the existing networks and also connect the non-networked part of the organization to a corporate network. Also assume some segments presently may have their own servers. Ethernet and token-rings are used in integrated enterprise network spread across three floors. Currently, there are two token-rings in the third floor that work at two different speeds (4 and 16 Mbps) and routing is done by the company's router on the first floor. This router also provides access to WAN.

Reorganize the existing networking using switches to tie multiple technologies together and speed up the traffic. Adding a switch in each floor is permitted. Illustrate your solution.

Problem 6.3

Assume the following campus scenario. The networking spans three multistorey buildings each with three floors internetworked via mixed wiring topologies. Wiring in each floor runs to repeaters and bridges located in a communication closet on the first floor of the building. Further, in each communication closet, a router ties together the three buildings through a 100 Mbps FDDI backbone.

It is observed that the existing routers are not able to cope with constant data rushing from the users in each building. The routers en route cause excessive latency and congestion on each segment as well as on the backbone itself. Deployment of intensive applications by the users is also posing heavy bandwidth demands.

The company also wants upgrading of the existing networking so as to include graphic stations (CAD/CAM), and video conferencing.

Describe a solution to augment the existing networking by including switches in each floor of the three buildings. The revised network should still have the 100 Mbps FDDI backbone ring. Illustrate your answer.

6.5   Concluding Remarks

Today's network design and/or network expansion efforts can be divided into two major steps: Logical layout of the network and physical development of network products. In the logical design phase, the network topology is conceived and its hierarchy is decided. At the network-layer level, the designer looks into the options for a feasible addressing model and selects bridging, switching, and routing protocols including security and management design considerations.

The next phase of physical design starts with ATM product selection consistent with the logical topology conceived. Associated devices are to be carefully chosen to implement the conceived network into a practical reality. A top-down network design is recommended, which facilitates a structured performance assessment of the test network and allows optimizations of features such as traffic management, QOS considerations, and switching mechanism.

In general, understanding ATM switching functions and knowing the performance aspects of ATM network products are crucial for a successful design and implementation of ATM networking. Chapter 8 offers additional information in regard to pragmatic design efforts towards ATM networking.

Bibliography

[6.1]   T. Viswanathan: *Telecommunication Switching Systems and Networks* (Prentice-Hall of India Pvt. Ltd., New Delhi, India: 1995)

[6.2]   M. de Prycker: *Asynchronous Transfer Mode – Solution for Broadband ISDN* (Prentice-Hall, International (UK) Ltd., London, England: 1995)

[6.3]   D.E. McDysan and D.L. Spohn: *ATM Theory and Applications* (McGraw-Hill Co., New York, NY: 1998)

[6.4]   R. Händel, M. N. Huber, and S. Schröder: *ATM Networks Concepts, Protocols, Applications* (Addison-Wesley, Publishing Co., Workingham, England: 1994)

[6.5]   R.O. Onvural: *Asynchronous Transfer Mode Network: Performance Issues* (Artech House, Inc., Boston, MA: 1995)

[6.6]   D. Hill: *The Switching Book I & II* (Xylon Corp., Calabasas, CA: 1996)

[6.7]   C. Gadecki and C. Heckart: *ATM for Dummies* (IDG Books Worldwide, Inc., Foster City, CA: 1997)

[6.8]   J. Walrand and P. Varaiya: *High Performance Communication Networks* (Morgan Kaufmann Publishers, Inc., San Francisco, CA: 1996)

[6.9]   E.R. Coover: *ATM Switches* (Artech House, Inc., Norwood, MA: 1997)

[6.10] T.M. Chen and S.S. Liu: *ATM Switching Systems* (Artech House, Inc., Norwood, MA: 1995)

7

ATM: OPERATIONS, ADMINISTRATION, MAINTENANCE, AND PROVISIONING

Who does housekeeping?

"The goal of transmission management is to move bits efficiently and effectively (between end-entities)"

R.Panko
Business Data Communications and Networking, 1997

7.1   Introduction

ATM management: A briefing

The management of ATM networks is a method by which a centralized administrative authority monitors the behavior of the network. Such management efforts per se have not been exclusively considered by the CCITT, inasmuch as there are already well-established local standards for the management of public switching systems, and, hence, the public ATM networks are expected to comply with those standards.

However, the ATM Forum has focused its efforts on devising a management framework for private ATM implementations (Fig.7.1), which cooperates with the management aspects of local PSTN in supporting the ATM. The ATM Forum UNI standards specified in its first draft refer to an *interim layer management interface* (ILMI). ILMI allows two management entities in the nodes that are directly connected by a UNI to manage that UNI, through a symmetric, peer-to-peer protocol. Based upon a contribution from Hughes LAN systems, ATM Forum adopted the ILMI concept, which has become subsequently responsible in defining the so-called ILMI *management information base* (MIB).

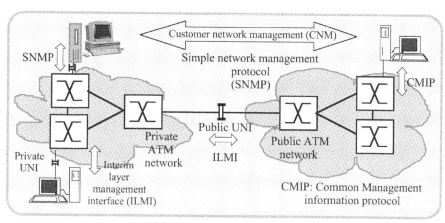

Fig. 7.1: Management of ATM networks

In order to understand the underlying aspects of the ATM management briefed above, it is necessary to consider the general considerations and basic issues of telecommunication network management. The following paragraphs offer relevant details [7.1-7.5].

7.1.1 Functional aspects of network management

ISO specifies the functional areas of network management as follows:

- *Configuration management:*   Exercising control on, collecting data from, and providing data to the managed objects for the purpose of the continuous operation of interconnection services
- *Fault management:*   Detection, isolation, and correction of the abnormal operation of network resources
- *Performance management:*   Evaluating the behavior of managed objects and assessing the effectiveness of the communication involved
- *Security management:*   Protecting the privacy and integrity of the managed objects
- *Accounting management:*   Imposition of tariff and billing the users for making use of managed objects.

In reference to the ATM network, the above management functions can be specifically identified as efforts towards the specified tasks enumerated below:

Configuration management:
This ATM management function involves the following:

- ✓ Creation and detection of VP/VC connections
- ✓ Ascertaining the connection status
- ✓ Determination of the number of active connections at an interface
- ✓ Determination of the maximum number of connections that can be supported at an interface
- ✓ Determination of the number of pre-configured connections at an interface
- ✓ Configuring the number of VPI/VCI bits supported
- ✓ Configuring and determining the status of interface address information.

Fault management
Fault management in ATM context refers to the following *notifications*:

- ✓ Inability to establish ATM connection
- ✓ ATM connection failure
- ✓ Multiple/simultaneous failures
- ✓ Adjacent peer failure to UNI.

Also, fault management involves supporting *operation, administration, and maintenance* (OAM) of fault management flows.

Performance management
This assesses network hardware, software, and transmission media. Essentially it does the following tasks:

- ✓ Checking whether an ATM connection complies with QOS requirements
- ✓ Checking on traffic contract violations

✓   Supporting OAM performance management flows.

Security management

In order to protect a network's integrity, the following security efforts are specified:

✓   Information authentication
✓   Controlled access to network resources
✓   Check on the integrity of data received
✓   Confidential aspects of information to be released
✓   Nonrepudiation: Verification on sender/receiver getting the sent data.

Accounting management

This performs the recording of the following for the purpose of billing:

✓   Connection QOS
✓   Connection bandwidth
✓   Connection duration
✓   Number of cells successfully transmitted and received
✓   Number of cells received with error
✓   Number of cells violating the contract.

7.1.2   Concept of OAM&P

Telecommunication network management is a gamut representing a complex set of triangles with technology, performance, customer needs, and economics depicting the vertices of each of them as illustrated in Fig. 7.2.

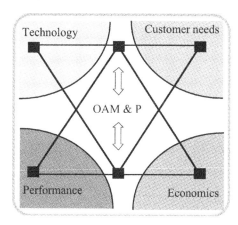

Fig. 7.2 Telecommunication network management

Enclaved within this set of triangles is the *operations, administrations, maintenance, and provisioning* (OAM&P) whose constituent functions can be briefly stated as follows:

■   *Operations*:   It functions around the clock within the infrastructure of the network carrying out such performances as needed to execute the designated operations of the network

■   *Administration*: This refers to designing/conceiving the network, processing orders, assigning addresses, and keeping track of network usage versus tariffs imposed

- *Maintenance*: Should any part of the network fail to function as needed, preventive maintenance and post-failure repairs are warranted for continuous network functioning

- *Provisioning*: It is the task of installing, setting up, updating parameters, verification of performance, and dismantling of the network.

In short, network management is a method of controlling a complex data network so as to maximize its efficiency, productivity, and reliability by means of the four functions indicated above.

User – OAM&P – network

The OAM&P functions involve processes across the user-to-network or vice versa (user ↔ network). The associated tasks can be summarized as topdown efforts as shown in Fig. 7.3.

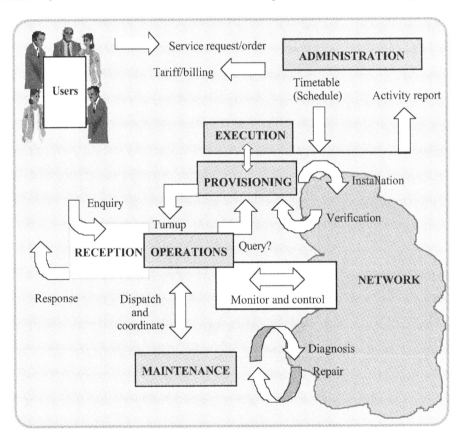

Fig. 7.3: OAM&P functions across user ↔ network

Operations: User ↔ network

As defined above, a network operation involves the instant-by-instant performance monitoring of a network. This effort identifies those instants and events for which it would require necessary intervention so as to pull the network back into compliance as specified towards ideal performance objectives set by the designers. The functional aspects of operations in a network are as follows:

✓ Informing the organization of the past performance history of the network. This information on the past performance includes service failures and corrective measures undertaken

✓ Watching for faults and invoking corrective and maintenance operational commands. These commands include issuing corrective controls for fault removal and resolving customer complaints

✓ Effectively and exhaustively using network capabilities in an optimum manner

✓ Coordinating the efforts between administration, maintenance, and provisions at all phases of a connection.

Maintenance: User ↔ network

As indicated earlier, maintenance is the combination of all technical and corresponding administrative actions, including supervision actions, intended to retain an item in, or restore it to a state in which it can perform a required function. An overview of maintenance considerations is presented in Table 7.1:

Table 7.1: <u>Overview of maintenance</u>

Maintenance actions	Description
Performance considerations	The functions of a managed entity are monitored by continuous or by periodic checking of functions. As a result, information on the maintenance event is obtained and conserved
Defect and failure detection via preventive maintenance	Any malfunctions or projected malfunctions are detected/ ascertained by continuous or periodic checking. Hence, maintenance event information and status on various alarms are produced
System protection via redundancy	Influence of the failure of a managed entity is minimized by blocking or via change-over to other entities. Thus, the failed entity is excluded from operation
Failure and performance information	Failure information is passed on to other management entities. That is, alarm indications are given to other management planes. Response to a status report request is also given.
Fault localization for post-failure maintenance	Determination by internal or external test systems of a failed entity if failure information is insufficient

Normally maintaining a network may involve unanticipated changes. Maintenance does not, however, include changes "not instigated via the administrative design or service provisioning process". Examples of maintenance efforts are:

✓ Replacing failed equipment
✓ Trouble-shooting physical circuit problems
✓ Changing the interface cards etc.

Administrations: User ↔ network

In a telecommunication network, administration implies efforts of persons performing necessary planning. This task may include:

✓ Conceivable upgrading of the network with additional features and functions
✓ Elimination of obsolete hardware and software elements

✓   Designing of a network (private, public, or hybrid version) and/or migration to state-of-the-art network technology
✓   Procurement and installation of network constituents
✓   Administrative planning to stage operations, maintenance, and provisioning activities
✓   Analysis of the traffic, subscriber usage, and billing.

Provisioning: User ↔ network

This is a superimposed effort to make changes on the administratively planned networking. It is an activity to make the necessary changes as per a set of rules and procedures or by using the knowledge-base derived from past provisioning experience. The changes being provisioned can be of routine type or refined ones. An example of provisioning is updating a vendor switch software.

Performance testing such as measuring the BER, ATM *loop-back call testing*, throughput determination, delay assessment via pinging etc., also come under the purview of provisioning.

OAM&P hierarchical levels

The OAM&P levels match the ATM networking layers. Further, OAM&P flows are bidirectional. For example, suppose a VPC is to be monitored by supervising cells sent out at one end-point and mirrored at the other end-point. OAM&P evaluates it at the sending end and a response is made feasible if required.

7.2   OAM&P Architectures

7.2.1 Reference configuration

In reference to the OAM&P architectures deployed, the essential sections of a BISDN/ATM networking are:

- Customer network (CN)

- Customer premises network (CPN)

- Subscriber premises network (SPN).

Pertinent to the above, the reference configuration of the B-ISDN UNI (same as 64 kbps ISDN UNI) is illustrated in Fig. 7.4.

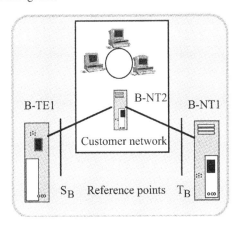

Fig. 7.4: Reference configuration of the B-ISDN UNI
(B-NT1: Broadband network termination 1; B-NT2: Broadband network termination 2; and B-TE1: Broadband terminal equipment)

In Fig. 7.4, the *functional groups* are: B-NT1, B-NT2, and B-NTE1, and the *reference points* are T_B and S_B. The functional aspects of these interfaces are as follows:

- B-TE2: Broadband terminal with a standard interface

- B-NT1: This performs only line transmission termination and related OAM&P functions.

- B-NT2:

  ✓ PBX (Private branch exchange)
  ✓ LAN (Performing multiplexing and switching of ATM cells)

- CN: This covers areas where users have access to the public network via their terminals. It is a part of the telecommunication network located at the user side of B-NT1.

Note:

- Physical interfaces may or may not take place at T_B and S_B. If realized, they must comply with certain standards

- Interface between CN and public network is usually done at the reference point T_B; CN coincides with the function group B-NT2.

7.3   Customer Networks

7.3.1   *Types of customer networks*
The types of customers implicitly refer to the categories of CNs specified on the basis of:

1. Environment (such as residential, business)
2. Number of users (topology).

Residential environment

This includes a domestic line for a family, which requires broadband facilities mainly for entertainment (such as video-on-demand etc.). This could be a kind of uniform facility offered to a single flat/house with no special switching capabilities within the residential CN.

Small business environment

This is similar to residential CNs but facilitated in an office/shop plus residential place such as in home-based businesses. In such cases, the facilities provided for the office/shop and broadband facilities extended to the residential sector should be segregated. This requires *internal switching*. That is, a small business environment may need internal switching at the CN level.

Medium/large business environment

This refers to a large office, commercial place, or a factory environment. Here, the broadband facility may not be required inasmuch as the entertainment support may not be warranted. On the other hand some interactive services such as the following may have to be facilitated.

  ✓ Telephony
  ✓ Video conferencing
  ✓ High-speed data transmission with internal switching capability.

The service is provided typically to 10-100 users spread across a few floors in a complex.

Large business environment

This environment can be characterized as follows:

    ✓   Number of users may exceed 100

    ✓   Networking may spread over a distance up to 10 km

    ✓   Again, switching facility is required.

Factory environment

In this case, the size of the CN may be in the order that it corresponds to a medium or a large business environment. Here again, an intercommunication switching facility may be required.

7.3.2 Requirements on customer networking

Service requirements

    ✓   Service mixing

    ✓   Supporting mixed services.

These services depend on customer type and customer specific characteristics such as bit rate, minimum and maximum delays, and CDV.

Structural requirements

    ✓   Flexibility

    ✓   Modularity

    ✓   Reliability

    ✓   Physical performance

    ✓   Cost.

Flexibility

It is the ability of the CN to cope with the changes in the telecommunication system. The subdivisions of flexibility are as follows:

- *Adaptibility*

  Requirement which measures how the CN deals with changes which do not alter the global scale of the CN (that is, warranting a new wiring). This requirement is important in the terminal area of residential/small business environments

- *Internetworking*

  It describes how the CN can interface to other networks. It has been vastly studied in LANs and WANs

- *Expandability*

  Requirement specification on the mode of growth, namely, introduction of new services, increasing bit rate to be supported, installing new terminals, or expanding the size of the CN

- *Mobility*

  It is the ability to interchange terminal interfaces. This requires a universal interface.

Modularity

This is the provision of a flexible structure. That is, a network is not to be limited to a few applications only

Reliability

Sensitivity of CNs to errors such as bit errors, terminal failures, and user-induced errors. Reliability in crucial CN locations (such as in hospitals) is normally beefed up via redundancy. That is, the terminal equipment is duplicated within the large CNs

Physical performance

This refers to an optimum use of the physical medium. It includes the following: Coding efficiency, cable-length, and cost-effectiveness of the hardware involved

Cost

This is an important requirement of accepting the CN. Normally, the residential CNs are kept low in cost, and with other CNs, when expanded, the cost is increased incrementally.

7.3.3 Physical configurations of realizing CNs
Star configuration

As illustrated in Fig. 7.5, each terminal is directly connected to the B-NT2 by a dedicated line. The B-NT2 can either be a centralized system or a LAN-like structure where terminals are connected to a common medium via medium adapters.

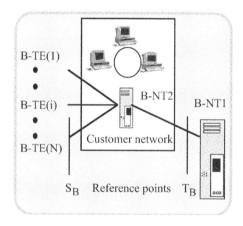

Fig. 7.5: Classical star configuration

Dual-bus configuration

Here, the terminals are directly connected to a common shared medium (Fig.7.6). These represent certain versions approved by ITU-T. The motivation for recommending these versions are:

✓   Simplicity in deployment
✓   Economy
✓   Plus-point considerations of B-ISDN.

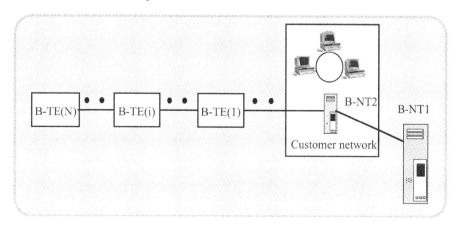

Fig. 7.6: Dual-bus configuration

Shared medium configuration

- This configuration allows (Fig. 7.7) extensions by adding a new terminal easily. (If a star configuration is used, the extension costs are high, as it may need an additional or a larger multiplexer.)

- Further, this configuration includes a medium access function. This is supported by the generic flow control protocol of the ATM layer. That is, the flow control protocol provides an orderly and fair access of the terminal to the shared medium by supervising the cell streams and assigning capacity to the contending terminals on a per-cell basis.

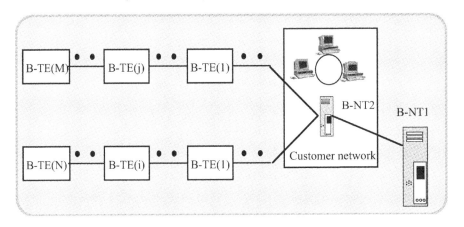

Fig. 7.7: Shared medium configuration

7.4   ATM Management Standards

The state-of-the art ATM management standards refer to the functions of the many existing management considerations of OAM&P. These management standards include the following:

7.4.1   Standards on ATM management

OSI: This corresponds to architectures based on *open system interface*, which defines the functions of *fault, configuration, accounting, performance, and security* (FCAPS)

ITU-T: This standard defines a physical and *local telecommunication management network* (TMN) architecture. It lays out a common structure for managing transmission voice and data networks. TMN is a standardized version of interoperable network management for implementations such as in the European RACE ATM technology. It is based on: OSI standardized *common management information service elements* (CMISE) and the associated *common management information protocol* (CMIP) for the Q-3 interface in the TMN. It has been advocated by ITU-T and ANSI.

OMG&TINA-C: These address ATM networks as objects and perform management from an object-oriented systems point of view

NMF: This is the *Network Management Forum* standard, which broadens the OSI functional model to reflect how providers may build the management systems

ATM Forum: The relevant standard defines a network management architecture corresponding to a structure for future *management information base* (MIB) and its related definitions.

TINA-C: This *Telecommunication Information Networking Architecture Consortium* provides methods for realizing an integrated management of all parts of a communication network.

7.5   Network Management Protocols

ATM networks are large and extensive. Therefore, network management is an essential support function of an ATM network. Such functions enable the network managers a look through the operations and performance of the ATM networks. Most ATM switches include the *simple network management protocol* (SNMP) described later for interoperation within the network management systems. (Some carrier-class switches may also use proprietary interfaces to facilitate proprietary management systems.)

7.5.1   Network management types

ATM management: Vendor-specified

The switches sold in the market may come with an option to purchase the management system designed to work with them. These switches have several built-in enhancement features towards network management (such as graphical user interfaces, and advanced LAN management for ATM workgroup switches). The vendor-specified management, however, requires a select group of equipment to make the network.

ATM management: Model-based

There are five defined interfaces of the ATM network management model. These cover the entire ATM environment, both at the LAN and WAN levels extending into private as well as public networks. They are the outcomes of the efforts due to the *Network Management Forum* (NMF) and the *Internet Engineering Task Force* (IETF).

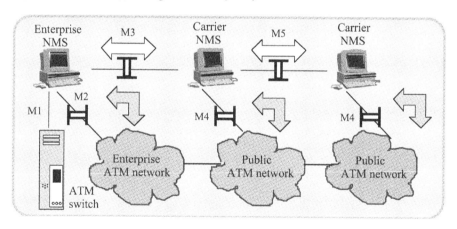

Fig. 7.8: ATM management: Model-based

The specifications with respect to the model depicted in Fig.7.8 primarily include the interface requirements, the logical management information base with fault, performance, and configuration objects. In Fig.7.8 the M3 interface is SNMP-based, while M4 includes both SNM and CMIP (described below) for carrier management.

ATM management: Cell-based

ATM network elements are designed to share the management information over a set of preassigned VCs. This is done via OAM cells, which enable ATM network devices to gather end-to-end statistics and reduce the number of needed MIBs. The defined OAM cells are:

    ✓   *Alarm indication signal* (AIS) cells
    ✓   *Far-end reporting failure* (FERF) cells.

Network switches generate AIS cells to communicate failures of switches or VCs. FERF cells communicate failure of one-half of a full-duplex ATM connection. The ATM Forum has also defined a special OAM *loopback cell*, which is useful to test new installations and network troubleshooting.

ATM management: Via monitoring

Two ATM Forum-specific management schemes refer to:

✓ ATM monitoring (AMON)
✓ Remote monitoring (RMON).

AMON supports replicating a VC ATM stream through a nonintrusive, real-time external monitoring device. RMON provides for embedded probes in ATM equipment so as to capture real-time information pertinent to the network operation.

Unique management issues concerning ATM

As defined earlier, network management includes the global aspects of OAM&P. It manages the planning, organizing, monitoring, and controlling of activities and resources within a network so as to provide a level of service that is acceptable to users. At the same time the associated cost is profitable to the network providers. When implementing these attributes specific to ATM, there are certain unique issues vis-à-vis the network management as listed below:

✓ Standards are still being developed
✓ Development of such standards are not in pace with technological changes
✓ ATM/B-ISDN is targeted to support older/existing legacy systems concurrent to supporting new functions and a variety of applications
✓ ATM includes all complexities and problems involved in LAN, MAN, and WAN, along with unknown and anticipated problems/complexities of those services being planned
✓ ATM aims at point-to-multipoint connections and multiplicity of application/services.

There are two options for ATM network management as illustrated in Fig.7.9. They are known as *centralized* and *distributed* architectures.

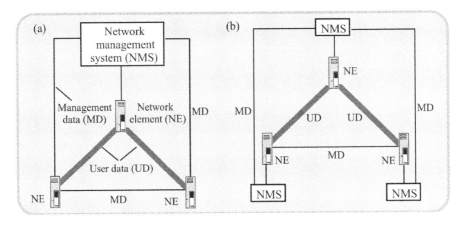

Fig. 7.9: Centralized and distributed architectures; (a) Centralized and (b) distributed

ATM network management frame work

The three general areas covered by the management of ATM networks are:

- ✓ *Interface management* including UNI, DXI, and LAN emulation
- ✓ *OAM frame work/layer management*
- ✓ *Global network management* for total management of ATM networks and services.

Interface management

This deals with the exchange of information at the interface level. It is used for configuration and alarms of ATM interfaces. It also facilitates the interfacing of different devices.

Layer management (OAM)

This is intended for continuity and loopback testing of a segment or at end-to-end VC/VP levels. It allows checking a user circuit through the network.

Within the scope of network management functions and services that interact to provide the necessary network management tools and control, the exclusive set of diagnostic and alarm reporting mechanisms defined by the ITU-T using special purpose cells is known as the *operation, administration, and management* (OAM) cells. The associated functions of OAM includes fault management, continuity-checkings, and performance measurement. These are indicated in the *ITU-T I.610 (1995) Recommendation on B-ISDN operation and maintenance principles and functions.*

OAM is the critical aspect of ATM networking. Its hierarchical levels refers to OAM flows associated with the physical and ATM layers. As indicated above, the functions of these flows refer to: Performance monitoring, failure detection, system protection, fault localization, and forwarding of failure information.

The ATM layer part of OAM functions are specified within the ATM cell itself. The physical layer part is indicated in the overhead structure of SONET/SDH.

Global management

This is concerned with the configuration of an ATM network using one or more switches and controlling/monitoring of ATM devices. It offers a top-down total management.

System constituents of ATM management frame work

The ATM *network management system* (NMS) is comprised of:

- ✓ Management station (manager)
- ✓ Agent
- ✓ MIB
- ✓ Network management protocol.

Manager: A *manager* serves as an interface for the *administrator* with the NMS. It converts the administrator's instructions to monitoring and control data required by the network elements. Relevant services involved are: Data analysis, fault recovery, and database to hold details on managed elements

Agent: An *agent* is a *network management entity* (NME). It is a software that performs network management related efforts; for example, collection of data on resources and storing related statistics locally. An agent responds to the commands of the administrator (manager)

MIB: This is the *management information base*. It is a collection of objects representing a particular aspect of a managed effort. For cxample, managers perform a monitoring task by retrieving the value of a particular object or change the settings of a network by modifying the value of the corresponding object

NMP: This *network management protocol* carries out the communication between the manager and the agents. The two NMPs used are:

     ✓   IETF specified *simple network management protocol* (SNMP)
     ✓   ISO specified *common management information protocol* (CMIP).

A network management protocol provides a means of communication by facilitating the following:

     ✓   Reading and updating the attributes of managed objects
     ✓   Requiring the managed objects to carry out specific functions
     ✓   Reporting the results stemming from managed objects
     ✓   Evolving and destroying the managed objects.

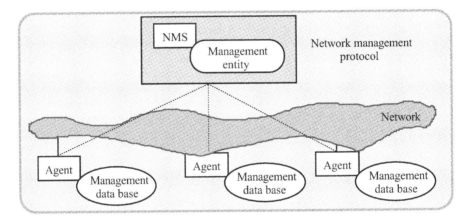

Fig. 7.10: Manager/agent model

The two standard management protocols, which employ the manager/agent model, are as follows:

7.5.2 *SNMP - based network management systems (NMS)*

Simple network management protocol (SNMP) defined in IETF (RFC-1157) refers to the *Internet network management protocol*. It provides a means to monitor status and performance as well as to set configuration parameters. SNMP uses *management information base* (MIB) as a data structure.

SNMP is widely used in TCP/IP networks. It represents a connectionless protocol. Its agents reside within the managed devices and are designed to function with minimal system resources. That is, SNMP will not let the devices slow down significantly in facilitating the NMS.

The functions of an agent in SNMP are:

     ✓   Gathering of data about the number and type of error messages received, number of bytes and packets processed by the device, the maximum queue length, etc.
     ✓   Storing the collected data in the MIB, which resides in the device itself.

The processing of collected data is done by the management application, which communicates with the managed devices via SNMP.

Essentially, SNMP is an application layer oriented protocol. It interfaces with the user datagram protocol at the transport layer. It can span across heterogeneous networks.

The basic services offered by SNMP are:

✓ Fetching of variables
✓ Storing of the variables fetched.

There are five SNMP commands that can be identified as follows:

✓ *Get*: This refers to fetching a value of a specified variable
✓ *Get-next*: This request fetches an anonymous variable (whose exact name is unknown). This is useful for going up the steps across a table
✓ *Get-response*: This is a response to a fetch command with the required data
✓ *Set*: This command stores a value via a particular variable
✓ *Trap*: This is concerned with sending a message in response to a particular event.

SNMP commands and responses contain an authentication (*community name*) to identify a set of management systems that is permitted to execute SNMP commands.
The essential parameters of SNMP are listed below:

✓ Physical layer type: DS-3, SONET, medium (UTP, fiber, coax etc.)
✓ Number of ATM VPs/VCs allowed, their active status, and the extent of each address field used
✓ Statistics on ATM cells received, sent or dropped
✓ Traffic descriptors (SCR, BT, QOS etc.)
✓ Setting of MIB parameters covering VPs/VCs on the UNI.

7.5.3 *Common management information protocol* (CMIP)

This protocol was developed by ISO. Unlike the SNMP, CMIP is designed to offer generic solutions to overall network management. However, this protocol is quite complicated and is largely used in service provider networks. The complication refers to the much involved command structures of CMIP (in comparison to those of SNMP). The CMIP is more comprehensive and versatile due to its inherent complexity.

The ISO specified management contains three entities:

✓ Layer management entities (LME)
✓ System management application entities (SAME)
✓ CMIP protocol.

LMEs

These operate at each OSI layer, monitoring the operation of the system at that specific layer. It does not govern the performance of the system as a whole. Such a function is vested with SMAEs.

SMAEs

Each managed object/device has its own SMAE. This assimilates the information collected by the LMEs and produces an overall view of the device functioning with respect to its networking functions.

CMIP

The SMAEs of different networks communicate with each other via CMIP. This enables the system administrators to gather and analyze data about the functioning of the network in its global output.

Management information base (MIB)

MIB defines objects in terms of primitives such as strings, integers, and bit-maps. It allows a simple form of indexing. Each object has a name, a syntax, and an encoding. MIB objects are defined as structures that allow organizational ownership of sub-addresses to be defined. MIBs can be regarded as types of information that can be accessed and manipulated in ATM interfaces, end systems, switches, and networks.

The information gathered by the agents is vital for network management since all activities are based on it. Therefore, MIB standards have been evolved to specify exactly what information is to be gathered and how it is to be stored.

These standards define network management variables and their meanings. The variables refer to the *structure of management information* (SMI). The functions of SMI are:

  ✓ Restricting the types of variables allowed to exist in the MIB
  ✓ Specifying the rule for naming them.

As per SMI, the MIB variables are defined as *ISO abstract syntax notation 1* (ASN.1). This has a readable notation for documentation and an encoded notation for communication protocols.

SMI enables defining the contents of the SNMP MIB. The structure is defined by the ISO branch of the global object identifier name-space tree (ITU-T/ISO).

ATM interface management

This is required for the configuration and securing status information on ATM interfaces. The interfaces defined are: UNI, DXI, B-ICI, and LAN emulation UNI (LUNI).

7.6 ATM Forum ILMI

7.6.1 *The UNI ILMI MIB structure*

ILMI is adopted for configuration and status information at a specific UNI. That is, ILMI allows ATM devices to exchange fault and performance management information over a UNI interface. It does not, however, provide security or accounting management functions. It simply supports the exchange of management information between the UNI management entities related to both ATM and physical layer parameters. The communication between adjacent UNI management entities so that one UNI management entity can access the MIB information associated with its adjacent UNI management entity takes place on the basis of SNMP commands. These commands are transported over AAL-5 and do not use IP or user datagram protocol (in contrast with original SNMP provisions). The ATM Forum framework does not advocate an agent on either side of the UNI. Instead, it has defined a UNI management entity that uses SNMP over AAL-5. The UNI ILMI MIB is depicted in Fig. 7.11.

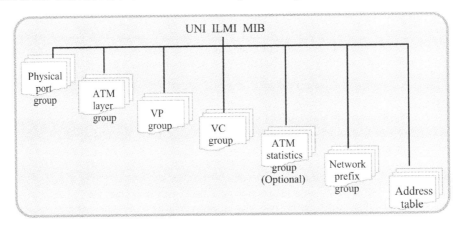

Fig. 7.11: The UNI ILMI MIB structure

The constituents of the MIB structure in Fig.7 have the following contents:

Physical port table:	Configuration status information (port index) address, transmission type such as OC-3, DS-3, etc., physical media type, and operational status
ATM layer group:	Maximum number of VPCs and VCCs, supported paths, path indicator bits, private/public port type
VP group:	Path indicator, QOS class supported in each direction, VP operational status
VC group:	Similar information as VP group but at VC level
Network prefix group:	Prefix on ATM address
ATM statistics group:	Counts on idle cells, cell-drops etc
Address group:	List of addresses associated with the UNI

Thus, the underlying aspect of the ILMI standard is the SNMP, which has now become the de facto standard for multi-vendor network management. In summary, ILMI defines an MIB for UNI, together with a protocol stack-SNMP running directly over AAL-5 (or, optionally, AAL-3/4), and a reserved VPI/VCI. The peer-to-peer management entities adopt this protocol stack to manage a single UNI. An ATM switch supports such multiple management entities and associated MIBs, one for each UNI. The ILMI is a stand-alone management element of an entire network. Further, the ILMI consists of both public and private switches, as well as the UNI.

In practice, the private and the public ATM switches are likely to support their own management agents and MIBs. These MIBs would contain information on the operation of the whole switch, not just restricted to individual UNIs. The associated management agents are to be monitored and controlled by *network management systems* (NMS), in both private and public networks.

In general, the protocols that have been adopted to communicate between the management agents and the NMS may vary. But, the private networks are very likely to use the SNMP. Further, the public networks will most probably go for the OSI *common management and information protocol* (CMIP), which is more common in the public network. The flow of communication between public and private NMSs is described by what is known as the *customer network management* (CNM). It provides users some limited access to information concerning the public UNI. It also enables control by the users on the public UNI. Such efforts are also to be based on SNMP as the CNM support for the SMDS and frame relay.

In reference to the management PDUs, either from SNMP or CMIP, they may be carried on reserved VPI/VCI values. This is done by using the ATM network itself, or via *overlay management networks*. The manner in which a particular switch is managed is generally proprietary. The use of publicly known MIBs and common protocols, however, help to ensure that a single NMS can manage a multi-vendor network. The prevailing indictions on the early adoption of industry standard protocols such as SNMP for the management of ATM networks is regarded as an encouraging development in the ATM management issues.

As indicated above, the ATM Forum-defined ILMI is based on SNMP. A default value of VPI = 0, or VCI = 16 has been chosen for the ILMI because CCITT/ITU has already reserved VPIs 0 through 15 (that is, the first 16) for future standardization. Alternatively, another VPI/VCI value can be manually configured identically on each side of the UNI for ILMI application. Use of this method is undesirable since it is not automatic.

The ILMI is specified to operate over AAL-3/4 or AAL-5 as a configuration option with support for a connectionless *datagram-oriented transport-layer protocol* belonging to the IP suite (UDP/IP) configurable option. Therefore, in order for the *UNI management entities* (UMEs) to interoperate, the AAL (either 3/4 or 5) and the higher layer protocol (either UDP/IP or Null 0) must be chosen.

Fig. 7.11 defines the structure of the ILMI MIB. Each interface has a set of object groups indicated by the branches in the tree. The physical layer group has an additional two groups for common and specific objects (or attributes). The interface index for accessing the data associated with a particular UNI interface is always a part of the object index. In the default interface index, zero means that the SNMP command refers to the UNI interface over which the SNMP message was received. Other interfaces can be referenced by supplying nonzero interface indices in SNMP requests. As a basic form of security access control, the network implementation should ensure that only interface indices applicable to a particular user UME can be accessed. Some object groups, such as VPC and VCC, have multiple entries per interface and have additional indices.

Two versions of the ILMI MIB have been specified in ATM Forum UNI specification version 2.0 and version 3.0. The version 3.0 MIB is backward compatible with the version 2.0 MIB.

The ILMI uses the standard systems group by reference, which supports items such as identification of the system name, and the time that the system has been up. The systems group also provides standard TRAPs, such as when a system is restarted or an interface failure is detected.

Each object in the group of the ILMI MIB tree is briefly addressed below. For more details, refer to the ATM forum UNI specification version 3.0.

- The physical layer group is *required*, *read only* and has an index, namely, the *interface index*, which refers to the following objects:

  ✓ An interface address
  ✓ A *port type identification*, such as DS-3, STS-3, 100 Mbps, or 150 Mbps Fiber is specified from a defined list
  ✓ A *media type identification*, such as coax, single-mode fiber, or multimode fiber specified from a defined list
  ✓ The *operational status* is specified as a choice from the following: *Other, in service, out of service, or loopback*
  ✓ Other *specific (adjacency) information* is specified to another MIB, such as the standard DS3 MIB.

- The ATM layer group is *required*, *read only*, and has an index, namely, the *interface index*, which refers to the following objects:

  ✓ An integer that specifies the *maximum number of VPCs*, which can be defined on this interface
  ✓ An integer that specifies the *maximum number of VCCs*, which can be defined on this interface
  ✓ An integer that specifies the *number of configured VPCs*, which can be defined on this interface
  ✓ An integer that specifies the *number of configured VCCs*, which can be defined on this interface
  ✓ An integer that specifies the *maximum number of VPI bits*, which are active on this interface
  ✓ An integer that specifies the *maximum number of VCI bits*, which are active on this interface
  ✓ ATM device type
  ✓ ATM public/private interface type indicator
  ✓ UNI/NNI signaling version
  ✓ ILMI version.

- The ATM statistics group is *optional*, *read only*, and has an index, namely, the *interface index*, which refers to the following objects:

  ✓ The total number of *received cells* on this interface
  ✓ The total number of *transmitted cells* on this interface that contains user data
  ✓ The total number of *dropped received cells*, which were dropped for one of the following reasons: Invalid header, HEC detected header error, or not configured.

- The network prefix group is *required* at the private UNI and *optional* at the public UNI *read/write*. The network prefix group has an index, which is the *interface index* plus the *network prefix*. It refers to the following objects:

  ✓ A variable length string of between 8 and 13 bytes defining the *network prefix* as either the NSAP or E.164 formats in the ATM Forum version 3.0 UNI specification

  ✓ A *network prefix status* that can have the value of either valid or invalid.

- The user part ATM address (or *ATM address*, for short) group is *required* at the private UNI and optional at the public UNI, *read/write*, and has an index, namely, the *Interface Index* plus the *ATM address*, which refers to the following objects:

  ✓ An *ATM address* of 8 bytes defining the low-order bytes of the NSAP signaling address format in the ATM Forum version 3.0 UNI specification

  ✓ An *ATM address status* that can have the value of either valid or invalid.

Fig. 7.12 illustrates the SNMP message flows associated with the address registration portion of the ILMI MIB. The *network prefix group* resides in the ILMI MIB on the *user side* of the UNI interface, while the *address group* resides on the *network side* of the UNI interface. The address group is not applicable for native E.164 addressing since the 8-byte network prefix completely specifies the address. All 20-byte *ATM end-system address* (AESA) formatted addresses have a 13-byte network prefix.

Address registration

This is done as a part of ILMI. It is the major aspect of the automatic configuration of *private network node interface* (PNNI) reachability information. This is required when using ATM SVC networking.

The underlying principle behind address registration is that it allows the network to communicate to the user the valued address prefixes for a particular logical ATM UNI. Subsequently, the user registers complete addresses by suffixing the address prefix with the *end system identifier* (ESL) and the *selector byte* (SEL) fields. The user may also register with each address connection scope.

By the above procedure, the user is relieved of manually configuring large numbers of user addresses. ILMI also facilitates source authentication, since the originating switch may screen the calling party address information element in the call setup phase message by comparing it against the set of registered address prefixes.

Registration occurs at initialization time (that is, a *cold start trap*), or whenever a prefix or address is to be added or deleted. At the time of initialization, the address and prefix tables are initialized to empty status. One side registers a prefix or address with the other side by first *SETTING* the address value and its status to be valid. The other side transmits a *RESPONSE* with either a *no error* or a *bad value* parameter in order to indicate the success or failure, respectively, of the registration attempt made. If a SET message is sent and no *RESPONSE* is received, then the SET request should be retransmitted. If the SET message is received again, then a RESPONSE indicating *no error* is returned. If a SET request is received, which attempts to change the status of an unregistered prefix or address to invalid, then a RESPONSE with a "*NoSuchName*" error is returned.

The address registration scheme via ILMI MIB is shown in Fig. 7.12.

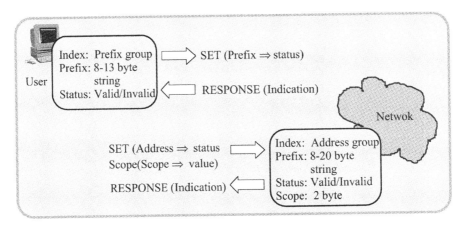

Fig. 7.12: Address registration via ILMI MIB

The function of the service registry MIB information part of the ILMI enables a general purpose service registry to locate ATM network services (like LAN emulation configuration server and ATM name server).

ATM Forum AESA formats

AESA formats are specified as per ATM Forum UNI version 4.0. The constituent parts of each address are as follows:

Initial domain part (IDP)
This consists of authority and format identifier (AFI) and initial domain identifier (IDI)

AFI field: *Identifies the format for the remainder of the address (1 byte field)*
IDI field: *Identities the network addressing authority responsible for the assignment and allocation of the AESA domain specific part.*

Domain specific part (DSP)
This consists of lower order DSP (LO-DSP) and higher order DSP (HO-DSP) fields. DSP is made of a field length of 20 bytes minus the size of IDP.

LO-DSP: *This contains an end system identifier (ESI) and a selector byte (SEL). Both ESI and SEL are identical for all IDI formats.*
HO-DSP: *ATM Forum UNI 3.1 has combined the routing domain (RD) and area identifier (AREA) fields into a single HO-DSP.*

There are three ATM-Forum defined AESA formats known as:

    ❑   *Data country code (DCC) AESA format*
    ❑   *International code designator (ICD) AESA format*
    ❑   *E.164 AESA format.*

These are illustrated in Fig. 7.13.

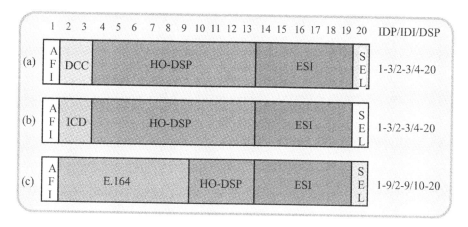

Fig. 7.13: AESA formats: (a) DCC; (b) ICD and (c) E.164

Group addresses

These are defined (Annex 5/ATM Forum UNI 4.0 specification) with respect to assigning a group address to more than one UNI port within a network.

When a user places a point-to-point call to a group address, then the network routes the call to the port "closest" to the source associated with the specified group address. That is, the network sets up the connections to any of several possible destinations. Hence, this capability is known as "anycast" signaling.

7.7   IETF AToMMIB

7.7.1 The ATM management information base

The scope of *ATM management information base* (called the AToMMIB) covers the management of ATM PVC-based interface, devices and services. Managed objects are defined for ATM interfaces, ATM VP/VC virtual links, ATM VP/VC cross-connects, AAL-5 entities, and AAL-5 connections supported by ATM end systems, ATM switches, and ATM networks.

The AToMMIB uses a grouping structure similar to the ILMI described earlier to collect objects referring to relating information and provide indexing. The AToMMIB defines the following group:

- ✓ ATM interface configuration
- ✓ ATM interface DS-3 PLCP
- ✓ ATM interface TC Sublayer
- ✓ ATM interface virtual link (VPL/VCL) configuration
- ✓ ATM VP/VC cross-connect
- ✓ AAL-5 connection performance statistics.

The *ATM interface configuration* group contains ATM cell layer information and the configuration of local ATM interfaces. This includes information such as the port identifier, interface speed, number of transmitted cells, number of received cells, number of cells with uncorrectable HEC errors, physical transmission type, operational status, administrative status, active VPI/VCI fields, and the maximum number of VPCs/VCCs.

The *ATM interface DS-3 PLCP* and *the TC sublayer group* provide the physical layer performance statistics for DS-3 or SONET transmission paths. This includes statistics on the bit error rate and errored seconds.

The *ATM virtual link* and *cross-connect groups* allow management of ATM VP/VC virtual links (VPL/VCL) and VP/VC cross-connects. The virtual link group is implemented on end-systems, switches, and networks, while the cross-connect group is implemented on switches and network only. This includes the operational status, VPI/VCI value, and the physical port identifier of the other end of the cross-connect.

The *AAL-5 connection performance statistics group* is based upon the standard interface MIB for IP packets. It is defined for an end-system, switch, or network that terminates the AAL-5 protocol. It defines objects such as the number of received octets, the number of transmitted octets, the number of octets passed on the AAL-5 user, number of octets received from the AAL-5 user, and the number of errored AAL-5 CPCS PDUs.

AToM MIB specifies M1, M2, and some of the M3 management interface. It describes how to configure a network and addresses of various AALs and switch-related management. It uses the SNMP version 2 format.

The associated framework needs an IP agent to reside on either side of the interface being managed. The manager gets management information from the agents using SNMP over UNI datagram protocol or IP.

The AToM MIB specific information defined for managing UNI/ATM devices within an ATM network and/or cross-connect nodes include the following:

- ✓ ATM interface configuration
- ✓ DS-3 interface
- ✓ Transmission convergence sublayer
- ✓ ATM traffic parameters
- ✓ ATM VP/VC links
- ✓ ATM VP/VC cross-connects
- ✓ AAL-5.

Example 7.1
Describe the VPC/VCC failure management/reporting procedure envisaged in ATM.

Solution
It is well known that a physical connection in a telecommunication system may experience a failure. Similarly, the virtual connections (VPC/VCC) may also fail due to:

- ❑ Failure of the physical link that supports VPC/VCC
- ❑ Invalid VPC/VCC translation
- ❑ Failure in delineating the ATM cells from the payload of underlying physical links.

When a VPC/VCC failure is detected, an *alarm indication signal* (AIS) must be forwarded to the local network management system (to initiate restoration/repair) and to various nodes along the failed connection.

The AIS to the nodes along the failed connection is done by generating an AIS at the intermediate node, (which has detected a failure) and passed on to alert the upstream nodes of the failure; alternatively a *far-end received failure* (FERF) signal is generated by the node terminating the failed connection to alert the upstream nodes of a failure in the downstream side.

Fig. 7.14 illustrates the proliferation of alarm indications.

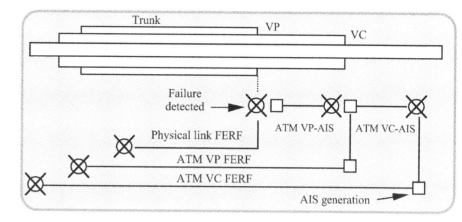

Fig. 7.14: Proliferation of AIS

Example 7.2

What is meant by VPC/VCC OAM *cell loopback* capability?

Solution

The OAM cell loopback capability permits operations information to be inserted at one location along a VPC/VCC connection and returned (or looped back) at a different locale without interrupting the service.

This is carried out as follows: An OAM cell is inserted at any accessible point along the virtual connection, namely, at a local end-point or intermediate point with instructions in the OAM cell payload for the cell to be looped back at one or more other identifiable points along the connections.

Four different possible cases of VPC/VCC OAM cell loopback testing are as follows (Fig. 7.15):

❑　　Case (1)

　　　OAM cell is inserted and looped back within the same single network

❑　　Case (2)

　　　This is an end-to-end OAM cell loopback that extends across multiple networks

❑　　Case (3)

　　　The OAM cell is inserted at the edge of one network and looped back at an end-point of the virtual connection

❑　　Case (4)

　　　This is an end-to-end loopback in which an OAM cell is inserted at an intermediate point with necessary instructions for it to be loopback at both end-point locales in series.

OAM cell loopback enables functions such as preservice connectivity verification, fault sectionalization, and on-demand cell-delay measurements.

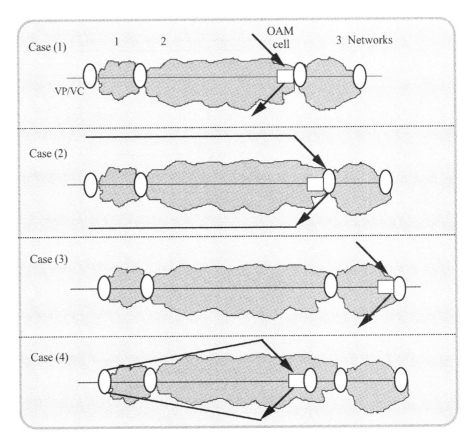

Fig. 7.15: OAM cell loopback case (1) to case (4)

7.8 Concluding Remarks

The OAM&P considerations in reference to networks are mainly specified to define a comprehensive approach in order to realize an effective and reliable teletransfer communication. Particular to ATM, such considerations account for the correlatory attributes of network configuration and traffic performance under broadband situations. The heterogeneous traffic and increasing consumer applications pose a lasting demand on OAM&P aspects of ATM only to be met with evolving standards, foreseeable regulations, and growing technology in the perpetual framework of information technology.

Bibliography

[7.1]   D. Minoli and T. Golway: *Planning & Managing ATM Networks* (Manning Publications, Inc., Greenwich, CT: 1997)

[7.2]   D. Ginsburg: ATM – *Solutions for Enterprise Internetworking* (Addison-Wesley Longman, Ltd., Harlow, England: 1999)

[7.3]   R.O. Onvural: *Asynchronous Transfer Mode Network – Performance Issues* (Artech House, Inc., Boston, MA: 1995)

[7.4]   M. Toy: *ATM Development and Applications – Selected Readings* (IEEE Inc., Piscataway, NJ: 1996), Chapter VII

[7.5]   D.E. McDysan and D.L. Spohn: *ATM Theory and Applications* (McGraw-Hill, Inc., New York, NY: 1998)

8

ATM NETWORKING: IMPLEMENTATION CONSIDERATIONS

ATM: Is it "ready or Net"?

"(ATM has a) vaunted ability to deliver QOS"

K. Dillon
"Effortless Expansion for ATM Networks"
in *Data Communications*, November 1998

8.1 Introduction

8.1.1 ATM networking: The past, the present, and the future

Since the mid-1990s, ATM networking has been wooed by technology-driven companies as a multiservice architecture. It has been conceived and used to support a variety of telecommunication including frame-relay, native LAN, and FDDI interconnections. These are essentially ATM-based technologies, which have entered the telecommunication networking scenario as forerunners to a host of more ATM services being deployed by user organizations.

The role of ATM technology in modern telecommunications can be viewed in the context of broadband demand, anticipated requirement of existing and futuristic networks, and the expected evolutions in teletransfer techniques.

Today, ATM is regarded as a major team player of information technology in meeting the bandwidth demands of corporations and institutions. Its bandwidth scalability is expected to mediate the anticipated traffic growth versus the quest for more and more network capacity. Such bandwidth needs are being facilitated via fiber optics, but "managed" by ATM efficiently so as to cater to the needs of multiple users and applications [8.1-8.12].

Apart from being an efficient bandwidth manager, ATM technology also bears a variety of features and plus-points conducive to meeting the expectations of evolving trends in telecommunications as discussed in earlier chapters. Its logical level management blended with fiber-optics-based physical level transmissions render ATM networking a comprehensive conception of the Gbps rate communication platform, offering varying levels of service guarantees.

Within the unlimited scope that prevails, ATM enjoys today, in its outlay and conception, strategies relevant to the networking aspects of:

- ✓   Changing the existing networks into ATM-centric
- ✓   Native ATM implementations.

In such implementations, who could be considered the patrons of ATM networking? In the existing networking environment and in the foreseeable applications, the users/patrons of ATM are:

- ✓   Commercial enterprises
- ✓   Governmental agencies
- ✓   Educational institutions
- ✓   Medical facilities
- ✓   Industrial/manufacturing sectors
- ✓   Domestic users.

The needs of these customers are many. The associated applications of ATM are kaleidoscopic. They may represent a potpourri of voice, data, video, telemetry, signaling, paging, fax, ISDN, conferencing, multimedia, and other information-objects created and manipulated by a set of new and emerging classes of applications.

In the years to come, ATM will become an inevitable and integral part of nearly every telecommunication networking user environment. ATM-enabling, migration-to-ATM, and deploying native ATM are and will be, the desiderata of telecommunications. They would represent *"intelligent networks"* dealing with service classes, which would support almost any unpredictable information flow.

Described in the following sections are ATM devices, the art of ATM networking, and design considerations of practical implementations.

8.1.2 ATM LAN

As the ATM concept was brought into practice, an imminent application that was foreseen was implementing ATM LAN. Generally, the first generation LAN is comprised of CSMA/CD and token-ring LANs, and the second generation is the FDDI. ATM LAN, therefore, refers to the third generation of LANs. It is designed to provide aggregate throughputs and real-time transport guarantees that are needed for wideband, high-speed, and multimedia applications. The typical requirements of this third generation LAN are:

✓ Supporting multiple, guaranteed classes of services. For example, live video may require a guaranteed bandwidth connection and a file transfer program that utilizes a background class service warranting a low BER transmission

✓ Providing a scalable service throughput. (That is, the throughput may grow with respect to host and aggregate capacity requirements.)

✓ Facilitating a seamless internetworking between LAN and WAN technology.

ATM LAN, in general, has been conceived and the relevant designs have been focused to cope with all the requirements as above cohesively.

ATM LAN: Definition

ATM LAN implies the use of ATM as a data transport protocol anywhere within the local premises.

Type of ATM LANs

✓ *Gateway-based ATM LAN*: This refers to a LAN system that uses an ATM switch that acts as a router and traffic concentrator for linking a premises network complex to an ATM LAN

✓ *Backbone ATM switch-based ATM LAN*: This is a LAN system comprised of a high-performance multimedia set of workstations and other end-systems, which connect directly to an ATM switch.

Apart from the aforesaid "pure" systems, a hybrid of two or more types of such networks can also be used to make an ATM LAN. Typical ATM LAN configurations are illustrated in Fig. 8.1.

The backbone ATM LAN is designed to include links to the outside world and the switches used must have buffering capacity for speed conversion compatible for mapping the data rate of LAN to ATM. Further, these switches must perform protocol conversion (for example, MAC protocol of LAN to ATM cell stream).

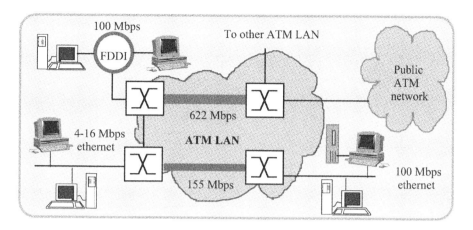

Fig. 8.1: ATM LAN configurations

An ATM LAN configuration yields for scalability by adding the following:

    ✓   Higher capacity
    ✓   Higher number of switches with more throughput in switches
    ✓   Higher data rate of trunks between switches.

The backbone of ATM LAN is a simple system, but it does not address all of the needs of local communication. For example, the end-systems (workstation, servers etc.) may still remain attached to shared-media LANs with the corresponding limitations on data rate imposed by the shared medium.

A better approach than ATM backbone architecture is to use hubs as depicted in Fig. 8.2. Here, an ATM hub includes a number of ports that operate at different data rates and use different protocols. A hub contains several rack-mounted modules, with each module containing ports of a given data rate and protocol.

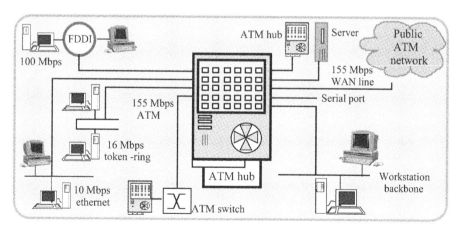

Fig. 8.2: ATM LAN hub configuration

The differences between the backbones of ATM LAN and ATM LAN hub configurations refers to the way in which individual end-systems are handled. In the ATM hub, the end-system has a dedicated point-to-point link to the hub; and each end-system has a communication hardware and software to interface with a particular type of LAN.

ATM hubs (Fig. 8.2) are often employed in a hierarchical manner to concentrate access for many individual users to a shared resource such as a server or a router as shown in Fig. 8.3.

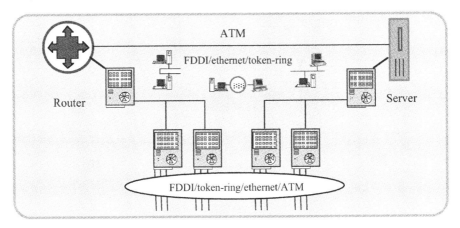

Fig. 8.3: ATM LAN architecture: Interface and functions of a hub-based system

In ATM LANs, ATM adaptation is rendered via *ATM interface cards* on the end-systems. Such cards support a wide range of industry standard interfaces, such as UTP, DS-1, DS-3, 100 Mbps (4B/5B), 140 Mbps (4B/5B), 155 Mbps fiber channel, OC-3, or even OC-12. They may also typically come equipped with their own microprocessors. Hub cards have a broad range of functionality, including PVCs and SVCs, multicast (point-to-point) and broadcast, AAL-3/4 and AAL-5 processing, and guaranteed QOS; and they can support TCP/IP and ATM application programming interfaces (APIs), and are SNMP MIB and CMIP compliant. ATM adapters are designed to support a variety of system buses, such as EISA, ISA, VME, etc.

Why does a user need an ATM interface card on the end-system?

The threshold of ethernet switching is 10 Mbps provided to each user and integration of applications (such as phone, computer, and video) would, however, require the capacity of ATM in the LAN environment. Hence, providing ATM to the desktop is accomplished by means of an interface card.

Other local devices of ATM LAN

Other local ATM devices include *ATM multiplexers/concentrators*, *ATM bridging devices*, and *ATM channel service units/digital service units* (CSU/DSUs). Fig. 8.4 depicts relevant interfaces and functions of these devices. An ATM multiplexer takes multiple, often lower-speed ATM interfaces and concentrates them into higher speed ATM trunk interfaces. A bridging device takes a bridgeable protocol, such as ethernet or token-ring, and connects it over an ATM network. This enables the user devices to feel as if they were on the same shared medium, or segment, as shown in the Fig. 8.4. The CSU/DSU takes the frame-based ATM DXI interface over a *high-speed serial interface* (HSSI) and converts it into a stream of ATM cells.

A switch can perform multiplexing; however, multiplexers are designed exclusively and are usually less expensive than switches because they have fewer functions to perform. Examples of ATM multiplexers or concentrators include the following:

A *bridging device* that encapsulates a bridged protocol, such as ethernet, emulating the bridging functions of the encapsulated protocol. These functions refer to *self-learning* and *self-healing capabilities*.

The CSU/DSU performs the conversion from an HSSI DTE/DCE interface operating up to 50 Mbps utilizing the frame-based ATM DXI protocol to an ATM UNI interface.

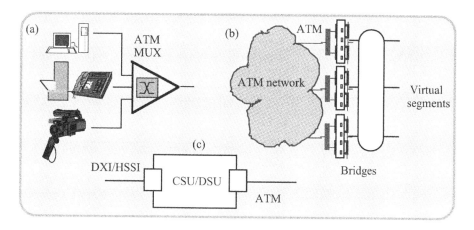

Fig. 8.4: (a) ATM multiplexers, (b) bridges, and (c) CSU/DSUs

ATM software

The associated software architecture of the ATM environment is illustrated in Fig. 8.5. In general, the software architecture includes the following constituents:

- A control processor software to perform:

    ✓   ATM cell processing
    ✓   AAL processing (for example, AAL-3/4, AAL-5).

- A host computer software to function as:

    ✓   A device driver, which performs:

        ⇒   ATM signaling and addressing including proprietary protocol considerations
        ⇒   Functions of an SNMP agent.

- An interface to upper level protocol modules in the operating system

        ⇒   TCP/IP, and proprietary *application programming interface* (API).

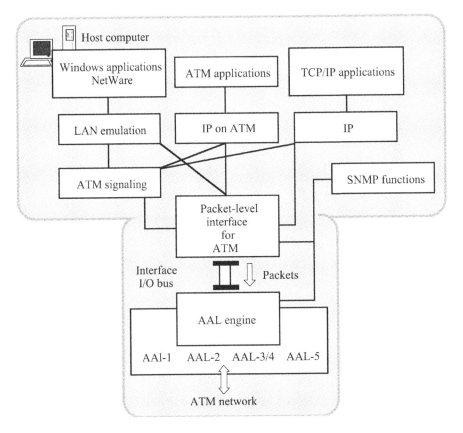

Fig. 8.5: ATM software architecture

An ATM LAN example

Due to its scalability in data rates and network throughput, much of the demand for ATM LAN technology has been influenced by computer-based applications that require more bandwidth than what the conventional LAN schemes can offer. One such example is a campus local area network, wherein the network often has to handle the rising user population as well as transmitting heterogeneous data streams such as voice, video, and data.

In a particular project at a university campus, for example, a high-speed ATM-based network can be introduced to serve one or two campuses. The switches implemented can operate at speeds ranging from Mbps to Gbps and each may support several hundreds of interfaces, which are connected to several workstations and servers. This network can also be connected to remote sites. The first phase of the project could be primarily an experimental stage, serving as a foundation for further developments. The second phase can be focused on creating the necessary components for the ATM campus-wide network.

In the ground-zero stage of the project, the feasibility of the technology can be demonstrated. Furthermore, it can be applied to a variety of high-speed applications. A network can be built consisting of a few multi-port broadcast packet switches. The switching network design may consist of a *copy network* (CN), a *routing network* (RN), and *broadcast translation circuits* (BTCs). Typically, a cell entering the system has several new fields added to it. These fields contain information required for processing a cell within a switch. The copy network makes copies of multipoint cells. Once the cell-routing field is selected, the cell is sent to the final destination. To make the ATM network "scalable" and efficient, a signaling system consisting of a communication protocol and network control software can be also developed.

Observations made from the results of the ground-zero stage may indicate certain merits and demerits of the ATM LAN implementation at the campus level. Typical observations may correspond to:

- Advantages of the implemenattion, such as:

  ✓ Conceivable minimum cell loss and cell delay characteristics
  ✓ Effective reduction in manufacturing costs as a result of the simple packaging adopted.

- Disadvantages of the implementation, such as:

  ✓ In spite of the possibility of reduced manufacturing costs, a tendency towards increased cost resulting from marketing and engineering strategies
  ✓ Considerable time involved in testing the network prior to its commission.

An ATM LAN based on a network of switches and links to hosts should be designed such that the bandwidth of the network increases with the number of hosts, unlike conventional LANs such as FDDI. An ATM LAN should meet the standards of existing networks, some of which can be listed as follows:

  ✓ Support for existing protocols such as TCP/IP
  ✓ Easily reconfigurable so that connection/disconnection of the hosts may require no administrative intervention
  ✓ Reliability and robustness of the network playing a vital role
  ✓ Perform efficiently so that the existing paths are used to their highest capacity before access is denied to a host
  ✓ Support network management frameworks currently being used.

Besides the generic LAN requirements as above, the ATM network is also expected to display the following capabilities:

  ✓ Support multiple classes of services
  ✓ Provide full bandwidth multicasting
  ✓ Have an application programming interface since ATM offers some unique capabilities such as bandwidth selection
  ✓ Support standard adaptation layers.

Hardware considerations

Each connection in an ATM LAN is essentially a "point-to-point" link between switches and hosts. Consequently, two kinds of hardware need to be designed:

  ✓ *ATM interface*: This allows the host to connect to the network. Programmable I/O interface cards have to be designed accordingly
  ✓ *Local ATM switches*: These switches serve as nodes for the network. They have two basic functions, namely:

    ⇒ Virtual connection management performed by software
    ⇒ Cell-routing, performed by hardware.

Software considerations

The relevant software considerations vis-à-vis ATM LAN include the following:

✓ Switch control software functions to perform the "switch side" capabilities of the signaling protocol
✓ Device driver software functions to implement adaptation layer functions
✓ Applications programming interface functions invoking the ATM application software
✓ Software capabilities of SNMP agents for network management and collecting information/statistics on the network's usage.

Signaling protocol

To cope with the evolving aspects of ATM standards on setting up/removing virtual channels on demand, simple protocols for ATM network signaling for ATM LAN environments have been considered and designed on a proprietary basis. Standard signaling protocols, though currently not available, can still be applied once they are developed, since the network can handle multiple signaling schemes. The proprietary designs evolved have been kept simple with supporting features of both conventional LANs and ATM LANs; also, the protocol messages are specified clearly.

ATM LAN applications

ATM LANs can be used for a variety of applications such as:

✓ *Multimedia applications*: Contrary to the conventional LAN schemes like FDDI whereby the performance degrades with increasing network load, ATM LANs depict a consistent performance regardless of any multimedia communications among the end-systems
✓ *Distributed computing*: Time taken to transmit and receive a cell over the ATM LAN is very low, around 25 microseconds. Proprietary ATM LAN switches and interface cards offer better performance compared to conventional LANs. It is also possible to interconnect ATM LANs when wide-area ATM services are provided.

The trends in fusing LAN and ATM call for engineering development and research efforts largely pertinent to traffic management problems posed to network planners. For example, suppose a LAN Internetwork for a location has to be planned. It may be required to forecast the network applications for the near future. Such forecasting is required to foresee the telecommunications needs and plan the system so as to enhance the existing network capability.

A typical assignment considering an ATM LAN expansion plan, for example, may include (and not be limited to) the efforts to provide for:

✓ Digital telephones
✓ Administrative e-mails/file-transfers
✓ Video telephone (picture quality)
✓ Video conferencing at desktop level (video, text, and voice)
✓ Accessing remote database, including multimedia, etc.
✓ Interworking with other campuses.

8.1.3 Designing an ATM network

Designing an ATM internetworking can be viewed essentially in its prospects of being deployed as follows:

✓ ATMs in LAN environments
✓ ATMs in WAN environments
✓ ATMs in public networks.

In order to understand and appreciate the underlying considerations of using ATM technology in the aforesaid environments, it is essential to first study the network expansion principle, as described below.

The networking expansion preliminaries refer to formulating a planning stage design-concept on the implementation profile as well as on hardware requirements based on traffic forecast considerations. As an illustration, the following example, pertinent to a LAN Internetwork can be considered to understand the implementation philosophy.

Example 8.1

Suppose a network planner of a LAN Internetwork is required to forecast the requirements and capacity of an existing network being expanded to meet certain anticipated new applications and needs.

Consider the following scenario of applications and discuss their expansion feasibility vis-à-vis compatible networking:

- ❑ Administrative e-mail
- ❑ Document distribution
- ❑ Text (file) transfer
- ❑ Basic telephony
- ❑ Browsing a data base.

Solution

In the expansion plan, two stages of expansion are assumed. The first stage may refer to a near-term strategy and the second stage is to be done after the full implementation of the first stage.

Application	Present status	Near-term expansion (Stage 1)	Stage-2 expansion
E-mail	10 Mbps ethernet	10 Mbps ethernet	100 Mbps FDDI ring
Document distribution	File transfer via ethernet	50 Mbps upgraded wiring	Heavy-duty serve access via 100 Mbps FDDI ring
Telephony	Basic T-1 service	Basic T-1 service	Upgraded T-3 service to support video telephony
Data base browsing	Via ethernet access	Via ethernet access	Upgraded database, which may include multimedia information access via the FDDI ring.

Any networking expansion strategy would involve, in general, the following considerations in respect of existing and future deployments:

- ✓ End-entities and customer premises equipment (CPE)
- ✓ Connections depicting the physical media, which are used in the intra- and/or interconnections across the network
- ✓ Information rate in bps

✓ Service class
✓ Management and protocol issues
✓ Access specifics.

In an expansion scenario, the network planner should perform a structured analysis of the existing systems and the expansion being considered. Relevant points of procedural interest and necessary predesign documentation are as follows:

- Description of customer's applications

  ✓ Name of each application that the customer runs over the network
  ✓ Type of the application such as database, multimedia, e-mail, CAD/CAM etc.
  ✓ Potential number of users who access each application
  ✓ Number of servers or hosts that enable the required applications
  ✓ Specific remarks on scalability and migration trend from an application.

- Description of network protocols

  ✓ Name of each protocol on the network
  ✓ Type of the protocol, such as client/server, session-layer etc.
  ✓ User profile using each protocol
  ✓ Number of hosts/servers deploying each protocol
  ✓ Specific remarks on scalability concerns and migration trends from a protocol etc.

- Description of identifiable bottlenecks

  ✓ A bottleneck on the network refers to the condition caused by a traffic that does not have a source or destination on the local network segment (nonlocal traffic)
  ✓ Identification of network segments in terms of the number of segments and their logical names
  ✓ Determination of the percentage of traffic on the source and the destination, both being local
  ✓ Determination of the percentage of traffic on the local source and a nonlocal destination
  ✓ Determination of the percentage of traffic on the nonlocal source and a local destination
  ✓ Determination of the percentage of traffics on the source and a destination both being nonlocal.

- Description of business constraints

  ✓ Understanding of the corporate structure and the associated information flow
  ✓ Identification of mission-critical data and/or information technology aspects
  ✓ Existing policy of the corporation on vendors, licensed applications, protocols, and platforms
  ✓ Proprietary considerations, if any
  ✓ Administrative constraints on local networking. For example, the registrar's office in a university dealing exclusively with student records, or firewall protection to a segmented part of a network

    ✓   Identification of the business customer's competitive profile and the related implication of the involvement of information technology.

- Description of existing network availability aspects

  ✓ Knowing the current network downtime and *mean time between failure* (MTBF) factors
  ✓ Locating the fragile and failure-prone segment of the network
  ✓ Determining the loss of revenue to the customer resulting from networking failure and downtimes
  ✓ Identifying the causes for the failures encountered and listing possible solutions in the changes to be made.

- Description of network performance

  ✓ Listing results of the host-to-host performance characteristics of the network
  ✓ Identification of possible improvements to the performance.

- Description of existing network

  ✓ Gathering statistics on network segments using tools like protocol analyzer, network monitor, and network management tools
  ✓ Collection of the following information (per segment per day):

    ⇒ Total megabytes handled
    ⇒ Total frames handled
    ⇒ Total number of physical layer errors (such as CRC errors) observed
    ⇒ Total number of MAC layer errors (such as collisions, token-ring soft error and FDDI ring operations)
    ⇒ Total number of multicast and broadcast frames handled.

  ✓ Computation of average and peak network utilization

- Description of network utilization

  ✓ Finding average network utilization on an hourly basis by configuring the monitoring tool. Typical network utilization may include the following:

    ⇒ Relative network utilization
    ⇒ Absolute network utilization
    ⇒ Average frame size
    ⇒ Multicast/broadcast rate.

- Description of the status of the routes

  ✓ Identification of interfaces
  ✓ Identification of buffers
  ✓ Identification of processes
  ✓ Characterization of router performance per a specified time in terms of:

    ⇒ CPU utilization
    ⇒ Input/output queue drops
    ⇒ Missed/ignored packets.

- Description of existing network management systems and tools

  - ✓ Documenting the type of platform and network management tools used
  - ✓ Gathering of results and past status.

- Description of the status of existing internetwork

  - ✓ Checking the network's health
  - ✓ Ascertaining the current thresholds on:

    - ⇒ Type of traffic
    - ⇒ Applications
    - ⇒ Internetworking devices
    - ⇒ Topology
    - ⇒ Criteria on acceptable performance.

  - ✓ Prognostic aspects on network's performance such as:

    - ⇒ Extent of saturation (in terms of a specified percentage of network utilization) on shared ethernet segments (up to 70%), WAN links (up to 70%)
    - ⇒ Response time (≤ 100 ms)
    - ⇒ Extent of broadcast/multicast traffic (limited to 20%)
    - ⇒ Extent of CRC error in units of 106 bytes of data
    - ⇒ Extent of collision (< 0.1% of the packets) in ethernet and token-ring segments
    - ⇒ Ethernet of at least one ring operation per hour on FDDI rings
    - ⇒ Extent of other-utilizations of routers
    - ⇒ Extent of output queue drops (<100/hour) and input queue drops (< 50/hour)
    - ⇒ Extent of buffer misses (< 25/hour)
    - ⇒ Extent of ignored packets (< 10/hour) on any interface.

8.1.4   Network design: A hierarchical approach

The hierarchical design approach of internetworks refers to modeling the design strategy via layers. For example, the OSI model allows the design of two computers to communicate via an internetworking on a layer-by-layer basis. It provides for a modular concept that allows creating a design element, which can be modified as necessary. The hierarchical approach reduces the complexity of design. Also, it facilitates fault diagnostics on a localized basis.

The hierarchical network design can be done via three strata as illustrated in Fig. 8.6. In reference to Fig. 8.6, the functional attributes of each structure are as follows:

Core stratum
This is the backbone part of the network enabling high-speed considerations. It should offer high reliability, facilitate redundancy and fault tolerance, adapt itself swiftly to changes, cause low latency, bear minimal slow filtering and other processes, offer good manageability, and spread across a well-defined, but limited perimeter of operation.

Distribution stratum
This refers to the demarcation boundary between the access and core layers of the network. Its functions include policy aspects (such as ensuring that a traffic from a network will be directed to a specified interface), security, address, workgroup access, domain definitions on broadcasting/multicasting, routing between VLANs, media translations (for example, between ethernet and token-ring), redistribution between two different routing protocols, and specifying the boundary between static and dynamic routing protocols.

Access stratum

This is a layer that enables a user to access a local segment in a network. It supports both switches and shared bandwidth LANs in a campus environment. In home/office environments, this stratum governs access to remote sites into corporate networks using WAN technologies such as ISDN, frame-relay, and leased lines.

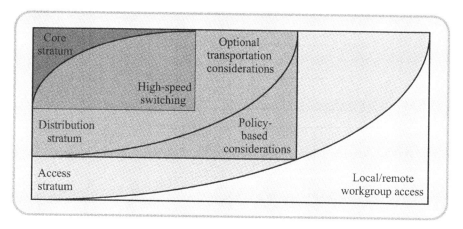

Fig. 8.6: Hierarchical network design

In reference to ATM networking, the aforesaid stratified network architecture is illustrated in Fig. 8.7.

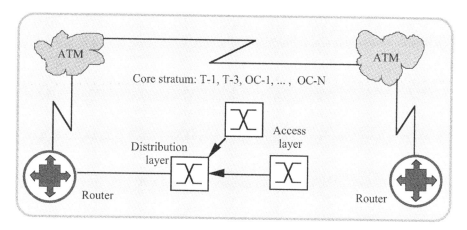

Fig. 8.7: ATM network topology

Consistent with the above considerations, the design perspectives of implementing ATM networking architecture are indicated in the following sections.

8.2   ATM Architecture Alternatives

As a part of ATM networking architectures, the CPE deployed represents vendor products compatible for ATM networks. They are either premises-based or intended for public/private networks. Their use in the ATM networking can be identified in the different environments enumerated below in reference to relevant architecture alternatives.

8.2.1 ATM in LAN environment

Commensurate with the existing vendor products of CPEs such as hubs and routers (described in Chapter 7), an ATM networking in LAN environments can be realized in different configurations. The CPE, in this case is essentially a multiprotocol routing device with ATM capabilities (especially in traffic management). A typical ATM routing used in a backbone architecture is illustrated in Fig. 8.8 as an example.

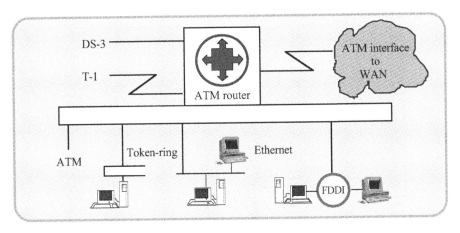

Fig. 8.8: ATM router based backbone architecture

In Fig. 8.8, the interfaces to the router may include all the LAN interfaces, such as *digital subscriber interface* (DSI), *digital exchange interface* (DXI), *high-speed serial interface* (HSSI) etc.

The architecture alternatives of ATMs that allow interoperability with various network environments are as follows:

✓ Router-centric
✓ ATM-centric
✓ Router plus ATM backbone architecture
✓ LAN emulation (ATM Forum LANE 1.0 1995)
✓ Classical IP over ATM (CLIP)
✓ Multiple protocols over ATM (MPOA)
✓ Voice and telephony over ATM (VTOA)
✓ Internetworking frame-relay (FR) and ATM
✓ Interworking frame-relay (FR) and ATM.

Router-centric architecture

Here, a set of routers form the *backbone architecture* and the ATM switches constitute *access vehicles*. The routing is performed by the backbone and local switching functions are carried out by the ATM switches as illustrated in Fig. 8.9.

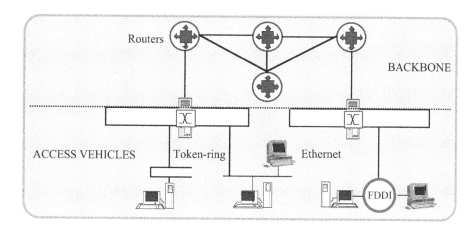

Fig. 8.9: Router-centric architecture

ATM-centric architecture

In Fig. 8.10, the ATM switching nodes depict the backbone and the routers enable the access. This ATM-centric view is considered as the provisioning architecture and is expected to be deployed in the public network.

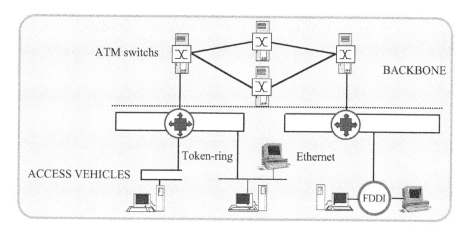

Fig. 8.10: ATM-centric architecture

Mixed architecture

This refers to a combined strategy of blending routing and ATM-backbone architectures. This is used as a transition stage across router-based and ATM-based backbones. These are popular in enterprise networking involving multiple locations. A typical mixed architecture is depicted in Fig. 8.11.

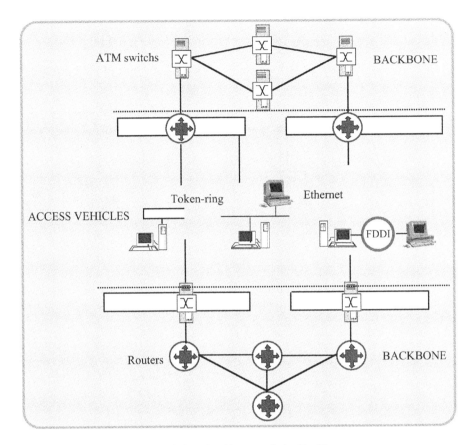

Fig. 8.11: A balanced architecture of mixed backbones

8.2.2   LAN emulation (LANE)

LANE is a technology (designed as a standard by ATM Forum) that network designers can deploy to internetwork the legacy LANs (such as ethernet and token-rings), with ATM-attached devices. That is, LANE is a technology that allows an ATM network to function as a LAN backbone. In this context, the ATM network must provide for multicast and broadcast support, address mapping (MAC-to-ATM), SVC management, and a usable packet format.

The need for LANE relies on the fact that the existing vast local area networking adopted for data transport be made to coexist in a network design that opts to use ATM devices. LANE uses ATM to provide virtual interconnection where an intelligent peripheral address encapsulation occurs. Here, the router-generated IP addresses are mapped onto ATM cells and transported across the ATM network using one of the ATM VCCs, thus facilitating a LAN connectivity.

LANE uses MAC encapsulation (OSI layer 2) because this approach supports the largest number of existing OSI layer 3 protocols. The culminating result is that all devices attached to an *emulated LAN* (ELAN) appear to be over a bridged segment. Hence, whether the protocol is Apple-Talk or IPX, it has the same performance features as the other, as in a traditional bridged environment. A typical LANE network is illustrated in Fig. 8.12.

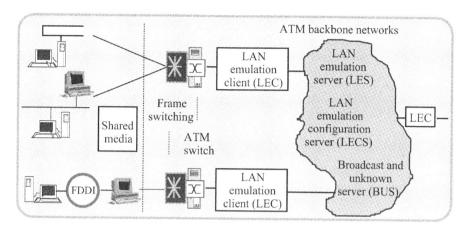

Fig. 8.12: ATM LANE network

Emulated LANs (ELANs)

ELAN refers to a logical network defined by the ATM Forum LANE specification comprising both ATM and attached legacy LAN end-stations. It represents the ATM segment of the virtual LAN (VLAN). (Note: VLAN is described below).

With the LAN emulation protocol architecture, it is possible to set up a number of logically independent, emulated LANs. An emulated LAN can support a single MAC protocol of which two types are currently defined:

❑   *Ethernet/IEEE 802.3*
❑   *IEEE 802.5/token-ring.*

Possible combinations obtainable from an emulated LAN are:

❑   *End-systems on one or more legacy LANs*
❑   *End-systems attached directly to an ATM switch.*

End-system on an emulated LAN should have a unique MAC address. Communication between end-systems on different emulated LANs is possible only through bridges or routers.

VLANs ..

The VLAN is defined as a network architecture that allows geographically distributed users to communicate as if they were on a simple physical LAN by sharing a simple broadcast and multicast domain. ATM Forum LANE supports VLANs, which in essence, specify a logical rather than physical grouping of devices. The devices are grouped using switching management software so that they can communicate as if they were attached to the same wire, when in fact they might be located on a number of different physical LAN segments. Because VLANs are based on logical instead of physical connections, they are highly flexible.

A VLAN consists of an ELAN segment along with legacy LAN segments. VLAN represents a logical collocation of users to share resources of distributed LAN-based server and application resources that may emerge and spread across multiple sites. Since VLAN enables the users on different physical LAN media to communicate with each other, it would appear to the users that they are all physically on the same LAN medium. That is, as far as the resources are concerned, they will all appear local to LAN users, notwithstanding the fact that they are located in different LAN segments. Thus, VLANs are created in an ATM by using VCCs to define different groups of users, each of which appears to use a different logical connectionless server (CLS) function for communication. Each virtual LAN can be made from a different set of users at different locales. Further, necessary connectivity to WAN is also facilitated in VLAN implementations.

VLAN may support multiple MAC and network protocols. It is not constrained by a specified MAC protocol (such as IP over ethernet) to be used by every user.

An example of VLAN implementation is depicted in Fig. 8.13. Suppose there are multiple users as shown located physically at different sites. All these users can access a server placed in a different location under the pretext that all these users share the same physical local ethernet or FDDI.

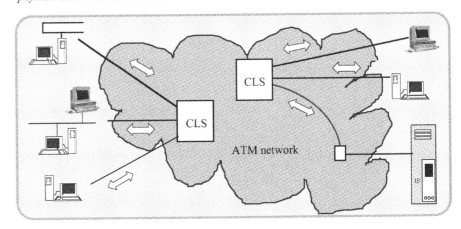

Fig. 8.13: VLAN implementation (CLS: Connectionless server function)

VLANs have been built with some proprietary switching hubs to switch between user groups. Typically, proprietary inter-switch link (ISL), IEEE 802.10, and ATM LANE are adopted to implement LANE. In a single profile, VLAN may connect two workgroups via LAN switches acting as LECs. That is, each workgroup maps to the physical ports on the LAN switches. This is called static VLAN mapping. In a dynamic (port mobility) system, a user moving from one location on a campus may attach to an unassigned port at a new location.

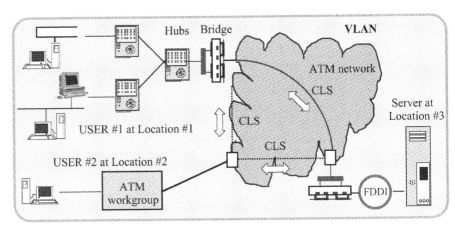

Fig. 8.14: ATM LANE: Virtual LAN emulation

ATM LANE: VLAN considerations

The realization of VLAN via ATM LANE is illustrated in Fig. 8.14. The LANE architecture in Fig. 8.14 refers to the bridging method (like transparent bridging and/or source routing bridging).

Problem 8.1

Draw a schematic to illustrate how the functions of a backbone LAN can be simulated in an ATM network by using a set of VCCs that provide a function equivalent to a backbone LAN based on VLAN/CLS considerations.

LANE implementation

Consider the interoperability of end-systems on a variety of interconnected LANs. End-systems attached directly to one of the legacy LANs implement the media access control (MAC) layer appropriate to that type of LAN. End-systems attached directly to an ATM network implement the ATM and AAL protocols. As a result, the following compatibility considerations arise:

- ✓ Interaction between an end-system on an ATM network and an end-system on a legacy ATM
- ✓ Interaction between an end-system on a legacy LAN and an end-system on another legacy LAN of the same type (for example, two IEEE 802.3 networks)
- ✓ Interaction between an end-system on a legacy LAN and an end-system on another legacy LAN of a different type (for example, an IEEE 802.3 network and an IEEE 802.5 network).

The most general solution in such situations refers to using a router. The router operates at the level of the *Internet protocol* (IP). Here, all the systems implement IP, and all the networks are interconnected with routers. As a result, the following can be assumed:

- ✓ If data are to travel beyond the scope of an individual LAN, they are directed to the local router. There, the logical link control (LCC) and MAC are removed and IP is implemented. Then the resulting IP protocol data unit (PDU) is routed across one or more other networks to the destination LAN, where the appropriate LCC and MAC are invoked
- ✓ Similarly, if one or both of the end-systems are directly attached to an ATM network, the AAL and ATM layers are stripped off and added to an IP-PDU.

Though the abovesaid router-based approach is effective, it adds overhead and causes delay at each router processing. Such delays can be significant in large networks.

An alternative approach is to convert all end-systems so that they operate directly on ATM. This is done with a seamless technology covering local and wide area components. However, with millions of ethernet and token-ring nodes added on shared-media LAN environment, one shot upgrading/conversion as above may not be affordable. As such, the ATM Forum created a specification for the coexistence of legacy LANs and ATM LANs. As stated earlier, this is known as the *ATM LAN emulation*.

Objectives of ATM LAN emulation

LANE is evolved to enable the existing shared-media LAN nodes to interoperate across an ATM network and to interoperate with devices that connect directly to ATM switches. Specifically, ATM LAN emulation defines the following:

✓ The method by which end-systems on two separate LANs of the same type (same MAC layer) can exchange MAC frames across an ATM network

✓ The way in which an end-system on a LAN can interoperate with another end-system emulating the same LAN type and attached directly to an ATM switch

✓ ATM Forum has not yet specifically addressed interoperability between end-systems on different LANs with different MAC protocols.

ATM LAN emulation protocol architecture

LANE protocol architecture specifies the interaction of an ATM attached system with an end-system attached to a legacy LAN. As stated before, an end-system attached to the legacy LAN is unaffected. It is able to use the ordinary repertoire of protocols including the MAC protocol specific to this LAN and LLC running on the top of the MAC. Thus, the end-system may run TCP-IP over LLC and various application-level protocols on top of that. But, these different application-level protocols are unaware that there is an ATM network underneath them. The essence of this strategy is to use a bridge-device namely, an ATM-LAN converter.

The bridge logic of ATM-to-LAN conversion refers:

✓ Capability of converting MAC frames to and from ATM cells. This is a major function of ATM LAN emulation module

✓ As per ATM Forum specification, AAL-5 segments the MAC frames into ATM cells and reassembles incoming ATM cells into MAC frames

✓ For outgoing ATM cells, the ATM-LAN conversion bridge connects in the usual fashion to an ATM switch as part of an ATM network

✓ A host on a legacy LAN exchanges data with a host attached directly to an ATM network

✓ To accommodate this exchange, the ATM host must include a LAN emulation module that accepts MAC frames from AAL and must pass up the contents to an LLC layer. That is, the host is emulating a LAN because it can receive and transmit MAC frames in the same format as the distant legacy LAN

✓ As far as the end-systems on the legacy LAN are concerned, the ATM host is just another end-system with a MAC address

✓ Thus, the entire LAN emulation process is transparent to the existing systems implementing LLC and MAC.

LAN emulation via clients and servers

LAN emulation is based on client/server architecture. The technique being addressed via client/server approach would, however, face the following hurdles:

✓ Devices attached directly to ATM switches and to ATM-to-LAN converter systems have ATM-based addresses. Hence, there is a burden to perform translation between these addresses and MAC addresses

✓ ATM deploys a connection-oriented protocol involving virtual channels and virtual paths. Therefore, the connectionless LAN MAC protocol has to be supported over this connection-oriented framework

✓ Multicasting and broadcasting on a shared-medium LAN is normally achieved with ease. But a question is often posed how this capability can be carried over into the ATM environment.

Clients and servers

Client: *A workstation in a LAN that is set-up to use the resources of a server. It is a node or software program that requests services from a server*

Server: *In a telecommunication network, servers are the trunks or the service process such as the call center agents, which fulfill the users' service requests. It is a node or a software program that provides services to clients. In a LAN, servers are devices that provide specialized services such as a file, print, and modem or fax pool.*

The solution to problems of LANE implementation via the client/server approach as suggested by ATM Forum is as follows:

- Client should operate on behalf of devices that are attached to legacy LANs and that use MAC addresses, and the responsibilities of the client are:

  ✓ Adding its MAC entities into the overall configuration
  ✓ Dealing with the tasks pertinent to translation between MAC addresses and ATM addresses.

- Responsibilities of the server

  ✓ Integrating MAC entities into the overall configuration
  ✓ Managing all of the associated tasks such as finding addresses and emulating broadcasting.

In general, servers can be either separate components or parts of ATM switches. LAN emulation service has three types of servers:

  ✓ LAN emulation configuration server (LECS)
  ✓ LAN emulation server (LES)
  ✓ Broadcast and unknown server (BUS).

These distinct versions enable efficient operation and reduced communication burden. The specific tasks of these three servers are as follows:

  ✓ *LAN emulation client (LEC)*: This sets up control connections to LAN emulation servers and data connections to other clients. Also, it maps MAC addresses to ATM addresses
  ✓ *LAN emulation configuration server*: This assists a client in selecting an LES
  ✓ *LAN emulation server (LES)*: LES performs initial address mapping and accepts clients
  ✓ *Broadcast and unknown server (BUS)*: This performs multicasting.

The LEC that resides in the end-stations provides a *LANE user-to-network interface* (LUNI) (a standardized interface between an LE client and an LE server within the scope of LANE service). The LUNI emulates MAC level services for ethernet token-ring to upper layers in the end-stations. These MAC level services include address resolution, data forwarding, and control functions. The LES enables the control coordination function with an ELAN. It performs a registration function and also maintains a table of all registered active members of an ELAN. LECs register the MAC addresses they represent with the LES and the relevant address resolution (for LECS) refers to maintaining a MAC and/or a route descriptor to the ATM address conversion table.

The address resolution function at LES is resolved at LES itself or it is forwarded to other clients through BUS for necessary response. The BUS does the following functions for ELAN:

- ✓ Broadcasting/multicasting
- ✓ Address resolution for unknown clients
  (Unknown server function).

The BUS does broadcasting/multicasting to the members of the ELAN using a MAC address whenever a LEC indicates. The first unicast exchange between two LECs is done though BUS if the destination's LEC has not registered by then with the LES. The BUS server broadcasts the first message so that the unknown LEC will receive it and respond, thereby indicating its ATM address. Upon receiving this LEC address, a direct VCC patching is done between the two LECS.

The BUS component of LANE does the administrative functions. It manages the assignment of LECs to a particular set of ELANs. A LEC can be a part of more than one ELAN. As indicated above, by furnishing the ATM address of its LES server, a LEC becomes an active member of the ELAN with the support rendered by the BUS. The request by a LEC to join a particular ELAN is done as above only after the BUS validates such a request in terms of the specified policies and configuration database.

The above scheme is a logical strategy independent of the physical premises of LECs. That is, LECs may reside at diverse locations and yet can become a part of an ELAN. (This is in contrast with legacy LANs wherein the membership for the LAN is dictated by its physical location.)

LUNI

The interface between LECs and LESs does the following sequences of LAN emulation indicated below:

- ✓ *Initialization*: A client obtains the ATM address of the LES for that emulated LAN (in order to join that emulated LAN). For example, the client may establish a VCC to the *LAN emulation configuration server* (LECS)
- ✓ *Configuration*: Upon a connection being established between the client and the LECS, the client can engage in a dialogue with the LECS
- ✓ *Joining*: Through this dialogue, the client obtains the necessary information (such as LEC's ATM address) to join an emulated LAN. Then it proceeds to setting the LES via a JOIN REQUEST providing information pertinent to its MAC address, LAN type, maximum frame size, a client identifier, and a *proxy indication* on an ATM address. If LES is prepared to accept this client a JOIN RESPONSE is returned.
- ✓ *Registration and BUS initialization*: This refers to a client setting up a data connection to the BUS
- ✓ *Data transfer*: This phase indicates the ability of the client to send and receive MAC frames.

Proxy indication

To indicate whether a client corresponds to an end-system attached directly to an ATM switch or is a LAN-to–ATM converter supporting end-systems on a legacy LAN.

LANE protocol interface

In reference to the LUNI functions indicated above, the LANE protocol interface structure can be illustrated as in Fig. 8.15.

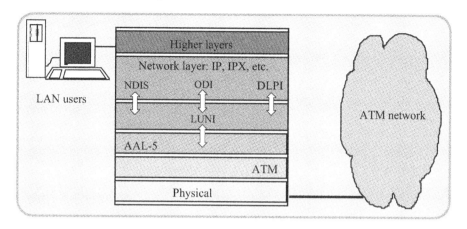

Fig. 8.15: The protocol interface of LANE implementation
(NDIS: Network driver interface specification; ODI: Open data-link interface; DLPI: Data-link provider interface and LUNI: LANE user-to-network interface)

LANE connectivity

Pertinent to the protocol interface of LANE shown in Fig. 8.16, the LANE connectivity specified between LECs and LAN emulation services are:

✓   Switched VCC, which requires ATM signaling function
✓   Permanent VCC, which handled by layer management.

Further, the VCCs can be categorized as *control VCCs* and *data VCCs*. As the names imply, the control VCCs support the control traffic while the data VCCs carry the information-bearing, encapsulated ethernet/token-ring frames. Each control/data VCC is set dedicated to a single ELAN.

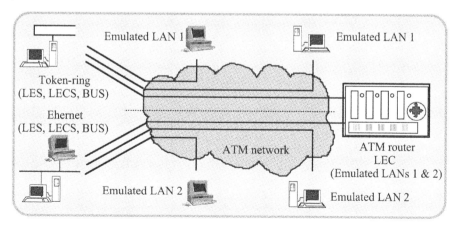

Fig. 8.16: Interconnectivity in LANE

Control VCCs

The control VCCs are established between an LEC and a LES during the establishment of the LEC initialization phase. It is strictly used for carrying the LE-address resolution protocol traffic and control frames. There are three control VCCs as indicated below along with their respective functions:

Configuration direct VCC:	This is a bidirectional point-to-point control connection between an LEC and an LECS. It is established via *broadcast low layer-information* (B-LLI) signaling during the LEC connection phase. It is done to get information such as the address of the LES. The B-LLI signaling implies that the connection carries the LE control packet formats. Once the LEC connection to ELAN is established, the configurations direct VCC may be optionally retained for further configuration inquires throughout the connection session, and an LEC may also use this connection to inquire about other LECs during its participation in the ELAN
Control direct VCC:	This is a bidirectional point-to-point control connection between LEC and LES. It is established during the LEC initialization phase. Unlike configuration direct VCC, maintenance of this connection is mandatory throughout the participating of the LEC in the ELAN
Control distribute VCC:	This is a unidirectional point-to-point or point-to-multipoint control connection from LEC to a set of LECs. This is an optional control connection that the LES may establish during the initialization of an LEC. If it is established, this connection is maintained throughout the participation of the LEC in the ELAN.

Data VCCs

There are three versions of data VCCs as named below together with their functional requirements:

Data direct VCC:	This is a birectional point-to-point data connection between LECs intended for exchanging unicast data. If an LEC wants to send a packet to another LEC for the first time and if it does not have the ATM address of this destination LEC, the source LEC issues a LE-address resolution protocol request to BUS to get this address. In response to this request, when the ATM address is obtained, the source sets up a data direct VCC to the destination LEC and this connection is used throughout the data exchange session
Multicast send VCC:	This is a birectional point-to-point data connection between LEC and BUS. An LEC establishes this connection via the same procedure as for the data direct VCC. When the LEC wants to send multicast data to multiple elements of the ELAN, it uses this VCC to pass the multicast to the BUS for distribution. The LEC may also use this VCC to send initial unicast data to an unknown destination through the BUS. The LEC maintains this connection during its participation in the ELAN and accepts data from this VCC, which the BUS may use to send data on a response back to the LEC
Multicast forward VCC:	This is a unidirectional point-to-point or point-to-multipoint data connection set by the BUS to the LEC during the initialization of the LEC for the ELAN sign up. The main function of this VCC is to distribute data from the BUS to the member LECs of the ELAN. The LEC accepts this VCC and maintains it during its participation in the ELAN. The BUS may opt to forward data either over multicast send or over multicast forward VCC and takes care to see that the LEC is not receiving duplicate data from both VCCs.

In order to understand the LANE implementation, it is first necessary to define the salient aspects of LANE:

✓ LANE is a means to seamlessly interconnect legacy LANs with new, high-performance, local area ATM networks

✓ LANE protocols enpowers ATM hosts and server applications to interwork with devices already residing on existing legacy LANs

✓ In LANE implementation, ATM cells are encapsulated in native LAN frames via the *cells-in-frames* (CIF) concept

✓ ATM standard (IETF REC 1483) defines a means of multiplexing multiple protocols over a single ATM VCC.

Implementation of LANE involves considerations pertinent to:

✓ Hardware and software of emulated LAN

✓ Components and connections involved in LANE

✓ Procedural aspects of LEC in automatic configuration, initialization, joining a virtual network, registration, broadcast, and data transfer

✓ Optimal LANE capabilities.

Hardware and software of LANE

LANE is an attempted wedlock between connection-oriented point-to-point ATM transmissions and inherently connectionless, shared-medium broadcast-capable LAN communications. The consummation is mediated by LANE through emulation of a broadcast medium.

The extent of broadcast traffic would, however, limit the capacity of the emulated LAN to that of the slowest interface. Broadcast would also limit the overall size of an ELAN.

Example 8.2

Suppose the backbone support to LANE implementation is OC-3 and 6% of the total traffic is of broadcast type.

Estimate the number of ethernet stations that can be supported by an ELAN if implemented.

Solution

6% of total OC-3 capacity = $(6/100) \times 155.52$ Mbps

$$\approx 10 \text{ Mbps}$$

That is, a single ethernet station can be supported to an extent of 10 Mbps by the OC-3 backbone. Hence, the number of stations on the ELAN is restricted to a minimum of a 500 and a maximum of 2000.

The design and implementation of LAN are arrived at by providing optimal overall throughput of shared medium LANs by fastening together a broadcast server and fast connection switching technique. A workstation with an ATM NIC card plus LANE software can join an existing LAN "plug and play" operation by supporting ethernet and/or token-ring source route bridging as well as VLAN considerations.

In short, LANE hardware/software elements enable computers running the same applications on legacy LANs (like ethernet) to connect directly onto ATM-enabled systems via high-performance ATM networks. Fig. 8.17 illustrates relevant considerations.

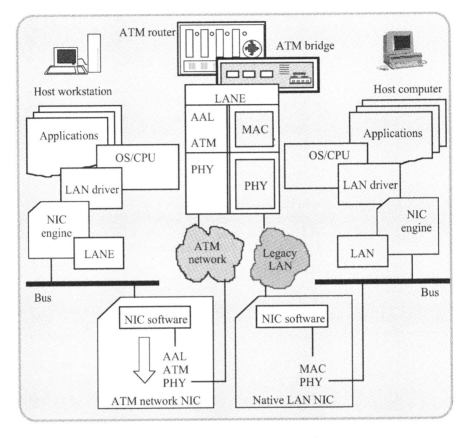

Fig. 8.17: LANE and legacy LAN interworking

The constituent parts and their functions in Fig. 8.17, which illustrate a seamless interworking of ATM LANE and legacy LAN devices, are as follows. Essentially LANE enables legacy applications to run unaltered using an existing software device driver interface on a host computer/workstation by means of an ATM NIC.

- ✓ ATM NIC: This runs a set of applications lodged on a specific operating system (OS). The information from the host is encapsulated into ATM cells, which traverse on an ATM network to an ATM compatible bridge/router
- ✓ LAN driver software: This is a part of the OS [for example, *Microsoft network driver interface specification* (NDIS)]
- ✓ LANE software: This is provided at the NIC/OS section so as to interface with the ATM NIC via NIC driver software located in the host
- ✓ ATM compatible bridge/router: This maps the incoming ATM cells to a legacy LAN MAC and physical layer
- ✓ LAN NIC: This card on the legacy LAN side interfaces to LAN NIC driver software in the host that enables an identical interface to the LAN driver software of the OS.

Thus, applications running on the ATM NIC and native LAN NIC networked workstations will not find any difference in terms of software functions.

Operational flow and connection management in LANE
Operation flow

Pertinent to the sequence involved in LANE, the operational flow can be specified by a flow chart as in Fig. 8.18.

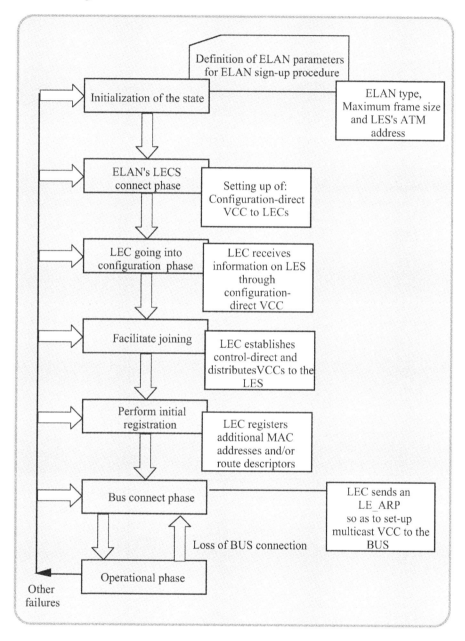

Fig. 8.18: Operational flow in LEC

LANE connection management

The management of the LANE environment refers to the following:

- VCCs

  - ✓ PVCs set up manually using the ATM layer management entity
  - ✓ SVCs set up via ATM UNI signaling protocol.

- LANE elements

  - ✓ LEC
  - ✓ LES
  - ✓ BUS.

- *Data direct VCC via SVC*: Call setup/release procedure

  - ✓ SETUP, CONNECT, CONNECT_ACK: Same as in SVC setup
  - ✓ READY_IND, READY_QUERY: LANE-specific sequences.

The sequences of call set-up are illustrated in Fig. 8.19.

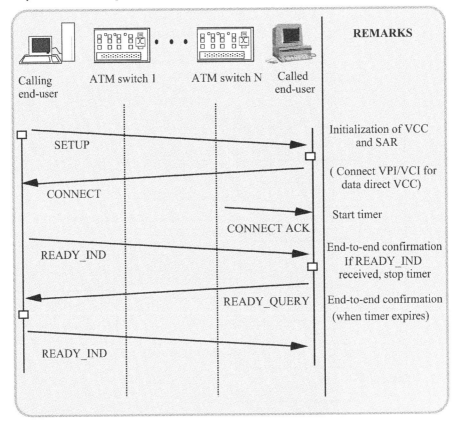

Fig. 8.19: CALL SETUP procedure in LANE

In summary, LAN emulation provides a transparent bridging of traditional LANs over an ATM backbone. It also allows LAN-attached end-stations, (such as personal computers and workstations), to interoperate with ATM-attached devices such as database servers. With LAN emulation, network managers can apply ATM to network hot-spots while protecting their committed investment in existing equipment and applications.

There are proprietary approaches available to LAN emulation but the ATM Forum has specified a vendor-independent solution. The LAN emulation (LANE) specification defines

LANE clients and LANE servers (see Fig. 8.20). LANE clients reside in ATM-attached bridges, routers, switches, and end stations. They translate frames into cells (and vice versa) and steer connectionless LAN traffic onto the correct ATM virtual circuits. LANE services, which reside in the network or in dedicated servers, group attached devices into emulated LANs and assist LANE clients with address translation and broadcast distribution.

The ATM Forum has designed LAN emulation carefully so that LANE clients and ATM interfaces can be added to existing devices, like LAN servers, without disrupting installed applications or network operating systems. Adherence to the LANE specification protects user investment in application hardware and software and ensures interoperability in multivendor internetworks.

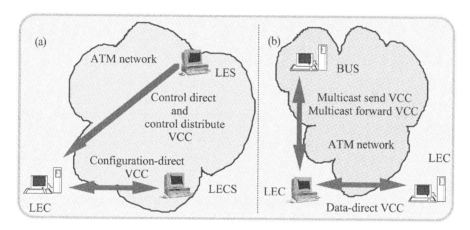

Fig. 8.20: LANE clients and servers: (a) Control VCCs and (b) data VCCs

The internetwork architectures that take advantage of switch-based building blocks are being marketed. Some of them are designed to relieve specific bottlenecks, others try to revitalize the entire internetwork.

8.3   Classical IP (CLIP) over ATM

To extend the use of ATM in the IP-based LAN environment IETF RFC 1577 prescribes what is known as *classical Internet protocol* (CLIP) and *address resolution protocol* (ARP) over the ATM. The prime objective of CLIP is to take advantage of higher wire speed offered by ATM technology, at the same time protecting the interests of IP-based applications by facilitating eventual migration of IP-based legacy networks (LANs) to ATM-based networks.

IP and ARP

Internet protocol (IP) is a protocol originally developed by the Department of Defense (DoD) to support the internetworking of dissimilar computers across a network.

Address resolution protocol (ARP) refers to the set of procedures and messages in a communications protocol that determines which physical network address (MAC) corresponds to the IP address in the packet.

The CLIP as defined in RFC 1577, specifies how a "classical" IP subnetwork should be implemented on top of ATM cell switching (see Fig. 8.21). This approach uses AAL-5 to segment IP packets into ATM cells. It also specifies how virtual circuits can be used to construct a sub-network and how IP addresses can be translated to virtual circuit identifiers.

The specification defines a standard method for exploiting the scalability of ATM in today's internetwork. Classical IP over ATM integrates ATM LANs into IP internetworks, but ignores non-IP protocols, hybrid frame/cell LANs, and multicast applications. In contrast, ATM LAN emulation establishes a standard for multiprotocol switched internetworking and hybrid frame/cell LANs as well.

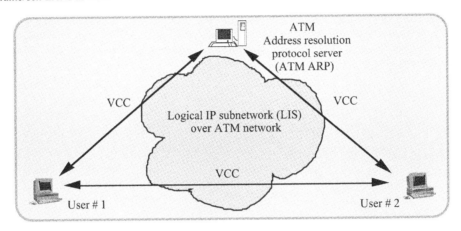

Fig. 8.21: CLIP

Classical IP over ATM is described in IETF RFC 2225. Functionally, CLIP allows IP traffic to be carried over ATM networks by resolving destination IP addresses into ATM addresses, which are then used for routing the traffic over the ATM network to the destination. The mechanism for performing this address resolution is called *ATM address resolution protocol* (ATMARP).

The basic components of a classical IP over ATM are:

- ✓ ATM-attached stations: These may be *end stations* (ES), *LAN access devices*, and, a *logical IP subnet* (LIS), which is formed by a group of these stations
- ✓ ATM switches
- ✓ ATMARP servers: These are typically implemented in an ATM switch. Each LIS must have an ATMARP server, which will resolve the IP addresses into ATM addresses. A single ATM switch may act as an ATMARP server for multiple LIS, but the switch must be a member of each LIS.
- ✓ Router: When hosts on different LISs wish to communicate, all traffic must be sent to a router, which forwards the traffic to the destination. The router must have membership in each LIS for which this service is to be performed.

For traffic crossing purposes, a LIS boundary is used as an intermediate router. This traffic is a carryover from RFC 1122, which defines the requirements for communication between Internet hosts. For this reason, this is called a "classical" procedure.

Merits of the classical IP

- ✓ Routers provide firewalls that prevent unwanted access to a LIS
- ✓ Network administrators are allowed to configure ATM networks using the same models as those used for legacy networks.

Demerits of the classical IP

   ✓ Guaranteed end-to-end QOS is preserved.

  The resolution of an IP address to a corresponding ATM address is the basic concept that all methods supporting classical IP over ATM have to deal with. This implies that all ES must posses both an IP and an ATM address in a classical IP over ATM subnetworks. The RFC 1577 document specifies that classical IP over ATM is a direct replacement for the "wires" interconnecting the ES, LAN segments, and routers in an LIS. There are several protocol methods employed in the deployment of classical IP over ATM. To name a few: *Inverse address resolution protocol* (InARP), the ATMARP, and the interconnecting LIS.

8.3.1 *Signaling standards in classic IP over ATM*

  The RFC 1755 document specifies the details relevant to hosts and routers required to achieve interoperations when employing the SVC capabilities (as specified in RFC 1577). It refers to the use of the UNI 3.0/3.1 information elements (IEs) in the SVC implementation of classical IP over ATM.

- AAL parameter IE

   ✓ AAL-5, Max SDU size 65,535 bytes, Null SSCS

- Broadband low layer information (B-LLI) IE

   ✓ Logical link control (LLC)
   ✓ Layer 3 protocol in IP (L3)

- ATM traffic descriptor IE

- Broadband bearer capability IE

- QOS parameter IE

- Called/calling party address

- Called/calling party sub-address (optional)

- Transit network selection (optional).

The fields indicated above are mandated IEs unless otherwise specified. For example, in case end stations use the AAL parameter and B-LLI information elements (at call setup time) so as to establish the correct mappings for higher layer protocols, the address and sub-address fields may be either of the network service access point format or the E.164 number format.

8.3.2 *Next hop resolution protocol (NHRP)*

  The classical model for IP over ATM specified in RFC 1577 requires that all communication between distinct LISs occur through a router. In view of this constraint, networks can be designed more efficiently (as an alternative strategy) via ATM "short-cut" routers, which enable a direct communication between nodes connected to the same ATM network (but being part of two different LISs). This would allow bypassing throughput bottlenecks at intermediate router hops. The continued use of ATM internetworks increases the possibility that two ESs may connect to the same ATM network. Besides, ATM can be expected to provide better services with respect to voice, video, and real time data applications than any legacy connectionless IP routers.

  Studies done by the IETF over the years in evaluating protocols designed to address these issues has led to a solution known as the *next hop resolution protocol* (NHRP) in subnetworks such as ATM, frame-relay, and SMDS. The NHRP facilitates different LISs on the same nonbroadcast multiaccess (NBMA) network to decouple the local versus the remote forwarding decision from the addressing convention that defines the LISs. Therefore, the NHRP permits systems to directly interconnect over a NBMA network, independently of the addressing based on conventions such as traffic parameters or QOS characteristics.

8.3.3   IP multicast over ATM

Present ATM standards have not defined a service comparable to the addressing features present in SMDS IP with a capability that allows one address to broadcast to all other addresses in the group effectively, in emulating a LAN. Such an effort obviously will be useful in the exchange of LAN topology updates, ARP messages, and information required in routing protocols.

The basis on which multicasting/broadcasting can work is defined in the IETF RFC 2022, which specifies the means to implement *IP multicast over ATM*. It defines two methods for implementing ATM multicast via,

       ✓   Multicast server (MCS)
       ✓   Full mesh point-to-multipoint method.

The IETF choice of AAL-5 to transport IP over ATM has significant consequences on the design of multicast services. One of the inherent aspects of AAL-5 is that all the cells from a packet must be transmitted sequentially on a single VCC. This implies that in reference to a point-to-multipoint connection, the transmission could be strictly from the root to each of the leaves. However, if the leaves have to transmit to the root, then the cells interleaved from multiple packets and arriving at the root would result in AAL-5 SAR failures leading to loss of packet data.

In the *multicast server approach* as shown by Fig. 8.22, all nodes are joined into a particular multicast group and a point-to-point connection is established thereof with the *multicast server* (MCS). The MCS can have a point-to-multipoint connection as shown in Fig. 8.22. As a second option, it may emulate a broadcast connection by transmitting packets in the reverse direction on every point-to-point connection. The MCS receives packets from each of the nodes on the point-to-point connections and then retransmits them on the point-to-multipoint connection. This design ensures that the serialization integrity aspect of AAL-5 is met with. That is, all cells of an entire packet are transmitted prior to cells from any other packet being sent.

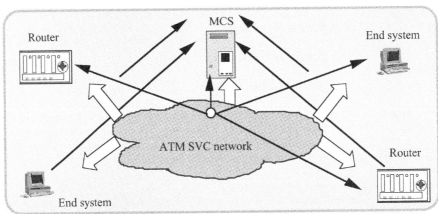

Fig. 8.22: MCS based ATM multicast option

In the *full mesh point-to-multipoint* connection approach, illustrated in Fig. 8.23, a point-to-multipoint connection between every node in the multicast group is established. Therefore, every node is capable of transmitting and receiving from every other node in the system. The number of connections that is required by this network, however, poses a complexity in configuring the PVCs for such a network. Hence, it is an existing notion that a point-to-multipoint SVC capability is rather essential to the practical deployment of a multicast over an ATM network. The implementation of the multicast capability has advantages as well as disadvantages.

The point-to-multipoint mechanism requires each node to maintain a specified number of connections for each group. This places a connection burden on each of the nodes, as well as on the ATM network. On the other hand, the multicast server mechanism requires at the most two connections per node. This approach requires that only the server supports a large number of connections. Hence, the multicast server mechanism is more viably expandable in terms of being able to dynamically change its multi-cast group membership. Nevertheless it presents a potential bottleneck as well as a single point of failure in ATM networks.

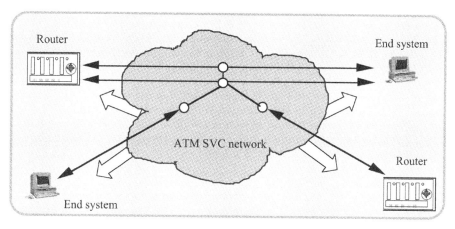

Fig. 8.23. Full mesh point-to-multipoint ATM multicast option

8.3.4 *IP over ATM*: State-of-the-art

The emergence of the network layer switching technique has created competition for the LAN switching technique. Such a competition is reflected in terms of the associated network performances and price ranges. Network switching is an operation aimed at the backbones of networks and it performs routing and switching functions only once and not for every packet that transmits over the network to a common destination.

When the IP and ATM committees were developing the protocol standards for IP broadcast over ATM, manufacturers were already building solutions due to the vast growth in IP internetworking. A good number of these manufacturers contributed by publishing their works as IETF RFCs in addition to issuing public proclamation that their approach would be an open non-proprietary solution. The first of such companies was Ipsilon Networks. The approach of Ipsilon Networks was to place its *IP switching over ATM* with an efficient protocol. The Toshiba Corporation brought about a similar approach with its *cell switch routers* for efficiently interconnecting classical IP over ATM. The most recent contender in this arena is Cisco Systems with its *tag switching* architecture, which not only works with ATM but also on a large number of legacy LANs. Yet another approach came into being, known as (IBM's) *aggregate route-based IP switching* (ARIS). As more and more contenders entered the scene, the IEFT formed a *multiprotocol label switching* (MPLS) workgroup to sort all these various standards and come up with a common industry standard.

Ipsilon's IP switching technique

In early 1995, Ipsilon Networks suggested that "ATM had already lost the battle for the desktop" to *fast ethernet*. It stated that LANE technology is nothing more than LAN switching technology and that the software burden of complex ATM protocols have made the proposed IP over ATM budget-intense. Besides it publicly questioned the directions of the ATM Forum and the IETF - the two organizations forming what is known today as the Inter-networking Operating Committee (IOC) in regard to the structure of implementing IP over ATM. Ipsilon pointed out the duplication of functions, the difficulties in implementing LANE, NHRP, MPOA, and the entire process of IP over ATM, as growing problems within the related scope of the technology.

The company then published key consideration of their protocol in the Internet RFCs in 1953, 1954, and 1987. This strategy placed all aspects of their protocols at the disposition of any manufacturer. It included algorithms available to hosts and routers as well as source codes free of charge. Their approach classified traffic into either short- or long-lived flows. With the advent of new components, the IP-switching approach applies only to long flow-cycles (as in FTP, long Telnet sessions, http, and extended Web multimedia sessions). The IP switching handled all short-lived sessions such as interactive traffic DNS, e-mail, and SNMP in the same manner IP routers handle them in the present.

The Ipsilon Network's IP switching architecture has a set of components, interfaces, and protocols of the relevant IP switching architecture. Its IP switch consists of two logical components: An ATM switch and an IP switch controller. Any ATM switch with a switch controller capable of making and breaking hundreds of virtual channel connections per second can be part of an IP Switch. Ipsilon also developed software for IP switch controllers, which connect an ATM UNI to the ATM switch, and multiplexes ATM cells at the VCC level to an IP router and Ipsilon switch controllers. ATM UNIs connect the IP switch to a set of upstream and downstream nodes. The nodes (namely, the hosts or routers) connect to interfaces on the IP switch so as to transfer IP data packets. These nodes should enable adding labels to IP packets. Further, they do the multiplexing of various flows of IP packets associated with the labels onto different VCCs. These nodes then interface to the IP switch controller through *Ipsilon's flow management protocol (IFMP)* as defined in the IETF RFCs, 1953 and 1954.

Toshiba's cell switching router (CSR)

Toshiba's *cell switching router* (CSR) is defined in its vendor proprietary IETF RFC, 2098. It defines another proposal for handling IP over ATM networks. The *cell switching router* (CSR) has ATM cell switching capabilities in addition to conventional IP datagram routing and forwarding.

IP datagrams are normally forwarded through hop-by-hop paths using the CSR routing function, in the same manner as in the IP switching approach. It is the role of the routing function to automatically recognize the long-lived flows and to either assign, or establish shortcut ATM paths precisely as with the IP switching model. On the other hand, the CSR adds new concepts as well. Primarily, it proposes to handle more than the IP protocol. Its second function is to permit shortcut connections to be preconfigured or established through interaction with RSVP. The CSR in addition proposes setting up shortcut routes that may bypass several routers. It also employs a *flow attribute notification protocol* (FANP). Besides all this, the CSR also implements standard IP routing protocols, the ATM Forum PNNI protocol, and ATM signaling.

Toshiba's CSR proposal places a strong emphasis on meeting application specific QOS and bandwidth specifications, even over non-ATM networks. Besides, the CSR refers to the interconnections of classical IP over ATM, MPOA, and switched IP networks. One of the CSR objectives is to internetwork these multiple varieties of IP over ATM at greater throughput than any interconnection methods in vogue.

The Cisco System's tag switching

The Cisco System's tag switching architecture was announced in 1996. Details on this development can be found in IETF RFC, 2105.

Tag switching is a high-performance, packet-forwarding technique based on the concept of label (header) swapping. Swapping headers at the intermediate nodes leads to realizing an end-to-end connection. Inasmuch as ATM VCC switching produces a special case of general header swapping using the VPI/VCI fields in the cell header, the switch/routers know whether to switch cells or assemble them and route the resultant packets based upon the information derived from the tag distribution protocol (TDP).

A set of *tag edge routers* at the boundaries of an ATM network provide network layer services and apply tags to packets. The tag switches/routers at the core of the network switch tagged packets or cells based upon the tags determined through the information piggybacked onto the standard routing protocols or via Cisco's *tag distribution protocol* (TDP). The tag switch/routers and tag edge routers implement standard network routing protocols, such as *open shortest path first* (OSPF) and *border gateway protocol* (BGP).

IBM's aggregate route-based IP switching (ARIS)

IBM's aggregate route-based IP switching (ARIS) is defined as a multicast distribution tree rooted at its egress point, and traversed in reverse. The egress refers to a unique identifier (depicting an egress router IP address). The IBM ARIS makes effective use of its route tree to determine the forward path as the reverse direction along such a minimum spanning tree.

The flows from the leaves of the spanning tree back towards the root merge at several points. The ARIS system operates over frame switched networks as well as cell-switched networks. In order for it to work over a cell-switching network, the ARIS requires ATM switches capable of VC merging in order to support larger networks. The VC merging capability of ATM switches permits the *ARIS integrated switch routers* (IRS) group cells from individual AAL-5 PDUs from different inputs and switches them onto a shared VCC on its path back from the common egress point.

The ARIS implements an IP forwarding table that includes a reference to the point to the multi-point switched path determined by an explicitly specified egress point on the IP network.

The IETF multiprotocol label switching (MPLS)

The IETF in its interest to build an expandable Internet backbone established the *multiprotocol label switching* (MPLS) to provide one common specification to the industry. The present framework combines a set of functional specifications from the proprietary approaches described previously. Below is a list of major requirements from the MPLS framework document and their sources.

> ✓ Existing approaches have adopted a short, fixed length layer 2 switching label designed to achieve lower cost, higher performance packet forwarding than the traditional router technology
> ✓ The conceived MPLS should support simple-cast as well as multicast
> ✓ There is a need for expansion (growth) on the order of N streams to establish best traffic efforts as required by the IBM ARIS strategy
> ✓ The MPLS should carry the ability to support RSVP and flow control recognition from Ipsilon's IP switching and Toshiba's CSR proposals
> ✓ It must further support all topology-driven protocols as those defined by Cisco's tag switching
> ✓ The label switching strategy should maintain compatibility with existing legacy IP routing protocols and coexist with devices not capable of supporting MPLS
> ✓ It must specify operation in a hierarchical network and be capable of supporting large-scale internetworks. The document defines the concept of stacking multiple labels in front of one packet to achieve this goal
> ✓ It must be independent of any specific data link technology. It must optimize all particular data link networks such as frame-relay, ATM and MPLS.

The IETF document on MPLS compares its approach with the other two approaches for constructing large internetworks, namely, directly interconnected routers, and routers connected by a full mesh of virtual channel connections over a core ATM network.

8.4   Multiprotocol over ATM (MPOA)

This is a standard issued by the ATM Forum to formalize running the multiple network layer protocols over ATM. It enables connectivity ATM LANEs and ATM emulated LANs from different subnetworks of an IP-network using a *cut-through routing*. (The cut-through routing implies the following: MPOA allows devices attached to an ATM network to communicate directly with each other, even if they are on two different subnetworks, as opposed to communicating through a router.) MPOA is a result of feedbacks received from the implementations of LANE and CLIP. One of the deficiencies observed in LANE and CLIP strategies is that it is still a necessity to adopt classical routers to carry traffic between the logical subnetworks created by relevant protocols even if the logical subnetworks resided over the same

ATM network. But such routers could cause excessive latency and choke the traffic inasmuch as each and every packet has to be individually processed by these routers. Further, these routers regenerate the packets from the ATM cells for the purpose of interpreting the packet header (for routing); this causes additional overhead concerns.

The router concept could be deployed in the traditional LAN environment wherein the traffic generated in the LAN segments is IP-based and the extent of such traffic traversing the routers is small (almost 20 to 25%).

With the advent of changing traffic patterns imposed by higher bandwidths in the LAN backbone, the ATM has to deal with this change and this could possibly augment the bottleneck at the routers. The MPOA is designed to eliminate such bottlenecks.

The essential features of MPOA can be summarized as follows:

- ✓ While LANE describes how LAN frames are bridged over ATM and ignores network-layer protocols, MPOA specifies how traffic should be routed over ATM and takes network layer protocols into account
- ✓ Multiprotocol over ATM refers to a standard that defines how routers, LAN switches, and hosts running multiple layer 3 protocols may optimize forwarding paths and multicasting across ATM, at the same time taking advantage of ATM QOS capabilities
- ✓ MPOA is a model for multiprotocol layer 2/3 networking across ATM, and defines various servers, protocols, and encapsulations.
- ✓ MPOA is an evolution of LANE plus classical ATM
- ✓ It can be considered as an evolution of LANE now including layer 3 routing plus LANE's layer 2 bridging
- ✓ It offers the benefits of a fully-routed environment across ATM in terms of bandwidth and service classes
- ✓ It separates routing and switching functions
- ✓ It is a unified overall internetworking architecture applied to layer 3 across ATM
- ✓ MPOA adds more functionality for connecting ATM LANs as compared to the more fundamental LANE standards.

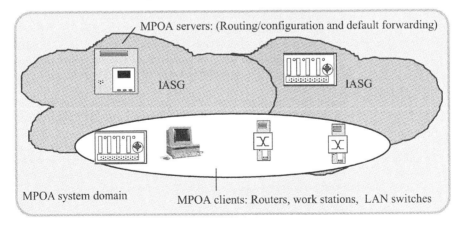

Fig. 8.24: The architecture of MPOA system

Definitions

❑   *IASG:*   *Internet address subgroup. It is an equivalent to a nonoverlapping range of internetwork layer addresses summarized by a layer 3 routing protocol. It is a basic unit of organization within MPOA*

❑   *LIS:*   *Logical IP subnetwork where all devices have direct connectivity with each other and adhere to the IP subnet model*

❑   *LAG:*   *Local address group. It is a set of ATM connected systems with direct connections with each other.*

(The LIS, LAG, or other layer 3 subnet within MPOA, are not necessarily of a MAC broadcast domain.)

MPOA scenario and strategy refer to the following:

✓   Protocols in use: IP, IPX, and DECnet
✓   Multiprotocol transport across ATM and internetworking with LANE be facilitated economically.

8.4.1 Deployment of MPOA across the ATM network

The implementation of MPOA is consistent with the following considerations:

✓   Existing LAN segments be connected to the ATM network via LAN switches acting as *edge device functional groups* (EDFGS)
✓   Hosts to act as ATM-attached *host functional groups* (AHFGs)
✓   Existing LANE domain be connected to the same backbone, integrating it thereby into the MPOA environment
✓   As a part of MPOA, a route distribution service be deployed between the *route server functional group* (RSFG) and the EDFGS/AHFGs
✓   A router to provide necessary MPOA server functions, including: LASG *coordination functional group* (ICFG), *default forwarder functional group* (DEEG), *route server functional group* (RSFG), and *remote forwarder functional group* (RFFG).

8.4.2 MPOA domain

This domain contains a number of devices operating as MPOA servers and clients. The *MPOA client* (MPC) is an entity that resides either in a LAN switch or an ATM-attached switch. It is nearly analogous to the LEC in LANE, but is capable of layer-3 operation. The *MPOA server* (MPS) is responsible for client registration, address resolution, and routing.

MPOA splits the traditional role of the router into two roles by off-loading packet forwarding from the router to the hosts and edge devices. The ability to separate packet forwarding from other router functions allows a more flexible implementation of the two components. Enabling packet forwarding in a separate device from the router is known as *out-through routing* (as stated earlier). This is a unique characteristic of MPOA. In ATM networks, which cover large geographical areas MPOA provides an efficient VLAN platform.

8.4.3 MPOA implementation

MPOA involves the following operational steps:

✓   Configuration: This refers to each MPOA device retrieving its configuration information from LANE LECs as they come online

✓	Discovery:	This operation allows MPC and MPS to recognize other online MPOA devices
✓	Target resolution:	This is an extended version of NHRP resolution request protocol to identify the ATM address of the destination end-point
✓	Contention management:	This operation creates, maintains, and terminates control and data VCCs
✓	Data transfer:	In the MPOA environment this occurs either over the default router path or over the shortcut VCC. The degree of shortcut VCCs is the measure of efficient unicast data transfer in an MPOA environment.

In the configuration operation, MPCs and MPSs get their configuration data from LANE LECS as a default option. Next, the discovery operation allows the MPOA devices to discover each other through LANE LE-ARP dynamically. The target resolution enables finding the ATM address of a destination with a particular IP address. Creating, maintaining, and terminating control and data VCCs are the efforts of the connection management.

8.4.4   IETF protocol of MPOA

MPOA uses two IETF protocols as a part of its architecture. They are:

✓   Next hop resolution protocol (NHRP)
✓   Multicast address resolution server (MARS)/MAC.

The NHRP refers to a protocol intended to optimize routing access to a network where direct access from any member of one LIS to any other member of any other LIS is possible. (Such a network is known as a nonbroadcast multiple access network or NBMA wherein broadcasting is not possible due to scaling issues.) The NHRP where optimizing the routing access NBNA networks allow end-systems to open VCs across LIS boundaries.

The MARS is a registry within the MPOA for ATM network service access point (NSAP) to layer 3 multicast address mapping. MARS is intended to resolve IP addresses to a single ATM point-to-multipoint address that reaches an entire group of users.

8.5   Voice over ATM (VTOA)

Voice band traffic refers to and includes the speech, fax data, modem data, and recorded audio (music etc.). There are two approaches that can be used for the transport of such *voice traffic over ATM* networks:

✓   Under constant bit rate (CBR) traffic conditions, using AAL-1 to support circuit emulation service (CES)
✓   Under real-time variable-bit-rate (rt-VBR) traffic conditions, using AAL-2 or AAL-5 adaptation.

There are three prescribed standards that have been addressed by the ATM Forum:

✓   AAL-1: Circuit emulation
✓   AAL-1: Structured circuit emulation
✓   AAL-2: Variable bit rate voice over ATM.

Circuit emulation allows the user to establish an AAL-1 ATM connection to support a circuit, such as a full T-1 or E-1, over the ATM backbone. *Structured circuit emulation* establishes an AAL-1 N × 64 kbps circuit, such as a fractional T-1 or E-1 over the ATM backbone. Finally, the *VBR voice over ATM* uses an AAL-2 connection to provide highly efficient voice over ATM support.

The ATM Forum specifications on VTOA are listed in Table 8.1.

Table 8.1: <u>ATM Forum specifications on VTOA</u>

Specification	Document	Date
• Circuit emulation service 2.0	af-vtoa-0078.000	Jan, 1997
• Voice and telephony over ATM to the desktop	af-vtoa-0083.000	May, 1997
• (DBCES) dynamic bandwidth utilization in 64 kbps time-slot trunking over ATM using CES	af-vtoa-0085.000	July, 1997
• ATM trunking using AAL1 for narrow band services v1.0	af-vtoa-0089.000	July, 1997
• Voice and telephony over ATM to the desktop	af-vtoa-0083.001	Feb, 1999
• ATM trunking using AAL2 for narrowband services	af-vtoa-0113.000	Feb, 1999
• Low speed circuit emulation service	af-vtoa-0119.000	May, 1999
• ICS for ATM trunking using AAL-2 for narrowband services	af-vtoa-0120.000	May, 1999

Circuit emulation service 2.0

This specification defines a *structured* and *unstructured service* over an ATM CBR permanent virtual circuit with options for synchronous or asynchronous clock recovery. Basically, it implies the definition of a *"circuit"* traffic over ATM networks. The working concept is that since ATM is essentially a packet (rather than a circuit-oriented transmission technology), it is expected to emulate circuit characteristics in order to provide good support for CBR traffic.

Dynamic bandwidth utilization in 64 kbps time-slot trunking over ATM

The strategy here is to specify a method for enabling *dynamic bandwidth CES* (DBCES) in an ATM network. It is based on detecting which time-slots of a given TDM trunk are active and which are inactive. When an inactive state is detected in a specific time-slot, the time-slot is dropped from the next ATM structure and the bandwidth it had been using can thereof be reutilized for other services. This method may be applied via any method of time-slot activity detection, such as *channel associated signaling* (CAS) and *common channel signaling* (CCS).

Low speed circuit emulation service (LSCES)

The LSCES-IS specifies point-to-point unstructured CES for low-speed interfaces such as V.35 and EIA-449. Many applications use these kinds of interfaces, for example, video conferencing.

8.5.1   Voice and telephone over ATM to the desktop

The relevant document indicates two versions. It specifies the ATM Forum's interoperability agreement for supporting voice and telephony over ATM to the desktop.

ATM trunking using AAL-1 for narrow band services v1.0

This scheme defines the trunk signaling for the interworking of B-ISDN to N-ISDN, as well as ISDN CCS support and interpretation. Basically, it provides additional efficiency in the use of the ATM backbone and narrowband access resources, augmenting the basic service CES definition. Specifically, it allows the allocation of resources on an ad hoc basis, depending on expected or actual traffic loading. It permits traffic to and from multiple end-points (PBXs) to share those allocated facilities and allows calls to be routed to the desired destination facilities, which eliminates the need to dedicate specific narrowband channels to every destination.

ATM trunking using ALL-2 for narrow band services

This method brings AAL-2 support to the former specification, allowing voice compression, releasing the bandwidth whenever the voice application does not require it, that is, whenever the talker is silent or when the call has completed routing and switching narrowband calls on a per-call basis.

ICS for ATM trunking using AAL-2 for narrowband services

This is an appendix to the specification for the ATM trunking using AAL-2 for narrowband services. In order to evaluate conformance of a particular implementation, it is necessary to have a statement on those capabilities and options that have been implemented. Such a statement is called an *implementation conformance statement* (ICS).

From these specifications' evolution, it is clear that CES has been fairly standardized and has actually been implemented by several manufacturers. But still it has performance problems and it does not take full advantage of the switching efficiency of ATM. There are actually two main reasons for these shortcomings:

✓ The implementation of an emulated circuit inside an ATM network requires more bandwidth due to the ATM cell overhead. For example, an E-1 (2048 kbps) requires 2308 kbps of ATM bandwidth

✓ The T-1/E-1 circuit is configured as a point-to-point ATM permanent virtual circuit; that is, per definition, this circuit is always active even if no data is being transmitted.

Despite the above, AAL-1 is still useful in cases where interoperability is necessary or when there is no AAL-2 support, and whenever the ATM network provides the functional equivalent to a *digital cross-connect system* in TDM networks. Consequently, the natural tendency is to use AAL-2 and an approved specification has been released in 1999 in support of this strategy.

In order to analyze VTOA implementation, the associated adaptation layers and signaling functions are briefly reviewed below:

A review on AAL…

ATM adaptation layer

The ATM adaptation layer (AAL) as discussed in Chapter 4, performs functions required by the user and control and management planes. It supports the mapping between the ATM layer and the next higher layer. The adaptation layer has been subdivided into two sublayers: Convergence sublayer (CS) and segmentation and reassembly (SAR).

ATM adaptation layer 1 (AAL-1)

The ATM adaptation layer 1 provides the following services:

❑ *Transfer of information with a constant bit rate*
❑ *Transfer of timing information between source and destination.*
❑ *Transfer of structure information between source and destination.*

The drawbacks of AAL-1 are that the bandwidth is used when there is no traffic and there is no standard mechanism for compression, silence detection, idle channel removal, or CCS. Further, a single user of the AAL is supported, and attempts to reduce the delay would require considerable bandwidth.

ATM adaptation layer 2 (AAL-2)

The AAL-2 ITU-T definition was developed to overcome the allocating bandwidth problem in CES-based voice transmission (structured or unstructured). Further, this adaptation provides the flexibility that allows minimization of delay for voice applications.

Voice compression over the ATM WAN presents a problem concerning the necessary time required to fill an ATM cell with voice samples. In the case of 64 kbps (using A-law or μ-law), it takes 6 ms (48 × 8 bits = 384 bits), but if CS-ACELP is used, it will take 48 ms to fill a cell.

There are four solutions proposed:

❑ *Filling the cell as if it has a full 64 kbps channel and losing that bandwidth*
❑ *Use of shorter cells that may, however, result in changing the end-systems and switches*

❑   *Multiplexing multiple channels in a single cell in a TDM fashion that provides better efficiency but cannot handle variable length voice samples*

❑   *Using a layered cell, which includes a header for sample and allows the sample to get any place in the cell, even allowing cell boundary crossing.*

Consequently, AAL-2 was defined to provide a bandwidth-effective transmission of low-rate, short, and variable packets in delay sensitive applications. It enables the support for both VBR and CBR applications. The VBR services enable statistical multiplexing for a higher-layer that will perform voice compression, silence detection and suppression, and idle channel removal. AAL-2 allows multiple user channels on a single ATM virtual circuit and permits varying traffic conditions for each individual user or channel.

The most important features of AAL-2 are:

❑   *Efficient bandwidth usage through variable bit rate ATM traffic class*

❑   *ATM bandwidth reduction support for voice compression, silence detection/suppression, and idle voice channel selection.*

❑   *Multiple voice channels with varying bandwidth on a single ATM connection.*

AAL-2 is divided into two sublayers, namely, the common part sublayer and the service specific convergence sublayer. It does not have a SAR sublayer.

AAL-2 common part sublayer

The CPS has two components, a CPS packet and a CPS PDU. The CPS functions are:

❑   *Identification of the AAL users*

❑   *Assembling/disassembling the payload associated with each individual user*

❑   *Error correction*

❑   *Interfacing the service specific conversion sublayer.*

Multiplexing is allowed at the AAL layer using multiple connections that can be associated to a single ATM layer connection. The AAL-2 user selects the QOS provided by AAL-2 through the choice of the AAL-SAP used for data transfer.

The CPS is defined on an end-to-end basis as a concatenation of AAL-2 channels, where each channel is a bidirectional virtual channel established over a PVC, SVC or SPVC.

AAL-2 service specific convergence sublayer

The SSC sublayer is the interface between the AAL-2 CPS and the higher layer applications of the individual users. This sublayer has several definitions for different applications such as mobile telephony, fax services, etc.

ATM adaptation layer 5 (AAL-5)

AAL-5 has been defined to provide a more simple protocol for VBR, than AAL-3/4. The common part supports VBR traffic, both connection oriented and connectionless. The protocol eliminates the overhead of AAL-3/4 leaving a 48-byte payload available for the user.

Most data protocol works over AAL-5, with the exception of SMDS/CBDS that works over AAL-3/4. Even the ATM control plane makes use of this layer. AAL-5 is also used while transporting MPEG-2 video across ATM.

Signaling...

There are two basic needs for signaling on public or private networks, namely, user-to-network and between the internal elements of the network. For traditional voice systems (over public or private networks), signaling refers to the mechanism necessary to establish a connection, to monitor and supervise its status, and to terminate it by using the switching fabric designed for that particular network.

Signals can be classified as follows:

- ❑   *Topologically identified*
- ❑   *Functionally specified*
- ❑   *Physically implemented.*

The topologically identified feature of signaling refers to where the signaling is dispositioned. It is categorized as:

- ❑   *Customer line signaling (user-to-switch)*
- ❑   *Interoffice trunk signaling (switch-to-switch)*

Functionally specified characteristic of signaling denotes the purpose of signaling such as the following:

- ❑   *Supervising (busy, idle line, origination, or termination)*
- ❑   *Addressing (routing and destination signals)*
- ❑   *Alerting (advising the addressee of an incoming call)*

Physically implemented aspect of signaling depicts the way the signaling is implemented. The relevant versions of signaling are:

- ❑   *In band-signaling, namely, residential loops and channel associated signaling (CAS). It places supervisory and address instructions in the same stream as the actual user information.*
- ❑   *In facility out-of-band (for example, ISDN D channel) signaling*
- ❑   *Out band (CCSS7) signaling.*

Channel associated signaling (CAS)
    The CAS uses some of the bits of the T-1/E-1 frame to transmit signaling information. Supervision is sent using channel-associated signaling. Address information is sent as PCM-encoded tone signaling.
    CAS does not require a separate signaling channel. The same ATM VCC that carries voice information also transports the signaling information.
    An example of CAS signaling message exchange sequence is indicated below:

- ❑   *The source PABX generates a signaling request*
- ❑   *The interworking function (IWF) responds with a confirmation*
- ❑   *The IWF then issues a request for connection to the destination IWF for a 64 kbps channel*
- ❑   *Once this VCC is established, the seizure, the wink, and dialed digits may proceed across the ATM backbone to the destination IWF*
- ❑   *The IWF establishes a connection with the remote PABX*
- ❑   *The PABX sends an answer message, which is forwarded back across the network.*

Common channel signaling (CCS)

Signaling information for all the channels is transmitted on a separate time-slot. Supervision and address information may be sent using CCS. Examples are as follows: E-1: 30th channel for signaling, ISDN (2B+D) or CCSS7.

Under CCS, a 64 kbps signaling channel is always maintained between IWF or between IWF and a voice signaling entity within the network, while the actual voice channels VCCs are established on demand. Multiple voice channels may share a single VCC, resulting in resource optimization.

Under CCS, in the absence of any established VCCs, the IWFs, and the message exchange follow the steps shown below:

❑   The source PABX generates a narrowband-signaling request (N-SETUP)
❑   The IWF responds with a CALL PROCEEDING and a connection request (B-SETUP) across the ATM network for the signaling channel
❑   The destination responds with a confirmation (B-CONNECT)
❑   Source IWF establishes one or more 64 kbps VCCs for the voice traffic
❑   The narrowband signaling messages can go across the IWF signaling VCC.
❑   Destination IWF forwards a N-CONNECT to the remote PABX
❑   Remote PABX responds with a N-CONNECT to the source PABX.

If both IWF signaling VCC as well as the 64 kbps voice VCC were already established, only the N-CONNECT and N-SETUP messages would be exchanged.

ITU-T signaling system 7 (SS7)

The goal of SS7 is to provide an international standard for common channel signaling suitable for stored program control exchanges, computers and PBXs. It operates on digital networks with DS-0 channels, and it accommodates low speed analog links if desired. It meets the requirements of call control, signaling for telecommunications services such us POTS, ISDN, and circuit-switched data transmissions services. The system includes redundancy of signaling information.

Trunk signaling methods

❑   d.c. (direct current) signaling (on-hook/off-hook)
❑   In-band pulse signaling
❑   Out-of-band pulse signaling
❑   Internal digital signaling.

Loop side signaling methods

❑   E&M
❑   CAS
❑   DTMF
❑   d.c. signaling.

8.6   Circuit Emulation Service

8.6.1   Service description

Circuit emulation allows the user to establish an AAL-1 ATM connection to support a circuit such as a full T-1 or E-1, over the ATM backbone. Specifically, the following types of CBR service have been defined:

✓   Structured DS-1/E-1 N × 64 kbps (fractional DS-1/E-1) service
✓   Unstructured DS-1/E-1 (1.544 Mbps, 2.048 Mbps) service
✓   Unstructured DS-3/E-3 (44.736 Mbps, 34.368 Mbps) service
✓   Structured J-2 N × 64 kbps (fractional J-2) service
✓   Unstructured J-2 (6.312 Mbps) service.

8.6.2 Structured and unstructured services

The *structured service* is the closest approximation to a wireline transmission. It is a point-to-point circuit emulation over PVC. A problem associated with this service is that the entire bandwidth has to be dedicated to this service (no knowledge of internal framing).

In the *unstructured N × 64 service*, the network manager can allocate the number of channels required and it is possible to group channels to go to different destinations from a single T-1/E-1 interface. (This is not possible with CCS.) ATM performs as an existing TDM and cross-connect system. It does not support SONET/SDH. One problem involved is the delay for cell construction (6 ms) and the need to wait for a cell to be full before sending it to the network. The recommendation is to use at least 6 channels and adopt a strict delay budget so as to avoid echo cancellation.

The *structured DS-1/E-1/J-2 N × 64 service* is modeled after a *fractional DS-1/E-1/J-2 circuit*, and is useful in the following situations:

- ✓ The N × 64 service can be configured to minimize ATM bandwidth by only sending the time-slots that are actually needed
- ✓ The N × 64 service provides clocking to the end-user equipment, so that it fits into a fully synchronous network environment
- ✓ Because it terminates the *facility data link*, the N × 64 service can provide an accurate link quality monitoring and fault isolation for the DS-1/E-1 link between the IWF and the end-user equipment.

The *unstructured* DS-1/E1/J-2 service provides transparent transmission of the DS-1/E-1/J-2 data stream across the ATM network and is modeled after an asynchronous DS-1/E-1 leased private line. It allows for the following situations:

- ✓ End-user equipment may use either the standard (SF, ESF, G.704 or JT-G.704) or non-standard framing formats when end-to-end communication of the *facility data link* or *alarm states* is important
- ✓ When timing is supplied by the end-user DS-1/E-1/J-2 equipment and carried through the network, the end-user equipment may or may not be synchronous to the network.

The *unstructured DS-3/E-3 service* provides basic DS-3/E-3 CES and allows for the following situations:

- ✓ A standard or a non-standard framing may be used by the end-user DS-3/E-3 equipment
- ✓ End-to-end communication of P-Bit, X-Bit, and C-Bit channels is provided
- ✓ Timing is supplied by the end-user DS-3/E-3 equipment and carried through the network. The end-user equipment may or may not be synchronous to the network.

8.6.3 ATM virtual channel requirements

The ATM virtual channels requirements are basically concerned with the PCR calculation for all service rates mentioned above.

Signaling

ATM and TDM signaling have no mapping specification. It takes the ATM signaling from the signaling specification of UNI 3.1 and corresponding SVC is optional for the IWF. Addresses and identifiers are defined here for SVCs. Also the SETUP signaling message parameters are defined, for example, forward PCR, backward PCR, quality of service parameters, and AAL-1 parameters. For each one of these parameters a specific code definition is established.

Call initiation procedures

In these procedures, the sequence of events that an IWF needs to implement for the automatic initiation of a SVC is suggested. For the initiation procedures, there are two modes of operation namely, the *passive* (wait for a call) mode and *active* (call periodically if a call is not taking place) mode. The operator can control the mode using SNMP.

Management

The IWF has two types of interfaces: ATM interface and CBR interface. A definition for the SNMP MIBs necessary to monitor and control both IWF interfaces is provided.

8.6.4   Dynamic bandwidth utilization

Service description (64 kbps time-slot trunking over ATM)

The dynamic bandwidth CES (DBCES) has been defined as an improvement over the CES 2.0 in which, as mentioned before, the entire BW has to be dedicated to this service (no knowledge of internal framing).

DBCES specifies a method for enabling dynamic bandwidth utilization in an ATM network based on detecting which time-slots of a given TDM trunk are active and which are inactive. When an inactive state is detected in a specific time-slot, the time-slot is dropped from the next ATM structure and the bandwidth it had been using may be reutilized for other services. This method may be applied by adopting any method of time slot activity detection, for example, CAS and CCS.

IWF

In this case, the interworking functions perform the following functions:

- ✓ Circuit emulation services (CES), and structured DS-1/E-1 N × 64 kbps service (as per ATM Forum af-vtoa-0078.000 section 2)
- ✓ Time-slot activity detection
- ✓ Dynamic structure sizing (DSS) of the AAL-1 structure, which correlates with the active time slots in the TDM to ATM direction
- ✓ Recovering the active time-slots from the AAL-1 structure, in the ATM to TDM direction, and placing them in the proper slots in the TDM stream
- ✓ Placing the proper signals (for example, ABCD) in each of the time-slots of the recovered TDM stream.

ATM virtual channel requirements

The consideration here is that the actual PCR will depend on the number of active channels in opposition to the simple CES case where the bandwidth utilization is fixed.

Signaling

Again UNI 3.1 SVC support is optional for DBCES. The call/connection control procedures of UNI 3.1 apply. The SETUP message is defined in a similar way as in specification for CES.

Call Initiation procedures
Same as CES 2.0

Management
Same as CES 2.0

8.6.5   Low Speed circuit emulation service

Service description

The low speed CES (LSCES-IS) specifies point-to-point unstructured CES for low speed interfaces like V.35 and EIA-449. The purpose of this specification is to provide an interoperability reference for ATM equipment that needs to interface with several pieces of equipment, which handle applications such us video conferencing, low speed multiplexers, WAN interfaces, encrypted data, and telemetry. In other words, the LSCES interface should emulate a point-to-point EIA-449 circuit or a V.35 circuit.

IWF

Two timing modes are available to the equipment attached to the LECES interface: *Synchronous*, in which the LSCES interface supplies the timing and may be traceable to the primary reference source, and *asynchronous*, in which the timing information is transported using SRTR or adapting timing.

ATM virtual channel requirements

Peak cell rate (PCR) is calculated for V.35 and EIA-449

Signaling

Signaling procedures correspond to those of UNI 4.0

Call initiation procedures

These apply to the initiation procedures for establishing a SVC between two circuit emulation processes.

Management

SMNPv2 compliant.

8.6.6   ATM trunking using AAL-1 for narrowband services V1.0
Service description

This specification defines the capabilities of the interworking function to provide a means for the interconnection of two narrowband networks through an ATM network via ATM trunks. An ATM trunk is defined here as one or more ATM VCs that carry a number of 64 kbps narrowband channels and associated signaling between a pair of IWFs.

The IWF's main function is to adapt the narrowband input interface into one that could be carried on an ATM network. The IWF for ATM trunking using AAL-1 provides a call-by-call switched service to the narrowband network. For the support of out-of-band signaling, the IWF terminates the narrowband signaling and transports all narrowband messages in a separate signaling VCC to the remote IWF. For the support of CAS, the IWF terminates the narrowband signaling, and signaling information is transported in the same ATM virtual connection that carries the voice information. The services available to the narrowband network are independent of those in the ATM network.

IWF

IWF refers to the following: *Signaling termination* (extraction and inserting of narrowband signaling), *call handling* (call setup and release from the narrowband equipment), *switching* (any channel to any ATM trunk), and *multiplexing* (channel combination from multiple sources into the same trunk).

Signaling

There are two types of signaling definition used depending on the interface considered: The narrowband interface signaling is done by means of N-ISDN or CAS signaling. The IWF-IWF signaling is Q.931 or CAS (E&M and DTMF).

8.6.7   ATM trunking using AAL-2 for narrowband services
Service description

The ATM trunking using AAL-2 for narrowband services presents a solution for an efficient transport mechanism to carry voice, voice-band data, circuit mode data, frame mode data, and fax traffic. Voice transport includes support for compressed voice and non-compressed voice together with silent removal.

There are two alternatives for trunking switched and non-switched efforts. In the first case, the narrowband channel can take any AAL-2 channel within a VCC between IWFs; the second case addresses the situation in which there is a permanent correspondence between a narrowband channel and the AAL-2 channel and the VCC designated for its support. Two possible applications have been described in the specification: Access trunking to the PSTN and PBX to

PBX trunking. Both cases involve voice signals crossing the ATM network and the IWF being able to concentrate both based-voice data or compressed voice.

IWF

The IWF provides options for voice band data (a regular dial-up modem signal) , fax data, circuit mode data for N × 64 kbps, DTMF information through DTMF packets, and frame mode data through SAR functionality. The salient IWF functions and their descriptions are shown in Table 8.2.

Table 8.2: IWF functions

IWF Function	Description
Multiplexing/demultiplexing	To combine and recover individual narrowband channels
Switching	Any narrowband channel connected to any AAL-2 channel
Signaling termination	Receive/insert signaling from the narrowband interface into both the TDM narrowband and the ATM interface
Call handling	Interpret call setup and release from the narrowband interface
SSCS user functions	Voice codecs for compression, fax demodulation, etc.
AAL-2 SSCS	User information formatting for transport in AAL-2 channels
AAL-2 CPS	For multiplexing AAL-2 connections into ATM cells
SAAL	SSCF+SSCOP+AAL-5 CPCS

Signaling

In order to support the two modes of trunking mentioned before, there are two types of narrowband signaling that the IWF has to be capable of: Signaling termination (for switched trunking) and signaling without termination (for non-switched trunking). The first one carries the control information via either CCS using AAL-2 or AAL-5, or via CAS with DTMF. For the second case, the control information signals are transported either using CAS with DTMF or CCS using an AAL-5 VCC.

Voice encoding

The support for voice encoding is application-dependent. IWF's may support any ITU-T voice-encoding algorithm.

Idle channel suppression

Based on the ABCD signaling bits or CCS messages, the IWF can detect an idle channel and allocate the corresponding bandwidth to other applications or channels. This is called *idle call state termination*. Another method is called *idle code detection*, where the IWF is capable of analyzing certain patterns that indicate when a channel is idle.

Silence detection and removal

During non-idle call periods there are periods of silence on a transmission path, either because the other party is speaking or due to the regular silence regimes in speech. Improvements in bandwidth efficiency can be gained by not transmitting during those silent intervals.

Echo cancellation

For voice applications, the combination of delay and echo causes impairment to speech quality. Delay is introduced into the end-to-end connection by a variety of factors including packetization, compression algorithms, physical transmission time, and switching of ATM cells in the network and queuing. As described in Chapter 2, echo is caused by hybrids used to go from a four-wire to two-wire circuit and acoustical feedback at the end user's terminal (especially when the user places a telephone receiver down on a hard surface). IWF may support echo cancellation.

8.6.8 *Voice and telephony over ATM: AAL-1 or AAL-5 option*
Service description

The ATM Forum's specifications (af-vtoa-0083.0000 and af-vtoa-0083.0001) specify the definition of an interoperability agreement for supporting voice and telephony over ATM to the desktop. In the specifications, the functions that a native ATM terminal and an internetworking function reare quired to provide are defined, and the three types of attachments are indicated, namely,

✓   B-TE (*broadband terminal equipment*) to private or public B-ISDN
✓   Private B-ISDN to public N-ISDN
✓   Private B-ISDN to private N-ISDN.

The B-TE can establish a voice communication through a B-ISDN on a per-call basis. The traffic supported is 64-kbps PCM-encoded voice (G.711). Each voice call uses one VCC. The B-TE may invoke supplementary services on a per-call basis. A B-TE has either a public UNI or a private B-ISDN. A B-TE attached to a private B-ISDN has the option of either using AAL-1 or AAL-5. However, a B-TE attached directly to a public B-ISDN may have to support the AAL-1 specifications for voiceband signal transport to conform to the public B-ISDN offerings.

IWF

The IWF converts the voice traffic on the ATM side (B-ISDN) to voice traffic on the narrowband side (N-ISDN). On both sides the traffic resides in the 64 kbps PCM encoded voice. One ATM VCC is mapped to one N-ISDN channel dynamically on a per-call basis. For this purpose, the IWF also maps the B-ISDN signaling information (on VC = 5) to the N-ISDN signaling information (on the D channel).

Signaling

Interworking between the N-ISDN basic call signaling protocol and the B-ISDN basic call/connection control signaling protocol is achieved by fully terminating each protocol in the IWF.

8.7   Interworking Frame-Relay (FR) and ATM

Because of its attractive pricing structure, FR has been favored as a corporate backbone networking. However, the quality of service (QOS) and support for very high-speed trunks are two things (obtainable with ATM) that the frame-relay does not offer. While frame-relay is suited for applications including LAN internetworking, SNA migration, and remote access and other applications, such as broadcast video and server farm support, it may be better suited for ATM networks. Hence, there is a niche in getting the best of both technologies.

Basically there are two options for connecting frame-relay and ATM: One can run a frame across an ATM backbone or simply connect the frame-relay and ATM networks directly. The defined schemes that make such internetworking possible are *frame-relay/ATM service interworking*, and *frame-based user-to-network interface* (FUNI).

Frame-relay/ATM interworking is a solution that provides users with low cost access to high-speed networks. Ratified by both the ATM and Frame Relay Forums, the *frame-relay/ATM PVC interworking implementation agreements* (IAs) provide a standards-based solution for interworking between existing or new frame-relay networks and ATM networks without any changes to end-user or network devices.

Technical functions supporting end-to-end communications in a frame-relay/ATM network are specified by the IWFs generally located at the switch where the two services meet. The IWF is primarily responsible for mapping various parameters and functions between frame-relay and ATM networks. These include the following:

✓   Frames or cells are formatted and delimited as appropriate
✓   Discard eligibility and cell loss priority are mapped
✓   Congestion indications are sent or received appropriately (frame-relay's forward explicit congestion notification is mapped to ATM's explicit forward congestion notification)
✓   DLCI to VPI/VCI mapping is performed
✓   Data encapsulation methods are converted (in the case of service interworking).

PDU formatting and delimiting is accomplished by remapping the FR protocol stack into an ATM stack at the ingress of the network for segmentation and transport over AAL-5 and vice versa at the egress. The IWF also supports traffic management by converting ATM and frame-relay traffic conformance parameters, supporting PVC management interworking via status indicators and providing upper layer user protocol encapsulation.

8.7.1   Network interworking

Basically network interworking permits two frame-relay devices to be linked transparently across an ATM network. Neither of these frame-relay devices knows that there is an ATM network in the middle. The ATM cloud is essentially transparent to the devices at the subscriber end and switches are basically tunneling devices, meaning that they carry the header and payload data as they are. This means that multiprotocol LAN traffic (and other higher layer procedures) are transported transparently, just as they would be over leased lines as illustrated in Fig. 8.25.

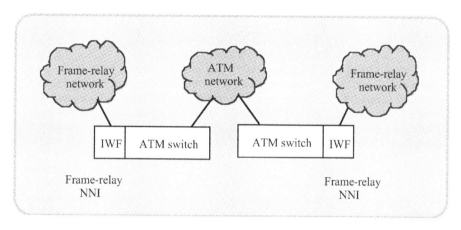

Fig. 8.25: ATM and frame-relay internetworking

Typically, network interworking is used in situations where a network manager uses the ATM public network in lieu of a leased line, to connect two or more frame-relay (FR) networks. The network interworking function is typically integrated into the ATM and/or frame-relay switches at the point where the two networks meet. From the perspective of frame-relay networks, the networking/interworking interface actually appears to be a standard frame-relay NNI (network-to-network interface).

The IWF enables all those mapping and encapsulation functions necessary to ensure that the service provided to the frame-relay CPE is unchanged by the presence of an ATM transport. As shown Fig. 8.26, the *frame-relay service specific convergence sublayer* (FR-SSCS) is a part of the FR/ATM IWF protocol stack. The FR-SSCS uses a PDU format identical to Q.922 core minus the CRC-16, FLAGs, and zero bit insertion. The SAR and CPCS (common part convergence sublayer) of AAL-5 in conjunction with the ATM-layer ATM *user-to-user indication* (end of PDU) provides segmentation and reassembly (frame delimiting) for FR-SSCS PDUs. The AAL-5 CPCS CRC-32 provides error detection over the FR-SSCS PDU. The discard eligibility (DE) field in the Q.922 core frame is copied unchanged into the DE field in the FR-SSCS PDU header and mapped to the ATM *cell loss priority* (CLP) bit of every ATM cell generated by the segmentation process of that frame. The frame level FECN is mapped to the cell level EFCI.

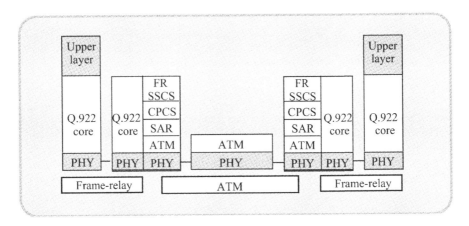

Fig. 8.26: FR/ATM IWF protocol stack

Each frame-relay PVC can be carried over individual ATM PVCs (many-to-one multiplexing), or all of the frame-relay PVCs can be multiplexed onto a single ATM PVC (one-to-one multiplexing). This is accomplished by using the *data link connection identifier* (DLCI) value at the FR-SSCS sublayer. The values used here have to be agreed upon between the two ATM end systems, because they otherwise have no significance outside of their originating networks. Multiplexing multiple frame-relay PVCs is more cost effective than using multiple leased lines.

8.7.2   Service interworking

Service interworking is defined in the *Frame-Relay Forum's FRF.8 specification*. It allows a frame-relay device to communicate with an ATM device. Either of these devices does not realize that the other is a "foreign" technology. Instead of tunneling frame-relay traffic across the ATM net, service interworking relies on protocol conversion and there is no communication with a distant frame-relay network. This is in contrast to network interworking in which FR frames are transported over ATM and processed at the FR-SSCS within the CPE. This means that the CPE must be configured to interoperate with the distant FR network.

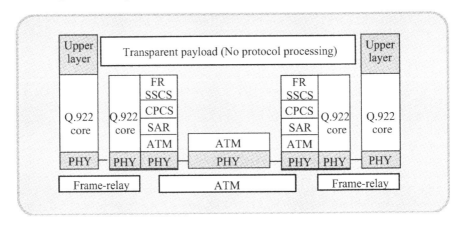

Fig. 8.27: Internetworking function for TPS

Fig. 8.27 shows the protocol stack for the interworking function, for *transparent protocol support*. The protocol stack uses a "null" SSCS for describing the interworking function. Within the interworking function, this SSCS provides interfaces using standard primitives to the UNI Q.922 DL core on one side, and to AAL-5 CPCS on the other. The FR frame is mapped into an AAL-5 PDU; the flags of the FR frame, inserted zero bits, and CRC-16 are stripped. The Q.922

DL core frame header is removed and some of the fields of the header are mapped into the corresponding ATM cell header fields. The service IWF maps the frame-relay DLCI to the ATM VPI/VCI, the FECN bit maps to the PT field, and the DE bit maps to the CLP bit.

For *transparent protocol support*, when encapsulation methods other than RFC 1490 or 1483 are used, the IWF forwards the data unaltered. Transparent mode is used when the terminal equipment on one side of the IWF uses the encapsulation method of the terminal equipment on the other side.

Service interworking using the protocol translation mode is shown in Fig. 8.28. The FR multiprotocol encapsulation procedures (RFC-1490) are not identical to the ATM multiprotocol encapsulation procedures (RFC-1483 and RFC-1577). When providing FR/ATM Service Interworking for multiprotocol routers, it is necessary for the IWF to convert the multiprotocol protocol data unit headers from FR to ATM and vice versa. Translation mode also supports the interworking of routed and/or bridged protocols (for example, ARP translation).

Service interworking is the interface implementation of choice if there is a need to route IP traffic. In order to pass IP traffic transparently, service interworking must also handle translation of the upper layer (IP) encapsulation technique. Inasmuch as this service is based on protocol conversion it is well suited for this translation: Frame-relay uses IETF RFC 1490 to encapsulate IP traffic while ATM uses RFC 1483.

In using the service networking, the frame-relay PVC culminates where the ATM PVC begins, and vice versa. Invariably, there is a one-to-one correspondence of frame-relay and ATM PVCs — unlike network interworking. This is, however, disadvantageous to frame-relay networks with short average frame lengths because they incur extra overhead on ATM networks.

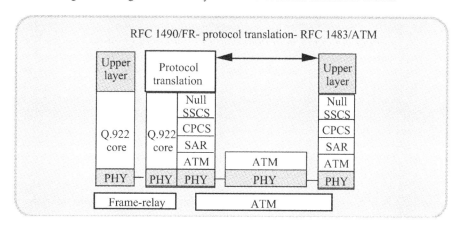

Fig. 8.28: Service IW using the protocol translation mode

Both network interworking and service interworking can be external to the network. In practice, they are integrated into ATM and/or frame-relay switches at the point where the networks handshake. Since service interworking directly connects frame-relay networks to ATM backbones (LAN or WAN), it facilitates an unmasked migration path to ATM.

Another merit of using service interworking is that network managers can maintain their investment in legacy network equipment. For example, they can install ATM at high-bandwidth data centers and connect branch sites using frame-relay, without having to modify the frame-relay services at the remote site. This is true when these branch offices need only the data services and do not require the QOS guarantees of the ATM for voice and video. If the WAN linking the branch offices to these data centers is dominated by frame-relay technology, then service interworking is the logical choice.

8.7.3   FUNI

Network interworking and *service interworking* are the two more common strategies in networking setups largely based on frame-relay. But in WANs (where ATM already dominates) it is more appropriate to deploy ATM access devices at all branch offices. This is an ideal scenario for FUNI (*frame-based user-network interface*). It allows CPE without ATM hardware to have cost-effective access to ATM WANs through the use of a frame-based format defined by the ATM Forum. This is done via firmware upgrades in routers, for example.

FUNI warrants FUNI-compatible software in the equipment at the user-end and a complementary frame-based interface. Also, it needs FUNI software in the switch to which the user equipment connects. Users send FUNI frames to these carriers that offer FUNI service. Within the switch interface of the carrier, the frames are segmented into cells and are relayed across the ATM WAN backbone. Cells coming from the network are reassembled into frames and sent to the user. A significant advantage expense-wise is the fact that hardware costs of segmentation and reassembly are transferred from the user equipment to the switch where it can be shared across a large number of users. Since frames of up to 2000 bytes in length are sliced up at these centralized switches, more efficient use is made of the access line bandwidth.

FUNI can be regarded as an improvement over another interface for interworking into ATM, namely, the ATM DXI (*data exchange interface*). In ATM DXI, the data is segmented on the user side. The DXI standard requires an external DXI-enhanced CSU (*channel service unit*) to convert frames before they are sent over an access line and DXI software in the user equipment. It essentially provides a segmentation and reassembly function in an external piece of equipment, which is located at the customer premises.

The DXI-enhanced CSU has always been a significant concern in respect of the cost involvement, which has, hence, been rendered unnecessary using FUNI. The consequence is that DXI sends cells over the WAN access lines whereas FUNI sends frames and therefore bandwidth utilization is less efficient than the FUNI. Also FUNI supports $N \times 64$-kbps rates, while the lowest speed supported by ATM DXI is DS1/E1.

There is a method akin to accessing the ATM network over a frame interface. It refers to SMDS-DXI (Switched Multi-megabit Data Service). It is in fact a variation of the ATM-DXI making use of a different encapsulation over the DXI.

The major benefit of ATM FUNI is that it is *ATM ready*; it can provide for signaling, QOS and OAM support. Although limited to VBR services, it uses the same schemes as cell-based ATM UNI in the following areas:

- ✓   Upper layer multiprotocol encapsulation and address resolution
- ✓   Traffic parameters
- ✓   ILMI
- ✓   ATM OAM functions
- ✓   SVC signaling
- ✓   VPI/VCI multiplexing.

DXI and FUNI allow frame-based access to an ATM network, while frame-relay allows frame-based access to a frame-relay network. FUNI is not designed to provide direct interoperability between ATM users and frame-relay users. That is, one cannot simply connect frame-relay user equipment to an ATM network's FUNI and expect it to function in that context. Other than having a frame structure similar to frame-relay and the ability to operate on the same type of hardware as frame-relay, FUNI has little in common with frame-relay.

Further, the entire ATM services are not available over FUNI. FUNI mandates the support of AAL-5 (in the ATM switch), while AAL-3/4 is optional. Services requiring the use of other ATM adaptation layers are not supported over FUNI. Also, support of some ATM QOS classes like ABR is not possible.

For a SVC to be established, it does not matter whether the destination is another FUNI, a DXI, or an ATM UNI. When the FUNI user sends traffic to the ATM network, it is not a concern whether the traffic terminates at another FUNI or at an ATM UNI. All of these service considerations are transparent.

FUNI frame formats

DXI, FUNI, and frame-relay are similar in their frame structures. The DXI and FUNI header within the frame are identical to each other, but are different in respect of the frame-relay header.

A FUNI PDU is almost identical to a frame used in frame-relay transmissions. It has a flag, header, payload, frame check sequence, and another flag. The header contains a 10-bit frame address. Six FUNI address bits are mapped into the ATM VCI and the remaining four get mapped into the VPI. This allows for a limited form of VPC service, supporting 16 VPCs or 1024 VCCs. The two *frame identifier* (FID) bits determine whether the FUNI frame contains either user information (that is, data, signaling, or ILMI) or an OAM cell. The *congestion notification* (CN) bit maps to the *explicit forward congestion notification indication* (EFCI) payload type value in the ATM cell header. The *cell loss priority* (CLP) bitmaps to the corresponding bit in the ATM cell header. The 0 and 1 bit in the header serve as an address extension function.

8.7.3   Bridging ATM and frame-relay

There are still a number of technical burdens to be considered when attempting to bridge the two technologies of FR and ATM. A consensus-based ideal solution has not been yet reached. A persisting issue refers to the mapping of frame-relay traffic to ATM, which presents a challenge for both network managers and service providers. Other major considerations are frame size distribution, traffic shaping and policing, QOS support, addressing differences, and incompatible signaling functions.

In service interworking, the percentage of ATM cells with mostly empty payloads increases as the size of frame-relay frames decreases. In the worst-case scenario, for example, it might take two 53-byte ATM cells to transport one 64-byte frame-relay frame, which would add up to 50% more overhead. This then adds to higher bandwidth requirements for the ATM implementation versus the equivalent frame-relay implementation. This may not be a typical scenario, but it shows how knowing the expected frame size distribution is important, especially for the service providers before they make any QOS guarantees.

Traffic shaping, (see Chapter 5) refers to the controlling of the traffic load offered to the network, in order to minimize congestion. It smoothes out traffic flow and reduces cell clumping, which results in a fair allocation of resources and reduced average delay time. Traffic shaping takes place on the transmitting interface, in conformance with the traffic contract and uses a form of the leaky-bucket algorithm known as token bucket.

Traffic policing as indicated in Chapter 5 refers to the regulation of the flow of data at the receiving end or at intermediate nodes, whereby cells that exceed certain performance levels are discarded or tagged. In ATM, receiving interfaces employ traffic policing in the form of the UPC, a way to compare the behavior of the VC to what it has been informed to expect by the traffic contract. If a large number of cells arrive at a receiving port of a switch, it can either mark the CLP bit in the ATM header for later discard, or drop the cells until congestion clears. For this function, a GCRA leaky-bucket paradigm is adopted.

Quality of service is another issue that still remains to be studied and implemented in some way using frame-relay. ATM was conceived with QOS in mind. On the other hand, the frame-relay was not, because it was initially developed as a data-only service. Now, the frame-relay may have to provide some sort of QOS, because it is being used more and more for time-sensitive traffic like voice and video.

In the context of ATM, the two classes of service, namely, CBR and ABR lack standards-based frame-relay equivalents. For these service classes to be mapped onto frame-relay, the possible techniques required to translate the signaling information still remain as open questions.

A standard for translating addresses between both environments also needs to be implemented. The frame-relay addressing is based on the E.164 and X.121 specifications defined by the ITU. ATM addressing supports E.164, as well as three forms of ATM end-system addresses (AESAs) defined by the ATM Forum. E.164 addresses are obviously easy to translate between both environments. AESAs, on the other hand, are not. Most ATM customer premises equipment supports AESAs. Current efforts by the respective technical committees include developing FRESAs (frame-relay ESAs), frame-relay equivalents to AESAs.

Signaling is another area that has to be brought into a common platform while implementing FR and ATM jointly. Not only are the signaling protocols to be translated, but also the functions they perform (which are not always mirrored in the other environment) need to be redefined. For example, the ITU's X.76 spec defines frame-relay SVC signaling over an NNI. This specification forms the basis of the Frame-Relay Forum's FRF.10 recommendation and it defines such value-added services as reverse charging, transit network identification, and closed user groups. On the other hand, ATM Forum's counterpart specification, namely, PNNI has no such equivalent. Similarly, ATM has a strong set of OAM functions for circuit loopbacks and continuity checks, while frame-relay does not.

8.8 TCP/IP on ATM

Transmission control protocol/Internet protocol (TCP/IP) is one of the two protocol architectures which forms the basis of interoperable communication standards. (The other is the OSI reference model.) TCP/IP in essence is an interoperable architecture model and OSI is a standard model for classifying communications.

Implementation of TCP/IP in end-systems of a communication model is illustrated in Fig. 8.29. Its layer functions are as follows:

- ✓ Application layer: This contains the logic needed to support the various user applications such as file transfer.
- ✓ Network access layer: It is concerned with the exchange of data between an end-system and the network to which it is attached.
- ✓ In the case where two devices are attached to different networks, IP procedure provides a routing function to let data traverse multiple interconnected networks. This protocol is available both in end-systems as well as in routers.
- ✓ TCP protocol enables a transmission control. That is, it provides a control so that when different applications exchange data, all such data arrive at the destined application reliably and in the same order in which it is sent. TCP is an embedment of a transport layer (or host-to-host) layer shared by all applications.
- ✓ Physical layer: This is the physical interface between a data transmission device (for example, a workstation, computer etc.) and a transmission medium or network.

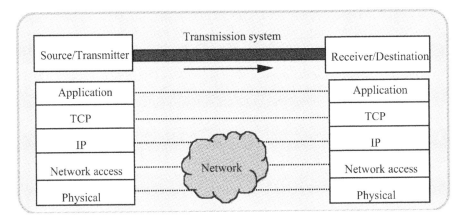

Fig. 8.29: TCP/IP protocol structure

For ATM to be adopted as a feasible infrastructure to facilitate data traffic, it has to able to interwork with legacy network equipment and with the protocols used by these network components. The common protocols used in legacy are the *transmission control protocol* (TCP) and *Internet protocol* (IP) indicated above. The underlying concept behind TCP/IP support on ATM can be understood by delving into the following considerations:

    ✓   The way ATM addressing is done among switching elements within an ATM network and PNNI protocol for routing within ATM networks
    ✓   Implementing Internet protocol (IP) over the ATM network
    ✓   Supporting transmission control protocol over the ATM network.

8.8.1   *Internet structure and facilitation of its support on ATM*

The Internet traffic is made of two parts namely, the downward traffic towards the end-user and the upward traffic commencing from the user side.

In the downward direction of Internet traffic supported by ATM, each protocol module adds its header information to create ethernet frames. Then these ethernet frames travel to the destination, which is an ethernet port of one of the ATM switches, namely, the *network management system controller* (NMS). The NMS controller segments the ethernet frame into 48 octet cells in the AAL layer. The ATM layer adds the ATM header and passes the cells to the ATM switch. In the upward direction, the ATM cells are reassembled into an ethernet frame. The receiving ethernet driver passes it upward to either the ARP module or IP module.

Addressing within ATM networks

In order to carry IP traffic over ATM networks, the first task is to resolve the destination IP address into an ATM address. Then, a route should be established to forward the traffic to the destination. If a PVC is used, then the connection has already been established and data transmission may begin. On the other hand, if a SVC is used, signaling procedures should precede data transmission.

ATM addresses are commonly referred to as *network service access point* (NSAP). The format used is specified in (network-host) form, where the network parameter identifies a particular subnetwork while the host parameter identifies a particular end-system on the subnetwork. The NSAP addressing plan specifies each address to an initial domain part (IDP) followed by a domain specific part as discussed in Chapter 4.

8.8.2   *Routing within ATM networks*

ATM routing occurs in private or public networks by different procedures. Routing in public networks has not been subjected to standardization, but within the private networks the ATM Forum has established a standard called the *private network-network interface* (PNNI). The PNNI aims at providing interoperability in multivendor networks.

The PNNI is a hierarchical routing protocol that includes the following components:

    ✓   *ATM switching nodes*: Identified by an ATM address
    ✓   *Peer groups*: A logical grouping of ATM nodes
    ✓   *Border node*: An ATM node that has a link to a node in another peer group
    ✓   *Horizontal link*: A logical link between ATM nodes within the same peer group
    ✓   *Outside link*: A logical link between ATM nodes within different peer groups
    ✓   *Peer group leader*: An ATM node that is responsible for aggregation of peer group information.

The functions of the PNNI routing protocol include embedding topological information in hierarchical addressing to specify the routing details. Such information is used recursively at each hierarchical level. The PNNI functions can be specifically enumerated as follows.

- *Enabling the discovery of neighbors and link status*: An ATM node first discovers its neighboring nodes by executing a "Hello protocol". It involves sending a "Hello packet" to any neighboring nodes that might be on the far end of the link. The Hello packet includes a node ID, ATM end system address, peer group ID and a Port ID

- *Procedure synchronization of topology databases*: Relevant databases that provide information that allows the node to compute a route to any reachable ATM address

- *Flooding of PNNI topology state elements (PTSE) to other peer group members*: The flooding process involves a node sending PNNI topology state packets to each neighboring node

- *Electing the peer group leader (PGL)*: The PGL is responsible for aggregating information within the peer group and flooding that information to other PGLs

- *Making a summary of topology state information*: The PGLs summarize peer group information prior to flooding to the higher-level peer group

- *Constructing of the routing hierarchy*: Pertinent to the exchange of details as above, a PNNI-capable switch receives a SETUP message across a UNI, it determines a route, and then appends a designated transit list (DTL). The DTL has a list of ATM addresses to be transited along the path to the requested connection endpoint. As each switch processes the SETUP message, it removes the information element containing its address from the DTL and passes the message on to the next node

- *Specifying the route*: The selection of the "best" route is decided by two parameters:

  ✓ *Metrics*:   Entities that determine the suitability of the path based upon parameters whose values are meaningful when applied over the entire end-to-end path
  ✓ *Attributes*:   Parameters, which decide whether a given link or a node is suitable for inclusion in a route based upon parameters whose values may be associated with a specific link or node.

The PNNI has the ability to make routing decisions within the ATM networks based upon QOS parameters.

Classical IP over ATM: Revisited

As indicated earlier, classical IP over ATM allows IP traffic to be supported by ATM networks by resolving destination IP addresses into ATM addresses, which are then used for routing the traffic over the ATM network to the destination. The underlying considerations to perform this address resolution are called the *ATM address resolution protocol* (ATMARP). The relevant aspects of classical IP over ATM are described in RFC 2225. It is based on RFC-1483 for encapsulation and RFC-1577 for address resolution.

The components of implementing a *classical IP over ATM* are:

    ✓   ATM-attached stations: These may be end stations and/or LAN access devices. A *logical IP subnet* (LIS) is formed by groups of these stations
    ✓   ATM switches
    ✓   ATMARP servers: Typically, these are implemented in an ATM switch. Each LIS must have an ATMARP server, which can resolve IP

addresses into ATM addresses. A single ATM switch may act as an ATMARP server for multiple LIS, but the switch must be a member of each LIS

✓ Router: Whenever hosts on different LIS want to communicate, all traffic should be sent to a router. This router forwards the traffic to the destination. The router must, however, have membership in each LIS for which this service is to be performed.

Merits of the classical IP:

✓ Routers stand as firewalls and prevent unwanted access to an LIS
✓ Network administrators can configure ATM networks using the same models as used for legacy networks.

Demerits of the classical IP:

✓ Classical IP prevents guarantees on end-to-end QOS.

Communication between stations in the same LIS

Suppose a station X1 communicates with another station X2 over the ATM network and both stations are in the development LIS.

X1 has an IP address to X2 but it does not know the ATM address of this station. Then, for SVC service the transaction proceeds as follows:

When X1 does not find the ATM address of X2 in the IP-to-ATM address translation table, it sends an ATMARP request to the ATMARP server. The ATMARP server sends a reply to X1 containing the ATM address of X2. Based upon this reply, X1 then establishes a direct connection to X2 using ATM signaling and X1 and X2 exchange data. After the transaction is over, the connection is retained for a while to allow subsequent traffic to be immediately exchanged between the two stations without the latency of another connection setup.

Communication between stations in different LIS

In this case, suppose a node Y1 in the development LIS wishes to communicate with a node Z2 in the production LIS. The relevant transaction for a SVC service proceeds as follows:

Suppose Y1 does not find the ATM address of Z2 in the IP-to-ATM address translation table. It then sends an ATMARP request to the ATMARP server. Inasmuch as the destination IP address belongs to a station in another LIS, this communication must go through a router. The ATMARP server sends a reply with the router ATM address. Y1 then establishes a direct connection to the router using ATM signaling and it sends data to the router. If the router does not know the ATM address of Z2, it will issue an ATMARP request to the ATMARP server using the IP address of Z2. The ATMARP server for the production LIS sends an ATMARP reply to the router containing the ATM address of Z2. The router then establishes a direct connection to Z2 via ATM signaling and the router forwards the data to Z2.

ATMARP packets under classical IP over ATM are encoded in AAL-5 protocol data units using LLC/SNAP encapsulation. Classical IP over ATM is concerned only with the maintenance and resolution of IP and ATM addresses, and not with subsequent connection establishment between end stations, or encapsulation rules used for data transfer over the connection.

Next hop resolution protocol (NHRP): Revisited

In reference to classical IP over ATM, whenever two end stations are in different LISs, the traffic should travel over two IP hops, namely, one from the source to the router, and the other from the destination. The two stations, residing on the same network, could be connected directly via an SVC. Such a route, which crosses LIS boundaries, is called a *cut-through route*.

The *next hop resolution protocol* (NHRP) supports the cut-through routing. In NHRP, the IP to ATM address resolution queries are sent from a *next hop client* (NHC) to a *next hop server* (NHS) using an NHRP request. The process is as follows: Suppose a sending node S1 does not find the ATM address of a receiving node R2 in its IP-to-ATM address translation table. Then S1 sends an NHRP resolution request to the NHS for the LIS containing the IP address for R2. The NHS sends a NHRP resolution reply to S1 containing the ATM address of R2. S1 establishes a direct connection to R2 using ATM signaling, and S1 and R2 exchange data.

In the event of the first NHS being not able to provide a reply, the request can be forwarded to the second NHS. This scenario for a NHRP service refers to the following: Suppose a node C1 sends a NHRP resolution request to NHS1 for the ATM address of D2. NHS1 forwards the NHPR resolution request to NHS2. Then NHS2 sends an NHRP resolution reply to C1 containing the ATM address of D2 enabling C1 to establish a direct connection to D2. Subsequently C1 and D1 would exchange data.

If NHS2 is not in a position to resolve the request, the request will be sent to yet another NHS. This process will continue following an IP route to the destination until a NHS, which is capable of resolving the NHRP resolution request is found.

After the IP address is resolved to an ATM address by the serving NHS, a NHRP resolution reply is sent to the originating NHC. This reply can be sent directly from the serving NHS to the originating NHC, if a direct SVC/PVC exists. Otherwise, the reply can trace a reverse path adopted by the NHRP request.

Another situation may involve the use of a router in a multihop connection between an ATM station and a station that is not directly connected to the ATM networks. For example, suppose a node K1 sends an NHRP resolution request to the NHS requesting the ATM address of a node L2. The NHS forwards the NHRP resolution request to a router that can reach L2. Since the router is aware that L2 is not on the ATM network, it will send an NHRP resolution reply to K1 containing its own ATM address. K1 then establishes a direct connection to the router using ATM signaling. Hence, K1 and the router can now exchange data. The router assembles the ATM cells into IP packets and forwards them over the legacy network to legacy LAN router. The legacy LAN router then forwards the IP packets to L2.

Multicast IP over ATM networks

In addition to the point-to-point connection between IP hosts, ATM should be able to support IP multicast. The solution offered in ATM Forum UNI 3.0. /3.1 uses a *multicast address resolution server* (MARS) to enable multicast between IP hosts on the same LIS. The MARS is the multicast equivalent to the ATMARP server of classical IP over ATM. A MARS responds to address inquiries using class D IP addresses by returning a list of ATM addresses associated with a given multicast IP address.

ATM multicast IP configuration

A configuration that can be used for an ATM multicast IP is a mesh of directly connected IP hosts. Alternatively, the configuration can use a point-to-point connection from each IP host to a multicast server (MCS). The MCS maintains a point-to-multipoint broadcast connection to each host. In this case, the MCS may become a performance bottleneck limiting ATM multicast performance in comparison to the multicast mesh. However, the multicast mesh requires a greater number of VCs.

The MARS usually communicates to the hosts over a point-to-multipoint cluster control VC. This cluster refers to a set of ATM hosts that participate in direct ATM connections to achieve multicasting among the group. When a MCS is employed, the MARS maintains a server control VC, which is used to inform the server about current cluster membership.

The MARS does not have a direct impact in multicasting. Its main function is to maintain the mappings between IP addresses and ATM addresses. The MARS maintains two types of address maps, namely;

✓   *Host map*, which contains a mapping between an IP address and a set of ATM addresses for each ATM host in the cluster

✓   *Server map*, which contains a mapping between an IP address and the ATM addresses for the MCSs associated with the cluster.

TCP over ATM

TCP facilitates a connection-oriented, flow-controlled, block data transfer protocol for the transmission of files and data stream. TCP guarantees complete, error-free sequenced data delivery over a "best effort" datagram delivery service but makes no guarantee regarding delay. It uses a sliding window to mange data flow. But it does not use explicit rate control and has no explicit congestion notification either.

The TC protocol adjusts its data rate to available bandwidth and recovers from data loss and out-of-sequence delivery.

TCP sliding window flow control

This protocol allows the window size to vary as a function of time. Each acknowledgement specifies the number of octets that have been received and also includes a window advertisement to indicate how many more octets of data the receiver is ready to accept. The functions of TCP sliding window protocol are as follows:

- ✓ If the window advertisement indicates a larger window size than the previous acknowledgment, the sender increases the size of its sliding window in response
- ✓ If the window advertisement indicates a smaller window size than the previous acknowledgment, the sender reduces the size of its sliding window in response.

This protocol enables flow control, reliable data transfer, and allows TCP to adapt to the available bandwidth on the network because of its variable window. However, the sliding window protocol does not offer a flow control mechanism between the source and intermediate nodes. This mechanism is asserted only between communicating endpoints. The intermediate nodes can be congested depending on the amount of data being received. To correct this condition TCP uses a congestion control mechanism.

Under congestion, data transit time across the network increases all the intermediate nodes. As a result, due to queueing, the buffers may overflow. Hence data may be lost.

TCP window size

TCP can be regarded as a "*stop and wait*" protocol, sending a window of data and then waiting for acknowledgements of previously sent data before sending additional data. To maximize the network throughput the "*bit pipe*" must be kept full. This corresponds to the TCP window size being large enough to continuously transmit data for a period of time equal to the RTT, where RTT is the *round trip time*. The RTT measures the elapsed time from when data is sent from sender to when sender acknowledges receipt of that data. The bandwidth-delay product determines the minimum TCP window size meeting this condition, namely, TCP window size > link bandwidth × RTT.

For TCP window sizes greater than the bandwidth-delay product, there must be sufficient capacity among the queues within the network routing/switching nodes to absorb the excess packet traffic. The upper limit on the TCP window size with no packet loss is: TCP window size < link bandwidth × RTT +Queue. The original TCP specification had the window size as 64 kilobytes. In broadband links, the bandwidth-delay production normally exceeds this 64 kilobytes limit.

The TCP segment is the unit of transfer in TCP communications and consists of a header, followed by data. These segments have a variable size; however, the end entities must agree upon a TCP *maximum segment size* (MSS) that is to be used. The network has a *maximum transfer unit* (MTU), which defines the maximum IP frame size. A typical implementation sets the MSS to be some multiple of 512 to ensure that the segment length is less than the default MTU value of 576 octets. In ATM networks the MTU defines the maximum size of the AAL PDU that is passed by the IP protocol layer across the AAL-SAP. Within ATM, TCP segments must be fragmented into ATM cells.

TCP performance over ATM networks and QOS considerations

Due to the nature of TCP/IP traffic, the unspecified bit rate (UBR) and the available bit rate (ABR) are the ATM QOS classes most frequently studied for this application. AAL-5 is the ATM adaptation method used. The performance of TCP is strongly influenced by the interaction with the ATM QOS. TCP can run on an ATM-attached host, or on a host attached to a legacy network that interconnects to an ATM network through a router serving as an edge device.

Packet flow is controlled between communication end-points based upon TCP mechanisms. UBR provides no traffic management feedback with network congestion being indicated by lost or excessively delayed acknowledgements that are detected by the TCP source only. Congestion may occur at the edge device or within the ATM network. With an ABR control loop operating within the TCP control loop, ABR would push the congestion outside of the ATM network to the edge devices.

For strictly TCP/IP-oriented data applications, ATM provides no inherent performance benefits over packet networks. ATM may even introduce additional complexity in terms of large required switch buffers for URB and the need for coordination between TCP and ATM flow control mechanisms when ABR is used.

8.9   ATM-Centric Asymmetric Digital Subscriber Lines (ADSLs)

The Internet has been swift in bringing society under the wings of the Information age. Surfing the World Wide Web (WWW), brings a wealth of information within the reach of anybody.

The growing interests and churning possibilities that the Internet showers into the world of communications are, however, hampered by the access speed. Today, Web pages appear to devour time to its limit in loading graphics and compressed video images on the screen of the computer. The bottleneck that chokes timely Web interaction is the copper that supports the Internet traffic and the bandwidth associated with the information disseminated.

Analog modems used in practice have come to the edge of their possible applications and modern data rate is rarely sustained with POTS lines. With millions of access lines being connected with 19-26 gauge copper twisted pair cabling, the bulging quest for more bandwidth has to come over this "twisted" media. Other technologies such as CATV and satellite are competing with the existing POTS lines and the telcos need a business model to make their move on the chessboard of the communication world. That is, a model that does not include millions of dollars of upgrades on the existing infrastructure.

In order to slide away from the PSTN and analog signaling, the change to digital became necessary. This resulted in *Digital subscriber lines* (DSL), which have been available for years but have never found many consumer type subscribers. The technology has however, matured today to include *asymmetric digital subscriber lines* better known as "ADSL". This "hot" technology is a cost-effective means to bump up the bandwidth by a factor of 20 plus by developing a technical solution for the copper loop from the telco's central office (CO) to the customer.

The boost in bandwidth conceived with ADSL is mainly in one direction, and that is the reason why it is called *asymmetric*. That is, the downstream rate (the direction of incoming data from the Web) is much greater than the upstream rate (pertinent to the interaction with Web pages or uploading of data to corporate servers that do not need the maximum data rate). The physical layer made of twisted copper lines has this new technology to provide the next level of Web throughput but one has to consider the OSI model based data link layer protocol for improved bandwidth as well.

ATM and ADSL technology combined can offer several avenues of technological traits and fascinating possibilities for the telecommunication information superhighway. For example, subscribers can enjoy the video advantages of the Web such as multimedia and video-on-demand. The other beneficiaries of ATM over ADSL are as follows:

Governmental agencies

    ✓   Department of Defense
    ✓   National archives and libraries
    ✓   Public information management services

✓   Distance learning programs.

Health and hospital domains

✓   Patient records database
✓   Patient billing and bookkeeping
✓   Video based consultation and patient welfare tracking
✓   Distance medicine and prescription filling
✓   Residential homecare
✓   Remote teaching of medical sciences
✓   Continued education on medical advances.

Finance management

✓   Trading rooms and banking operation
✓   Network consolidation
✓   Critical site backup vis-à-vis disasters
✓   Insurance operations.

Manufacturing sector

✓   Distributed database management of engineering operations
✓   Supplier/vendor database access sharing on products
✓   Just-in-time ordering
✓   Personnel training
✓   Product training and maintenance
✓   Customer support services.

Educational environment

✓   Distance learning
✓   Virtual classrooms
✓   Home-study sessions.

Originally the ISDN service was conceived as a solution for low-end computer data transfer applications of the PSTN. The data from ISDN connections traverse the same telcos switching infrastructure as normal voice communication. However, such telcos nodes are already overburdened with intense voice traffic sought by the growing subscriber population. The long Web connect times rob resources needed by the normally short-duration phone calls found on the PSTN. The telcos, hence have been looking for other services to relieve "burgeoning data traffic" on its PSTN and have become anxious to ease the already overcrowded network. Hence, ADSL conceived.

Considering the teletraffic burstiness models of ISDN, ADSL, and cable (CATV), ISDN dictates a full Central Office (CO) upgrade instead of line-by-line upgrade as new subscribers decide to opt for the service. The investment by the telcos has always been a hard business model to portray in a saleable framework. ADSL does not force the telcos into an all or none business model situation. A step-by-step upgrade to the COs is possible with ADSL. This process moves forward commensurate with the customer demands. A rival to ATM over ADSL to be considered is held by the traditional cable television (CATV) companies. Broadband digital services offered by the cable companies seem feasible since their coaxial lines already pass over 90% of the homes in the United States and about 65% consume the service. But this service is based on one-way analog equipment not meeting the requirement for two-way digital traffic. Hence another large investment will be needed if the upgrading of the cable system is attempted. There are alternatives available to this highly expensive route, such as using a telephone line for the upstream or return data. This is similar to the asymmetric operation of ADSL but adds new additional costs to the subscriber — probably not cost-effective for the average Internet user. Cost comparison would be similar to ISDN subscription rates and probably not as accessible in certain business areas such as business parks and office buildings.

The following plus points can be indicated for personal broadband services in the form of DSL against ISDN, cable and/or satellite digital systems:

✓ Reusing the existing twisted pair infrastructure avoids investing in new fiber optic distribution systems and distribution nodes

✓ Since DSL uses the existing twisted pair infrastructure, service can be supplied to virtually any subscriber with phone service.

✓ Costly central office switch upgrades as with ISDN can be avoided since DSL traffic bypasses switches entirely

✓ DSL implementations can share the same phone line providing the subscriber's voice service

✓ DSL can provide a dedicated high speed digital two-way access channel for each subscriber rather than using a shared media distribution architecture

✓ The investment by the telcos can be recovered by a graceful tariff raise profile.

The built-in wiring infrastructure around the world, made of loops of twisted pair copper, indicates ADSL as the choice connectivity. ADSL with its higher data rates and with the ATM protocol offering video, text, and data with QOS, brought the personal broadband service into reality in lieu of other technologies.

The *Universal ADSL Working Group* (UAWG) made of PC makers and service providers push for ADSL, which has a maximum bandwidth of 8 Mbps but also a *mass market, consumer oriented version of ADSL*. This *more affordable* version of ADSL is known as *G.Lite*. The G.Lite specification is an extension of the ANSI standard T1.413 ADSL. It is going through the ITU process of standardization to gain interoperability. The two main documents for physical layer hardware of G.Lite are:

✓ G.992.1: Asymmetric digital subscriber line (ADSL) transceivers, G.DMT

✓ G.992.2 Splitterless ADSL transceivers, G.Lite.

Mandatory requirements, recommendations, and options are the contents of these documents. G.Lite is expected to bring a technology with a single, simple solution for personal broadband service. The G.Lite efforts place a momentum on the DSL *"deployment and availability of this service by reducing complexity and driving down over all costs"*. In the relevant scheme, the main electronic component eliminated is the splitter that has to be installed on the subscriber's premises. The cost of this equipment plus installation fees detracts from an inexpensive and easily maintained Internet connection.

ADSL was standardized in late 1997 and G.Lite was standardized in September 1998 (or June 1999?). G.Lite recommends supporting up to 1.536 Mbps downstream (data to end-user) and 512 kbps upstream (end-user to ISP). That is, about 25 times the bandwidth currently realizable with analog modems. Also G.Lite is interoperable with and able to migrate up to a full-rate ADSL known as G.DMT. Some of the major differences between G.DMT and G.Lite are:

▪ *ATM-only transport*: G.DMT has the option to allow for data transmission using bit serial synchronous transfer mode (STM) as an alternative to ATM cell transport. G.Lite specifies the sole use of ATM cell transport

▪ *Fast retrain*: Fast retrain is based on the concept of multiple stored profiles. It is a must for G.Lite, since line conditions may vary, due to G.Lite's splitterless implementation. G.DMT has no provision for fast retraining

- *Power savings*: Power management depicting the full rate and full power, sleep, and idle, is aimed at power saving considerations. Power savings should encourage mobile, battery-powered laptops, and green computer users to rapidly adopt G.Lite. However, since G.DMT has no provision for power management, a potential incompatibility exists

- *ATU-R loop timing*: ATU-R transmission of timing, synchronized to the ATU-C, is the classic method of timing a network. This is supported by both G.DMT and G.Lite

- *Reduced overhead framing mode only*: G.DMT supports four framing structures. G.Lite only supports framing structure #3, using the interleave data buffer definition.

The splitterless operation specified in the G.Lite specification may cause signaling problems that must be overcome with considerable signal-processing. The appropriate processing technique needed includes interleaving, use of Reed-Solomon forward error correction and trellis encoding.

The available hardware of ADSL lays the foundation of a new and improved data connection to world wide resources. But the Web networking protocol of the present times must be upgraded to ATM. The advantages of using ATM as the layer 2 protocol over ADSL are:

- ✓ Protocol transparency
- ✓ Support of multiple QOS classes and capability to guarantee levels or QOS
- ✓ The fine-grained bandwidth salability of ATM
- ✓ Evolution to different xDSL members.

8.9.1 *ATM based broadband home access system*

The ATM Forum defines the ATM access system and home network for residential broadband network, *residential broadband architectural framework* (July 1998). The *ATM RBB reference architecture* defines five functions including networks, terminations, and equipment as depicted in Fig. 8.30. They are:

- ✓ Core ATM network
- ✓ ATM access network
- ✓ Access network termination
- ✓ Home ATM network
- ✓ ATM end system.

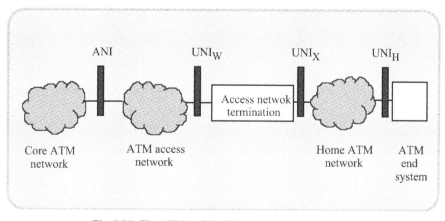

Fig. 8.30: The ATM residential broadband architecture

705

The interfaces shown in Fig. 8.30 are specific to the access network technology, access network termination, home network, and ATM end system. They support a cell-based UNI, or optionally, a frame-based UNI, for ATM transport between these elements.

Core and access ATM networks

The core ATM network consists of one or more ATM switches, ATM network management, and servers which provide network-specific functions.

The *ATM access network* consists of an *ATM digital terminal* (ADT) and *access distribution network*. The ADT provides the *access network interface* (ANI) to the core ATM network. This interface must be in compliance with existing ATM standards with no variations to the ATM physical and ATM layers specified. The ATM Forum has considered several ATM access Networks. To name a few:

  ✓ ATM over hybrid fiber coax (HFC)
  ✓ ATM over passive optical network based access networks
  ✓ ATM over asymmetric digital subscriber line (ADSL)
  ✓ ATM over very high speed digital subscriber line (VDSL).

The ATM over *passive optical network* (PON) access network is defined to interface ATM to fiber to the home (FTTH) and *fiber to the building/curb* (FTTB/C). The optical access network can be configured in a variety of ways such as ring, point-to-point, or point-to-multipoint. In the PON access network, passive optical splitters are used to share across several terminal/*optical network units* (ONU) over multiple subscribers. Then, a medium access control (MAC) is also required to allow the terminals or ONUs to access the PON access networks in the upstream direction.

Inasmuch as ADSL is designed to make use of the existing copper telephone lines to transmit a high bit rate (Mbps) to subscribers, the ATM over ADSL is one of the good choices to delivery high information rates such as video-on-demand to subscribers. ADSL facilitates a high bit rate downstream channel (up to 6 Mbps) and a lower bit rate upstream channel (up to 640 Kbps) depending on loop length and noise.

The ATM Forum defines an ATM over ADSL reference model in which the ATM access network is called ADSL-ADT. It consists of the ATM switch and/or concentrator and the *ADSL transceiver unit-central office* (ATU-C). In the access network termination, the ADSL-NT typically consists of the ATM layer functions, *ADSL transceiver unit - remote terminal* end (ATU-R), mux/demux, and an interface to the HAN. The ADSL Forum has defined the interfacing standard for ATM and ADSL.

8.9.2   Very high speed digital subscriber line (VDSL).

This is a transmission system designed to support very high bit rates over twisted-pair coppy wires. VDSL can transmit bit rate, in asymmetric mode, up to 52 Mbps in the downstream direction and up to 6.4 Mbps in the upstream direction or up to 32 Mbps in both directions in the symmetric mode. However the service distance of VDSL only 1 - 4.5 kft due to high bit rate transmission. Fiber optical network is used to extend the service area.

The VDSL-ADT (*VDSL ATM digital termial*) may consist of VDSL transceiver unit-central office (VTC-C), POTS splitter to segregate the POTS and VDSL channel, concentrator and/or switch, mux/dmux, interface to some kind of an optical access network and interface to the core ATM network.

The home ATM network

The *home ATM network* (HAN) connects the access network termination and the ATM end system. It consists of two functionl groupings, namely *home distribution device* and *home distribution network*.

The home distribution device is optional, and it performs cross-connecting, switching and/or multiplexing of ATM virtual connections between the user-network interface and one or more ATM end systems. The home distribution network can be implemented with a single point-to-point link, with a star configuration or with a shared media tree and branch topology. There are three media types indicated for the home distribution network, 100 Ohm Cat 5 TP, 120 Ohm Cat 5 TP and *plastic optical fiber* (POF). The nominal maximum reach between nodes is 50 m.

ASDL technology

In essence, ADSL is a new conception towards providing high-speed connectivity. It uses an advanced signal modulation technique called *discrete multi-tone* (DMT); it can transmit data over normal existing copper telephone lines at rates 100 times faster than current 56 kbps modems. ADSL can deliver over 6 Mbps downstream and 640 kbps upstream (depending on the loop length and noise conditions) on the telephone lines while simultaneously preserving lifeline POTS.

The ADSL Forum has defined the technical guideline for architectures, interfaces, and protocols for the telecommunications networks incorporating ADSL transceiver. In a typical ADSL configuration, the equipment on the right-hand side is located on the customer premises. Devices which the end-user accesses, such as a personal computer, television, and ISDN terminal equipment connect to a premises distribution network via service modules. These service modules employ either STM, ATM, or packet transport modes. An existing twisted pair telephone line connects the user's ADSL modem to a corresponding modem in the public network.

ADSL downstream transport

From the ANSI T1.413 DMT ADSL standard, the logical data channels on an ADSL link can have up to four simplex downstream channels (AS0, AS1, AS2, and AS3; shown in Table 8.3) that are causes of asymmetrical throughput, and three full-duplex channels (control channels LS0, LS1, and LS2). The logical channels are sometimes called *bearer channels*. For all logical data channels, the actual rates must be multiples of 32 kbps. The purpose of having the different logical channels is to allow ADSL to be flexible enough to support many different applications.

The transmission rate depends on the noise environment in the loop and the class of operation. Therefore, some of these channels can be activated in the initialization process with possibly a few different optional transmission throughputs. The four classes of channels and throughput combinations are shown in Table 8.3. The classes are explained in the following,

Table 8.3: <u>ADSL logical data channels</u>

Channel	Type	Allowable rates	Remarks
AS0	Downstream/ simplex	0-8.129 Mbps	This channel is most commonly used as the lone downstream channel
AS1	Downstream/ simplex	0-4.608 Mbps	
AS2	Downstream/ simplex	0-3.072 Mbps	
AS3	Downstream/ simplex	0-1.536 Mbps	
LS0	Duplex	0-640 kbps	This channel is most commonly configured as an upstream simplex channel. May have different rates in each direction.
LS1	Duplex	0-640 kbps	May have different rates in each direction
LS2	Duplex	0-640 kbps	May have different rates in each direction

- *Class 1*: The maximum simplex throughput is 6.144 Mbps for up to four channels (AS0, AS1, AS2, and AS3). The maximum duplex channel throughput is 640 kbps for up to two channels (LS1 and LS2) and the control channel (LS0) throughput is 64 kbps

- *Class 2*: The maximum simplex throughput is 4.608 Mbps for up to three channels (AS0, AS1, and AS2). The maximum duplex channel throughput is 608 kbps for up to one channel (LS1 or LS2) and the control channel (LS0) throughput is 64 kbps

- *Class 3*: The maximum simplex throughput is 3.07 Mbps for up to two channels (AS0, and AS1). The maximum duplex channel throughput is 608 kbps for up to one channel (LS1 or LS2) and the control channel (LS0) throughput is 64 kbps

- *Class 4*: The maximum simplex throughput is 1.536 Mbps for up to one channel (AS0). The maximum duplex channel throughput is 160 kbps for one channel (LS1) and the control channel (LS0) throughput is 16 kbps, which is not implemented in the control channel slot but through the synchronization overhead.

Table 8.4: <u>Channel options by transport class for optional ATM channel rates</u>

Transport class:	1	2	3	4
Downstream simplex bearers:				
Maximum capacity (in Mbps)	6.144	4.608	3.072	1.536
Bearer channel options	1.536	1.536	1.536	1.536
(in Mbps)	3.072	3.072	3.072	
	4.608	4.608		
	6.144			
Maximum active sub-channels	Four (AS0, AS1, AS2, AS3)	Three (AS0, AS1, AS2)	Two (AS0, AS1)	One (AS0 only)
Duplex bearers:				
Maximum capacity (in kbps)	640	608	608	608
Bearer channel options	576			
(in kbps)	384	384	384	
	160	160	160	160
	C(64)	C(64	C(64	C(64
Maximum active sub-channels	Three (LS0, LS1, LS2)	Two (LS0, LS1 or (LS0, LS1)	Two (LS0, LS1) or (LS0, LS2)	Two (LS0, LS1)

Data bits, sent between ATU-C and ATU-R are organized in frames and are gathered into superframes. A superframe consists of 68 ADSL frames. Some frames have special functions. Frames 0, 1, and 34 are used to carry error control information and indicator bits for managing the link. One ADSL superframe is sent every 17 milliseconds. Each ADSL frame has two parts, fast data and interleaved data parts. The fast part is protected by forward error correcting (FEC). The throughput data rate that is sent in the fast part is faster than in the interleaved part due to the delay in the interleaving process. So if the data is delay sensitive, the data should be sent in the fast part.

ADSL can support a wide variety of high-bandwidth applications. That is, it can deliver broadband services, such as high-speed Internet access, telecommuting, virtual private networking, and streaming multimedia content, to homes and small businesses.

Compared to the OSI reference model, ADSL is a technology that is positioned at the physical layer. ADSL is then a physical layer protocol that could also be positioned at the ATM physical layer. It would then be possible to utilize CBR, VBR and ABR services as specified with ATM. This forms the basis for ATM transportation over ADSL.

An ATM-ADSL reference model has been defined by ADSL Forum. According to it, the ATU-C and ATU-R interface to the twisted copper pair via splitters. The splitters separate the ADSL signal and the POTS signal. The ATU can be considered as the ADSL modem or line interface equipment. The task is now to pass ATM traffic from the ATM core network via the Access Node to the terminal equipment on the *premises distribution network* (PDN). The *core network interface element* in *access node* (AN) or so-called *digital subscriber line access multiplexer* (DSLAM) provides the interface between the AN and the core ATM Network. The VPI/VCI translation and higher-layer function provide the Multiplexing/Demultiplexing of the VCs between the ATU-Cs and the core network interface on a VPI and/or VCI basis. The ATU-C can support both "Fast" (TC-F, *transmission convergence-fast*) and "Interleave" (TC-I, *transmission convergence-interleave*) channels ("Fast" means lower latency path and higher bit error rate and "interleave" means higher latency path and lower bit error rate).

Broadband network termination (B-NT) performs the interface between the twisted pair cable and the PDN or terminal equipment. The ATU-R provides terminating/originating the transmission line. The B-NT may contain VPI/VCI translation functions to support multiplex/demultiplex of VCs between the ATU-R and the PDN/TE interface element on a VPI and/or VCI basis.

Access node

The *access node reference model* consists of these three planes; data, control, and management. The core ATM network interface element function contains interfacing layers, ATM and PHY layers, between the access node and the core ATM network via V interface. The ATM core networks interfaces may include the physical interface to n × DS-1, DS-3, and SONET/SDH. The PHY function performs ATM layer functions such as UNI and traffic management. Examples of such functions are multiplexing among different ATM connections, cell rate decoupling, cell discrimination, cell loss priority, traffic shaping, policing, and congestion control.

VPI/VCI translation and higher-layer function perform cell routing based on a VPI/VCI to appropriate ATU-C and to the "Fast" or "Interleaved" path of modem in the downstream direction. For the upstream direction, this function block combines/concentrates the data stream from ATU-C to form a ATM cell stream to the core ATM network.

ATU-C performs the CO ADSL PHY layer function to support ATM transport which includes the ATM transmission convergence (TC) functions for both of "Fast" and "Interleaved" paths. Also this functional block performs ATM transport protocol specific functionality such as idle cell insertion, header error control generation, cell payload scrambling, bit timing ordering, cell delineation, header error control verification, and cell rate decoupling.

The VP/CE multiplexer function is part of the ATM layer. It performs the combination of cell stream from the "Fast" and "Interleaved" buffers into a single ATM cell stream in downstream direction. In another direction, upstream, this function block performs cell routing based on VPI/VCI to the "Fast" and "Interleaved" paths. The ATM layer performs the ATM functions the same as the one in the access node including VPI/VCI translation. PDN/TE interface element function performs the physical layer interface between the B-NT and the PDN or TE over the T or S interface.

Channelization

For ATM systems, the ATM data stream uses different VPs and/or VCs for the channelization of embedded different payloads. Then, the basic requirements for transport ATM cells over ADSL are for at least one ADSL channel downstream and at least one ADSL upstream channel. In the ANSI T1.413 standard, for the transport of ATM over ADSL, the ADSL modem uses the AS0 channel downstream and the LS0 channel upstream for the single latency class while AS1 and LS1 are reserved for dual latency. The ADSL modem can use both "Fast" and "Interleaved" paths for transport of ATM.

Dynamic rate change

Comparing the bandwidth of an ADSL link and ATM, the ADSL link is much lower than typical ATM line interfaces. Also the bandwidth of the ADSL link may vary during connections due to a noisy environment. The ADSL Forum has considered this case for transport ATM traffic over an ADSL link. Especially for UBR traffic, most systems send at the line rate they can see on their local NIC interface, which may be much higher than the ADSL link further down the line. This causes difficulties, as the end-systems may not be aware of this restriction when shaping their outgoing ATM cell traffic. Excess cells sent by the client or the server may be lost due to buffer overload in the ADSL-NT or ADSL-LT. Such cell loss causes massive packet loss at the AAL5 and PPP layers, resulting in aggressive throttling of the TCP connection.

Studies performed by ADSL Forum on dynamic rate change behavior has led to *dynamic rate adaptation* (DRA) and *dynamic rate repartitioning* (DRR) at the physical layer. Both DRA and DRR are performed during normal operation without losing the ADSL connection. However, DRA and DRR are optional functions.

Dynamic rate repartition

Dynamic rate repartition is a specific function that reallocates bandwidth between the "Fast" and "Interleaved" channels without changing the total aggregate bandwidth of the link. The rate repartition may be performed with a maximum of 125 ms of service interruption.

Since the data rate over the ADSL link depends on the physical environment or other factors such as noise, the rate may vary from time to time. The variable data rate of the ADSL link causes problems to the network, which spans network management, traffic management, signaling, and operation management functions.

During the rate change transition, the data rate on the ADSL link may be zero up to three seconds. Then the physical layer notifies the ATM layer when any dynamic rate change occurs.

In dynamic rate adaptation, an ADSL link initially is set up with the maximum contract aggregate bit-rate and the minimum contract aggregate bit-rate. This means that the transport of guaranteed services such as CBR can be up to the minimum contract aggregate bit-rate and transport of non-guaranteed services such as UBR can be up to the maximum contract aggregate bit-rate. Whenever the actual aggregate bit-rate is lower than the minimum contract aggregate bit-rate, a link failure alarm is generated from the physical layer to the ATM layer.

8.10   Wireless ATM

To understand why *wireless ATM* is a "hot topic" today, it is pertinent to understand the market drivers behind the scene. These market drivers indicate the existing cooperation between standards organizations, which has led to the agreements on wireless broadband ATM architectures on their performance. Foreseeing the comforts of wireless ATM-based teletraffic, several vendors and education institutions have cooperated in realizing several prototype networks. The Magic WAND, RACE II, and RDRN are a few to mention.

The quest for large bandwidth coupled with the ever-increasing mobility of today's society are the precursors for wireless ATM. It has been estimated that the need and market for wireless ATM can jump about 100% by 2005.

The conceivable applications of wireless ATM refer to the following locales: The workplace, home, and "wireless to the curb." The workplace application will allow multimedia, bandwidth-intensive applications to run on mobile devices. Video conferencing, access to multimedia databases, and Internet access will not be limited by the number and location of LAN ports and wiring. Mobile devices might also include ATM-ready PDAs. Workplace applications are likely to be the first applications implemented. Home-based applications might also include:

- ✓ Portable enhanced television services, including video-on-demand, home-shopping, banking, games etc.
- ✓ Wireless in-home networking and Internet access for traditional PCs and mobile network computers
- ✓ High quality audio distribution throughout the home, allowing more flexible positioning of audio components, cordless telephony, and replacement of miscellaneous legacy wiring systems.

The "wireless to the curb" application of wireless ATM differs from the first two applications in that both the base station and subscriber devices would be fixed.

On the consideration of possible uses for a wireless ATM network, there are several constraints that must be looked into when the standards are developed. The primary factor is the QOS. Other hurdles to be faced are:

- *Access schemes* in reference to protocols that can overcome high error rates and noise inherent in radio systems

- *Reliability and availability* specified by coverage areas, fading temporary outages, error detection, and correction

- *Service ubiquity* posed by wireless reaching hard downtown areas, carriers' responsibility to acquire licenses, acquiring rooftop space, and deploying access stations

- *QOS mobility* to ensure consistent QOS and handoff as a user wanders.

- *Applications* made to overcome the limitations inherent in wireless transmissions.

Standards on wireless ATM are still emerging. Several organizations have been working on standards from many different perspectives. Both European and North American standardization efforts are on trial. The standards cover a wide range of topics pertaining to both additions to the ATM Protocol standards for communication on air (both mobile and fixed) and enhancements to wireless standards for usage and globalization.

The ITU-T has launched its International Mobile Telecommunications - 2000 (IMT-2000) initiative "to provide wireless access to the global telecommunication infrastructure through both satellite and terrestrial systems, serving fixed and mobile users in public and private networks" (ITU-IMT, 1999). Study Group 11 of the IMT met as recently as July 1999 to address *signaling for broadband and multimedia networks and services*. The IMT-2000 is part of the larger *wireless access systems* (WAS) effort of the ITU, which is addressing the following concerns:

- *Question ITU-R 77/8*: Adaptation of mobile radio communication technology to the needs of developing countries

- *Question ITU-R 125/9*: Point-to-multipoint radio systems

- *Question ITU-R 140/9*: The use of mobile-derived technologies in fixed wireless access applications

- *Question ITU-R 215/8*: Frequency bands, technical characteristics, and operational requirements for wireless access local loop systems

- *Question ITU-R 212/8*: Radio local area networks for mobile applications

- *Question ITU-R 142/9*: Radio local area networks (RLANs).

The WAS has already established several ITU-R recommendations regarding wireless access.

Working within the framework of IMT-2000 is the UMTS Forum, (which was founded in 1996). It is a non-profit organization with representation from over 180 member organizations. This group is working with other standards organizations including the ITU, ETSI, GSM Association, and ANSI to develop the Universal Mobil Telecommunication System (UMTS). The UMTS Forum met August 9–12, 1999 for a workshop on 3G Mobile Broadband.

The ETSI has also established a standardization project for *broadband radio access networks* (BRAN). Prior to the establishment of the BRAN project ETSI had released *functional specification EN 300 652 high performance radio local area networks* (HIPERLAN). The BRAN project will enhance the HIPERLAN functionality to HIPERLAN type 2 and begin working on standards for wireless access and broadband interconnect.

At present there are 60 documents in various stages of approval under the BRAN project. The BRAN project has also been working closely with the ATM Form, IEEE, and the ITU-R to avoid duplication of effort and ensure a cohesive set of standards.

The ATM Forum (ATMF) has also begun looking into standards for wireless ATM. In 1996 the wireless ATM working group (WATM) was formed. In part the WATM working group will address:

✓ Radio access layer protocols including radio physical layer, medium access control for wireless channel errors, data link control for wireless channel errors, and wireless control protocol for radio resource management

✓ Mobile ATM protocol extensions including hand-off control (signaling/NNI extensions, etc.), location management for mobile terminals, routing considerations for mobile connections, traffic/QOS control for mobile connections, and wireless network management.

Currently there is also a specification in draft form: Wireless ATM Capability Set 1. The IEEE Communications Society, which is responsible for the 802 series of LAN/WAN standards has two relevant standards, which are currently in draft status:

LAN/MAN (802) Unapproved Drafts	
P802.11a/D5.0	Wireless MAC and PHY specifications: High speed physical layer in the 5 GHz band (draft supplement to 802.11-1997)
P802.11b/D5.0	Wireless LAN MAC and PHY specifications: Higher speed physical layer (PHY) extension in the 2.4 GHz band

In addition, there are smaller groups such as the Wireless Broadband Association and the Delson Telcom Group's Task Force — Wireless Mobile ATM (TF-wmATM). These are small independent organizations working on generic solutions to wireless ATM Access. The TF-wmATM focuses on implementation issues and has held several conferences.

8.10.1 WATM: Architecture

In this configuration the wireless base stations act as an interface into the switched wireline ATM network. This network is based on the wireless ATM reference architecture. This reference architecture defines new interfaces for the ATM protocol stack. The wireless user interface, or interface "W" UNI deals with handover signaling, location management, wireless link and QOS control. The *radio access layer*, or "R" RAL, governs the signaling exchange (that is, channel access, data link control, etc.) between the WATM terminal adapter and the mobile base station. The *mobile network interface*, or "M" NNI, governs signaling exchange between the WATM base station and a mobile capable ATM switch, as well as mobility-related signaling.

Considering the WATM protocol stack, the bottom three layers make up the radio access layer (RAL). The RAL is required to support the new wireless specific protocols that will define the physical and access layers. At the physical layer there are multiple competing solutions. Currently the two front runners are CDMA and TDMA but new technologies are on the way. It is likely that there will be more than one answer. The MAC relates to the user's ability to access the wireless media in order to guarantee the QOS. The base station is the coordinator for channel access, and is aware of the wireless resource utilization. The MAC layer will likely have to support multiple physical layer protocols.

The data link control (DLC) protocol will have to support new error handling to account for wireless channel errors. The DLC will be responsible for transmission and acknowledgement of frames, frame synchronization and retransmission, and flow control.

8.10.2   Wireless ATM prototype networks

Currently there are several proof-of-concept projects on WATM. Generally, a consortium of vendors and educational institutions operates these projects. The *Magic Wireless ATM Network Demonstrator* (WAND) and the *rapidly deployable radio networks* (RDRN) are two such projects to mention.

The Magic WAND is a project under the auspices of the advanced *communications technologies and services* (ACTS) research program of the European community. The WAND project is a consortium of the following six vendors and seven educational institutions.
The Magic WAND project is a broad project with three main goals:

■   To specify a wireless, customer premises access system for ATM networks that maintain the service characteristics and benefits of the ATM networks to the mobile user

■   To promote the standardization, notably in ETSI, of wireless ATM access as developed in this project

■   To demonstrate and carry out user trials with the selected user group and test the feasibility of a radio based ATM access system.

The Magic WAND project chose the 5 GHz frequency band for the demonstrator while conducting studies on the 50 Mbps operation in the 17 GHz frequency band. The project covered design, implementation, and testing with actual subscribers. The user trial will be conducted with user groups in a hospital and a mobile workplace environment. The medical scenario is meant to fully test the mobile multimedia capability of the Magic WAND. The mobile office scenario is meant to test legacy systems running in a wireless ATM environment.

8.10.3   Rapidly deployable radio networks (RDRN)

The RDRN project is funded by the Information Technology Office (ITO) of the DARPA and is carried out at the University of Kansas. The goal of the project is to: Create architectures, protocols, and prototype hardware and software for a high speed network that can be deployed rapidly in areas of military conflicts or civilian disasters where communication infrastructures are lacking and or destroyed (for example, in Desert Storm, Bosnia, Hurricane Andrew, Los Angeles earthquake).
The University of Kansas developed a wireless ATM architecture based on rapid deployment. Their rapidly deployable radio network (RDRN) architecture does not require a wired ATM backbone and is comprised of three overlaid networks:
A low power, omnidirectional network which supports location determination, switch coordination, and management functions. A "cellular-like" network system to support multiple end user's access to the ATM network. A multiple beam radio network with high capacity to support switch to switch communications.
Although wireless ATM is still in its infancy, expectations are high. The worldwide standards organizations, international educational institutions, vendors, and service providers are all working towards a standard WATM offering in the foreseeable future. When brought to fruition WATM will be able to provide today's mobile society with high speed, high quality mobile access from virtually anywhere.

8.11   Applications of ATM Technology

An overall consideration of migration towards ATM implementation in an existing networking refers to meeting the upcoming needs of network applications plus their QOS requirements. Such migration efforts rely on several different types of ATM service models being deployed in practice. There is no single method to achieve an optimized working implementation of ATM strategy vis-a-vis the technology considerations and user requirements. However, indicated in the following sections are general principles and some examples towards the migration to ATM consistent with typical applications, technology perspectives, and existing networking profile.

8.11.1 Classical applications

Migration to ATM in a "classical" sense refers to intermingling of traditional LAN and WAN methodologies and ATM concepts. Such an approach has restricted uses mostly to tailor-made enhancement strategies, such as upgrading file transfers, information retrievals and remote terminal access capabilities.

File transfer programs or *file transfer protocols* (FTPs) operable within the OSI environment are usually vendor-specific. When large volumes of file transfers are needed, upgrading the system to be ATM-based would facilitate faster file transfers. For example, classical T-1/E-1 leased lines may need a few hours to complete a file transfer whereas an ATM-based system may require only a few minutes for the same effort.

Migration to an ATM-based upgrading in such cases is quite justifiable though the initial cost of implementation may be significant. Also the host system should be prepared for a dedicated reception of the upgraded file transfers during the session of transfer.

Terminal access programs (TAP) such as Telnet are universal versions and offer remote access to windowed environments (like X-Remote). Such windowed environments can be characterized as follows:

- ✓ High-bandwidth requirement with significant file transfer time
- ✓ Text-based applications involved correspond to large volume transfers
- ✓ Combinations of text, still-images, and video conform to using multiple windows.

Information retrieval and multicasting: There has been a phenomenal growth in multicasting over the Internet in the recent times. In the relevant efforts, the end-systems packetize voice, images, and video, and broadcast them for multiple recipients. In these applications, the associated synchronized voice and video file presentations for electronic white-boards are also increasing. Existing PCs offer support for such multicasting with Internet protocols (IP) for workgroups, campus users, metropolitan, and global subscribers. However, IP limits such operations since overloading may occur in the event of traffics concerning high-fidelity video, user-generated conference, even at the campus level when expansion plans are introduced. As a result, upgrading from transmissions at 500 kbps to 1.5 to 4 Mbps may invariably be warranted. Therefore, the Internet shift to ATM as a backbone technology becomes imminent to support such multicasting strategies.

Further, multicasting has been growing exponentially in data distribution tasks based on the wide usage of Web servers accessed via graphical front-ends such as Netscape^{TM}, Mosaic^{TM} etc. in business transactions and home-page accessing.

Home-page

The first screen seen by remote users when accessing a Web site. This page can be seen linked to additional screens, data bases, images repositories, other sites, and even incoming video.

In the retrieval phase (that is, when a user retrieves information at a Web site or opening a home page etc), the host resources must be judiciously used without sacrificing bandwidth considerations. That is, multicasting and remote access of wideband information should be properly integrated on a real-time basis. This would justify the deployment of ATM.

Locators and resolvers

Traditional *domain name servers* (DNS) in Internet systems support resolving host names against IP addresses and various locator systems for Internet users (For example, X-500). Conventional DNS use text-only and are static. Future expansions would, however, require DNS to accommodate extensive texts so as to provide information on a large-scale basis dynamically (such as facilitating a global directory). Relevant strategies may again require ATM support.

New classes of ATM applications

Classical ATM applications are directed at upgrading the existing non-ATM systems with the inclusion of ATM concepts. Hence, they are essential *"ATM-enabled"* systems. There are, however, entirely new and totally ATM-based systems envisioned in the upcoming trends and developments especially in the computing and video fields.

Server farm implementation

File servers (such as Netware/Windows etc) are deployed in central facilities due to considerations of maintenance, data integrity, and security. In general, file services are accessed by end-users from local area servers as well as from wide area servers. In such efforts, these servers and end-users must be linked via high-speed connectivity. ATM offers this feasibility with sufficiently low latency characteristics while supporting relevant LAN traffic. Success of such implementations would require the following:

- ✓ Highly reliable service-provider file services
- ✓ Bridging of different workgroups
- ✓ An ATM switch acting as a gateway between, for example, switched multi-megabit data service (SMDS) and a LAN-based ATM network
- ✓ Need for backup servers in case of disaster recovery situations

  ⇒ Bandwidths offered by ATM can facilitate such backup processes efficiently
  ⇒ An ATM system will also enable continuous availability of such backup systems or for those user needs on an ad hoc basis (with a fee levied for such services).

Collaboration service

Collaboration is specified by and refers to connection of computers and access to generated data. The connection of computers (for example, the parallel operation of multiple computing systems) calls for an environment, either campus-based or wide-area based, with multiple and parallel operation of systems. Such an operation would require:

- ✓ Enormous data exchange in real time
- ✓ Implementation of high-speed transmission.

Originally, non-ATM high-speed links (such as DS-3) have been used for such applications. But, the upcoming needs dictate multiple supercomputers to operate in parallel over a wide area covering a national network of research laboratories, universities etc. at speeds extending into the Gbps regime.

In reference to collaboration on access to generated data, the real-time access is not critical. But an abundance of data (such as visual sequence downloading, storing, redisplaying etc.) will be the part of associated efforts. In such situations, ATM deployment would reduce access/downloading time whenever real-time access of extensive data is needed.

Residential/domestic applications

Use of ATM-based networking advocated in residential applications involves, a single-link to support voice, video and data, unidirectional video distribution (video-on-demand), bidirectional voice (plus video), video telephony, and unidirectional data from provider-operated data bases and Internet gateways.

ATM Forum suggestions for residential applications of ATM networking: Video-on-demand, interactive games, video telephony, healthcare, home-based business, information-on-demand, distance learning, wireless/PCs with ATM, security/surveillance etc.

8.11.2   ATM implementation: Examples

In the strategies towards migration to ATM-enabled and/or ATM based new systems involve first an appraisal of the following:

    ✓   Present type of communication system/networking
    ✓   Types of applications
    ✓   Bandwidth requirements
    ✓   QOS requirements
    ✓   Budget/cost-effectiveness
    ✓   Expansion scenario
    ✓   Engineering feasibilities.

In reference to ATM implementation, there are a number designated applications, which can be grouped as follows:

    ✓   Classical
    ✓   LANE-based
    ✓   MPOA-based
    ✓   Frame-relay access-based
    ✓   SMDS access-based.

Commensurate with the categories indicated above, the following subsections illustrate some typical examples of ATM implementation strategies.

8.11.3   *Migration to ATM-based Systems: Examples*

Example 8.3

Suppose the existing campus multiprotocol FDDI backbone is overloaded due to the following reasons:

    ❑   An increase in user population
    ❑   Deployment of new applications
    ❑   Introduction of new and more powerful end-systems.

As a result, the FDDI, ethernet and/or token-ring are found to be inadequate to meet the enhanced congestion. Further, due to increased bandwidth requirements and application needs possibly, a set of QOS parameters are to be met with. Suggest a migration strategy by which implementing, a continued support of existing multiprotocol applications would prevail. Further, as far as possible, the investment in the existing hardware should be conserved.

Solution

ATM can be deployed as a new switched backbone to interconnect routers supporting IP, IPX, and AppleTalk. This classical approach can be based on PVCs with the following considerations:

    ❑   Recommending the use of an ATM backbone interconnecting the various routers (for example use of RFC 1483 and LLC/SNAP encapsulation for multiprotocol)
    ❑   Logical link control/simple network management protocol
    ❑   Multiprotocol encapsulation over AAL-5 in support of permanent VCs and switched VCs.

The existing networking environment has multiple layer 3 protocols and hence, the ATM migration being planned should be created to address layer 3 protocols. That is, the ATM should enable these address maps in the routers for layer 3 protocols (such as Internetwork packet exchange (IPX) and AppleTalk as well as the other two protocols in use). The number of PVCs required is not much since almost all users may connect to

LAN ports on the routers. The proposed effort should also allow bridging to support nonroutable protocols as explained below:

RFC 1483 encapsulation (or any other ATM end-system capable of bridging) would enable a router-based initiation of point-to-point broadcast capable VCC or a point-to-multipoint capable VCC upon which the bridged traffic can be transmitted.

Switches and routers deployed must support SNMP for management and the ILMI (which facilitates interfacing an end-system and an ATM switch for status and configuration reporting as well as for registering/deregistering ATM address). The ILMI is used between the switches and the ATM-connected routers and workstations. Also, a connectivity to corporate WAN (PSTN) can be established via serial interface on one of the routers.

Relevant PVC parameters of the system to be migrated in are:

- Cell-loss priority (CLP): 0 (high priority) or 1 (low priority)
- Peak cell rate (PCR): Specified in kbps
- Average cell rate (ACR)/Sustainable cell rate: Specified in kbps and not to exceed 50% of the PCR
- Burst tolerance (BT)/Maximum burst size: Expressed in multiples of 32 cells
- Adaptation (AAL): AAL-5 (all end-system) and AAL-3/4 (some routers reserved for SMDS/CBDS)
- Encapsulation: LLC/SNAP (RFC 1483), MUX (RFC 1483), NLPID (RFC 1490), QSAAL (signaling), and SMDS (RFC 1209)
- Maximum transmission unit (MTU): Set at 9188 octets maximum.

In the existing networking, data flows over in the existing fiber-based ethernet backbone, connecting the various buildings with the campus computer center. Users share LAN segments, while servers are on shared or dedicated segments. These segments are connected via routers to the ethernet backbone. As the volume of data transfer increases due to new applications and greater use of networking, the backbone becomes congested and all users suffer. Some servers require higher-speed interfaces than that provided by even a dedicated ethernet.

In the migrated structure, the ethernet backbone is replaced with ATM, interconnecting existing routers. Some servers are now connected directly to ATM, while some users are provided with dedicated ethernet, via LAN switching. These steps introduce the necessary bandwidth when needed, while requiring no changes in networking paradigms.

Example 8.4

Today's campus networking (for example, a University LAN) is depicted in Fig. 8.32. It consists of:

- Bus-based workgroup LAN (10BaseT/UTP: Z_0 = 100 ohms; maximum 100 meter run)
- Ring-based workgroup LAN (16 Mbps/STP: Z_0 = 150 ohms; maximum 100 meter run) (*Note*: Z_0: Characteristic impedance of the cable)
- A file-server operating at 10 Mbps: 10BaseT/UTP
- A simple engineering workstation operating at 10 Mbps: 10BaseT/UTP.

It is planned to incorporate the following as a migration strategy towards ATM enabling:

(a) Bus-based LAN is retained for basic PC applications
(b) Additional multimedia-based workgroup hosts are to be added
(c) In addition to the existing file-server, a video server is to be included

(d)   The existing engineering workstation is to be upgraded to accommodate multimedia applications

(e)   The public network is to be accessed

(f)   With the upgraded features, this campus networking is to be linked to an off-campus center (via public network). This off-campus center is also assumed to facilitate the user applications up to 100-155 Mbps rates

(i)   Draw a block-schematic indicating the inclusions of the above features in the campus networking via an ATM-enabled approach

(ii)   Tabulate the data rate requirements, maximum link-lengths in the new networking, and the type of cabling needed in each cases.

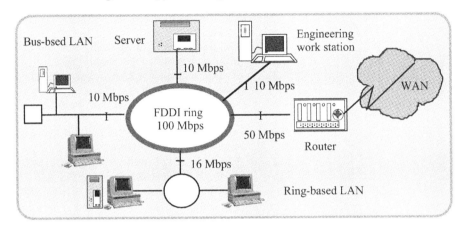

Fig. 8.32: Today's LAN environment

Solution

(i)   Block-schematic of the conceived networking towards expansion is presented in Fig. 8.33.The existing setup (Fig. 8.32) is expanded to include the features being contemplated. Essentially, the following governing factors are to be noted:

❑   Network expansion and interworking have the basis pertinent to the type of network and the rate of transmission being supported as illustrated in Fig. 8.33

❑   ATM LAN: ATM Forum specifies ATM LAN as an HSLAN or as a backbone. It can be supported in bus-based (10 Mbps) or ring-based (16 Mbps) topology. It constitutes the legacy networking in the campus as depicted in Fig. 8.33

❑   Expansion thereof can be done in a simple form as shown in Fig. 8.33 using interconnections of smaller ATM switches together in a distributed manner

❑   Multimedia workstations, broadband video servers etc. can be connected directly on ATM switches as well as the existing LANs (Bus, ethernet or FDDI).

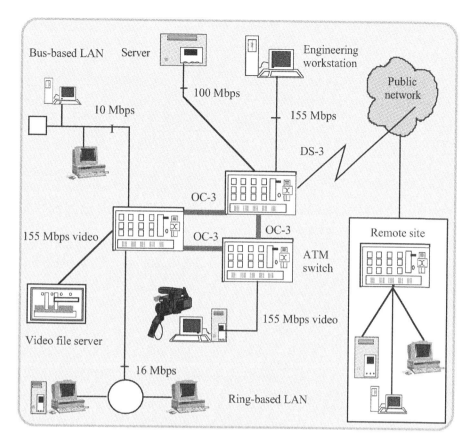

Fig. 8.33: Expanded network

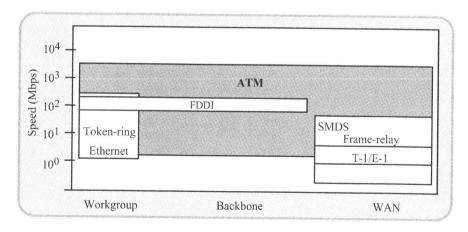

Fig. 8.34: Network versus speed

(ii)   Relevant to Fig. 8.33 the details are as follows:

Data rates	Cabling	Remarks
10 Mbps	10BaseT (UTP)	IEEE STD: 802.3I
16 Mbps	STP	Z_o = 100 ohms
		Z_o = 100 ohms or 150 ohms
100 Mbps	Multimedia fiber	For private UNI only such as FDDI physical layer with a 4B/5B line code/125 Mbaud physical rate.
		Block coded:
		• For private UNI: 8B/10B line code −194 Mbaud. Frame structure 10 ATM cell for every 26 ATM cell
155 Mbps	Multimode fiber	• SDH/SONET framing for both private and public UNI
		• Limited to 100 meter links
		• 12 ATM cells mapped into 125 µsec
	UTP	
44.736 Mbps	DS-3/multimode fiber/coax	

Interfaces
Comply fully with ATM Forum, ITU-T, and ETSI specifications
Interface types

❑   SONET/SDH STS-3c/STM-1 155 Mbps multimode fiber
❑   TAXI 100 Mbps multimode fiber
❑   SONET/SDH STS-3c/STM-1 155 Mbps single-mode fiber
❑   DS-3 over coaxial cable
❑   E-3 over coaxial cable
❑   STS-3c/STM-1 over unshield twisted pair category-5 cable (UTP-5)
❑   STS-1 (55 Mbps) over unshield twisted pair category-3 cable (UTP-3).

Physical layer	Data rate	Mode	Connector
STS-3c/STM-1	155 Mbps	Multimode fiber	SC
TAXI 4B/5B	100 Mbps	Multimode fiber	MIC (FDDI style)
STS-3c/STM-1	155 Mbps	Single-mode fiber	SC
STS-3c/STM-1	155 Mbps	UTP-5	RJ-45
DS-3	45 Mbps	Coaxial cable	BNC
E-3	34 Mbps	Coaxial cable	BNC

Connection support

❑   AAL: AAL-1/2   Voice, video
                   AAL-5   Other services/ traffic types
❑   Priority connections/levels and PVC, SVC, etc.

Each connection through the ATM switch can be labeled as either high priority (requiring low cell-delay variation) or low priority (tolerant of cell-delay variation). High-priority connections can typically be used for voice traffic, while low-priority connections can usually handle data traffic. Connections can be either permanent virtual connections (PVCs) or switched virtual connections (SVCs). PVCs are set up through the serial port, with parameters stored in nonvolatile memory for retention following a power failure or reset. By comparison, SVCs are set up by ATM end-stations using ATM signaling protocols to communicate with the switch.

Functional aspects of the connections

- ❏ Supporting both permanent and switched virtual circuits
- ❏ Supporting virtual channel (VC), virtual path (VP), point-to-point, and point-to-multipoint connections
- ❏ Elimination of single points of failure through fully integrated support for ATM Forum V3.0 Q.2931 UNI signaling
- ❏ Supporting ATM point-to-point connections per interface and point-to-multipoint connections per switch
- ❏ Allowing construction of multi-switch networks via NNI standard support.

Feature	ATM-Forum UNI 3.1 Q.2931 Version 1	Future Q.2931 versions
Call Call topology	Single connection • Bidirectional point-to-point • Unidirectional point-to-multipoint	Multiple connections • Bidirectional point-to-poi • Unidirectional point-to-multipoint • Multipoint-to-multipoint
Point-to-mutlipoint setup	• Sequential setup • Parallel setup	• Sequential setup • Parallel setup • Atomic setup • Leaf-initiated joining
Dynamic calls	• Dynamic add/drop of parties	• Dynamic add/drop of parties • Dynamic change of traffic contract • Dynamic add/remove one of a call
Traffic description	• Peak cell rate • Sustainable cell rate • Maximum burst size • QOS classes	• Peak cell rate • Sustainable cell rate • Maximum burst size • QOS classes • Attribute-value pairs
Traffic contract negotiation		• User-network • User-user

Connection topologies

- ❏ Point-to-point
- ❏ Point-to-multipoint
- ❏ Multipoint-to-point.

Bandwidth symmetry

(a) Symmetric/asymmetric bilateral
(b) Unilateral asymmetric
(c) Asymmetric.

QOS parameters

Relevant PVC parameters are:

Call loss priority (CLP)	0 (high priority) or (low priority)
Peak cell rate (PCR)	Value in kbps
Average cell rate (ACR)	Value in kbps should not exceed % that is the
(sustainable cell rate)	PCR
Burst tolerance (BT)	
(Maximum burst size)	In multiple of 32 cells
Adaptation (AAL)	AAL-5 (all end-systems): AAL-3/4 (some routers for SMDS/CBDS only) LLC/SNAP (RFC 1483), MUX (RFC 1483), NLPID (RFC 1490), QSAAL (signaling), SMDS (RFC 1209)
MTU (maximum transmission units)	9188 octets maximum

The design details can be tabulated in the following format:

Source	Destination	VCC#	VCI	VPI	AAL	Encap	PCR (kbps)	ACR (kbps)	BT (× 32 cells)
Main campus	Off campus	1	0	60		LLC SNAP	80000	40000	3
Engineering Workstation . etc.									

Problem 8.1

In reference to Example 8.4, assuming the use of RFC 1483 and LLC/SNAP encapsulation for multiprotocol transports,

(a) Decide and tabulate the ATM adaptation layer (AAL#) accessing each of the hosts/workgroups involved
(b) Indicate the appropriateness of PVC and SVC for the accesses of (a)
(c) Indicate the signaling VPIs/VCIs, if a specific work-group accesses another work-group within the campus, and both within as well as the off-campus
(d) Tabulate typical QOS parameters: CLP, PCR, ACR, BT.

Example 8.5

A university campus has the following telecommunication and internetworking infrastructure:

- ❑ Basic telephone facility with an exchange within the campus
- ❑ Legacy LAN support on 10BaseT in the administrative building exclusive to serve the administration, finance, purchase, and personnel departments
- ❑ A high-speed LAN on FDDI that supports the large-sized files of registry, admissions, and other student-related offices spread across the campus
- ❑ A legacy LAN supporting eight nonengineering/science colleges
- ❑ Cat 5 cabled LAN with servers in the engineering and science colleges with multimedia terminals (at least one per college)

❑   A media center to support applications such as full-motion video, high resolution graphics, audio, and text data

❑   A research center internetworked with the science and engineering colleges. Also, it is planned to connect it to an adjacent Defense Research Complex to interoperate with certain applications of the Defense Research Complex (DRC). Assume that these applications at DRC are of special categories with protocol structures of their own.

An expansion plan is conceived to upgrade the interoperation between the individual LANs and also extend the entire internetworking of the campus to get accessed to an off-campus center of the university within the metropolitan area. Further, the campus is to be facilitated WAN access to communicate nationwide on high-speed at least at an OC-3 level.

Suppose a migration to ATM is undertaken to realize required bandwidth/speed, low-latency and reliable connectivity.

(a) Tabulate the combined, multiple protocols, interfaces, speeds, and transport media use

(b) List the following for the various links: QOS class, CLP, PCR, ACR, BT and AAL #

(c) Indicate whether users belong to (virtual) LANs based on logical grouping or physical locations. Why?

Solution

Project summary

The purpose of this design is to migrate a few of the 10BaseT individual networks existing among the different colleges and divisions on a university campus to an ATM switch based network. The engineering and science colleges, the media center, and the research center will be the divisions that form the essential part of this migration. Further, as part of this migration, a new division, namely the Defense Research Complex, has to be connected to the Research Center using ATM technology.

The enhanced network should be also capable of providing WAN access to communicate nation-wide at the OC-3 level.

A relevant proposal on conceivable networking is as follows: Each of the divisions previously mentioned will have an ATM switch and hub to satisfy the needs for the users and servers within that LAN. A powerful ATM switch can be installed at a central location in the campus to provide backbone connectivity to all the different ATM networks. VLANs will be defined per each division to improve data speed.

Since the campus offers a basic telephone facility with an exchange within the campus, it will allow the use of a private ATM network instead of using the public ATM network.

Existing environment: Summary

❑   There is a 10BaseT legacy network in the administrative building exclusive to serve the administration, finance, purchase and personnel departments

❑   Currently, there is a high-speed FDDI ring to support large sized files of registry, admissions, and other student-related offices spread across the campus

❑   Eight non-engineering/science colleges have legacy LANs

❑   Engineering and science colleges have a CAT-5 cabled LAN that serves the multimedia terminals

❑   The media center supports applications such as full motion video, high-resolution graphics, audio, and text data

❑   The research center is interconnected with the science and engineering colleges

❑   There is a basic telephone facility with an exchange within the campus.

Network requirements

The proposed ATM migration should cover the following considerations:

- ❏ The engineering and science colleges need to be migrated to ATM technologies. Backbone connectivity among these to colleges and research centers is still required
- ❏ The research center should be connected to the Defense Research Complex to interoperate with certain applications of special categories with protocol structures of their own.
- ❏ Due to the high data rate in the media center, this division needs to be migrated to ATM switch technology
- ❏ All the divisions that will have ATM implemented should be provided with WAN access to communicate nationwide
- ❏ The administrative building and the non-engineering/science colleges will keep their existing legacy networks.

The itemized details of the migration planning/design proposed are indicated below.

ATM switch design parameters	Engineering and science colleges	Research center
AAL supported	AAL-1: To support regular data transfer. (Some burstiness could be expected). AAL-2: To support minor multimedia variable rate compressed video	AAL-1: To support regular data transfer
Multicast/Broad-cast support	Multicast or broadcast is not needed for the multimedia applications or the regular data transmission	There is no multicast or broadcast involved on the data transfers for this department
VPI/VCI support	VPI/VCI support. The signaling default VPI = 0, VCI = 5 will be used for everything (VC assignment, removal and checking). No metasignaling cells will be used to negotiate on signaling VCI and signaling resources.	VPI/VCI support. The signaling default VPI = 0, VCI = 5 will be used for everything (VC assignment, removal and checking). No metasignaling cells will be used to negotiate on signaling VCI and signaling resources.
Bandwidth symmetry	Asymmetric/ bilateral. The central ATM switch will handle a higher bandwidth (to connect to the WAN).	Asymmetric/ bilateral. The central ATM switch will handle a higher bandwidth (to connect to the WAN).
QOS parameters	Data ratio 1:1 • Cell loss priority (CLP): 0 high, 1 low • Peak cell rate (PCR): 1.5 Mbps • Average cell rate (ACR): 1.5 Mbps • Burst tolerance (BT): 2 (There is VBR and CBR traffic).	Data ratio 1:1 • Cell loss priority (CLP): 0 high, 1 low • Peak cell rate (PCR): 1.5 Mbps • Average cell rate (ACR): 1.5 Mbps • Burst tolerance (BT): 2 (There is VBR and CBR traffic).
Connection technology	• Point-to-point • PVCs	• Point-to-point • PVCs

ATM switch design parameters	Defense Research Complex	Media center
AAL supported	AAL-1: To support regular data transfer, AAL-5: To support encapsulation of proprietary protocols for specific applications. ATM switch will work as a gateway on this division	AAL-1: To support regular data transfer. AAL-2: To support variable rate compressed video and audio (delay sensitive)
Multicast/Broadcast support	Application and data are confidential. Data is very sensitive; request will be sent on a one-to-one basis. No multicast or broadcast is supported.	Each user will request different video, graphic, audio, or text data from the servers. No multicast or broadcast will be supported.
VPI/VCI support	VPI/VCI support. The signaling default VPI = 0, VCI = 5 will be used for everything (VC assignment, removal, and checking). No metasignaling cells will be used to negotiate on signaling VCI and signaling resources.	VPI/VCI support. The signaling default VPI = 0, VCI = 5 will be used for everything (VC assignment, removal and checking). No metasignaling cells will be used to negotiate on signaling VCI and signaling resources.
Bandwidth symmetry	Asymmetric/ bilateral. The central ATM switch will handle a higher bandwidth (to connect to the WAN).	Asymmetric/ bilateral. The central ATM switch will handle a higher bandwidth (to connect to the WAN).
QOS parameters	Data ratio 1:2 • Cell loss priority (CLP): 0 high, 1 low • Peak cell rate (PCR): 1.5 Mbps • Average cell rate (ACR): 1.5 Mbps • Burst tolerance (BT): 2 (There is VBR and CBR traffic).	Data ratio 1:3 • Cell loss priority (CLP): 0 high, 1 low • Peak cell rate (PCR): 1.5 Mbps • Average cell rate (ACR): 1.5 Mbps • Burst tolerance (BT): 2 (There is VBR and CBR traffic).
Connection technology	• Point-to-point • PVCs	• Point-to-point • PVCs

Hardware considerations
ATM switch to be provided at the colleges and departments.

❑   *Size*: 2.5 Gbps

❑   *Blocking factor*
Non-blocking or with a probability of blocking of 1×10^{-10} (virtually non-blocking).

❑   *Buffer capacity*
The larger the buffer, the less the traffic will be dropped, but the greater delay.

❑   *Buffering method*
Input, this ATM switch processes video request from the multimedia stations.

❑   *Switching delay*
The delay measured in µs as the total one-way through the switch. For this backbone switch the delay should be less than 40 µs.

❑   *Interfaces*
One OC-1.
One 155 Mbps interface, to connect to the ATM hub/servers.

ATM switch to be provided at the central administration

-     *Size*: 5 Gbps

-     *Blocking factor*
  Non-blocking or with a probability of blocking of 1×10^{-10} (virtually non-blocking).

-     *Buffer capacity*
  The larger the buffer, the less the traffic will be dropped, but the greater the delay. For the OC-3 link, the buffer capacity will be 65,535 cells with a delay of 186 ms.

-     *Buffering method*
  Input and output, this ATM switch services the ATM campus backbone as the WAN connection to communicate nation wide.

-     *Switching delay*
  The delay measured in µs as the total one-way through the switch. For this backbone switch the delay should be less than 20 µs.

-     *Interfaces*
  One OC-3/STM-1.
  Five OC-1, to connect to the different ATM colleges/department.

Links

Each of the different ATM colleges and departments (engineering and science colleges, research center, Defense Research Complex, and the media center) will connect to the central administration ATM switch via OC-1 links. This bandwidth provides a cost-effective solution for the backbone connectivity.

The central administration ATM switch will have also an OC-3 link to connect the campus nationwide.

Cabling

Each of the engineering and science colleges was already using CAT-5 cables. This cable can support speeds upto 100 Mbps, which constitutes a good speed to support the applications for these colleges.

Existing fiber optic cable will be used to connect the backbone on the engineering and science colleges and the research center. New multimode fiber optic cable will be run from the Defense Research Complex and the media center to the central administration ATM switch.

The existing legacy networks in the administration building and non-engineering colleges will not be migrated to ATM switching.

New servers

Security servers

The Defense Research Center may bear confidential/classified information. The access to the data stored on these servers, therefore, will have to be restricted only via a security server (primary and secondary). The security should run special authentication and authorization software that can use the DES III mechanism. This method provides a secure way to access the servers with the proper privileges.

Each user will have a DES card. This card will run an algorithm that will synchronize with the same algorithm running at the server. Each time a user requests an access to a specific database, it will be challenged and the DES card will generate a number. The customer will maintain a fixed PIN number for that specific card and the PIN number has to be entered after the challenge has been generated.

The use of dynamic passwords will assure that only the person who has the card and knows the PIN number can access the classified information.

Further, the users can be assigned with privileges on a group basis. The use of security servers on a group-wide basis will maintain the integrity to the restricted access to different levels of information. Each time a packet is generated from the user, the security servers will check through the access lists defined for the user in order to allow the packet or reject it.

Implementation schemes (Figs. 8.35 and 8.36)

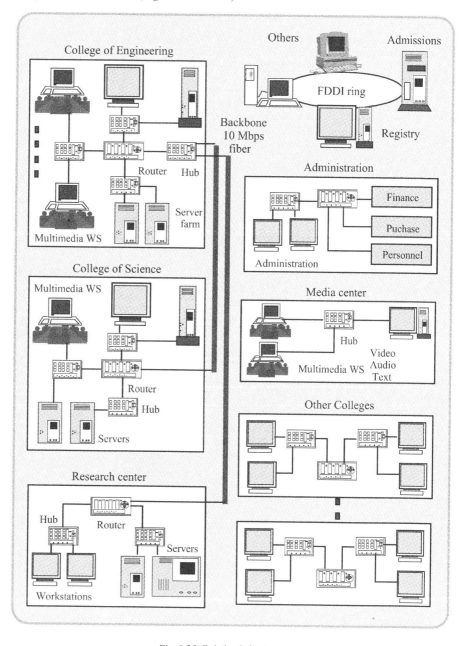

Fig. 8:35: Existing infrastructure

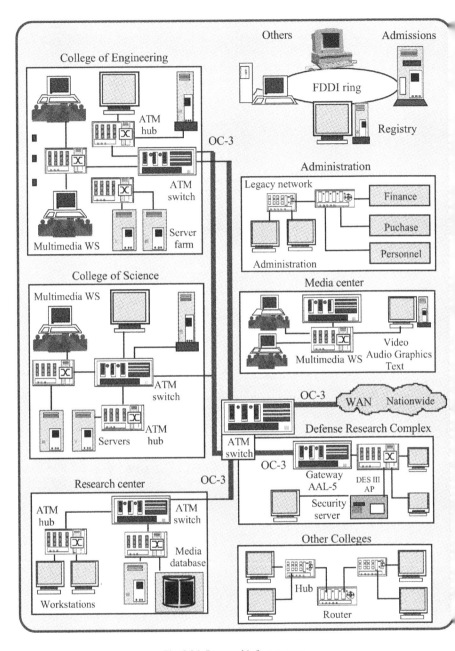

Fig. 8.36: Proposed infrastructure

Example 8.6

Suppose a campus networking is on an IP backbone and its existing structure is overloaded due to the following reasons:

- ❑ A hike in user population
- ❑ Deployment of new applications
- ❑ Upgrading of end-systems to more powerful versions.

Propose a migration strategy to accommodate the congested traffic by still retaining the IP environment.

Solution

The suggested strategy towards migration is as follows: Since backbone is IP only, SVCs can be deployed. The ATM backbone-based system can be used via IP. That is, ATM can be deployed as a new switched backbone to interconnect the routers.

Redundant ATM ARP servers can be deployed to provide address resolution between IP and ATM address. It allows the use of SVCs without requiring the use of preconfigured address maps. Further, optional PVCs can be created within the network for routing updates, and PVCs could be configured for signaling and the ILMI. The *maximum transmission unit* (MTU) can be set to a specified extent throughout the network.

Problem 8.2

Shown in Fig. 8.37 is a projected campus network for a large university. Typical distances involved are also shown in Fig. 8.37. End-entities are comprised of one or many/all of the following at each node:

Broadband terminal
Engineering/science workstations
Physician's workstation
Ethernet
Image-database

Suggest, with a block-schematic and relevant tables (on traffic parameters, cabling, types of VC, etc.), an initial phase of implementation of ATM networking compatible for the applications indicated.

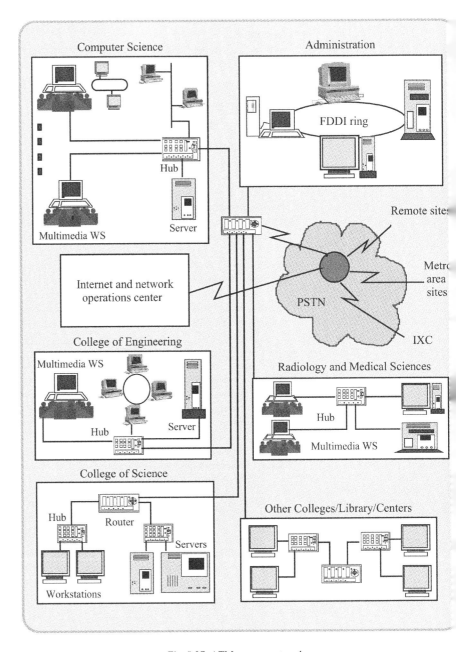

Fig. 8.37: ATM campus network

Note: Approximate distances – (a) Between Computer Science Department and other colleges/library/centers locations: 2 miles; (b) Between College of Engineering and other colleges/library/centers locations: 1 mile, and (c) From the locations of Colleges of Engineering, Science and other colleges/library/centers to Radiology/Medical Sciences buildings: 6 miles.

Problem 8.3

Using the relevant concepts indicated perform the following migration to ATM-enabled/based networking for the scenario described below:

In a growing city, an integration-expansion of networking is planned. The major participants of the plan are: A medium-sized university campus (with about 10 colleges/20,000 students), and the local municipality. The individual requirements are as follows:

University campus

Administration: Academic and nonacademic departments: Secured computational facility, e-mail/Internet and voice (telephony).

Colleges/departments: Voice and table-top computers to staff/faculty, e-mail/Internet (certain departments with multimedia services).

Engineering design center: This has multimedia, CAD/CAM and graphic facilities

Academic computer center: Mainframe facility.

Library facilities: Comprehensive database.

The present networking at the university campus consists of legacy LAN, ethernet and FDDI backbone. All the buildings are within a radius of 3 miles. The upgrading required is as follows: Expansion of multimedia across the campus and campus library extension to municipal library for exchange of database contents.

Municipal complex

City hall: Administration, public works, records center — all located within a 1/2 mile radius.

Public library, police department, fire department: Situated in different parts of the city within metropolitan networking limitations.

Present networking of the municipal complex consists of legacy LAN (ethernet, star-connected to cityhall complex. The university is within a distance of 5 miles from the municipal building.

The upgrading required is: Expansion of a multimedia facility at the municipal library, a more comprehensive record storing, and a provision for limited access to records to public. Necessary firewalls on secured information (such as in the police department) may be facilitated.

Describe a plan/design for the expansion planned.

Example 8.7

Suppose a corporation has its offices situated at multiple locations, interconnected by public PVC-only ATM service. The following are assumed:

- ❑ Expansion requires implementing SVC-based ATM backbones at the multiple (say, three) sites
- ❑ Bandwidth requirements indicate that these sites be interconnected via an ATM WAN
- ❑ Service providers offer only a PVC-based ATM.

Explore a possible migration to deploy ATM in the systems.

Solution

The implementation strategy would warrant a method to "tunnel" the SVC signaling information across the ATM WAN since the service-provider offers an ATM which is PVC -based.

As a part of the ATM service implementation, the customer can be provided with CPE ATM switches, which are capable of VP switching. ATM switches set in the core of the network should perform VP switching as well.

The following are to be implemented:

❑   Configuration of the necessary VPCs between the switches across the ATM backbone

❑   Remapping signaling requests normally carried on VPI = 0; VCI = 5 to VPs of VPI other than 0

❑   Within a given VPC, establishment of VCCs on demand by the signaling channel.

In Fig. 8.38, the circuit from Miami to Atlanta is remapped from VPI = 3 to VPI = 9 between the 2 locations, while the signaling VCI is preserved as it is. The solution accommodates end-to-end SVCs although the intervening ATM WAN service supports only PVCs. It is further assumed that multiple signaling channels are supported over a physical interface in Fig. 8.38. Otherwise, each CPE switch would require two physical interfaces into the ATM WAN.

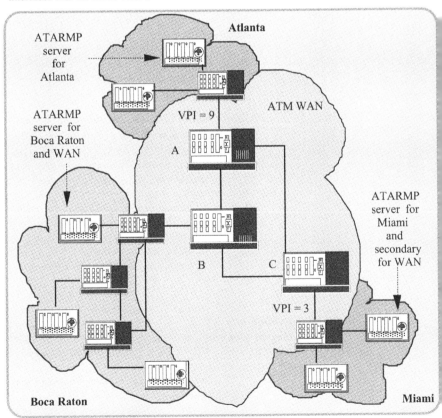

Fig. 8.38: Proposed migration

Note: A, B, and C are ATM switches located within the core of the network.

Example 8.8

Suppose an existing system operates on a frame-relay backbone. It is required to connect non-ATM capable routers to ATM networking. While deploying the frame-based interface to ATM, the investment in non-ATM capable routers should be preserved.

Assume that the currently existing ATM-based network is as in the previous example (Example 8.7). A fourth site is required to be connected to the ATM WAN and it does not warrant the use of high-speed ATM connection. Design appropriate migration considerations.

Solution

The additional (fourth site) connection does not require the use of high-speed ATM connection. Therefore, ATM cell overhead for this connection is an uneconomical burden at E-1/DS-1 data rates. Hence, a frame-based interface is to be implemented into the network.

The main consideration in a migration effort is using a frame-based ATM interface. The frame-based UNI (FUNI) has vendors' support on its standardization. Its interface to ATM can support signaling as well as required QOS (traffic-shaping). It provides necessary SAR for connection to the ATM switch. That is, relevant support can enable segmentation and reassembly (SAR) at the SAR sublayer where PDUs are segmented and reassembled/rebuilt from ATM cells.

Procedure

- By way of an enhancement to the system described in the previous example, a router can be added to the ATM WAN at Orlando. This router connects to an ATM WAN service at 2 Mbps. Hence, a standard serial interface, (such as X.21) can be used. Alternatively, in a scenario where higher speeds are expected, the HSSI can be adopted

- At Orlando, the IP address is configured to the router interface. Inasmuch as the router connects directly to the ATM WAN, it is given a WAN backbone address

- The encapsulation is set to FUNI, configuring PVCs with RFC 1483 SNAP. The individual PVCs are to be mapped within the Orlando router to the IP address of the destination router

- The distant ends are configured for each of the PVCs at the three original router sites. The other connections throughout the network may rely on SVCs. But they must still be defined as PVCs. Further, the encapsulations used at each end of the VCC should be identical. For example, a PVC defined as LLC/SNAP at one router must be identical to that specified at the destination side.

Typical PVCs configured are as follows:

Source	Destination	Number of VCCs	VPI/VCI	AAL #	Encapsulation type	PCR/ACR	BT
Orlando	Miami Branch	11	1/60	5	LLC/SNAP	2000/1000	2
Orlando	Boca Raton Branch	12	1/70	5	LLC/SNAP	2000/1000	2
Orlando	Atlanta branch	13	1/80	5	LLC/SNAP	2000/1000	2

- As in Example 8.7, the new PVCs are configured at each of the implemented ATM switches

733

❏   Connectivity between the new router and each of the original routers as well as with any other end-systems connected to the network should be verified

❏   Though the FUNI, which supports SVCs is implemented, due to the requirement dictated by dynamic routing, (which would require a PVC in any case) the SVCs are not implemented. With static routing, or if the network grows to include several other nodes, the use of SVCs would be logical and desirable. Then the required parameters can be configured across this interface, along with the signaling and ILMI PVCs, QOS parameters, and the NSAP address of the ATMARP servers. This part of the procedure maps very closely to Example 8.6, where one may develop maps for the SVCs based on the same QOS parameters as those used for the PVCs indicated above.

Example 8.9

Consider the following:

❏   A company specializes in helping oil exploration corporations locate the prospective sites of mineral oil deposits

❏   This company does subterranean probe-drilling and collects seismic data from various survey sites via appropriate instrumentation. Such data are at terabyte levels

❏   Data-processing condenses these collected data to about 50 gigabyte, high resolution images which are supplied to the customer-corporations for analyses and interpretations towards prospective oil strikes

❏   Analyses of these data-sets are computationally and graphically intensive

❏   Conventional delivery media are magnetic tapes. But, spinning these tapes, writing on them, and reading them would take days to weeks

❏   Hence, ATM-based networking is considered as an alternative fast-delivery medium.

Existing set-up and networking requirements towards expansion:

❏   User-side workstations: High-end SGIs, mid-range desktop users/servers

❏   Legacy LAN workgroups/ethernet

❏   High-speed LAN (for example, HiPPI) links a few supercomputers, which have logged-in data sets

❏   Expansion envisaged to include:

(a)   Company-wide internal networking with ease of configuration and use

(b)   Interoperability and performance

(c)   A combination of HiPPI and ATM

(d)   Distance (link-lengths) involved within the premises: < 100 meters

(e)   Service extension to customers across the city as well as between cities with reduced cycle times of delivery of services.

Develop a block-schematic to indicate the concept planning that includes the expansion strategy contemplated as above. Indicate, as necessary, the types of switches (ATM switches) needed, interfaces warranted, cable-types, data-rates, and any special feature that can be included.

Solution

The concept planning is illustrated in Fig. 8.39. It provides a company-wide internal networking including legacy LAN. Hence, interoperability is made feasible. The networking combines HiPPI and ATM: HiPPI LAN supports the supercomputing (data-processing) system and is linked to high-end desk-tops.

Workstations are SGIs, Indigos, Crimsons, and high-end Sparcs (for example, servers provided for could be Sparcs.)

ATM backbone configuration (as shown in Fig. 8.39) allows interworking of existing ethernet, low-speed workstations etc., (which may be needed for conventional company office environment), apart from facilitating high-end data crunching workstations.

Interfaces

- ❑ Internal premises cabling: Cat-5 UTP (155 Mbps support ~ 100 meter links)
- ❑ OC-3 for extension to MAN
- ❑ Other details on priority levels, PVC, SVC, signaling QOS parameters etc. should be stipulated.

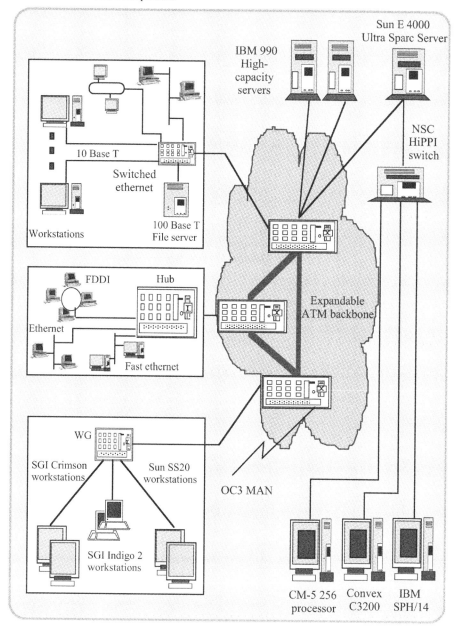

Fig. 8.39: ATM backbone configuration

Example 8.10

In bank-based derivative trading, the bank's derivative units should obtain accurate, crucial, minute-to-minute reports quickly and at any time of the day via telecommunication networking, so as to cope with the volatile, international financial marketing (that fluctuates rapidly).

Suppose, in a particular bank, such reports are available in the existing system with the following infrastructure:

- ❑ Availability of the reports only twice a day
- ❑ Lack of minute-to-minute information on derivative transaction has led the bank to potential losses
- ❑ Current network deploys fast ethernet via 100BaseT/FDDI.

Consider an expansion plan that facilitates:

- ❑ Availability of the reports any time of the day
- ❑ Reduced bottlenecks
- ❑ Dealer transaction exposure requests are provided rapidly at desk-top level
- ❑ Expandability to handle more dealer requests
- ❑ Interoperability.

Develop a block-schematic that describes an ATM-based networking in the bank premises to include the aforesaid planning strategies.

Solution

Extension of the current network (of fast ethernet via 100BaseT/FDDI) to incorporate the desired strategies is indicated in Fig. 8.40.

Note:

In Fig. 8.40, the high-speed data/text transmission is of primary importance. No video service needed. Other considerations are:

- ❑ Low latency
- ❑ Reliable connectivity
- ❑ LAN emulation provides transparent support for multiple network protocols over an ATM backbone
- ❑ Users belong to logical groups, rather than physical location
- ❑ Multicast support
- ❑ Consistent with the above features, 100 Mbps rate, low BER, high BT (with relevant PCR and ACR) services are to be provided
- ❑ Interfaces should accommodate the multiple protocols involved
- ❑ Multicasting feature should be considered and reliable signaling thereof should be provided for
- ❑ CAC/congestion reduction requests.

In summary, in the present network ethernet and token-ring LANs are connected via routers to an FDDI backbone. Central servers are also connected to the FDDI ring. New applications and increasing use of the network result in congestion on the backbone and sub-optimal access to servers. The user is not ready to migrate fully to ATM, and is comfortable with FDDI as a technology.

In the migration pursued, FDDI switching is deployed at the core, interconnecting existing routers. Servers also connect to these redundant FDDI switches. Switches use ATM internally, but only FDDI is visible to the user. This solves the current networking problem while providing a path for future ATM deployment.

In the existing infrastructure, the dealers connect via hubs to redundant routers located in the building. Servers are connected directly to routers. The introduction of new applications and hardware requires a greater bandwidth than that provided by hubs and exceeds the throughput of central routers. Network managers also require greater control over workgroups and membership.

In the conceived migration, LANE is deployed via an ATM backbone, along with LAN switching to most users. Hubs are preserved where user bandwidth does not justify dedicated ethernet. Switches are connected to the redundant ATM core. Routers interconnect VLANs and provide LANE services. Some servers are connected directly to the ATM backbone.

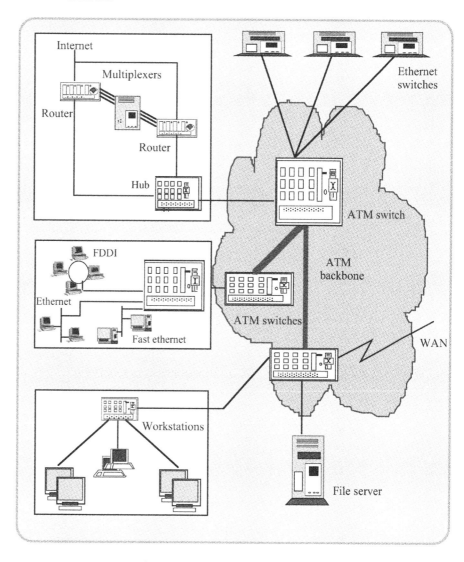

Fig. 8.40: Implementation of Example 8.10

In summary, the current setup uses separate voice and data networks to interconnect sites. As data volume grows, networking charges become excessive. Users wish to rationalize networking charges by connecting to the new ATM offering, but require high WAN reliability.

In the migration contemplated, voice and data services are provided by a nationwide ATM multiservice offering. CPE is a hybrid service multiplexer/router, with backup provided via ISDN for both voice and data traffic. Networking charges are reduced by combining access over a single service.

Example 8.11

Develop an ATM network-planning design in respect of a school district administration support service using the following information:

- ❑ The county school administration has an exclusive building complex
- ❑ There are three senior high schools, three junior highs, and ten elementary schools managed by the county
- ❑ Consistent with the present status on networking the required expansion is as follows:

    - ATM-based LAN/WAN, as necessary
    - All sites are connected through the PSTN cloud
    - All sites to have access to a public library media center which has documentary archives available online through CD-ROMs located in optical disk libraries
    - PC multi-media terminals can be used as necessary
    - Availability of Internet-ready Web site browsing and e-mail facility
    - Facility to get network access from home by the district staff and school features. (Assume a reasonable population).

- ❑ Your design should describe and/or include the following:

    - Overall infrastructure: Existing and proposed
    - Sharing resources across platforms
    - Bus-based LAN being retained for basic PC applications at each site
    - Multimedia based workgroup hosts can be added as needed
    - File-servers and/or video servers can be added as necessary
    - Public network is accessed and used
    - Block-schematic of the features included as above in the networking implemented
    - Tabulation of data rate requirements, maximum link-lengths in the network proposed, and the type of wiring needed
    - Multicasting/broadcasting requirements
    - Types of connections to be supported: VC, VP, PVC, SVC, point-to-point, point-to-multipoint etc.
    - Types of signaling protocols supported
    - CLP, PCR, ACR, and BT requirements.

Note: 1) Assume that only the senior high schools have multimedia workstations to access public library video on ATM.

2) All other networking can be conventional (legacy) LAN/WAN.

Solution

Project summary

The purpose of this design is to develop an ATM network design for a school district administration, support service for three senior high schools, three junior high schools, and ten elementary schools.

Currently, the only connectivity available from any of the mentioned schools with the county school administration is performed through dedicated phone-lines.

Communication is established from the different schools PCs or terminals to the main host: AS/400 at the central administration site.

The senior high schools will be the only schools with multimedia workstations. These schools will have access to a public library media center documentary archives and also will be provided Internet access for Web browsing and e-mail. Due to the traffic volume, the senior high schools will be migrated to an ATM environment. This design proposed will make use of ATM hubs and switches on the mentioned schools.

Since the junior high schools or the elementary schools do not have different bandwidth requirements, their existing legacy systems will not be modified or migrated to ATM.

All the senior high schools will connect to the central administration through a public ATM network. The cost of using a public network is lower than using a private data network (that is, switches and trunks can be shared across multiple customers reducing the cost) and due to the trunks having speeds higher than any access line speed, there is a lower delay and loss.

A new service will be implemented for remote access by the district staff and school teachers. The solution proposed will be ISDN to the home. This will provide a cost-effective solution for those that need to dial remotely. For the central site, B-ISDN is supported by the public ATM network.

Networking requirement
Services
New services to be added to the schools are as follows:

- ❑ All senior high schools will have access to public library media center documentary archives located in optical disk libraries
- ❑ All senior high schools will have Internet access for Web browsing and e-mail services
- ❑ School teachers and district staff will be provided access from home
- ❑ No changes will be implemented to existing legacy networks for junior high schools or elementary schools.

Bandwidth
Data
For the junior high schools and/or elementary schools the workstations and terminals will be still attached to the existing legacy environment and will not be migrated to an ATM hub/switch. Data will continue to be transferred from the PCs and terminals to the AS/400 on the central administration through the phone-lines. Bandwidth will remain the same.

For the senior high schools, data will be transferred from the legacy network (all the PCs and terminals connected through a 10BaseT hub) to the ATM backbone switch. The new OC-3 links will be used to transfer the data and the new video (see next section) frames. Data bandwidth is presented to represent not a considerable portion of the traffic.

Video
For this environment, a new form of information will be transferred from the public library to the multimedia workstations on the senior high schools video traffic. The bandwidth requirements for video distribution is 16 Mbps per user.

Since users can request videos-on-demand, the traffic is unidirectional and there might be more than one request per user at certain times and more than one user requesting videos in the same LAN. Two scenarios need to be considered:

- ❑ The multimedia workstations could be attached to the legacy network (10BaseT hub) or,
- ❑ Multimedia workstations will be connected to an ATM hub.

For the first situation, if more than one multimedia workstation is requesting video traffic, overload, data loss, and delays, could be presented in the LAN (due to the bandwidth required to transfer a video). The second option provides the requested throughput at a low cost (the cost per port on an ATM hub is less expensive than the cost of an ATM backbone switch port). The currently proposed design will implement the second option, which uses an ATM hub for the multimedia workstations.

Hardware considerations
Multimedia workstations
Physical interface rate: 25 Mbps Desktop 25
Physical media type: SM Fiber
System bus: PCI

AS/400 servers
The AS/400 servers can be connected directly to the ATM backbone switches on the central administration. This will ensure the highest throughput for all data requested from the different sites. The interface proposed is used for both the private and the public UNI.
Physical interface rate: 155 Mbps
Physical media type: UTP-5: 155 Mbps can be transported over copper pairs
System bus: S-bus

WEB servers – proxy servers and multimedia servers
All these servers will have a high data transfer request; the interface proposed should be able to provide 155 Mbps of throughput. Since the public network is being used, the proposed interface is the same as the previous AS/400 servers.

Physical interface rate: 155 Mbps
Physical media type: UTP-5
System bus: S-bus

ATM switch
The following are the minimum hardware characteristics for the ATM switches to be used on this design.

Senior high schools
Size: 2.5 Gbps

Blocking factor
Non-blocking or with a probability of blocking of 1×10^{-10} (virtually non-blocking).

Buffer capacity
The larger the buffer, the less the traffic will be dropped, but the greater delay. For the OC-3 link, the buffer capacity will be 65,535 cells with a delay of 186 ms.

Buffering method
Input, this ATM switch processes video requests to the video public library.

Switching delay
The delay measured in μs is the total one-way through the switch. For this backbone switch, the delay should be less than 40 μs.

Interfaces
One OC-3/STM-1.
10BaseT, to connect to the legacy network.
One 155 Mbps interface, to connect to the ATM hub/servers.

Central administration
Size: 5 Gbps

Blocking factor
Non-blocking or with a probability of blocking of 1×10^{-10} (virtually non-blocking).

Buffer capacity
The larger the buffer, the less the traffic will be dropped, but the greater delay. For the OC-3 link, the buffer capacity will be 65,535 cells with a delay of 186 ms.

Buffering method
Input and output locations. The ATM switch services the video request from the senior high schools and performs requests to the video public library.

Switching delay
The delay measured in µs as the total one-way through the switch. For this backbone switch, the delay should be less than 20 µs.

Interfaces

-      One OC-3/STM-1
-      One 155 Mbps interface, to connect the AS/400 host/proxy server
-      A 10BaseT port to connect the firewall (that filters all the traffic from the Internet)
-      B-ISDN support, for remote ISDN users.

Links
Central administration
        Assuming that all the senior high schools are located in the same city (5 to 50 km), there will be a MAN type of network. The rate required is 155 Mbps between the schools and the central administration site; this can be achieved through an OC-3 link. Taking some statistical information, one may expect the following typical workstations traffic characteristics:

Data

-      There will be an average of 28 messages per second
-      5% of the messages are between 10 kbit and 11 Mbit
-      25% of the messages are between 1 kbit and 16 kbit
-      70% of the messages are less than 1 kbit.

This will result in an average traffic of approximately 1.5 Mbps per workstations.

Video
The video rate required is about 6 Mbps
A MAN operating at 155 Mbps and an acceptable occupancy level of 60%, allows 30 workstations processing data and 8 multimedia workstations doing video transfers to be attached to the MAN, assuming that 50% of the workstations are simultaneously active. This number of workstations is very reasonable and cost-effective for the solution and the speed of the link proposed.

Central administration-public library
        Since only senior high schools will have access to the videos on the public library, and assuming that 30% of the multimedia workstations will be accessing the videos, that is, 30% of the total bandwidth for each of the three senior high schools, an approximately bandwidth of 140 Mbps will be required between the central

administration and the public library. Using the optical carrier speed hierarchy, an OC-3 will be needed (155.02 Mbps). This bandwidth will be enough to support the requirement for video transfers.

Home to schools

A 64 kbps (128 kbps using both B channels) ISDN link will be enough for the district staff and the school teachers. Implementing ISDN from the home is cost-effective for the type of information being transferred. No video will be transferred.

Cabling

For the local 155 Mbps links from the ATM switches to the servers, Cat 5 twisted pair cable will be used. The servers will have a RJ-45 connector.

For the OC-3 links from the senior high schools to the central administration and from this site to the public library, the cabling will be multimode fiber. The framing will be SDH/SONET and the connector SC.

All the legacy networks (10BaseT) type of connections will use Cat 5 patch cables and the connector will be RJ-45.

Interfaces

The following are the interfaces to be used by the ATM switches:

- ❑ For the servers, the interfaces will be STS-3c/STM-1. This provides 155 Mbps over Cat 5 twisted pair copper cable
- ❑ For the OC-3 links, the interfaces will be SONET/STS-3c/STM-1 at 155 Mbps over multimode fiber optic cable.

New servers
Proxy server

Each of the senior high schools will have a proxy server. This server will retain the most used WEB pages in order to eliminate traffic on the OC-3 links from the senior high schools to the central administration. The server will be connected directly to the ATM backbone switch at each location. There will be a central proxy server located in the central administration building. All the remote proxy servers will update their records to the central proxy server after hours.

Firewall

A firewall server will be added for the connection to the Internet. This server will run specialized software to filter and block traffic from the Internet. This server will protect all the internal networks for the school administration for software or TCP/IP port attacks. The "clean" or internal side of the firewall will be connected to the ATM backbone switch through a 10BaseT data rate. The "dirty" or external side of the firewall will be connected to an ethernet port on a router. The router will have a T-1 connection to the Internet.

Multicasting/broadcasting requirements

The user video requests will be done from the multimedia workstations to the servers. There is no requirement for multicasting or broadcasting. Each multimedia station makes individual requests to the servers.

WEB browsing traffic will be also generated from each of the PCs to the different WEB sites. No multicasting will be present.

Connections supported
Connection topology

Point-to-point: Each school will be connected to the central administration via the ATM backbone switch. This is a point-to-point connectivity.

Connection type

    Permanent virtual circuits (PVCs): All the schools will be connected to the central site on a permanent basis through the OC-3 links.

Bandwidth symmetry

    Asymmetric/bilateral: Each of the schools will have an OC-3 link to the central administration building. On the central site, an OC-3 will be used. More traffic will be handled on the central administration side than on the senior high school side.

    From the public library to central administration, the traffic will be also asymmetric; the link will be OC-3. Requests will be made from the senior high schools to the multimedia server at the public library.

AAL

-     AAL-1 will be used to support data transfers, assuming the bit rate to be constant.
-     AAL-2 will be used to support VBR video transfer.

Signaling protocol supported

-     Routing cells through the network will be done through VPIs and VCIs. The signaling default VPI = 0, VCI = 5 will be used for everything. (VC assignment, removal, and checking).
-     No metasignaling cells will be used to negotiate on signaling VCI and signaling resources.

QOS parameters
Cell loss priority (CLP)

    Since ATM utilizes the CLP capability on user request, network resources are allocated to CLP = 0 (high priority) and CLP =1 (low priority) traffic flows. The cell with the highest priority is the least to be discarded. A value of 1 in the CLP field means that the cell has low priority; in other words, it may be selectively discarded during congested intervals in order to maintain a low loss rate for the high-priority CLP = 0 cells.

Peak cell rate (PCR)

    The PCR traffic parameter for a connection is defined at the physical layer SAP. Since the PCR is a mandatory source traffic parameter, it applies to connections supporting both CBR and VBR services. For this model, the PCR will be defined at 155 Mbps.

Average cell rate (ACR)

    The average cell rate is about 50% of the total bandwidth, approximately 75 Mbps on both directions, from the multimedia workstations to the video servers and in the other way.

BT
Can be set up to 1.

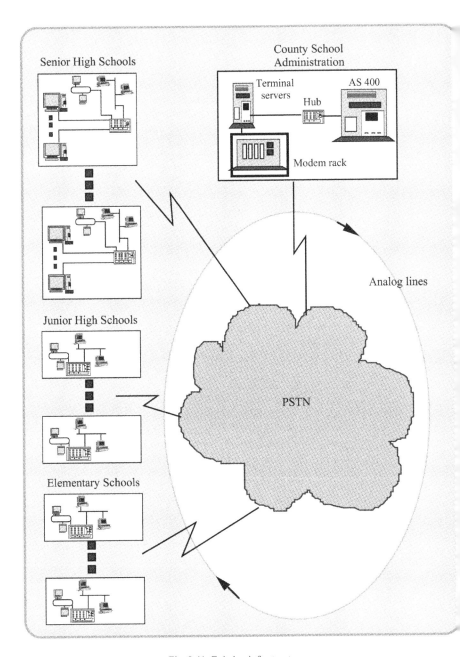

Fig. 8.41: Existing infrastructure

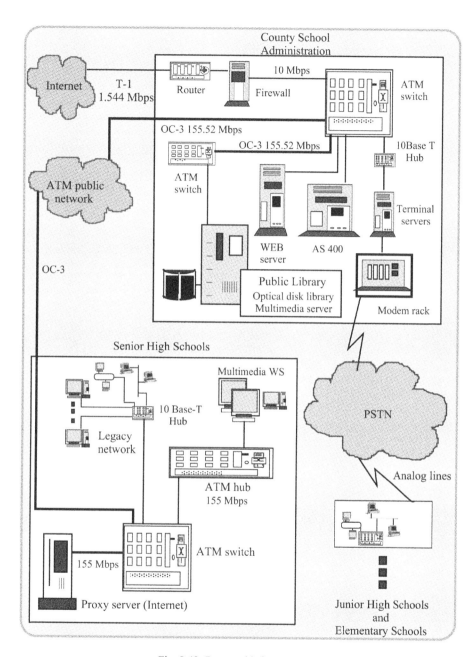

Fig. 8.42: Proposed infrastructure

Example 8.12

Currently, a city library has its documentary archives (texts and photography) complied using the state-of-the-art information technology system. The complied data-sets are stored in a minicomputer (say, IBM AS/400) and are available to library-users (within the library premises) at desk-top PCs via legacy LAN networking.

An expansion plan is contemplated to facilitate the following:

- Upgrading of documentary archives with each document to feature a one-minute audio-visual abstract
- The above upgraded documents are to be on several online CD-ROMs located in optical disk libraries
- These digitized data-sets are to be made available to library-users (within the premises) on standard PC multimedia terminals. The data sets on documented archives (tests, video clips and fixed images plus voice-commentary) accessed at the PCs are to be completely "transparent", i.e., the users should have the impression that they are looking at local data when, in fact, it comes from the server.

(a) Draw a block-schematic of the networking with the expansion strategy indicated. (Assume 2 IBM AS/400 servers and a total of 30 PC terminals (for library users are available in the library)

(b) Assume bit-rate requirement at 155 Mbps. The digitized data-sets can be made available to a branch library located:

(i) Close to the main library (say within 10 km);
(ii) Within the metropolitan areas of the city; and
(iii) In a suburban town about 50 km from the main library.

In each of the above cases, indicate the extension methodologies to be pursued.

(c) Construct a table indicating the following:

(i) Bandwidth and number of ATM interfaces to be supported by the ATM switches being considered
(ii) Types of AAL(s) and traffic types to be supported
(iii) Types (with reasons) of the priority levels to be provided
(iv) Multicast/broadcast requirements
(v) Types of cabling within (and outside) the premises
(vi) Types of interfaces (SONET/SDH, DS, ... etc.) to be supported
(vii) Types of connections to be supported: VC, VP, PVC, SVC, point-to-point, point-to-multipoint, etc.
(viii) Types of signaling protocols to be supported
(ix) Tabulate typical QOS parameters to be managed: CLP, PCR, ACR, BT
(x) Any other special feature you may foresee.

Solution

Current environment

Applications:

- Text and photography.

Hardware premises

- One AS400 server
- Desk-top PCs
- Legacy LAN networking.

Expanded environment

Applications:

- ❑ Text
- ❑ Video clips
- ❑ Fixed images
- ❑ Voice documentaries.

Hardware premises:

- ❑ Two AS/400 w/on-line CD-ROMs and optical disk
- ❑ 30 Multimedia PCs.

User requirements

- ❑ Data sets on documented archives (texts, video clips, and fixed images plus voice commentary) accessed at the multimedia PC's are to be completely transparent.

Analysis

File transfer programs, that is, text data, remain the same. The key point here is that new forms of traffics, such as video and voice, are added. Thus, the multimedia PCs will be receiving a variety of traffic, namely video, voice, and data. All this traffic requires synchronization and as per user request, low latency or delay.

It is important to note that the video and voice data is one way (on demand) and the environment may allow multiple windows open at the same time requesting different video clips.

A summary of multimedia application bandwidth requirements are:

Application	Bandwidth (Bits/second/User)
Images database retrieval: Pictures	28,800
Voice annotated text	96,000
Voice annotated image	86,000
Video distribution	6,000,000

A legacy LAN could handle this kind of traffic but it will incur latency problems when more than one client or multimedia PC on the same legacy LAN requests video clips from the AS/400 servers. In addition, a mix of data, voice, and video will make things worse because there is no prioritization whatsoever making the legacy LAN overloaded and incurring heavy delays. A solution is to migrate the multimedia PCs legacy LAN to an ATM LAN. The ATM NIC possibilities for the multimedia PCs are:

Physical interface rate	Physical media type	System bus
25 Mbps desktop 25	SM fiber or copper cables	EIAS/PCI
51 Mbps SONET	MM fiber or copper cables	EISA/PCI

The ATM Forum recommendation on ATM-to-desktop is 52 Mbps. It is a low cost interface and permits already installed copper cables to be used. Nevertheless, since our multimedia PCs only generate video clips on demand (one way), a 25 Mbps interface will suffice and reduce implementation costs.

All the 30 multimedia PCs will be connected to an ATM-ready hub. This will eliminate the bottleneck problem and will achieve real-time information transfer and scalable throughput.

Further, it is necessary to connect the two AS/400 servers directly to an ATM switch because the host systems should be prepared to transmit video clips to possibly more than 30 users. That is, a high throughput from these servers is warranted. A 155 Mbps connection is, therefore, recommended. The NIC card is characterized by:

Physical interface rate	Physical media type	System bus
155 Mbps SDH	UTP-5 (< 100 m)	S-bus

The ATM-ready hub will also connect to the ATM switch.

The branch office and the offices within the metropolitan areas of the city and suburban towns can be connected via a metropolitan area network (MAN) or wide area network (WAN) service provider. A guarantee of 155 Mbps is required. Hence, the UNI must be set to OC-3. At the other end of the MAN/WAN cloud, the connection has to be made using OC-3 to other ATM switches.

Fig.8.43 depicts the block-schematic of the networking expansion strategy.
Note: One can also use a direct fiber link between the branch office close to the main library such an SM fiber at 155 Mbps using STM-1 but the cost will be much higher.

Summary of the expansion strategy is as follows:

Location	Distance	Strategy
Branch office	~ 10 km	OC-3
MAN area	< 50 km	OC-3
Suburban town	~ 50 km	OC-3

The characteristics for the ATM switches selected in the implementation can be summarized as indicated below:

ATM switch Design feature	ATM backbone switch
Bandwidth	ATM campus (CA) switch supporting: • Up to 5 Gbps
# ATM interfaces	ATM interfaces: • 2 or more OC-3 interfaces • 2 or more 155 Mbps interfaces.
AAL supported	AAL-2: To support variable rate compressed video and time sensitive audio: • Bit rate is variable • Timing must be preserved end-to-end • Connection oriented. ALL-1: To support regular data transfer (maybe some burstiness) • Bit rate is constant • Connection oriented • Timing preserved end-to-end. AAL-5 is not required since one assumed user is not running any other variable length PDU's
Priority levels	High priority: • Voice/video: require low cell delay variation Low priority: • Data transfer: can tolerate cell delay variation
Multicast/broadcast requirements	User application does not require multicasting or broadcasting. Assume each individual multimedia PC makes independent requests and requests are only generated by the end user not the server. See connection types and signaling below.
	155 Mbps links

Cabling	• Cabling: UTP • Limited to 100 m • RJ-45 connector OC-3 155 Mbps links • Cabling: Multimode fiber • SDH/SONET framing • SC connector
Physical interface	• SONET/STS-3c/STM-1 155 Mbps multimode fiber • STS-3c/STM-1 155 Mbps over UTP-5
Connection types	Bandwidth symmetry: • Asymmetric/bilateral (more BW from the main office to the network) Connection topology • Point-to-point • PVCs • No SVC is required for this specific application
Signaling protocols	• VPI/VCI support • Single signaling VC support • Signaling for VC assignment • Signaling for VC removal • Signaling for VC checking • No metasignaling is required for this specific application

QOS	Source: AS/400
	Destination: Multimedia PC's
	• Cell loss priority (CLP)
	    0-high priority
	    1-low priority
	• Peak cell rate (PCR)
	    Approximately 155 Mbps
	• Average cell rate (ACR)
	    Approximately 70 Mbps
	• Burst tolerance (BT)
	    6
	Source: Multimedia PC's
	Destination: AS/400
	• Cell loss priority (CLP)
	    0-high priority
	    1-low priority
	• Peak cell rate (PCR)
	    Approximately 2 kbps
	• Average cell rate (ACR)
	    Approximately 1 kbps
	• Burst tolerance (BT)
	    0

continued ...

ATM switch Design feature	ATM switch 1, 2, and 3
Bandwidth	ATM campus (CA) switch supporting: • Up to 5 Gbps
# ATM interfaces	ATM interfaces: • 1 or more OC-3 interfaces • 1 or more 155 Mbps interfaces.
AAL supported	AAL-2: To support variable rate compressed video and time sensitive audio: • Bit rate is variable • Timing must be preserved end-to-end • Connection oriented. AAL-1: To support regular data transfer (may be some burstiness) • Bit rate is constant • Connection oriented • Timing preserved end-to-end. AAL-5 is not required since one assume user is not running and other variable length PDU's
Priority levels	High priority: • Voice/video: required low cell delay variation Low priority: • Data transfer: can tolerate cell delay variation
Multicast/broadcast requirements	User application does not require multicasting or broadcasting. Assume each individual multimedia PC makes independent requests and request are only generated by the end user not the server. See connection types and signaling below.
Cabling	OC-3 155 Mbps links • Cabling: Multimode fiber • SDH/SONET framing • SC connector
Physical interface	• SONET/STS-3c/STM-1 155 Mbps multimode fiber
Connection types	Bandwidth symmetry: • Asymmetric/bilateral (more BW from the network to the branch offices) Connection topology • Point-to-point • PVCs
Signaling protocols	• VPI/VCI support • Single signaling VC support • Signaling for VC assignment • Signaling for VC removal • Signaling for VC checking • No meta-signaling is required for this specific application

QOS	Source: AS/400
	Destination: Multimedia PC's
	• Cell loss priority (CLP)
	    0-high priority
	    1-low priority
	• Peak cell rate (PCR)
	    Approximately 155 Mbps
	• Average cell rate (ACR)
	    Approximately 70 Mbps
	• Burst tolerance (BT)
	    6
	Source: Multimedia PC's *continued ...*

Destination: AS/400
- Cell loss priority (CLP)
  0-high priority
  1-low priority
- Peak cell rate (PCR)
  Approximately 2 kbps
- Average cell rate (ACR)
  Approximately 1 kbps
- Burst tolerance (BT)
  0

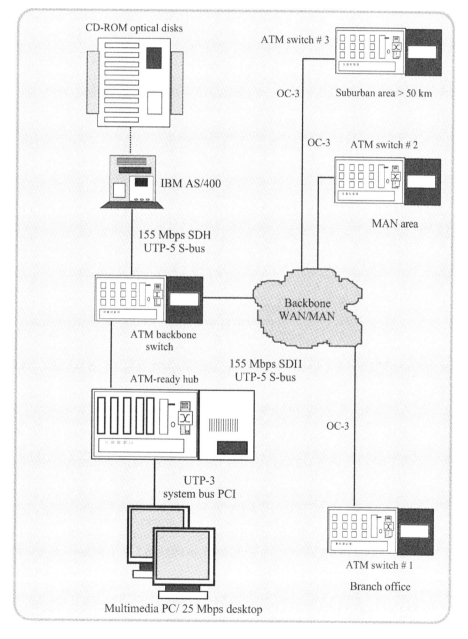

Fig. 8.43: City library networking

Note: Internal premises cabling - Cat 5 UTP for 155 Mbps support (limited to 100 m) and extension to WAN/MAN - OC-3 for 155 Mbps support

Example 8.13

Suppose there are sites required to be linked, which do not share common network addresses, and routing has to be done across an ATM WAN at least suboptimally across LIS boundaries. For example, multiple corporations (or a single organization with multiple addresses) may need to be connected to an ATM network. In such cases, a direct connectivity across the ATM network is preferable instead of "jumping through" routers due to addressing constraints involved.

Suggest a migration strategy towards ATM implementation.

Solution

NHRP, which optimizes routing across NBMA (non-broadcast multiple access) networks can allow end-systems to open VCs across LIS boundaries. That is, NHRP provides for address resolution across multiple IP networks, at the same time can enable short-cut routing across ATM networks. This allows two end-systems on separate IP networks to establish direct VCCs across the ATM network.

This strategy is preferred in the present case (as compared to using ATM ARP) because ATM ARP is limited to a single network and in using it, connectivity is forced to exist from the ATM cloud in facilitating a connection from one IP network to another.

A mesh of NHRP servers within the network can be used to provide the address resolution. (Note that NHRP is limited to IP protocols only.) Suppose various sites assumed to be located within an NHRP environment are, however, geographically separated. Enterprise or CO access/CO class switches can be used to interconnect (the assumed) four sites. These switches can implement congestion avoidance and the associated buffer mechanisms are adequate to support application traffics across the wide area.

Example 8.14

It is required to share product details and service profile information among the employees of a corporation and offer relevant training across the enterprise. The implementation is to be based on an ATM network. Develop a relevant networking strategy.

Solution

The proposed strategy would involve the deployment of IP multicasting across the ATM backbone using network layer multicasting techniques and ATM point-to-multipoint VCCs. This strategy is also suitable for the metropolitatn/WAN where multiple video channels can be delivered to subscribers. (This may not be the case when the video-on-demand (VOD) application is warranted, because multicasting infers that a number of users may wish to receive a given video feed at any instant in contrast with a situation wherein any user may be permitted to view an individualized program as in VOD.)

Clients of the required service can be attached to routers and the PIM (*protocol independent multicast* — an Internet protocol, which provides for efficient routing of multicast data across an internetwork) manages the network layer multicasting.

Thus ATM, is deployed here as high-speed backbone supporting IP-layer multicasting.

Example 8.15

The problem is to facilitate a guaranteed QOS for multimedia traffic across an interconnecting network comprised of both non-ATM and ATM segments. What is a feasible strategy?

Solution

The proposed strategy is the deployment of *resource reservation protocol* (RSVP), which allows PCs and workstations to request QOS from a network. Such RSVP requests may be mapped on ATM signaling requests to end-systems and routers within the network.

The routers at the periphery of the ATM cloud can remap from the integrated service QOS to ATM QOS.

In the example illustrated in Fig. 8.44, two sets of end-systems are connected to the network. The first set implements RSVP within the networking stack. The second set, not capable of generating and responding to RSVP requests, can however, use static RSVP parameters as originally configured in the routers along the path of connectivity.

In the example shown, a multicast flow using audio/video application is assumed.

If this network does not include ATM at the core, it amounts to making only the static RSVP entries within the routers closet to the sources and receivers. These routers would then emulate end-system characteristics within the cloud. However, since the ATM is introduced into the network, the RSVP parameters must be mapped into ATM QOS.

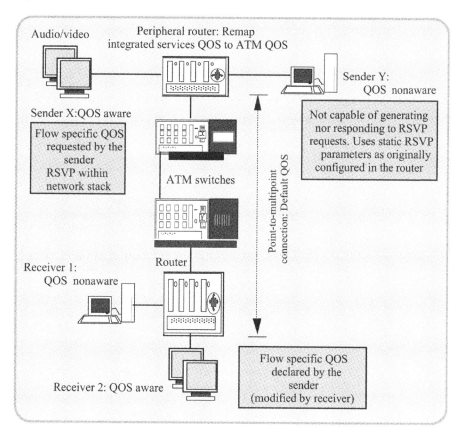

Fig. 8.44: Implementation scheme for Example 8.15

Example 8.16

A present setup has two parallel networks one assigned for mainframe traffic and the other for IP. It is necessary to combine these two networks for cost-effectiveness, at the same time preserving the class of service for the transaction-oriented main frame traffic. What is a possible recourse?

Solution

Deployment of ATM networks to provide a high-speed backbone for the interconnection of IBM main frames for use in disk mirroring is suggested.

The data from the *advanced peer-to-peer networking* (APPN), an IBM's internetworking architecture adopted to replace *system network architecture* (SNA), data is to be carried across the ATM in native format (that is, IP encapsulated).

APPN transport and direct IBM channel interconnect take place at routers acting as network nodes. IBM's *high performance routing* (HPR) would facilitate supporting APPN session transmission priority and *class of service* (COS). It also can facilitate subsequent mapping of APPN session priority and COS to ATM QOS. Since this architecture does not support SNA, any currently used SNA has to be upgraded to APPN. Alternatively, LANE/IP encapsulation should be implemented.

Example 8.17

Suppose the OAM scenario changes due to staff moves, adds and changes in small organizations, and no requirement for separate VLANs arises. However, additional bandwidths may be required to users as well as in the backbone due to the deployment of new applications. Develop a migration strategy via ATM Forum LANE.

Solution

Here LANE can be deployed on a per-port basis. It refers to the deployment of ATM in the backbone with LAN switching to the existing hubs and to some end-systems. Virtualization of the workgroup via LAN can be considered for greater mobility.

The migration suggested is for a small network where security between end-systems is not required. It can be a single VLAN implementation across ATM and refers to a port-based VLAN. That is, a single LAN port on a LAN switch/router is adopted. No configuration on routing function is needed in such single VLANS.

The router would require a connectivity to an external network. The example shown in Fig. 8.45 is for ethernet LAN-ELAN. (It can be used for token-ring also.)

Configuration refers to all ports on all LECs' members of the same VLAN. That is, each LAN switch implements only one LEC process. This single LEC corresponds to a single LECS, LES, and BUS located on a router. ILMI, as usual, will be part of LANE, with the LECs learning their network prefixes from the ATM switch. Network service access point (NSAP) address of the LECS is configured by the network administrator, and the address of the LES and the BUS are accessed from the LECS.

Note on port mobility: Suppose there are two workgroups (WGs) — say, planning and engineering a LAN switch port may dynamically identify either of these WGs. A user moving from one location on a campus, for example, may attach to an unassigned port at a new location. Depending on the level of security involved, the assigned port at the new location will take the identity of the user's WG automatically based on a MAC or network layer address.

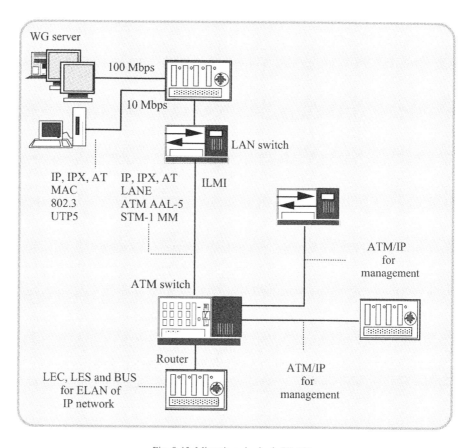

WG server

100 Mbps

10 Mbps

LAN switch

IP, IPX, AT   IP, IPX, AT   ILMI
MAC   LANE
802.3   ATM AAL-5
UTP5   STM-1 MM

ATM/IP
for
management

ATM switch

Router

LEC, LES and BUS
for ELAN of
IP network

ATM/IP
for
management

Fig. 8.45: Migration via single VLAN

Example 8.18

Suppose a small number of users need VLAN capabilities. Further, the user requires a larger bandwidth at the workgroup (WG) core, at the same time supporting the existing applications. Though there may be a need to partition the network, the user does not wish to introduce routing at this stage. Indicate a migration strategy by deploying VLANs across the ATM, based on ATM Forum LANE.

Solution

The migration can be based on the deployment of the LANE domain, servicing multiple VLANs based on filtering. That is, the virtual LAN assignment would require a network layer address and/or an identity based on MAC. Alternatively, the VLAN assignment can be done on the basis of the switch capacity to filter source and destination addresses (facilitated by filter off-sets within the data frame).

Filtering can be done at both the MAC (layer 2) and network (layer 3) layers to define identity (membership) within a VLAN. This filtering is used in conjunction with routing between VLANs to maintain internetworking. The MAC based filtering allows subdividing single (layer 2) VLAN. With this partitioning, the network routing is avoided. The LANE can be deployed on a per-port basis emulating routing between VLANs. The proposed architecture of the LANE domain is equivalent to a bridge network. The LANE domain contains the LAN switches. Each switch can act as the LECS, LES, and BUS for the ELAN.

Example 8.19

Growth considerations of an organization warrants multiple VLANs based on functions involved. Relevant OAM is expected to be expensive in realizing the movements within the organization, which would require supporting end-user protocols and applications. Suggest an implementation strategy of VLANs across an ATM based on ATM Forum LANE.

Solution

Original WG is divided into subgroups based on functions involved. Multiple VLANs can be facilitated to these (multiple) subgroups. Here LAN scaling and security are warranted. A LANE domain can be configured to contain multiple ELANs connected via a router. Each ELAN would require an LES, a BUS, and one or more LECs. A single LECS is sufficient for the entire LANE domain. Port mobility is introduced at the LECs whereby a single port may be assigned to multiple ELANs. For example, a single file server can be connected to a single physical port on a LAN switch while serving multiple VLANs. The same applies to high performance workstations connected directly to the ATM switch and implementing LEC software.

Example 8.20

Network growth may require some systems to serve multiple WGs. That is, some servers would require access from multiple VLANs. This requirement may include: Central e-mail, Web servers, and other resources of common interest throughout the organization. That is, all WGs would get access to these resources. Develop a migration effort built upon LANE considerations by attaching control servers to multiple VLANs.

Solution

An ATM connected server (central server) may be connected to a LAN switch ethernet (10 or 100 Mbps) or token-ring interface. This can be done two ways: (i) The server can be placed on its own VLAN with all traffic routed to the system. (High performance routing), or alternatively (ii) a single LAN switch ethernet or token-ring interface can be allowed to service multiple VLANS.

In essence, the migration would involve multiple VLANs on some LAN switch ports. This system would require LANE capable NICs.

Example 8.21

Suppose there is a requirement to interconnect a LAN in a WG to a classical ATM WAN. It is assumed that LANE is deployed currently throughout a campus and ATM has to be now extended across MAN/WAN. Describe a compatible migration strategy.

Solution

The suggestion is to deploy an internetworking between the LANE and classical domains. That is, the strategy should involve deploying devices capable of internetworking the two domains.

All traffic leaving the campus can be sent through a router connected to both the campus ATM backbone and the ATM MAN/WAN. This routing is required for both inter VLAN connectivity as well as for interworking to other data services.

Example 8.22

Suppose the protocols in use are IP, IPX, and DECnet and it is required for this multiprotocol to be transported across an ATM backbone and the relevant internetworking with LANE be facilitated economically. Suggest a migration strategy.

Solution

The migration may involve deployment of MPOA across the ATM network. Suppose the classical approach with ATM ARP for IP and manual configuring the address maps for IPX an DECnet are adopted. Then, the overhead involved will be considerable. Hence, MPOA at the same time incorporating VLANs is proposed. MPOA essentially is conceived on its associated client/server functions. It would require ATM switches, end switches, and routers. MPOA permits the distribution of routing functions by incorporating some layer 3 functionality in the LAN switches. This would help deploying scalable VLANs. Existing LAN segments can be connected to the ATM network via an LAN switch acting as *edge device functional groups* (EDFGs). Hosts can act as *ATM-attached host functional groups* (AHFGs). The existing LANE domain can be connected into the same backbone, integrating it thereby into the MPOA environment. MPOA domain is based on two Internet address subgroups (IASGS) and MPOA clients (EDFGs and AHFGs) are members of one or both of these groups. The same considerations apply to IPX and DECnet. Each of the IASGS can be associated with a VLAN.

As a part of MPOA, a router distribution service can be deployed between the route server functional group (RSFG) and the EDFGs/AHFGs. The router provides the necessary MPOA server functions, including, IASG coordination functional group (ICFG), *default forwarder functional group (DFFG), route server functional group* (RSFG), and *remote forwarder functional group (RFFG)*.

Example 8.23

Currently, a large enterprise uses FR to connect branch offices to a central site at its headquarters. It is necessary to interconnect some of the branch locations via ATM. Such an interconnect should facilitate interworking of sites connected via FR with the newly envisaged ATM connections. Indicate a possible strategy and describe the implementation profile.

Solution

In the scenario indicated, a service provider offers FR service and the enhanced traffic conditions are such that the existing core switches may not handle the traffic load. It is necessary to preserve the present investment on FR but extend the network's performance by providing a long-term scalability matching the traffic load enhancements (in the core).

The migration strategy can be based on two options:

❑   Installing high performance FR
❑   Deploying an ATM core.

The existing FR backbone can be redeployed to the periphery of the network, thereby inserting a new, higher bandwidth, ATM core. Then, most of the users will still be connected to the existing FR switches or concentration nodes. But those requiring DS-1/E-1 and above may have to be accommodated at the ATM core using the FR-UNI, if the switch supports this, or FR encapsulation across ATM. Thus, the new high speed backbone that interconnects FR nodes overcomes the additional overload on existing FR backbone. For the service internetworking between the ATM and FR networks proposed, IWF does the conversion between DLCIs (used across FR) and the VPI/VCI fields (of the ATM). If FR has SVC services, the FR NSAP addresses can be directly mapped to ATM NSAP addresses at the IWF.

Example 8.24

Suppose a service provider wants to rationalize existing data services by facilitating as many such services as possible over a common ATM infrastructure. This requires decommissioning the specific hardware and network management of the existing distributed queue dual bus (DQDB) service. The rationale behind this effort is that current SMDS based on single-service equipment may result in higher cost, and the bandwidth is limited to DQDB architecture.

Explain a method to replace this DQDB-based SMDS service with ATM service.

Note: *DQDB* is a networking technology based on the transmission of cells (slots) on two buses. It is defined in IEEE 802.6 and is used for non-ATM SMDS networks.

Solution

Deployment of SMDS/CBDS service across an ATM structure relies on the *connectionless service function* (CLSF), which is a hardware module part of an ATM switch. (It maintains an address table and forward data to the destined address, and it multicasts or broadcasts on an ad hoc basis.) The relevant strategy is illustrated in Fig. 8.46 SMDS over ATM makes use of AAL-3/4 or AAL-5.

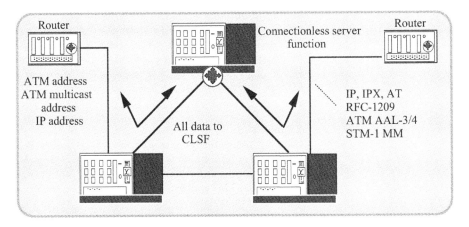

Fig. 8.46: Implementation scheme for Example 8.24

In the existing network, a public service provider maintains separate data networks for leased lines, frame-relay, and possibly SMDS. This requires separate switching systems, trunks, management systems, and user interfaces. The provider wishes to rationalize the service offering while allowing interworking between the various services.

In the migration being pursued, an ATM backbone integrates multiple services, dedicated circuits, frame relay, and SMDS now share a single hardware platform and backbone, resulting in cost savings. Some sites not justifying ATM are still provided with frame-relay. The backbone is also capable of integrating voice services.

Example 8.25

Suppose some sites of an enterprise are on SMDS and the others are on ATM-based SMDS. It is necessary to facilitate a transparent end-to-end connectivity between these two sites. This scenario refers to a subscriber being currently on the DQDB-based SMDS service, and new installations are to be based on SMDS over ATM so as to evolve a seamless interworking between the two networking technologies. Suggest a methodology of implementation.

Solution

The problem is to enable an interworking between users connected to an ATM-based SMDS and an already existing SMDS network based DQDB (802.6). This method should provide transparency between users (in interconnectivity) on both networks.

For ATM users, IWF between the two network plays no part in the end system configurations. Gatewaying between ATM and DQDB-based SMDS could be a plausible solution.

Example 8.26

Suppose there is a requirement to combine voice traffic over ATM WAN in order to economize the cost of operation. Suggest a technique to replace the existing, dedicated lines or TDMs, by an ATM network.

Solution

The user wants to replace the existing TDM network with ATM combining WAN data, voice, and video across the same network. This can be done via the deployment of ATM circuit emulation service (CES) across the enterprise network. That is, the technique should involve deploying ATM backbone with switches capable of servicing data, voice, and video. The ATM Forum CES allows connecting PABXs and video CODECs.

Example 8.27

Considering the methods of providing the delivery of video and data services to users over the existing residential infrastructure, suppose it is necessary to deliver enhanced services (such as VOD) to residential subscribers. Indicate how to deploy ATM as a multiservice backbone to the home.

Solution

A public service provider can implement VOD or a service close to VOD (near-VOD/staggercasting) across the existing PSTN or cable infrastructure. The service offered should permit select preencoded and/or live video feeds. Eventually voice service must be supported.

Using ATM as an infrastructure for the delivery of VOD refers to direct encapsulation of MPEG-2 over AAL-5 and physical delivery possibilities may include the following:

- ❑ FTTC
- ❑ FTTH
- ❑ ASDL
- ❑ VDSL
- ❑ HFC.

All of the above can be supported by ATM. At the video source (head-end), the incoming NTSC/PAL/SECAM video stream can be digitized in real time by an MPEG-2 encode. (Alternatively, previously encoded and compressed video content from a video server can be adopted). The resulting video stream (MPEG-2 transport stream) is then encapsulated into AAL-5 by means of suitable server architectures. (For example, 500 movies in compressed format can be placed in disk storage with two terabyte storage space). At AAL-5, the video is distributed to subscribers over point-to-multipoint VCCs. Each channel is a separate point-to-multipoint VCC at about 4 Mbps (assuming S-VHS format) including 384 kbps for audio.

Subscribers wishing to see a particular channel may choose the proper VCC via a point-to-point control VCC connected to a set-top box with the server. Between the ATM switch and set-top box, the ATM VCCs are modulated for transport over the residential access network. The set-top box demodulates the signal and splits the analog video coming from regular analog TV channels. The compressed video (of VOD) is decompressed and used.

In summary, in the existing system, the cable provider is limited to a traditional video offering due to hardware and infrastructure limitations. Users rely on PTT/telco for voice and data services. The cable provider wishes to provide both voice and data services across the infrastructure.

In the migration proposed, the cable provider deploys ATM-capable set-top boxes, integrating all services over a single infrastructure. An ATM switch at the headend combines video, voice, and data services.

Example 8.28

It is contemplated to provide a data overlay service for Internet connectivity such that this new service should integrate with the existing cable infrastructure. It should also provide necessary bidirectional bandwidth to all users. Develop a strategy to provide this data service across the existing cable infrastructure.

Solution

Metropolitan data outlay network on existing HFC infrastructure can be considered. It refers to deploying ATM at the head-end of the network and ethernet services to users. The data is sent modulated across existing cable plant.

Use of multiple distribution hubs provides scalability up to or more than 200,000 subscribers.

Problem 8.4

A convention center organizes a showcase of exhibitions that provides a forum for the world's leading graphics developers to share the least in computer-based graphics technology.

An internetworking is setup in the center that has an infrastructure to support applications, namely, full-motion video, high-resolution graphics, audio and data.

The natural choice being the ATM networking, the (ATM) backbone switches implemented, offer high-bandwidth, low-latency, and reliable connectivity.

- The networking also includes LAN connections within the convention center
- It facilitates WAN connections to a site within the city via OC-3 fiber
- Internet facility is provided to the participants in the center.

Assume LAN emulation strategy and develop a concept networking that represents the facilities implemented in the convention center as above.

Tabulate the combined multiple protocols, interfaces, speeds and transport media used

Indicate whether users belong to (virtual) LANs based on logical grouping or physical locations. Why?

List the following for the various links: QOS class, CLP, PCR, ACR, BT, and AAL #

Perform a power throughput analysis for the OC-3 optical link between the convention center and a remote site (WAN) assuming the following data:

Link distance: 40 km
Fiber: Single-mode type with attenuation 0.25 dB/km
Wavelength: 1550 nm

Source: LD with power output: 5 dB
Detector: APD with sensitivity at OC-3c speed for BER = 10^{-10}: -25 dB
Splices: 4 per km; splice loss: 0,125 dB/splice
Connectors: 2; connector loss: 0.5 dB/connector
System loss: (Due to aging and temperature degradation): 4 dB

Are there any repeaters needed? If so, what is the repeater spacing?

8.12   Concluding Remarks

ATM: The tempo and the hype on this new technology of the information highway have set a carpet that rolls into the new millennium. The scope for this technology is immense and it has grown exponentially in a niche that holds new devices and specialized tasks. ATM, in fact, has opened a gateway to the information highway of this new millennium.

Bibliography

[8.1]   S.V. Kartalopoulos: *Understanding SONET/SDH and ATM – Communications Networks for the Next Millennium* (IEEE Inc., Piscataway, NJ: 1999)

[8.2]   D.E. McDysan and D.L. Spohn: *ATM Theory and Applications* (McGraw-Hill, Inc., New York, NY: 1998)

[8.3]   D. Ginsburg: *ATM – Solutions for Enterprise Internetworking* (Addison-Wesley Longman, Ltd., Harlow, England: 1999)

[8.4]   S. Schatt: *Understanding ATM* (McGraw-Hill, Inc., New York, NY: 1996)

[8.5]   D. Teare: *Designing Cisco Networks* (Cisco Press, Indianapolis, IN: 1999)

[8.6]   J. Martin, K.K Chapman and J. Leben: *Asynchronous Transfer Mode: ATM Architecture and Implementation* (Prentice Hall PTR, Upper Saddle River, NJ: 1997)

[8.7]   W. Stallings: *High-Speed Networking – TCP/IP and ATM Design Principle* (Prentice-Hall, Inc., Upper Saddle River, NJ: 1998)

[8.8]   M. Sexton and A. Reid: *Broadband Networking: ATM, SDH and SONET* (Artech House, Inc., Norwood, MA: 1997)

[8.9]   T.M. Chen and S. Liu: *ATM Switching Systems* (Artech House, Inc., Norwood, MA: 1995)

[8.10]   A. Gillepsive: *Access Networks – Technology and V5 Interfacing* (Artech House, Inc., Norwood, MA: 1997)

[8.11]   K. Sato: *Advances in Transport Network Technologies* (Artech House, Inc., Norwood, MA: 1996)

[8.12]   D. Hill: *The Switching Book, I & II* (Xylon Corp., Calabasas, CA: 1996)

Glossary of Networking Terms: Definitions and Acronyms

Terminology		Definition
A		
Access control field	ACF	The first byte in ATM header (802.6)
Access node	AN	-
Access rate	AR	Speed of a channel accessed into a backbone network
Acknowledgement	ACK	A message or control bytes in a protocol acknowledging the reception of a transmitted packet. ACKs can be separate packets piggybacked on reserve traffic packet also known as positive acknowledgement
Active line state	ALS	A possible status of FDDI optical fiber
Adaptive differential pulse code modulation	ADPCM	A type of voice compression in PCM technique that typically uses 32 kbps
Additive increase rate	AIR	An ABR service parameter that controls the rate at which the cell transmission rate increases
Address	-	An identifier of a source or destination in a network such as IPv4, E.164, NSAP, and IPv6
Add/drop multiplexer	ADM	A process in which a part of the information carried by telecommunication system is dropped at an intermediate node or some information is added at a node for subsequent transmission
Address prefix	-	A string of the high-order bits of an address used in routing protocols procedures
Address registration	-	A method of using the ILMI where a network communicates valid address prefixes to the user, and the user registers specific AESAs with the network. LANE clients (LECs) provide address information to the LAN emulation server (LES) using this protocol
Address resolution protocol	ARP	A way to resolve a destination address for a lower-layer protocol (e.g., ATM or MAC) from a known address for another higher-layer protocol (e.g., IP). It lets the routers to adjust between different protocols/domains
Administrative unit	AU	The payload plus pointers of SDH
Advanced Research Project Agency	ARPA	U.S. agency that created ARPANET packet network
Agent	-	Representative software residing in a managed network device that reports MIB variables through SNMP
Alarm indication signal	AIS	A keep-alive signal that represents an OAM function used in fault management for physical and ATM layers. A system transmits AIS upstream if it detects a fault. The downstream end system responds with a remote defect indication (RDI). AIS is also known as blue signal
Allowed cell rate	ACR	ACR is the current rate in cell/s at which a source is allowed to send. It is an ABR service parameter
Alternate mark inversion	AMI	A line coding where 0 (space) is no voltage and successive 1s (marks) are pulses of opposite polarity
Alternate routing	-	Selecting other paths if an attempt to set up a connection along a previously selected path fails
American National Standards Institute	ANSI	A U.S. standard making body and a member of ISO
American standard code for information interchange	ASCII	A code based on 7 bits plus parity

American wire gauge	AWG	The U.S. standard designator of the wire size. e.g., 24 AWG 0.150 ohm/kft, 24 AWG 0.195 ohm/kft
Analog	-	Signals, which are continuous functions of time (like voice) that possess an infinite number of values, in contrast with digital signals that have a discrete number of values
Anycast	-	Routing calls in a network to the closest ATM end-system, which is assigned the anycast address within a specified scope. Anycast is defined in the ATM Forum's UNI 4.0 and PNNI 1.0 specifications that supports load-balancing
Application programming interface	API	A software module that separates OS from application. It represents a set of functions used by an application program to access to system capabilities (e.g., packet transfer or signaling)
Asymmetric digital subscriber line	ADSL	A type of XDSL technology
Asynchronous transfer mode	ATM	A type of framing recommended for B-ISDN. It is used for information transfer in a high-speed connection-oriented transmission with asynchronous (statistical) multiplexing and switching of multiple applications (like voice, video, and data) specified in international standards. It uses fixed-length 53 byte cells to support the multiple types of traffic. It is asynchronous in the sense that cells carrying user data need not be periodic and are multiplexed statistically (asynchronously) on a first-in first-out (FIFO) basis except upon contention in which case, a priority schedule is specified
Asynchronous transmission	-	The transmission of data through start and stop sequences without the use of a common clock
ATM adaptation layer	AAL	It represents two sublayers intended to segment large PDUs into ATM cells. Relevant adaptation follows a set of internationally standardized protocols and formats so as to support circuit emulation, packet video and audio, and connection-oriented and connectionless data services. These protocols map higher layer user traffic into a stream of ATM cells and convert the received cells back to the original form at the destination. The two sublayers are the segmentation and reassemble (SAR) sublayer and the convergence sublayer (CS). • AAL-1: Concerned with constant bit rate (CBR) traffic such as digital voice and video. It supports applications sensitive to both cell loss and CDV. It emulates TDM circuits • AAL-2: Supports connection-oriented services with variable rate that have timing transfer and limited delay variation requirements. Used with time-sensitive, variable bit rate traffic such as packetized voice or video • AAL-3/4: Intended for both connectionless and connection-oriented variable rate services. Originally, the standards defined two distinct protocols but later merged them into a single AAL. Adopted primarily to support SMDS • AAL-5: Used towards connection-oriented and connectionless variable bit rate data services without timing transfer requirements. Examples are typical bursty data traffic found in LANs. Commonly used for data applications

ATM address resolution protocol	ATMARP	IETF FRC 1577 defines ATM ARP as a means to map IP addresses to ATM hardware addresses. It functions like the conventional ARP on a LAN when mapping network-layer addresses to MAC addresses
ATM end system address	AESA	It is a 20 byte long, NSAP-based address. It has a hierarchical structure that supports scaling to global networks. Its initial domain part (IDP) is assigned by the network and the domain specific part (DSP) is assigned by the end user
ATM Forum	-	International consortium founded in 1991 by Cisco, NET/ADAPTIVE, Northern Telecom and Sprint. Recommends for ATM
ATM management entity	ATMM	A management entity in an ATM device
ATM service categories	-	A set of categories that define QOS parameters, traffic parameters and the use of feedback pertinent to the services supported on ATM links. Currently defined categories are CBR, rt-VBR, nrt-VBR, UBR and ABR
AToM management information base	AToM MIB	Defined in IETF RFC 1695, AToM MIB allows SNMP net management systems to monitor and configure ATM devices. AToM MIB reports on device or traffic status and accepts configuration commands
Attenuation	-	Reduction of signal magnitude or signal loss, usually expressed in decibels
Authority and format identifier	AFI	It is a part of a network level address header. It represents the first byte of an ATM address that defines the number assignment authority and the format of the remainder of the address. The ATM Forum currently defines formats for international code designator (ICD), data country code (DCC), and E.164
Automatic repeat request	ARQ	It is an error correction scheme for data links used with a CRC. It involves an automatic request for retransmission
Available bit rate	ABR	An ATM service category in which the network delivers limited cell loss if the end user responds to flow control feedback. The ABR service does not control cell delay variations

B

B1	-	SOH byte carrying BIP-8 parity check in SONET transmissions
B2	-	LOH byte carrying BIP parity check in SONET transmissions
B3	-	POH byte carrying BIP parity check in SONET transmissions
B channel	-	Bearer channel representing a DS-0 for user traffic. In ISDN bearer service, it is a channel that can carry either voice or data at a speed of 64 kbps
Backbone	-	A network interconnectivity several other networks as the primary path for traffic between those networks
Backward explicit congestion notification	BECN	A procedural convention in frame-relay for a network device to notify the user source device of network congestion
Backward explicit congestion notification cell	BECN cell	A type of resource management (RM) cell. Either the network or the destination may generate a BECN RM-cell to throttle the ABR source in binary mode
Backward sequence number	BSN	The sequence number of a packet that is expected next (in SS7)
Bandwidth	-	The extent of transport resource available to pass information, measured in Hz for analog transmission (i.e., frequency passband) and bits per second (bps) for digital transmission

Bandwidth balancing	BWB	A method to reduce an access at a station to a transmission bus so as to improve fairness (802.6)
Bandwidth-on-demand	BoD	A dynamic allocation of line capacity to busy users
Bandwidth management service	BMS	An AT&T offering mimicking a private network with equipment located at the CO
Basic rate access	BRA	Access facilitated in ISDN (2B + D) loop
Basic rate interface	BRI	In a given local loop, an ISDN access interface type composed of two B channels, each at 64 kbps, and one D channel at 16 kbps (2B+D) rate
Basic system reference frequency. (Formerly Bell-SRF)	BSRF	Stratum 1 level clock source at 8 kHz
Basic telecommunication access method	BTAM	An older IBM mainframe communication software
Basic transmission unit	BTU	A frame structure of SNA
Baud rate	-	A set of discrete signals sent once per signaling interval at the physical layer. If the baud in each signaling interval represents only one bit, then the baud rate equals the bit rate. Some transmission schemes may send multiple bits per signaling interval, in which case bit rate $\neq$ baud rate
Beacon	BCN	A set of frames sent down stream by a station on a ring when upstream input is lost (802.5)
Beginning of message	BOM	A protocol indicator contained in the first cell of an AAL3/4 PDU. It is a type of segment (cell) that starts a new MAC frame prior to continuation and end of messages in SMDS
Beginning tag	BTag	A field in header of a frame whose value is required to make ETag
Bellcore client company	BCC	One of the RBOCs that owns Bellcore
Bell system practice	BSP	A format prescribed for documents concerning the equipment used by telcos
Best effort	-	A representative QOS class in which no traffic parameters are specified and no guarantee is given on traffic delivery. The unspecified bit rate (UBR) service category is an example of best-effort services
Bill of material	BOM	List of all constituent parts (hardware and software) of an assembly
Binary coded decimal	BCD	A 4-bit expression for zero ('0000') through nine ('1001')
Binary mode ABR	-	A specific type of available bit rate (ABR) that interworks with older ATM switches, which only supports the explicit forward congestion indication (EFCI) function in the ATM cell header. Binary mode uses the congestion indication (CI) bit in a resource management (RM) cell in the reverse direction to control the rate of the source
Binary synchronous communication	BSC	A half-duplex protocol
Bipolar violation	BPV	Two successive pulses being of same polarity in certain line coding schemes
Block check code	BCC	A CRC or a similar algorithm, which calculates a number that finds the transmission errors
B-ISDN inter-carrier interface	B-ICI	A specification defined by the ATM Forum for the interface between public ATM networks in support of user services (e.g., CRS, CES, SMDS, or FR) transport on multiple public carriers
Bit	-	A unit of information measure of binary signal elements

Bit compressed mux	BCM	Multiplexer M44 adopted for ADPCM
Bit error ratio (rate)	BER	It is a ratio of errored bits to total bits transmitted
Bit error rate test	BERT	A test to evaluate the BER
Bit per second	bps	Number of bits transmitted per second. A transmission rate in digital communication
Bit-stuffing	-	A technique used in an asynchronous system
Bit-interleaved parity	BIP	A parity check, which conglomerates all the bits in a block into units (such as a byte) and performs a parity check for each bit position in a group. In this BIP-x error checking method, each of x bits is parity of every x^{th} bit in a data block. (x = 8 in SONET; 16 in ATM)
Bridge	-	A network entity interconnecting local area networks at the OSI data link layer, filtering and forwarding frames according to MAC addresses
Broadband	-	Bandwidths in excess of DS-3 (45 Mbps) or E-3 (34 Mbps)
Broadband connection oriented bearer	BCOB	A part of SETUP message containing information that indicates the bearer service requested by the calling user. Types include A (see CBR) and C (see VBR) which may include the specific AAL and its parameters for use in interworking. BCOB-X defines a service where AAL, traffic type, and timing are transparent to the network.
Broadband digital cross-connect system	B-DCS	A switching system adopted in DACS, OC-1, STS-1, DS-3 and higher rates only
Broadband high layer information	B-HLI	A part of SETUP message containing information that is used by end systems for compatibility checking
Broadband intercarrier interface	B-ICI	An interface between ATM networks
Broadband inter-switching system interface	B-ISSI	An inter-switching that exists, for example, between ATM nodes
Broadband integrated services digital network	B-ISDN	A network standard developed by ITU-T supporting an integrated, high-speed transmission, switching and multiplexing of data, audio, and video, and ATM is specified as a suitable transport mode
Broadband ISDN user part	B-ISUP	A protocol defined in ITU-T Q.2761 for use between carrier networks operating between origin, destination, and transit nodes. It interoperates with the narrowband ISUP using the same SS7 message transfer part (MTP) protocol
Broadband low layer information	B-LLI	Information element in a SETUP message identifying the layer 2 and layer 3 protocols used by an application
Broadband terminal equipment or adapter	B-TE or B-TA	An equipment categorized for use in B-ISDN, which includes end systems and intermediate systems. TAs convert from other protocols while TEs operate in a native ATM mode
Broadcast	-	Transmission sent to all addresses on the network or subnetwork
Broadcast and unknown server	BUS	A component of the LAN emulation strategy that receives all broadcast and multicast MAC packets as well as MAC packets with an unknown ATM address. The BUS transmits these messages to every member of an emulated LAN (ELAN)
Buffer allocation size	BAsize	Number of octets specified in SMDS frame
Burstiness	-	A source traffic characteristic parameter defined as the ratio of maximum rate to the average rate
Burst tolerance	BT	An entity proportional to the maximum burst size (MBS). It measures conformance checking of the sustainable cell rate (SCR) as part of the usage parameter control (UPC) function

Byte	-	A set of 8 bits; an octet
Byte interleaved	-	Sets of 8 bits each, for example, from STS-1 they are placed in sequence in a multiplexed or concatenated STS-N signal

C

Cable modem	-	A modem operating over coaxial line of cable TV providers
Caching	-	A method of replication in which information learned in a previous transaction is adopted to process a late transaction
Campus network	-	A set of LAN segments used in a campus
Carrier sense multiple access/ collision detection	CSMA/CD	An access methodology of a protocol employed in ethernet where stations (NICs) first listen to the bus and transmit only when the bus is free. If the transmitting stations detect a collision, they retransmit the frame after a random time-out
Carrier sensing area	CSA	An area defined by a local loop length out of a CO. It refers to the creation of DS-0 capabilities in a remote area of the exchange/CO by using subscriber carrier to establish a remote wire center
Category-x cable (x = 1, 2, ...5)	Cat-x cable	Voice and/or data grade twisted pair of copper wires Cat-1: Unspecified Cat-2: 1 Mbps cable for low-speed data circuits Cat-3: 16 Mbps cable for 10Base T and 4 Mbps token-ring Cat-4: 20 Mbps cable for 10Base T and 16 Mbps token-ring Cat-5: 100 Mbps cable for high-speed data-grade copper technology. It has about 3 twisted/inch
Cell	-	The packetized ATM protocol data unit (PDU) used in transmission, switching and multiplexing. It is a fixed-length 53 octet packet comprised of a 5-octet header and a 48-octet payload
Cell delay variation	CDV	An ATM UNI traffic parameter that specifies a QOS parameter and measures the difference in delay between successive cell arrivals. Standards define one-point and two-point measurement methods for CDV
Cell delay variation tolerance	CDVT	A tolerance parameter used in the usage parameter control (UPC) specification of the peak cell rate (PCR). It effectively defines how many back-to-back cells may enter the network for a conforming connection
Cell error ratio	CER	A QOS specific parameter quantified in terms of the fraction of cells received with errors at the destination
Cell header	-	The header part of an ATM cell. It is a 5-octet part of the cell that defines control information used in processing, multiplexing and switching cells. It contains the following fields: GFC, VPE, VCI, PT, CLP, and HEC
Cell information field	CIF	The 48-byte payload in each ATM cell
Cell-loss priority	CLP	It is a signaling bit in ATM cell. It is an one-bit field in the cell header that indicates the relative discard priority. The network may discard nonconforming cells by designating a CLP set to 1 during periods of congestion so as to preserve the cell loss ratio (CLR). (CLP = 1: Low priority and CLP = 0: High priority)
Cell-loss ratio	CLR	A QOS parameter negotiated at the connection setup and specifies the ratio of lost (i.e., not delivered) cells to the total cells transmitted
Cell-misinsertion rate	CMR	The number of misinserted cells (namely, those that were never sent by the source) divided by a specified time interval. It is a QOS parameter

Cell relay function	CRF	A function of an ATM device operating at the virtual path or channel (VP or VC) level to translate the VP and VC identifiers (VPI and VCI) between an input port and an output port on an ATM device
Cell relay service	CRS	A service offered by carriers in guaranteeing delivery of ATM cells to end-users according to standards and specifications
Cell transfer delay	CTD	A QOS parameter quantifying the cumulative delay between two measurement points for a specific virtual connection
Cellular radio advanced mobile phone system	AMPS	A cellular radio system operating at 825 – 845 MHz and 870 – 890 MHz over a serving area divided into cells with frequency reuse in nonadjacent cells
Central office terminal	COT	Equipment at CO end of digital loop or line
Central processor	CP	CPU that runs network under center-weighted control
Center weighted control	CWT	A central process that runs network-wide functions (in contrast to nodes, which do local functions)
Central office	CO	A telephone company switching office (exchange) providing an access to end-entities of the network and its services. It is also known as a central wire center or a central exchange
Centrex	-	A service offered by local exchange carriers in lieu of PBX
Channel associated signaling	CAS	A method adopted in telephony to associate on-hook and off-hook conditions with a TDM signal representing voice channels. (e.g., Robbed bit signaling uses the LSB of a 8-bit PCM sample to signal telephony channel state in North American telephony)
Channel service unit/data service unit	CSU/DSU	A customer service unit used to interface CPE to the WAN/leased line. It is a digital-to-digital interface. This may be a separate device, or built into a switch or a multiplexer port card. The DSU function provides a data communications equipment (DCE) interface to a user data terminal equipment (DTE). The CSU functions support the standard WAN interface to connect low-speed digital device to high-speed digital highway
Channel unit	CU	A unit that performs the function of interfacing the multiplexer to a specific type of input signal
Circuit emulation service	CES	It is an ATM Forum specification defining emulation of time division multiplexed (TDM) circuits (e.g., DS-1 and E-1) over ATM adaptation layer 1 (AAL-1). CES operates over VCCs, which support the transfer of timing information. Typically, CES operates at a CBR. CES operates in an unstructured (i.e., a 1.544 Mbps DS-1) or a structured mode (i.e., N × 64 kbps). CES may also support channel associated signaling (CAS)
Circuit switching	CS	A process of connecting one circuit to another. In telecommunication, it is a connection-oriented technique based on either time- or space-division multiplexing and switching providing minimal delay. It uses TDM, rather than packets
Circuit-switched channel	CSC	A circuit-switched connection
Circuit-switched digital capability	CSDC	An AT&T version of Sw56
Circuit-switched public data network	CSPDN	Circuit-Switched network supported by public carriers
Class of service	COS	A specific service type (as in ATM)

Class 5 exchange	-	An end-office depicting the lowest-level switch in the PSTN hierarchy. It is also known as the local exchange
Classical IP over ATM	CLIP over ATM	Adaptation of address resolution protocol (ARP) to facilitate ATM over a single logical IP subnetwork (LIS) defined by IETF FRCs 1483, 1577, and 1755
Clear-to-send	CTS	A lead on interface indicating that the DCE us ready to receive the data
Client	-	A work-station in a client/server LAN
Client/server	-	An ambient for a LAN wherein the PCs serve either as a workstation (client) or as a server
Clock	-	A controlled timing source of information used in synchronous transmission systems
Closed user group	CUG	-
Code mark inversion	CMI	A line code or line signal for STS-3
Code division multiple access	CDMA	A spread-spectrum technique wherein broadcast frequency changes rapidly in a (encoded) pattern known to the receiver
Coder-encoder	CODEC	A unit that converts analog voice to digital and back
Coding violation	CV	A transmission error in SONET section
Collision domain	-	A section of a LAN medium where collision are detected
Common carrier	-	A telco that offers telecommunication services to the general public as part of the PSTN
Common channel signaling	CCS	A system in which signaling is completely separate from switching and information transmission i.e. signaling is done over a channel different from the one, which carries the voice and data
Common communication subsystem	-	Level 7 application services in SNA
Committee for European Electrotechnical Standardization	CENELEC	-
Committee for European Standardization	CEN	-
Committed information rate	CIR	A frame-relay terminology that defines the rate the network commits to deliver frames
Common management information protocol	CMIP	A part of OSI management interface standard of ITU-T supporting administration, maintenance, and operation (OAM&P) information functions
Common (network) management information service	CMIS	A service that runs on CMIP (OSI)
Common part AAL	CPAAL	Common part of AAL that may be followed by a number to indicate type
Community antenna television	CATV	Cable TV
Common part convergence sublayer	CPCS	An AAL convergence sublayer (CS) that passes primitives to an optional service specific convergence sublayer (SSCS). AAL-1, AAL-3/4, and AAL-5 standards have a CPCS function, which is mandatory
Compression	-	A method of conserving bandwidth and/or storage by reducing the number of bits required to represent information
Computerized branch exchange	CBX	A computer-based PABX

Concatenation	-	Linking together various data structures. (e.g., Concatenated STS-Nc, and concatenated VT)
Congestion	-	A queueing-theoretic specified condition in which the line up traffic demand exceeds available network resources (i.e., bandwidth or buffer space) for a sustained period of time
Congestion collapse	-	A condition where re-transmissions ultimately result in markedly reduced effective throughput. Adequate buffering, intelligent packet discard mechanisms, or available bit rate (ABR) flow control in ATM networks are preventive strategies to avoid congestion collapse phenomenon
Congestion control	-	A technique involving resource allocation and traffic management mechanism so as to avoid or enable recovery from congestion situations
Congestion indication	CI	A bit in a resource management (RM) cell that indicates congestion. The receiving end sets this bit in response to cells received with the EFCI bit set in the cell header
Connection admission control	CAC	An algorithmic procedure adopted to determine whether an ATM device accepts or rejects an ATM connection request. Generally, CAC algorithms accept calls only if enough bandwidth and buffer resources exist to establish the connection at the requested quality of service (QOS) without impacting connections already in progress
Connection end-point	CE	(specific to ATM)
Connection end-point identifier	CEI	(specific to ATM)
Connectionless network	CLN	A network in which packet address is unique and network routes entire traffic involved
Connectionless protocol	CNLLP	-
Connectionless service	CNLS	-
Connectionless network service	CLNS	A service defined by OSI for the transfer of information between users employing globally unique addresses (e.g., LAN, IP, and SMDS)
Connection-oriented network	CON	A network that defines a single-path per logical connection
Connection-oriented network service	CONS	A service where the network should first establish a connection between the source and destination prior to data transfer. (e.g., N-ISDN, X.25, frame relay, and ATM)
Constant bit rate	CBR	One of the possible ATM service categories. It supports applications like voice and video requiring a constant bit rate, constrained CDV, and low CLR connection
Consultative Committee International for Radio	CCIR	Sister group to CCITT, which eventually became a part of Telecommunication Standardization Bureau (TSB) in 1993
Consultative Committee International for Telegraphy and Telephony	CCITT	Now, referred to as the ITU-T and merged as a part of TSB since 1993
Contention	-	A competitive condition arising when two or more stations attempt to transmit at the same time using the same transmission channel
Continuous ARQ	-	A sliding window ARQ, which eliminates the need for a transmitting device to wait for ACKs after a block of data
Continuation of message	COM	A protocol indicator contained in any cell that is not the first or last cell of an AAL-3/4 PDU. A segment that lies between BOM and EOM (ATM, SDMS)
Control plane	C-plane	Out-of-band signaling system for U-plane

Convergence layer PDU	CS-PDU	Information plus new header and trailer to make packet that is segmented into cells or SUs
Convergence sublayer	CS	In the layered ATM protocol architecture the higher layer portion of the AAL that interfaces with the lower layer segmentation and reassembly (SAR) portion of the AAL. The CS sublayer interfaces between ATM and non-ATM formats. The CS may be either a common part (CPCS), or service specific (SSCS)
Crankback	-	A mechanism to partially release a connection establishment attempt, which encounters a failure back to the last node that made a routing decision. This mechanism allows PNNI networks to perform alternate routing in a standard way
Crosstalk	-	An impairment perceived in a circuit due to the invasion of signals carried by neighboring circuits
Current cell rate	CCR	A resource management (RM) cell field defined by the source indicating its transmission rate. When the transmitting end sends a forward RM cell, it sets the CCR to the current allowed cell rate (ACR). No elements in the network are allowed to adjust this value
Customer information control system	CICS	An IBM mainframe communication software with database
Customer network management	CNM	A capability that allows public network users to monitor and manage their services. CNM interfaces for monitoring physical ports, virtual channels, virtual paths, usage parameters, and QOS parameters have been defined by ATM Forum
Customer premises equipment	CPE	Equipment that resides in and is operated at a customer site
Customer premises	CP	Premises (at the user-end) as opposed to CO
Customer premises node	CPN	Same as CPE
Cyclic redundancy check	CRC	An algorithm implemented in a cyclic shift register that computes a check field for a block of data. The sender transmits this check field along with the data so that the receiver can either detect errors, and in some cases even correct errors in a bit stream

D

Data	-	Raw information represented in digital code, including voice, text, facsimile, and video
Data bus	-	A physical circuit path (typically in a computer), which supports the data traversing from one point to another
Data communication	-	Transfer of data over telecommunication facilities
Data communications channel	DCC	An overhead connection in D types in SONET management
Data communications (or circuit termination) equipment	DCE	A device function existing as an access for the user data terminal equipment (DTE) to a network. It interfaces the DCE to the PSTN (e.g., A modem, CSU/DSU, or switch port). Gender of interface on modem
Data country code	DCC	ATM Forum-specified, one of three authority and format identifiers (AFI). The ISO 3166 standard defines the country in which the NSAP-based address is registered
Data encryption standard	DES	An encryption standard, which is moderately difficult to break

Data exchange interface	DXI	A serial port protocol for SNMP for any speed. It is a frame-based interface between legacy CPE (e.g., router) and an external ATM CSU/DSU via a high speed serial interface (HSSI). Initially developed to support early introduction of ATM, the frame-based UNI (FUNI) provides a similar function without the expense of an external ATM CSU/DSU
Data flow control	DFC	Layer 5 of SNA
Data link	DL	-
Data link connection	DLC	A single logical bit stream in LAPD/layer 2
Data link control	DLC	Level 2 control of trunk to adjacent node (SNA)
Data link connection identifier	DLCI	Address in a frame (I.122). It is a frame-relay address designator locally significant to the physical interface for each virtual connection endpoint
Data link layer	DLL	An OSI model (layer 2), which establishes, maintains, and releases data-link connections between end points. It facilitates error detection and optional correction between adjacent network devices when operating over a noisy physical channel. (e.g., Frame-relay, ATM, and the LAN LLC and MAC sublayers)
Data service unit	DSU	A device in digital transmission. It adapts the physical interface on DTE device to a transmission facility such as T-1 or E-1
Data terminal equipment	DTE	A device functioning on transmitting data to, and/or receiving data from, a DCE (for example, a terminal or printer). Gender of interface on terminal or PC
Data terminal unit	DTU	Newbridge TA for data
Datagram	-	Under IPv4, datagram is a format of a packet including IP header, layer 4 header, plus data. It is the basic unit of transmission, switching, and multiplexing in a connectionless network service (CLNS). A datagram service does not guarantee delivery. A network may route datagrams along different paths; therefore datagrams may arrive in a different order at the destination than that generated by the source
Decapsulation	-	An inverse operation of encapsulation
Description	-	A reverse operation on encryption
Decibel	dB	$(1/10)^{th}$ of a bel; $10 \log_{10} (a/b)$ where (a/b) is the ratio of powers
Decibel referenced to 1 mw	dBm	Decibel specified in reference to 1 mw power (measured at 1004 Hz across 600 ohm load)
Decibel referenced to 1 mw – TLP specified	DBm0	Power-level measured at a point of a transmission system expressed in dB and referenced to the TLP level at that point
Decibel referenced to noise	dBrn	Power level specified relative to noise level = (dBm + 90)
Decibel referenced to noise (C-weighted)	dBrnC	dBrn, when the power passed through a C-weighted audio filter (that matches human earing response)
Defence ARPA	DARPA	Formerly known as ARPA
Degraded minute	DGM	Duration over which BER is between 10^{-6} to 10^{-3}
Delay dial start dial	DDSD	A protocol specified by a start-stop format for dialing into a CO switch
Delta channel	D channel	The ISDN out-of-band (16 kbps or 64 kbps, depending on BRI or PRI, respectively) signaling channel which carries the ISDN user signals or can be used to carry packet-mode data
Delta channel signaling system 1	DSS 1	An access protocol for switched connection signaling (Q.931 and ANSI T1S1/90 – 214)

Demultiplexing	-	Segregating multiple input streams from the multiplex stream
Dense wavelength division multiplexing	DWDM	Fiber-optics based multiplexing method in which a dense set of voice channels are placed each on a frequency corresponding to a wavelength of IR spectrum
Deregulation	-	A shift in telco business from a regulated monopoly to a competitive nonregulated market
Derived MAC protocol data unit	DMPDU	A 44–octet segment of upper layer packet plus cell header/trailer (802.6)
Designated transit list	DTL	An information element (IE) defined in the ATM Forum. PNNI signaling protocol used in a SETUP message that conveys information about an end-to-end source route from source to destination
Destination address	DA	An address block of a 6-octet value uniquely identifying an endpoint as defined by the IEEE media access control (MAC) standards
Destination service access point	DSAP	An address field in the header of LLC frame. It identifies an user within a station address (layer 2)
Destination signal transfer point code	DPC	In SS7, DPC is a level 3 address in a signaling unit of signal transfer point
Dial pulse	DP	Rotary dialing in lieu of dual tone multifrequency touch-tone dialing
Differential Manchester code	DMC	A line-code of pulse pattern that has a transition at the center of each bit (for clocking), and transition (none) at start of period for 0 (1); (802.5 specified)
Differential pulse code modulation	DPCM	A voice compression algorithm used in adaptive differential PCM
Digital	-	As opposed to continuous analog representation, an entity (such as voltage) having discrete values, 0s and 1s, (+ 1, 0, – 1) etc.
Digital access and cross-connect system	DACS (DCC/DXC)	A switching device of space-division and/or time-slot type that electronically interconnects at various PDH and SDH multiplexing levels for TDM signals
Digital access signaling system	DASS	Protocol adopted in UK for ISDN D-channel
Digital channel banks	D3	Third generation channel bank (24 channels on a single T-1C)
Digital cross-connect system	DCC or DXC	Same as DACs.
Digital data communication message protocol	DDCMP	-
Digital data service unit	DSU	A unit that converts RS-232 or other terminal interface to line coding for local loop transmissions
Digital data system	DDS	A network that caters dataphone digital service
Digital Equipment Corporation	DEC	-
Digitally multiplexed interface	DMI	Interface (AT&T) specified for 23, 64 kbps channels and a 24th for signaling
Digital multiplex system	DMS	-
Digital network architecture	DNA	A network architecture of DEC
Digital private networking signaling	DPNSS	PBX interface of CCS
Digital service unit	DSU	Same as CSU/DSU
Digital signal	DS	An electrical or optical signal that varies in discrete steps, that is in digital encoded format

Digital signal 0	DS-0	The world-wide standard on physical interface for digital transmission using TDM operating at 64 kbps. PDH standards map 24 DS-0s into a DS-1 or 32 DS-0s into an E-1
Digital signal 0A	DS-0A	Digital signal level 0 with a single rate adopted channel
Digital signal 0B	DS-0B	Digital signal level 0 with multiplex channels, subrate multiplexed in digital data system format
Digital signal 1	DS-1	North American standard physical transmission interface operating at 1.544 Mbps. The payload rate of 1.536 Mbps may carry 24 DS-0 channels, each operating at 64 kbps. Colloquially know as T-1. In CCITT countries DS level 1 refers to 2.048 Mbps
Digital signal 1A	DS-1A	A proposed designation for 2.048 Mbps in North America
Digital signal 1C	DS-1C	Two T-1s used largely by telcos internally
Digital signal 2	DS-2	Four T-1s. Uncommon in the United States, and mostly used in Japan
Digital signal cross connect	DSX-1	Cross-connect at level 1 part of DS-1 specification
Digital signal processor	DSP	Application-specific chip optimized for fast computations
Digital signal 3	DS-3	Constituted by 28 T-1s in the North American standard 44.736 Mbps digital transmission physical interface. The payload rate is approximately 44.21 Mbps. Used to carry digital data, or multiplexed to support 28 DS-1s
Digital subscriber line	DSL	An access line into subscriber premises from the CO
Digital subscriber signaling number 2	DSS-2	Generic name for signaling at a B-ISDN User-Network Interface (UNI) defined in standards by the ITU-T. N-ISDN standards define DSS-1 signaling between a user and a network
Digital signal cross connect, level Z	DSX-Z	Part specified at DS levels Z = 0A, 0B, 1, 1C, 2, 3 etc.
Direct access storage device	DASD	SNA – specific
Direct connect card	DCC	A data interface module on a T-1 bandwidth manager
Direct inward dial	DID	A CO directing a call to a specific extension on PBX
Discard eligibility bit	DE	Refers to a bit in the frame relay header which, when set to 1, indicates the particular frame is eligible for discard during congestion conditions
Distributed queue dual bus	DQDB	An IEEE 802.6 MAN protocol standard based upon 53-byte slots that support connectionless and connection-oriented, isochronous integrated services. The physical DQDB network has two unidirectional buses configured in a physical ring topology to access MANs, at T-1, T-3, or faster
Domain specific part	DSP	Represents the low-order portion of an ATM end system address (AESA) assigned by an end-user. It contains the end system identifier (ESI) and the selector (SEL) byte
Dual attachment station	DAS	A workstation that attaches to both primary and secondary FDDI MAN rings. This configuration enables the self-healing capability of FDDI
Dual tone multifrequency	DTMF	Touchtone dialing as opposed to dial-pulse/rotary dialing
Duplex	DX	Two-way audio channel bank plug without signaling
Dynamic alternate routing algorithm	DARA	-

E

E.164	-	An ITU-T-defined numbering plan for public voice and data networks containing up to 15 BCD digits. Numbering assignments currently exist for PSTN, N-ISDN, B-ISDN, and SMDS public networks. The ATM Forum 20 byte NSAP-based ATM end systems address (AESA) format encapsulates an 8-byte E.164 address as the domain specific part (OSP)
Early packet discard	EPD	A smart packet-discard mechanism congestion control technique that drops all cells from an AAL-5 PDU containing a higher layer protocol packet
Echo cancellation	-	A technique adopted to isolate and filter unwanted echoed (reflected) signal energy, commonly resulting from misinserted hybrids that convert 2-wire to 4-wire circuits in telephony
Echo return loss	ERL	-
Emulated local area network	ELAN	ATM Forum specified logical network on LAN emulation (LANE) comprising both ATM and legacy attached LAN end stations
Encryption	-	Security-specified algorithm applied to alter the appearance of data
Encapsulation	-	A method of wrapping data in a particular protocol header
End of message	EOM	A cell type carrying the last segment of a message. A protocol indicator used in AAL-3/4 that identifies the last ATM cell in a packet
End of text	ETX	A control byte
End node	EN	A node depicting a limited capability access device
End tag	ETag	A field in the trailer of a frame whose value should match that in beginning tag (BTag)
End of transmission block	ETC	A control byte in binary symmetric communication (BSC)
End system identifier	ESI	A field of 6 bytes in the domain specific part (DSP) of an ATM end system address (AESA)
Ending delimiter	ED	An unique symbol to mark the end of a LAN frame (e.g., TT in FDDI)
Enhanced service provider	ESP	A company that delivers its product over the phone
Enquiry	ENQ	A control byte that requests a repeat transmission or control of line
Enterprise network	-	A large and diverge internetwork connecting major points in an enterprise
Equal access	-	A key provision of the 1984 modified Final Judgement, which forces the BOCs to provide (1 +) access to toll for all interexchange carriers
Error checking code	ECC	Usually, a pair of bytes in a frame or packet derived from data to let the receiver test for transmission error
Error correction	EC	A process to check packets for errors and send again if required
Errored second	ES	A 1 second interval containing 1 or more transmission errors
Escape	ESE	An ASCII character
Ethernet	-	A protocol used in LANs. It is an IEEE 802.3 and ISO set of standards defining operation of a LAN protocol at 10, 100, and 1,000 Mbps. Typically, ethernet uses the CSMA/CD media access control (MAC) protocol that limits user data to 1500 byte. Ethernet operates over coax, twisted pair, and fiber optic physical connections

European digital signal level 1	E-1	A standard of the European CEPT-1 on digital channel operating at 2.048 Mbps. The payload of 1.92 Mbps may be employed transparently, or multiplexed into 32 individual 64 kbps channels. The E1 frame structure supports 30 individual 64 kbps digital channels for voice or data, plus a 64 kbps signaling channel and a 64 kbps channel for framing and maintenance
European digital signal level 3	E-3	A standard on European CEPT-3 for digital transmission service at 34.368 Mbps. It supports transmission at a payload rate of 33.92 Mbps, or it carries 16 E-1s when used in the PDH multiplexing hierarchy
European digital signal level 4	E-4	A standard on European CEPT-4 for digital physical interface at 139.264 Mbps. It operates at a payload rate of 138.24 Mbps, or carries four E-3s when operating in multiplexing mode
European Telecommunications Standards Institute	ETSI	The primary telecommunications standards organization in Europe, which coordinates telecommunication policies
Explicit congestion notification	ECN	A network warning terminals of congestion by setting bits in frame header (I.122). In frame-relay, the use of either FECN and BECN messages notifies the source and destination of network congestion, respectively
Explicit forward congestion indication	EFCI	A one-bit field in the ATM cell header payload type indicator (PTI). Any ATM device may set the EFCI bit to indicate a congested situation, for example, when buffer fill crosses a pre-set threshold
Explicit forward congestion indication mode	EFCI mode	One of the congestion feedback modes permitted in the available bit rate (ABR) service
Explicit rate	ER	A mode of available bit rate (ABR) service where switches explicitly indicate within a specific field in backward resource management (RM) cells the rate they can support for a particular connection
Exterior link	-	A link that crosses the boundary of the PNNI routing domain. The PNNI protocol does not run over an exterior link

F

Facimile	FAX	-
Facility assignment control system	FACS	A control system that facilitates a telco to manage its outside plant (local loops)
Far-end alarm	FEA	Alarm/status identification by repeating bit C-3 in DS-3 format
Far-end bloc error	FEBE	Alarm signal indicated by count of bit interleaved parity (BIP) errors received. ATM uses Z2 byte in SONET LOH
Far-end crosstalk	FEXT	Crosstalk perceived at the far-end (from the source) in a susceptible line
Far-end receive failure	FERF	An alarm signal in ATM
Far-end reporting failure	FERF	Indicates same condition as remote defect indication (RDI), but specifies more information about the failure
Fast ethernet	-	100 Mbps ethernet
Fast packet	-	A term used for advanced packet technologies like frame relay, DQDB, and ATM
Feature group	-	A series of services rendered by BOCs to provide access to IXCs. Feature group A gave customers access to IXCs, 7/10 digit telephone number plus personal ID etc.
Federal Communication Commission	FCC	Regulatory body in the United States, which regulates communications

Fiber distributed data interface	FDDI	An ANSI defined fiber-optic LAN operating at 100 Mbps using a dual-attached, counter-rotating ring topology and a token passing protocol
	FDDI-II	An FDDI standard with the additional capability to carry isochronous traffic (voice/video)
Fiber optics	FO	Thin fibers of glass or plastic carrying a transmitted light beam generated by a light emitting diode (LED) or a laser diode
Fiber optic central office terminal	FCOT	-
Fiber in the loop	FITL	Fiber optics from CO to customer premises (CP)
Fiber optic test procedure	FOTP	-
Fiber optic terminal system	FOTS	A mux or CO switch interface
Fiber-to-the-curb	FTTC	Fiber-based local loops from CO to just outside customer premises
File transfer access and management	FTAM	An OSI layer 7 protocol for LAN interworking (802)
File Transfer Protocol	FTP	A capability in the Internet protocol suite supporting transfer of files using the transmission control protocol (TCP).
Firewall	-	A router/remote access sever used as a security buffer between connected networks
First in/first out	FIFO	Traffic/device wherein prioritization is not supported
Fixed round-trip time	FRTT	The total of fixed and propagation delays from the source to the farthest destination and back used in the ABR Initial cell rate (ICR) calculation
Flag	F	The bit pattern '01111110' signaling the beginning or end of an HDLC frame.
Flow control	-	Method to ensure that a transmitting entity does not overwhelm a receiving entity with data
Flow of OA&M cells	F1	Over a SONET section (ATM) at level 1
	F2	Over a line
	F3	Between path terminating equipment (PTEs)
	F4	For metasignaling and VP management
	F5	Specific to a logical connection on one VPI/VCI
Foreign exchange	FX	Not the nearest central office. A line from a CO/PBX to beyond its normal service area
Foreign exchange office	FXC	An interface at the end of a private line connected to a switch
Foreign exchange subscriber/ station	FXS	An interface at the end of an FX line connected to an end-entity (such as telephone)
Format identification	FID	Bit C-1 in DS-3 format identifying M13/M28 signal
Forward error correction	FEC	A method used towards error correction by transmitting redundant information along with the original data. The receiver uses the redundant FEC data to reconstruct lost or errored portions of the original data. FEC does not employ retransmissions to correct errors
Forward explicit congestion notification	FECN	A convention used in frame relay for a network device to notify the user (destination) device of network congestion
Fractional T-1	FT-1	A colloquial reference to the transmission of N × 64 kbps TDM service on a DS-1 for 1 < N < 24. The name is based on the fact that the service provides a fraction of the T-1
Frame	-	A commonly used named for the data link layer PDU
Framing	F	Bit position in TDM frame where known pattern repeats
Framing bit	FB	-

Frame assembler/ disassembler	FAD	Functions like packet assembling/disassembling, but for frames
Frame alignment signal	FAS	Bit or byte used by receiver to locate TDM channels
Frame check sequence	FCS	A field in an X.25, SDLC, or HDLC frame, which contains the result of a CRC error detection and correction algorithm
Frame control	FC	A field intended to define the type of frame (in FDDI)
Frame mode bearer service	FMBS	FR on ISDN (now known as FR bearer service FRBS)
Frame-relay	FR	A WAN networking standard (specified by ANSI/ITU-T) for switching frames between end users. It operates at higher speeds with less nodal processing than X.25. ATM borrows many concepts from frame-relay (I.122, T1.617)
Frame-relay protocol data unit (I.122)	FPDU	-
Frame-relay switch/service	FRS	-
Frame-based user-network interface	FUNI	A protocol using a modified DXI header defined for access to ATM service at speeds ranging from 64 kbps up to DS-1/E-1 rates. In addition to DXI capabilities, FUNI adds support for OAM and RM cells
Frequency division multiplexing	FDM	A method of facilitating multiple simultaneous connections over a single high-speed channel by using individual frequency passbands for each connection used
Full duplex	-	Simultaneous bidirectional transmission of information over a common medium

G

Generic call admission control	GCAC	It is a congestion preventive procedure where a parameter is controlled via a technique defined in PNNI to allow a source to estimate the connection admission decisions along candidate paths to the destination
Generic cell rate algorithm	GCRA	An algorithm that decides how an ATM entity measures and controls negotiated service usage. It is a reference model of the ATM Forum for defining cell-rate conformance in terms of certain traffic parameters and tolerances (i.e., PCR, CDVT < SCR, BT, MBS). It is often referred to as the leaky-bucket algorithm
Generic flow control	GFC	It is the first half byte in the ATM header at UNI. That is, a 4 bit field in the ATM cell header with local significance to allow a multiplexer to control two levels of priority in attached terminals. There are only a few implementations that support GFC
General switched telephone network	GSTM	CCITT terminology replacing PSTN after the privatizations of the 1990s
General system for mobile communication (Group special mobile)	GSM	Part of CEPT working on cellular European-based digital mobile radio (DMR) in which a special A/D – D/A converter (codec) in the transreceiver digitizes the subscribers' voice signals
Global positioning system	GPS	Satellite systems, which report exact time
Gigabit ethernet	-	1 Gbps LAN technology (IEEE 802.3z)
Gigabits per second	Gbps	10^9 bits per second
Goodput	-	A measure of actual data successfully transferred from senders to receivers. In ATM: cell/s throughput of an ATM switch, if that switch experiences cell-loss leading to incomplete/ unusable frames arriving at the receiving end

Government OSI profile	GOSIP	A suite of protocols mandated for U.S. federal and UK contractors
Graphical user interface	GUI	-
Ground start	GS	A simple analog phone interface implementation
Group address	GA	Address intended for multicasting
Group – X facimile standard	G3 G4	Group 3, analog facimile standard at/up to 9.6 kbps Group 4, digital facimile standard at 56/64 kbps
Guaranteed frame rate	GFR	A service category proposed by the ATM Forum oriented toward frame-based, instead of cell-based, QOS and traffic parameters

H

Halt	H	A line state symbol in FDDI
Half duplex	-	Transmission of information over a common medium, in both directions such that information travels in only one direction at any time as determined by a control protocol
Handshaking	-	Two protocol entities synchronizing during connection establishment
Head of bus	HOB	A station and function that generates cell or slots on a bus (DQDB)
Header check sequence	HCS	A CRC on header fields only and not on information field. (It refers to header error control (HEC) in ATM)
Header error control	HEC	An error checking code in ATM cell for header only and not for data. A 1-octet field in the ATM cell header containing a CRC checksum on the cell header fields. HEC is capable of detecting multiple bit errors or correcting single bit errors
Header extension	HE	A 12-octet field for various information elements (SMDS)
Header extension length	HEL	A number of 32-bit words in HE (802.6)
Hello packet	-	Packets used by networking devices for neighbor discovery and recovery – an indication that the device is still in operation
Hertz	Hz	Unit for frequency of electrical signals representing cycles per second
Heterogeneous LAN management	HLM	An OSI network management system developed by IBM
Hierarchical network design	-	Designing scalable campus/enterprise network topologies using layer modular base
High bit rate digital subscriber line	HDSL	A Bellcore standard to carry DS-1 over local-loops without repeaters
High capacity digital device	HCDS	A Bellcore T-1 specification
High capacity multiplexing	HCM	Six channels of 9600 in a DS-0
High capacity voice	HCV	8 or 16 kbps scheme
High definition TV	HDTV	TV with double resolution image capability. An end-entity candidate for broadband networks
High density bipolar 3-zeros	HDB3	A line-cable for 2 Mbps lines replacing zeros with bipolar violations (CEPT)
High level data link control	HDLC	A layer-2 full-duplex protocol. It is a widely-used ISO and ITU-T standardized link layer protocol standard for point-to-point and multi-point communications. (e.g., Frame-relay, SNA, X.25, PPP, DXI, and FUNI protocols all use HDLC)
High performance routing	HPR	A method of dynamic call-routing in the PSTN
High-speed peripheral parallel interface	HiPPI	A computer channel simplex interface clocked at 25 MHz, corresponding to 800 Mbps when 32 bits wide or 1.6 Gbps when 64 bits wide
High-speed serial interface	HSSI	An interface at 600 or 1200 Mbps
HOP	-	A node-to-node information movement

Host	-	An end-station in a network
Horizontally-oriented protocol structure	HOPS	A proposed protocol structure for high performance interfaces at broadband rates
Hot-potato routing	-	A type of routing used in switching networks. The switch passes the message on as quickly as possible, often by routing to the line that has least traffic at the time
Hybrid ring control	HRC	A TDM sublayer at the bottom of data-link (2), which splits FDDI into packet and circuit-switched parts
Hub	-	A collecting point of several lines. It can serve as a repeater node
Hub-and spoke topology	-	A star topology
Hybrid fiber/coax system	HFC	A system offered by cable-network service providers to connect CATV networks to the service providers high-speed fiber-optic network
Hyper text transfer protocol	HTTP	A protocol in the TCP/IP suite operating over TCP that implements home pages on the World Wide Web (WWW)

I

Idle state	I	Line state in FDDI
Idle line state	ILS	A state depicting the presence of idle codes on optical fiber lines (FDDI)
Implementation agreement	IA	An agreement that refers to a subset of CCITT/ISO standards without options to ensure interoperability
Infrared	IR	Optical spectrum of wavelength greater than and adjoining the wavelength of red
Indication	IND (OSI)	-
Initial address message	IAM	The call request packet of SS7
Initial MAC protocol data unit	IMPDU	-
IMA control protocol	ICP	A protocol structure that enables inverse multiplexing over ATM (IMA) to combine multiple cell streams at the receiving end. The ICP cells enable endpoints to negotiate and configure, synchronize, and coordinate multiple physical links
Implicit congestion notification	-	An indication of the queueing status on congestion, which is performed by upper-layer protocols (e.g., TCP) rather than network or data link layer protocol conventions
Information element	IE	An information field of variable length within a signaling message defined in terms of type, length, and values according to various standards and specifications
Initial cell rate	ICR	The rate at which a source is allowed to start up following an idle period. It is established at connection set up and is between the MCR and the PCR
Initial domain part	IDP	The higher-order part of an ATM end-system address (AESA) assigned by a network
Input/output	I/O	(As in I/O ports)
Institute of Electrical and Electronics Engineers	IEEE	The U.S.-based, worldwide engineering professional body that prescribes standards for the electronics industry as necessary (e.g., IEEE LAN standard, namely, IEEE 802 services)
Integrated access and cross-connect system	IACS	An AT&T box that contains the digital access and cross-connect system (DACS) and multiplexing functions via packet-switching fabric
Integrated digital loop carrier	IDLC	A combination of remote data terminal, transmission facility, and IDT to feed voice plus data into a CO switch

Integrated digital terminal	IDT	An M24 function in a CO switch to terminate a T-1 line from remote data terminal
Integrated local management interface	ILMI	A SNMP-based PVC management protocol defined by the ATM Forum. It provides interface status and configuration information. It also supports dynamic ATM address registration
Integrated services digital network	ISDN	A digital network service that can support voice and collectively
	ISDN-UP	The user part of ISDN protocol from layer 3 and above for signaling for users (SS7)
Intelligent network	IN	-
Inter carrier interface protocol	ICIP	A protocol of connection between two PSTNs
Inter-exchange carrier	IXC or IEC	A long distance public switching network carrier providing connectivity between LATAs
Interface	I/F	The boundary between two adjacent protocol layers (i.e., link-to-network) in the OSI architecture. Per ITU-T, it is a physical connection between devices
Interim inter-switch signaling protocol	IISP	A protocol employing UNI-based signaling for switch-to-switch communication in private networks as specified by the ATM Forum. In contrast with PNNI, IISP relies on static routing tables and makes support for QOS an alternate routing optional
Intermodulation distortion	IMD	Undesirable modulation products
Intermediate bit rate	IBR	Bit rate between 64 and 1536 kbps denoting the fractional T-1 rates
Intermediate distribution frame	IDF	-
International code designator	ICD	One of three authority and format identifiers (AFI) defined for ATM end-system addresses (AESA)
Inter-office channel	IOC	A part of T-1 or other line between COs of the IXC
International Communications Association	ICA	An association of users group
Internet control message protocol	ICMP	A protocol that reports host errors detected in a router by IP
Interworking function	IWF	The conversation process between frame-relay and X.25, frame-relay, and ATM, etc.
Interworking unit	IWU	A protocol converter between packet formats like frame-relay and ATM
International Standards Organization	ISO	An international organization for standardization, based in Geneva (of which ANSI is a member) that establishes voluntary standards
International Telecommunications Union	ITU	A UN agency previously known as the CCITT. The ITU telecommunications standardization sector (ITU-T) defines standards for N-ISDN, frame relay, B-ISDN, and a variety of other transmission, voice, and data-related protocols
International standard	IS	-
Interswitching system interface	ISSI	An interface between nodes in public network and not available to CPE. (e.g., SMDS to B-ISDN)
Internet activity board	IAB	The body that defines LAN standards such as single network management protocol (SNMP)
Internet Engineering Task Force	IETF	Organization responsible for standards and specification development for TCP/IP networking and adopts requests for comments (RFCs)

Internet Protocol	IP	Network layer basis for TCP, UDP etc. It is a connectionless datagram-oriented network layer (layer 3) protocol containing addressing and control information in the packet header. It allows nodes to independently, yet consistently, forward packets towards the destination. Coexisting routing protocols discover and maintain topology information and for routing packets across a wide range of network topologies and lower layer protocols
Internet telephony	-	A generic term for techniques that run voice over IP networks
Internet protocol version 6	IPv6	A replacement of current IPv4
Interoperability	-	An interoperation capability of multiple, dissimilar vendor devices, and protocols to operate and communicate using a standard set of rules and protocols
Inverse multiplexing over ATM	IMA	A protocol defined by ATM protocol that distributes a stream of cells over several physical circuits (typically DS-1s or E-1s) in a round-robin fashion at the transmitter. The receiving end then reassembles these streams back into their original order
ISDN gateway	IG (AT&T)	-
Isochronous	-	A clocking capability supporting a consistent, timed access of network bandwidth for delay-sensitive transmission of voice and video traffic
Isochronous service data unit	ISDU	An upper layer packet from TDM or circuit-switched service (802.6)

J

Jitter	-	An impairment to a transmission signal in time or phase/delay, accumulation of which may introduce errors and loss of synchronization. Jitter may also be used to denote CDV
Joint Photographic Experts Group	JPEG	A part of ISO that defines the digital storage format for still photographs
Joint Technical Committee 1	JTC 1	A technical committee of IEC and ISO

L

Label switching	-	This is same as call address switching in which the data packets are relayed through a network by switching to an output port based upon the received label and swapping the label value prior to transmission on the output port. (e.g., frame relay, ATM, and MPLS)
LAN emulation	LANE	A standard specified by the ATM Forum enabling ATM devices to interwork seamlessly with legacy LAN devices using either the ethernet or token-ring protocols. Higher layer protocols are supported without change over LANE on ATM-based systems. LANE employs a number of functional elements in performing this role
LAN emulation client	LEC	A software residing in every ATM device, such as workstations, routers, and LAN switches. It bears an ATM address and performs data forwarding, address resolution, connection control, and other control functions
LAN emulation configuration server	LECS	It is a LANE component involved with the configuration of emulated LANs (ELANs) by assigning LECs to specific LANE servers (LES). Network administrations control the combination of physical LANs to form VLANs
LAN emulation network node interface	LNNI	A interface (NNI) operation between the LANE servers (LES, LECS, BUS)

LAN emulation server	LES	A server that takes the ATM and MAC address registrations from LECs. It then provides the MAC address to ATM translation via the LANE address resolution protocol (LE-ARP)
LAN emulation user-to-network interface	LUNI	A standard specified by the ATM Forum for LAN emulation on ATM networks. It defines the interface between the LAN emulation client (LEC) and the LAN emulation server components
Layer	-	A specific level in a protocol vertical hierarchy
Layer 2 PDU	L2 - PDU	A fixed-length cell (protocol data unit) in SMDS
Layer 3 PDU	L3 - PDU	A variable length (packet at OSI level 3)
Layer management	-	It is a network management function, which provides information about the operations of a given OSI protocol layer
Layer management	LM	A control function for protocol
Layer management entity	LME	The process that controls configuration etc. (802.6)
Layer management interface	LMI	The interface between the LME and network management systems in DQDB systems. A software at each OSI layer in SMDS (802.6)
Leaf initiated join	LIJ	ATM Forum 4.0 specified procedure on signaling specification that allows leaf nodes to dynamically join and leave a point-to-multipoint connection using special messages
Leaky-bucket	-	A widely used term to describe an algorithm used for conformance checking of cell flows against a set of traffic parameters. The leak rate of the bucket defines a particular rate, while the bucket depth determines the tolerance for accepting bursts of cells
Least significant bit	LSB	The lowest order bit in the binary sequence of a numerical value. It is the position in the byte that has the lowest value
Linear predictive code	LPC	A voice encoding technique
Line build out	LBO	Deliberate insertion of loss in a short transmission line to make it to act like a long line with greater attenuation
Line-terminating equipment	LTE	A unit capable of either originating or terminating a SONET OC-N signal
Line state unknown	LSU	A possible indication from FDDI reporting the line state monitored
Link layer service access point	LSAP	A logical address of boundary between layer 3 and LLC sublayer in 2 (802)
Link access procedure	LAP	Layer 2 protocol for error checking between 2 devices
Link access procedure balanced	LAPB	A high-level data link protocol for data sent into X.25 network, etc.
Link access procedure (for) D-channels	LAPD	A variant of LAPB for ISDN D-channels
	LAPD +	LAPD protocol for other than D-channels, for example B-channels
Link access procedure modem	LAPM	A part of V.42 modem standard
Link identifier	LI	Address consisting of VPI/VCI in ATM
Local access and transport area	LATA	Geographically constrained telecommunication area, within which a local carrier can provide communications services as defined by a regulatory agency
Local area network	LAN	A MAC-level data and computer communications network confined to short geographic distances. (e.g., ethernet)

Local area transport	LAT	A DECnet protocol for terminals
Local channel	LC	The local-loop
Local exchange carrier	LEC	A carrier depicting an intra-LATA communication service provider (such as an RBOC or an independent telephone company); telco
Local exchange node	LEN	A CO switch of LEC
Local loop back	LLB	-
Local management interface	LMI	A set of communications standards at user device-to-network level used in ATM DXI and frame relay systems
Logical connection identifier	LCI	A short address in the connection-oriented frame
Logical channel number	LCN	A form of address in a packet
Logical group node	LGN	A lower level peer group in the PNNI hierarchy
Logical link	-	A logically conceivable abstract representation of the connectivity between two (logical) nodes used in PNNI. This includes one or more physical circuits and/or virtual path connections
Logical link control	LLC	IEEE 802.2 specified sublayer standard that interfaces with the MAC sublayer of the data link layer in LAN standards. It carries out error control, broadcasting multiplexing, and flow control functions. It facilitates a common interface for all IEEE 802.X MAC protocols to user systems
Logical node	-	A logically conceivable abstract representation of a peer group or a switching system as a single point in PNNI identified by a unique string of bits
Logical unit	LU	An upper level protocol in SNA
Loop start	LS	Analog phone interface
Loss of frame	LOF	A situation or condition at the receiver or a maintenance signal transmitted in the PHY overhead indicating that the receiving equipment has lost the frame delineation. LOF is used to monitor the performance of the PHY layer
Loss of frame count	LOFC	Number of LOFs
Loss of pointer	LOP	A SONET error condition
Loss of signal	LOS	No received data or the incoming signal is not present

M

Main distribution frame	MDF	A large wire-rack at the CO for housing low-speed data and voice cross-connects
Management information base	MIB	OSI-specified description of a structured set of objects (e.g., integers, strings, counters, etc.) accessible via a network management protocols like SNMP and CMIP
Management AAL	MAAL	-
Management service	MS	(of SNA)
Master/slave	M/S	A relationship specific to protocols where the master entity issues command and the slave entity responds
Maximum burst size	MBS	Number of cells that may be sent at PCR without exceeding SCR. It is an ATM-specific traffic parameter that represents the maximum number of cells that can be transmitted at the connection's peak cell rate (PCR) such that the maximum rate averaged over many bursts is no more than the sustainable cell rate (SCR)
Maximum packet life time	MPL	Number of hops permitted before a packet is discarded
Mean time between service outages	MTBSO	-

Medium	-	The physical means of transmission representing a channel for information transmission (e.g., twisted-pair wire, coaxial cable, optical fiber, radio waves, etc.)
Medium access control	MAC	The lower sublayer of the OSI data link layer specific protocol (defined by the IEEE). It controls workstation access to a shared transmission medium. The MAC sublayer interfaces to the logical link control (LLC) sublayer and the physical layer. (e.g., are 802.3 for ethernet CSMA/CD, and 802.5 for token-ring
Medium access control address	MAC address	A control to identify uniquely via a 6-octet value of an endpoint and which is sent in an IEEE LAN frame header to indicate either the source or destination of a LAN frame
Medium access control primitive data unit	MAC PDU	(806.2 specific)
Medium access control service point	MSAP	A logical address (up to 60 bits) of the boundary between MAC and LLC sublayers (802)
Medium access control service data unit	MSDU	A data packet in LAN format, which may be long and variable prior to segmentation into cells
Media access unit	MAU	A device that is attached physically to a cable (802.3 layer 1)
Medium dependent interface	MDI	A link between a medium access unit and a cable (802 layer 1)
Medium interface connector	MIC	A dual-fiber equipment socket and cable plug used in FDDI
Megabit per second	Mbps	10^6 bits per second
Mesh	-	A network topology in which devices are organized with many, often redundant interconnections. Such interconnections are placed strategically between network nodes
Message identifier	MID	A sequence number whose value is used to identify all slots or cells that make up the same PDU in DQDB and AAL-3/4
Message handling system	MHS	OSI store and forward protocol
Message transfer part	MTP	A set of connectionless SS7 signaling protocol stack used within carrier networks at lower layer 3
Message toll service	MTS	A normal dial-up phone service
Metropolitan area network	MAN	Typically 100 Mbps transmission on a network, which operates over metropolitan area distances and number of subscribers. A MAN can carry voice, video, and data. (e.g., DQDB and FDDI)
Minimum cell rate	MCR	An available bit rate traffic descriptor that specifies a rate at which the source may always transmit traffic
Modified final judgement	MFJ	A court decision that led to a split of AT&T in 1984
Modulator-demodulator	Modem	A data communication equipment (DCE) that converts serial DTE data to a signal suitable for transmission over an analog telephone line. It also converts back the received signals into serial digital data for delivery to the DTE
Most significant bit	MSB	Position in byte with largest value. That is, the highest order bit in the binary representation of a numerical value
Moving Pictures Expert Group	MPEG	Standards set for compressing video. (e.g., MPEG1 is a bit-stream standard for compressed video and audio)
Multiple access unit	MAU	A hub device in a LAN (802.5)
Multicast	-	A connection type with the capability to disseminate information to multiple destinations on the network
Multicast address resolution server	MARS	Multicasting based address resolution comparable to the address resolution protocol (ARP) specified for classical IP over ATM. MARS resolves IP addresses to a single ATM multicast address that defines a cluster of end points

Multichannel voice frequency	MCVF	FDM based bank of voice transmissions
Multi-drop data bridging	MDDB	A digital bridging of PCM coded modem signals
Multi-frame alignment	MFA	A code in time slot 16 of E-1 to mark the start of superframe
Multimedia	-	A collective means of presenting a heterogeneous set of information such as text, data, images, video, audio, graphics to the user in an integrated fashion
Multi-peer multicast	MPMC	An N-way mutual broadcasting of messages
Multimode fiber	MMF	An optical fiber with 50- to 100-μm core diameter through which the light propagates along multiple paths or modes. Since the signal may travel different distances for each mode, pulse dispersion limits the transmission distance when compared with single-mode optical fiber
Multiplexer	MUX	A device to combine several input traffics
Multiplexer functions	M13	Multiplex function between DS-1 and DS-3 levels
	M24	Multiplexer function between 24 DS-0 channels and a T-1
	M44	Multiplexer for placing 44 ADPCM channels on a single T-1
	M48	ADPCM multiplexer to place 48 ADPCM channels on a single T-1 with signaling in each voice channel
	M55	ADPCM multiplexer that places 55 voice channels (in 5 bundles) on E-1
Multiplexing	-	A combining technique for multiple streams of information to share a physical medium
Multipoint-to-point	-	A one direction connection topology where multiple leaf nodes transmit back toward a root node. The intermediate nodes should provide a means to merge these streams, for example, by using separate VCCs on a multipoint-to-point VPC
Multiprotocol encapsulation over ATM	-	This protocol is defined in IETF RFC 1483, and allows multiple higher-layer protocols, such as IP or IPX, to be routed over a single ATM VCC using the LLC/SNAP MAC header
Multiprotocol over ATM	MPOA	A protocol specified means to route ATM traffic between virtual emulated LANs, bypassing traditional routers. MPOA is ATM Forum-specified and utilizes the next hop resolution protocol (NHRP) to determine short-cut paths between IP-addressed ATM systems attached to a common ATM SVC network

N

Narrowband-integrated services digital network	N-ISDN	An access at T-1 or less. It is a predecessor to the B-ISDN. It includes the standards for operation at the PRI rate and below. N-ISDN interworks with the traditional telephony standards
Network	-	A collective representation of telecommunication systems with autonomous devices connected via physical media providing a means for communications
Network interface card	NIC	A board that facilitates network communication capabilities to a computer system
Network interworking	-	A trunking arrangement of frame relay connections over an ATM network defined in the frame relay forum's FRF.5 implementation agreement
Network layer	-	Layer 3 of the OSI RM
Network management	-	Governance of the process of involving the operation and status of network resources (e.g., devices, protocols)
Network management system	NMS	A set of equipment and software used to monitor, control, and provide a communications network

Network node interface	NNI	An interface between ATM switches defined as the interface between two network nodes
Network node interface (or network-to-network interface)	NNI	A standard of ITU-T specified on an interface between nodes in different networks. The ATM Forum distinguishes between two NNI standards, one for private networks called PNNI and another for public networks called the B-ICI
Network parameter control	NPC	A traffic management technique performed at the NNI for traffic received from another network
Network service access point	NSAP	An open-system interface specified format defining the octet network address used to define an ATM end system address (AESA)
Next hop resolution protocol	NHRP	A protocol defined by IETF for routers to learn network-layer addresses over NBMA networks like ATM. A next hop server (NHS) responds to router queries regarding the next best hop toward the destination
Node	-	A point of interconnection within a telecommunication network
Non-broadcast multiple access	NBMA	A link layer network not intended to perform broadcasting, but, it has multiple access points (e.g., frame relay and ATM)
Non-real time variable bit rate	nrt-VBR	An ATM service category for traffic compatible for packet data transfers. It may experience significant CDV. The traffic parameters are PCR, CDVT, SCR, and MBS
Nrm	-	An ABR service parameter that specifies the maximum number of cells a source may send for each forward RM cell

O

Octet	-	Telecommunication networking term designating a field 8 bits long. It is synonymous with byte used in computer science
Office channel repeater/office channel unit	OCR/OCU	A channel service unit at CO level
Off premises extension	OPX	A line from PBX to another site
Open channel unit data port	OCU-DP	A channel bank plug I/O to 4-wire local loop and channel service unit on customer premises
Open network architecture	ONA	A FCC plan for equal access to public networks
Operations, maintenance, and administration parts	OAMP	The upper layer 7 protocol in SS7
Operator number identification	ONI	-
Operating system	OS	Main software to run a CPU
Open shortest path first	OSPF	A LAN routing protocol defined for IP that uses the Djikstra algorithm to optimally determine the shortest path so as to minimize some delay, and it does not just loop between nodes. It also defines a reliable topology distribution protocol based upon controlled flooding
Open systems interconnection reference model	OSI RM	ISO specified a seven-layer model defining a suite of protocol standards for data communications. (e.g., X.25 and CMIP are examples)
Operations and maintenance	OAM	A set of measures adopted for diagnostic and alarm reporting as defined by the ITU-T using special purpose cells. Fault management, continuity checking, and performance measurement (PM) are functions of OAM
Operations, administration, maintenance, and provisioning	OAM&P	Management functions and services that interact to provide necessary management tools and control in a telecommunications network

Operational support system	OSS	A support system used by telco to provision, monitor, and maintain facilities
Operating telephone company	OTC	Same as LEC
Optical carrier	OC	A carrier in the fiber transmission of SONET hierarchy. It indicates an optical signal
Optical carrier level N	OC-N	A level specified for the optical carrier signal in SONET, which results from converting an electrical STS-N signal into optical pulses. (e.g., OC-1: 51.84 Mbps; OC-3: 155.52 Mbps etc.)
Optical fiber	OF	-
Optical power received	OPR	Power received by a fiber optic termination
Optical return loss	ORL	-
Optical time domain repletometry	OTDR	A tester to locate breaks in optical
Order wire	OW	A DS-0 in overhead intended for voice path to support maintenance
Organizationally unique identifier	OUI	An IEEE 802.1a specified 3-octet field in the subnetwork attachment point (SNAP) header. It identifies administering organization for the 2-octet protocol identifier (PID) field in the SNAP header. Together the OUI and PID identify a particular routed or bridged protocol
Out of cell delineation	OCD	A receiver in search mode for cell alignment in ATM
Out of frame	OUF	A multiplexer searching for framing bit pattern
Out of synchronization	OUS	A state of multiplexers, which are out of synchronization. (They cannot transmit data)

P

Packet	-	A entity representing an organized and ordered group of data and/or control signals transmitted through a network. A packet is made by slicing a larger message
Pad length	PL	Number 0 to 3 of octets of 0s added to make the information field a multiple of 4 octets (802.6)
Packet assembler disassembler	PAD	Assembling serial data into packet formats and vice versa
Packet data group	PDG	12 octets FDDI frame not assignable to circuit-switched connections
Packet layer protocol	PLP	S protocol at layer 3 like X.25
Packet-switching	-	A technique of switching that segments user information into fixed or variable units called packets prefixed with a header. The network operates on the fields in the header in a store-and-forward manner so as to reliably transfer the packet across the network to the destination
Packet-switched network	PSN	A switched network that handles packets
Packets per second	PPS	-
Packet-switched data network	PSDN	-
Page counter control	PCC	(of SMDS)
Page counts modulus	PCM	SMDS header field
Page reservation	PR	SMDS header field
Partial packet discard	PPD	Discarding of all remaining cells of an AAL-5 PDU (except the last one) after a device drops one or more cells of the same AAL-5 PDU
Parts per million	PPM	
Path control	PC	Level 3 in SNA for network routing

Path overhead	POH	A set of bytes in SDH/SONET for channels carried between switches over multiple lines and through data communication channels
Path terminating equipment	PTE	SONET nodes on ends of logical connections
Payload	-	The nonheader part of 48 octet in 53 octet ATM cell, that contains either user information or OAM data as specified by the payload type indicator (PTI) and VPI/VCI fields in the cell header
Payload type indicator	PTI	An indicator specified by a 3 bit field in the ATM cell header that identifies whether the cell contains user data, OAM data, or resource management data
Peak cell rate	PCR	The maximum rate at which a source can transmit cells according to the leaky-bucket algorithm with a tolerance parameter defined by CDV tolerance. PCR is a traffic parameter
Peer group	-	A collection of nodes within a PNNI network sharing a common address prefix at the same hierarchical level
Peer group leader	PGL	An elected node of its peers to represent the peer group in the next level of the hierarchy
Performance loop back	PLB	Loop back done at point of ESF performance function at customer premises equipment
Performance monitoring	PM	A function in ATM
Permanent virtual channel connection	PVCC	A virtual channel connection (VCC) permanently specified for network management functions
Permanent virtual connection	PVC	A virtual path or channel connection (VPC or VCC) indicated for indefinite use in an ATM network by a network management system
Permanent virtual path connection	PVPC	A virtual path connection (VPC) permanently indicated for network management functions
Phase locked loop	PLL	An electronic circuit that recovers clock time from the data stream
Physical layer	PHY	A physical medium forming the bottom layer of the ATM protocol stack, which defines the interface between ATM traffic and the physical medium. It consists of two sublayers: the physical medium-dependent (PMD) sublayer and the transmission convergence (TC) sublayer
Physical layer convergence protocol	PLCP	A part of physical layer 1 of the OSI model specified as a protocol within the TC sublayer defining a framing structure over DS-3 facilities based upon the 802.6 DQDB standard
Physical layer medium dependent	PMD	Lower sublayer of the physical layer that defines how to transmit and receive cells over the transmission medium
Physical link signaling	PLS	Part of layer 1 that does encoding and decoding (e.g., Manchester coding/IEEE 802)
Physical medium attachment	PMA	Electrical driver for specific LAN cable in media access unit but separated from physical link signaling by the attachment unit interface
Physical unit	PU	SNA protocol stack that provides services to a node and to less intelligent devices attached to it
Plain old telephone system	POTS	Classical residential type telephone service
Plastic optical fiber	POF	Fiber for a very short distance haul
Plesiochronous digital hierarchy	PDH	A nearly-synchronous time division multiplexing (TDM) technique for carrying digital data over legacy high-speed transmission systems

Point-to-multipoint	-	A connection topology transmitting unidirectionally with one root node to two or more leaf nodes
Point-to-point	-	A unidirectional or bidirectional connection topology with two end entities only
Point-to-point protocol	PPP	Nonproprietary multiprotocol serial interface on routers for WAN links
Policing	-	A watching mechanism commonly used to refer to usage parameter control (UPC)
Port	-	Physical interface with respect to a computer, multiplexer, router, CSU/DSU, switch, or other device
Postal, Telephone and Telegraph	PT&T	An authority controlling telecommunications
Preamble	PA	A period of usually steady signal ahead of a LAN frame to specify timing, reserve a cable etc.
Premises distribution system	PDS	The voice and data wiring inside a customer office
Presentation layer PDU	PLPDU	(of OSI)
Presentation services	PS	Level 6 of SNA
Primary market area	PMA	A metro area as served by MAN
Primary out	PO	Fiber optical port that sends light into the main fiber ring (FDDI)
Primary rate interface	PRI	A narrowband N-ISDN access interface type operating on a DS-1 physical interface organized as 23 bearer (B) channels operating at 64 kbps and one data (D) channel operating at 64 kbps (23B + D). The European version of N-ISDN operates on an E-1 interface organized as 30 B channels and one D channel
Primary reference clock	PRC	GPS controlled rubidium oscillator used as a bottom stratum level of timing
Primary subset identifier	PSI	Part of address in network level header (Manufacturing automatic protocol)
Private automated branch exchange	PABX	An electronic PBX
Private branch exchange	PBX	An enterprise located circuit-switch that connects telephony-based equipment and provides access to the public switched telephone network (PSTN)
Private network	-	An exclusive network providing intra-organizational connectivity only. Typically implemented via interconnecting switches, multiplexers, PBXs, or routers via leased line
Private network-network interface	PNNI	A private network node interface, which provides interoperability in multi-vendor networks
Private network-network interface topology state element	PTSE	Information specific to PNNI about available bandwidth, QOS capability, administrative cost, and other factors used by the routing algorithm
Private network-network interface topology state packet	PTSP	Procedures for PDU and flooding used by PNNI to reliably distribute topology information throughout a hierarchical network
Propagation delay	-	Time taken for data to travel over a network end-to-end
Protocol	-	A set of rules and procedural conventions formally governing the formatting and sequencing of message exchange between two communicating systems

Protocol data unit	PDU	A set consisting of control information and user data exchanged between peer layer processes. It also refers to the message that contains information residing in the header and trailer fields associated with the particular layer
Protocol identification	PID	A code allocation by CCITT to identify specific protocols
Public data network	PDN	A packetized data network
Public network	-	A telecommunication network operated by a service provider that shares resources across multiple users
Public service/utility commission	PSC/PUC	A regulatory agency of telecommunication at state level
Public switched digital service	PSDS	Switched 56 k intra LATA service
Public switched telephone network	PSTN	The telecommunications network accessed by conventional telephones, PBX trunks, modems, facsimile machines etc
Pulse code modulation	PCM	A method of converting an analog signal (such as voice) into a digital stream representing digitized samples of the original analog signal following the Nyquist sampling theorem
Pulse per second	PPS	Speed of rotary dial pulses

Q

Q.n	-	Protocols in signaling system 7 (e.g., Q.921 – CCITT recommendation for level 2 protocol in SS7; Q.931 – CCITT recommendation for level protocol in SS7)
Quadrature amplitude modulation	QAM	A method of converting digital signals into analog signals for transmission over a telephone network
Quantizing noise	-	The A/D (and subsequent D/A) effects by a quantization process produces discrepancies between the input and output analog waveforms. The resulting error is the quantization noise
Quality of service	QOS	ITU-T/ATM Forum-specified set of parameters and measurement procedures to quantify loss, errors, delay, and delay variation
Quality of service class	-	ATM Forum (UNI 3.1) specified one of five broad groupings of QOS parameters being largely replaced by the notion of ATM service categories and specific QOS parameters
Queue	-	A ordered set of messages, packets or cells lined up and awaiting service at some processor, like multiplexer or switch
Queue arbitrated	QA	Position of packet traffic that contends for bandwidth (DQDB)
Queued packet synchronous exchange	QPSX	An old name (of Australian origin) for distributed queue dual bus (DQDB)
Queueing theory	-	A branch of statistical theory adopted to study queues, of their arrival and departure characteristics, servicing involved, contention, if any etc.
Queueing delay/time	-	The delay/time involved in awaiting to send/receive a message while being serviced
Quiet	Q	A line state in FDDI

R

Radio frequency	RF	Spectrum of EM frequency intended for wireless transmissions
Radio frequency interface	RFI	Undesirable interference caused by the external radio frequency (RF) radiations on a telecommunication system
Rate decrease factor	RDF	A decremental factor required by an ABR source on its transmission rate if congestion occurs

Rate increase factor	RIF	An incremental factor required by an ABR source on its transmission rate if the received resource management cell indicates no congestion
Read only memory	ROM	A programmable memory holds relatively static data
Ready state	-	A condition at the DTE/DCE interface that indicates that the DTE is ready to accept an incoming call and the DCE is ready to accept a call request
Real time-variable bit rate	rt-VBR	A category of VBR in an ATM service that supports traffic and requires transfer of timing information via constrained CDV. It is suitable for carrying packetized video and audio
Real time protocol	RTP	A part of the IP suite, that conveys sequence numbers and time-stamps between packet video and audio applications (Application: Internet telephony)
Received data	RD	A lead on electrical interface
Redundancy checking	-	A performance (in codes) depicting a calculation on received data and comparison of results with redundant codes to check for certain processing or transmission errors
Redundant code	-	Additional bits (parity) added codes for error checking (detection/correction) proposes
Reference point	R	Interface reference point in the ISDN model to pre-ISDN phones and terminals
Regional Bell Operating Company	RBOC	One of about 22 local service telephone companies, which resulted from the break-up of AT&T in 1984
Region Bell Holding Company	RBHC	One of the seven "baby Bells"
Registered jack	RJ	A connector for UNI (e.g., RJ-11 for telephone and RJ-11 for T-1)
Regulatory agency	-	An agency that controls common carrier tariffs or frequency spectrum allocation, EMI/RFI conformance etc.
Reject	REJ	A logical link control frame that acknowledges received data units while requesting re-transmission from specific errored frame (layer 2)
Release	REL	Release specifying signaling packet on disconnect (SS7)
Release complete	RELC	A packet to acknowledge disconnect (SS7)
Remote access control facility	RACF	A security program of SNA
Remote access memory	RAM	A voltalite memory chip
Remote alarm indication	RAI	(of ATM)
Remote data base access	RDA	A service element of OSI
Remote defect indication	RDI	One of the OAM function types used for fault management
Remote fiber terminal	RFT	-
Remote monitoring MIB	RMON	For remote configuration and data retrieval from a network monitoring device, RMON provides the database
Remote terminal unit	RTU	-
Request	REQ	(of OSI)
Request disconnect	RD	A secondary station unnumbered frame requesting the primary station for a disconnect (DISC) (layer 2)
Request for comment	RFC	Documents that are modified (as per comments) and then adopted by IETF as Internet standards
Research and Advance Communications in Europe	RACE	A program of Europe intended to develop broadband systems
Residual time stamp	RTS	Control information in ATM to support CBR service
Response	RSP	(of OSI)

Resource management	RM	A designated set of cells that communicates information about the state of the network like bandwidth availability and impending congestion to the source and destination
Resource reservation protocol	RSVP	IETF protocol to support different classes of service for IP flows such as guaranteed QOS and controlled load service classes
Return to zero	RZ	Signal pauses at zero voltage between each pulse when making zero crossings
Reuse address resolution protocol	RARP	An Internet protocol to let a diskless workstation learn its IP address from a server (TCP/IP)
Reverse address resolution protocol	RARP	Protocol in TCP/IP stack that allows a diskless station to determine its IP address when its MAC address is known
Ring	-	A common bus network topology in a closed-loop form
Ring	R	One of the conductors in a standard telephone twisted pair, 2 wire local loop. It is connected to the "ring" part of a phone plug of the DTE-to-DTE side of a 4-wire interface
	R1	The ring (R-level) of the DCE-to-DTE side of a 4-wire interface
Robbed bit signaling	RBS	(in PCM)
Round-trip time	RTT	The resultant time-delay incurred in a transmission of a message from a source to a destination, the generation of a response by the destination node, and the subsequent transmission of the response back to the source entity
Router	-	An OSI model specified device operating at layers 1 (physical), 2 (data Link), and 3 (network)
Routing	-	A technique of finding the most efficient circuit path for a message end-to-end transmission
Routing table	-	A table at a node on a message-switched network indicating the preferable route on an outgoing line for each destination
RS	-	A prefix designated by Electronic Industries Association (EIA) for data communication serial connections. (e.g., RS-232, RS-232c, etc.)

S

Scalability	-	Capacity of a network to keep pace with the change and growth of service and architecture
Secondary out	SO	A fiber optic port that sends light into the secondary fiber ring (FDDI)
Section	-	A part of transmission facility in SONET depicting the section between the network element and regenerator
Section overhead	SOH	A set of bytes in SDH specified for channels supported through repeaters between line terminations like data communications channels or switch
Section terminating equipment	STE	A SONET repeater
Segment	-	The payload (user data) portion of the slot in DQDB
Segmentation and reassembly sublayer	SAR	The protocol vertical hierarchy representing the lower half of the common part AAL for types 1, 3/4, and 5
Segmentation and reassembly protocol data unit	SAR-PDU	A segment of CS-PDS with additional heading (and possibly a trailer). (e.g., An ATM cell)

Segmentation Unit	SU	Information field of L2-PDU ≤ 44 octets of a L3-PDU (SMDS, 802.6)
Selector	SEL	A byte reserved for multiplexing by the end user and represents the low order 1 octet field in the domain specific part (DSP) of a 20 octet ATM end system address (AESA). The network does not use this field for routing decisions
Self-healing	-	A process of automatically restoring service by a network in reference to the remote traffic around a failed link or network element
Self-healing ring	SHR	A topology that survives on one failure in line/node (802.6 etc.)
Sequenced information	SI	A LAP-D frame type
Sequence number	SN	An order in the transmission of frames/cells within a channel or a logical connection
Sequence number protection	SNP	CRC and parity calculation on sequence number field in the header (ATM/AAL-1)
Server message block	SMB	A LAN client-server protocol
Service access point	SAP	A locale where services of a lower layer are available to the next higher layer. It is named according to the layer providing the services. (e.g., the interface point between the PHY layer and the ATM layer is called the PHY-SAP)
Service control point	SCP	CPU that is linked to SS7, which supports carrier services
Service data unit	SDU	The part of information unit transferred across a service access point (SAP)
Service information octet	SIO	A field in message signal unit (MSU) used to distinguish individual users in SS7
Service interworking	-	Service interworking provides a method of transparently interworking frame relay and ATM end users and its function refers to converting between semantically equivalent fields in the frame relay and ATM headers (FR Forum FRF.8)
Service specific connection-oriented protocol	SSCOP	It is a part of service specific convergence sublayer (SSCS) portion of the signaling AAL (SAAL). It refers to an end-to-end protocol that provides error detection and correction by retransmission, status reporting between the sender and the receiver, and guaranteed delivery integrity. (UNI and NNI signaling protocols use the same SSCOP layer)
Service specific convergence sublayer	SSCS	This refers to a portion of the convergence sublayer (CS) that is dependent upon the specific higher layer protocol
Service specific coordination function	SSCF	ITU-T defined different functions at the UNI and NNI as regard to signaling ATM adaptation layer-(SAAL) that resides between the SSCOP and the signaling application
Serving office	SO	A control office where an IXC has a POP
Session layer	-	Layer 5 of OSI
Session layer PDU	SPDU	PDU, OSI-specified
Set asynchronous balance mode	SABM	A connection request between HDLC controllers or LLC entities (layer 2)
	SABM - extended	SABM with optional 16 bit control fields
Set number response mode	SNRM	Unnumbered command frame (layer 2)
Severely errored second	SES	An interval in which BER exceeds 10^{-3} > 319 CRC errors in ESF, frame slip, or alarm is present
Shaping	-	A smoothing technique used by the source to modify bursty traffic characteristics to match specified traffic parameters, like PCR, CDVT, SCR, and MBS

Shielded twisted pair	STP	A twisted pair of copper wires with a jacket shielding used for longer distance and/or higher-speed transmission than the unshielded counter part. STP is also less sensitive to electrical noise and interference than unshielded twisted pair (UTP)
Shortest path first	SPF	A LAN routing protocol to minimize delays, not just hops between nodes
Signaling ATM adaptation layer	SAAL	A specific set of protocols residing between the ATM layer and the signaling application
Signaling connection control part	SCCP	An upper layer 3 protocol (SS7)
Signal ground	SG	The second lead to balance E-lead in voice tie line in E&M (ear and mouth) signal circuit
Signaling information field	SIF	Payload of a signaling packet or message signal unit (SS7)
Signaling indication processor outage	SIPO	An outage alarm of processor failure that receives signaling packets (indications) (SS7)
Signaling link selection	SLS	A field in routing label of SU that keeps related packets on the same path to presence delivery-order (SS7)
Signal-to-noise ratio	SNR	-
Signaling return loss	SRL	-
Signaling system number 7	SS7	A common channel signaling standard used in telephony and ISDN and defined by ITU-T. It provides NNI signaling and network intelligence utilized in telephony. Replaces CCIS in ISDN
Simple main transfer protocol	SMTP	(TCP/IP)
Simple efficient adaptation layer	SEAL	The name original assigned for AAL-5
Simple mail transfer protocol	SMTP	Internet protocol offering e-mail services
Simple network management protocol	SNMP	This is an Internet network management protocol. It provides a means to monitor status and performance as well as set configuration parameters
Simplex	-	A traffic of information supports one-way transmission on a medium
Signal frequency	SF	A type of on/off hook signaling within telcos
Single-attachment stations	SAS	A station attached to only the primary ring of the FDDI. It does not provide self-healing
Single-mode fiber	SMF	A fine cored (8- to 10-μm) core diameter optical fiber with a single mode of propagation. It is used for higher speeds or longer distances as compared with multimode optical fiber
Single segment message	SSM	A frame short enough to be carried by one cell
Slot	-	A specified space depicting the basic unit of transmission on a DQDB bus
Soft permanent virtual channel connection	SPVCC	VCC provisioned at ATM end points using the PNNI MIB. The ATM network employs signaling message with a unique SPVCC information element (IE) to dynamically establish and restore the intermediate portions of the connection
Soft permanent virtual path connection	SPVPC	VPC provisioned at the ATM end points using the PNNI MIB. The ATM network employs signaling message with a unique SPVPC information element (IE) to dynamically establish and restore the intermediate portions of the connection
Source address	SA	Field in frame header (802)
Source routing	-	A routing strategy scheme in which the originator determines the end-to-end route. (e.g., token-ring and PNNI)

Source routing transparent	SRT	A variation of source routing combined with signaling tree – algorithm for bridging (802)
Start of header	SOH	Control byte in BSC
Static route	-	A route specified configured and entered in a routing table
Station	-	A concept point representing an addressable logical entity or an actual physical network entity capable of transmitting, receiving, or repeating information
Statistical time-division multiplexer	STDM	-
Store-and-forward switching	-	Frame switching technique in which frames are totally processed a priori to onward transmission out of the appropriate port
Status	S	Signaling bit in CMI
Stream	ST	A network layer protocol for very high-speed connections
Structured data transfer	SDT	An accurate clock (at source and destination) controlled AAL-1 data transfer mode which supports transfer of N × 64 kbps TDM payloads
Subfield	SF	(SNA)
Subnetwork	-	In IP networking, a network chasing a particular subnet address
Subscriber loop carrier	SLC	A digital loop system
Subscriber-network interface	SNI	An interface depicting DQDB user access point into an SMDS network
Subrate digital data multiplexing	SDM	A digital data system service to place multiple low-speed channels in a DS-0
Subsystem number	SSN	Local address of SCCP user (SS7)
Super frame	SF	A frame made of 12 T-1 frames
Supervisory frame	S	Command bearing frame at LLC level
Sustainable cell rate	SCR	An ATM-specific traffic parameter that characterizes a bursty source. It defines the maximum allowable rate for a source in terms of the peak cell rate (PCR) and the maximum burst size (MBS). It is measured in terms of the ratio of the MBS to the minimum burst interarrival time
Sustained information rate	SUS	Audible signal preceding an announcement by the network to a caller
Switch fabric	-	A collective attributes and central function of a switch, which buffers and routes incoming PDUs to the appropriate output ports
Switched fractional T-1	SWIFT	A telco service defined by Bellcore and includes full T-1
Switched multimegabit data service	SMDS	A metropolitan area network conceived as connectionless service offered using the DQDB protocol or HDLC framing
Switched multimegabit data service interface protocol	SIP	Three levels of protocol which define the SMDS SNI user information frame structuring, addressing, and error control
Switched virtual channel connection	SVCC	A virtual channel connection established and released dynamically via control plane signaling
Switched virtual connection	SVC	A virtual logical connectivity between endpoints established by the ATM network on demand when signaling messages received from the end user or another network
Switched virtual path connection	SVPC	A signaling specification (ATM Forum 4.0) defining a VPC established and released dynamically via control plane signaling
SW56	-	Switched 56 kbps digital dial-up service

Synchronous	-	Highly accurate clock controlled transmission, switching, and multiplexing systems as in SONET/SDH
Synchronous	Sync	-
Synchronous allocation	SA	Time allocated to FDDI station to send sync frames (802.6)
Synchronous character	SYN	16h ASCII
Synchronous data link control	SDLC	A protocol defined by IBM for use in SNA environments. It is a bit-oriented protocol
Synchronous digital hierarchy	SDH	A standard for the physical layer of high-speed optical transmission systems similar to SONET in terms of transmission speeds. It differs from SONET in overhead functions (SDH is defined by ITU-T)
Synchronous optical network	SONET	A Bellcore and ANSI defined North American standard for high-speed fiber-optic transmission systems. It is similar to the ITU-T SDH, but incompatible
Synchronous residual time stamp	SRTS	Defined in AAL-1 SRTS is a mode for transparently transferring timing from a source to a destination. It requires that both source and destination have accurate clock frequencies
Synchronous transfer mode	STM	A communications method defined by ITU-T that transmits a group of time division multiplexed (TDM) streams synchronized to a common reference clock. In contrast to ATM, STM systems reserve bandwidth according to a rigid hierarchy regardless of actual channel usage
Synchronous transfer module-n	STM-n	A basic unit of SDH (synchronous digital hierarchy) defined in steps of 155.52 Mbps. The variable n represents integer multiples of this rate. (e.g., STM-1 at 155.52 Mbps, STM-4 at 622.08 Mbps, and STM-16 at 2.488 Gbps)
Synchronous transmission	SYNTRON	Byte aligned format for electrical DS-3 interface
Synchronous transfer signal	STS-n	A basic unit of the SONET multiplexing hierarchy defined in steps of 51.84 Mbps. (e.g., STS-3 at 155.52 Mbps, STS-12 at 622.08 Mbps, and STS-48 at 2.488 Gbps). STS is viewed as an electrical signal to differentiate it from the SONET optical signal level designated as OC-N
Synchronous transport module (n concatenated)	STM-nc	A set of SDH standard transmission rates, which treat the payload as a single concatenated bit stream. ATM uses this mode of SDH transmission. (e.g., the payload rate of STM-1c is 149.76 Mbps, and STM-4c is 599.04 Mbps)
Synchronous transport signal level nc (n concatenated)	STS-nc	It is a SONET mode that concatenates the synchronous payload envelopes (SPE) to deliver a single high-speed bit stream to the higher level user. ATM uses this mode of SONET transmission. (e.g., the payload rate for STS-3c is 149.76 Mbps, and STS-12c is 599.04 Mbps)
System application architecture	SAA	A compatibility scheme for communications among IBM computers
Systems management application process	SMAP	All functions at layer 7 and above to monitor and control network (SS7)

T

T (interface)	T	An interface between NT-1 and NT-2 in ISDN
T (delimiter)	T	An ending delimiter (802.6)
T1 (data multiplexer)	T1 DM	A multiplexer that brings DS-0 Bs together on a DS-1 (Telco phone)

T1 (standard)	T1	The standards committee responsible for transmission issues in the United States. Its European counterpart is ETSI and in Japan, TTC
T-1 (transmission)	-	Digital DS-1 system supported on a four-wire transmission/repeater for operation at 1.544 Mbps
T-3 (transmission)	-	Digital DS-3 system
Target token-rotation time	TTRT	An expected or allowed period for token to circulate once around ring (802.4 and 802.6)
Tandem switching system	TSS	An intermediate switch that interconnects circuits from the switch of one telco CO to the switch of second telco CO in the same exchange
Task oriented procedures	TOPS	Telco document for equipment operation and maintenance
Technical and office protocol	TOP	(for LANs)
Technical subcommittee (for standards setting)	TSC	-
Technical subcommittee	T1D1	- of T1 for basic rate U interface
	T1E1	- of T1 for subscriber network interface
	T1M1	- of T1 for network management system and operation support system
	T1Q1	- of T1 for ADPCM, voice compression etc.
	T1S1	- of T1 for ISDN bearer
	T1X1	- of T1 for SONET and SS7
Technical advisory	TA	A Bellcore standard in draft form before becoming a technical reference (TR)
Technical reference	TR	A final Bellcore standard
Technology systems	TSY	A Bellcore group renamed network technology (NWT)
Telecommunications	-	An electrical/optical transmission, switching, and multiplexing of voice, video, data, and images
Telecommunication Association	TCA	-
Telecommunication device for the deaf	TDD	A teletype machine/terminal with modem for dial-up access
Telecommunications Industry Association	TIA	A successor to EIA
Telecommunications management network	TMN	A support network to run a SONET network
Telecommunications Standardization Bureau	TSB	A bureau setup by ITU in 1993 as a result of the merger of CCITT and CCIR
Telecommunications Standardization Sector	TSS	A variant of TSB
Telecommunications Technology Committee	TTC	A Japanese standard-making body
Teletypewriter	TTY	An U.S. counterpart of teleprinter
Teletype writer exchange	TWX	A switched service (of originally Western Union) separate from Telex
Terminal access controller access control system	TACACS	An authentication protocol that provides remote access authentication and related services like event logging
Terminal adapter	TA	An adapter that matches ISDN formats to the existing interfaces like V.35, R-232
Terminal channel	TC	Same as local loop
Terminal equipment	TE	A support for native ISDN or B-ISDN formats without a TA
Terminal end-point identifier	TEI	A subfield in the second octet of LAP-D address field
Terrestrial digital service	TDS	T-1 and DS-3 service of MCI

Test command	TEST	An LLC UI frame to create loopback (layer 2)
Terminal interface unit	TIU	CSU/DSU or NT1 for SW56 service that handles dialing
Throughput	-	A measure depicting rate of information arriving at and passing through a specific point in a network
Time assigned speech interpolation	TASI	An analog voice compression compatible to DS-1 and statistical multiplexing of data
Time division multiplexing	TDM	A digital technique depicting the method of aggregating multiple simultaneous transmissions (circuits) over a single high-speed channel by using individual time-slots for each circuit
Time division multiple access	TDMA	Stations, which take turns sending in bursts via LAN or satellite
Time-domain reflectometry	TDR	A device used on transmission lines to locate problems on physical layers
(Time) measurement interval	T	Bc/CTR in seconds (Committed burst/Committed information rate)
Timing monitoring system	TMS	-
Time-slot	-	A set of bits in a serial bit stream determined by framing information using the TDM paradigm (e.g., in DS-1 and E-1 TDM systems, an 8 bit time slot transferred 8,000 times per seconds delivers a 64 kbps channel)
Time-slot interchange	TSI	A technique of temporarily storing data bytes so that they can be sent in a different order than received: A method to switch voice or data
Tip	T	One of the conductors in a standard twisted-pair, 2-wire local loop, or one of the DTE-to-DCE pair of 4-wire interface
	T1	T-lead of the DCE-to-DTE pair in a 4-wire interface
To be determined	TBD	(in unfinished technical standards)
Token	-	A designated marker indicating the station's right to transmit that can be held by a station on a token-ring or FDDI network
Token-ring	-	One of the LAN protocols standardized in IEEE 802.5 that uses a token-passing access method for carrying traffic between network elements. Token-ring LANs operate at either 4 or 16 Mbps
Total user cells	TUC	A count kept per VC while monitoring. It represents a field in OAM cell
Traffic contract	-	A contractual agreement between the user and the network regarding the expected QOS provided by the network subject to user compliance with the predetermined traffic parameters such as PCR, CDVT, SCR, MBS
Traffic descriptor	-	A set of generic traffic parameters that capture the intrinsic traffic characteristics of a requested ATM connection. (e.g., UPC, CAC, PCR, SCR, MBS, and ABR
Traffic shaping	-	A queueing method to limit bursty surges in the traffic, which may otherwise congest a network
Trans-Atlantic Telephone	TAT	(Applied to cables as TAT-8)
Transmission control	TC	Control in level 4 (SNA)
Trunk conditioning	TC	Insertion of various signaling bits in A and B positions of DS-0 during carrier failure alarm connections
Transaction capabilities application part	TCAP	Lower layer of SS7

Transmission control protocol	TCP	A popular and widely used layer 4 transport protocol to reliably deliver packets to higher layer protocols like FTP and http. It performs sequencing and reliable delivery via retransmission inasmuch the IP layer does not perform these functions. TCP dynamically maximizes the usage of bandwidth due to its dynamic windowing feature
Transmission control protocol/Internet protocol	TCP/IP	A widely used combination of network and transport protocol developed over the past decades internetworking. Also known as the Internet protocol suite
Transmission convergence	TC	Specific to ATM, a sublayer of the PHY layer responsible for transforming cells into a steady flow of bits for transmission by the physical medium dependent (PMD) sublayer
Transmission group	TG	One or more links between adjacent nodes (SNA)
Transmission level point	TLP	In reference to gain/loss in a voice channel, the measured part at a point minus TLP specified at that point refers to power at 0 TLP site
Transaction processing	TP	Task of a terminal on-line with a host computer
Transmit data	TD	-
Transport layer	-	Layer 4 of OSI
Transport level interface	TLI	(for UNIX)
Transport protocol of class N (N = 0 to 4)	TP-N	OSI layer 4
	TP-0	Connectionless TP (ISO 8602)
	TP-4	Connection-oriented TP (ISO 8073)
Transports protocol data unit	TPDU	(of OSI)
Transport processing service element	TPSE	(of OSI)
Transfer service	TS	(of OSI)
Transport service data unit	TSDU	(of OSI)
Twisted pair copper wire	-	The classical telephone transmission medium, consisting of 22 to 26 AWG insulated copper wire. TP can be either shielded (STP) or unshielded (UTP)

U

Unassigned cell	-	A dummy cell inserted by the transmission convergence (TC) sublayer when the ATM layer has no cells to transmit. This function decouples the physical transmission rate from the ATM cell rate of the source
Unavailable second	UAS	A time period during which BER of a line has exceeded 10^{-3} for ten consecutive seconds until next available second starts
Uncontrolled not ready	UNR	A signaling bit in CMI
Unshielded twisted pair	UTP	A twisted pair of copper wire without jacket shielding. It is used for short distances since it is more subject to electrical noise and interference than shielded twisted pair (STP)
Unicast	-	A message that is sent to a single network node with a specified address (unicast address)
Unified network management architecture	UNMA	An umbrella software of AT&T
Universal asynchronous receiver transmitter	UART	An interface chip for serial asynchronous port
Universal data protocol or user datagram protocol/Internet protocol	UDP/IP	A transport layer

Universal synchronous/ asynchronous receiver transmitter	USART	An interface chip for sync/async data I/O
Universal service order code	USOC	-
Universal coordinated time	UTC	The ultimate global time difference
Unix operating system	UNIX	An operating system developed in 1969 at Bell laboratory
	UNIX 4.3 BSD	UNIX Berkeley Standard Distribution
	UNIX system V Release 4.0	Developed by AT&T
Unspecified bit rate	UBR	A service category of ATM where the network makes a best effort to deliver traffic. It has no QOS guarantees with only a single traffic parameter, PCR
Usage parameter control	UPC	Commonly used in public ATM service, UPC denotes a set of actions taken by the network to monitor and control traffic offered by the end user
User datagram protocol	UDP	In reference to an IP suite, a connectionless datagram-oriented transport-layer protocol
User-to-network interface	UNI	ITU-T/ATM Forum defined interface point between end-users and a network

V

Value added network	VAN	A packet-switched network with access to databases, protocol conversion etc.
Variable bit rate	VBR	The varying rate of transmission generic to sources that transmit data intermittently. The ATM Forum divides VBR into real-time and non real-time (rt-VBR and nrt-VBR) service categories in terms of support for constrained Cell delay variation (CDV) and cell transfer delay (CTD)
Variable quantizing level	VQL	A voice encoding method
Variance factor	VF	A metric used in PNNI GCAG computed in terms of the aggregate cell rate on a link
Vector quantizing code	VQC	A voice compression technique that runs at 32 and 16 kbps
Very plain old telephone service	VPOTS	A POTS with no switching, automatic ring down (ARD) etc.
Very small aperture terminal	VSAT	A small size satellite disk (under 1m diameter)
Via net loss	VNL	A loss measure related to TLP
Video display terminal	VDT	-
Video-on-demand	VOD	A special service that enables users to remotely choose and control the playback of a video
Virtual channel	VC	Same as VCC
Virtual channel connection	VCC	An ATM-specific concatenation of virtual channel links (VCLs) connecting endpoints which access higher-layer protocols. VCCs are unidirectional
Virtual channel cross-connect	VCX	A device to switch ATM cells on logical connections
Virtual channel identifier	VCI	A identifier of 16-bit value in the ATM cell header, which in conjunction with the VPI value identifies the specific virtual channel on the physical circuit
Virtual channel link	VCL	An ATM network specific reference configuration between two connecting points where the VPI and VCI values are either modified or removed. (e.g., every ATM device is a VCL)
Virtual circuit	VC	A logical circuit conceived to enable a reliable service between two end entities in a network

Virtual container	(VC-n)	A cell of bytes carrying a slower channel to define a path in SDH; VC-n refers to Ds-n, n = 1 to 4
Virtual LAN	VLAN	A group of devices on a single or multiple LANs, which are configured to communicate as if they are attached to the same wire, though they are in reality located at different LAN segments. The configuration is enabled by a management software
Virtual LAN trunk	-	A stand-alone physical link supporting more than an VLAN
Virtual LAN trunk protocol	VTP	A switch-to-switch and switch-to-router VLAN management protocol, which exchanges VLAN configuration changes as they are made to a network (A Cisco product)
Virtual path	VP	Same as VPC
Virtual path connection	VPC	An ATM-specific concatenation of virtual path links (VCLs) connecting virtual path termination (VPT) endpoints. VPCs are unidirectional
Virtual path connection identifier	VPCI	An identifier used in signaling messages to logically associate a real VPI with a particular interface
Virtual path identifier	VPI	An identifier value in the ATM cell header that identifies the specific virtual path on the physical circuit. The VPI field is 8 bits long at the UNI and 12 bits long at the NNI
Virtual path link	VPL	An ATM network specific reference configuration in ATM networks between two connecting points where the VPI value is either modified or removed. (e.g., every ATM device is a VPL)
Virtual private network	VPN	A networking in which protocols and processes enable an organization to get interconnected securely of its private network to a public network or the Internet
Virtual source/virtual destination	VS/VD	A mode of available bit rate service that divides flow control into a series of concatenated flow control loops, each with a virtual source and destination
Virtual telecommunications access method	VTAM	SNA protocol and host communications program
Virtual telecommunications network service	VTNS	-
Virtual tributary	VT	A SONET-specific lower level time division multiplexing structure
Virtual tributary envelope	VTE	The real payload plus path overhead within a VT in SONET
Voice frequency	VF	300 to 3300 (- 4000) Hz
Voice grade	VG	Related to the basic analog phone line
Voice grade private line	VGPL	An analog private telephone line
Voice on IP	VoIP	Transmission of telephone calls on IP
Voice response unit	VRU	Automated method to deliver information and accept DTMF inputs
V.n	(n = 1 to 6, 10, 11, 13, 15, 16, 19–22, 22 bis, 23–26, 26 bis, 27, 27 bis, 27 ter, 28-32, 35-37, 40, 41, 50-56)	Data communication standards

W

Wavelength division multiplexing	WDM	A multiplexing scheme to place information at different wavelength carriers in an OF
Wait before transmit positive acknowledgement	WACK	A control sequence of DLE plus second character such as 30 ASCII, 6B EBCDIC
Wide area network	WAN	A networking covering a large geographic region. Usually, WANs employ carrier facilities and/or services
Wideband channel	WBC	One of 16 FDDI subframes of 6.144 Mbps accessible to packet and/or circuit connections
Wideband digital cross-connect system	W-DCS	3/1 DACS intended for OC-1, STS-1, DS-3 and below including T-1
Wide area telephone service	WATS	-
Window	-	A protocol data structure concept of establishing an optimum number of frames or packets that can be outstanding (unacknowledged) before the source transmits again. (e.g., X.25, TCP, and HDLC)
Windowing	-	A flow-control mechanism used when sending data using windows
Wire speed	-	The maximum (theoretical) throughput of a network or circuit
Wiring closest	-	A room designed for wiring a data or voice networks
World Wide Web	WWW	A large network of Internet servers. It provides hypertext and other services to terminals, which run client applications like browsers

XYZ

X-Digital subscriber line	XDSL	A digital subscriber line designated by its variety, X. (e.g., ADSL and HDSL)
X.n	-	Data communication recommendations level 3 protocol to access a packet switched network as defined by CCITT. (e.g., X.25)
X-off	-	Transmission off depicting ASCII character from receiver to stop sender
X-on	-	Transmission on, meaning ok to resume sending
eXpress transfer protocol	XTP	A simplified protocol for low processing purposed for broadband networks
X-stream	-	A generic nature of four fully digital data services of British Telecom (namely, Kilostream, Megastream, Satstream, and Switchstream)
X-window system	XWS	For communication between X terminals and UNIX workstations, the window system provided for relevant tasking and graphing
Yellow alarm control bit	Y	Alarm control bit in sync byte of T1DM (y = 0 indicates alarm)
Younger committee	-	A UK committee that considered the problems of data protection and privacy (1972)
Zone	-	A logical group of network devices specific to AppleTalk
Zone information protocol	ZIP	Mapping of network numbers to zone name in AppleTalk
Zone information protocol storm	ZIP-storm	Several ZIP queries being sent on numerous network segment. This is an undesirable network event

SUBJECT INDEX